HANDBOOK
ON
SATELLITE
COMMUNICATIONS

Third Edition

HANDBOOK
ON
SATELLITE
COMMUNICATIONS

Third Edition

A JOHN WILEY & SONS, INC.
PUBLICATION

INTERNATIONAL
TELECOMMUNICATIONS UNION

For ordering and customer service, call 1-800-CALL-WILEY

Library of Congress Cataloging-in-Publication Data is available.

ISBN 0-471-22189-9

Printed in the United States of America

10 9 8 7 6 5 4 3 2 1

TABLE OF CONTENTS

FOREWORD TO THE THIRD EDITION

The origin of the "Handbook on Satellite Communications (FSS)" goes back to 1980, when Study Group 4 (SG 4) of the CCIR (now ITU-R) decided on the preparation of such a Handbook, the main purpose of which was to extend to developing countries the benefit of an adequate knowledge of what was, at that moment, a relatively new and highly promising telecommunication technique. More generally, in the spirit of the ITU's technical cooperation objectives, the Handbook was to "provide administrations and organizations with a tutorial document to assist them in the preparation of their programmes and in the education of their personnel".

The first edition was prepared by a SG 4 Ad-Hoc Group and was published in 1985. Following the success of this Handbook, under a decision of the XVIth CCIR Plenary Assembly, a new SG 4 "Handbook Group" was formed with the participation of a large panel of experts. This Group was to prepare an updated second edition of the Handbook in order to take account of the rapid technological evolution such as the increasing importance of digital transmissions and advent of ISDN, new satellite networks based on small sized, low-cost earth stations, etc. and also the general evolution of the telecommunication environment and, in particular, competition with fiber optics cables.

The second edition was published in 1988. Again, it proved to meet the needs of engineers, planners and operators, with a much wider audience than just those in developing countries, for which it was originally intended.

Following this second edition, the Handbook Group, under a new commission from ITU-R SG 4, published three Supplements on selected topics. Supplement 3 (1995), notably, is a handbook by itself on "VSAT Systems and Earth Stations" (VSAT, an acronym for Very Small Aperture Terminals, designates small sized earth stations, generally intended for business communication networks).

The preparation of this Edition 3 was decided during the meetings of Study Group 4 and of the Handbook Group in Bern (January 1995) with the purpose of accounting again with the evolution of satellite communications. However, it should be emphasized that it is not a minor revision of the second edition, but an entirely new, updated, document. While, of course, keeping some parts of Edition 2, it has been significantly augmented and many texts have been completely rewritten.

The rationale for this extensive revision and updating is that satellite communications are nowadays faced with rapidly changing perspectives in the market and technical environments, which needed to be accounted for in such a publication:

- *Market*: Trends towards more deregulation and towards integration in the Global Information Infrastructure (GII), increased demand for satellite TV broadcasting, new opportunities for direct data transmission (e.g. via Internet) and for satisfying telecommunication needs in developing countries, increased competition with fiber optics cables, etc.;

- *Technical*: Widespread growth of digital transmission with a high degree of hardware integration and with signal processing and control by software, compatibility with Broadband ISDN (B-ISDN) and Asynchronous Transfer Mode (ATM) operation, advent of non-geostationary Orbit (non-GSO) satellite systems and networks and subsequent needs for new orbit/frequency spectrum allotments and for utilization of higher frequency bands, compressed digital TV broadcasting, etc..

The preparation of this Edition 3 has been performed by the "Handbook Group" under the same working principles as for the previous editions, i.e.: periodic (annual) meetings, liaisons with other relevant ITU groups and autonomy of decisions under the control of Study Group 4. However, after two meetings (Geneva, March 1996 and January 1997), it was decided that the final drafting and the approval of the book would be conducted by correspondence, which, due to the very wide coverage of the book, proved to be a rather long process. It should also be noted that, as a consequence of its autonomy, the Handbook Group remains responsible for the text of the handbook, no formal review by Study Group 4 being required.

On behalf of ITU-R, I would like to acknowledge the efforts of the Handbook Group Members and, in particular, of J. Salomon (France) who chaired the Group continuously since its creation in 1980. As stated in the "Overview of Edition 3 and Acknowledgments" below, this recognition should be extended to all coordinators, contributors and authors, to the organizations which so kindly brought their support to the members and contributors, to the Study Group 4 and its Chairman, Dr. Ito, to SG4 Working Parties and to their Chairmen for their active cooperation and finally to the Radiocommunication Bureau Secretariat, especially at first with Mr. Deyan Wu, then with Mr. Jinxing Li, who, through their efficient assistance, facilitated the publication.

I anticipate that this Edition 3 will experience an even greater readership than the previous editions and will contribute to a better technical knowledge of all aspects of satellite communications and, therefore, to their development and progress.

Robert W. Jones
Director,
Radiocommunication Bureau

OVERVIEW OF THE THIRD EDITION AND ACKNOWLEDGMENTS

This Edition 3 of the "Handbook on Satellite Communications (FSS)" is a completely revised and updated version of Edition 2 (Geneva, 1988). While, hopefully, keeping the tutorial and didactic character of the previous edition, it takes into account the evolution of techniques and technologies, in particular as concerns the current preeminence of digital communications (including the last developments on B-ISDN, SDH, ATM, etc.) and the advent of the new systems based on non-geostationary Orbit (non-GSO) satellites. In this viewpoint, most texts of Edition 2 have been completely rewritten.

The Handbook is divided in Chapters, Appendices and Annexes. Appendices are included at the end of the chapters for giving more development on specialized subjects. The Annexes, at the end of the book cover subjects of general interest.

The Handbook is a self-standing document that is supposed to give a comprehensive description of all issues relative to satellite communications systems operating in the Fixed Satellite Service (FSS). However, for more details on such a specialized subject as the small earth stations and systems, known as Very Small Aperture Terminals ("VSATs"), the reader is referred to Supplement 3 to Edition 2 ("VSAT systems and earth stations", Geneva 1995).

Chapter 1 ("General", Coordinator: Ms. I. Sanz Rodriguez, Hispasat, Spain) gives a general summary of the Handbook content, by giving an overview of the operational, technical and regulatory current characteristics of satellite communications.

Chapter 2 ("Some basic technical issues", Coordinator: Chairman, Mr. J. Salomon, France) provides the reader with basis for a technical understanding the book, including data on quality, availability and link budgets.

Chapter 3 ("Baseband signal processing and multiplexing", Coordinators: Ms. A.Grannec, Mr. Chaumet and Mr. Duponteil, France Telecom), while keeping a few data on analog signals, is centered on digital signal processing and multiplexing, including modern techniques such as voice Low bit Rate Encoding (LRE), video signal compression (e.g. MPEG), Asynchronous Transfer Mode (ATM, see also in Chapter 8), Synchronous Digital Hierarchy (SDH). The subject of Forward Error Correction (FEC) is also introduced in length.

Chapter 4 ("Carrier Modulation Techniques", Coordinator: Dr. L.Kantor, Russian Federation), while again keeping some information on analog modulation, puts an emphasis on digital techniques, including description of digital modems (for continuous signal and for TDMA) and filters. In supplement to the subject of FEC dealt with in Chapter 3, the new technique of Trellis Coded Modulation (TCM), which combines error correction and modulation is introduced.

Chapter 5 ("Multiple access, assignment and network architectures", Coordinator: Mr. R. Zbinden ✝, Swiss Telecom) describes the various modes of multiple access (actually a specific and very useful feature of satellite communications): multiple access by Frequency Division (FDMA), by Time Division (TDMA), including its various derivations (FDMA-TDMA, Random access, etc.) and Code Division (CDMA). Other subjects are dealt with, such as Demand Assignment (DAMA), satellite network architectures and problems of intermodulation in non-linear amplifiers (in Appendix 5.2).

Chapter 6 ("Space segment", Coordinator: Mr.M.W.Mitchell, NTIA, USA), begins by a description of the various possible satellite orbits and of non- GSO satellites constellation design. A somewhat detailed description of the communication satellites and of their communication payload (including the case of transponders with on-board processing) is then given. Other issues, e.g. launching, positioning, station keeping, management of satellite operation and launchers, are also dealt with. Two examples of projected non-GSO satellite systems are described in Appendix 6.1.

Chapter 7 ("Earth segment", Coordinator: Mr. M. Sakai, NEC, Japan) gives a very comprehensive technical description of earth stations and of their equipment (Antenna system, Low Noise Amplifiers, High Power Amplifiers , Communication equipment). In particular, the most advanced types of communication equipment (TDMA, FDMA-TDM, etc.) are the subject of detailed developments. Separate sections are specifically devoted to Satellite News Gathering (SNG) and Outside Broadcasting (OB) earth stations and also to earth stations for direct reception of TV (including of multiple digital programs) and audio programs. The main characteristics of the earth stations standardized by Intelsat and Eutelsat are summarized in Appendix 7.1.

Chapter 8 ("Interconnection of satellite networks with terrestrial networks and user terminals" Coordinators: Mr. J. Albuquerque, then Mr. Young Kim, Intelsat). After recapitulating the problems of interconnection with conventional terrestrial networks, notably with public digital international networks (telephony and fax), this chapter deals with general considerations on protocols (generally based on the OSI model, see Appendix 8.2) for the interconnection with VSAT and other data networks, with user data terminal equipment, and, finally with ISDN and ATM networks (in supplement to Chapter 3). Other subjects considered in this chapter are: Digital Circuit Multiplication Equipment (DCME), Echo cancellation, signalling (with details in Appendix 8.1), Interconnection of television networks and TV distribution and direct broadcasting (analog and digital).

Chapter 9 ("Frequency sharing, interference and coordination" (Coordinator: Ms. A. Grannec, France Telecom). In this very substantial chapter, all questions relating to the frequency and orbit sharing by actual or potential satellite communication operators are dealt with. Of course this text is given for information and tutoring only. When planning or implementing new systems or new stations in the FSS, the reader must refer to the relevant ITU regulations and recommendations (and, in particular to the Radio Regulations (RR)) and other documentation. The Chapter has been prepared taking into account the issues of the last meetings of ITU, notably of the last World Radio Conference (WRC-1997). However, the reader should be advised that many texts analyzed in this Chapter could be revised by the next, incoming WRC-2000.

The main aspects worked on by these conferences result from the advent of the non-GSO satellite systems and of the need to allot them frequency bands, taking into account their possible interference with other radio communication systems of the various services, satellite (GSO or non-GSO) or terrestrial. The chapter is supplemented by a very complete table of the frequency allocations to the various satellite services (Appendix 9.1)

Annex 1 ("Propagation medium effects", Coordinator: Mr. K. Masrani, Radio Agency, UK) summarizes the different effects that contribute to the overall losses along the Earth-space paths. Simplified methods for estimating the signal attenuation are given (in view of the link budget calculations) and the effects of atmosphere on polarization are also described.

Annex 2 ("Typical examples of link budget calculations", Coordinator: Mr. J. Salomon, Chairman). This Annex illustrates by some typical examples the link budget calculations described in Chapter 2.

Annex 3 ("General overview of existing systems", Coordinator: Mr. J. Salomon, Chairman) provides information on the main characteristics of satellite communication systems under current operation throughout the world. However, because the included information is limited to those administrations/organizations which have actually supplied updated information in response to a questionnaire sent by ITU, the Annex should be considered as giving important and typical examples but is not claiming for completeness.

Finally, **Annex 4** ("Relevant ITU references", prepared by Mr. J. Li, Counsellor, ITU-R Secretariat) gives a list of the main relevant ITU publications (Radio Regulations, Recommendations from ITU-R and ITU-T, etc.) relevant to the subject of the Handbook.

In the framework of the ITU-R/Study Group 4, this new edition is the result of the collective work of the "Handbook Group". Due acknowledgements should be granted to all active participants and, primarily, to the Coordinators of the various parts of the book. There were also many invited contributors whose expertise was highly appreciated and which should be thanked for their participation. It should be emphasized that these acknowledgments should be extended to the administrations, organizations or companies which, through their kind and efficient support, permitted the works and efforts of these participants, coordinators and contributors.

A list of these administrations/organizations/companies is given below. Recognition and thanks to each of them should not be limited to the parts of the book they have directly prepared or coordinated, but to their general participation to the works of the Group, through their delegates.

- France Telecom (CNET), for Chapters 3 and 9;
- Hispasat (Spain) for Chapter 1;
- Intelsat for Chapter 8;
- JFC Informcosmos (Russian Federation) for Chapter 4;
- NEC (Japan) for Chapter 7;
- NTIA (USA) for Chapter 6;
- Radio Agency (UK) for Annex 1;
- Swiss Telecom for Chapter 5;
- Teledesic for § AP6.1-1;
- and Alcatel for the support of the Chairman and for expert contributions which were largely used in the preparation of many parts of the Handbook: Digital modulation techniques (MMr. C. Bertrand and Ch. Bergogne), Earth station antenna components (Mr. A.Bourgeois), Digital SCPC and IDR equipment (Mr. J. Bousquet and Mr. C. Bareyt), ATM (Mr. D. Cochet), Quality and Availability (Mr. J. P. Dehaene), TDMA equipment and Demand Assignment (Mr. E. Denoyer), Error correction: FEC and TCM (Mr. J.Fang), Digital modems (Mr. M. Isard), Satellite orbits (Mr. E. Lansard), Link budget examples (MMr. T. Ménigand and T. Gervaz), Earth stations for direct reception of digital TV (Mr. M. Part), SkyBridge (Mr. Rouffet);

Moreover, many thanks are also due to:

- SG4 Working Party 4A and to his Chairman for their constant and active support through liaison statements and transmission of relevant documents and for the direct comments and advises of their Chairman, Mr. A. Reed;

- SG4 Working Party 4B for the transmission of many useful documents;

- SG4 Working Party 4SNG for the direct preparation of § 7.9 on Satellite News Gathering;

- and to all relevant Administrations or Organizations which have supplied updated information on their satellite systems in the framework of Annex 3.

CHAPTER 1

General

1.1 Introduction

Satellite communications now form an integral part of our new "wired" world. Since their introduction in about 1965, satellite communications have generated a multiplicity of new tele-communication or broadcasting services on a global or regional basis. In particular, they have enabled a global, automatically-switched telephony network to be created. Although, since 1956 (TAT1), submarine cables began to operationally connect the continents of the world with readily available telephony circuits, only satellite communications have enabled completely reliable communication links for telephony, television and data transmission, to be provided over all types of terrestrial obstacles, whatever the distance or remoteness of the locations that have to be connected.

At present, about 140 000 equivalent channels[1] (June 1998) are in operation, accounting only for the INTELSAT system. Although the growth of the satellite part on intercontinental telephone highways has been somewhat slowed by the installation of new, advanced, optical fibre submarine cables, satellites continue to carry most of the public switched telephony traffic between the developed and the developing world and between developing regions and this situation should continue, thanks to the advantages of satellite systems and, in particular to the simplicity and the flexibility of the required infrastructure.

It can be predicted that the future of satellite communications will rely more and more on the effective use of their specific characteristics:

- multiple access capability, i.e. point-to-point, point-to-multipoint or multipoint-to-multipoint connectivity, in particular for small or medium density scattered data or voice/data traffic between small, low-cost earth stations (business or private communications networks, rural communications, etc.);

- distribution capability (a particular case of point-to-multipoint transmission), including:

  - TV programme broadcasting and other video and multimedia applications (these services are currently in full expansion);

  - data distribution, e.g. for business services, Internet wideband services, etc.;

- flexibility for changes in traffic and in network architecture and also ease of operation and putting into service.

---

[1] The term "equivalent channel" is used in lieu of "telephone circuit" to indicate that this channel may carry any type of multimedia traffic (voice, data, video).

A major factor in the recent evolution of satellite communications is the nearly complete replacement by digital modulation and transmission of the previous conventional analogue technique, bringing the full advantage of digital techniques in terms of signal processing, hardware and software implementation, etc.

It should also be emphasized that satellite communications will no longer be based, as was the case up till now, only on satellites in geostationary orbit (GSO), but also on the utilization of non-geostationary (non-GSO) satellite systems, with lower altitude (low-Earth or medium-Earth orbit (LEO or MEO)) satellites. These new systems are opening the way to new applications, such as personal communications, fixed or mobile, innovative wideband data services, etc.

This Handbook attempts to bring together basic facts about satellite communication as related to the fixed-satellite service (FSS). It covers the main principles, technologies and operation of equipment in a tutorial form with the intention of assisting administrations and organizations in the preparation of satellite programmes and in the education of their personnel who have technical or general responsibilities related to satellite communications.

The technical characteristics of the satellites and of the earth stations currently implemented in most of the international, regional and national satellite communication systems are summarized in Annex 3 of this Handbook.

1.2   Historical background

Table 1.1 below summarizes some of the main events which have contributed to the creation of a new era in worldwide communications. It is of interest to note that only 11 years passed between the launching of the first artificial satellite (USSR's Sputnik) and the actual implementation of a fully operational global satellite communications system (INTELSAT-III) in 1968. Note that Table 1.1 tentatively anticipates some future operations.

1.3   Definition of satellite services

This section deals with the official ITU definitions relative to the various satellite services currently in operation.

1.3.1   Fixed-satellite services FSS

According to the Radio Regulations (RR No. S1.21), the FSS is a radiocommunication service between given positions on the Earth's surface when one or more satellites are used. These stations located at given positions on the Earth's surface are called earth stations of the FSS. The given position may be a specified fixed point or any fixed point within specified areas. Stations located on board the satellites, mainly consisting of the satellite transponders and associated antennas, are called space stations of the FSS (see Fig. 1.1).

At the present time, with very few exceptions, all links between a transmitting earth station and a receiving earth station are effected through a single satellite. These links are comprised of two parts, an uplink between the transmitting station and the satellite and a downlink between the satellite and the receiving station. In the future, it is envisaged that links between two earth stations could use two or more satellites directly interconnected without an intermediate earth station. Such a link between two earth stations using satellite-to-satellite links would be called a multi-satellite link. The satellite-to-satellite links will form a part of the inter-satellite service (ISS).

TABLE 1.1

Satellite communications historical background

1929:	The Problem of Space Flight. The Rocket Engine, by Hermann Noordung, describes the concept of the geostationary orbit.
1945 (May):	In a visionary paper, Arthur C. Clarke, the well-known physicist and author, describes a world communication and broadcasting system based on geosynchronous space stations.
1957 (4 Oct.):	Launching of the Sputnik-1 artificial satellites (USSR) and detection of the first satellite-transmitted radio signals.
1959 (March):	Pierce's basic paper on satellite communication possibilities.
1960 (Aug.):	Launching of the ECHO-1 balloon satellite (USA/NASA). Earth-station to earth-station passive relaying of telephone and television signals at 1 and 2.5 GHz by reflection on the metalized surface of this 30 m balloon placed in a circular orbit at 1 600 km altitude.
1960 (Oct.):	First experiment of active relaying communications using a space-borne amplifier at 2 GHz (delayed relaying communications) by the Courier-1B satellite (USA) at about 1 000 km altitude.
1962:	Foundation of the COMSAT Corporation (USA), the first company specifically devoted to domestic and international satellite communications.
1962:	Launching of the TELSTAR-1 satellite (USA/AT&T) (July) and of the Relay-1 satellite (USA/NASA) (December). Both were non-geostationary, low-altitude satellites operating in the 6/4 GHz bands.
1962:	First experimental transatlantic communications (television and multiplexed telephony) between the first large-scale, pre-operational earth stations (Andover, Maine, USA, Pleumeu-Bodou, France and Goonhilly, UK).
1963:	First international regulations of satellite communications (ITU Extraordinary Radio Conference). Initiation of sharing between space and terrestrial services.
1963 (July):	Launching of SYNCOM-2 (USA/NASA), the first geostationary satellite (300 telephone circuits or 1 TV channel).
1964 (Aug.):	Establishment of the INTELSAT organization (19 national Administrations as initial signatories).
1965 (April):	Launching of the EARLY BIRD (INTELSAT-1) satellite, first commercial geostationary communication satellite (240 telephone circuits or 1 TV channel). First operational communications (USA, France, Federal Republic of Germany, UK).
1965:	Launching of MOLNYA-1 (USSR), a non-geostationary satellite (elliptical orbit, 12 hours revolution). Beginning of television transmission to small size receive earth stations in USSR (29 Molnyas were launched between 1965 and 1975).
1967:	INTELSAT II satellites (240 telephone circuits in multiple access mode or 1 TV channel) over Atlantic and Pacific Ocean regions.
1968-1970:	INTELSAT III satellites (1 500 telephone circuits, 4 TV channels or combinations thereof). INTELSAT worldwide operation.
1969:	Launch of ATS-5 (USA/NASA). First geosynchronous satellite with a 15.3 and 3.6 GHz bands propagation experiment.
1971 (Jan.):	First INTELSAT IV satellite (4 000 circuits + 2 TV channels).
1971 (Nov.):	Establishment of the INTERSPUTNIK Organization (USSR and 9 initial signatories).
1972 (Nov.):	Launching of the ANIK-1 satellite and first implementation of a national (domestic) satellite communications system outside the USSR (Canada/TELESAT).

1974 (April):	WESTAR 1 satellite. Beginning of national satellite communications in the USA.
1974 (Dec.):	Launching of the SYMPHONIE-1 satellite (France, Federal Republic of Germany): the first three-axis stabilized geostationary communications satellite.
1975 (Jan.):	Algerian satellite communication system: First operational national system (14 earth stations) using a leased INTELSAT transponder.
1975 (Sept.):	First INTELSAT IVA satellite (20 transponders: more than 6 000 circuits + 2 TV channels, Frequency reuse by beam separation).
1975 (Dec.):	Launching of the first USSR geostationary Stationar satellite.
1976 (Jan.):	Launching of the CTS (or Hermes) satellite (Canada), first experimental high-power broadcasting satellite (14/12 GHz).
1976 (Feb.):	Launching of the MARISAT satellite (USA), first maritime communications satellite.
1976 (July):	Launching of the PALAPA-1 satellite. First national system (40 earth stations) operating with a dedicated satellite in a developing country (Indonesia).
1976 (Oct.):	Launching of the first EKRAN satellite (USSR). Beginning of the implementation of the first operational broadcasting satellite system (6.2/0.7 GHz).
1977 (June):	Establishment of the EUTELSAT organization with 17 administrations as initial signatories.
1977 (Aug.):	Launching of the SIRIO satellite (Italy). First experimental communication satellite using frequencies above 15 GHz (17/11 GHz).
1977:	ITU World broadcasting-satellite Administrative Radio Conference (Geneva, 1977) (WARC SAT-77).
1978 (Feb.):	Launching of the BSE experimental broadcasting satellite for Japan (14/12 GHz)
1978 (May):	Launching of the OTS satellite, first communication satellite in the 14/11 GHz band and first experimental regional communication satellite for Europe (ESA: European Space Agency).
1979 (June):	Establishment of the INMARSAT organization for global maritime satellite communications (26 initial signatories).
1980 (Dec.):	First INTELSAT V satellite (12 000 circuits, FDMA + TDMA operation, 6/4 GHz and 14/11 GHz wideband transponders, Frequency reuse by beam separation + dual polarization).
1981	Beginning of operation in the USA, of satellite business systems based on very small data receive earth stations (using VSATs).
1983:	ITU Regional Administrative Conference for the Planning of the Broadcasting-Satellite Service in Region 2.
1983 (Feb.):	Launching of the CS-2 satellite (Japan). First domestic operational communication satellite in the 30/20 GHz band.
1983 (June):	First launch of the ECS (EUTELSAT) satellite, (9 wideband transponders at 14/11 GHz: 12 000 circuits with full TDMA operation + TV. Frequency reuse by beam separation and by dual polarization).
1984	Beginning of operation of satellite business systems (using VSATs) with full transmit/receive operation.
1984 (April):	Launching of STW-1, the first communication satellite of China, providing TV, telephone and data transmission services.
1984 (Aug.):	Launching of the first French domestic TELECOM 1 multi-mission satellite: 6/4 GHz, telephony and TV distribution; 8/7 GHz, military communications; 14/12 GHz, TVRO and business communications in TDMA/DA.
1984 (Nov.):	First retrieval of communication satellites from space, using the space shuttle (USA).
1985 (Aug.):	ITU World Administrative Radio Conference (WARC Orb-85) (1st session on utilization of the geostationary orbit).
1988 (Oct.):	ITU World Administrative Radio Conference (WARC Orb-88) (2nd session on utilization of the geostationary orbit).
1989:	INTELSAT VI satellite (Satellite-Switched TDMA, up to 120 000 circuits (with DCME), etc.)
1992 (Feb.):	ITU World Administrative Radio Conference.
1992 (Feb.):	Launching of the first Spanish HISPASAT-1 multi-mission satellite: 14/11-12 GHz distribution, contribution, SNG, TVRO, VSAT, business services, TV America, etc.; 17/12 GHz, DBS analogue and digital television; 8/7 GHz governmental communications.

Up to 1996:	9 INTELSAT VII-VIIA satellites
1997 -1998:	INTELSAT VIII satellites
1998 onwards:	Launching of various non-geostationary satellites and implementation of the corresponding MSS systems (Iridium, Globalstar, etc.) and FSS systems (Teledesic, Skybridge, etc.).
1999:	First INTELSAT K-TV satellite (30 14/11-12 GHz transponders for up to 210 TV programmes with possible direct to home (DTH) broadcast and VSAT services).
2000:	INTELSAT IX satellites (up to 160 000 circuits (with DCME)).

Inter-satellite links (ISLs) of the ISS may be employed to provide connections between earth stations in the service area of one satellite to earth stations in the service area of another satellite, when neither of the satellites covers both sets of earth stations.

A set of space stations and earth stations working together to provide radiocommunications is called a satellite system. For the sake of convenience, a distinction is made in the particular case of a satellite system, or a part of a satellite system, consisting of only one satellite and the associated earth station which is called a satellite network.

The FSS also includes feeder links, i.e. links from an earth station located at a specific fixed point to a space station, or vice versa, conveying information for a space radiocommunication service other than for the FSS. This category includes, in particular, uplinks to the satellites of the broadcasting-satellite service (BSS) and up and downlinks between fixed earth stations and satellites of the mobile satellite service (MSS). Definitions of BSS and MSS are given in § 1.3.2. below.

All types of telecommunications signals can be transmitted via FSS links: telephony, facsimile (fax), data, video (or a mix of these signals in the framework of integrated services data networks (ISDN)), television and sound programmes, etc.

The latest communications satellite generations, operating in FSS frequency bands are equipped with high-power transponders, which makes it possible to implement broadcasting services direct to the general public for individual reception (direct-to-home (DTH) applications) through very small receiving antennas (television receiving only (TVRO)), and for community reception (professional applications and domestic applications).

1.3.2 Other satellite services

Other satellite services listed below use different frequency band allocations than the FSS. These services are not directly covered by this Handbook although similar techniques are employed and many commonalities exist between these services and those of the FSS.

Mobile-satellite service (MSS): According to the Radio Regulations (RR No. S1.25), this is a radio-communication service between mobile earth stations and one or more space stations, or between mobile earth stations by means of one or more space stations (see Fig. 1.2). This includes maritime, aeronautical and land MSSs. Note that, in some modern systems the earth stations may consist of very small, even hand-held, terminals.

Broadcasting-satellite service (BSS): This is a radiocommunication service in which signals transmitted or retransmitted by space stations are intended for direct reception by the general public using very small receiving antennas (TVROs). The satellites implemented for the BSS are often called direct broadcast satellites (DBSs). The TVROs needed for BSS reception should be smaller than the ones needed for operation in the FSS. The direct reception shall encompass both individual reception (DTH) and community reception (CATV and SMATV) (see Fig. 1.3).

Some other satellite services which are mainly oriented to very specific applications are: radiodetermination-satellite service, radionavigation-satellite service, meteorological-satellite service, etc.

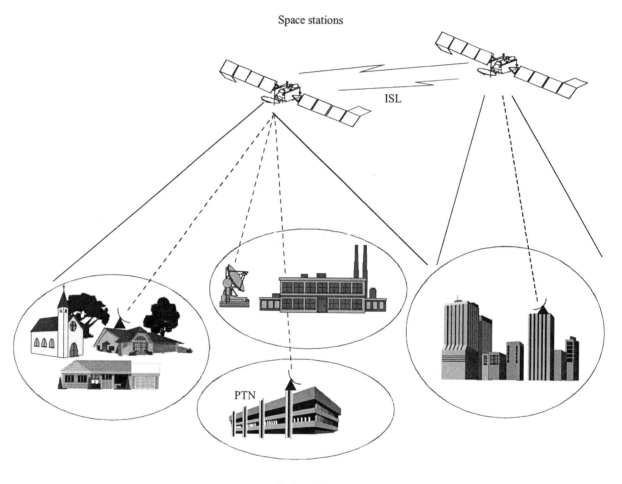

(A possible inter-satellite link (ISL) with another satellite is represented)

PTN: public telephone network

Sat/C1-01

FIGURE 1.1

Generic illustration of FSSM

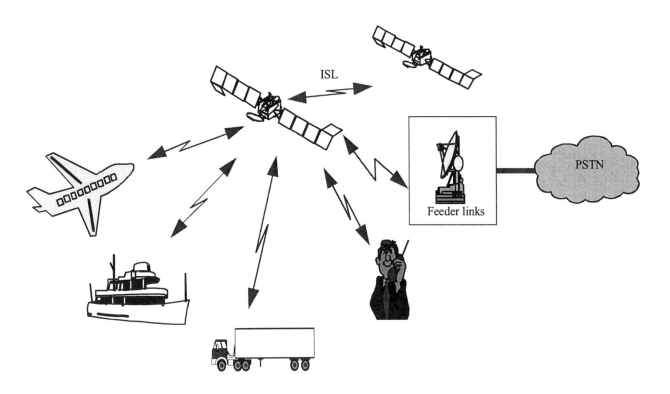

(A possible inter-satellite link (ISL) with another satellite is represented)

PSTN: public switched telephone network

Sat/C1-02

FIGURE 1.2

Generic illustration of mobile-satellite services

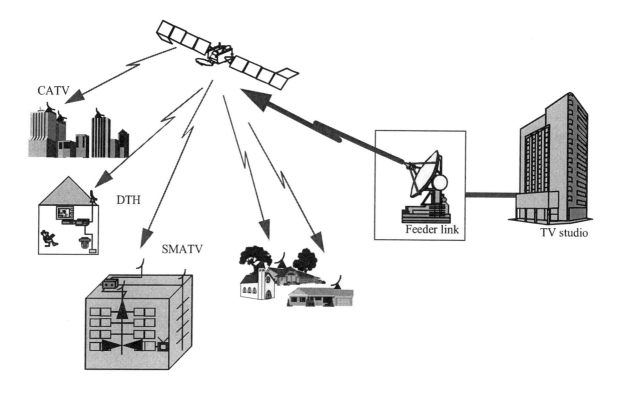

CATV: Cable television network
SMATV: Satellite master antenna TV
DTH: Direct-to-home TV

Sat/C1-03

FIGURE 1.3

Generic illustration of broadcasting-satellite services

1.4 Characteristics and services

1.4.1   Main components of a communication satellite system

1.4.1.1   The space segment

The space segment of a communication-satellite system consist of the satellites and the ground facilities providing the tracking, telemetry and telecommand (TTC) functions and logistics support for the satellites.

i)   The satellite

The satellite is the core of the network and performs all the communication function in the sky using active elements. It is composed of an assembly of various telecommunication subsystems and antennas. The satellite is also fitted with service equipment to provide the following functions:

- bus structure,

- power supply,

- attitude control,

- orbit control,

- thermal control,

- telemetry, telecommand and ranging.

The telecommunication equipment is composed of transponders. There are different kinds of transponders: transparent transponders and on-board processing (OBP) transponders. The most widely used are transparent transponders. Transparent transponders perform the same functions as radio-relay repeaters; they receive transmissions from the Earth and retransmit them to the Earth after amplification and frequency translation. The antennas associated with these transponders are specially designed to provide coverage for the parts of the Earth within the satellite network.

OBP transponders means they are capable of performing one or more of the three following functions: switching (in frequency and/or space, and/or time), regeneration and baseband processing. At present OBP transponders are not widely used but in the future some services will benefit from this technology.

Telecommunication satellites are based on the technologies and techniques used by most other artificial satellites. The repeater (transponder) technology is, however, specific to this type of satellite and is derived from that of terrestrial telecommunication equipment. Certain components, such as solar cells and travelling wave tubes, are tailored to satellite applications. Other components are derived from standard production items, but have been specially selected and subjected to manufacturing checks and final space quality control tests. The ability of a production line to produce components of space quality is checked by a process known as "space qualification".

At the present time, most communication satellites with a few exceptions describe a circular orbit on the equatorial plane at an altitude of about 36 000 km, resulting in a 24 h period of revolution round the centre of the Earth. They are thus synchronous with the Earth's rotation and appear relatively motionless in relation to a reference point on the Earth's surface. This characteristic enables the satellite to provide permanent coverage of a given area, which simplifies the design of earth stations, since they are no longer required to track satellites moving at considerable angular velocities and makes more efficient use of the radioelectrical spectrum and orbital resources. They are consequently located on the geostationary-satellite orbit (GSO) and are designated as geostationary satellites.

Geostationary satellites are generally placed in orbit in two stages:

a)   first a launcher places the satellite in an elliptical transfer orbit (perigee typically about 200 km, apogee about 36 000 km, period 10.5 h);

b)   secondly in order to reach the geostationary-satellite orbit, the satellite uses an auxiliary motor which is fired approximately at the apogee of the transfer orbit so as to make it circular. In addition thrust of the apogee motor is so directed that by the time it finishes its burn the orbital plane coincides with the equatorial plane.

After this operation apogee has taken place, the satellite is allowed to slowly drift until it reaches the vicinity of the desired longitudinal location. Corrections are then finally made to position the satellite accurately on the stipulated longitude.

Although this Handbook is mainly oriented to GSO satellites, there are other systems using non-GSO satellites. Up till now, such systems were mostly dedicated to military or governmental purposes. At present, however, several non-GSO systems are under construction or projected for operation in the MSS for mobile services using hand-held terminals (Iridium, Globalstar, Odyssey, ICO, etc.) and also in the FSS, e.g. for business data networks (Teledesic, Skybridge, etc.).

Most of these projected non-GSO systems use low-Earth orbit (LEO) satellites, designed to operate at altitudes between 400 and 1 500 km, but a few others use medium-Earth orbit (MEO) satellites orbiting at an altitude between 7 000 and 12 000 km. Of course, these satellites being no longer seen as fixed by an earth observer, a given network needs multiple satellites, the number of satellites being an inverse function of their altitude. The design of these systems is clearly derived from the architectures of terrestrial mobile communication cellular networks: The principle of operation is that, from his personal terminal, a subscriber anywhere in the operating zone, should be able to communicate through an overhead satellite via a link to another satellite which is located at that moment overhead his correspondent in an operating zone anywhere else in the world.

Some advantages of these systems are: lower propagation delay which is desirable, especially for telephony services, higher elevation angle, a very useful feature for operation in urban areas, etc. Of course, there is a price to be paid: lower orbits imply a (much) greater number of satellites in the system and can make it more difficult to get compatibility with other GSO services. Also LEO and MEO systems imply complex operation and network management, specially when a global capability is aimed at. However, the satellites should be smaller and their launching cheaper.

ii)   Tracking, telemetry and telecommand

These subsystems are used for carrying out from the ground the following operations for the logistics support of the satellites:

- tracking the position of the satellite (angular position, distance) and determining attitude while it is being placed in orbit and on station and then throughout its life to supervise operation and transmit correction instructions;

- telemetering of various on-board functions;

- command of various on-board functions;

- supervision of telecommunication functions, especially of the carriers in the various transponders.

The latter operation is used to check the functioning of the network and to ensure that emissions from different earth stations comply with specifications (power, frequency, etc.).

These operations are performed by means of a special earth station and are usually centralized at a network control centre. For some communication modes, this centre and other specialized stations also control synchronization, demand assignment, etc.

1.4.1.2   The earth segment

The earth segment is the term given to that part of a communication-satellite system which is constituted by the earth stations used for transmitting and receiving the traffic signals of all kinds to and from the satellites, and which form the interface with the terrestrial networks.

An earth station includes all the terminal equipment of a satellite link. Its role is equivalent to that of a terminal radio-relay station. Earth stations generally consist of the following six main items:

- the transmitting and receiving antenna, with a diameter ranging from 50 cm (or even less in some projected new systems) to more than 16 m. Large antennas are usually equipped with an automatic tracking device which keeps them constantly pointed to the satellite; medium-sized antennas may have simple tracking devices (e.g. step-track), while small antennas generally have no tracking device and although normally fixed, can usually be pointed manually;

- the receiver system, with a sensitive, low noise amplifier front-end having a noise temperature ranging from about 30 K, or even less, to several hundred K;

- the transmitter, with power ranging from a few watts to several kilowatts, depending on the type of signals to be transmitted and the traffic;

- the modulation, demodulation and frequency translation units;

- the signal processing units;

- the interface units for interconnecting with terrestrial networks (with terrestrial equipment or directly with user equipment and/or terminal).

The size of this equipment varies considerably according to the station capacity.

1.4.2   Frequency bands

Most FSS links involve transmission from an earth station to a space station (uplink, U/L) and retransmission from the space station to one or several earth station(s) (downlink, D/L). Since large channel capacities and consequently large bandwidths are required, very high radio frequencies are implemented. These frequencies are subject to allotment by ITU.

Historically, bandwidths around 6 GHz (U/L) and 4 GHz bands (D/L) have been commonly paired and many FSS systems still use these bands (which are often called the "C-Bands"). Military and governmental systems traditionally used 8 and 7 GHz bands (the "X-Bands"). A number of systems also operate around 14 GHz (U/L) and 11-12 GHz (D/L) (the "Ku-Bands"). In the future, due to the saturation of these bands, the 30 GHz and 20 GHz bans (the "Ka-Bands") should be more and more implemented although they are subject to high meteorological attenuation.

Other frequency bands have also be allotted to the feeder links.

All these systems are capable of using a common antenna for transmitting and receiving since the ratio of the U/L to the D/L frequencies is no more than 1.5. Another advantage of this arrangement is that the propagation conditions are relatively similar and that the polarizations are likely to be correlated on both links.

Table 1.2 summarizes the main frequency bands used in FSS along with typical applications.

A detailed Table of Frequency Allocations to the FSS, BSS, MSS and ISS is given in Appendix 9.1 of Chapter 9 of this Handbook.

TABLE 1.2

Summary of frequency bands used in the FSS for GSO

Frequency bands (GHz)			Typical utilization
Current denomination	Up path (bandwidth)	Down path (bandwidth)	
6/4 (C-Band)	5.725-6.275 (550 MHz)	3.4-3.95 (550 MHz)	National satellites [Russia: Statsionar and Express International (Intersputnik)]
	5.850-6.425 (575 MHz)	3.625-4.2 (575 MHz)	International and domestic satellites. At the present the most widely used bands: Intelsat. National satellites: Westar, Satcom and Comstar (USA), Anik (Canada), Stw and Chinasat (China), Palapa (Indonesia), Telecom (France), N-Star,(Japan)
	6.725-7.025 (300 MHz)	4.5-4.8 (300 MHz)	National satellites (FSS Plan, RR Appendix S30B)
8/7 (X-Band)	7.925-8.425 (500 MHz)	7.25-7.75 (500 MHz)	Governmental and military satellites
13/11 (Ku-Band)	12.75-13.25 (500 MHz)	10.7-10.95 11.2-11.45 (500 MHz)	National satellites (FSS Plan, RR Appendix S30B)
13-14/11-12 (Ku-Band)	13.75-14.5 (750 MHz)	10.95-11.2 11.45-11.7 12.5-12.75 (1 000 MHz)	International and domestic satellites in Region 1 and 3. Intelsat, Eutelsat, Russia (Loutch), Eutelsat Telecom 2 (France), DFS Kopernikus (Germany), Hispasat-1 (Spain)
		10.95-11.2 11.45-11.7 12.5-12.75 (750 MHz)	International and domestic satellites in Region 2. Intelsat, Anik B and C (Canada), G-Star (United States), Hispasat-1 (Spain)
18/12	17.3-18.1 (800 MHz)	BSS bands	Feeder link for BSS Plan
30/20 (Ka-Band)	27.5-30.0 (2 500 MHz)	17.7-20.2 (2 500 MHz)	International and national satellites: various projects under study (Europe, United States, Japan), N-Star (Japan), Italsat (Italy)
40/20 (Ka-Band)	42.5-45.5 (3 000 MHz)	18.2-21.2 (3 000 MHz)	Governmental and military satellites

1.4.3   Different types of system

An overview of satellite systems currently in operation is given in Annex 3 to this Handbook. Telecommunication satellites were first used to set up links over very long distances, as the financial penalty of using satellites for short distances when compared to conventional methods was too large. The first links were therefore intercontinental.

Space technology has since made considerable advances which have resulted in a substantial cost reduction. At the same time, new financing facilities have become available to potential operators, who are now able to lease all or part of the capacity of a satellite, rather than purchasing it, with a consequent reduction in financial risk. Moreover, it is now possible to take out insurance for launching, which amounts to sharing the financial risks among several users. All these factors have promoted the development of regional and even national applications. Many countries with large territories which had no telecommunication networks, have been able to rapidly acquire complete networks thanks to satellites. Countries which already had networks have taken advantage of the favorable characteristics of satellite system, to supplement them and develop new services.

From the point of view of coverage the FSS systems may be divided in two broad categories:

1.4.3.1   Global coverage systems

Global coverage systems are satellite communications systems designed to operate all over the world. They primarily carry international traffic, although they can also be used to provide regional and domestic services. The main global coverage system operator is INTELSAT, an international organization.

Apart from INTELSAT, other global coverage systems currently in operation are INTERSPUTNIK, also an international organization and PANAMSAT and ORION, which are private companies.

Since 1965, satellites have grown in complexity and performance. Most earth stations are fitted with relatively large diameter antennas (i.e. more than 12 m in the 6/4 GHz band) and with sophisticated equipment (automatic tracking devices, very low-noise receivers, high-power transmitters). Although simpler earth stations with smaller diameter antennas (3.5 to 12 m) are acceptable for low and medium traffic capacities, and are generally necessary for end-user premises installations, the number of large antennas should continue to expand to meet world traffic demand and television quality standards. Advances in satellite technology have, and will continue to drive the size of earth stations downward while still continuing to operate in a bandwidth limited mode in the satellites and thus making the most efficient use of the space segment. It should also be noted, that INTELSAT's space segment charges per circuit have been declining steadily as the number of system users constantly rises.

Global systems are also used to provide regional and domestic (national) systems as explained below.

1.4.3.2   Regional and national systems

A regional communication satellite system provides international communication between a group of countries which are in geographical proximity or which constitute an administrative or cultural community. Examples of regional satellite systems include the ARABSAT (League of Arab States), EUTELSAT (European Telecommunications Satellite Organization), TELEX-X (Sweden, Norway and Finland), PALAPA (Association of Southeast Asian Nations) and HISPASAT-1 (Spain, Europe and America's Spanish speaking areas). An overview of some existing regional systems is included in Annex 3.

A national communication satellite system is one which provides telecommunications within a single country. This type of system is used by a large number of countries where satellite systems are economically competitive with terrestrial systems.

In fact; satellite systems are especially well suited for the provision of telecommunication services over large or dispersed service areas. Where countries cover vast territories and have formidable natural obstacles (dense forest, mountain ranges, large stretches of desert, archipelagos), a scattered population and a rudimentary infrastructure, satellites can provide a means to rapidly establish a telecommunication network which meets the following specific requirements:

•   high-quality and low-cost links, especially with rural areas;

•   extension of television services to all communities within the country.

Systems of this type are often set up by leasing or purchasing one or more available transponders in an existing satellite. However when the traffic increases, the implementation of a dedicated satellite system may prove profitable. In fact, there is, at present, a lot of such national systems in operation (see Annex 3). These systems often include a large number of earth stations (between several tens and several thousand), so that the earth segment share in the total cost of the system predominates. The aim is therefore to simplify earth stations as far as possible, in order to reduce their purchase and operating costs. The trend is, and will increasingly be, to use simplified local stations which can be mass-produced at low cost (especially for low-traffic telephone services and for television programme reception). These stations have small diameter (e.g. 3 to 5 m), fixed pointing antenna, low-power transmitters (several watts) and compact equipment for telecommunication and onward connection to the terrestrial network. The construction of these stations must be particularly rugged and reliable, their operation should require no permanent local staff and their design, similar to that of a radio-relay system, should take local conditions into account as far as possible (general environment, availability of primary energy sources, etc.). In this field, very-low cost earth stations for rural telephony are becoming available.

The use of similarly designed stations is also increasing for new telecommunication services, *in particular* for business and administrations (telephone, data, videoconference, etc., see § 1.4.5.3 below.). These stations may be installed directly on the user's premises or in the immediate vicinity of a group of users, in order to reduce as far as possible the length of terrestrial feeder lines. In this framework, very popular services, particularly in the 14/12 GHz band, makes use of "microstations" or "very small aperture terminals" referred to as VSATs[2].

---

[2] For more details see "VSAT Systems and earth stations". Supplement 3 to the 2nd Edition of the Handbook on satellite communications. ITU Radiocommunication Bureau.

It should be noted, however, that such simple earth stations can be introduced only into networks using relatively high-power satellite transponders.

1.4.4 Basic characteristics of satellite telecommunications

This Handbook is devoted primarily to the fixed-satellite service. However there is a clear trend toward convergence between the technologies used, and even the applications served, in the FSS and in the MSS and BSS services (the feeder links for the MSS and BSS are definitely included in the FSS services). In consequence, this Handbook includes some *issues* related to the MSS and the BSS.

The most specific characteristics of satellite communications are described below.

1.4.4.1 Coverage

Satellite links allow for communication between any points on the Earth's surface, without any intermediate infrastructure and under conditions (technical, cost, etc.) which are independent of the geographical distance between these points, provided they are located within the satellite coverage area[3].

In the case of a GSO satellite, the points to be served must be situated, not only in the region of the Earth, visible from the satellite[4], but also within the geographical areas covered by the beams of the satellite antennas: these areas are called the coverage areas of the communication satellite system[5] (see Fig. 1.4). The satellite antenna beams can be "shaped" to form specific coverage areas tailored to the region to be served.

Since the satellite is located at a great distance from the Earth (35 786 km vertically above the so-called "sub-satellite" point, the very high free space propagation loss (e.g. about 200 dB at 6 GHz) should be offset at the earth stations by:

• high-gain (i.e. large diameter, high performance) antenna with low susceptibility to noise and interference (the antenna is used for both reception and transmission);

• high-sensitivity receiver (i.e. with a very low internal noise);

• powerful transmitter.

---

[3] In future, links of the inter-satellite service may permit the connection of points which are not covered by the same satellite.

[4] Round an axis of revolution passing through the centre of the Earth and the satellite, this area falls within a cone with the satellite at its apex and apex angle of 17.3° which also corresponds to a cone with the centre of the Earth at its apex and an apex angle of 152.7°. These conditions describe an area from which the satellite is visible from the earth stations at an angle of at least 5° above the horizontal direction.

[5] It should be noted that the satellite is not necessarily located vertically above points within these coverage areas: it has been shown that it is often convenient to cover a limited area of the Earth (in the case of regional or national telecommunication systems) by an oblique beam from the satellite.

On the other hand, the requirement to be met by the earth station depends directly on the performance of the satellite transponder. In particular, the smaller the area to be served (and hence the land coverage area the greater can be the directivity of the beam of the antenna connected to the transponder and the effective radiated power[6] of the satellite which results in lower earth-station performance requirements (with consequently lower station cost). It is therefore desirable that the coverage area should be only just large enough for the region to be served.

1.4.4.2   Multiple access

An outstanding operational feature used in FSS telecommunication is multiple access. Multiple access is the ability for several earth stations to transmit their respective carriers simultaneously to the same satellite transponder.

This feature allows any earth station located in the corresponding coverage area to receive carriers originated by several earth stations through a single satellite transponder.

Conversely, a carrier transmitted by one station into a given transponder can be receive by any earth station located in the corresponding coverage area. This enables a transmitting earth station to group several carriers into a single-destination carrier.

The most commonly used types of multiple access are **frequency division multiple access (FDMA)** and **time division multiple access (TDMA)**.

In FDMA, each earth station is allocated a specific frequency (with the necessary bandwidth) for the emission of a carrier forming part of a multiple-destination multiplex. Each of the corresponding stations must receive this carrier and extract the channels intended for it from the baseband.

FDMA is often associated with frequency division multiplex (FDM).

However, FDMA can also be combined with other types of multiplexing and modulation, especially time division multiplexing (TDM) with digital modulation, which generally uses phase shift keying (PSK). The INTELSAT business service (IBS), the EUTELSAT business service (SMS) and INTELSAT intermediate data rate (IDR) carriers are examples of this type of modulation.

For links involving sporadic, low-density traffic, FDMA can be used without multiplexing, i.e. each carrier is modulated by only one telephone channel. This technique is known as single channel per carrier (SCPC). The modulation employed is generally either analogue FM or digital phase shift keying.

---

[6] This power is called e.i.r.p. (equivalent isotropically radiated power).

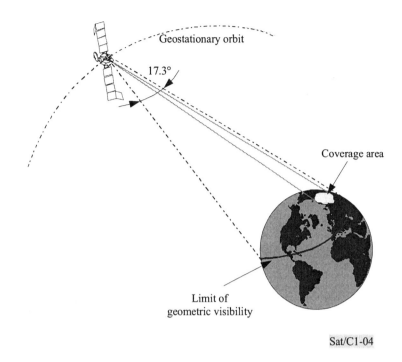

FIGURE 1.4

Coverage by a geostationary satellite

Another very important type of multiple access is time division multiple access (TDMA). The INTELSAT system includes several TDMA networks operating at a transmission rate of 120 Mbit/s. The EUTELSAT (Regional) and Telecom I (France) satellite systems were designed to use TDMA.

In TDMA, each station is periodically allocated, on the same carrier and within a "frame", a period of time (a "burst"), during which it emits a digital signal forming part of a multiple-destination multiplex. Each corresponding station receives this burst and extracts from it the digital channels intended for it. Multiplexing associated with TDMA is effected by time division (TDM), the telephone (or data transmission) channels being themselves coded digitally (e.g. PCM). The carrier is generally phase shift keyed by the digital signal.

TDMA was the first major digital modulation system implemented in the INTELSAT system, where space segment capacity is allocated on the basis of sequential time shared use of the entire transponder bandwidth. Each INTELSAT TDMA network comprises four reference stations per network including the back-up stations, a number of traffic terminals and one satellite. The reference stations provide network timing, and control the operation of traffic terminals. The traffic terminals operate under the control of a reference station, and transmit and receive bursts containing traffic and system management information. The reference stations and the traffic terminals are interconnected by the satellite.

There are other types of multiple access, such as spread spectrum multiple access (SSMA), and more particularly, code division multiple access (CDMA).

In CDMA, the transmitted signals are not discriminated by their frequency assignment (as in FDMA), nor by their time slot assignment (as in TDMA), but by a characteristic code which is superposed on the information signal. At present, CDMA is generally reserved for specific applications, e.g. for some personal communications systems.

Multiple access can also result from various combinations of FDMA and/or TDMA and/or CDMA and can be performed or changed in the satellite by on-board processing (OBP).

In any case, multiple access processes can also be classified into two categories, referring to their assignment mode:

- pre-assigned multiple access (PAMA), in which the various channels are permanently allocated to the users;

- demand assigned multiple access (DAMA), in which a transmission channel is assigned only for the period of a call (telephone call, data packet, etc.). The great majority of satellite telecommunication system use DAMA, but in the case of sporadic traffic varying in time, the concentration properties of the DAMA process considerably enhance the efficiency of the communication satellite system.

Questions relating to multiple access are discussed in detail in Chapter 5 of this Handbook.

1.4.4.3   Distribution

Multiple access, as described above, generally relates to two-way links and especially to duplex telephone links. When the links are one-way and more precisely, when the information signal is emitted by a specific earth station towards stations which are often assigned for reception only (generally numerous and scattered throughout the coverage area), the distribution (broadcasting) capability of the satellite is used. This capability is particularly useful for television (transmission of television programmes) and for certain data transmission services (e.g. data banks). In some cases, it may be useful for the *receiving* stations to have a capability to transmit service channels, reserve channels or selection request channels, etc.

1.4.4.4   Frequency reuse and bandwidth utilization

A telecommunication satellite constitutes a very wide band high traffic capacity relay: the primary energy sources on the satellite can supply a large number of transponders [e.g. 50 transponders with a power of 5 to 10 W on the INTELSAT-V satellite]. These transponders share the total effective bandwidth: in the frequency bands now most widely used (6/4 GHz and 14/11 GHz), the available bandwidth is 500 MHz and the bandwidth of each repeater is usually about 40 or 80 MHz. It has become current practice to reuse the available bandwidth several times, thus considerably increasing the total effective bandwidth. This frequency reuse can be effected by two procedures, which are mutually compatible:

- frequency reuse by beam separation: the same frequency bands are transmitted by the satellite antennas using different transponders by means of directional and space-separated radiated beams;

- frequency reuse by polarization discrimination (also known as dual polarization frequency reuse): the same frequency bands are transmitted by the satellite antennas through different transponders using two orthogonal polarizations of the radio-frequency wave;

- Figure 1.5 shows an example of the way in which the 500 MHz frequency band (at 6/4 GHz) is reused four times in the INTELSAT-V systems. Thus the total effective bandwidth of the INTELSAT-V satellite, 2 590 MHz, is distributed as follows:

  - 6/4 GHz band: 375 MHz bandwidth reused four times and 125 MHz is used twice (for global beam);

  - 14/11 GHz band: 420 MHz bandwidth reused twice (by beam separation).

In the future, the total effective bandwidths available *could be* further increased:

- by increased reuse of frequencies;

- by using the expanded bands allocated by WARC-79; or

- by using higher frequencies, in particular the 30/20 GHz bands (providing an available bandwidth of 3.5 GHz).

For example, the INTELSAT-VI satellite reuses parts of the 6/4 GHz (WARC-79) bands six times, while the Japanese satellite CS-2 uses 30/20 GHz bands.

The high traffic capacity of a communication-satellite system can be used in a flexible manner, ranging from very high capacity (e.g. 960 or even more telephone channels)[7] point-to-point or point-to-multipoint services, to services with low-density and very sporadic traffic. There are only a few specific constraints, connected with the hierarchy of terrestrial transmission networks (for both analogue and digital multiplexers).

Some systems have been specifically designed for incorporating transponders corresponding to different communications services ("multi-mission payload"). This is the case, for example, for the HISPASAT 1A and 1B satellites, which includes the following payload: 18 FSS transponders for telecommunication services over Spain, Europe and North, Central and South America (14/11-12 GHz, double linear polarization with frequency reuse), five BSS transponders for TV distribution and broadcasting (17/12 GHz, circular polarization) and two transponders in the 8/7 GHz band for governmental services.

---

[7] A circuit is a two-way voice link involving telephone connection between two users. A satellite telephone channel is the part of the satellite transponder bandwidth and power utilized for a one-way telephone link. Therefore two satellite telephone channels are needed for transmission of one circuit. A satellite telephone channel is often called a half circuit.

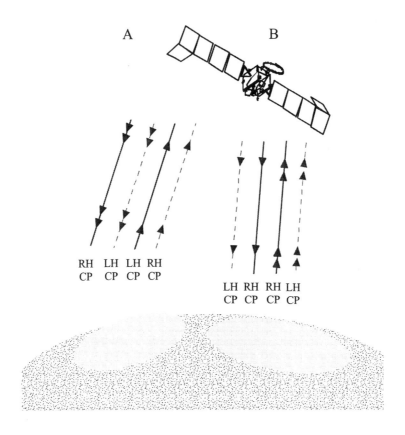

Between lines with one arrow and lines with two arrows:
Double frequency reuse by beam separation. The same frequency bands are used, the first time for transmission from land area A and reception in land area B (lines with one arrow) and the second time for transmission from land area B and reception in land area A (lines with two arrows).

Between solid and broken lines:
Double frequency reuse by dual polarization. The same frequency bands are used, the first time for transmission by left-hand circular polarization (LHCP) and reception by right-hand circular polarization (RHCP) (solid lines) and the second time for transmission by right-hand circular polarization (RHCP) and reception by left-hand circular polarization (LHCP) (broken lines).

                                                                                                              Sat/C1-05

FIGURE 1.5

Diagram showing quadruple frequency reuse in the INTELSAT-V system

1.4.4.5   Propagation delay

An important feature of satellite links is the propagation delay. In the case of GSO systems, owing to the distance of the geostationary satellite from the Earth, the propagation time between two stations via the satellite can reach approximately 275 ms.

During telephone calls, the round-trip propagation time is approximately 550 ms and the use of echo control devices, such as echo suppressor or the more efficient echo cancellors, is essential to avoid unacceptable deterioration of the subjective transmission quality. Moreover, since the present ITU-T recommended limit for one-way delay is 400 ms, double-hop connections, i.e. those using two satellite links, should not be used except under the most exceptional circumstances (ITU-T Recommendation G.114). However, recent tests on double-hop circuits have shown a marked improvement in the subjective performance when the circuits are equipped with echo cancellers equipment (as opposed to previously used echo suppressors).

Note that double-hop connections are especially liable to occur in the international telecommunications of a country whose national traffic is itself routed (at least in part) via a satellite system[8].

ITU-T Recommendation E.171 – The International Telephone Routing Plan – establishes the routing principles with respect to the use of satellites in international connections. This recommendation states that it is desirable in any connection to limit the number of international circuits (satellite or terrestrial) in tandem for reasons of transmission quality. However, in practice, the large majority of international telephone traffic is routed on direct circuits, i.e. a single international circuit between international switching centres.

NOTES

- The one-way delay figure of 400 ms is currently under active review in ITU-T. A figure of 500 ms is being given consideration for telephony as a result of recent subjective tests in the United States of America. Some telecommunication services are not affected by propagation delay, even in the case of multiple hops. Examples of these are television or broadcasting services and certain types of data services.

- Inter-satellite links should contribute to avoiding double hops, but this is not a current practice in GSO systems.

- Propagation delay may give rise to problems and even incompatibilities in the choice of signalling methods, especially where compelled signalling methods are concerned. ITU-T Recommendation Q.7 should be referred to in this connection. Similarly, satellite propagation delay in all media creates constraints in signalling and routing procedures for certain data transmission networks, especially in packet-switched transmission systems, if correct compensation measures are not adhered to.

- Of course, delays are greatly reduced in the new non-GSO systems, in particular those implementing low-Earth orbits (LEOs). For example, the propagation time between two earth stations via satellite orbiting at 1 000 km is generally less than 10 ms.

1.4.4.6   Flexibility and availability

Satellite telecommunications have other particularly interesting operational characteristics:

---

[8] They may even be triple hops, if two countries in international communication each route their national traffic via a satellite.

- round-the-clock availability for 365 days a year with service continuity generally exceeding 99.99%;

- rapid installation and bringing into service of earth stations, irrespective of the distance and accessibility of the area to be served;

- great flexibility for changes of services and traffic plans and for all changes in the Earth segment (introduction of new stations, increased traffic capacity, etc.).

To begin with, the implementation of the ground segment of a satellite network is relatively simple because the number of physical installations is minimal. To install a satellite network, a planner need only consider the sites where service is required. In comparison, the installation of an optical fibre cable system or of microwave links requires first that the right-of-way be secured from organizations such as governments, utility companies, and railroads. Hundreds or even thousands of sites must be provided with shelter and power (and even access roads in the case of terrestrial microwave). After the entire system is installed and tested, all the equipment must be maintained to assure continuous service. Even then, one outage along the route could put the chain out of service until a crew and equipment can arrive on the scene to effect repairs.

Finally, the time to implement satellite networks and add stations has been reduced from one to two years to one to two months. In contrast, a terrestrial fibre network is like a major highway project and can take years to design and construct.

1.4.5   Range of services provided

Almost anything that can be classified as telecommunications can be carried by satellite communications. A possible classification of satellite telecommunications services, independent of international, regional or national (domestic) designations, is as follows:

- telephony/facsimile (fax), etc.;

- television, video and audio;

- data transmission and business services;

- integrated services digital network (ISDN);

- emergency communications;

- cable restoration services.

Communications satellites have steadily supplemented older means of communication to the point that they now represent a mature technique. At the same time satellites have, due in large part to their unique characteristics, opened many opportunities for services which were previously economically unattractive, very difficult to implement or near the limits of technology. Satellites can be tailored to meet specific needs (e.g. voice, data, television, or combinations thereof) using either analogue or digital techniques over either wide or relatively small geographic areas. Digital techniques offer the opportunity for sophisticated signal processing and efficiently intermixing various services or providing completely new services e.g. business services.

1.4.5.1 Telephony, fax, etc., services

i) Service implementation

Telephony, fax and various other low bit-rate data transmission services, which were previously based on analogue transmission, are nowadays systematically implementing digital technologies. The use of time division multiplexed digital carriers, especially when combined with such techniques as adaptive differential pulse code modulation (ADPCM), low bit rate encoding (LRE) and digital speech interpolation (DSI) (e.g. with digital circuit multiplication equipment (DCME)) can provide increased traffic capacity, i.e. larger numbers of channels on such carriers.

The use of broadband integrated service digital networks (B-ISDN) including asynchronous transfer mode (ATM) and synchronous digital hierarchy (SDH) concepts will even increase the flexibility and the capacity of the satellite links and will allow mixing of telephony and fax transmission with other services (data, video, etc., see § 1.4.5.5 below.).

In the international area, global systems such as INTELSAT operate as international common carriers of traffic between major land masses and/or individual countries. This provides interconnection between various domestic terrestrial networks and allows orderly, systematic flow of normal telephone traffic across national boundaries.

Direct routing, via routing and diversity:

The "hierarchical" World Routing Plan, as originally conceived by the International Telecommunication Union (ITU), has been completely revised since the introduction of satellite systems. International satellite systems enable the "direct" interconnection of various countries' international telephone switching centres instead of making use of a structured network hierarchy (designed for wholly terrestrial systems) to step through the international network from origin to destination. This "short circuiting" effect of satellite systems also reduced the need for "via routing"[9] to a minimum, as each country obtained earth stations enabling a direct connection to correspondents a third of the world away.

Global planners have always considered terrestrial systems (including submarine cables) and satellite systems to be complementary and have established a large number of terrestrial/satellite paths as a means of establishing direct access circuits. In a few instances where this seemed impracticable, double-hop satellite circuits have been established.

Another feature of satellite systems is the ability to offer diversity routing for the purpose of service protection. Diversity is possible by access to multiple earth stations in different countries from a single point (provided via routing capability through a third country in the event of an earth station or local national routing problem) and through second or third satellites in the same coverage area. The INTELSAT network frequently offers diversity of two or more paths either by the use of satellites operating in the same region, or by the use of satellites operating in the two regions with

---

[9] The term "via routing" refers to the use of an intermediate or transit country or switch by an originating country to reach the destination country.

common coverage areas. This system diversity capability is also used for the restoration of traffic routed from other media that are periodically out of service, e.g. submarine cable systems.

Regional telephone services can be provided by an international satellite system as well as by a dedicated regional satellite system. The service provided is similar to the international service in the fact that communication links between PTTs and/or common international carriers entities may exist, however, a direct tie into an international network does not normally exist.

Domestic telephone services can be provided by an international satellite system, by regional satellite systems or by a dedicated domestic satellite system. Domestic telephone services can extend from multichannel links between major communication centres that interface with the terrestrial networks, private networks between industrial or government groups, to thin-route and rural telephony links between isolated locations. The nature of the service can range from conventional telephone services provided in regions where a terrestrial telecommunication infrastructure does not exist, to specialized services that supplement or parallel an extensive terrestrial network and satisfy the unique requirements of the business, government or private sectors.

ii)   Fax, data, etc.

The general conditions for implementing facsimile (fax), telegraphy, telex and other low (or medium) bit rate data transmission services in the framework of international, regional and domestic satellite communications systems are very similar to those given above for telephony.

Fax is at present the largest non-voice service operating on the world's public switched telephone network (PSTN). Facsimile is a high resolution raster scanning image transfer system that uses PSTN modem technology for its communications. Groups 1 and 2 (for analogue in PSTN); Group 3 (digital in PSTN) and Group 4 (digital in ISDN) were standardized by ITU-T for facsimile terminal equipment.

Up to now, due to the relatively low extent of the ISDN, the major market for fax machines remains with Group 3 equipment.

Another service which is currently stimulating much interest is the Internet service. Like fax service, Internet uses PSTN modem technology for its communications. Internet is a worldwide network connecting personal computers (PCs). At present several millions of PCs are connected. It is based on TCP/IP protocol and includes services like: e-mail, news transfer, remote terminal, file transfer, etc.

1.4.5.2   Television/video and audio services

i)   TV, video and audio distribution

The distribution of television is a major service provided by the FSS and BSS. This ranges from conventional television programming (entertainment, news, special events), educational/instructional programmes to teleconferencing applications. The availability of satellite links for TV distribution eliminates the distance/cost relationship associated with terrestrial delivery of signals. In some instances satellites provide the only available, or cost effective means of signal distribution. The applications include the processing of satellite news gathering (SNG, see below), the distribution of TV programming or video information between locations for re-broadcasting over terrestrial stations or for redistribution through cable networks (CATV), etc.

In addition, slow scan or broadband TV signals can be used in support of teleconferencing or educational activities, where audio and visual information is required to enhance effective communication. Educational applications can range from traditional classroom situations, health care, and agricultural training, to specialized instructional information for the handicapped, etc.

In a similar way, the FSS and BSS can also provide distribution of radio programmes (e.g. high fidelity audio and stereo).

It is also possible to implement broadcasting services direct to the general public for individual reception (DTH applications) through very small receiving antennas (television receiving only TVRO), and for community reception (professional application, CATV, SMATV, direct domestic applications DTH). In this area, digital TV transmission offers new possibilities, as explained below.

ii)   Digital television broadcasting

Advanced digital techniques for picture and sound coding, channel coding and modulation, have proven their efficiency and reliability. Multimedia applications and communications using digital video technology are already becoming a reality. The availability of digital techniques, at low cost, are therefore the key for the present introduction of satellite multi-programme television services ("bouquets").

During 1990 some experimental projects showed that the digital video compression system known as "motion compensated hybrid discrete cosine transform coding" was highly effective in reducing the transmission capacity required for digital television. Before that, digital television broadcasting was thought to be impractical to implement.

Digital television is now a reality and is opening broad possibilities to offer new services to the users in addition to conventional TV services.

Most of the systems use ISO/IEC MPEG-2 for audio and image coding. MPEG-2 video is a family of systems, each having an arranged degree of commonality and compatibility. It allows four source formats, or levels, to be coded, ranging from Limited Definition (about today's VCR quality) to full HDTV each with a range of bit rates. In addition to this flexibility in source formats, the MPEG-2 system allows different profiles. Each Profile offers a collection of compression tools that together make up the coding system. A different Profile means that a different set of compression tools is available.

Communication satellites have and will have an outstanding role in the provision of these new services. There are already systems operating with digital television (DVB, etc.). All of them use a very similar network architecture and source coding (MPEG-2). Some of them are flexible systems which are designed to cope with a range of satellite transponder bandwidths (26 to 72 MHz for the DVB-S system).

Medium and high power satellites operating in the FSS/BSS frequency bands, are the ideal means for the rapid introduction of these new services, allowing maximum exploitation of the transponder capacity. A significant evolution from the analogue to the digital world, can be foreseen by the end of the century, with the progressive integration of the services provided by satellite channels and terrestrial digital networks. The need for harmonization and commonality in source coding and multiplexing techniques, for use on the various delivery media, is the key factor for the evolution towards this scenario.

Without doubt, it can be anticipated that the opportunities for new services, the start of the multimedia era and the increasing demand for broadcasting bouquets will lead to a high demand for satellite capacity in the coming years.

iii)   Satellite news gathering

Satellite news gathering (SNG) allows organizations and broadcasters to pick up events, whenever they happen, and to deliver their picture and sound to the studios for editing and/or broadcasting. SNG systems range widely in capacity, weight and cost depending on the characteristics of the service they must provide. Immediacy, availability of infrastructure and quality are elements to be considered when determining which kind of SNG system to be used.

Broadly speaking they can be classified in two groups:

- SNG Trucks and Trailers: They are transportable earth stations mounted on a truck, trailer or van. They are able to deploy antennas in the range of 3 m (C band) and 2 m (Ku band);

- SNG Fly Away: These are free standing transportable stations which can be shipped by air on regular flights. The antennas are usually foldable for easy transport and range from 1.2 m to 1.8 m (Ku band).

The new digital video compression techniques will provide further advantages for digital SNG services. With 8 Mbit/s data rate for digital video coding it is possible to use 0.75 m to 1.2 m SNG terminals.

SNG earth stations are described in some detail in Chapter 7 (§ 7.9).

1.4.5.3   Data transmissions and business services

Commencing in the 1970s, significant advances in technology have resulted in computers that are faster, simpler and less expensive. As a result, they are now capable of providing real-time access to production and economic data that are important for decision-making. The availability of new digital technology, and the interest in improving productivity have resulted in an increasing demand from industrial, financial and administrative organizations for a number of new services that were impractical or excessively costly. The geographical distribution of computer facilities within an organization or administration, that need to "talk together", can be overcome through the use of high-speed satellite links.

Also, in the framework of the recent explosive development of the Internet, satellite communications could offer to this new medium the advantages of direct user-to-user broadband links.

In consequence, it is expected that data transmission will offer new market opportunities to satellite communications.

The propagation delay characteristic of satellites could pose a problem to early protocols. Some of the protocol error controls require a distant end response to each transmitted event, which, with a satellite hop in each direction, results in a total delay of about 500 ms. On terrestrial circuits, the same round trip takes tens of milliseconds.

Although the propagation delay can easily be compensated for, and protocols meeting the newest international standards are satellite-friendly, throughput performance may inadvertently be impaired by the use of a wrong protocol. The future trends with non-GSO satellite networks should improve this situation by reducing this delay.

The advancement and convergence of the technologies of telecommunications and data processing is often referred to as "telematic services" and includes:

- sharing of computational capacity between facilities (e.g. one or several centralized and/or distributed computers);

- real-time backup and recovery/continuation of processing during periods of failure in processing systems;

- establishment of high speed communications links for providing rapid exchange and switching of information between major modes in data processing networks (with particular application to data communications in the packet switching mode);

- simultaneous updating of a number of distributed processing centres and outlying locations;

- data bank transfer;

- information broadcasting and data distribution (with or without return channels for conversational/interactive processes);

- facsimile transmission extending from simple low bit-rate facsimile to high-definition transmission of printed matter, manuscripts, photographs, drawings, etc. for such applications as Internet, electronic mail services, remote printing, etc.;

- teleconferencing facilities extending from the most simple audio conferences with visual aids to more sophisticated image transfer aided conferences and up to full videoconferences between multiple users. It should be noted that teleconferencing facilities may include transportable earth stations.

The large scale development of satellite data communications, and, more generally, telematics, has been allowed for in the planning of most satellite systems. National or domestic satellite systems in the United States were the first to provide these business services in the mid to late 1970s. Business services have become international in scope with the advent of the INTELSAT business service (IBS). Additionally, the EUTELSAT business service (SMS) and TELECOM 2, HISPASAT (DVI), etc. for European users, have been designed with characteristics fully compatible with the INTELSAT IBS, so as to enable interworking between systems (see Annex 3 for more details).

Also, as explained below, private dedicated networks are currently major actors in the field of satellite data communications.

Private networks and VSATs

VSAT (very small aperture terminal) is a well recognized and popular term for designating very small, low cost earth stations which are directly connected to the users.

Most important applications of VSAT systems are in the form of private, closed user-group communication networks in which the remote VSATs are directly installed in the premises of each remote user. Most VSAT networks operate with a central, larger, earth station ("the hub"), which distributes, controls and/or exchanges the information towards or between the remote VSATs.

The range of VSAT applications is continuously expanding. Some examples are listed below:

•   One-way applications; Data distribution: Distribution of information, in the form of digital signals, from the Hub to all subscribers (data broadcasting), or to a limited number or subscribers in the network (data narrowcasting), e.g. for news, press releases, weather information bulletins from meteorological agencies to airports, announcement displays, computer program remote loading, audio distribution, video distribution, etc.

•   One-way applications; Data collection: This architecture is used in the reverse direction, i.e. from VSATs towards the Hub, for data collection purposes. Some applications are for meteorological or environmental monitoring, pipeline, electric power network monitoring from unattended stations, etc. However, in most cases, such an operation needs some form of control and management from the central facility, which means that the VSAT should implement a receive function and not really be considered one-ways.

•   Two-way applications: VSAT networks are nowadays commonly used for diverse types of two-way data transmission, in particular for all types of interactive or inquiry/response data interchange and file transfer. Some examples of applications are: financial, banking, insurance (for file transactions and bulk transfers from local agencies toward the central processing facility), point-of-sale operations, credit card verification, management and technical assistance, reservation operations for airlines, travel operators and hotels, electronic mail services, etc.

More details on VSAT technology, applications and services will be found in: "VSAT Systems and earth stations", Supplement No. 3 to the Handbook (ITU, 1995).

1.4.5.4   Integrated services digital network (ISDN) services

The concept of an integrated services digital network (ISDN) was introduced by ITU-T in the 1970s on the basis that "... ISDN might be the ideal worldwide communication network for the future". ISDN will upgrade the existing telecommunications network and provide fully digital, end-to-end (up to the subscriber) services. It will also allow the introduction of a standard, out-of-band signalling, so that voice and data services can be integrated on the same network.

ISDN is being planned and implemented internationally to be a single global switched network providing all services through standard customer interfaces. Narrow-band ISDN (N-ISDN) is based on synchronous transfer mode (STM) and on a standardized 64 kbit/s access for all services with data rates from 64 Kbit/s to 2 Mbit/s.

The main feature of the N-ISDN concept is the support of a wide range of voice and non-voice applications in the same digital network. A key element of service integration for an N-ISDN is the provision of a range of services using a limited set of connection types and multipurpose user-network interface arrangements. N-ISDNs support a variety of applications including both switched and non-switched connections. Switched connections in an N-ISDN include both circuit switched and packet-switched connections.

Although some obstacles had to be overcome for ensuring compatibility and interworking of satellite links with ISDN procedures, protocols and ITU-T Recommendations (specially as concerns bit error ratio and delay), satellites have proven their capability for facilitating the rapid implementation of ISDN services on an international scale. For example, INTELSAT satellite facilities provide ISDN services in three ways: time division multiple access services (TDMA); intermediate data rate (IDR) carriers; and super IBS. All of these satellite services are capable of providing better than 1×10^{-7} error ratio capabilities and system availability in excess of 99.96% of the year.

Broadband ISDN (B-ISDN) was promoted by ITU-T from 1988 with the approval of the first ITU-T Recommendation I.121 with the objective to integrate all voice, data and video services.

It is based on:

- the synchronous digital hierarchy (SDH), which replaces the previous plesiochronous digital hierarchy (PDH), with more flexibility and much higher bit rates[10];

- the asynchronous transfer mode (ATM) as a transport mechanism[11].

  Of the many elements in the ISDN which are already standardized by ITU-T, satellite systems are primarily affected by the bit error ratio (BER) and availability performance which it is necessary to achieve to meet its share of the overall BER requirements to the user. However, thanks to the recommendations prepared by ITU-R, a transmission quality comparable to fibre optics can be obtained on satellite links (see Chapter 2, § 2.2) whenever necessary to allow their utilization in B-ISDN networks.

1.4.5.5   Emergency communication services

During times of natural disaster, civil disturbance or serious accidents, normal terrestrial-based communication facilities are frequently overloaded, temporarily disrupted or destroyed. The availability of satellite communication facilities ensures that one element of the system remains isolated from terrestrial-based disruptions, i.e. the satellite or space segment. Through the deployment of small transportable earth terminals to the emergency location, communications can be established and the process of restoring the necessary services (communications, aid, food/water distribution, etc.) assisted.

For example, since 1984 INTELSAT has provided, with increasing frequency, short-term communications with remote locations under emergency conditions or facilities for reporting special events.

---

[10] In particular, SDH allows a transparent multiplexing process. For example, a 64 kbit/s channel can be accessed directly from the highest SDH multiplex hierarchy. The SDH transmission multiplex bit rates currently extend from 255 Mbit/s (basic frame or STM1) to 2.4 Gbit/s (STM16). See Chapter 3, § 3.5.3.

[11] ATM is a data transmission technique somewhat similar to packet-switched systems, but with fixed data length packets (called cells). See Chapter 3 (§ 3.5.4).

In these cases very small, readily transportable earth terminals, have been used to provide temporary television (at 14/11-12 GHz band) and/or audio only communications.

In similar conditions, but under less dramatic circumstances, communication in the FSS are ideally suited for establishing temporary links for satellite news gathering (SNG) (see above §1.4.5.2 iii)) and links with locations hosting special events that require unusual communications facilities and a high instantaneous traffic capacity, e.g. international meetings, sporting events, etc.

1.4.5.6   Cable restoration services

Traditional cable restoration service is designed for restoring services via satellite during outages of telecommunication cables. The cable restoration services are provided by INTELSAT on an occasional use as well as on a longer term basis when there is a planned outage.

INTELSAT currently offers three different cable restoration services, each aimed at the restoration of a different type of size of cable:

*   wideband submarine cable restoration, for the restoration of digital optical fibre cables;

*   cable repair service, for the planned restoration of medium capacity (up to 4 000 circuits) cable systems when extensive repair is needed;

*   cable circuit restoration, for the restoration of telecommunication cables on an occasional use basis.

1.5 Regulatory considerations and system planning

1.5.1   Introduction

This section summarizes the various matters that have to be resolved before establishing and operating satellite telecommunications links.

It should be stressed that this section will primarily deal with the various regulations set up by ITU. It will not describe the regulations particular to each country (or group of countries). Of course, any operator wishing to establish a satellite system or network in a given country must comply with the particular regulations and standards in vigour and refer to the regulatory agency of this country (e.g. the FCC in the United States of America) to get the authorizations for launching, implementing and operating its system or network. It should be noted, however, that following the worldwide trend of "deregulation", the previous restrictions to satellite system operation have been considerably lightened or even suppressed in many countries (e.g. in Europe, under the auspices of the CEPT, etc.)[12].

Also, this section will not deal with the standards which are applicable to the equipment and to the modes of operation with which a satellite system should comply. On a global basis, these standards take the form of ITU (-R and -T) Recommendations (see Annex 4 to this Handbook), but others are

---

[12] Particular regulatory and also standardization problems relative to VSATs are dealt with in "VSAT Systems and Earth Stations" (Supplement 3 to the Handbook), Chapter 5.

established by standardization agencies in various countries, or group of countries (e.g. by the FCC and other organizations in the United States of America, by ETSI in Europe, etc.) and often also by various technical or industrial groups (e.g. MPEG for digital television). Finally, other standards have also been established by the satellite operators themselves, e.g. by INTELSAT and EUTELSAT.

The different cases to be considered here are:

- international or national links;

- utilization of an existing operational satellite system (space segment) or establishment of a new dedicated satellite system.

Subsection 1.5.2 deals with the general problems, mainly of a regulatory nature, to be dealt with in each of these cases.

Setting up a new satellite system, which may be either a regional system with the participation of a group of countries, or a purely national (domestic) system, is obviously much more difficult than using an existing system: in fact, the decision to implement a new satellite system usually results from a long-term process, which may be preceded by the phases outlined below:

- utilization of the space segment of an existing satellite system, usually by leasing space segment capacity;

- preliminary economic and technical studies of the validity and profitability of a new system considering the traffic growth and the possible need for new telecommunication services;

- technical and operational preliminary experiments e.g. by using an existing satellite, if available, or even by launching an experimental or pre-operational satellite.

In this framework, the progressive evolution of the earth segment should be considered. For example, earth stations implemented in the first phase should remain usable as a part of the new system earth segment.

Cases may also occur where the implementation, from the outset, of an entirely new satellite system may be necessary e.g.:

- if existing satellites cannot carry the envisaged services;

- if new technologies are better adapted to the specific requirements, etc.

Subsection 1.5.3 is devoted to the problems of planning new satellite systems.

1.5.2   Global regulations considerations

This subsection deals with various considerations, particularly regulatory ones, that an organization may encounter when bringing into service and operating FSS telecommunication links.

These considerations are of two types:

a) Considerations relating to rules for the internal operation of a satellite system: internal (administrative and technical) system regulations, types of traffic, traffic capacity, technical characteristics of space and earth stations, etc. Other problems relate to internal protection against mutual interference between space stations, earth stations and terrestrial radio stations in the service area of the satellite system concerned.

b)   Considerations relating to external interference protection, i.e. protection of the system against interference from space and earth stations belonging to other satellite systems and from other terrestrial radiocommunication systems and, conversely, protection of these other systems from interference by the system.

The four basic cases of satellite communication links that may arise are given below:

*   existing satellite system[13] using international (including regional) links;

*   existing satellite system using national links;

*   new satellite system[14] using international (including regional) links;

*   new satellite system using national links.

It should be noted that in the case where the technical parameters of an existing satellite system are changed substantially, the existing system would be categorized as a new satellite system.

1.5.2.1   Existing satellite systems

In the case of an existing system, the new user establishes his telecommunication links by installing his own stations in accordance with the internal operating rules of the system concerned. Problems of type b) above will generally already have been solved and the new user will only have to take into account considerations of type a).

1.5.2.1.1   International (including regional) links

An overview of the main international and regional satellite systems now in service (or being brought into service) are summarized in Annex 3.

For each of these systems an international organization, composed of the countries which are signatories of the constituent agreements, is responsible for managing the system or, more frequently, its space segment.

To establish direct telecommunication links within systems such as these, the government of the concerned country has to contact, either directly or through its telecommunication administration or a telecommunication organization under its jurisdiction, the international organization concerned, which will inform it of:

*   the telecommunication services offered and the technical conditions to be fulfilled;

*   the financial conditions and charging methods;

---

[13] Existing satellite system: a system which has already been coordinated according to ITU procedure and which has already been registered by the Radiocommunication Bureau (BR) (as explained in the following sections).

[14] New satellite system: a system which is in a planning stage and/or the coordination of which has not yet been completed according to ITU procedure.

- the regulatory procedures required to establish the earth stations as a part of the established space system. These procedures are required to:

  - avoid unacceptable interference to other users;

  - insert the required links in the general traffic plan and in traffic capacity forecasts.

It should be noted that, in any case, the above-mentioned procedure with the international organization concerned does not replace ITU procedures for registration with the Radiocommunication Bureau (BR) of new telecommunication facilities, as given in § 1.5.2.2.

1.5.2.1.2 National links

There are two possibilities for setting up national links in the framework of existing satellite systems:

a) If the territory where the user wishes to set up links is already served by an autonomous satellite system adapted to these links, the new user shall apply to the administration or the body owning or managing the system. Significant countries operating satellite systems are listed below (see Annex 3):

- Australia/New-Zealand (Optus Communications/AUSSAT);
- Argentina (NAHUELSAT);
- Brazil (SBTS);
- Canada (TELESAT);
- China (STW, CHINASAT 1, ASIASAT);
- France (TELECOM);
- Germany (KOPERNIKUS);
- India (INSAT);
- Indonesia (PALAPA);
- Italy, (Telecom Italia, ITALSAT);
- Japan (N-STAR);
- Korea (KOREASAT);
- Spain (HISPASAT);
- United States of America (SATCOM of RCA, COMSTAR of AT&T, WESTAR of Western Union, SBS, GSTAR of GTE, etc.);
- Russian Federation (Molnya-3, Statsionar, Loutch).

b) On the other hand, if there is no autonomous national system, consideration may be given to leasing (or even buying) part of the space segment (i.e. a certain transmission capacity) of an existing satellite system with a coverage area which includes the territory concerned.

In particular, INTELSAT offers space segment capacity leasing and selling possibilities for a wide range of services, including:

- public switched telephony;
- television and radio broadcasting;

- public switched data networks;

- private business networks.

It should be stressed that the method of leasing a certain transmission capacity in the space segment of an existing satellite system usually makes it possible for the user country, administration or entity to avail itself freely of this capacity (though with some restrictions referred to below) to design, execute and operate a complete national satellite telecommunication system, consisting of the leased part of the space segment and of its own earth segment. It should be noted, however, that:

i)   the organization owning or managing the space segment supplies the technical characteristics of the leased part and rules for its use. These rules ensure the compatibility of the leased part with the whole satellite system and protect the system from interference, overloading, etc.;

ii)   also in the case of national links, ITU-R procedures for registration with the BR of new telecommunication facilities, as given in § 1.5.2.2, have to be followed, after having established the technical characteristics of the national network.

These characteristics and rules must be taken into account by the user in choosing the parameters of the stations in his earth segment and the transmission characteristics of his radio carriers.

The total cost of the system will of course consist of investment costs (provision and installation of stations) and recurrent expenses, e.g. operational costs and the cost of leasing the space segment, Since the last-named costs generally depend directly on the capacity leased, optimization of the total cost is a difficult problem and must be considered along with the technical considerations mentioned above (station parameters, transmission characteristics). In solving this problem of optimization, attention must be given not only to the existing situation of the national network and traffic requirements, but also to their future development[15].

1.5.2.2   New satellite systems

When planning a new system, it is necessary not only to establish rules for internal operation and to examine the internal protection of the system against interference (type a)) problems, § 1.5.2), but to consider from the outset questions relating to mutual interference between existing systems and the planned new system (type b)) considerations, § 1.5.2).

The ITU Radio Regulations lay down procedures and limits with a view to avoiding interference harmful to the efficient operation of all telecommunication services including the FSS, BSS, MSS and ISS. These procedures and limits are dealt with in Chapter 9 of this Handbook, and only a few general points are made here.

---

[15] The following case is cited as an example only:
   A domestic communication network for sporadic low-density traffic can be established by using small local stations communicating among themselves in the FDMA-SCPC mode with preassigned carriers. When the traffic demand grows (owing to an increase in traffic density and/or the number of stations), a transfer from preassignment to demand assignment may be considered. This method might serve to meet increased demand, entailing a certain amount of expenditure for re-equipping stations, but not necessitating any increase in the capacity leased (and hence no increase in the cost of leasing the space segment).

The ITU Member States have established a legal regime which is codified through the Constitution/Convention, including the Radio Regulations (RR). These instruments contain the main principles and lay down the specific regulations governing the following major elements:

- frequency spectrum allocations to different categories of radiocommunication services (Article S9 of the Radio Regulations);

- international recognition of these rights by recording frequency assignments and, as appropriate, orbital positions used or intended to be used in the Master International Frequency Register (Article S11 of the Radio Regulations)

In general terms, the Radio Regulations distinguish between non-GSO and GSO satellite networks, which are subject to different regulatory regimes. Any GSO satellite network in any frequency band has to coordinate its planned use of orbit and frequency spectrum with any other GSO system likely to be affected for which notification was received at an earlier date by the ITU Radiocommunication Bureau (BR). Non-GSO networks are subject to coordination only in respect of certain specific space services in certain frequency bands identified by footnotes to the Table of Frequency Allocations (footnote with a reference to application of RR No. S9.11A).

The main regulatory procedures applicable to space systems are contained in Article S9 and in Appendix S5, of the Radio Regulations, which contain criteria for the identification of administrations with which coordination is to be effected. In line with the normal practice in connection with space services, Article S9 contains a two-step procedure consisting in the publication of simplified advance information on the planned network followed by coordination with systems likely to be affected, observing an order of priority determined by the date of submission of the coordination data to ITU. The same Article envisages the need for coordination of a planned space system (space and earth station) vis-à-vis other non-GSO or GSO systems as well as with terrestrial services sharing the same frequency band. The procedure for notification and recording of space network frequency assignments in the Master International Frequency Register (MIFR) is described in Article S11 of the Radio Regulations.

The Radio Regulations contain also provisions for sharing between space and terrestrial services. Above 1 GHz, Article S21, Sections I-III, aims to protect the GSO space services from the terrestrial emissions of fixed and mobile services, the transmitting stations of which must be pointed 2° away from the GSO and observe power level limits. With respect to the protection of terrestrial receiving stations vis-à-vis space stations, Sections IV and V respectively provide the minimum angle of elevation of earth stations and the limits of pfd from space stations. Article S22 indicates further the rules relating to space radiocommunications services (station keeping, pointing accuracy of antennae on geostationary satellites, earth station off-axis power limitations ...).

1.5.3   Implementation of new systems

This subsection attempts to provide general information, somewhat in the form of a reminder, on the various steps for installing and putting into service a major project for a new complete satellite system. The process should be greatly simplified in the case of a new network of earth stations operating on an existing satellite (e.g. a VSAT or a rural network).

1.5.3.1   Preliminary techno-economic studies

The first step is to assess the service requirements to be met by the satellite system taking into consideration the overall perspective of growth of the telecommunication network and economic, industrial, social and political factors.

Introduction of a satellite system for supplementing and augmenting long-distance telecommunications is usually planned with the following well-known advantages in view:

i)   coverage over a large area;

ii)   costs independent of distance and intervening terrain, ease of connecting locations in mountainous regions, islands, deserts, forests and swamps;

iii) speed of installation and capacity to provide short-term and emergency services using transportable terminals;

iv)   broadcast nature of satellite transmissions and facility in providing television, radio, teleconferencing, news, data and facsimile services;

v)   flexibility to adopt varying traffic demands and types of services;

vi)   capacity to provide media diversity for terrestrial cable and radio links.

A satellite system transmission plan should take into account these advantages and should be compatible with the national switching and transmission plan. For details of switching system characteristics, the ITU-T Handbook entitled "Economic and Technical Aspects of the Choice of Telephone Switching Systems" should be consulted. After assessing the total service requirements, allocation to various transmission media – cable, radio, optical fibre and satellite – would have to be undertaken. For this purpose, the ITU-T GAS 3 Manual on Transmission Systems – "Economic and Technical Aspects of the Choice of Transmission Systems" (Vols. 1 and 2) – should be of assistance.

A cost/benefit analysis may be carried out to compare a satellite network with a terrestrial equivalent. Satellite systems may be economical beyond several hundred kilometres though the cross-over distance could be lower if the advantages cited above are taken into consideration. System costs are assessed and compared on the basis of the present price of the system.

1.5.3.2   Initial planning

a)   Traffic projections

Satellite design life is around 10 to 14 years. Prior to the launch of the satellite a period of at least three to four years is required of preparing spacecraft specifications, developing request-for-proposals (RFP), RFP evaluation, and ordering and construction of the spacecraft. This means that during the initial planning of a dedicated satellite system, traffic projections have to be made for about 10 to 15 years. This means that these projections must include future applications (data, multimedia, etc.).

b)   Preliminary system engineering

From projections of traffic to be carried over a satellite network, a traffic matrix is prepared indicating the number of carriers, channel capacities and inter-connections required. Assessment of the number of satellite transponders required is an iterative process involving tentative assumptions regarding earth-station and satellite parameters to begin with and then optimizing the parameters in relation to overall system costs, channel capacity and performance required. Main parameters for spacecraft are transponder bandwidth, e.i.r.p. and G/T; and for earth stations they are G/T, antenna transmit and receive gain, system noise temperature and transmitted power. For the system, the choice of frequency band, modulation method and multiple access parameters, are important factors.

1.5.3.3 Detailed planning

Following the preliminary studies and development of the satellite system proposals, a planning organization is required to be set up for detailed centralized planning and implementation of the system at the headquarters of the administration concerned.

The planning group is responsible, for obtaining administrative, technical and financial approval for the scheme and for its execution, monitoring and control. Systematic planning is essential for cost-effective implementation of the satellite system. Network analysis, critical path method (CPM) and programme evaluation and review technique (PERT) are methods for carrying out a complex project. In these techniques, the project work is broken down logically into its component parts and these are recorded on a network model or diagram which is then used for planning and controlling interrelated activities needed to complete the project.

Planning of a satellite system may be broadly divided into two areas: i) space segment and ii) earth segment. These two segments may be handled by different organizations in some countries and procedures for coordination between the planning groups of these organizations should be established.

i) Space segment planning

The establishment of the space segment requires a major effort in analysis, planning and execution of work in a highly sophisticated technology area and involves the following steps:
- collection of customer requirements, service and mission requirements;
- preliminary studies regarding various options for spacecraft, spare philosophy, time schedules and system cost;
- orbit and spectrum requirements, selection of orbit slot and follow up of coordination procedures;
- selection of launch vehicle;
- preparation of request-for-proposal (RFP) for satellites and associated control facilities;
- evaluation of RFP and award of contract;
- finalization of launch vehicle agreement;
- establishment of satellite and launch vehicle contract monitoring.

ii) Earth segment planning

Planning of the earth segment would involve the following main steps:

- collection of service requirements and traffic matrix, finalization of transmission plan taking into consideration the specified space segment characteristics;

- preparation of the overall definition for the ground segment;

- finalization of earth-station equipment specification, preparation of request-for-proposals for ground segment, evaluation of RFP and ordering of equipment;

- selection of earth-station sites and electromagnetic compatibility (EMC) survey;

- notification and frequency clearance for selected sites;

- preparation of plans for building and award of contract for building construction;

- ordering of water and electric supplies and power plant.

1.5.3.4   Installation

Construction of spacecraft, supply of earth-station equipment, site selection and construction of approach roads and buildings are long lead time items and these activities are carried out concurrently.

Simultaneously, arrangements should be made for recruitment and training of staff for installation, operation and maintenance. Provision for training by manufacturers may be provided when ordering equipment.

The construction of buildings and installation of equipment could be a challenging task particularly in developing countries – due to lack of infrastructure, resources and skilled manpower. ITU-T GAS 7 Handbook on Rural Telecommunications provides valuable guidance for those undertaking work in rural and remote areas.

The installation group has to carry out the following activities:

- construction of buildings, civil works and provision of infrastructure;

- detailed engineering including drawing-up of layout plan for equipment and definition of the interfaces between the different items of equipment;

- monitor equipment supply orders and order for any shortcoming and installation materials;

- transportation of equipment and materials to site, physical inspection of items supplied and dispatch of reports to the manufacturers, as necessary, for damage and shortage;

- award of contract for installation of antenna and towers and supervision of work;

- installation and acceptance testing of equipment and power plant;

- commissioning of equipment, local line-up;

- assistance in the evaluation of the earth stations by a network operations control centre (NOCC) as per acceptance procedures. Acceptance testing includes antenna radiation patterns, G/T and e.i.r.p. parameters;

- line-up for operational links;

- integration of circuits into the terrestrial telecommunication network, trunk automatic and manual exchanges, etc., and coordination with other organizations for extension of circuits to users.

1.5.3.5 Operation and maintenance

The operation and maintenance group has the following responsibilities:

- establishment of procedures for coordination between agencies concerned with spacecraft operations, earth-station operations and the telecommunication network;

- laying down schedules for manning earth station and definition of duties;

- preparation of line-up procedures, operational procedures and preparation of test schedules for equipment as per manufacturer's recommendation;

- setting up of a network operations and control centre (NOCC) for monitoring and control;

- preparation of daily, weekly and monthly operational schedules as per transmission plan and service requirements and notification of these to a Spacecraft Control Centre and NOCC;

- establishment of central and regional repair centres, calibration facilities and fault control procedures.

CHAPTER 2

Some basic technical issues

2.1 Characteristics of a satellite link

2.1.1 The basic satellite link

Figure 2.1 shows, in its simplest form, a satellite link carrying a duplex (two-way) communication circuit: the earth station A transmits to the satellite an uplink (U/L) carrier wave (modulated by the baseband signal, i.e. by the signal from the message source transmitted by the user terminal) at radio frequency (RF) F_{u1} (e.g. 5 980 MHz). The satellite antenna and transponder system receives this carrier and, after frequency conversion from F_{u1} to F_{d1} (e.g. 5 980 MHz – 2 225 MHz = 3 755 MHz), amplifies and re-radiates it as a downlink (D/L) wave which is received by the earth station B. To establish the return link, B transmits a U/L carrier at another RF F_{u2} (e.g. 6 020 MHz) which is received by A at the converted D/L RF F_{d2} (e.g. 6 020 MHz – 2 225 MHz = 3 795 MHz).

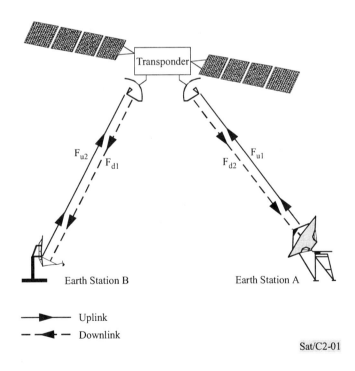

Sat/C2-01

FIGURE 2.1

The basic satellite link

Note that:

- more generally, the satellite is equipped with several transponders (the two links F_{u1}/F_{d1} and F_{u2}/F_{d2}, bearing the two half-circuits, may be transmitted through different transponders);

- in the most advanced types of satellites, the signal does not simply incur a frequency conversion in a transponder, but is subject to more complex operation, including demodulation/remodulation, baseband processing, etc. Such operations (called on-board processing, or OBP) are described in various parts of this Handbook (e.g. in § 6.3.4);

- the figure is, of course, independent of the satellite distance and type of orbit.

The link must be designed to provide reliable, good quality communication, which implies that the signal transmitted by the sending earth station must reach the receiving earth station at a carrier level sufficiently above the unwanted signals generated by various, unavoidable sources of noise and interference.

The quality of the communication, i.e. of the baseband message signal received by the user terminal, is derived from the received power-to-noise ratio, accounting for the modulation/demodulation and possibly encoding/decoding processes:

- In the case of analogue communications, frequency modulation is generally used and the communication quality is measured by the signal-to-noise ratio (S/N), which is derived from the carrier-to-noise ratio (C/N) at the receiver input and from the frequency modulation parameters (see below, § 4.1).

- In the case of digital communications, the communication quality is measured by the information bit error ratio (BER).

   The BER is derived from the carrier-to-noise-density ratio (C/N_0) (or from the (E_b/N_0)) at the receiver input and from the coding and modulation parameters. Here E_b is the energy per information bit and N_0 is the noise power spectral density (noise power per Hz, i.e. $N_0 = N/B$, where B is the RF noise bandwidth).

Based on (C/N), (C/N_0), (E_b/N_0), (S/N) or BER, and as a function of the link availability, the level of quality of the communication link or of the received message, in the various analogue or digital types of systems is the subject of a number of ITU (ITU-R and ITU-T) recommendations, as explained in detail below (§ 2.2).

The fundamental design factor of a satellite link is the calculation of the link budget, i.e. the calculation of the (C/N), (C/N_0) or (E_b/N_0) as a function of the characteristics of the satellite, of the earth stations and of the local environment and interference conditions.

Before going to link budget analysis and calculation (§ 2.3), it is necessary to review some basic notions on the antennas (because they constitute the earth-space interface), on the link noise and on the influence of the propagation medium.

2.1.2   Basic characteristics of an antenna

This subsection defines only the antenna parameters needed for link budget calculations. Other basic characteristics of an antenna such as radiation diagrams, beamwidths and polarization are described in Appendix 2.1.

2.1.2.1   Antenna gain definition

An isotropic transmitting antenna radiates a spherical wave with a uniform power $p_o/4\pi$ in any direction (θ, φ) of the surrounding space (p_o being the power available at the input of the antenna).

A directional antenna will radiate a power $p(\theta, \varphi)$ in the direction θ, φ (see Figure 2.1).

The definition of the gain of an antenna is:

$$g(\theta, \varphi) = \frac{p(\theta, \varphi)}{p_o/4\pi} \qquad (1)$$

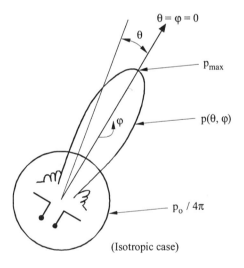

Sat/C2-02

FIGURE 2.2

Radiated power of a transmit antenna

p_o is also the total radiated power (in all directions) and it can therefore be expressed as:

$$p_0 = \int\limits_{0}^{2\pi} \int\limits_{0}^{\pi} p\,(\theta, \varphi) \sin \theta \; d\theta \; d\varphi \qquad (2)$$

The maximum of the gain function is:

$$g_{\max} = \frac{p_{max}}{p_o / 4\pi} \qquad (3)$$

The maximum gain, $g_{\max}$, is often called simply the "antenna gain", g, and is usually expressed as follows in decibels, and more precisely, in decibels over the gain of an isotropic antenna (dBi):

$$G = 10 \log g \;\; (dBi)$$

The definition of antenna gain is more straightforward if the antenna is considered during transmission, as in this subsection. However, according to the reciprocity theorem, the properties of an antenna are the same for transmission and for reception. For example, the beginning of this subsection could read, for reception: "An isotropic receiving antenna is able to receive uniformly from the same transmitter positioned in any direction (θ, φ) a power $p_o/4\pi$. A directional antenna will receive a power $p\,(\theta, \varphi)$ from the same transmitter positioned in the direction (θ, φ)...".

The reciprocity theorem holds for all antenna characteristics defined in this subsection. The following subsection is more easily explained if the antenna is considered during reception.

2.1.2.2   Antenna effective aperture and gain

If a radio wave, arriving from a distant source (in the present case – the satellite, see Figure 2.3) impinges on the antenna, the antenna "collects" the power contained in its "effective aperture area" A_e. If the antenna were perfect and lossless, this effective aperture area A_e would be equal to the actual projected area A (e.g. for a circular aperture $A = \pi D^2/4$). In practice, taking into account losses and the non-uniformity of the "illumination law" of the aperture:

$$A_e = \eta \cdot A$$

where:

$\eta =$ antenna efficiency $(\eta < 1)$

$(\eta = 1$ for a uniform "illumination law" and no losses, typical values are between 0.6 and 0.8).

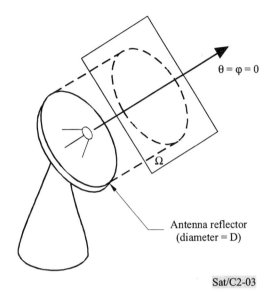

FIGURE 2.3

Effective aperture of a receive antenna

A very important relation between g_{max} and A_e in square metres is:

$$g_{max} = \frac{4\pi A_e}{\lambda^2}$$

(4)

where λ is the wavelength (m)

$$\lambda = \frac{c}{f}$$

c: RF waves velocity $= 3 \times 10^8$ (m/s)

f: radio frequency (Hz)

$$g_{max} = \frac{4\pi \eta A}{\lambda^2}$$

(5)

For a circular aperture (diameter = D in metres)

$$g_{max} = \eta \left(\frac{\pi D}{\lambda}\right)^2$$

(6)

Or, in dB:

$$G = 10 \log g_{max} = 9.94 + 10 \log \eta + 20 \log\left(\frac{D}{\lambda}\right) \text{dBi}$$

(7)

Equation (4) is given for the direction of maximum antenna gain (i.e. for $\theta = \varphi = 0$). However, it can be generalized as:

$$g(\theta, \varphi) = \frac{4\pi A_e(\theta, \varphi)}{\lambda^2} \tag{8}$$

$A_e(\theta, \varphi)$ being the effective aperture area in direction (θ, φ).

For an isotropic antenna ($g_{max} = g(\theta, \varphi) = 1$), the effective aperture is A_{iso} where:

$$A_{iso} = \frac{\lambda^2}{4\pi} \tag{9}$$

Therefore, the equations above may also be written as:

$$g(\theta, \varphi) = \frac{A_e(\theta, \varphi)}{A_{iso}} \tag{10}$$

$$A_{e\,max} = A_e(\theta = \varphi = 0)$$

$$g_{max} = \frac{A_{e\,max}}{A_{iso}} \tag{11}$$

2.1.3   Power radiated and received by an antenna

2.1.3.1   Power flux-density, equivalent isotropically radiated power and free-space attenuation

The power flux-density (pfd) is the power radiated by an antenna at a sufficiently large distance d, in a given direction.

For an isotropic antenna in free space conditions, the power p_e at the antenna input is radiated uniformly in any direction throughout a sphere centred on the antenna and the (pfd) is (in w/m^2):

$$(pfd)_i = p_e/(4\pi d^2) \tag{12}$$

The power flux-density (pfd) radiated in a given direction by an antenna having a gain g_e in that direction is thus:

$$(pfd) = p_e \cdot g_e/(4\pi d^2) \tag{13}$$

The product $p_e \cdot g_e$ is called the equivalent isotropically radiated power or e.i.r.p. This is an essential factor in the evaluation of the link budget. It has the dimension of power and is therefore given, when expressed in logarithmic units, by $(P_e + G_e)$ in dBW.

For a receiving antenna having an effective area A_e, the power reaching the antenna is equal to:

$$p_r = p_e \cdot g_e \cdot A_e/(4\pi d^2) \tag{14}$$

or, using the general equation (4),

$$p_r = p_e \cdot g_e \cdot g_r \cdot (\lambda/4\pi d)^2 \tag{15}$$

where:

> g_r: receiving antenna gain

> λ: wavelength

The value:

$$l = (4\pi d/\lambda)^2 \tag{16}$$

or, expressed in decibels:

$$L = 20 \log (4\pi d/\lambda) \tag{17}$$

represents the free-space attenuation between isotropic antennas.

The value of this free space attenuation is represented in Figure 2.4, as a function of frequency, with some distances (d) as a parameter. The distances shown are typical of the orbits of satellite systems currently in operation, or projected, as already explained in Chapter 1. For a more detailed description of satellite orbits, the reader is referred to Chapter 6, § 6.1:

- d = 36 000 km is the highly popular geostationary satellite orbit (GSO) distance (note that the altitude is more precisely 35 786.1 km). Due to the – approximately – fixed position of the satellite (as seen from an earth station), this was, up to now, the only distance practically implemented for civilian communication satellites either for international or for domestic traffic.

  NOTE – d = 36 000 km is (approximately) the geostationary orbit altitude. In fact, in the calculation of the free-space attenuation, the distance must account for the obliquity of earth station to satellite path. For example, with a 30° elevation, the actual distance is d = 38 607 km.

- d = 10 000 km is typical of medium earth orbit (MEO) satellites, also often known as intermediate circular orbit (ICO): this is the orbit chosen by some organizations for systems implementing very small, portable (even pocket size) earth stations (often called "terminals") in the framework of personal telecommunications (PTs) networks. This is the case, in particular, for the Inmarsat P project (which operates in the mobile-satellite service – MSS).

- d = 1 000 km is a typical value for low earth orbit (LEO) satellites. There are, at present, many "PT-like" system projects which are based on the implementation of LEO satellites operating either in the MSS or in the FSS.

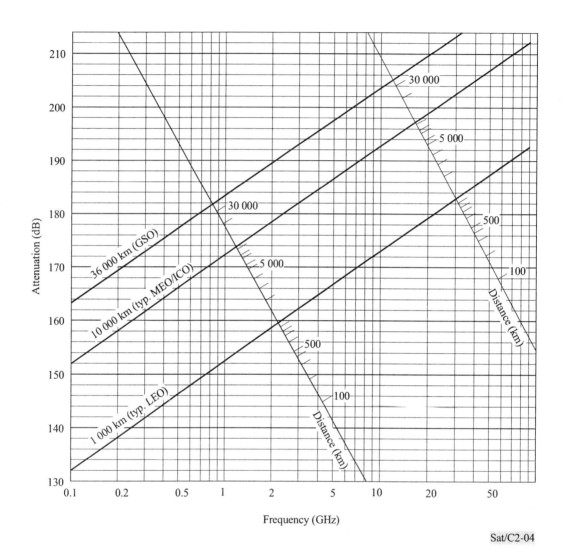

FIGURE 2.4

Free-space attenuation (L)

NOTE – Parallel lines can be drawn at the various distances by connecting corresponding points on the two distance scales.

One of the main reasons for the trend towards the utilization of MEO or LEO satellites for "PT-like" systems appears clearly on this figure: at 6 GHz, the attenuation is reduced from about 200 dB for GSO operation to less than 190 dB for MEO/ICO and to less than 170 dB for LEO. The reduction is even greater in the MSS allocated 1.5 GHz band (e.g. 156 dB for LEOs). Even in the 30 GHz band, the free-space attenuation, in conjunction with high-gain satellite antennas, can be compatible with the utilization of hand-held terminals (e.g. nominally less than 182 dB for LEOs; however an additional attenuation accounting for rain is to be provided).

2.1.3.2   Additional losses

In addition to the losses due to free-space propagation, received power calculations must also take into account, both in the uplink and downlink, miscellaneous losses as follows:

- atmospheric losses representing the attenuation due to propagation in the atmosphere and the ionosphere (see Annex I to the Handbook). This may vary from a few tenths of a decibel at 4 GHz to several tens of decibels at 30 GHz, according to local precipitation conditions and the elevation angle of the satellite;

- losses due to polarization mismatch at the antenna interface and to cross-polarization caused by propagation;

- losses due to antenna offset with respect to the nominal direction commonly referred to as pointing error losses. These may generally be expressed in terms of a change in antenna gain as a function of the off-boresight angle θ:

$$\Delta G \approx 12 \ (\theta/\theta_0)^2 \ \text{(dB)} \tag{18}$$

where:

$$\theta_0 \approx 65 \ (\lambda/d) \tag{19}$$

θ_0: is the half-power (-3 dB) beamwidth (in degrees)

d: being the (circular) antenna diameter.

G can also be read on the nomogram in § A.2.1.2 of Appendix 2.1;

- feeder losses representing the losses in the transmitting antenna feeder and between the receiving antenna and the receiver input. These losses are generally included in the e.i.r.p. on emission and in the station sensitivity on reception.

2.1.4   Noise power received on the link

Noise power at the receiver input is due both to an internal source (typical of the receiver) and to an external source (antenna contribution).

2.1.4.1   Noise temperature

In calculating link budgets the term N_0 (noise spectral density in W/Hz) or the term T (noise temperature) is used in preference to the noise power N, so that there is no need to specify the bandwidth B in which the noise is measured. The relation between these terms is:

$$N = k \, T \, B \quad \text{and} \quad N_0 = N / B \tag{20}$$

where k is Boltzmann's constant (1.38×10^{-23} Joule/Kelvin). T must be expressed in absolute degrees i.e. Kelvin (K) (temperature expressed in Kelvin = temperature expressed in Celsius degrees + 273).

N_0 and N are sometimes expressed in decibels as follows:

$$N = 10 \log k + 10 \log T + 10 \log B \qquad dB(W)$$

$$N_0 = 10 \log k + 10 \log T \qquad dB(W/Hz)$$

where:

$$10 \log k = -228.6 \text{ dB (Joule/Kelvin)} \qquad (21)$$

2.1.4.2   Noise temperature of a receiver

An ideal noiseless receiving amplifier would amplify an input noise not more than the input signal (i.e. with the same gain). Due to internal noise, an actual receiving amplifier will bring additional noise power.

The noise caused by a receiver is usually expressed in terms of an equivalent amplifier noise temperature T_R. It is defined as the temperature of a noise source (resistance) which, when connected to the input of a noiseless receiver, gives the same noise at the output as the actual receiver.

The receiver is actually composed of cascaded circuits and, more precisely of a few amplifying stages or other networks (such as the down-converter, etc.), each one having its own gain g_i and its noise temperature T_{Ri}. It can be easily demonstrated that, under such conditions the receiver noise temperature is:

$$T_R = T_{R1} + (T_{R2}/g_1) + (T_{R3}/g_1 g_2) + ...$$

This formula is important because it shows that the noise contributions of the successive stages are lessened by the total gain of the preceding stages. Consequently, the RF amplifier, called the low noise amplifier (LNA) must have a low T_{R1} and a high g_1.

Common values of T_R for the LNAs used in modern receivers are between 30 K and 150 K, depending on the frequency band and on the LNA design.

Note that, in small earth stations (receive-only, small stations for business communications, called VSATs, etc.), the LNA is generally included with the down converter (D/C) in a single unit called low noise converter (LNC) or low noise block (LNB).

Note also that the noise caused by the receiver is sometimes expressed by its noise Figure F (or by $F_{dB} = 10 \log F$), the relation between F and T_R being: $T_R = (F-1) T_0$, T_0 being, by convention, equal to a normal ambient temperature value of 290 K. In fact, since T_R is usually much less than 290 K, T_R is more practical to use than the noise figure in satellite communications.

2.1.4.3   Noise temperature of an antenna

The noise temperature of an antenna is the translating, in terms of noise temperature (according to formula (20)) of the collection, by the antenna of the external noise. Therefore, it can be expressed by the following integral:

$$T_A = 1/4\pi \iint g(\Omega) \cdot T(\Omega) \cdot d(\Omega) \qquad (22)$$

where:

$d(\Omega)$:   is the elementary solid angle in the direction Ω;

$g(\Omega)$ and $T(\Omega)$:   are the antenna gain and the equivalent noise temperature of the noise source in the direction Ω.

It should be noted that the noise contributions can arise either from actual noise sources (e.g. sun or radio stars) or from absorbent material (e.g. water vapour). The second case results from the blackbody theory. In fact any material radiates the energy it absorbs and can be considered as a blackbody of the same size with such a temperature that it would radiate with the same spectral density at the considered frequency.

External noise contributions may be divided into two categories:

a)   Noise from the terrestrial noise environment, due to:

   •   a1) the atmospheric attenuation of the sky (water vapour, fog, clouds, rain, oxygen);

   •   a2) the ground.

   In practical applications, this is, by far, the dominant contribution.

b)   Extraterrestrial noise from radio stars, the Sun, the Moon, etc. and from the 2.7 K background cosmic noise. This accounts generally for a very small contribution (a few K), except some situations such as the – non-operational – situation where the earth station antenna intercepts the Sun ($T_{SUN} \approx 10^4$ K; the Sun's interference is dealt with in Recommendation ITU-R S.733, Appendix 2 to Annex 1, and in § 7.2.1.2 of this Handbook).

The result of (22) is very different, whether the antenna is pointing towards the ground (satellite antenna) or towards the sky (earth station antenna).

In the first case, T_A can generally be taken as equal to the earth absorption temperature, i.e. about 290 K, since the main lobe of their radiation diagram necessarily intercepts the Earth's atmosphere and ground.

In the second case, due to contributions in category a), T_A is highly dependent on the antenna beam elevation angle. For a low elevation, the antenna collects rather high noise contributions from the atmosphere a1) (the lower the elevation, the longer the length of the rays in the atmosphere) and also from the ground a2), through that part of the radiation diagram which hits the ground (i.e. side lobes, including back lobes and, even possibly, a part of the main lobe in the case of the wide lobes radiated by small antennas).

For illustration, Figure 2.5 gives an example (from Recommendation ITU-R PI.372.6) of the sky noise temperature for an earth station antenna (for frequencies between 1 and 55 GHz, with the elevation angle as a parameter). This diagram is the result of a calculation using a model of the atmosphere. Contribution a1) only is included, in the case of a "clear sky", assuming various hypotheses, as quoted. Under these conditions, the sky noise temperature is called "brightness temperature" in Recommendation ITU-R PI.372.6. The figure is given simply to show the general trends and, in particular, the important increase around 23 GHz (due to resonance in the water vapour absorption). In practical cases and notwithstanding contributions a2) and b), noise temperature due to rain, fog, etc. needs to be added. For more data on the subject, refer to Chapter 7, § 7.2.1.

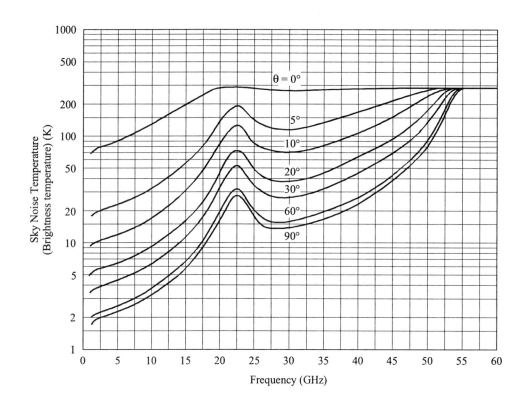

(Clear air, 7.5 g/m^3 water vapour) θ: elevation angle

(Source: Recommendation ITU-R PI.372-6)

Sat/C2-05

FIGURE 2.5

Noise temperature due to the atmospheric attenuation of the sky

2.1.4.4   Noise temperature of a receiving system

Figure 2.6 shows a practical receiving system, with an antenna with a noise temperature T_A and a receiver with a noise temperature T_R.

An attenuating section is inserted between the two parts. This section represents the losses (generally ohmic losses) in the antenna and in the feeder (i.e., the RF link, waveguide, coaxial or any other element).

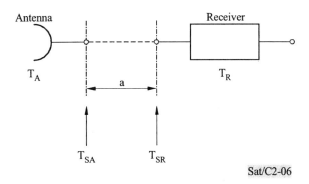

FIGURE 2.6

Noise temperature of a receiving system

Under these conditions, the total noise temperature of the system is:

$$T_{SR} = T_R + T_a\,(1 - 1/a) + T_A/a \tag{23}$$

or:

$$T_{SA} = T_A + T_a\,(a - 1) + aT_A \tag{24}$$

where:

a: is the attenuation, expressed as a power ratio (a ≥ 1, i.e. in decibels, $a_{dB} = 10 \log a$)

T_a: is the physical temperature of the attenuating section (generally taken = 290 K)

T_{SR}: is referred at the receiver input, which means that, in subsequent calculations, the receiver can be assumed to be noiseless

T_{SA}: is referred at the antenna output, which means that, in subsequent calculations, the attenuating section and the receiver can be assumed to be noiseless.

The formulas show that the contribution of the attenuating section can be quite significant (≈ 7 K per 0.1 dB loss). However, in the usual case where the attenuation remains low (a ≈ 1) and if the contribution of the attenuating section is included into T_A, then the receiving system noise temperature can be written simply:

$$T_S \approx T_{SR} \approx T_{SA} \approx T_A + T_R \tag{25}$$

2.1.4.5   Figure of merit of a receiving station

In the link budget calculations (§ 2.3), it will be seen that the signal-to-noise ratio for reception in a satellite link depends on the characteristics of the space (satellite) and earth stations as given by their respective figures of merit.

The figure of merit of a receiving station is defined as the ratio between the gain of the antenna in the direction of the received signal and the receiving system noise temperature. This gain-to-noise temperature ratio (G/T) is generally given for the maximum gain (g_{max}, see formula (2) in § 2.1.2.1) and expressed in decibels per Kelvin ($dB(K^{-1})$):

$$(G/T) = 10 \log g - 10 \log T_{SA} (dB(K^{-1}))$$

$$\approx 10 \log g - 10 \log T_S (dB(K^{-1}))$$

Earth stations G/T typical values range from 35 $dB(K^{-1})$ (main 4 GHz stations with a 15 to 18 m diameter antenna) to some 15.5 $dB(K^{-1})$ (12 GHz "VSAT" data transmission microstations with 1.2 m antenna). Space stations G/T at 6 GHz typically range from about -19 $dB(K^{-1})$ for a global beam antenna to -3 $dB(K^{-1})$ for a pencil beam (zone beam) antenna.

The G/T is a very important parameter of an earth station. The methods used for its measurement and the contributions to its noise temperature are the subject of Recommendation ITU-R S.733 and are described in Chapter 7 (§ 7.2.6).

2.1.5   Non-linear effects in power amplifiers

2.1.5.1   General

The transmission equipment of the satellite transponders and of the earth station generally includes stages exhibiting non-linear input-output characteristics. This is particularly the case for the output stages, i.e. the power amplifiers which can use microwave tubes (generally travelling wave tubes (TWTs) or also klystrons) or solid-state power amplifiers (SSPAs). Figure 2.7 shows a typical example of the behaviour of a satellite TWT near the saturation point, i.e. the maximum output power operation.

When multiple carriers are simultaneously transmitted at different frequencies in the bandwidth of the power amplifier, non-linearities cause intermodulation, i.e. give rise to unwanted signals, called intermodulation products.

Even if a single signal is amplified, unwanted effects appear:

- AM-AM conversion, i.e. distortion of the output versus input amplitude;

- AM-PM conversion, i.e. non-constant phase shift between input and output signals, as a function of the input power. This effect is particularly harmful in the case of digital transmissions where the information is generally carried by phase modulation.

In consequence, non-linearity in power amplifiers (satellite and earth stations) induces three types of undesirable effects:

i)   The generation of unwanted signals interfering with wanted signals. These unwanted signals can often be assimilated to white noise (interference noise) and account, in the link budget calculations, as a term $(C/N_0)_I$.

ii) A degradation of the overall BER performance in the case of digital transmission.

iii) A reduction of the output power, due to the need for operating the amplifier in a sufficiently linear region (i.e. with a "back-off"[1] below saturation) in order to decrease the above-mentioned effects.

These effects can be reduced by the use of linearizers, i.e. wideband predistortion circuits located at the input of the amplifier. Several types of linearizers have been developed and are actually utilized (see Chapter 7, § 7.4.5.2.3).

Effects i) and ii) will be discussed briefly below and in much more detail in Appendix 5.2 and in Chapter 7 (§ 7.4.5).

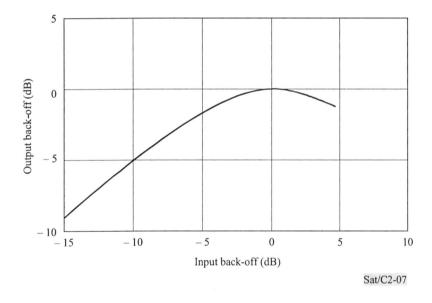

Sat/C2-07

FIGURE 2.7

Example of the input-output characteristic of a satellite TWT

---

[1] The input back-off is defined as the ratio between the input power at saturation (single carrier) and the actual operating input power. The definition of the output back-off is the same (replacing input by output). The inverse of these ratios may also be used in these definitions and, anyway, they are always converted to dBs.

2.1.5.2   Intermodulation

When several carriers (at frequencies f_1, f_2, etc.) are simultaneously amplified in a non-linear amplifier, which is the case in frequency division multiple access (FDMA), intermodulation products are generated at frequencies $f = m_1 f_1 + m_2 f_2 + ... + m_N f_N$ (m_1, m_2, ..., m_N being positive or negative integers). The integer $m = |m_1| + |m_2| + ... + |m_N|$ is called the order of the intermodulation product. In the usual case where the operating bandwidth is small, compared to the radio frequency, only odd-order products fall into the bandwidth and have to be considered.

Intermodulation product power decreases with the order and only third-order products and sometimes fifth-order products are to be taken into account.

The number of intermodulation products increases very quickly with the number of input carriers (for example, for 3 carriers, there are 9 products and for 5 carriers there are 50).

Figure 2.8 (a) shows the third-order products of a first type generated at frequencies $2f_1 - f_2$ and $2f_2 - f_1$ by two equal level input carriers f_1 and f_2. When N carriers are simultaneously amplified, all possible products of this type appear, at frequencies $2f_i - f_j$ (with i, j = 1, 2, ..., N).

Moreover, when $N \geq 3$, products of a second type appear at frequencies $\pm f_i \pm f_j \pm f_k$ (only one minus possible) and become dominant. Figure 2.8 (b) shows the third-order products of this second type generated at frequencies $f_1 + f_2 - f_3$, $f_1 - f_2 + f_3$ and $f_2 + f_3 - f_1$ by three equal level input carriers f_1, f_2 and f_3.

There are N (N–1) products of the first type and (1/2) N (N–1) (N–2) products of the second type (e.g. respectively 90 and 360 for N = 10).

It should be noted that if one of the carriers has an input level greater than the others, there is a "capture effect" which means that it is amplified with a higher gain.

Other data on intermodulation in power amplifiers are given in Chapter 7 (§ 7.4.5.2.1) and methods for calculating the level of the intermodulation products are detailed in Appendix 5.2.

In actual cases, the intermodulation products are modulated as are the input carriers (e.g. analogue FDM, FM modulated carriers). This reduces their power density and, as explained above, the spectral density of the intermodulation caused by multiple modulated carriers is often considered as noise and is accounted for by a term $(C/N_0)_{IM}$ in the link budget calculations.

Due to its interfering effects, the intermodulation caused by the satellite and earth station power amplifiers must be limited, both in the operating bandwidth of the system under consideration and outside this band (to protect other systems). In particular, the satellite operators (e.g. INTELSAT) generally specify a limit to the out-of-band intermodulation products caused by a particular earth station (e.g. 20 dBW/4 kHz).

To reduce intermodulation in multicarrier operation, the amplifier needs to be driven with a sufficient back-off: an output back-off of 3.5 dB (corresponding to an input back-off of about 8 dB) up to 10 dB is typical for satellite amplifiers. In the case of earth station HPAs, a greater back-off is usually required, in the range of 5 to 10 dB (output). However the situation can be improved by the utilization of linearizers (see Chapter 6, § 6.3.3.3 and Chapter 7, § 7.4.5.2.3).

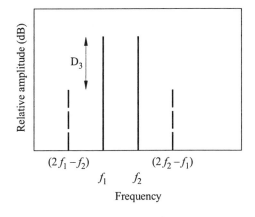

(a) First type: two equal level carriers

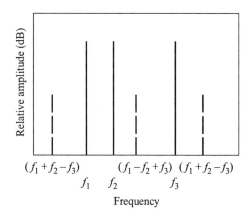

(b) Second type: three equal level carriers

Sat/C2-08

FIGURE 2.8

Third-order intermodulation products

2.1.5.3 Non-linear effects in digital transmission

If multiple digital carriers are simultaneously transmitted through a common satellite (and possibly earth station) power amplifier(s), they suffer the same intermodulation problems as described above.

This is the case, for example, of time division multiplexed (TDM), phase shift keying (PSK) modulated carriers with medium bit rates (e.g. 64 kbit/s to 8 Mbit/s or more), which are transmitted in FDMA through a common transponder (e.g. standard IDR and IBS INTELSAT and SMS EUTELSAT carriers).

This is also the case, for example, of medium bit-rate FDMA-TDMA carriers.

However a digitally encoded, phase modulated RF carrier, having a constant amplitude envelope should not suffer from AM-AM and AM-PM effects.

Moreover, amplifier non-linearity should, in principle, have no effect at all in the case of a single high bit-rate TDMA carrier (e.g. the classical 120 Mbit/s, 4-PSK modulated INTELSAT carrier) which is generally transmitted through a full wideband satellite transponder and through a dedicated earth station HPA.

However, even in this case, non-linearities must be taken into account in the transmission link (which comprises at least two non-linear circuits: the satellite and the earth station amplifiers). This is due to the conversion of amplitude modulation into phase modulation (AM/PM) resulting from non-linearity effects on signals distorted by frequency filtering at the amplifier input. For more details on the non-linearity effects and on the practical back-off levels to be used, see Chapter 7, § 7.4.5.2.2.

Note that the same considerations apply to CDMA carriers (see Chapter 5, § 5.4.7.1).

2.2 Quality and availability

2.2.1 Introduction

Transmission quality and availability are essential performance issues of all communication systems. Due to the random effects of the propagation environment (interference, rain attenuation, etc.) or equipment failures, it is not possible to guarantee perfect transmission. Therefore, any transmission performance can only be described in terms of the probability of achieving a given level of quality or availability (or, equivalently, by the percentage of time during which that level can be achieved). More precisely:

- the transmission quality, as explained at the beginning of this chapter (§ 2.1.1), is conventionally expressed in terms of a signal-to-noise ratio (S/N) in the case of analogue transmission or of a bit error ratio (BER) in the case of digital transmission. Usually there is a specification requiring that the operating (S/N) shall remain higher than (or equal to) a given limit or that the operating (BER) shall remain lower than (or equal to) a given limit during some given, specified, percentage of time. It is very important to note that the quality is only defined during the time where the link is considered to be available;

- the transmission link availability is generally related to a limit performance, called a threshold and characterized by (analogue transmission) an abrupt increase of noise (decrease of (S/N)) or (digital transmission) by a (BER) limit. When the performance degradation exceeds this limit for a given time duration, the link is considered to be no longer available. This threshold is determined by the capability of equipments to work properly in extreme conditions (e.g. equipment synchronization limitations at a low reception level).

In preparation for the next section (§ 2.3) on link budgets, this section deals with the quality and availability objectives to be attained in a satellite link.

In fact, transmission margins are needed in the link budget in order to comply with the quality and availability specifications, taking into account the statistical laws used to characterize the random events impairing the transmission, the most important being generally due to atmospheric effects.

It should be noted that quality and availability are not completely independent because either one or the other can determine the link performance and thus the minimum required received power (or C/N_0) to be obtained from the link budget.

This is schematically represented (for digital transmissions) in Figure 2.9 below. In fact, as a rather general rule, the performance is generally under the constraint of the availability in the 14/11-12 GHz band, while the constraint comes from the quality in the 6/4 GHz band, due to the much higher effect of rain attenuation in the former case.

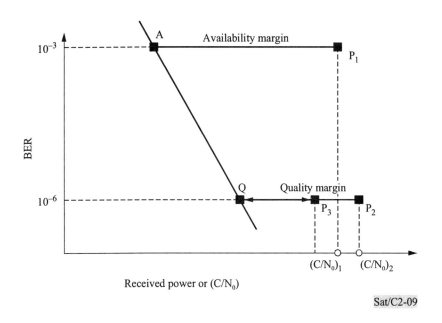

FIGURE 2.9

**Examples of received power (or C/N0) limitations by
availability or by quality performance constraint**

The diagram typically represents the performance curve of a satellite digital communication receiver, i.e. BER vs. C/N_0 for the modem including an FEC (forward error correction) codec and accounting for various implementation losses.

Using, as examples, the specifications of Recommendation ITU-R S.579 (see below § 2.2.3.1) and S.614 (see below § 2.2.3.3):

• point A represents the availability threshold (10^{-3}). An availability margin corresponding to the statistically possible atmospheric propagation effects (during no more than 0.2% of the worst month) is to be added, giving point P_1;

• point Q represents the quality threshold (e.g. 10^{-6}). A quality margin corresponding to the propagation effects (during no more than 2% of the worst month) is to be added, giving point P_2 or P_3 (depending on the local statistics).

In a first case (($C/N_0)_1$), the C/N_0 corresponding to P_1 is greater than the C/N_0 corresponding to P_3 and the link performance is determined by the availability constraint. In a second case (($C/N_0)_2$), the C/N_0 corresponding to P_2 is greater than the C/N_0 corresponding to P_1 and the link performance is determined by the quality constraint.

Of course, the conclusions are dependent on the slope of the performance curve, i.e. on the modem + FEC characteristics.

2.2.2   Quality of Service (QoS)

It is very important to note that the term quality is used throughout this chapter in a rather restricted sense and should not be confused with the more general concept of quality of service (QoS) which is becoming an ever more important issue in telecommunication and computer communication[2].

In fact, QoS covers a wide variety of notions depending on the application, service or user interest. Although its general definition remains rather broad, ITU-T has issued many specific recommendations on the subject. In these recommendations, the set of parameters is shared between a succession of layers describing the functions implied in the overall user-to-user communication process, similarly to the open system interconnection (OSI) reference model[3].

ITU-T Recommendation E.800 defines as follows the QoS and also another significant quality parameter, the network performance (NP): "The QoS is the collective effect of service performances which determines the degree of satisfaction of a user of the service. The NP is the ability of the network or network portion to provide the functions related to communication between users. The NP contributes to continuity of performance and to service integrity".

This definition of QoS is founded on the degree of satisfaction (GoS: grade of service) which remains a rather vague notion.

ITU-T Recommendation E.430 identifies a relationship between QoS and NP parameters. It gives a list of the ITU-T recommendations related to QoS and NP and covering all aspects of telecommunications in the case of a connection (access, information transfer and release).

It should be noted that, *a priori*, the notion of QoS is not of much interest to customers in a traditional public monopoly supply situation for telecommunication services. However, once a competitive market place is established, then the QoS may become very important, although it can be unfamiliar to the customer.

When applying QoS (and NP) notions to the (lower) layers (and to the network portions) which are covered by satellite communications, very diverse, but specific, performance parameters can be included, such as: transmission quality and link availability (as defined above in § 2.1.1) and also: time delay, system capacity and throughput, security and data protection, MOS (mean opinion score) evaluation of the voice quality (especially in the case of low bit-rate voice encoding by a vocoder) or of the video quality, etc. However, the latter parameters, which are not directly dependent on the link budgets, will not be considered further in this chapter.

In fact, the seamless integration of a satellite portion in a network (in particular in modern B-ISDN networks) should be evaluated on the basis of the QoS to be expected by the end users. This QoS should not be excessively affected or degraded, due to possible delays or specific bit error

---

[2]   Quality of service is the subject of many publications. See, for example: "Quality of service in telecommunications. Part I: Proposition of a QoS framework and its application to B-ISDN, and Part II: Translation of QoS into ATM performance parameters in B-ISDN", Jae-Il Jung, IEEE Communications Magazine, August 1996, pp. 104-117.

[3]   From the three lower layers (physical, data link, network) to the four upper ones (up to the application layer). For an overview of the OSI model, see Appendix 8.2 in this Handbook.

distribution introduced by the satellite link. A portion of the allowable degradation will be allocated to the satellite portion of the link.

In the case of ATM transport (taken as an important example), the traffic and congestion control functions should be able to satisfy sufficient QoS requirements for all foreseeable services. The basic QoS parameters that correspond to a network performance objective in satellite ATM networks may include cell loss ratio (CLR), maximum and mean cell transfer delay (CTD) (especially in real time applications) and cell delay variations (CDV)[4].

2.2.3   ITU-R and ITU-T recommendations

The main specifications relative to quality and availability can be found in ITU-R and ITU-T recommendations. The basic recommendations on communications are established by ITU-T. The parts relative to radio links (and, in particular, to satellite links) are translated by ITU-R in recommendations applicable to these links.

As general remarks concerning these recommendations, it should be noted that:

* they are primarily relative to public, international links;
* they are based on reference long-distance links which allocate a limited contribution to satellite paths;
* most of them are circuit-oriented;
* they are mostly adapted to permanent, transparent transmissions (e.g. an ISDN link).

IMPORTANT NOTE – It should be noted that all recommendations quoted below are subject to occasional revisions: this is marked by a supplementary revision number, e.g. 353-7. This revision number is not mentioned here. Moreover, there are many more details and important notes in the text of the recommendations. In fact, before any precise engineering calculation, the latest edition of the relevant ITU recommendation must be referred to.

2.2.3.1   Availability objectives

The unavailability of a continuous link is defined as the ratio, in percentage, of the unavailable time to the total time. The availability of a link is defined as: 100 – unavailability (%).

According to Recommendation ITU-R S.579, an FSS link set up between the ends of the hypothetical reference circuit or of the hypothetical reference digital path referred to in Recommendations ITU-R S.352 and S.521 should be considered to be unavailable if one or more of the following conditions exist at either of the receiving ends of the link for longer than 10 consecutive seconds[5]:

---

4   For more details, see, for example, "Satellite ATM networks: A Survey", I.F. Akyildiz and Seong-Ho Jeong, IEEE Communications Magazine, July 1997, pp. 30-43.

5   These 10 s are considered to be unavailable time. The period of unavailable time terminates when the same condition ceases for a period of 10 consecutive seconds. These 10 s are considered to be available time.

- in analogue transmission, the wanted signal entering the circuit is received at the other end at a level at least 10 dB lower than expected;

- in digital transmission, there is a break in the digital circuit (i.e. there is a loss of frame alignment or timing);

- in analogue transmission of a telephone channel, the unweighted noise power at the zero relative level point, with an integration time of 5 ms, exceeds 10^6 pW0;

- in digital transmission, the bit error ratio (BER), averaged over 1 second, exceeds 10^{-3}. However, in the new Recommendation ITU-R S.1062, the condition of unavailability is defined by the advent of 10 consecutive severely errored seconds (SES, see below, § 2.2.3.3 ii)).

Recommendation ITU-R S.579 provisionally stipulates that, in the FSS:

- the unavailability of a hypothetical reference circuit or digital path due to equipment should be not more than 0.2% of a year;

- the unavailability due to propagation should be not more than:

  - 0.2% of any month for a hypothetical reference digital path;

  - X% of any month for a hypothetical reference circuit (X being still under study, $X = 0.1$ being suggested).

2.2.3.2   Quality objectives, analogue transmission

Table 2.1 below summarizes the quality objectives for analogue telephony, according to Recommendation ITU-R S.353.

It should be noted that ITU recommendations generally refer to the percentages for "any month" or for "the worst month" (whichever is equivalent). This can be converted in yearly percentages. For example, it is generally assumed that:

- 10% of the worst month corresponds to 4% of the year;

- 2% of the worst month corresponds to 0.6% of the year;

- 0.2% of the worst month corresponds to 0.04% of the year.

TABLE 2.1

Quality objectives for analogue telephony (Recommendation ITU-R S.353)

Measurement conditions	Noise power at reference level
• 20% of any month (1 min. mean value)	10 000 pW0p
• 0.3% of any month (1 min. mean value)	50 000 pW0p
• 0.05% of any month (integrated value over 5 ms)	1 000 000 pW0p (unweighted)

NOTE – "Noise power at reference level" is the usual means of expressing the signal-to-noise ratio (S/N) for analogue telephony. It is expressed in pW (10^{-12} W) for a 0 dBmW reference signal. 10 000 pW0p is thus equivalent to S/N = 50 dB (the final "p" meaning: weighted by a psophometric filter).

Analogue television: For analogue television, the quality objectives for all long-distance TV transmissions (terrestrial and/or satellite) are given in ITU-T Recommendations J.61 and J.62.

These state that the signal-to-noise ratio (S/N, weighted) should be equal to or better than 53 dB for 99% of the time and 45 dB for 99.9% of the time. However, these quality objectives are generally associated with the requirements of systems used for broadcasting distribution networks and do not reflect the current design practice of satellite systems used for more general distribution particularly to small earth stations (TVRO: television receive only).

In fact, four modes of distribution of TV programmes by satellite are nowadays in very common usage, although they are not always separately defined in ITU-R recommendations:

• Primary distribution: TV programme distribution by an FSS satellite from a main earth station towards medium or small earth stations which feed local terrestrial broadcasting stations. In this mode, the (S/N) is generally not worse than 48 dB.

• Satellite news gathering (SNG): temporary and occasional transmissions with short notice of television or sound for broadcasting purposes, using highly portable or transportable uplink earth stations operating in the FSS. In this service, the (S/N) is generally not worse than 48 dB. This mode and the equipment used are detailed in Chapter 7 (§ 7.9).

• Direct reception from an FSS satellite by small or very small TVRO stations either for collective (CATV: Cable TV, SMATV: satellite master antenna TV) or individual (DTH: direct-to-home) utilization. In this mode, the subjective quality is considered as fully acceptable to the majority of viewers for (S/N) = 40 dB.

• Direct reception from a BSS satellite (DBS: direct broadcast satellite) by very small individual TVRO stations (note that this mode is outside the scope of this Handbook).

2.2.3.3   Quality objectives, digital transmission

The ITU recommendations on digital transmissions are currently evolving:

i)   Digital (PCM) telephony and 64 kbit/s ISDN

Tables 2.2 and 2.3 below summarize the quality objectives according to ITU-R recommendations on the subject, i.e. Recommendation ITU-R S.522 for digital (PCM) telephony and Recommendation ITU-R S.614 for a "hypothetical reference digital path (HRDP) in the fixed-satellite service (FSS) when forming part of a 64 kbit/s international connection in the integrated services digital network (ISDN)". The HRDP definition is specified in Recommendation ITU-R S.521[6].

The objectives quoted in the third column of Table 2.2 result from the translation, by Recommendation ITU-R S.614, of the impact of ITU-T Recommendation G.821 on the satellite part of the HRDP ISDN connection. ITU-T Recommendation G.821 expresses the objectives in terms of "errored intervals" and, more precisely, of "degraded minutes", "severely errored seconds" and "errored seconds". Table 2.3 below recapitulates the ITU-T objectives relating to the 64 kbit/s ISDN connection respectively for the overall end-to-end and for the satellite HRDP.

TABLE 2.2

**Quality objectives for digital telephony and 64 kbit/s ISDN
(Recommendations ITU-R S.522 and S.614)**

Measurement conditions	Digital (PCM) telephony (Rec. S.522) BER	64 kbit/s ISDN (Rec. S.614) BER
• 20% of any month (10 min. mean value)	10^{-6}	
• 10% of any month		10^{-7}
• 2% of any month		10^{-6}
• 0.3% of any month (1 min. mean value)	10^{-4}	
• 0.05% of any month (1 s mean value)	10^{-3}	
• 0.03% of any month		10^{-3}

In Recommendation ITU-R S.614, the translation is performed by using statistical models of the distribution of errors according to calculations which are detailed in Annex 1 to the recommendation. However, the objectives resulting from the calculations are not unique and other

---

[6] On the subject of HRDP and, more generally on satellite transmission in ISDN, see § 3.5.5.

"bit error ratio masks" (i.e. diagrams giving acceptable BER as a function of the percentage of total time) may be used by the designer as long as these masks satisfy ITU-T Recommendation G.821.

TABLE 2.3

**Overall end-to-end and satellite HRDP error performance objectives
for international ISDN connections**

Performance classification	Definition	End-to-end objectives	Sat. HRDP objectives
Degraded minutes	minutes intervals with BER > 10^{-6} (more than 4 errors/minute)	< 10%	< 2%
Severely errored seconds	second intervals with BER > 10^{-3}	< 0.2%	< 0.03%
Errored seconds	second intervals with one or more errors	< 8%	< 1.6%

ii)   Digital transmission at or above the primary rate

Following the issue, by ITU-T, of a new Recommendation (G.826) entitled "Error performance parameters and objectives for international constant bit-rate digital paths at or above the primary rate" (i.e. 1.5 Mbit/s), ITU-R prepared Recommendation ITU-R S.1062 ("Allowable error performance for an HRDP operating at or above the primary rate").

This recommendation translates ITU-T Recommendation G.826 into a set of specifications usable by satellite system designers by apportioning the performance of the satellite part.

It should be noted that the general purpose of these ITU-R Recommendations (S.614 and S.1062) is to maintain the role of the FSS satellite links as a major supplier of reliable international digital communications.

Consistent with ITU-T Recommendation G.821, the requirements of ITU-T Recommendation G.826 are given in terms of errored intervals (EI). However, the definitions of the parameters are different. In ITU-T Recommendation G.826, the EI are defined in terms of error blocks (EB) as opposed to individual bit errors, in order to allow the verification of the adherence to the performance requirements on an in-service basis. This has important consequences for satellite transmissions when errors tend to occur in groups (which is the case when forward error correction (FEC) is employed). Table 2.4 below recapitulates the ITU-T objectives respectively for the overall end-to-end and for the satellite HRDP[7].

---

[7] This corresponds to a 35% allocation of the end-to-end objectives to any satellite hop, independent of the actual distance spanned.

TABLE 2.4

**Overall end-to-end and satellite HRDP error performance objectives
for digital connection at, or above, primary rate**

Performance classification	Definition	End-to-end objectives		Sat. HRDP objectives	
Bite rate		1.5 to 5 Mbit/s	>15 to 55 Mbit/s	1.5 to 5 Mbit/s	>15 to 55 Mbit/s
Bits per block		2 000 - 8 000	4 000 - 20 000	2 000 - 8 000	4 000 - 20 000
Errored seconds (ES) ratio (ESR)	ES/t: • ES: 1 s with one or more errored blocks • t: available time during fixed measurement interval	0.04	0.0075	0.014	0.0262
Severely errored seconds (SES) ratio (SESR)	SES/t: • SES: 1 s with 30% errored blocks or 1 SDP • SDP: Severely disturbed period: 4 contiguous blocks or 1 ms with BER 10^{-2}	0.002	0.002	0.0007	0.0007
Background block error (BBE) ratio (BBER)	BBE/b: • BBE: an errored block not occurring as a part of an SES • b: total number of blocks during fixed measurement interval (excluding blocks during SES and unavailable time)	$3 \cdot 10^{-4}$	$2 \cdot 10^{-4}$	$1.05 \cdot 10^{-4}$	$0.7 \cdot 10^{-4}$

NOTE – Only two bit rates (>1.5 to 15, >15 to 55 Mbit/s) are shown as typical examples in this table. For the other possible bit rates (>5 to 15, >55 to 160, >160 to 3 500 Mbit/s), see Recommendation ITU-R S.1062.

In order to get compliance with the requirements of ITU-T Recommendation G.826, Annex 1 to Recommendation ITU-R S.1062 establishes a methodology for generating "bit error probability (BEP)[8] masks", i.e. diagrams giving acceptable BEP/α as a function of the percentage of total time (α being the average number of errors per block). The translation is performed by using statistical models of the distribution of errors. However:

• the mask depends on the bit rate;

• the calculation depends on various assumptions and does not result in a unique mask.

---

[8] Calculations in Annex 1 to Recommendation ITU-R S.1062 (and also to ITU-R S.614) are made in terms of bit error probability (BEP). However measurements are made in terms of bit error ratio (BER). In practice, with a sufficient number of BER measurements, BEP is obtained.

A mask for typical bit rates is given in Recommendation ITU-R S.1062 (see Figure 2.10 below). However, it should be emphasized again that the ITU recommendation, in its latest edition, must be referred to before any precise calculation is made.

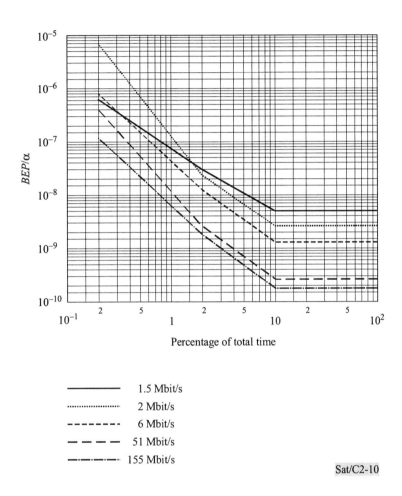

FIGURE 2.10

**Generated masks for satellite hop according
to Recommendation ITU-R S.1062**

iii)   SDH satellite transmission

Following the issue of new ITU-T Recommendation G.828 on error performance for international constant bit-rate synchronous digital paths, ITU-R is currently preparing a new recommendation on the "allowable error performance for a hypothetical reference digital path based on the synchronous digital hierarchies (SDH)".

The principles and methodology for translating parameters of ITU-T Recommendation G.828 in terms of bit error ratio (BER) or, more precisely, of bit error probability (BEP) in the satellite channel, are the same as those used as explained above for translating ITU-T Recommendation G.826 into Recommendation ITU-R S.1062. This methodology will be used to generate the necessary bit error probability (BEP) design masks. In order to fully comply with the requirements of ITU-T Recommendation G.828, the bit error probability (BEP) divided by the average number of errors per burst (BEP/α) at the output (i.e. at either end of a two-way connection) of a satellite hypothetical reference digital path (HRDP) forming part of an international connection based on SDH should not exceed these design masks during the total time (worst month).

iv)   ATM satellite transmission

As explained in Chapter 3 (§ 3.5.4) and in Appendix 3.4, the asynchronous transfer mode (ATM) is a specific packet-oriented mode which uses asynchronous time division transmission and switching of fixed size blocks (53 bytes long) called cells.

Following ITU-T recommendations on ATM, and, in particular, Recommendation I.356 (on B-ISDN ATM layer cell transfer performance) and Recommendation I.357 (on B-ISDN availability performance for semi-permanent connections), ITU-R is currently working on the translation of the performance objectives in terms applicable to connection portions including satellite links in order to issue recommendations on these subjects. This subject is dealt with in Chapter 8, § 8.6, to which the reader is referred.

Digital television

Similar considerations to the analogue TV case apply as well to the distribution of digital TV programmes. However, no specific recommendations on the performance requirements of digital primary distribution and SNG services exist for the time being. The performance objectives adopted by the operators are generally derived from the recommendations relevant to digital transmissions at or above the primary rate (ITU-T G.826, ITU-R S.1062).

The only available performance requirement appears to be provided by the European (ETSI) standard ETS 300 421. It is relevant to the DVB-S specification on digital broadcasting systems for television, sound and data services broadcasted by 11-12 GHz satellite systems. The indicated system performance objective is "quasi error free" operation, corresponding to less than one uncorrected error event per transmission hour, i.e. BER $\leq 10^{-10}$ or 10^{-11}, depending on user bit rate.

It seems to be generally assumed that the DVB standard will extend its applicability to other digital television delivery services, thus filling the missing areas relevant to QoS in digital television transmission applications.

2.2.4   Satellite operator standards

In addition to ITU-R recommendations, satellite operators such as INTELSAT and EUTELSAT have issued quality standards for their business services as expressed by Table 2.5 below.

TABLE 2.5

INTELSAT and EUTELSAT standards for IBS and SMS business telecommunications services

BER	Maximum percentage of total time during which BER is exceeded			
	INTELSAT (super IBS) EUTELSAT (high grade SMS)		INTELSAT (basic IBS) EUTELSAT (standard SMS)	
	Worst month	Year average	Worst month	Year average
10^{-3}	0.2%	0.04%	-	-
10^{-6}	2%	0.64%	3%	1%
10^{-7}	10%	4%	-	-
10^{-8}	Clear sky (INTELSAT only)		-	

NOTES

• The "super IBS/high grade SMS" columns give characteristics nearly identical to Recommendation ITU-R S.614 for 64 kbit/s ISDN links (with a supplementary slightly more stringent INTELSAT condition for clear sky).

• The "basic IBS/standard SMS" columns fix a quality objective (with a threshold at BER = 10^{-6}) but without any availability constraint).

2.2.5 Private corporate network operator performance requirements

Existing specifications listed above may be ill adapted to private corporate networks (such as VSAT networks) due to three specific factors:

• they feature direct end user connections which means that there is no need to account for other possible "last mile" connection performance deterioration;

• their quality/availability requirements can be highly dependent on the services offered (e.g. voice/data). In fact, direct consideration, by the user and the vendor, of the actual service requirements and of the transmission mode (e.g. packet) and protocols (e.g. ATM) could often lead to a better definition of the needed specifications;

• they are generally proposed in a highly competitive market environment (competition between satellite companies and/or even between satellite communications and other – terrestrial – communications means).

The two extreme cases of voice only and data only are given below as an example of the consequences of service offering on the performance requirements:

• Voice (telephony) services generally LRE (low bit rate encoding) with bit rates in the range of 2 to 8 kbit/s. Such vocoders generally include specific, powerful, source forward error correction coders (FEC) to protect the most sensitive bits. In fact, above a BER threshold which can be as high as 10^{-3} or even 10^{-2}, the subjective quality is defined more by the voice compression algorithm than by the link quality. Such applications are thus characterized by the judgement that any BER better than the availability threshold is considered acceptable.

• Data transmission services generally require much higher BER performance (10^{-6} or even 10^{-10}, e.g. for financial applications), which can be achieved:

- by providing for sufficient margin in the link budget;

- by increasing the available satellite e.i.r.p.;

- by implementing an uplink power control system, i.e. by increasing the power transmitted by the earth station in relation to atmospheric attenuation conditions;

- by using powerful FEC schemes, e.g. concatenated Reed-Solomon (RS) codes;

- by applying a convenient protocol for error correction in the transmission process (OSI lower layers), such as ARQ (automatic repeat request). Such a technique is normally applied in a random access technique like Aloha (see Chapter 5, § 5.3.6) where collisions, and therefore errored packets, cannot be avoided. The use of ARQ reduces the information throughput and introduces supplementary transmission delays which may impact on the quality perceived by the user but which can be accepted in applications where true real time is not a necessity. ARQ, which can also be implemented in the upper OSI layer (application), could be applied notably when error-free transmission (e.g. financial services) is required.

It is beyond the scope of this section to define quality and availability specifications for all types of services, but, as seen from the user's point of view, some characteristics which could be used to establish such specifications are listed below. Note that the notion of "quality of service (QoS)" and the related ITU-T recommendations could be referred to (see above § 2.2.2):

Telephony transmission

- The quality, when using LRE, is influenced both by the speech degradation due to the vocoder proper (and dramatic progress has been achieved in this field during recent years) and by transmission errors degrading the vocoder decoding process.

  In fact, below a sharp threshold, "clicks" appear in the voice flow. These clicks and other subjective voice quality criteria (such as intelligibility and speaker recognition) can be used to establish bit or packet transmission error specifications.

- The required availability could be based on a period (e.g. 10 seconds) with more than a given number of clicked seconds. The link blocking probability, due to possible traffic overload, is also to be taken into account when dimensioning the network capacity.

Data transmission

- The quality and availability specifications are generally based on the in-service detection of errored blocks (the blocks could be related to the transmission system proper, or, if possible, to the user blocks or messages).

- The transmission delay, due not only to the propagation, but also to the involved protocols, should also be specified.

- More generally, the blocking probability and the available information throughput should also be a part of the specification on quality. This specification could be based, for example, on the maximum duration of the transmission of a given message (this includes, in particular ARQ processes).

2.3 Link calculations

2.3.1   Introduction

Referring to Figure 2.1, the overall performance of a one-way link between two earth stations A and B depends on the characteristics of three elements: the uplink (A to satellite), the satellite transponder and the downlink (satellite to B). This section explains the calculation of the overall link budget for such a one-way satellite link. Of course, such a calculation can be directly extended to the case of multiple access links.

As explained above in § 2.1.1, the purpose of a link budget is to calculate the quality of a satellite communication:

- in the case of analogue, frequency modulation transmissions, this quality is evaluated by the signal-to-noise ratio (S/N);

- in the case of digital communications, this quality is measured by the information signal bit error ratio (BER).

However, it should be emphasized that, in practical applications, an inverse process is generally followed: for the transmission of a given signal between two earth stations (or even between two user terminals) with given availability and quality requirements[9], the final purpose of a link budget is to calculate the technical design parameters needed for the signal (type of modulation, error correction encoding, etc.) and for the earth station and, possibly, for the space station, i.e. the satellite (G/T, e.i.r.p., etc.). These technical parameters determine the type of equipment needed (type and size of antennas, power of the amplifiers, modems, codecs, etc.).

This § 2.3 is limited to the calculation of the factors (C/N_0), which do not postulate the choice of the transmission bandwidth (B) nor of the modulation and coding processes. It is only after the introduction, in Chapters 3 and 4, of the various coding and modulation techniques that the conversion of the (C/N_0) into (S/N) or into BER will permit evaluation of the final transmission performance. Some indications on the subject will be given at the end of the section.

Note also that this § 2.3 is only devoted to the basic formulas. Practical cases of link budget calculations will be given in Annex 2 to this Handbook.

Some basic relations between (C/N), (C/N_0), (C/T) and (E_b/N_0) are recalled in Table 2.6 below.

Important: In all formulas and calculations below, small letters (lower case) are used when numerical units are implied (with the following exceptions: T for the noise temperature, B for the bandwidth occupied by the signal, R for the digital information signal bit rate). Capital letters are used when decibels are implied.

---

[9]   As explained above in § 2.2, the required availability and quality are generally formulated in terms of percentages of time and result from the statistical behaviour of the propagation losses along earth-space paths.

TABLE 2.6

Relations between (C/N), (C/N$_0$), (C/T) and (E$_b$/N$_0$)

•	$(c/n) = (c/n_0) \cdot B^{-1}$	or, in dB: $(C/N) = (C/N_0) - 10 \log B$
•	$(c/n_0) = (c/T) \cdot k^{-1}$	or, in dB: $(C/N_0) = (C/T) - 10 \log k$

In the case of digital transmissions:

•	$(c/n_0) = (e_b/n_0) \cdot R$	or, in dB: $(C/N_0) = (E_b/N_0) + 10 \log R$

Therefore:

•	$(c/n) = (e_b/n_0) \cdot (R/B)$	or, in dB: $(C/N) = (E_b/N_0) + 10 \log (R/B)$

Where:

c: carrier power (w)

e$_b$: energy per information bit (J)

n: noise power (w)

n$_0$: noise power spectral density (w/Hz)
(noise power per Hz, i.e. $n_0 = n/B$)

k: Boltzmann's constant ($k = 1.38 \cdot 10^{-23}$ Joule/Kelvin or, in dB,
$10 \log k = -228.6$ dB/(J $\cdot$ K^{-1}))

NOTE – The ratio (R/B) depends on the symbol/bit rate in the modulation (e.g.: 1 for 2-phase phase shift keying (BPSK), 2 for 4-phase PSK (QPSK)), on the forward error correction (FEC) rate and on a bandwidth filtering factor.

For example, for QPSK modulation with FEC rate = 1/2, (R/B) ~ 2 $\cdot$ (1/2) $\cdot$ (1/1.2) ~ 0.8.

These considerations will be developed further in Appendix 3.2 and Chapter 4.

2.3.2   Uplink (C/N$_0$)$_u$

In accordance with the formulas in § 2.1.3.1, the power level received at the input of the satellite receiver is given by:

$$c_u = p_e \cdot g_{et} \cdot g_{sr}/l_u \text{ (w)} \tag{26}$$

where:

p$_e$: the output power of the earth station high power amplifier (HPA)

g$_{et}$: the earth station antenna transmit gain in the direction of the satellite, whence:

p$_e \cdot$ g$_{et}$: the equivalent isotropically radiated power of earth station (A) in the direction of the satellite, i.e.: (e.i.r.p.)$_e$

l$_u$: the free-space attenuation in the uplink (§ 2.1.3.1 and Figure 2.4)

g$_{sr}$: the satellite receiving antenna gain in the direction of the transmitting earth station A, including losses in the feeder between the antenna output and the receiver

The carrier-to-noise density ratio in the uplink is then given by:

$$(c/n_0)_u = (e.i.r.p.)_e \cdot g_{sr} /(l_u \cdot kT_u)$$
$$= (g/T)_s \cdot (e.i.r.p.)_e /l_u \cdot k^{-1} \qquad (27)$$

where:

T_u: equivalent noise temperature of the uplink at the satellite receiver input (K) (see § 2.1.4.4)

$(g/T)_s$: figure of merit of the space station (K^{-1}).

Formula (27) can be rewritten as follows:

$$(c/n_0)_u = (g/T)_s \cdot (\lambda^2 / 4\pi) \cdot (e.i.r.p.)_e /(4\pi d^2) \cdot k^{-1}$$

Here ($\lambda^2 / 4\pi$) is the effective aperture area of an isotropic antenna (§ 2.1.2.2, Formula (9)).

As $(e.i.r.p.)_e /(4\pi d^2) = (pfd)_u$, the power flux-density transmitted by the earth station antenna at the actual distance (d) of the satellite (§ 2.1.3.1, formula (12)), this figure is often included in the satellite specifications as the operating flux-density at the transponder input.

In consequence, another useful method of expressing formula (27) is:

$$(c/n_0)_u = (g/T)_s \cdot (\lambda^2 / 4\pi) \cdot (pfd)_u \cdot k^{-1} \qquad (28)$$

Formulas (27), (28) are often expressed in decibels (i.e. $(C/N_0)_u = 10 \log_{10}(c/n_0)_u$), as:

$$(C/N_0)_u = (G/T)_s + (E.I.R.P.)_e - L_u + 228.6 \qquad (29)$$

$$(C/N_0)_u = (G/T)_s + 10 \log (\lambda^2 / 4\pi) + (PFD)_u + 228.6 \ (dB \cdot Hz) \qquad (30)$$

Typical examples of L_u are: 199.75 dB at 6 GHz and 207.1 dB at 14 GHz (for a distance d = 38 607 km corresponding to a GSO satellite at 30° elevation). Typical examples of the effective aperture area of an isotropic antenna (in dB, i.e. $10 \log (\lambda^2 / 4\pi)$) are -37 dB(m^2) at 6 GHz and -44.37 dB(m^2) at 14 GHz.

2.3.3   Downlink $(C/N_0)_d$

The level of the carrier received at the input of the earth station receiver is given by:

$$c_d = p_s \cdot g_{st} \cdot g_{er}/l_d \ (w) \qquad (31)$$

where:

p_s: the output power of the satellite transponder amplifier

g_{st}: the satellite antenna transmit gain in the direction of the earth station, whence:

$p_s \cdot g_{st}$: the equivalent isotropically radiated power of the satellite in the direction of the receiving earth station, i.e.: $(e.i.r.p.)_s$

l_d: the free-space attenuation in the downlink (§ 2.1.3.1 and Figure 2.4)

g_{er}: the receiving earth station antenna gain, including losses in the feeder between the antenna output and the receiver.

Hence, the carrier-to-noise density ratio in the downlink is:

$$(c/n_0)_d = (e.i.r.p.)_s \cdot g_{er} /(l_d \cdot kT_d)$$
$$= (g/T)_e \cdot (e.i.r.p.)_s /l_d \cdot k^{-1} \tag{32}$$

where:

T_d: equivalent noise temperature of the downlink at earth station receiver input (K) (see § 2.1.4.4)

$(g/T)_e$: figure of merit of the earth station (K^{-1}).

Formula (32) is often expressed in decibels as:

$$(C/N_0)_d = (G/T)_e + (E.I.R.P.)_s - L_d + 228.6 \tag{33}$$

Typical examples of L_d are: 196.20 dB at 4 GHz and 205 dB at 11 GHz.

2.3.4   Link budget for a transparent transponder

The overall link budget calculation depends on whether the satellite is equipped with a conventional transponder or a regenerative transponder. In the former case, the role of the transponder is simply to amplify the uplink signal (with minimum distortion and noise). This is the reason why it is often called a transparent transponder[10].

In the latter case, the uplink (generally digital) signal from earth station A is demodulated in the transponder, then regenerated (often after implementing some decoding and baseband processing), re-modulated, amplified before being downlink transmitted to earth station B.

This subsection deals with transparent transponders while § 2.3.5 will deal with regenerative transponders.

2.3.4.1   Combined uplink and downlink $(C/N_0)_{ud}$

The total $(c/n_0)_{ud}$ of the link between the earth stations A and B, including only thermal noise contributions is the ratio of the signal power to the total thermal noise power, at the receiver input of B.

The signal power is: $c_{ud} = c_u \cdot g \cdot g_{st} \cdot g_{er}/l_d$, where c_u, g_{st} and g_{er} have already been defined in § 2.3.1 and 2.3.2 and where g is the transponder gain.

The noise spectral density is the sum of the uplink and downlink contributions, i.e.: $n_0 = n_{0u} \cdot g \cdot g_{st} \cdot g_{er}/l_d + n_{0d}$.

Therefore:

$$(c/n_0)_{ud} = \frac{c_u}{n_{0u} + n_{0d} \cdot l_d/(g \cdot g_{st} \cdot g_{er})}$$

---

[10] Sometimes also called a "bent pipe" transponder.

Now, since: $g = p_s/c_u$ and $c_d = p_s \cdot g_{st} \cdot g_{er}/l_d$

it follows that: $g = (c_d/c_u) \cdot l_d/(g_{st} \cdot g_{er})$

and, after simplifications:

$$(c/n_0)_{ud} = \frac{c_u}{n_{0u} + n_{0d} \cdot (c_u/c_d)}$$

or:
$$(c/n_0)_{ud}^{-1} = (c/n_0)_u^{-1} + (c/n_0)_d^{-1} \qquad (34)$$

This is an essential relation. However, since the satellite gain (g) has been assumed to be the same for the signal power and for the uplink noise, it is strictly valid only for a linearly operated transponder (including its power amplifier). But it can be shown that, even in a non-linear operation (near power amplifier saturation), it remains valid with a sufficiently good approximation.

Note that this relation is in numerical values, not in decibels. In the course of calculations, the (c/n_0)s are generally converted to (or from) decibels (dB), with $(C/N_0) = 10 \log (c/n_0)$, or reciprocally.

Of course relation (34) and its equivalent in dB can also be written in the same form by replacing the factors (C/N_0) by factors (E_b/N_0), (C/T) or (C/N).

The importance of relation (34) comes from the fact that it proves to be the general form for adding other (C/N_0) factors (or (E_b/N_0), (C/T), (C/N)). This will be shown below.

2.3.4.2   Other noise contributions to the link budget

As explained above in § 2.1.4.6, other noise contributions must be included in link budget calculations:

*   intermodulation products and other effects due to equipment non-linearities caused by multiple carrier operation in the satellite transponder (including, possibly, multiple paths) and in the earth station. As indicated in § 2.1.5.1 and 2.1.5.2, this intermodulation noise is generally assimilated to a white noise (intermodulation noise) and accounted for by a term $(C/N_0)_i$;

*   interference generated by the same satellite system or other systems. Again, this type of interference can generally be treated as non-coherent, quasi-white noise (interference noise).

An indicative list of such interference sources is given below.

Interfering emissions may come from the same satellite system in the form of:

*   transmissions in adjacent channels;

*   orthogonally polarized transmissions.

Interfering emissions may also come from other satellite systems and from terrestrial radio systems, in particular:

*   emissions received from an adjacent satellite by the side lobes of earth stations of the concerned system (downlink);

- reception, by the concerned satellite of off-axis emissions from earth stations operating towards an adjacent satellite (uplink);

- terrestrial emissions, notably radio-relay links operating in the same (shared) frequency band.

  However, the level of such emissions from other systems (intersystem interference) can be reduced by a careful system design and is constrained by:

  - the limits imposed by the Radio Regulations;

  - the limits imposed by various ITU-R recommendations, such as Recommendation S.524 (in the particular case of VSATs by Recommendations S.726, S.727 and S.728);

  - the various frequency sharing and coordination procedures (see Chapter 9).

- the most common method for dealing with interference is to include an additional noise contribution $(C/N_0)_p$ in the link budget. This is the simplest method to provide sufficient margins in the link budget. This may be employed when interference from other systems is sufficiently low, for example when the "apparent increase in the equivalent satellite link noise temperature" due to interference is lower than $\Delta T/T = 6\%$ (one interfering system) or 20% (several interfering systems). For more details on such situations, see Chapter 9.

It should be noted that situations can arise where this interference noise contribution becomes dominant in the link budget. This may be, for example, the case in rural communications systems, when very small earth station antennas are used. Such examples will be found in Annex 2 ("Typical examples of link budgets").

2.3.4.3   Total link budget $(C/N_0)_{total}$

Formula (34) above can be extended and generalized as follows to include all the other noise contributions and to derive the final result in the total (C/N_0) for the overall link budget from earth station A to earth station B:

$$(c/n_0)_{total}^{-1} = (c/n_0)_u^{-1} + (c/n_0)_d^{-1} + (c/n_0)_i^{-1} + (c/n_0)_p^{-1} \tag{35}$$

If more practical, the term $(c/n_0)_{pother}$ may itself be split into the various additional noise contributions, according to the same general type of relation:

$$(c/n_0)_p^{-1} = (c/n_0)_{p1}^{-1} + (c/n_0)_{p2}^{-1} + (c/n_0)_{p3}^{-1} + ...$$

Again, the (c/n_0)s are generally converted to/from (C/N_0)s (dB).

Through formula (35), an interesting case of link budget optimization can be derived: the intermodulation noise level can be calculated (see Appendix 5.2) as a function of the number of carriers, their relative level, their frequency distribution and the satellite amplifier (TWT) characteristic and operating point (see Figure 2.7 in § 2.1.5.1). There are two conflicting factors: increasing the TWT back-off (i.e. lowering the operating point and improving the linearity) reduces the intermodulation noise (i.e. increases the term $(C/N_0)_i$). But, at the same time, this reduces the usable power for the downlink and hence the term $(C/N_0)_d$. In consequence, when varying the TWT operating point, there is a maximum in $(C/N_0)_{total}$, which is an optimum for the link budget.

This is illustrated by the typical example of Figure 2.11 (note that, in this figure, the factors used are (C/T)s, not (C/N_0)s, but this does not change the conclusions).

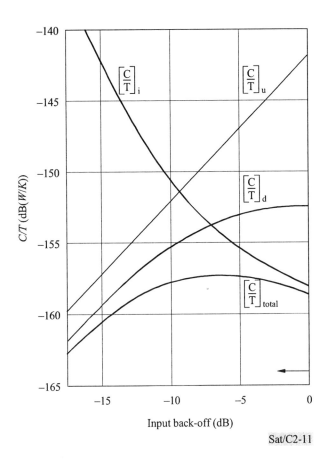

FIGURE 2.11

**Illustrative example of an FDMA
link budget optimization**

Transmission parameters

- 6/4 GHz band, 90 MHz bandwidth transponder

- FDMA

- 20 equal level carriers

- Each carrier: 30 telephone channels, FDM-FM multiplex (4.05 MHz bandwidth)

- Required quality: $(C/T)_{total} = -157.4$ dB(W/K) (this corresponds to $(S/N) = 51.4$ dB (see § 4.1))

NOTE – The $(C/T)_{total}$ represented on the figure corresponds to the worst case (i.e. the carrier which is the most affected by intermodulation products).

2.3.4.4   Atmospheric losses

In addition to the free-space attenuation, atmospheric losses must also be accounted for in link budget calculations.

These atmospheric losses, which are particularly significant at the higher frequencies (over 10 GHz), are mainly due to precipitation and depend therefore:

- on local meteorological conditions;

- on the requirements in terms of link availability and transmission quality. As explained above in § 2.2, these requirements are expressed as percentage of time. They can be considered as margins for accommodating operation at a given level of the meteorological conditions.

The calculation of the atmospheric losses is given in Annex 1.

In consequence:

- for the uplink, a term $-(A_p)_u$ may have to be included in formulas (29) or (30);

- for the downlink, a term $-(A_p)_d$ may have to be included in formula (33). Moreover, there is another consequence of a downlink atmospheric attenuation: as explained above in § 2.1.4.3, the antenna noise temperature increases which results in a decrease in the actual figure of merit $(G/T)_e$ of the earth station.

Note that, for a given percentage of time and for a given meteorological environment, the term $-(A_p)_u$ is generally greater than $-(A_p)_d$, due to the higher frequency band used on the uplink.

For a given total link availability requirement, there is an optimum sharing of the partial percentages of time of unavailability for the up and downlinks (for a given local meteorological environment). The calculation of this optimum gives the terms $-(A_p)_u$ and $-(A_p)_d$ to be added in the link budget.

Depending mainly on the demodulator performance, there are other circumstances where the constraint comes from the total link quality requirement (the total link availability requirement being then automatically obtained). In this case, the terms $-(A_p)_u$ and $-(A_p)_d$ are to be derived from the optimization of the partial percentages of time of quality threshold for the up and downlinks.

2.3.4.5   Remarks

i)   Power flux-density in multi-carrier operation

In formulas (28) or (30), the power flux-density $(pfd)_u$ transmitted by the earth station for the carrier under consideration is often calculated by referring to the maximum power flux-density $(pfd)_{s\,max}$, which corresponds to the satellite transponder saturation and which is included in the satellite communications payload performance characteristics. In the case of multi-carrier operation, this usable $(pfd)_{s\,max}$ is shared between all the carriers which are simultaneously transmitted through the transponder. However, some back-off is generally needed (especially in the case of multi-carrier operation) in order to operate the transponder with adequate linearity. Therefore, a general expression of the $(pfd)_u$ in formula (28) for the carrier under consideration is:

$$(pfd)_u = \frac{(pfd)_{s\,max}}{(bo)_{si}} - \sum (pfd)_{ui} \quad W/m^2 \tag{36}$$

In this expression, $\Sigma\,(pfd)_{ui}$ is the sum of the power flux-densities of the other carriers sharing the same transponder and $(BO)_{si} = 10\log(bo)_{si}$ is the input transponder back-off in decibels.

In the particular case of n equal carriers, this expression can be simplified and written directly in $dB(W/m^2)$ as follows:

$$(PFD)_u = (PFD)_{s\,max} - 10\log n - (BO)_{si} \qquad dB(W/m^2) \tag{37}$$

ii)   Earth station e.i.r.p.

Taking into account the relation between the transponder input and output back-offs (as shown for example in Figure 2.7, similar expressions can be used in formulas (32) or (33) for calculating the $(e.i.r.p.)_s$ or $(E.I.R.P.)_s$ of the carrier under consideration, i.e.:

$$(e.i.r.p.)_s = \frac{(e.i.r.p.)_{s\,max}}{(bo)_{so}} - \sum (e.i.r.p.)_{si} \tag{38}$$

$$(E.I.R.P.)_s = (E.I.R.P.)_{s\,max} - 10\log n - (BO)_{so} \qquad dBW \tag{39}$$

where $(e.i.r.p.)_{s\,max}$ is the satellite e.i.r.p. when the transponder is saturated by a single carrier (this is a part of the satellite performance characteristics, $\Sigma\,(e.i.r.p.)_{si}$ is the sum of the $(e.i.r.p.)_s$ of the other carriers and $(BO)_{so} = 10\log(bo)_{so}$ is the output transponder back-off in decibels).

Note that similar expressions can be used for calculating the earth station $(e.i.r.p.)_e$ in formulas (27) and (29) by reference to the saturated output power of earth station high power amplifier.

2.3.4.6   Conclusions

As explained above, the conversion of the (C/N_0) into (S/N) or into BER depends on the modulation (Chapter 4) and coding (Chapter 3) techniques implemented on the link:

If analogue, frequency modulation (FM) is used, the formulas of § 4.1.1.1 (FDM-FM), 4.1.1.2 (SCPC-FM) and 4.1.1.4 (TV-FM), permit conversion of (C/N_0) into (S/N) (or reciprocally).

If digital modulation is used, the conversion (C/N_0) into BER (or reciprocally) is dealt with in § 4.2.2 (PSK modulation, in particular Figure 4.2.1) and in § 3.3.5 and Appendix 3.2 as concerns forward error correction (FEC) coding.

It should also be mentioned that link budget calculations are now generally performed by computer calculations. Some available computer programs are listed in "Supplement No. 2 to the CCIR Handbook on Satellite Communications (FSS)" (Handbook ITU-R M1, Geneva 1993). However, most of the currently used programs remain proprietary.

2.3.5   Link budget for a regenerative transponder

On-board regeneration is generally associated with digital modulation. As explained above in § 2.3.3, the uplink signal from earth station A is demodulated in the transponder, then regenerated and transmitted to earth station B.

Therefore, the uplink is separated from the downlink. The features of on-board regeneration will be dealt with in Chapter 6. However, some characteristics and advantages which are pertinent to the subject of link budget are listed below:

i)   Formula (34) does not apply here. Because demodulation is applied in the transponder, it is necessary to convert directly $(C/N_0)_u$ in terms of bit error ratio (BER).

Therefore, notwithstanding the type of on-board modulation/demodulation and anticipating the results of Chapter 4, the up and downlink (BER)s will be used here. In fact, formula (34), which shows the addition of noise contributions from the uplink and the downlink, is to be changed with the addition of $(BER)_u$ and $(BER)_d$:

$$(BER)_{ud} = (BER)_u + (BER)_d \text{ [11]} \qquad (40)$$

Of course, the calculation of $(BER)_{total}$ may need the addition of other (BER) terms (just as in formula (40)). However, these miscellaneous noise contributions are generally minimized, as discussed below in ii) to v).

However, it should be noted at this point that formula (40) (regenerative transponder) can bring a net advantage in the link budget, compared to formula (34) (transparent transponder), as shown by the two case examples given below, both using coherent PSK modulation/demodulation with a requested $(BER)_{ud} = 10^{-4}$ (corresponding to $(E_b/N_0)_{ud} = 8.4$ dB[12]):

• Let us assume first that $(E_b/N_0)_u = (E_b/N_0)_d$. For a transparent transponder $(E_b/N_0)_{ud} = (E_b/N_0)_u - 3$ dB which means that $(E_b/N_0)_u = 11.4$ dB is requested (the same for $(E_b/N_0)_d$).

In the case of a regenerative transponder $(BER)_{ud} = 2 \cdot (BER)_u$ and the requested (E_b/N_0), both for $(E_b/N_0)_u$ and $(E_b/N_0)_d$, is only 8.8 dB[12], an advantage of 2.6 dB.

• In the second case example, it is assumed that, with a regenerative transponder, $(E_b/N_0)_u > (E_b/N_0)_d$, e.g. the difference is about 1.5 dB. Then, due to the steepness of the PSK performance function, $(BER)_u$ becomes much smaller than $(BER)_d$ and $(BER)_{ud} \sim (BER)_d$.

---

[11] In fact, since the uplink error bits are transmitted on the downlink, a more exact relation should be:

$$(BER)_{total} = (BER)_u \cdot [1-(BER)_d] + (BER)_d \cdot [1-(BER)_u]$$

However, the (BER) products are negligible.

[12] Refer below to Chapter 4, § 4.2.2 (PSK modulation, Figure 4.2-1) for PSK performance (BER vs. (E_b/N_0)). Of course, all values given here are theoretical and do not account for various degradations and implementation margins.

For example, if $(E_b/N_0)_u = 10$ dB, $(BER)_u = 4 \cdot 10^{-6}$ and $(E_b/N_0)_d = 8.5$ dB will be sufficient for getting the requested performance $(BER)_{ud} \sim 10^{-4}$. In the transparent transponder case, $(E_b/N_0)_d = 13.5$ dB should be needed.

Of course, if $(E_b/N_0)_d > (E_b/N_0)_u$, this is the same as changing the downlink advantage to an uplink advantage.

Such link budget advantages can be used, either in the space segment, to reduce the on-board amplifier sizing, or in the earth segment, to implement smaller earth stations, in terms of antenna or power amplifier size.

ii) Most regenerative transponders are associated with some kind of on-board processing (OBP) and/or satellite antenna switching (SS). The format of the downlink signal (multiple access, multiplexing, modulation) can be different from the uplink one[13]. Under such conditions, intermodulation is scarcely a problem.

iii) The effects of non-linear distortions (see above, § 2.1.5.3) in the two links do not accumulate. The on-board RF amplifiers are fed by signals which are virtually free from amplitude modulation and can therefore be operated very near saturation.

iv) Any on-board multi-path interference arising from RF coupling among different channels is generally eliminated.

v) However, interference from other satellites (or terrestrial radio systems) may need to be accounted for.

2.4 Earth coverage and frequency reuse

2.4.1 Earth coverage by a geostationary satellite

2.4.1.1 Elevation angle from the Earth

For most regions of the world, the geostationary-satellite orbit (GSO) provides a uniquely favourable location for communication satellites. A spacecraft in this orbit appears fixed relative to the surface of the Earth, and problems of earth station antenna tracking are minimized or eliminated. Also, the GSO altitude of about 36 000 km provides global Earth surface coverage. The principal limitation in coverage is the area above about 75° North or South latitude. Figure 2.12 shows the Earth's surface, as seen from a GSO satellite, with a pattern indicating the elevation angles of the line-of-sight.

Figure 2.13 gives the satellite elevation angle and distance as a function of the earth station's latitude and longitude (relative to the sub-satellite point longitude).

---

[13] For example, FDMA can be implemented on the uplink, with TDM on the downlink. Or an uplink TDMA can be reconfigured by so-called on-board TST (time-space-time) switching. These various processes and network architectures are described in Chapter 6 (§ 6.3.3) and Chapter 5 (§ 5.6) (see also: Supplement 3, "VSAT Systems and Earth Stations", ITU, Geneva 1994 § 6.3, pp. 185-188).

The precise computation of the azimuth angle α (measured clockwise from North) and the angle of elevation, ε, of a satellite in a circular equatorial orbit can be expressed by:

$$A = \text{arc tan} (\tan \theta / \sin \varphi)$$

$$\varepsilon = \text{arc sin} \frac{K \cos \varphi \cos \theta - 1}{\sqrt{K^2 + 1 - 2K \cos \varphi \cos \theta}}$$

Eliminating θ between these two equations leads to:

$$A = \text{arc cos} \left\{ \left[\frac{\tan \varepsilon + K^{-1} \sqrt{\tan^2 \varepsilon + (1 - K^{-2})}}{1 - K^{-2}} \right] \tan \varphi \right\}$$

or to:

$$A = \text{arc cos} \frac{\tan \varphi}{\tan [\text{arc cos}(K^{-1} \cos \varepsilon) - \varepsilon]}$$

The provisional azimuth angle A shall be modified as follows according to the location of the earth station on the Earth's surface:

$\alpha = A + 180°$ for earth stations located in the northern hemisphere and satellites located West of the earth station.

$\alpha = 180° - A$ for earth stations located in the northern hemisphere and satellites located East of the earth station.

$\alpha = 360° - A$ for earth stations located in the southern hemisphere and satellites located West of the earth station.

$\alpha = A$ for earth stations located in the southern hemisphere and satellites located East of the earth station.

where:

α: azimuth angle (measured clockwise from North)

K: orbit radius/Earth radius, assumed to be 6.63

ε: geometric angle of elevation of a point on the geostationary-satellite orbit

φ: latitude of the earth station

θ: difference in longitude between the earth station and the satellite

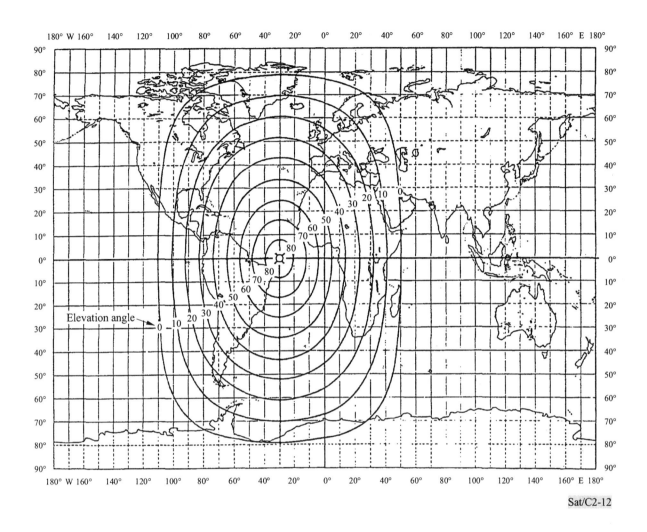

FIGURE 2.12

**World planisphere with a typical pattern showing the elevation
angle of the line-of-sight from an earth surface
point towards a GSO satellite**

NOTE – In this figure, the satellite is located, for illustration, at 30°W (sub-satellite point). If the pattern is copied on a transparency, it can be centred on any sub-satellite point on the Equator. The projection used in this figure (lines of equal longitude are parallel) distorts the actual geometry of the Earth, but offers a convenient method for determining GSO satellite coverage on the Earth's surface with a two-dimensional map.

There are practical limitations to the minimum usable elevation angle. Those depend on the local conditions: unobstructed field of view, atmospheric propagation conditions and, possibly, multipath effects.

As concerns propagation conditions, it should be kept in mind that they can be severely impaired at low elevation angles (see § 2.3.4.4 and Annex 1), especially at frequencies above 10 GHz.

Multipath effects are scarcely significant in satellite link operation (unlike terrestrial microwave links) because the earth station antenna half beamwidth is generally smaller than the elevation angle of the satellite direction (i.e. the angle of the main beam axis).

In summary, 6/4 GHz systems can often be used at elevation angles as low as 5°[14], while 14/12 GHz systems may require a minimum elevation angle of 10°.

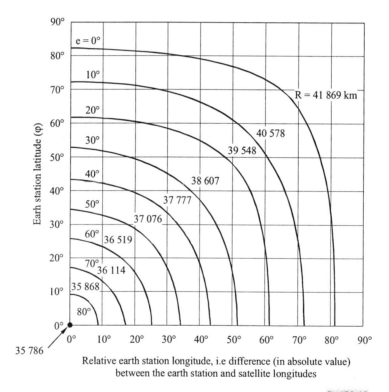

Sat/C2-13

FIGURE 2.13

**Elevation angle (e) and distance (R) of a GSO satellite,
as seen from an earth station**

---

[14] However, earth stations have been successfully operated at elevation angles lower than 5°. For example, earth stations have been implemented by Norway in the Spitzbergen islands (78° N) with a 1.7° elevation angle.

2.4.1.2   Types of earth coverages and of satellite antenna beams

Figure 2.12 shows a coverage pattern for a global circular beam satellite antenna centred directly over the sub-satellite point (17.4° beamwidth for a 5° elevation angle). In practice, this maximum earth coverage is only used for systems requiring global coverage. For example, Figure 2.14 shows how three global beam transponders can be implemented in order to cover the whole of the Earth's surface.

In most cases, satellite antennas are designed to radiate a much smaller effective beamwidth in order to increase the satellite e.i.r.p. in the service area, whatever the shape of this area. Such narrow-beam antennas should be designed so that the energy is concentrated towards this area and reduced to the lowest possible level outside. Therefore, in order to enhance orbit utilization, the objectives of the satellite antenna design should be to achieve the following characteristics (which can be mutually conflicting):

- the main beam pattern should model the service area (also called coverage area) as closely as possible. This means it should not be a simple circular or elliptical beam, but a beam with a special, complex shape, called a "shaped beam"[15];

- the radiation outside the service area (in particular the side lobes) should be minimized in order to reduce interference and ease the problems of coordination with other networks. Recommendation ITU-R S.672-2 specifies in detail the design objectives for the antenna gain (i.e. the radiation level) outside the coverage area. The annexes to this recommendation give precise descriptions of the gain contours of the beam in the coverage area;

- the antenna should be designed, as far as possible, to permit repositioning at other locations on the GSO, while keeping the required performance.

Remark on e.i.r.p. and G/T contours

The satellite figures of merit for transmission (downlink), i.e. the e.i.r.p. and for reception (uplink), i.e. the G/T, are represented on maps of the Earth as "iso-e.i.r.p." or "iso-G/T" contours for facilitating link budget calculations. In some cases, only the limiting contours at the skirts of the service area (beam edge), showing the minimum performance available inside this area, are represented. Of course, inside the service area, higher e.i.r.p. and G/T are available and this is often represented on the earth coverage figures by "iso-e.i.r.p." or "iso-G/T" step-by-step contours. This is shown below in Figures 2.15, 2.16, 2.18, 2.19 (iso-e.i.r.p.), 2.17 and 2.20 (iso-G/T).

The following types of coverages and of satellite antennas are most commonly used:

i)   Global coverage, as described above (Figures 2.12 and 2.14). The satellite antenna can then be a simple microwave circular horn.

ii) Circular or elliptical coverage, with beamwidths from about 5° to 10°. A focus-fed parabolic reflector is often used, but the utilization of offset-fed reflectors improves significantly the side lobe performance.

---

[15] "Circular", "elliptical" or "shaped" refers to the outline of the beam pattern in a plane normal to the main radiation axis (although, it should be recognized that the definition of such an axis could be ambiguous in the case of shaped beams).

iii) Narrow, simple (circular or elliptical) coverage, called spot beam coverage. An example is given in Figure 2.15 (INTELSAT VIIA spot beams).

Spot beams with a beamwidth as narrow as 0.5°, or even less, are feasible in the 14/11-12 GHz band (or higher frequency bands) but their utilization can be limited by the required pointing accuracy.

Spot beams are usually provided with a steering capability (with ground control). In advanced satellite designs, multiple spot beams or scanning spot beams can be implemented, generally in conjunction with an on-board switching matrix in order to cover the service area by a dynamic process (see Chapter 6, § 6.3.4).

Another possible utilization of multiple, narrow spot beams is to provide links to very small, even handheld, earth stations ("terminals") through cellular coverage of the Earth. However, most systems proposed for such a service (e.g. Teledesic and Skybridge) are based on non-geostationary (non-GSO) satellites (see below § 2.4.2). This is often also the case for many such systems operating in the MSS (e.g. Iridium, Globalstar, ICO, etc.).

iv) Shaped coverage which matches a specific service area (e.g. a particular country or territory), whatever its contour. Such shaped beam antennas are usually constructed with a reflector illuminated by multiple feeds connected to the transponders (on the receive or on the transmit side) through a "beam-forming network". Beam shaping results from the composition (i.e. the in-phase addition of the RF fields) of the spatially separated elementary patterns (or "beamlets") radiated by each feed.

Shaped coverages are implemented in most modern satellites and the antenna can be very complex when the contour of the service area is itself distorted. For example the Eutelsat II satellite antenna comprises 17 component feed horns, the INTELSAT V satellite 88 and the INTELSAT VI 147. Note that the smaller the beamlet width, the more accurately will the shaped beam be modelled to the service area (for more details, see Chapter 6). The figures below illustrate various typical examples of shaped coverages actually implemented with the Arabsat II satellites (Figures 2.16 and 2.17), with the INTELSAT VIIA satellites ("Hemispheric beams", Figure 2.18 and "zone beams", Figure 2.19) and with the Eutelsat II satellites (Figure 2.20).

The particular type of antenna to be used depends on frequency and on the available physical size and mass available on the launch vehicle payload. This size and mass determines the maximum dimension of the antenna (D) and of the (possibly multiple) feed assembly. Minimum feasible beamwidths (or beamlet widths) depend upon l/D (l being the wavelength). On the other hand, they also depend upon the precision with which the satellite altitude can be stabilized and controlled (see Recommendation ITU-R S.1064).

It should be noted that the sub-satellite point can be outside the service area. For a given beamwidth, advantage can be taken of the increased coverage offered by an oblique pointing of the satellite antenna. However, this advantage must be balanced against the requirement for a higher pointing accuracy and a lower elevation angle of earth station.

Other considerations, such as possible need for frequency reuse implementation (see below § 2.4.3), should also be accounted for.

Remark on satellite e.i.r.p.

The narrower the coverage, the greater will be the satellite antenna gain and therefore the e.i.r.p. (for a given amplifier power).

For example, on the 6/4 GHz transponders of INTELSAT VIIA satellites, the global beam has an e.i.r.p. of 29 dBW, to be compared with the zone beam e.i.r.p. of 36.5 dBW (beam-edge minimum figures). E.i.r.p. of 47 dBW or higher are not uncommon on 14/11 GHz domestic satellites (Australia, Canada, Eutelsat, Hispasat etc.).

High satellite e.i.r.p. permits simple, economical earth stations with small antennas to be implemented.

In the case of simple (circular or elliptical) beams, the antenna gain can be approximated by formulas (3) and (3*bis*) in Appendix 2.1. Note that, in consequence, the antenna gain and e.i.r.p. (for a given satellite amplifier power) does not depend on the frequency, but only on the coverage width.

Using this simple case, it can be shown that the antenna gain should be about 4 dB lower on the beam edge than at the beam centre[16]. As already explained, the utilization of shaped beam antennas can improve this situation by optimizing the gain inside the service area (i.e. getting a more uniform gain) and minimizing it outside.

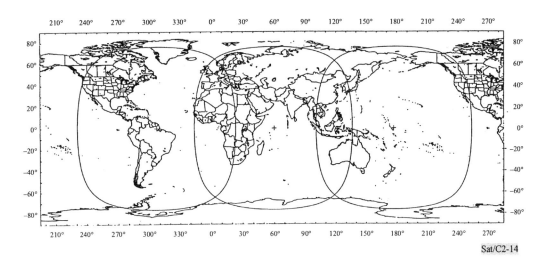

Sat/C2-14

FIGURE 2.14

World coverage by GSO satellites
(INTELSAT system with 3 global beams in the Atlantic, Indian and Pacific ocean regions.)

---

[16] The –4 dB figure corresponds to a theoretical optimum. This results from the natural shape of a simple antenna beam and can be illustrated by the following example: for a 5° circular service area (as seen from the satellite), and for a 5° wide antenna beam (measured at –4 dB level), the maximum gain at the beam centre is about 30 dBi and the gain at the coverage limit (±2.5°) is, of course, 26 dBi. Should the same service area be illuminated by a thinner beam (e.g. 4°), the beam centre gain will be higher (31.9 dBi in this example) but the coverage limit gain will be smaller (25.3 dBi). On the contrary, with a wider beam (e.g. 6°), these gains will be respectively 28.4 dBi and 25.7 dBi.

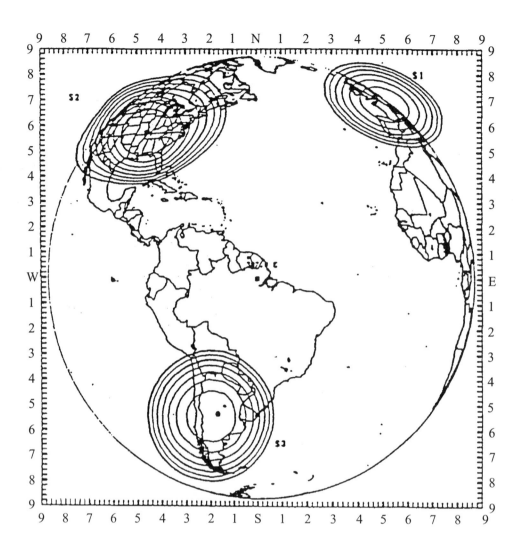

NOTE – (Satellite location: 307° E). 6/4 GHz spot beams transmit coverage. The nominal beam-edge e.i.r.p.$_s$ (dBW) are 45.0 (S1), 43.5 (S2) and 41 (S3). "Iso-e.i.r.p." contours are illustrated at 1 dB steps from the outer beam edge.

Sat/C2-15

FIGURE 2.15

INTELSAT VIIA typical satellite coverage

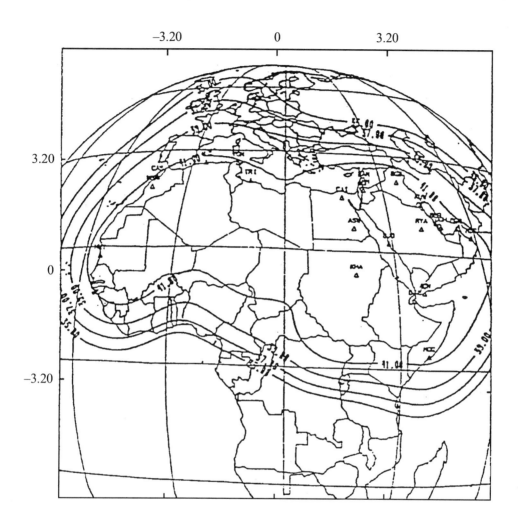

NOTE – 6/4 GHz transponder coverage (transmit, 4 GHz band: iso-e.i.r.p. contours. Right-hand circular polarization (RHCP) (frequency reuse)).

Sat/C2-16

FIGURE 2.16

Arabsat II typical satellite coverage

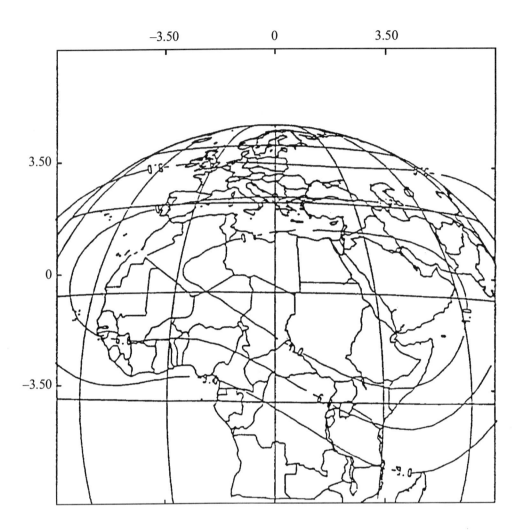

NOTE – 6/4 GHz transponder coverage (receive: 6 GHz band: iso-G/T contours. Left-hand circular polarization (LHCP) (frequency reuse)).

Sat/C2-17

FIGURE 2.17

Arabsat II typical satellite coverage

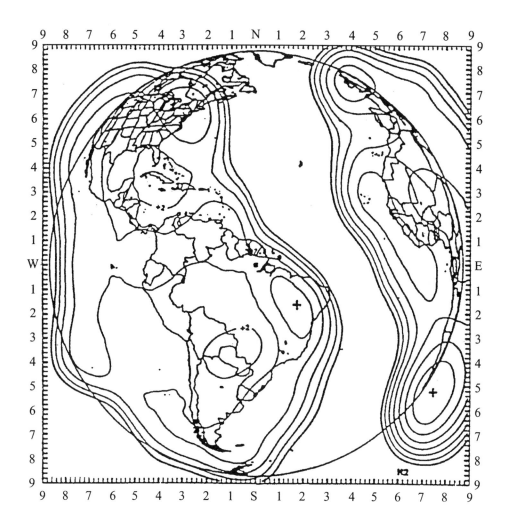

NOTE – (Satellite location: 307° E). 6/4 GHz "hemispheric" beams: 4 GHz band transmit coverage. The nominal beam-edge e.i.r.p.$_s$ (dBW) are 32.5 (H1 and H2). "Iso-e.i.r.p." contours are illustrated at 1 dB steps from the outer beam edge.

Sat/C2-18

FIGURE 2.18

INTELSAT VIIA typical satellite coverage

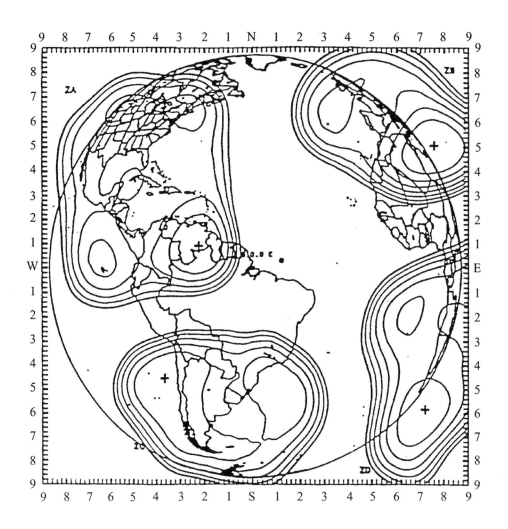

NOTE – (Satellite location: 310° E). 6/4 GHz "zone" beams: 4 GHz band transmit coverage. The nominal beam-edge e.i.r.p.$_s$ (dBW) are 33 (ZA, ZB, ZD) and 32.5 (ZC). "Iso-e.i.r.p." contours are illustrated at 1 dB steps from the outer beam edge.

Sat/C2-19

FIGURE 2.19

INTELSAT VIIA typical satellite coverage

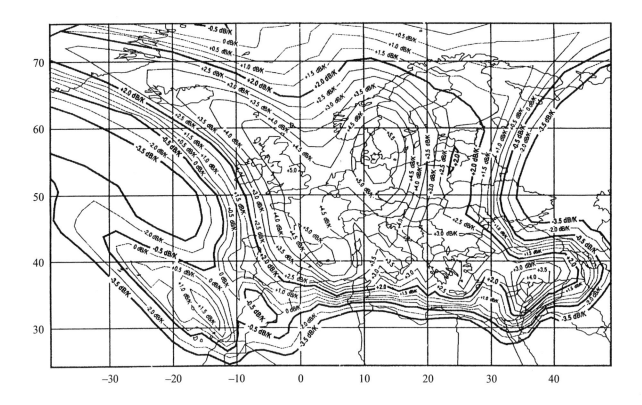

NOTE – (Satellite location: 16° E). 14/11-12 GHz: 14 GHz band receive coverage, "iso-G/T" contours.

Sat/C2-20

FIGURE 2.20

EUTELSAT II typical satellite coverage

2.4.2 Earth coverage by a non-geostationary satellite

This section refers to new systems based on non-geostationary satellite (non-GSO) systems. With the general purpose of providing very large numbers (maybe millions) of users with small – even hand-held – earth stations (terminals), such systems and concepts are currently the subject of very active studies, developments and of firm programs. Many of these systems are actually based on low-earth orbit satellites (LEOs), but other non-GSO systems, such as medium orbit – MEOs – and highly elliptical – HEOs – or even GSO satellite systems, have been and are proposed for the same services (see above § 2.1.3.1).

The considerations below are general and apply to many current projects, whether they pertain to the fixed-satellite service (FSS) or to the mobile-satellite service (MSS). In general, service to small mobile and hand-held terminals is provided via satellites in the MSS. Service to small, sometimes transportable earth stations is provided via satellites in the FSS.

Of course, non-GSO satellites are not seen from earth stations as fixed in the sky. For example, LEOs generally move in circular, low altitude, orbits at a relatively high speed (e.g. about 7 km/s for a 700 km altitude). In order to maintain the permanency of the service and of the communications, multiple satellites are regularly distributed around orbit planes. Moreover, in order to allow communications with various regions, multiple orbit planes are implemented. The complete set of satellites of all orbits forms the constellation implemented in the system.

These systems are generally based on cellular architectures similar to terrestrial mobile telephone networks. This means that the service area (which may extend over the whole surface of the Earth) is divided into cells. A given earth station communicates with a satellite via a satellite spot beam covering the cell in which the station is located.

However, in the case of LEOs, there is a significant difference in comparison with terrestrial systems: a given cell[17] is not serviced by a fixed satellite spot beam, but is regularly and permanently scanned by different transmit and receive spot beams corresponding to different satellites. In some systems, the satellite, when passing over a given cell, steers its antenna beam in order to keep it longer over the cell and to compensate for the satellite motion and for the Earth's rotation. Anyhow, and especially in the case of small cells, the continuity of the communications should rely on the soft call handoff techniques[18] commonly used in terrestrial cellular systems.

The instantaneous satellite footprint may cover a relatively large area (e.g. a 1 000 km diameter zone) but there are many possibilities: one cell can be associated with a single spot beam covering the whole footprint. In other proposed systems, the satellite footprint is divided into very small cells or microcells (e.g. 200 km, see below § 2.4.3.4 and Figure 2.22) corresponding to multiple thin spot beams.

The system can generally be connected to other networks (notably to public networks) through gateway earth stations. As concerns the connection of a communication from/to a given terminal to/from another terminal which can be located in another, distant, cell, i.e. serviced by another satellite, there are currently two philosophies: the more straightforward way is to transit through a gateway and, possibly, through two gateways connected by a terrestrial link. A more advanced process consists in connecting the concerned satellites through int-satellite links (ISLs).

Similar to terrestrial mobile systems, most satellite systems described above rely intensively on frequency reuse in the different beams and/or cells, as explained below (§ 2.4.3.4) More details on non-GSO systems will be found in Chapter 6 (§ 6.5).

2.4.3   Frequency reuse

A definition of the frequency reuse concept has been given in Chapter 1 (§ 1.3.3 and Figure 1.3).

---

[17] In this section, the term "cell" is associated with a given, fixed, earth zone. However, in some system descriptions, a cell is associated with a moving spot beam.

[18] Call handoff designates the transfer of a communication when the cell, or the satellite beam, servicing a given terminal is changing. Soft handoff is ensured when there is a complete continuity in the transfer, without any interruption in the communication.

The available spectrum in any given band is finite and the communication capacity of a satellite is determined accordingly. To increase satellite system capacity, methods of reusing the frequency spectrum have been developed and are actually implemented. As already explained, the two common methods are spatial isolation (also called beam separation) and polarization discrimination (also called dual polarization).

2.4.3.1 Spatial isolation frequency reuse

Spatial isolation frequency reuse on a given satellite is performed by implementing the same frequency band in different transponders and antenna beams towards different coverage areas. The antenna beams must be separated enough to provide sufficient isolation for avoiding harmful interference, a typical requirement being a 27 dB isolation. As illustrated in Figure 2.21, this means that, if an earth station in service area A receives a carrier destined for service areas B or C, this carrier must be at least 27 dB lower than its own destined carrier (at the same frequency).

The isolation may be limited by the cross-over relative gain level between two beams (between A and B in Figure 2.21) or by the side-lobe levels (e.g. between A and C). In the case of insufficient isolation, a possible solution is to introduce a supplementary isolation by implementing cross-polarized carriers between the too close beams (at transmit and receive).

As a simple example, nearly all current INTELSAT satellites use separate so-called "hemispheric beams" (see Figure 2.18) and "zone beams" (see Figure 2.19) in which spatial isolation frequency reuse is implemented. However, it is clear on the figures that, in these examples, there is a large angular distance between the various beams.

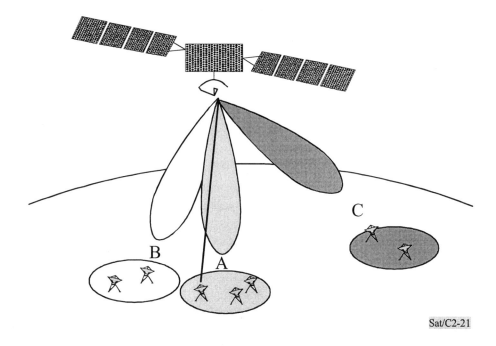

Sat/C2-21

FIGURE 2.21

Spatial isolation frequency reuse

NOTE – The transmit and receive beams may be radiated by a single satellite antenna (as represented) or by several antennas.

2.4.3.2   Polarization discrimination frequency reuse

Dual polarization frequency reuse on a given satellite is performed by implementing the same frequency band in two different transponders which serve the same coverage area, but with two orthogonal polarizations. The two polarizations may be linear (e.g. horizontal and vertical) or circular (right hand circular polarization: RHCP and left hand circular polarization: LHCP). On the satellite, there can be two different antennas, one for each polarization or a common antenna can be used. In the latter case, the (transmit or receive) antenna is connected to the transponder through a microwave coupler with two orthogonally polarized ports, called an ortho-mode transducer (OMT). On the earth stations, there is always a common antenna equipped with a transmit/receive OMT. Typical values of the discrimination between the two polarizations are between 30 and 35 dB.

Dual polarization frequency reuse techniques, which employ either circular or linear polarized transmissions, can be impaired by the propagation path through the Earth's atmosphere, resulting in interference between the two orthogonal polarized transmissions. This phenomenon, referred to as depolarization or cross-polarization, is induced by two sources: Faraday effect, which is caused by the Earth's magnetic field in the ionosphere, and precipitation, primarily rain or ice crystals.

Faraday effect causes a time-varying and frequency-dependent polarization rotation (not depolarization) in the orientation of the polarization plane. Peak values of Faraday rotation can be as high as 9° at 4 GHz and 4° at 6 GHz. If Faraday rotation is significant at the frequency of operation, a method of providing differential rotation of the planes of polarization must be provided at the earth station antenna. This is because the direction of rotation, as seen in the direction of propagation, is opposite for transmit relative to receive. Faraday rotation has a negligible effect on circular polarization and, in many cases for frequencies above 10 GHz, for linear polarization as well.

Rain-induced depolarization is caused by non-spherical rain drops, which produce a different attenuation and phase shift between orthogonal linear components of the signal which degrades the discrimination between either linearly or circularly polarized signals. Although differential phase shift is the primary cause of depolarization, differential attenuation also becomes important above 10 GHz.

Propagation models have been developed to predict depolarization impairments due to rain and other meteorological factors. In regions with very high rainfall rates (such as Central America and tropical zones), dual polarization frequency reuse operation could be subject to implementation problems. For more details, see Annex 1 (§ 3).

The application of dual polarization frequency reuse places additional requirements on the satellite and earth station antennas:

a)   As explained above, each part of the antenna must transmit (or receive) only RF fields according to its own polarization (e.g. LHCP) and must be isolated from the other polarization (e.g. RHCP): this requirement is called polarization discrimination (or polarization isolation). If polarization discrimination is not sufficient (e.g. less than 24 to 28 dB depending on other possible interference noise sources and including imperfect satellite and earth station antennas and also propagation medium effects), a signal at the same frequency will appear from the orthogonal polarization (also called X-polar) and will cause harmful interference at the normal (also called co-polar), wanted port.

b) This polarization discrimination requirement may impose constraints in the overall design of earth station antennas, including the components of the feed: duplexer and orthomode transducer (OMT)[19], polarizers (in the case of circular polarization), angular tracking system, primary source (radiating horn) and even on the reflector system (in particular mechanical symmetry of supporting struts), etc.

The transmit and receive radiation patterns of properly designed earth station antennas should exhibit a high polarization purity in the beam axis and in the main part of the beamwidth and even a sufficient purity (low X-Polar components) in the side lobes in order to avoid interference on the GSO.

Ultimately, polarization compensation devices and systems could be implemented in the earth station antenna and communication equipment. Such devices and systems have been developed but their practical application proves to be complicated.

Notwithstanding these implementation problems, due to the scarcity of orbit/spectrum resources, many satellite systems extensively use polarization discrimination frequency reuse.

Note that, even if dual polarization frequency reuse is not needed in an earth station at the beginning of operation, it could prove wise, at the programme planning phase, to provide for its future introduction. It must be also remembered that dual polarization must be accounted for in internetwork interference and coordination calculations.

More details on earth station dual polarization antenna design are given in Chapter 7, § 7.2.2.

2.4.3.3   Common application of both methods of frequency reuse

The two methods of frequency reuse described above are, in general, compatible and are actually often implemented simultaneously. In fact, this common implementation proves to be a powerful tool for augmenting satellite systems' traffic capacity and for improving orbit-spectrum utilization efficiency. It permits the provision of the equivalent of a bandwidth 2M x B on a satellite where an actual bandwidth B is allocated (M being the number of beams using the same frequency band by spatial isolation frequency reuse).

As an example, six-fold frequency reuse is implemented in the INTELSAT VI system.

However, a high degree of frequency reuse is not without cost: interference noise may eventually appear as the dominant factor in the link budgets and as the main limitation in satellite traffic capacity.

---

[19] Earth stations whose traffic plans provide for communicating simultaneously in both polarizations must be equipped with a four-port duplexer/OMT (two transmit and two receive ports).

2.4.3.4   Application of frequency reuse in non-geostationary satellite systems

Spatial isolation frequency reuse is systematically implemented in most of the non-GSO systems servicing very small terminals in the FSS or in the MSS. This permits the coverage of very large service areas with a minimum bandwidth.

In fact, as already explained in § 2.4.2, the service area is usually covered by a multiplicity of closely located and even intersecting satellite spot beams, which can correspond to the cells covering the Earth. In general, the various beams corresponding to cells or microcells around a given point on the Earth use different frequency bands in order to avoid mutual interference. Just as is done in terrestrial cellular systems, the mosaic figure of frequencies, i.e. the distribution of the frequency bands between the beams in the service area should be determined by the minimum angular distance required to ensure a sufficient isolation between beams using the same frequency band. The utilization of cross-polarization, i.e. of dual polarization frequency reuse, can also be envisaged to supplement isolation. A typical illustration of such arrangements is given in Figure 2.22.

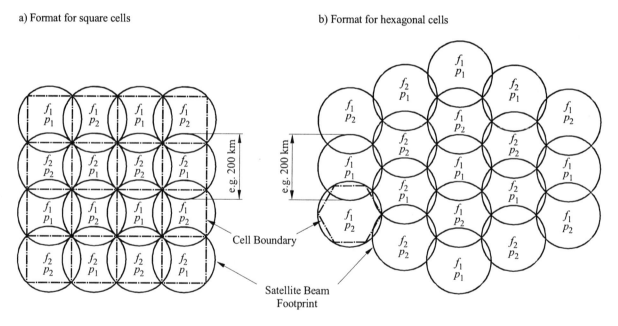

a) Format for square cells

b) Format for hexagonal cells

Area of Cell = e.g. 20 000 sq km

Area of Cell ≅ e.g. 26 000 sq km

f = frequency
p = polarization

Sat/C2-22

FIGURE 2.22

**Typical illustration of frequency and dual polarization
reuse in non-GSO satellite systems**

2.5 Other topics in orbit/spectrum utilization

2.5.1 Introduction

It is well known that the natural resources of the geostationary orbit and the frequency spectrum (this combination being called the orbit-spectrum resource) are in increasing demand. Consequently, when projecting new satellite communication services or increasing the traffic capacity of current systems, a careful appraisal must be made in order to optimize orbit-spectrum utilization to avoid causing interference to other systems and, more generally, to preserve as far as possible, these limited global resources. Considerable attention has been paid to this subject by international and national authorities in charge of telecommunications and by international bodies, especially the ITU-R.

NOTE – Although not formally included in the discussions of this section, the need for economy of use of spectrum also applies to the case of the non-geostationary (non-GSO) satellite systems.

Future utilization of the satellite orbits and frequency bands allocated to space services is considered in the framework of World Radiocommunication Conferences (WRCs) organized by ITU.

Assuming an average satellite spacing of 3° (for operation in a given band), only a few orbit positions remain available on the GSO.

The problems are particularly significant in the 6/4 GHz and 14/10-12 GHz bands for the portions of the GSO orbit serving regions with the highest volume of communications, e.g. the arcs from 49° E to 90° E (over the Indian Ocean), from 135° W to 87° W (serving North America) and from 1° W to 35° W (over the Atlantic Ocean).

The growth of traffic in these areas will have to be accommodated by the allocation of new frequency bands (or the extension of bands already allocated) or the use of higher frequency bands (e.g. 30/20 GHz) by a reduction of the satellite spacing and by a more efficient use of the bandwidths.

However, it should be recognized that the optimization of orbit/spectrum efficiency may result in an increase of system costs.

This section gives a short summary of the following subjects, which are important for efficient orbit/spectrum utilization:

*   modulation and coding techniques;
*   intersystem interference and orbit utilization;
*   on-board signal processing;
*   homogeneity between networks.

2.5.2 Modulation and coding techniques

Coding and modulation methods are discussed in Chapters 3 and 4. A brief preliminary discussion of the efficiency of analogue and digital modulation in terms of satellite capacity and interference immunity is given below.

For analogue systems using frequency modulation (FM), as the modulation index is increased, the capacity per satellite is reduced, but the baseband noise density due to interference at a given carrier-to-interference ratio also falls. This permits closer satellite spacing and, up to a point, results in an increase in the efficiency of use of the GSO.

For digital transmissions using phase-shift keying modulation (PSK), which are now predominant, interference immunity of a signal is increased as the number of phase conditions is reduced, again allowing closer satellite spacing, but the traffic capacity per satellite is then reduced. In fact, the utilization of the GSO tends to be optimized when the number of phases is 4 or 8.

In digital transmissions, the efficiency of spectrum utilization is also increased by the utilization of signal processing techniques, such as:

- low bit rate encoding (LRE) and digital circuit multiplication (DCM) for telephony;

- bit rate reduction techniques for video and television (e.g. MPEG standards) and for data transmissions.

However, forward error correction (FEC) coding which is generally needed to improve power efficiency, tends to reduce spectrum efficiency. Modern FEC processes can include both source coding (e.g. Reed Solomon (RS) codes) and channel coding (e.g. with Viterbi decoding) and even combined error correction and modulation processes (this is called trellis coded modulation (TCM) see Chapter 4, § 4.2.3.2).

2.5.3   Intersystem interference and orbit utilization

Intersystem interference and coordination is the subject of Chapter 9 of this Handbook.

The system elements that are likely to have the largest effect on space utilization efficiency for GSO networks sharing the same orbital arc and frequency band are:

- earth station antenna side-lobe characteristics;

- space station (satellite) antenna characteristics;

- satellite station-keeping and antenna pointing accuracy (see Chapter 6, § 6.3.1);

- cross-polarization discrimination;

- internetwork interference sensitivity.

Taking into account these elements, the geocentric separation between two GSO satellites with the same coverage area can be reduced to 3° or even less.

i)   Earth station antenna side-lobe characteristics

The radiation pattern of earth station antennas, particularly in the first 10° from the maximum gain axis and in the direction of the GSO, is one of the most important factors in determining the interference between systems using geostationary satellites. A reduction in side-lobe levels would significantly increase the efficiency of utilization of the orbit.

At present, most satellite system specifications call for complying with Recommendations ITU-R S.465 and 731 (Reference radiation patterns, co-polarized and cross-polarized, for use in coordination and interference assessment) and S.580 (Design objective for earth station antennas operating with GSO satellites).

Design problems for low side-lobe earth station antennas are dealt with in Chapter 7 of this Handbook (§ 7.2.1.1 and 7.2.2).

ii)   Space station (satellite) antenna characteristics

Satellite antennas should satisfy two conditions for optimizing orbit/spectrum utilization:

- in order to make the best use of their allocated orbit location, the radiation pattern should give the best fit coverage of the service area. This condition applies mainly to GSO satellites and can lead to the design of complex antennas, with multiple-feed, shaped beams;

- outside the service area, the side-lobe level, and more generally, the radiation level should be minimized in order to minimize intersystem interference.

Although work for the adoption of a reference radiation pattern for coordination purpose is not currently completed, ITU-R has issued a Recommendation (S.672) giving radiation patterns for use as a design objective in the FSS employing GSO satellites.

Design problems for high performance satellite antennas are dealt with in Chapter 6 of this Handbook.

iii)   Cross-polarization discrimination

The use of orthogonal linear or circular polarizations permits discrimination to be obtained between two emissions/receptions in the same frequency band from the same satellite. Typical values of on-axis polarization discrimination range from 30 to 35 dB for good antenna feed design. This augments the discrimination provided by the directional properties.

There are several means for taking advantage of polarization discrimination in view of optimizing orbit/spectrum utilization:

- The most common method is to implement systematically polarization discrimination frequency reuse. As already explained (Chapter 1, § 1.3.4.3), this is liable to multiply by two the traffic capacity and this can be combined with frequency reuse by beam separation to further augment this capacity.

- Another method where advantage could be taken of low side lobes and good polarization discrimination characteristics to increase traffic capacity, is to implement multiple beam frequency reuse antennas, or clusters of co-located satellites, in order to lay down multiple spot beams over the service area, instead of a single beam. Typically, adjacent beams would use different portions of the frequency band to avoid interference, but non-adjacent beams can reuse the same frequencies provided that the inter-beam interference is acceptable.

  Such systems can also be particularly well adapted to low earth orbit (LEO) non-GSO satellite networks (see § 2.4.3.4).

2.5.4   On-board signal processing

The implementation of on-board signal processing, regenerative transponders, can improve link budget parameters and thus enable lower transmission power. This would reduce interference with adjacent satellites and thus increase orbit utilization efficiency.

2.5.5   Homogeneity between networks

The most efficient orbit utilization would be obtained if all satellites using the GSO and which were illuminating the same geographical area and using the same frequency bands, had the same characteristics, i.e. if they formed a homogeneous ensemble. However, in practice, a perfectly homogeneous system will not exist, due to differences in the individual systems.

Consider two satellite systems A and B, using satellites having adjacent orbital positions. If A and B have widely different characteristics for satellite receiver sensitivity, downlink e.i.r.p. or associated earth stations, the angular spacing necessary to protect A against interference from B may differ from that necessary to protect B from A. In practice, the greater of the two angles must be selected. The extent to which this may represent an inefficient use of the GSO is dependent on many factors in the design of the satellite systems using orbital positions near those of A and B.

APPENDIX 2.1

Antenna radiation diagrams and polarization

In § AP2.1-2 some basic antenna characteristics such as antenna gain and aperture were defined. In this appendix, other important characteristics such as radiation diagrams, beamwidths and polarization are described.

AP2.1-1 Radiation diagrams and beamwidths

The directional properties of an antenna can be represented by its "radiation diagram" (or "radiation pattern"). This is shown in rectangular coordinates on Figure AP2.1-1, in a particular plane (e.g. a vertical plane through the maximum antenna gain axis).

The "half-power (–3 dB) beamwidth" (full beamwidth) is given by the important formula:

$$\theta_0 = k \cdot \frac{\lambda}{D} \tag{1}$$

where D is the diameter of the (circular) antenna aperture, λ is the wavelength and where k depends on the "illumination law" of the aperture. For a high efficiency earth station antenna $k \approx 65°$.

Formula (1) can be generalized for any relative power level (dB level). For example, the "10 dB beamwidth", i.e. the angle between directions where the power level (received or transmitted) is reduced to one-tenth, is given by:

$$\theta_{10dB} = k' \cdot \frac{\lambda}{D} \qquad (k' \approx 112°) \tag{2}$$

Relation (6) in § 2.1.2.2, is recalled here:

$$g_{max} = \eta \left(\frac{\pi \cdot D}{\lambda} \right)^2$$

where η is the antenna efficiency. Using this relation and formula (1) above, and assuming a typical efficiency $\eta = 0.65$, the maximum gain can be given in a very useful form by:

$$g_{max} = \frac{27\,000}{\theta_0^2} \tag{3}$$

Formula (3) can be generalized to non-circular, quasi-elliptically shaped antennas with an aperture of D1 and D2 axial dimensions (corresponding to half-power beamwidths θ_{01} and θ_{02} in the two main plans). In this case:

$$g_{\max} = \frac{27\,000}{\theta_{01} \cdot \theta_{02}} \tag{3bis}$$

Figure AP2.1-2 shows some common values of antenna gain and beamwidth. A nomogram for their calculation is given below in § AP2.1-2.

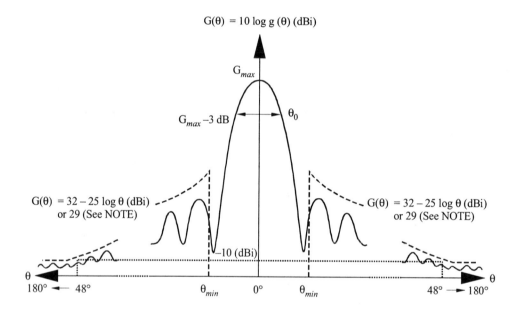

-------- : Reference diagram (Recommendation ITU-R S.465-5)

θ_{min} = 1° or 100 λ/D (whichever is greater)

NOTE – In Recommendation ITU-R S.580-4, the value becomes 29 as a design objective (with a different formulation).

Sat/AP2-011

FIGURE AP2.1-1

**Typical outline of the radiation diagram
of an earth station antenna**

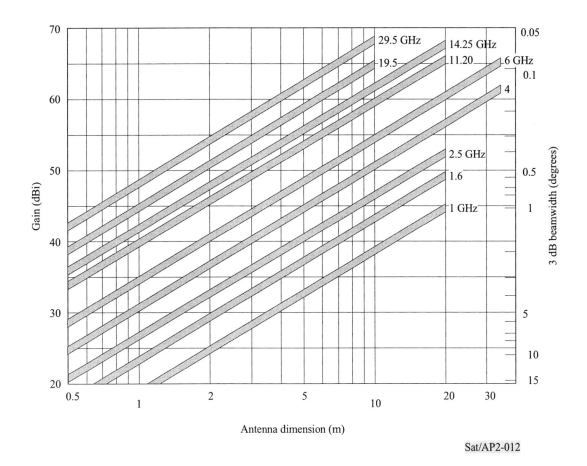

Sat/AP2-012

FIGURE AP2.1-2

**Antenna gain and 3 dB beamwidth
(for 0.6 to 0.8 antenna efficiencies)**

AP2.1-2 Nomogram for antennas with circular aperture

AP2.1-2.1 General

The antenna nomogram given in Figure AP2.1-3 enables the following antenna parameters to be simply estimated:

- gain versus frequency and diameter;
- allowable pointing error versus gain drop and nominal gain;
- gain drop versus pointing error and nominal gain.

The aperture efficiency is given by the ratio of the equivalent effective antenna area to the physical antenna area. It is defined at the throat of the antenna feed horn and hence does not include the losses of the feed system e.g. orthomode transducer, diplexer, filters and waveguide connections.

The typical errors of the approximation formula when used for beamwidth calculations lie within a range of approximately $\pm 15\%$.

AP2.1-2.2 Typical applications

i)   Gain versus frequency and diameter

- Connect the given frequency on scale f and diameter on scale d.
- The intersection of this line with the scale G_o indicates the antenna gain.

This gain can then be corrected for aperture efficiencies other than 0.7.

e.g.	Frequency f:	14 GHz
	Diameter d:	18 m
	Gain G_o:	66.9 dB for $\eta = 0.7$
	Gain G_o:	66.9 dB – 1 dB = 65.9 dB for $\eta = 0.55$

ii)   Antenna diameter versus gain and frequency

- Connect the given gain on scale G_o and frequency on scale f.
- The intersection of this line with scale d indicates the antenna diameter.

e.g.	Gain G_o:	60 dB
	Frequency f:	4 GHz
	Diameter d:	28.5 m

NOTE – The gain is normalized for $\eta = 0.7$.

iii)   3 dB beamwidth versus nominal antenna gain

- Connect the given nominal gain on scale G_o and the 3 dB point on scale ΔG.
- The intersection of the line with scale $\pm\theta$ indicates the half 3 dB beamwidth.

e.g.   Nominal gain G_o: 55 dB

          Gain drop ΔG: 3 dB

          $\pm\theta$ $\pm0.14°$

          3 dB beamwidth: $2 \cdot 0.14° = 0.28°$

iv)   Maximum pointing error versus gain drop and nominal gain

- Connect the given nominal gain on scale G_o and maximum gain drop on scale ΔG.
- The intersection of the line with scale $\pm\theta$ indicates the pointing error.

e.g.   Nominal gain G_o: 55 dB

          Maximum gain drop ΔG: 0.5 dB

          Maximum pointing error $\pm\theta$: $\pm0.058°$

v)   Gain drop versus pointing error and nominal gain

- Connect the given nominal gain on scale G_o and pointing error on scale $\pm\theta$.
- The intersection of the line with scale ΔG indicates the gain drop.

e.g.   Nominal gain G_o: 60 dB

          Pointing error $\pm\theta$: $\pm0.1°$

          Gain drop ΔG: 4.6 dB

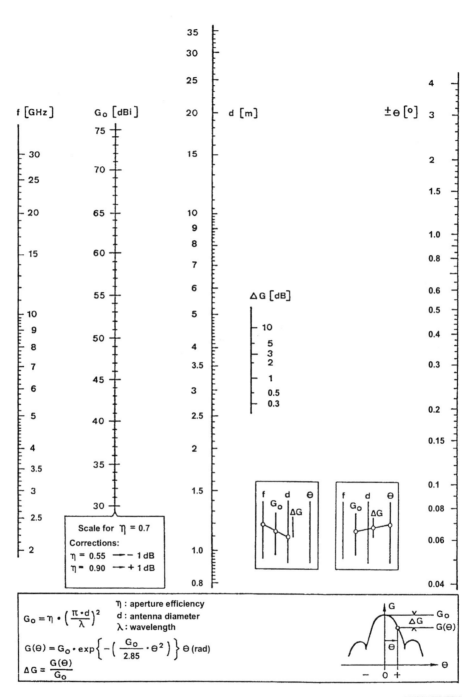

Sat/AP2-013

FIGURE AP2.1-3

Nomogram showing the relations between frequency, maximum gain, antenna diameter, beamwidth and gain drop due to pointing error

AP2.1-3 Antenna side lobes

Most of the power radiated by an antenna is contained in the so-called "main lobe" of the radiation pattern but some residual power is radiated in the "side lobes".

Conversely, due to the reciprocity theorem, the receive antenna gains and radiation patterns are identical to the transmit antenna gains and radiation patterns (at the same frequency). Therefore, unwanted power can also be picked up by the antenna side lobes during reception.

Side lobes are an intrinsic property of antenna radiation, and diffraction theory shows that they cannot be completely suppressed. However, side lobes are also due partly to antenna defects which can be minimized by proper design.

As explained in more detail in § 7.2, the recommended side-lobe characteristics of earth station antennas used in the FSS are defined in Recommendations S.465-5 and S.580-5 (see Figure AP2.1-1).

Moreover, the residual power radiated in the side lobes induces unwanted emissions from earth stations. These spurious emissions are subject to limitations specified by Recommendation ITU-R S.524-5 and, for the particular case of very small aperture terminals (VSATs), by Recommendation ITU-R S.728. As concerns satellite antennas (see § 6.3.1), their side-lobe characteristics are defined by Recommendation ITU-R S.672.

AP2.1-4 Polarization

The polarization of an RF wave radiated (or received) by an antenna is defined by the orientation of the electric vector (E) of the wave (Figure AP2.1-4).

This vector – which is perpendicular to the direction of propagation – can vary in direction and intensity during one RF period T (T = 1/f, where f is the frequency).

This means that, while travelling one wavelength during period T, the E vector not only oscillates in intensity but can also rotate.

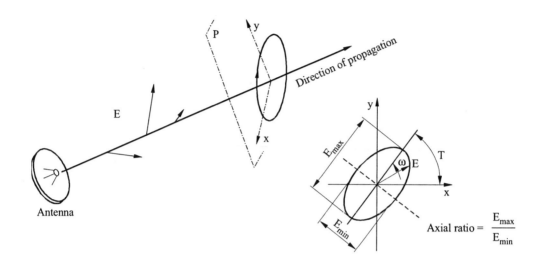

Sat/AP2-014

FIGURE AP2.1-4

Definition of RF polarization

In the most general case, the projection of the tip of the E vector on a plane (P) perpendicular to the direction of propagation describes an ellipse during one RF period: this is called elliptical polarization.

Elliptical polarization is characterized by three parameters:

*   rotation sense, as seen from the antenna (and looking in the direction of propagation): right hand (RH, i.e. clockwise) or left hand (LH, i.e. counter-clockwise);

*   axial ratio (AR) of the ellipse (voltage axial ratio);

*   inclination angle (τ) of the ellipse.

Most practical antennas radiate either in linear polarization (LP) or in circular polarization (CP) which are particular cases of elliptical polarization.

Linear polarization is obtained when the axial ratio is infinite (the ellipse is completely flat, i.e. the E vector only oscillates in intensity). Circular polarization is obtained when AR = 1.

For any elliptical polarization, an orthogonal polarization can be defined which has inverse rotation sense.

It can be shown that two orthogonally polarized waves are (theoretically) perfectly isolated. This means that an antenna can be equipped with two receive (or transmit) ports, each one being perfectly matched to only one polarization.

If, for example, a wave with a given polarization impinges on the antenna, the full power will be received at the matched port and no power at the other port. Therefore, the same antenna can receive (transmit) simultaneously two carriers with orthogonal polarizations at the same frequency: this is the basis for dual polarization frequency reuse systems (see § 1.3.4 vi) and § 2.4.3). Dual polarization frequency reuse is generally implemented, either with two linear orthogonal polarizations (e.g. horizontal and vertical) or with two circular orthogonal polarizations (LHCP and RHCP).

It can be shown that any elliptically polarized RF wave can be considered as the (vectorial) sum of two orthogonal components, e.g. of two linearly polarized waves (e.g. horizontal and vertical) or of two circular orthogonal polarizations (LHCP and RHCP).

An important characteristic of an antenna is its polarization purity. The RF wave radiated by an imperfect antenna contains a so-called co-polarized (wanted) component and a cross-polarized ("X-polar") (unwanted) component. Therefore, the polarization purity of an antenna can be described by its cross-polarization gain and radiation diagrams. This is shown on Figure AP2.1-5, which compares the co- and cross-polarization antenna diagrams, with a common reference, with the co-polarized maximum antenna gain.

The diagrams shown are theoretical. Note that, for practical antennas, diagrams are less regular and show no nulls, only minimums, but the important fact is that, generally, the cross-polarized diagram is minimum in the antenna axis and shows first lobes with maximum gains at a relative level of the order of -20 dB.

The polarization purity of an antenna, both at transmission and reception, is generally defined by its actual co- and cross-polarization diagrams. The cross-polarization isolation (XPI) is the ratio, in a given direction, of the co-polarized to the cross-polarized (X-polar) RF field amplitudes. It is generally evaluated in dB:

$$(XPI)_{dB} = 20 \log (XPI)$$

In Figure AP2.1-5: $(XPI)_{dB} = 10 \left[\log g_{co} (\theta) - \log g_x (\theta) \right]$

The most stringent limits of (XPI) are generally specified near the antenna axis (referring to Figure AP2.1-5, recall that, in actual antennas, there is not a null, but a minimum of the X-polar in the antenna axis).

For example, a specification could be: $(XPI)_{dB} \geq 30$ dB within the 1 dB "contour" (i.e. beamwidth) of the main (co-polar) antenna beam. Proper design can improve the polarization purity of an antenna, i.e. maximize (XPI). This may prove more difficult in the case of offset antennas[20].

---

[20] Offset antennas use a non-symmetrical parabolic reflector. This avoids radiation blockage by the antenna feed and improves most performances, in particular the co-polarized side-lobe level.

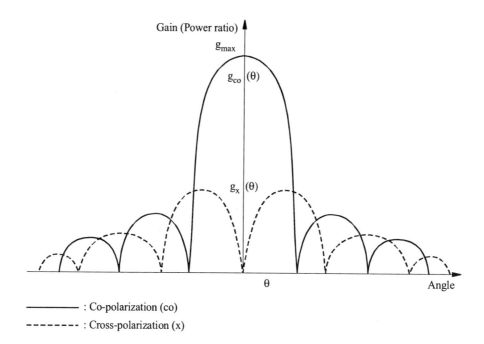

NOTE – The diagram shows absolute values: discontinuities at the nulls are due to amplitude changes (+ to –), i.e. phase reversals (0 to π). In most practical antennas, there are no nulls, but only minimums.

Sat/AP2-015

FIGURE AP2.1-5

Co-polarization and cross-polarization antenna diagrams (theoretical)

The importance of the polarization purity of an antenna comes from the fact that the unwanted, X-polar emissions can cause interference to other systems and that, at reception, unwanted components from other systems can cause interference to the (given) system. But, even more important, are the internal interferences in dual polarization frequency reuse systems.

This case is explained by Figure AP2.1-6.

In this very simplified figure (where, in particular, the satellite is not represented), the transmitting and the receiving earth stations are only represented by the "primary source" of their antennas (here a simple horn), a polarizer (in the case of circular polarization) and a so-called orthomode transducer (OMT).

The function of the OMT is to interface with the two orthogonal polarizations (here, LHCP and RHCP) of the dual polarization frequency reuse system, i.e. to connect two separate transmitters (through two transmit ports 1 and 2) and two separate receivers (through two receive ports 1 and 2), thus forming two links, able to operate in the same frequency band, one on LHCP, one on RHCP, with, theoretically, no coupling between these two links (for more details on primary sources, polarizers and OMTs, see, for example, Chapter 7, § 7.2.2). However, in practical systems where the polarization purity of each antenna is imperfect, there is some cross-coupling (X-polar), as shown on the figure by the components $(LHCP)_x$, relative to $(RHCP)_{co}$ and $(RHCP)_x$, relative to $(LHCP)_{co}$.

Note that in Figure AP2.1-6, the situation is rather complex, because the received X-polar components can be due both to the transmitting and to the receiving antenna (and even possibly to some cross-polarizations effects in the propagation medium).

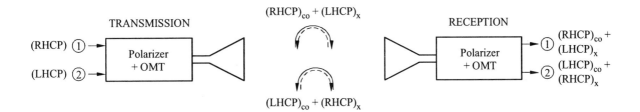

OMT: Orthomode Transducer
(RHCP):   Right Hand Circular Polarization
(LHCP):   Left Hand Circular Polarization
co: Co-polarization
x: Cross-polarization

NOTE – The same figure holds for a dual (linear) polarization frequency reuse system, if, for example, (LHCP) is changed to vertical and (RHCP) to horizontal.

Sat/AP2-016

FIGURE AP2.1-6

Transmission and reception of the co-polarized ((LHCP)co and (RHCP)co) and cross-polarized ((RHCP)x and (LHCP)x) components of two RF links of a dual (circular) polarization, frequency reuse system

In order to introduce the (XPD) definition, it will be assumed that the transmission antenna has a perfect polarization purity, as well as the propagation medium (i.e. the X-polar components represented in the middle of Figure AP2.1-6 are not present) and that the two RF waves (LHCP/RHCP or vertical/horizontal) are transmitted with equal amplitudes (i.e. two transmitters with the same power), then the cross-polarization isolation of the receiving antenna can be defined again as:

$$(XPI) = (RHCP)_{CO}/(LHCP)_X \text{ (at receive port 1), or:}$$
$$(XPI) = (LHCP)_{CO}/(RHCP)_X \text{ (at receive port 2) } [21]$$

---

[21] Note that, in these formulas, since the two RF transmissions are assumed to be perfectly polarized, the X-polar components $(LHCP)_x$ and $(RHCP)_x$ are only due to the imperfections of the receiving antenna. Because the co-polar and X-polar components appear at the same receiving port, they should be measured, either by transmitting the two tests signals at slightly different frequencies, or successively.

Remarks

- Again, the same relations hold for a dual (linear) polarization frequency reuse system, if, for example, (LHCP) is changed to vertical and (RHCP) to horizontal.

- Of course, the situation can be reversed for defining the (XPI) of the transmitting antenna (i.e. a perfect receiving antenna, etc.).

Now, if a single polarization (no frequency reuse) is transmitted by a perfect antenna, e.g. (LHCP), through a perfect medium, then the cross-polarization discrimination of the receiving antenna is defined as:

$$(XPD) = (LHCP)_{CO} \text{ (at receive port 1)} / (RHCP)_X \text{ (at receive port 2)}$$

In practice, both (XPI) and (XPD) are often confused and called simply isolation. In the case of nearly circular polarizations, the following formulas can be used:

$$\text{Isolation} = 20 \log \frac{(AR) + 1}{(AR) - 1} \qquad \text{dB}$$

where (AR) is the axial ratio of the elliptical polarization (as transmitted by the antenna from a single port). For $(AR) \leq \sqrt{2}$ (3 dB), a good approximation is:

$$(\text{Isolation})_{dB} = 24.8 - 20 \log (AR)$$

$$(AR)_{dB} = 17.37 \times 10^{-(\text{Isolation})dB/20}$$

The cross-polarizations characteristics of FSS antennas are specified in Recommendations ITU-R S.731 and, for VSATs, ITU-R S.727.

CHAPTER 3

Baseband signal processing and multiplexing

3.1 General

Satellite transmission systems are characterized by a particular combination of baseband processing, carrier modulation and multiple-access techniques. Although the two latter issues are developed in the following Chapters (4 and 5), the techniques are related as explained below.

The baseband signals used in the fixed-satellite service are usually subject to different kinds of processing. These may be inherent in the transmission methods used in the terrestrial interconnection links or they may be carried out just for the transmissions using the satellite link. In either case the baseband processing system affects substantially the performance of the satellite network.

Multiple-access techniques are concerned with methods by which a large number of earth stations can interconnect their respective transmission links simultaneously through a common satellite transponder.

Although multiple access methods use different concepts of baseband processing, multiplexing and carrier modulation, the techniques are closely linked, and it is often assumed that a certain combination of baseband processing, modulation methods and multiplexing techniques is inherent to a particular multiple-access technique.

Table 3-1 summarizes types of transmission systems by various combinations of baseband processing, multiplexing and modulation methods and further their adaptability to the four different multiple-access modes. In selecting these transmission systems, it is necessary to consider their advantages and disadvantages with respect to parameters such as fixed limits of satellite power, available frequency bandwidth and intermodulation products.

TABLE 3-1

Adaptability of baseband processing, multiplexing and carrier modulation methods to multiple-access modes

System	Baseband processing	Multiplexing	Carrier modulation	Adaptability to multiple access and assignment modes			
				FDMA	TDMA	PAMA	DAMA
FDM-FM CFDM-FM	None Companding	FDM	FM	X		X	
SCPC-CFM	Companding	None (single channel)	FM	X		X	X

System	Baseband processing	Multiplexing	Carrier modulation	Adaptability to multiple access and assignment modes			
				FDMA	TDMA	PAMA	DAMA
SCPC-PCM-PSK	PCM	None (single channel)	PSK	X		X	X
TDM-PCM-PSK	PCM	TDM	PSK	X	X		
					X	X	X
CDMA	PCM	None (single channel: Asynchronous CDMA)	PSK	X		X	X
		CDM (Synchronous CDMA)	PSK	X		X	

FDMA: Frequency Division Multiple Access
TDMA: Time Division Multiple Access
CDMA: Code Division Multiple Access (Spread spectrum)
FDM: Frequency Division Multiplex
CFDM: FDM with channel companding (SSBC)
TDM: Time Division Multiplex
CDM: Code Division Multiplex

PCM: Pulse Code Modulation
FM: Frequency Modulation
CFM: FM with syllabic companding (single channel)
PSK: Phase Shift Keying
SCPC: Single Channel Per Carrier
PAMA: Pre-Assigned Multiple Access
DAMA: Demand Assigned Multiple Access

3.2 Baseband signal processing – analogue

3.2.1 Voice and audio

Analogue processing of voice and audio signals can correspond to syllabic companding or voice carrier activation.

3.2.1.1 Channel-by-channel syllabic companding

The principle of companding is to reduce the dynamic range of the speech signal for transmission and to carry out the opposite operation at reception (i.e. the signal is compressed before and expanded after transmission). The term "syllabic" is used because the operating time constants of the compander are such that they can adapt to the syllabic variation of the speech signal.

The compander compression ratio is defined by the following formula:

$$R_c = \frac{n_e - n_{e0}}{n_s - n_{s0}}$$

where:

n_e: input level of speech signal,

n_{e0}: unaffected level,

n_s: output level,

n_{s0}: output level corresponding to an input level n_{e0}.

The preferred value of Rc, as given in ITU-T Recommendation G.162, is 2. In this case a 2 dB variation is reduced to 1 dB for transmission. The compandor expansion ratio is also defined in a similar manner.

The level of 0 dB is also recommended for the unaffected level of the compandor: it gives a higher gain in quality but increases the mean power of the channel (and the multiplex). However, since in the case of satellite transmission, power efficiency is of great importance, lower values are sometimes chosen. For example, a typical device with a –11 dBm0 unaffected level still subjectively provides about 10 dB of quality improvement (S/N) and does not modify the load (mean power) of the channel. The two operations of compression and expansion are carried out for each telephone channel, before multiplexing and after demultiplexing, respectively.

The companding operation results in an improvement of telephone channel quality, for two reasons:

- Objective reason

The noise power in the telephone channel is added only after compression and at a level well below the unchanged reference level. This results in a relative reduction of noise power at expansion in relation to the mean power of the speech signal (see Figure 3.1).

- Subjective reason

The additive noise in the telephone channel, always relatively low, is heard only during pauses in conversation. Since the effect of expansion is to reduce the noise power, which is well below the unchanged reference level, pauses in conversation become even more "silent", and this greatly improves the acceptability of the call. On the other hand, this effect is offset by the "hissing" accompanying the rise and fall of the noise level in the interval between periods of speech activity and inactivity.

The subjective reason is by far the most important, but figures for it can be given only on an experimental basis. In the interests of caution, the gain has been limited to about 13 dB, which means that for the same subjective telephone channel quality, the signal-to-noise ratio measured after demodulation (and before expansion) may be reduced by 13 dB.

The capacity of the FDM-FM carriers can thus be multiplied by a factor of 2 to 2.5.

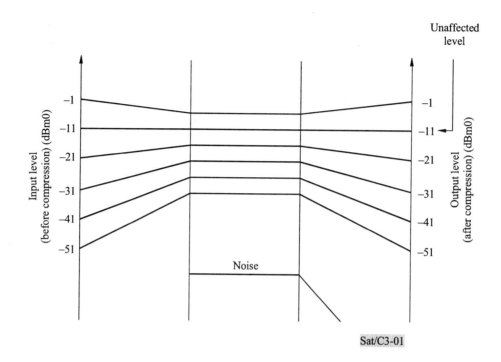

FIGURE 3.1

Effect of companding on noise level

3.2.1.2   Voice carrier activation

Voice activation is a method of achieving a bandwidth/power advantage through a device that withdraws the transmission channel from the satellite network during pauses in speech.

The basic parameters of the voice-operated switch are:

a)   the threshold level above which speech is assumed to be present (typically –30 to –40 dBm0 for a fixed threshold level switch);

b)   the time taken by the voice switch to operate on detection of speech (typically 6 to 10 ms);

c)   the hangover time during which the voice switch remains activated after the cessation of a speech burst on a speech channel (typically 150 to 200 ms).

The threshold level needs to be set fairly low to avoid missing large portions of the start of a speech burst and low speech levels. This makes the voice switch susceptible to high noise levels, thus reducing the net system advantage.

Voice activation has found practical application in SCPC systems where its use yields a power saving of about 4 dB.

3.2.2   Video and TV

3.2.2.1   Composite analogue TV signal

A TV analogue picture may be reproduced from four signals usually combined to a composite signal: the level of three fundamental spectrum components for each picture dot (red, green, blue) E_R, E_G, E_B, and a synchronization signal carrying the line pulses and the frame pulses. The composite video colour signals have been defined on the basis of a requirement for compatibility between the colour signal and the monochrome signal; this means that circuits in the monochrome receivers must correctly reconstitute (in monochrome) the pictures transmitted according to the colour system, while the colour receivers must operate correctly (in monochrome) when they receive monochrome TV signals. As a result of these conditions, the colour video signal has been made up from a luminance signal by adding to it the element required to transmit chrominance information.

The luminance signal is formed from E_R, E_G and E_B by linear combination. The amplitude of this signal ranges from 0.3 V (black) to 1 V (white). The rectangular synchronization signals with amplitude levels between 0 and 0.3 V are superimposed on this signal.

The two chrominance components are also linear combinations of E_R, E_G and E_B. These two chrominance components modulate a colour subcarrier whose spectral components are interlaced with the spectral components of the luminance signal. For NTSC and PAL systems, the two chrominance components modulate the amplitude of this subcarrier in quadrature. For the SECAM system, the two components modulate the frequency of this subcarrier sequentially.

ITU-T Recommendations J.61 and J.62 define the quality criteria for long-distance television transmission.

3.2.2.2   MAC TV signal

MAC, an acronym for multiplexed analogue component, is a television modulation method that has been developed for satellite television transmissions of both the broadcasting-satellite service (BSS) and the fixed-satellite service (FSS) where the satellite is used as an FSS link between, for example, a television studio and a cable-head distribution point.

In common with the PAL and SECAM systems, the electronic scan of one picture line across the television screen takes 64 μs. The video and audio components of the television signal must necessarily be compressed in time in order that they can be time multiplexed into one line period (see Figure 3.2). The MAC system retains the current 625 line 2:1 interlace structure so that it is compatible (with an appropriate decoder) with current TV receivers.

However, unlike traditional analogue composite TV formats (PAL, SECAM, etc.) the MAC system is based on time multiplexing of the luminance, chrominance and sound/data components. The video signal (luminance and chrominance) is analogue and the audio signal is digital.

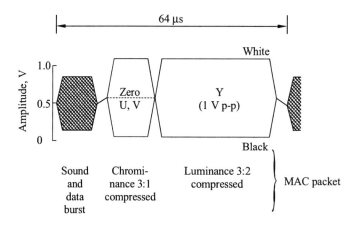

Y: luminance signal
U and V : colour difference signals
              (U = B – Y where B is the blue signal
              V = R – Y where R is the red signal)

Sat/C3-02m

FIGURE 3.2

A single MAC line

Currently there is no single internationally agreed MAC standard: in Europe three different types, namely C-MAC, D-MAC and D2-MAC, have been adopted by the EBU and the European Commission. The prefix (C, D or D2) denotes the type of sound/data modulation that is used. Details of each MAC type are provided in Table 3-2 below. Australia uses the B-MAC system for distribution of television programmes.

TABLE 3-2

MAC family types

MAC type	Data burst rate, kbit/s	Data burst type	Maximum number of audio channels	Intended usage
C-MAC	20-25	2-4 PSK	8	DBS
D-MAC	20-25	Duobinary	8	CATV
D2-MAC	10-125	Duobinary	6	DBS and CATV

The MAC system, when compared to existing TV formats, has the following advantages:

•   a better subjective picture quality than existing TV formats at all C/N ratios;

•   does not suffer from cross-colour or -luminance impairments because the chrominance and luminance signals are separated in time;

- does not use subcarriers, thus enabling the bandwidth of the component signals to be increased to provide an extended definition capability for future large screen displays;

- data can support multiple high-quality audio channels;

- is more efficient in transmitting blanking and synchronizing signals;

- can be readily encrypted.

Notwithstanding these advantages, MAC systems did not find many applications, due to the advent of digital TV systems.

3.2.3   Energy dispersal

3.2.3.1   Introduction

Energy dispersal is a scheme by which the spectral energy density of the modulated carrier can be reduced during periods of light loading at the baseband.

Energy dispersal is achieved by combining an energy dispersal signal with the baseband signal before carrier modulation. The use of energy dispersal is considered essential in the fixed-satellite service for the following reasons:

- spreading of energy is required to maintain the level of the interference to a terrestrial or a satellite system operating in the same frequency band;

- it is also useful to reduce the density of intermodulation noise produced in an earth station HPA or a satellite transponder operating in a multicarrier mode.

The practical techniques of energy dispersal differ depending on the nature of the baseband signal and the carrier modulation technique in use.

3.2.3.2   Energy dispersal for analogue systems

a)   Multichannel telephony systems

The easiest way to apply energy dispersal to this type of system is to add a dispersal waveform at a frequency which is below the lowest frequency of the baseband signal to be transmitted. The waveform may be sinusoidal or triangular, or may be a band of low-frequency noise. Among these, the triangular wave is considered to be the most practical from the viewpoint of the ease of implementation and its dispersal effect which has been shown to be within 2 dB of the dispersal obtained under full-load conditions.

Figure 3.3 shows an example of a triangular wave used as a dispersal signal. This is often referred to as "symmetrical triangular" because it has equal rise and fall times. Careful attention has to be paid to maintain the linearity of the waveform. Non-linearity such as roundings of the peaks will result in a deterioration of the dispersal efficiency.

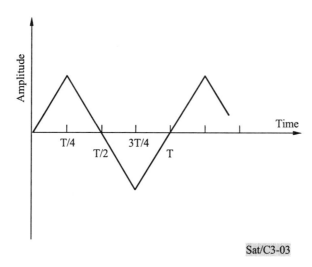

Sat/C3-03

FIGURE 3.3

Symmetrical triangular wave for use as a dispersal signal

b)   Television systems

The conventional energy dispersal waveform used for television systems is a symmetrical triangular wave at one-half the frame frequency in use. That is 30 Hz for NTSC employing 525/60 standards, and 25 Hz for PAL/SECAM employing 625/50 standards.

3.3 Baseband signal processing – digital

3.3.1   PCM, ADPCM

3.3.1.1   Pulse code modulation (PCM)

The first objective of coding is the digital representation of analogue signals which are intended to be transmitted by digital techniques.

In PCM, the analogue signal is sampled periodically at a rate equal to the Nyquist rate (i.e. twice the highest baseband frequency) and quantized in an agreed manner. In order to decrease the inherent error occurring in the process of quantization (quantization noise) the signal may be previously subjected to companding or nearly instantaneous companding.

ITU-T Recommendation G.711 specifies the PCM format to be used in international connections of voice signals. The channel sampling rate is 8 000 samples per second and each sample is coded into

8 bits giving a total bit rate of 64 kbit/s per speech channel (Figure 3.4). However, a coding standard of 7 bits per sample can be used in some systems (for example the INTELSAT SCPC system), which gives a bit rate of 56 kbit/s.

Some advantages of PCM are summarized in the following statements:

i)   PCM presents a small but definite performance advantage over FM at the lower signal-to-noise ratios.

ii) A PCM system designed for analogue message transmission is readily adapted to other input signals.

iii) PCM has the capability of regeneration and is very attractive for communication systems having many repeater stations.

PCM is often applied at voice channel level in connection with time-division multiplexing (TDM). Another use of PCM is the direct encoding/decoding of a frequency-division multiplexed (FDM) signal by the use of a transmultiplexer.

To provide improved performance for voice signals, the quantization steps are not evenly spaced: logarithmic compression and expansion are applied respectively at the transmit and receive sides. This is called instantaneous companding. The so-called A and μ laws, shown below, are functions defining this companding operation for voice PCM applications:

$$|V2| = \frac{\log(1 + \mu|V1|)}{\log(1 + \mu)} \qquad \text{(μ-law)}$$

and:
$$|V2| = \frac{A|V1|}{1 + \log A} \qquad 0 \leq |V1| \leq \frac{1}{A} \qquad \text{(A-law)}$$

$$|V2| = \frac{1 + \log(A|V1|)}{1 + \log A} \qquad \frac{1}{A} \leq |V1| \leq 1$$

where V1 and V2 are normalized input and output voltages. The parameters μ and A normally assume the values $\mu = 255$ and $A = 87.6$. Figure 3.5 illustrates these two companding laws for these specific values of μ and A.

Nearly instantaneous companding (NIC) is a source coding technique for reducing the bit rate in PCM. This is achieved by rescaling the quantizing range of a block of PCM samples according to the magnitude of the largest sample in the block. A control word is transmitted and used by the receiver to convert the received bit stream to the normal PCM format. It is expected that NIC will make possible the transmission of good quality speech at bit rates between 34 and 42 kbit/s.

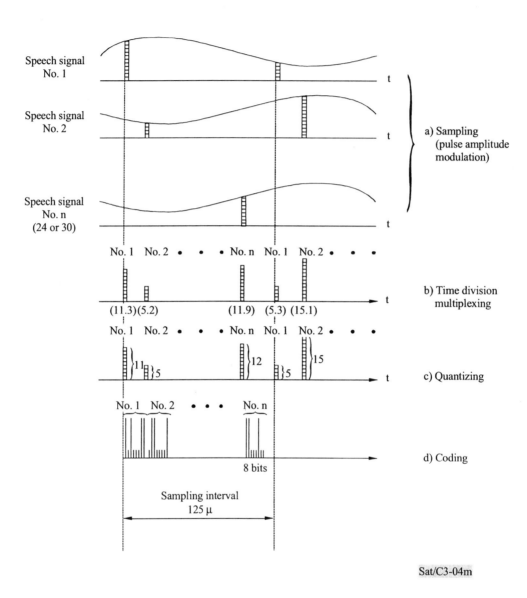

Sat/C3-04m

FIGURE 3.4

Transformation from analogue signal to PCM signal

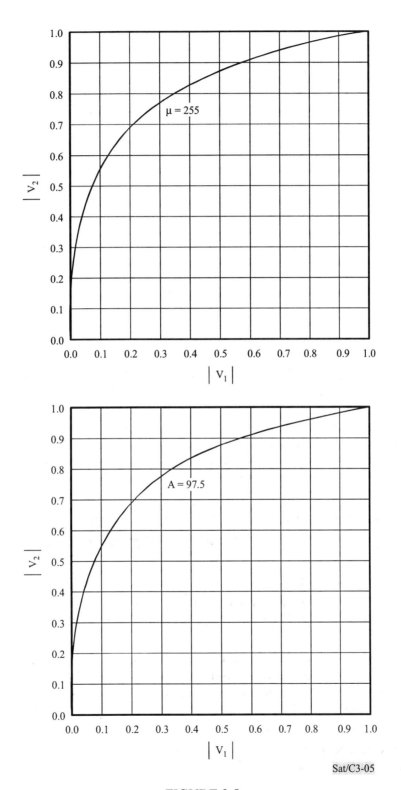

FIGURE 3.5

μ and A companding laws for PCM coding

3.3.1.2   Delta modulation (DM)

In delta modulation, the analogue signal is sampled at a rate higher than that for PCM (typically 24 to 40 kHz for good quality speech). Then, a one-bit code is used to transmit the change in input level between samples. DM is very simple to implement and is suitable for moderate quality communications (e.g. thin route traffic).

The effect of transmission errors in DM is not as serious as in PCM systems. This is because all the transmitted bits are equally weighted and are equivalent to the least significant bit of a PCM system.

The main drawback of DM systems is the difficulty of interconnecting them in a general way in an international network using analogue and digital links.

It is possible to improve the dynamic range of the encoder at the expense of increased circuit complexity by using an adaptative quantizer, wherein the step size is varied automatically in accordance with the time-varying characteristics of the input signal. In adaptive delta modulation (ADM) the variable step size increases during a steep segment and decreases during a slowly varying segment of the input signal.

3.3.1.3   Differential pulse code modulation (DPCM)

In DPCM, the difference between the actual sample and an estimate of it, based on past samples, is quantized and encoded as in ordinary PCM, and then transmitted. To reconstruct the original signal, the receiver must make the same prediction made by the transmitter and then add the same correction. Thus, DPCM, as compared to PCM, uses the correlation between samples to improve performance. In the case of voice signals it is found that the use of DPCM may provide a saving of 8-16 kbit/s (1 to 2 bits per sample) over standard PCM. For monochrome TV, a saving of about 18 Mbit/s (2 bits per sample) can be obtained.

3.3.1.4   Pulse code modulation and adaptive differential coding at 32 kbit/s (ADPCM)

The ADPCM coding principle combines two techniques: adaptive quantizing and linear prediction.

3.3.1.4.1   Adaptive quantizing

In PCM coding, the signal is logarithmically compressed in order to obtain a signal-to-quantizing noise ratio which is constant over a wide range of levels and which displays low sensitivity to the statistical characteristics of the signal.

For a signal of known statistical characteristics, it is possible to define an optimum quantizer of which the performance, in terms of signal-to-quantizing noise ratio for the signal, is appreciably better than that using logarithmic compression (up to 10 dB improvement in the S/N ratio). Unfortunately, the optimum quantizer performance deteriorates very rapidly as soon as the statistical characteristics of the signal to be quantized depart from that for which they were calculated.

Hence the idea of adaptive quantizing, in which the quantizing step is made variable in time as a function of the statistical characteristics of the signal to be quantized. The step adaptation is then selected in such a way as to approach at all times the performance of an optimum quantizer. Adaptation is effected by a summary statistical analysis of the quantized signal carried out by digitizing the quantizing noise.

Figure 3.6 explains the principle of a quantizer of this type, known as a feedback quantizer.

Figure 3.7 shows an example of a 3-bit adaptive quantizer. In these figures Δ_n is the instantaneous value of the quantizing step at time n. This value is adapted as a function of the instantaneous amplitude of the signal. The step variation rule makes use of the multiplicative coefficients of Figure 3.5 with $\Delta_{n+1} = \Delta_n \cdot M_n$. For example, if a signal 111 (strong signal) is transmitted, the quantizing step will be multiplied by M_4 (>1) at the following time. This process increases the dynamic range of the quantized signal.

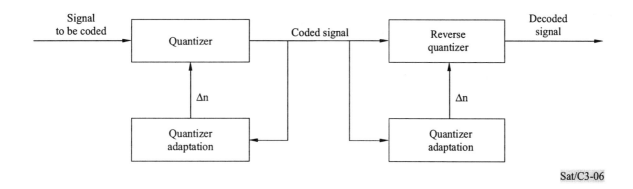

Sat/C3-06

FIGURE 3.6

Adaptive quantizer with feedback

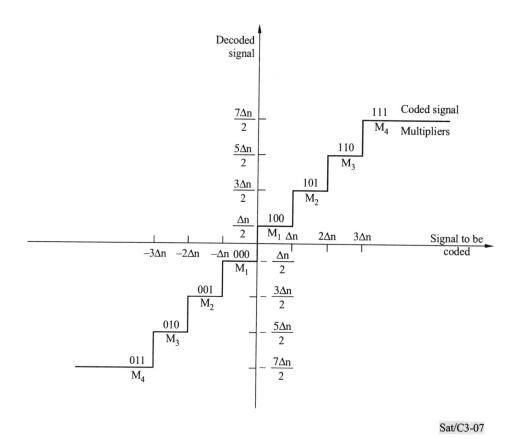

Sat/C3-07

FIGURE 3.7

Characteristic of an adaptive quantizer

3.3.1.4.2   Linear prediction

The successive samples of a speech signal (and more generally of a non-flat spectrum signal) display a degree of correlation. For numerous voice sounds, this correlation is very good. However, PCM-type coding processes the samples one after the other without allowing for any possible correlation. Linear prediction consists, in a sense, in extracting the redundancy associated with the correlation between successive samples.

When there is a high level of correlation between two successive samples, an initial application of this principle means transmitting the difference between them rather than the actual samples. Hence the term "differential", as used in ADPCM coding. More generally, linear prediction consists in calculating an approximation (or prediction) of a signal sample x_n from a few samples which precede x_n. Only the difference between the signal x_n and the predicted signal will therefore be transmitted. In the autoregressive models, the predicted signal is a linear combination of the preceding samples.

The best means of obtaining an accurate prediction is to adapt the coefficients of this linear combination to the short-term signal statistical characteristics, which also offer the advantage of ensuring compatibility with data transmission. In ADPCM applications, the predicted signal, as well as the predictor adaptation, are derived solely from the coded signal and not from the signal itself, so that the operation can be performed at both ends of the link.

3.3.1.4.3   ADPCM coding at 32 kbit/s

Figure 3.8 shows how adaptive quantizing and adaptive linear prediction are combined to produce an ADPCM coder and an ADPCM decoder. For 32 kbit/s coding, the quantizer has 16 levels (4 bits).

During study period 1981-1984, the ITU-T (former CCITT) defined a 32 kbit/s ADPCM coding algorithm based on the principles described above. The predictor is a 2-pole 6-zero adaptive filter. The quantizer is fitted with a simplified speech-data detector so that the adaptation speed can be reduced for steady-state signals, which enhances the quantizer performance for this type of signal. The algorithm is defined from signals coded in PCM at 64 kbit/s (A or μ-law), and described in ITU-T Recommendation G.721 "32 kbit/s ADPCM coding" which was revised in 1986 in order to improve its performance for FSK voiceband data modem signals at 1 200 bauds.

ADPCM coding at 32 kbit/s leads to a concentration ratio of 2. For example, this enables two CEPT 2.048 Mbit/s PCM links (see § 3.5.2) to be concentrated on just one 2.048 Mbit/s link. The use of these coding techniques could therefore double the capacity of satellite and submarine cable digital links.

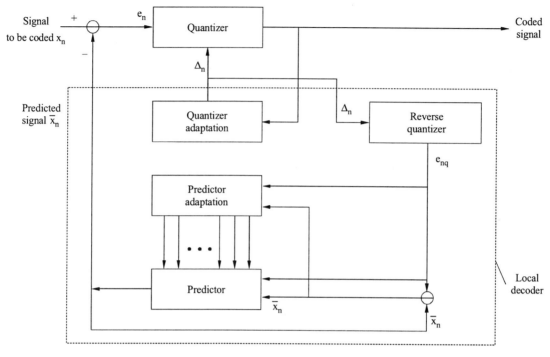

a) ADPCM coder

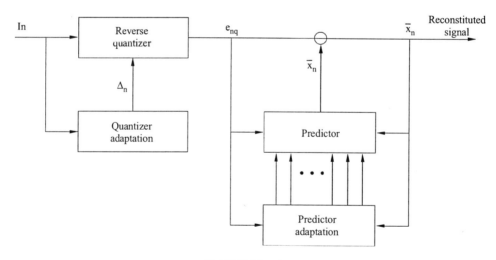

b) ADPCM decoder

e_n: difference signal
e_{nq}: quantized difference signal

NOTE - It can be verified that: $\bar{x}_n = \underline{x}_n + (e_{nq} - e_n)$.
The quantizing noise is included in $\bar{x}_n$ without amplification.

Sat/C3-08

FIGURE 3.8

Principle of an ADPCM coder and decoder

3.3.1.4.4   Degradation due to low bit-rate coding

This concentration technique brings about a signal degradation due to the addition of quantizing noise. The quality obtained with the ITU-T algorithm is roughly equivalent to that of 56 kbit/s PCM coding. For example, one ADPCM coder-decoder pair, providing a return to the analogue signal, is equivalent to three and a half 64 kbit/s PCM coder-decoder pairs in sequence.

Furthermore, the resistance to transmission errors is greater than with classical PCM coding. However, the algorithm only allows data transmission up to 4 800 bit/s.

3.3.1.4.5   Wideband speech coding within 64 kbit/s [Ref. 3-3][1]

The 300-3 400 Hz bandwidth, requiring a sampling frequency of 8 kHz, provides a speech quality referred to as toll quality. This is sufficient for telephone communications, but for emerging applications such as teleconferencing, multimedia services, and high-definition television, an improved quality is necessary. By increasing the sampling frequency to 16 kHz, a wider bandwidth, ranging from 50 Hz to about 7 000 Hz can be accommodated.

In 1986, the ITU-T defined a new algorithm for a wideband (50 Hz to 7 kHz) speech channel within 64 kbit/s. The technique involves the splitting of the wideband speech into two sub-bands (0-4 kHz and 4-7 kHz), and uses quadrature mirror filters (QMF) which avoid aliasing, and ADPCM in each sub-band. The ADPCM algorithm used in each sub-band is similar to the 32 kbit/s algorithm defined in ITU-T Recommendation G.721.

The bit rate allocated to the lower sub-band can be made variable from 32 to 48 kbit/s, corresponding to an overall bit rate of 48 to 64 kbit/s. For 48 or 56 kbit/s applications, the speech channel can be complemented by an auxiliary data channel of 16 or 8 kbit/s, multiplexed with the speech channel to form the overall 64 kbit/s channel.

The coder can operate in three different modes:

i)   64 kbit/s audio transmission with no data transmission;

ii)   56 kbit/s audio transmission with 8 kbit/s data transmission;

iii) 48 kbit/s audio transmission with 16 kbit/s data transmission.

The description of this algorithm is given in ITU-T Recommendation G.722 "7 kHz – audio coding within 64 kbit/s" (1986).

The main applications foreseen for this type of coding are for digital audioconference circuits and commentary channels for broadcasting. In the future, it could be used in the ISDN for wideband loudspeaker telephone circuits. Although the algorithm was first designed and optimized for 7 kHz speech applications, the algorithm can also provide good results for 7 kHz music signals. For speech applications, the level of quality is such that the coded speech at 64 kbit/s can be only hardly distinguished from the original 7 kHz speech without bit-rate reduction in formal listening tests.

---

[1]   Indicates references to Bibliography at the end of the Chapter.

Recent works on wideband speech coding have been focusing on coding algorithms below 64 kbit/s; for example:

- the MPEG-AUDIO layer III audio coding algorithm which results in a sub-band coder with 22 sub-bands that operates at a rate between 16 and 32 kbit/s;

- the CELP coding algorithm which is described in section 3.3.2 and Appendix 3.1.

3.3.2 Voice low rate encoding (LRE) [Ref. 3-4]

Studies are being performed in numerous countries in an attempt to reduce the bit rate for telephone signals.

This essentially means resorting to more sophisticated coding techniques than 32 kbit/s ADPCM coding; speech coding algorithms are being developed with two main interdependent objectives:

- a reduction of the bit rate assigned to a speech channel without degrading quality. This objective is of particular importance for radio systems which are limited in capacity (in terms of the number of channels or, alternatively, of the frequency bandwidth used);

- an enhancement of the quality of the signal transmitted without unreasonably increasing the bit rate; this approach is a key factor in the development of services such as teleconference, audioconference and videoconference and, more generally, new multimedia services (see § 3.3.1.4.5 above).

Speech coding technologies are grouped into two classes:

- waveform coding techniques, attempting to approximate the speech signal as a waveform, were the first to be investigated and are widely embodied in standards; for example, adaptive differential PCM (ADPCM) coding was standardized in 1984 (ITU-T Recommendation G.721) for bit rates around 32 kbit/s, and uses two signal processing techniques: prediction and adaptive quantizing (see § 3.3.1.2 above);

- parametric vocoders, attempting to model and extract the pertinent parameters from the source, and transmit them to the decoder where they are used to reconstruct a waveform in order to produce a sound similar or close to the original. The drawback when using purely parametric techniques consists in their sensitivity to the environment of the source (especially background noise) and the deterioration of the natural qualities of the voice.

At coding bit rates below 16 kbit/s and specifically the bit-rate range 9.6 to 4.8 kbit/s, the waveform coding method can provide reproduced speech which is highly intelligible. At coding bit rates in the range 4.8 kbit/s and below, source coding (vocoder) methods can provide acceptable quality.

The only way to achieve significant bit-rate reduction at present is to use hybrid techniques combining parametric and waveform coding principles.

The quality of the different coding techniques is measured subjectively by the MOS (mean opinion score) scale. Its rating has a five-point scale with categories, i.e. 1 (bad), 2 (poor), 3 (fair), 4 (good) and 5 (excellent). At higher bit rates the SNR (signal-to-noise ratio) can also be used. Today, CELP (code excited linear prediction) techniques, first introduced in 1985, take full advantage of the implementation of enhanced signal processing equipments and offer higher performance levels both in terms of quality and throughput (bit-rate) reduction.

Most low bit-rate coding procedures are proprietary to the developer. However, some are already standardized, especially for mobile communications, such as:

- IMBE (improved multiband excitation) for INMARSAT-M and Optus (Australia) mobile satellite services: 4.8 kbit/s (6.4 kbit/s with FEC);

- VSELP (vector sum excited linear prediction) for digital mobile radios in the United States: 7.95 kbit/s (13.0 kbit/s with FEC);

- VSELP (vector sum excited linear prediction) for digital mobile radios in Japan: 6.7 kbit/s (11.2 kbit/s with FEC);

- RPE-LTP (regular pulse excited long with term prediction), a hybrid coding – "analysis-by-synthesis" – technique, was standardized in 1988 by ETSI (the European Telecommunication Standards Institute) for the pan-European digital mobile radio system (GSM). It provides at only 13 kbit/s (22.8 kbit/s with FEC) a quality slightly lower than normal telephony, while being insensitive to transmission errors, always present in large quantities in mobile communications;

- ITU-T Recommendation G.728 contains the description of an algorithm for the coding of speech signals at 16 kbit/s using low-delay code excited linear prediction (LD-CELP);

- ITU-T Recommendation G.729 contains the description of an algorithm for the coding of speech signals at 8 kbit/s using Conjugate-Structure Algebraic-Code-Excited Linear-Prediction (CS-CELP).

3.3.3 Data transmission [Ref. 3-1 and 3-2]

Data transmission networks are classified according to the switching technique they implement.

Circuit switching

In a circuit switching network, each circuit has its own multiplexing and switching resources allocated to it during the whole duration of the communication. Apart from voice calls, fax (facsimile) calls are the dominant application with file transfer and computer network access. Fixed data rates between 2.4 kbit/s and 9.6 kbit/s are commonly available for reasonably fast transmission of facsimile.

Packet switching

In a packet switching network, data to be transmitted are structured in packets, and multiplexing and switching resources are affected to each communication circuit but only for the duration of the packet. This technique is particularly well suited to interactive applications; typical protocols used are the X.25, SNA/SDLC or TCP/IP protocols.

A network based on packet switching ensures three different functions:

i)   packet routing;

ii) integrity of the information contained in the packets, ensured by error detection on the packets and their retransmission if the case arises;

iii) permanent control of information flows by a window mechanism to avoid network congestion.

A window mechanism specifies the number of packets that can be sent from a transmitter to a receiver before any reverse acknowledgement is required.

Message switching

This technique is based on the reception, storage and retransmission of messages by the network; a connection between sender and recipient is therefore not required. The time needed to forward a message will depend on the amount of messages being simultaneously processed and on the data transfer rate along the information path.

The occurrence of new needs (e.g. high bit-rate data transfer over long distances) has led to the implementation of enhanced packet data transmission techniques, including:

- Frame relay, which supports data transmission over a connection-oriented path, enabling the transmission of variable length data units over an assigned virtual connection. At present Frame Relay networks authorize a length up to 4 096 bytes, to be compared with the size of 128 bytes typically encountered in current implementations of packet switching.

- Cell relay, which concerns the asynchronous transfer mode (ATM) technology for the B-ISDN and the distributed queued dual bus (DQDB) technology for access and information transfer in metropolitan area networks (MANs); § 3.5.4 and Appendix 3.IV provide a description of the ATM technique.

- Wideband packet technology (WPT) which provides efficient transport of information in the form of packets at transmission rates up to 150 Mbit/s, and uses state-of-the-art digital signal processing techniques to compress speech, voiceband data, facsimile and digital data to provide efficient transport of these signals, thereby significantly increasing the capacity of digital transmission facilities.

This technology has been approved by ITU-T and is documented in Recommendations G.764 and G.765. WPT uses circuit emulation protocols that make the compression equipment appear transparent to the user. With the advent of WPT, various types of signals can be integrated onto the same packet link or digital transmission facilities (e.g. fibre optic and metallic cables – under sea or terrestrial – digital radio and satellite).

Equipment that employs WPT is referred to as packet circuit multiplication equipment (PCME), as opposed to digital circuit multiplication equipment (DCME, see below § 3.3.7.2) which uses circuit technology.

The link level protocol to which the packet conforms is defined in ITU-T Recommendations Q.921 and Q.922. These procedures define how the bits in a frame should be interpreted. The type of packet (speech, voiceband data, facsimile, digital data, signalling, or control data) is identified by the protocol identifier in the packet header, as defined in ITU-T Recommendations G.764 and G.765. The applicable ITU-T Recommendations are:

- G.764   Voice packetization – packetized voice protocols
- G.765   Packet circuit multiplication equipment
- Q.921   ISDN user-network interface. Data link layer specification
- Q.922   ISDN data link layer specification for frame mode bearer services.

The packet circuit multiplication equipment (PCME), i.e. the terminal equipment, should be able to correctly classify the input signal into one of the following classes for the application of the most appropriate signal processing technique:

- Speech

- Voiceband data at less than 1 200 bit/s

- Voiceband data at 1 200 bit/s FSK and up to 2 400 bit/s

- Voiceband data at 4 800 bit/s but less than 7 200 bit/s

- Voiceband data at 7 200 bit/s or greater.

Through correct signal classification, transmission bandwidth is utilized efficiently resulting in higher average bits per sample for speech and better subjective performance especially under relatively high loading conditions.

For voiceband data at speeds up to 144 kbit/s, the PCME uses the 40, 32, 24-kbit/s fixed ITU-T Recommendation G.726 ADPCM algorithm. The choice of algorithm depends on the coding type assigned to the different classes discussed above.

The PCME uses a facsimile demodulation/remodulation algorithm compliant with ITU-T Recommendation G.765 to compress Group 3 facsimile resulting in up to 9:1 compression ratio. The transmitting PCME demodulates the V.29 or V.27 signal back to its 9.6, 7.2 or 4.8 kbit/s baseband signal, packetizes it and transmits each packet to the receiving PCME.

To compress digital data transported in a 64 kbit/s channel, the PCME uses two circuit emulation protocols, 1) DICE (digital circuit emulation) and 2) VDLC (virtual data link capability), which are described in ITU-T Recommendation G.765.

Using the DICE service, the PCME can interface to 64 kbit/s digital data channels or other channels of specific formats. It can remove specific bit patterns (e.g. idle codes) and this reduces transmission cost.

3.3.4 Video and TV signal

The rapid development of source coding techniques and their implementation with the help of integrated circuits leads to the existence of many new services which are based on picture coding techniques. Applications are to be found not only in the field of broadcasting but also in the area of interactive audiovisual communications. With the techniques available today, a high data rate reduction factor is achievable together with a high quality of the decoded picture. Therefore, efficient use of existing networks is possible and the costs of picture transmission are significantly reduced.

The signal can be digitized by means of a PCM coding system, which samples the signal and quantizes each sample.

The possibility for bit-rate reduction and hence for achieving the target bit rates, depends to a large extent on the service for which they are intended: for videophone and videoconference, the camera is fixed and looks at one or more individuals who are relatively immobile; on the other hand, broadcast television pictures are quite different in character: movement of camera, zoom, changes of perspective, etc.

For example, in the case of broadcast television distribution circuits, the third level of ITU-T transmission hierarchies has been fixed as the objective. This corresponds to 34 Mbit/s in Europe, 32 Mbit/s in Japan and 44 Mbit/s in the United States. In the case of videoconferencing, the more restricted range of signals to be transmitted and the different concept of picture quality means that the first level of ITU-T transmission hierarchies may be used, namely 2.048 Mbit/s in Europe and 1.544 Mbit/s in Japan and the United States. Even lower bit rates may be contemplated for the videophone service.

For the transmission of a complete service, the picture signals may be accompanied by other signals. For broadcast television, these signals can be associated with the programme (such as high-quality sound channels), control signals, signals for additional services, etc. All these signals have to be multiplexed so as to produce a single bit stream.

The presence of transmission errors generally calls for the use of an error detecting/correcting code in order to protect the transmitted signal. By these means, a transmitted bit-error ratio can be reduced from around $1.0 \cdot 10^{-6}$ to values between $1.0 \cdot 10^{-8}$ and $1.0 \cdot 10^{-9}$. This improvement is necessary because bit-rate reduction systems enhance the sensitivity of the signal to transmission errors. Therefore, the protection method and the multiplexing structure cannot be selected independently of the bit-rate reduction method when the transmission system is designed.

For the reason mentioned above, the required picture quality may change according to the service involved. The quality constraint is stricter for broadcast television than for videophone or videoconference. In order to assess the relative performance of the different coding systems for a given application, one must be clear about the criteria used for comparison.

The first to be considered is the information compression rate, which is defined as the ratio of the transmission rate in PCM to the mean transmission bit rate after compression.

The second point to be considered, which is just as important in the case of broadcast television, is the picture quality after it passes through the transmission chain. ITU-R's objective is to prevent the reception by the viewer of a lower-quality picture than that produced by an analogue transmission chain. This quality is evaluated by means of subjective tests.

The complexity of the processes must also be limited as much as possible, since complexity gives rise to problems of feasibility and cost of the coding and decoding equipments. Moreover, the existence of a transmission channel between transmitter and receiver, causing transmission errors, raises the problem of the sensitivity of the decoding system to these errors and their effect on the quality of the signal received by the final observer. This important problem is covered by specific studies.

Appendix 3.III provides the description of selected digital video compression techniques. The corresponding compression rates are given in Table 3-3 below for illustrative purposes.

TABLE 3-3

Some selected video compression examples

	Throughput (Mbit/s)	Compression rate
Theoretical	216	–
Achieved	165	1
Differential PCM	140	1, 2
Professional	34	5
MPEG 1	115	140
MPEG 2	4 to 10	15 to 40

The reader is referred to Appendix 3.III for a description of selected digital video compression techniques such as the MPEG 2 (Moving Picture Experts Group 2) and the DVB (Digital Video Broadcasting) standards:

The MPEG 2 system and video specifications are dealt with in the following Recommendations:

- ITU-T Recommendation H.222 "Information technology – generic coding of moving pictures and associated audio information: systems";

- ITU-T Recommendation H.262 "Information technology – generic coding of moving pictures and associated audio information: video".

The MPEG standard may be used for transmission of video signals over VSAT systems.

Based on top of the MPEG 2 standard, the DVB standard can essentially be seen as a complete package for digital television and data broadcasting. In particular the concept has been adapted to satellite digital television transmission (DVB-S).

3.3.5 Error control coding

3.3.5.1 Introduction to channel coding

The two fundamental problems related to reliable transmission of information [Ref. 3-8] were identified by C.E. Shannon:

- the use of a minimal number of bits to represent the information given by a source in accordance with a fidelity criterion;

- the recovery as exact as possible of the information after its transmission through a communication channel in the presence of noise and other interferences.

The first problem is usually identified as a problem of efficient communication to which the source coding provides most practical solutions. The second is a problem of reliable communication to which the channel coding, known also as error control coding, is the basic technique.

Shannon proved that, by properly encoding, these two objectives can always be achieved, provided that the transmission rate R_b verifies the fundamental expression $H < R_b < C_{cap}$, where H is the source entropy, and C_{cap} the channel capacity.

The source entropy, H, is the measure of the real quantity of information emitted by the source. The channel capacity, C_{cap}, is the measure of the maximum information quantity that two parties can communicate without error via a probabilistically modelled channel. For example, the capacity of the additive Gaussian white noise (AGWN) channel can be expressed as $C_{cap} = W\log(1+C/N)$ (bit/sec), where W is the information transmission bandwidth and C/N is the carrier-to-noise ratio of the received signal.

The capacity of the channel is independent of the coding/modulation scheme used. Shannon's channel coding theorem exactly stated that, for a given signal-to-noise ratio, the error probability can be made as small as desirable, provided that the information rate R_b is less than the capacity C and that a suitable coding is used.

As indicated by Figure 3.9, the capability of channel coding to decrease the error probability is related to the introduction of extra bits in the information bit stream generated by the source coding, and thus, by an increase of the transmit bit rate. The inverse of this increase in percentage is the code rate r_c (<1) which can be seen as the average quantity of information carried by a channel binary symbol. The code rate is high when much larger than 1/2, and low when small. The symbol rate (or modulation speed), R_S, is related to the modulation and expressed in Bauds or symbols/sec. The information transmission rate, R_b, measures the information quantity in bits transmitted in units of time, it is denoted sometimes simply by bit rate. If an M-ary modulation is concerned, all these quantities are related by $R_b = r_c R_s \log_2 M$. The product of r_c and $\log_2 M$ is the number of bits per modulation symbol, called effective rate, R_{eff}, in bits/symbol, sometimes useful in system studies.

The bandwidth occupancy, W, of a coding and modulation scheme is determined by the modulation symbol rate, R_S. The code rate has an influence but through the combination with the modulation. According to the Nyquist criterion, $W=R_S$ is the minimum bandwidth necessary to transmit R_S symbols per second, called the Nyquist bandwidth. In practice the bandwidth occupancy will be slightly larger, an empirical rule being $W = 1.2 R_S$, so that $W = 1.2 R_b/R_{eff}$.

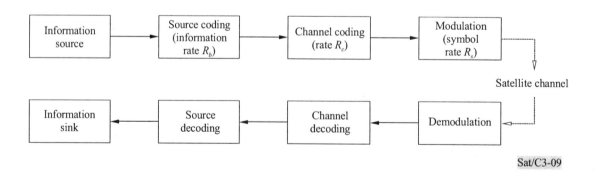

FIGURE 3.9

Information processing in a digital communication system

In satellite communications, channel coding is especially interesting because of the severe power and bandwidth limitations. This double constraint is constantly growing tighter because of the new requirements for advanced services. Besides the progress in multiple access schemes, resource assignment algorithms, signal processing techniques and advanced error control coding provide the most efficient means to realize highly reliable information transmission.

3.3.5.2 The two basic error control coding techniques

Two error control coding strategies exist: forward error correction (FEC), and automatic repeat request (ARQ).

The FEC corrects the errors at the reception without use of any information feedback to the transmit side. No additional delay is introduced except the processing time. A powerful FEC system may require a quite sophisticated implementation. The ARQ technique can be much simpler in hardware and achieve very good performance. Its drawbacks are the necessity of a return channel and the variable transmission delay.

Table 3-4 below summarizes the three main error control coding techniques.

TABLE 3-4

Comparison of the error control coding techniques

Coding technique	Through-put	Packet buffer	Performance	Delay	Return channel	Decoding	Bit rate
FEC	High	No	Medium/high	Short	No	Complex	High/very high
ARQ	Low	Yes	Very high	Long	Yes	Simple	Low/medium
Hybrid (HEC)	Medium	Yes	High	Medium	Yes	Medium	High

The ARQ technique, based on error detection coding and a retransmission protocol, is well adapted to the situations where a two-way channel is available. Typical examples of such systems can be encountered in a computer data network using satellite links. However, it is worthy to notice that the ARQ and improved ARQ, as well as HEC techniques, are widely used in modern digital communication and storage systems.

In a network environment, these two strategies are of particular interest. It is a judicious work for a design engineer to choose one of them or to combine them in a complex system to find the best solution. While error control coding with an ARQ strategy could generally be considered as an integrated part of the higher layer protocols (link and representation layers), the FEC is essentially a physical layer technique, mainly used in transmissions over a one-way channel. This is why the ARQ will not be further developed in this chapter. Interested readers are referred to reference [Ref. 3-9].

3.3.5.3   FEC (forward error correction)

The basic FEC techniques used in satellite communication systems can be classified into two categories:

- Convolutional codes, with trellis-coded modulation as extensions, presented in § 4.2.3. The convolutional codes can be further divided into three classes:

  - high rate orthogonal codes (or majority decodable codes) with threshold decoding;

  - short and medium memory codes and their associated punctured codes with Viterbi decoding using soft decision;

  - long memory codes with sequential decoding.

- Block codes, including both binary and non-binary, with algebraic decoding. In this category, the most successfully used are the Reed-Solomon (RS) codes. The block coded modulations are another generalization of this second category, when the codes are jointly optimized with modulation schemes. Block codes can be divided into the following classes:

  - BCH (Bose-Chaudhuri-Hocquenghem) codes decodable by algebraic decoding, including the Reed-Solomon codes as a non-binary subclass in BCH codes;

  - cyclic codes used for error detection;

  - simple codes such as the Hamming codes, Reed-Muller codes and certain particularly good ones: Golay, QR (Quadratic Residual), etc.

Some binary block codes such as BCH codes have been used in TDMA systems. The block codes are well adapted to time slot structures, and TDMA links are not severely power limited so that hard decision block decoders provide quite good performance. The main drawback of the block codes is that the soft decision is not as easily applicable as for the convolutional codes. However, soft decoding of block codes could be potentially interesting for satellite transmissions if a good decoding algorithm could be implemented. This remains an important research topic in FEC.

On the other hand, convolutional coding with soft decision Viterbi decoding is now a standard technique for satellite communications. This technique benefits mainly from the use of the soft decision, making full use of the channel information to improve the performance by about 2 dB with respect to the hard decision. A very simple definition of the soft decision can be made in terms of the quantization level at the output of the matched filter of the demodulator. If this quantization is of 1 bit, no reliability measure can be obtained from it and a hard decision is concerned. A soft decision must have a quantization level higher than 1 bit, from 2 bits up to infinity (unquantized). A reliability measure is available from the soft decision that is useful to improve the performance. This issue has been treated in § 4.2.3; some practical problems related to the soft decision will be further discussed in Appendix 3.2.

The association of both techniques results in an even more powerful FEC scheme: the concatenated coding system. This powerful FEC scheme has been introduced into satellite transmission systems in recent years, for a considerable increase of the service quality without a sensible expansion of bandwidth. While the inner code, with Viterbi decoding, can correct a large part of random errors and very short error bursts, the residual errors at the outputs of the Viterbi decoder tend to be grouped in bursts. Using a properly chosen interleaving that cuts the error bursts into shorter ones, a high rate Reed-Solomon code can be used as the outer code in order to correct most of these dispersed error bursts to achieve a very low bit-error rate. The introduction of concatenated coding and trellis-coded modulation into satellite communications are the most remarkable events in this domain.

3.3.5.4 Performance of error correcting coding techniques

An FEC system can improve the quality of a digital transmission link and this improvement can be appreciated from the following two aspects:

- a bit-error rate reduction, closely related to the service quality criterion; or

- a saving in the E_b/N_0 (or in C/N_0) to be considered in the link budget.

The E_b/N_0 or C/N_0 saving is often called the coding gain, expressed in dB. It can be expressed as the difference in E_b/N_0 or in C/N_0, at a certain BER value, of the coded system and the reference non-coded one. In the comparison between different transmission schemes, E_b/N_0 is usually used because it is independent of the coding scheme.

$$\Delta G = (E_b/N_0)\text{ref} - (E_b/N_0)\text{cod} \ [dB]$$

The merit of a coding system can also be appreciated in terms of the savings in C/N_0 and C/N, the carrier-to-noise density ratio, and the carrier-to-noise ratio, their relations to E_b/N_0 are given as:

$$[C/N_0] = [E_b/N_0] + 10 \log R_b \ [dB] \qquad [C/N] = [E_b/N_0] + 10 \log R_b - 10 \log W \ [dB]$$

A coding gain in E_b/N_0 means in general a gain in C/N but the coding gain in C/N depends on the bandwidth expansion with respect to the reference system. It is however possible to have a coding gain without a bandwidth expansion using trellis-coded modulation (TCM), § 4.2.3.

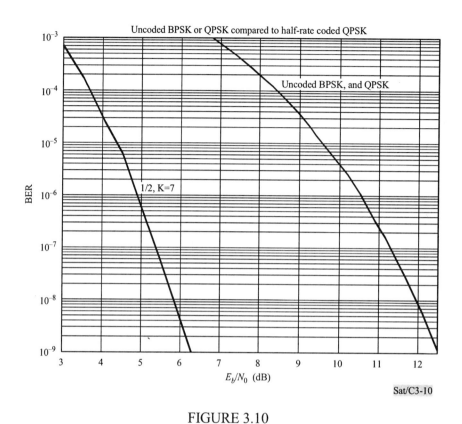

FIGURE 3.10

Merit of channel coding in a digital transmission system

The E_b/N_0 coding gains are nearly 5.2 dB and at least 6 dB for the bit-error ratios (BER) of 10^{-5} and 10^{-7} respectively. When compared to uncoded BPSK there is no bandwidth expansion and the C/N coding gains are also nearly 5.2 dB and at least 6 dB. If uncoded QPSK is chosen as reference, a bandwidth expansion of a factor of 2 exists, and the gains in C/N become respectively 8.2 dB and 9 dB.

Channel coding is a very efficient technique to improve transmission performance. It decreases the necessary E_b/N_0 (and thus the transmit power) to achieve a given BER objective, reducing the cost of other system components such as RF elements and antennas.

3.3.5.5   Typical parameters of error correcting techniques

The following Table 3-5 presents a short list of FEC techniques along with their performance.

TABLE 3-5

Typical techniques of FEC and their performance/complexity

Code	Decoding technique	Gain (BER=10^{-5})	Gain (BER=10^{-8})	Bit rate	Complexity
Convolutional codes	Threshold decoding	1.5-3.0	2.5-4.0	Very high	Low
Convolutional codes	Viterbi decoding (soft decision)	4.0-5.0	5.0-6.5	High	High
Convolutional codes	Sequential decoding (hard decision)	4.0-5.0	6.0-7.0	High	Low
Convolutional codes	Sequential decoding (soft decision)	6.0-7.0	8.0-9.0	Medium	Low
Concatenated codes (convolutional & RS)	Viterbi inner and algebraic outer	6.5-7.5	8.5-9.5	High	Medium
Concatenated codes (short block & RS)	Soft inner and algebraic outer	4.5-5.5	6.5-7.5	Medium	High
Short block linear codes	Soft decision	5.0-6.0	6.5-7.5	Medium	High
Block codes (BCH & RS)	Algebraic decoding (hard decision)	3.0-4.0	4.5-5.5	High	Medium

Remarks:

- With the above-mentioned classical techniques, a higher coding gain is obtained generally with a lower rate code, and consequently with more bandwidth expansion.

- The complexity may be evaluated by the number of computations in the decoding algorithm but must take into account commercial considerations, such as the cost of development. A VLSI decoder chip may be considered as cheap when developed in large quantities (for example Viterbi and RS decoders).

- A specific code should be evaluated in association with a modulation scheme. For example, a binary 1/2 convolutional code and its higher rate punctured ones must be associated with a QPSK modulation. (For a definition of punctured codes, see Appendix 3.2.) An RS code is best concatenated to an inner code which is well adapted to a suitable modulation scheme. See also Appendix 3.2 for a generality about the Reed-Solomon codes.

- A single RS code is not well matched directly to digital modulation. See § 4.2.3 for a more detailed treatment of joint optimization of channel coding and digital modulation.

- Some recent developments such as turbo-codes [Ref. 3-22] seem very attractive for satellite communications. These new codes are constructed by a parallel concatenation of short convolutional codes. The turbo-codes are decoded by cascaded Viterbi decoders with soft outputs. Extremely good performances have been shown by simulation and the first VLSI decoder chips have been tested. They will not be studied here, because the techniques are not yet technologically mature and much R&D effort remains necessary for their practical application.

3.3.5.6 Error correcting coding in existing satellite transmission systems

Table 3-6 presents some typical FEC applications in earlier and current satellite communication systems. More details are available in Appendix 3.2.

TABLE 3-6

Examples of the use of FEC on satellite digital links

Satellite system	Access mode	Info. bit rate	Type	R	Length	Distance	Decoding algorithm	Gain or improvement in BER
Intelsat	FDMA/ SCPC	48 kbit/s 56 kbit/s	self-orthogonal convolutional	3/4 7/8	80 384	5 5	threshold	10^{-4} to 5.10^{-9} 10^{-4} to 5.10^{-8}
		≥4.8 kbit/s	shortened Hamming	14/15	120	4		
		105 Mbit/s	BCH (128, 112)	7/8	128	6	table look-up	2.5 dB for 10^{-5}
Intelsat	TDMA	assignment channel	extended Golay (24, 12)	1/2	24	8		
Telecom 1	TDMA	nx64 kbit/s	Hamming	4/5	40	4	table look-up	10^{-6} to 10^{-10}
		8 Mbit/s 34 Mbit/s	convolutional	2/3	5	5	Viterbi	4.5 dB for 10^{-5}
Telecom 1	FDMA	48 kbit/s data	convolutional (Idaware)	3/4	3	3	threshold	10^{-4} to 10^{-11} burst errors
Eutelsat SMS Intelsat IBS	FDMA	nx64 kbit/s	convolutional	1/2	7	10	Viterbi	
Intelsat IDR	FDMA/ SCPC	nx64 kbit/s 1.5/2-8 Mbit/s up to 45 Mbit/s	convolutional & punctured	1/2 3/4 7/8	7	10 5 3	Viterbi	
Intelsat IDR NG	FDMA/ SCPC	nx64 kbit/s 1.5/2-8 Mbit/s	RS(219, 201) + convolutional & punctured	1/2 3/4 7/8	7	19(RS)	Viterbi Algebraic	
		up to 45 Mbit/s	RS (219, 201) + 2/3 PTCM 8PSK		7	19(RS)	Viterbi Algebraic	
SBS	TDMA	48 Mbit/s	Shortened Reed-Solomon (204, 192)	16/17	204	13	Algebraic	10^{-4} to 10^{-11}
TVSAT	TDM	900 kbit/s	BCH (63, 44)	44/63	63	8	Algebraic	
Inmarsat Std B	SCPC	300/ 600 bit/s	convolutional	1/2	7	10	Viterbi	+ interleaving
Inmarsat Std C	SCPC	9.6/ 16 kbit/s	punctured	3/4	7	5	Viterbi	Audio quality for 10^{-3}

Remarks:

- "Length" means the block length for the block codes, but the constraint length for convolutional codes. For self-orthogonal codes, "Length" is the constraint length of the code, it is also the so-called "decoding constraint length", denoted K=m+1, where m is the number of memory elements in the encoder. See [Ref. 3-9] and Appendix 3.2 for threshold decoding of the self-orthogonal codes.

- "Distance" means the minimum Hamming distance for the block codes, but the free distance for the convolutional codes. In particular, dmin=N–K+1 for the Reed-Solomon codes.

- "Viterbi" means 3-bit soft decision decoding by the Viterbi algorithm.

- "Algebraic" means hard decision algebraic decoding based on the Berlekamp-Massey iteration or the extended Euclid iteration procedures [Ref. 3-9].

- "Gain" is the E_b/N_0 coding gain measuring a certain BER.

- In concatenated coding systems, an interleaving between the inner and the outer codes is often applied. For example, up to 2 dB can be obtained as a supplementary gain using an appropriately chosen interleaving over a system without any interleaving.

- "Intelsat IDR" means the Intelsat IDR Standard IESS-308. The use of Reed-Solomon outer code is optional, such that the existing equipments remain compatible. In addition, the pragmatic trellis-coded 8PSK modulation (see § 4.2.3) is introduced.

- This table is far from exhaustive and may be partially obsolete. Concatenated coding with Reed-Solomon codes and trellis-coded modulation schemes are entering into the satellite transmission domains. Trellis-coded modulation (TCM) is dealt with in detail in Chapter 4 (§ 4.2.3). Furthermore, the "turbo-codes" may have a very interesting perspective in future satellite systems.

3.3.5.7 System design with error correcting coding

The engineering design of a FEC system may be an iterative process because of the large number of interactive factors to be treated. These factors are:

- the E_b/N_0 (C/N_0) value and the necessary margin used in the link budget analysis;

- the bandwidth expansion;

- the bit-error rate objective;

- the processing delay;

- the multiple access scheme, in particular the data format constraints and the multiple access interference analysis;

- the modulation and the demodulation schemes, in particular the synchronization threshold;

- the phase ambiguities resolution possibility with the code transparency;

- the availability of the decoder devices;

- the implementation complexity, the cost, and the development delay and economic efforts, etc.

The synchronization or demodulation threshold is the minimum signal-to-noise ratio above which the carrier recovery algorithm can work normally. Below that threshold the carrier can no longer be recovered and cycle slips appear. It is useless to use an over-dimensioned, too powerful FEC scheme, if it puts the required E_b/N_0 lower than the demodulation threshold. This demodulation threshold can be used as a lower bound of the E_b/N_0 value, for a FEC scheme. Upper bounds exist for general coding and modulation schemes using the technique of Chernoff and union bounds, explicit performance bounds have been derived for some popular codes such as binary convolutional codes using their transfer function, see [Ref. 3-10] and Appendix 3.2 for more details.

For coding systems, the most useful criterion to evaluate is the bit-error ratio (BER). Sometimes, symbol error rate (SER), and frame error rate (FER) could be more adequate, depending on the applications. For detection problems, missing detection probability and false alarm probability are of particular importance.

The design of a reliable communication system is strongly related to the use of a precise and realistic link budget. For this purpose, when a performance evaluation is needed, the most practical way may be a computer simulation. The actual computer simulation packages can evaluate rapidly the performance of a fairly complicated system, including the channel filtering, non-linear amplification, ACI (adjacent channel interference) and CCI (co-channel interference), phase noise, modem imperfections, etc.

The performance evaluation must be carried out in several steps. The first one, also the most important one, is to establish a proper mathematical model of the system. The model should be as simple as possible, but it must represent as closely as possible the main features of the system. A trade-off should be found between a complete and complex, and a simple and inaccurate, modelling of the transmission link. The second step consists in the computation by the most appropriate method for the model: analytical, semi-analytical or simulation.

Finally, it is reasonable to state that a well-designed FEC technique realizes a trade-off in system complexity between baseband digital processing and the RF hardware. A saving in C/N_0 can simplify considerably the RF part, reducing the antenna diameter, the power flux-density, etc. at the price of the FEC digital processing (essentially in the decoder) and of bit-rate increase.

It can be foreseen that more and more sophisticated FEC schemes will be utilized in future satellite communication systems. A partial list of references on coding theory and techniques is provided in the Bibliography at the end of this Chapter.

3.3.6   Encryption

Encryption is a means by which data, during transmission, are protected from being decoded by unauthorized users. The encryption is usually applied to the pure information pulse train on a channel-by-channel basis before it is subjected to multiplexing or error correcting coding.

There are many methods currently available to realize the encryption function. As an example, the additive cipher method which is recommended for use with the INTELSAT Business Service (IBS) is explained below.

In this method, encryption and decryption are achieved by the bit-by-bit modulo-2 addition of an identical pseudo-random sequence (called key stream sequence) to the data pulse train at the transmit and receive ends of the link. Synchronization of the encryption sequence at both ends of the link is achieved by initializing the key stream sequence every multiframe. This is signalled to the receive end using the encryption control channel which is included in the framing bits.

The key stream sequence in use is identified by a key identification which is numbered in the range of 0-255. The key is renewed after a certain period of time (e.g. 24 hours) to keep the encryption effective. The key change is signalled by a new key identification number sent via the encryption control channel and is implemented without a break in transmission.

3.3.7   Other signal processing techniques

3.3.7.1   Speech interpolation

Digital speech interpolation (DSI) techniques are based on the fact that, in a normal two-way conversation, each speaker actually uses the telephone circuit for only about half the total time taken up by conversation. Furthermore, gaps between syllables, words and phrases add to the idle time, so that a normal one-way telephone channel is active, on average, for only 35 to 40% of the time during which the circuit is connected.

Accordingly, if the same transmission channel can be assigned to different speakers on a voice-activated basis, the transmission channel is better utilized. This brings a significant gain in the number of conversations which can be routed over the same channel in a given period, and leads to a gain in traffic capacity (e.g. 2.5 times for 40% activity). The theoretical speech interpolation gain is defined by the ratio between the number of speakers (input trunks) and the number of transmission channels (bearers) required to service them.

In view of the large proportion of telephone channels used alternately for speech transmission and for transmission of data with a higher activity factor than that of speech, the practical gain generally achieved at the present time is approximately 2.

3.3.7.2   Digital circuit multiplication equipment (DCME)

The function of the digital circuit multiplication equipment is to concentrate a number of input digital lines (trunks) onto a smaller number of output channels (bearers), thereby achieving a higher efficiency of utilization of the link. This is quantified by the "circuit multiplication gain" which is defined as the ratio of the number of input channels over the number of DCME output channels.

The digital circuit multiplication equipment usually refers to both the low-rate encoding (LRE) and digital speech interpolation (DSI) functions where the combined use of DSI (see § 3.3.7.1) and LRE by ADPCM (see § 3.3.1.4) gives a circuit multiplication gain of about 5. The actual gain depends on the traffic loading and, in particular, the percentage of voiceband data calls being carried on the link. Physically, a DCME may interface with several 2.048 Mbit/s or 1.544 Mbit/s primary multiplex bit streams on the trunk side, and one 2.048 Mbit/s or one 1.544 Mbit/s bit stream on the bearer side (for digital multiplex signals see § 3.5.2).

ADPCM contributes to the gain by lowering the bit rate of the incoming 64 kbit/s PCM channel. Although various encoding algorithm and rates are possible, ITU-T recommends the following ITU-T-defined ADPCM algorithm for use with DCME:

	ADPCM rate	Bits per sample
a) Speech	32 or 24 kbit/s	4 or 3 bits (ITU-T G.721 and G.723)
b) Voiceband data	40 kbit/s	5 bits (ITU-T G.723)

a)   Speech transmission

The speech signal is normally coded at 32 kbit/s using the G.721 algorithm before being connected to an available 4-bit bearer channel. When, during a period of high peak load, an input speech channel turns active and there is no 4-bit bearer channel available, then a 3-bit overload channel is created by stealing least significant bits (LSBs) from three 4-bit bearer channels which are carrying speech. These 3-bit channels (both created and bit stolen) then undergo 24 kbit/s coding using an algorithm optimized for speech signals. This algorithm (called G.723) is derived from ITU-T Recommendation G.721 by modifying quantizing thresholds and their corresponding tables.

During this operation, speech quality is degraded due to a reduction in the signal-to-noise ratio of around 4 dB compared to the 32 kbit/s coding defined in ITU-T Recommendation G.721. However, this increase in quantizing noise is considered less annoying, subjectively, than the effect of speech clipping which is more evident when overload channels are not used. Experience shows that when circuit multiplication gain is greater than five, the subjective quality of speech is substantially better than that obtained by fixed bit-rate coding. The change in bit rate does not affect system operation as long as the changes in the quantizer are synchronized at both ends of the link.

b)   Data transmission: see § 3.3.3

c)   Load control.

The creation of overload channels permits a higher circuit multiplication gain but introduces some speech quality degradation. In order to meet the speech transmission performance requirements, the DCME traffic load should be controlled in such a manner that the average speech encoding rate does not remain below a predetermined value (typically 3.6 to 3.7 bits/sample) for an extended period of time. When this limit is reached, a built-in dynamic load control (DLC) function of the DCME will signal to the switch that no additional calls should be placed on the trunks served by the DCME. If the DLC function is not implemented, the traffic should be engineered to take into account the very rare occurrences of large peak loads which, in turn, would result in lower achievable multiplication gain.

3.3.7.3   Energy dispersal for digital systems (scrambling)

Energy dispersal used with digital signals is usually referred to as scrambling. The principle of scrambling is to randomize the transmission pulse train irrespective of the incoming information pulse train, thereby reducing the peaks of the carrier spectrum density. There are two general methods to achieve this, the pseudo-random scrambler and the self scrambler.

a)   Pseudo-random scrambler

In this method, the output of a pseudo-random code generator is modulo-2 added to the information pulse train to generate a pseudo-random transmission pulse train. At the receive end, the same pseudo-random code pattern is modulo-2 added to the received pulse train to recover the original information pulse train. Figure 3.11 illustrates an example of this type of scrambler/descrambler.

The pseudo-random code generator of the receive unit needs to be synchronized to that of the transmit unit; for this reason this kind of scrambler is usually used with those systems which include a frame alignment signal to enable such synchronization to take place (e.g. TDMA, INTELSAT Business Service).

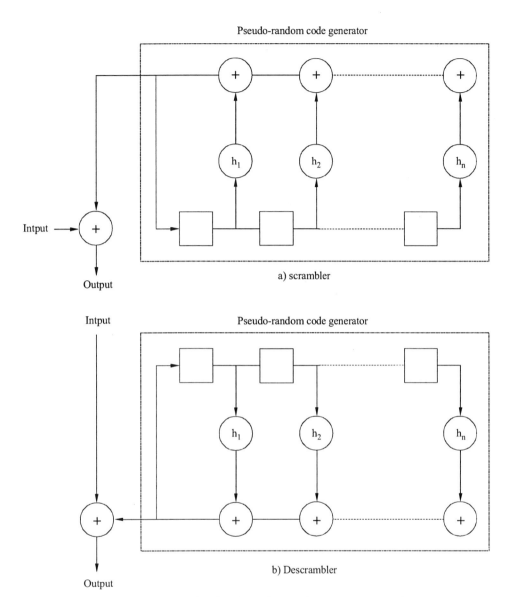

a) scrambler

b) Descrambler

NOTE: h_1, h_2 ... h_n represent the shift register internal connections which determine the code sequence.

Sat/C3-11

FIGURE 3.11

Pseudo-random scrambler

b)   Self scrambler

In this method, a shift register having a feedback loop, is installed at the transmit end to randomize the transmission pulse train. At the receive end, a shift register having a feed-forward loop is used to recover the original information pulse train. A typical circuit arrangement for this type of scrambler/descrambler is illustrated in Figure 3.12.

This type of scrambler does not require synchronization between the transmit and receive units, and is therefore suitable for use with a continuous mode of data transmission where no knowledge of frame alignment signal, etc., is assumed. However there is a disadvantage in that a single error in the transmission pulse train would result in r bits of errors, if r is the number of the stages of the shift register. Consequently, the scrambler is usually placed before the error correcting encoder at the transmit end, and the descrambler before the error correcting decoder at the receive end.

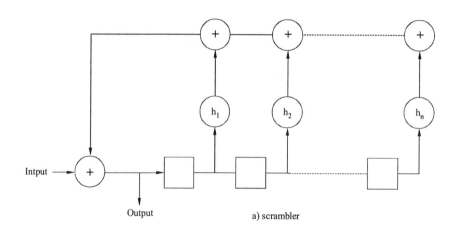

a) scrambler

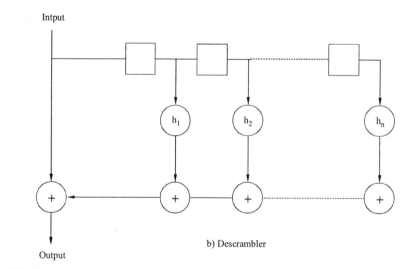

b) Descrambler

NOTE: h_1, h_2 ... h_n represent the shift register internal connections which determine the code sequence.

Sat/C3-12

FIGURE 3.12

Self scrambler

3.4 Multiplexing – analogue

Multiplexing is the reversible operation of combining several information-bearing signals to form a single, more complex signal. The signals that are combined in a multiplexer usually come from independent sources, such as subscribers in a telephone network. Prior to multiplexing, each signal travels over a separate electrical path, such as a pair of wires or a cable, whereas the multiplexed signal can be transmitted over a single communication medium of sufficient capacity. The reversibility of the multiplexing operation permits recovery of the original signals which often have different final destinations at the receiving end of the transmission link. This inverse operation whereby the original signals are recovered is called demultiplexing.

Multiple access may be considered as a special kind of multiplexing in which two or more signals, each from a different location, are sent by radio to a single transponder. The term "multiplexing", however, is usually restricted to describe those other situations where the signals to be combined arrive over electrical circuits. The most commonly used techniques for both analogue multiplexing and multiple access are based on the common principles of frequency division.

In a frequency division multiplex (FDM) system, a number of channels are arranged side by side in the baseband, and thus can be accommodated by a single wide bandwidth transmission system. The actual multiplexing process is as follows (Figure 3.13):

a)   each telephony channel transmits audio frequencies from 0.3 to 3.4 kHz, the baseband signals being in the form of single-sideband (SSB) signals with suppressed carriers at 4 kHz spacing;

b)   twelve telephone channels are frequency-converted to compose a basic group in the frequency range from 60 to 108 kHz;

c)   five basic groups are again frequency-converted to compose a basic supergroup in the frequency range from 312 to 552 kHz;

d)   frequency-conversion of this basic supergroup can be realized to multiplex many further telephone channels.

FDM is used for terrestrial communications, and conventionally the multiplexed telephone channels are arranged at baseband frequencies above 60 kHz. However, in satellite communications, one basic group is arranged in the frequency band of 12 to 60 kHz so as to use the baseband frequency bandwidth more effectively.

At the receiving earth station, the FDM signal is demultiplexed through a sequence of filtering and SSB-demodulation steps. Separation of supergroups, and individual channels by filters is possible with minimal impairment because of guardbands left in the FDM signal. It should be recalled that the voice signal occupies only 3 100 Hz of the 4 kHz channel. An important technical requirement of FDM concerns the accuracy and coherence of SSB carrier frequencies, which are usually derived from stable master oscillators.

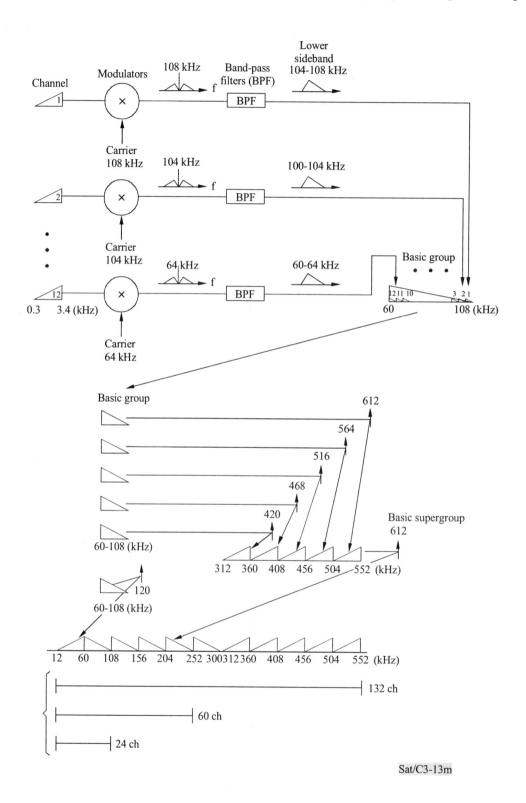

FIGURE 3.13

Frequency arrangement of frequency division multiplex (FDM) telephony

3.5 Multiplexing – digital

3.5.1   General

Digital transmissions take advantage of another technique, called time-division multiplexing, in which the bit streams arriving in parallel at a network node (for example a satellite communications terminal or an intermediate switching centre) and having the same destination are time-division multiplexed (i.e. interleaved) into a single serial bit stream for transmission on a single RF carrier.

As in an FDM system, a TDM signal is formed through a sequence of multiplexing steps. The resulting multiplex arrangement is called a hierarchy.

3.5.2   Plesiochronous digital hierarchy (PDH) [Ref. 3-1]

PDH transmission systems provide a very cost-effective means of transporting a large number of telephone circuits, due to reductions in the cost of integrated circuits and advances in optical technology.

The current PDH evolved mainly in response to the demand for plain voice telephony. The availability of cheap transmission bandwidth led to the proliferation of new, non-voice, telecommunication services, most of them aimed at the business customer. New service offerings resulted in new requirements to be met by the network, such as flexibility of providing new connections or dynamically distributing the capacity.

As shown in Figure 3.14 there are three kinds of digital hierarchy recommended by ITU-T. This figure also shows other bit rates, non-standard in the sense of the Recommendation, which are actually used mainly because they were introduced two decades ago: 274 Mbit/s in North America, 97 and 400 Mbit/s still used in Japan, and 560 Mbit/s, corresponding to a 4 · 140 Mbit/s multiplexed signal, widely used in long-distance networks in Europe.

The first order of digital hierarchy, which is often called the primary group, is either 1.544 Mbit/s or 2.048 Mbit/s. The 1.544 Mbit/s stream (used in North America and Japan) contains twenty-four 64 kbit/s channels and additional 8 kbit/s for framing and signalling. The 2.048 Mbit/s stream (used in the rest of the world) contains thirty 64 kbit/s channels carrying traffic and two additional 64 kbit/s channels (the respective time slots are No. 0 and No. 16) carrying framing and signalling information. For international working, a clock accuracy of better than or equal to $1.0 \cdot 10^{-11}$ is required with these primary digital bit streams.

Second and higher order bit streams are formed by multiplexing several lower order bit streams (called tributaries). However, the bit streams to be multiplexed may have slightly different bit rates. Even if these bit streams are nominally at the same bit rate, they have imperfect clock stabilities and hence are not exactly synchronous with one another or with the earth terminal clock. Furthermore, some of these bit streams may have arrived via a satellite link in a multi-hop transmission where the satellite-earth station propagation time is changing because of satellite motion. Consequently, although many of these bit streams may have the same nominal clock frequency they are not necessarily synchronous. In order to time-multiplex these bit streams, one or all of the bit streams should be buffered and have "stuff" bits added to synchronize them (justification).

For interworking between countries using 1.544 Mbit/s and 2.048 Mbit/s systems, a different hierarchy is recommended (called interworking hierarchy).

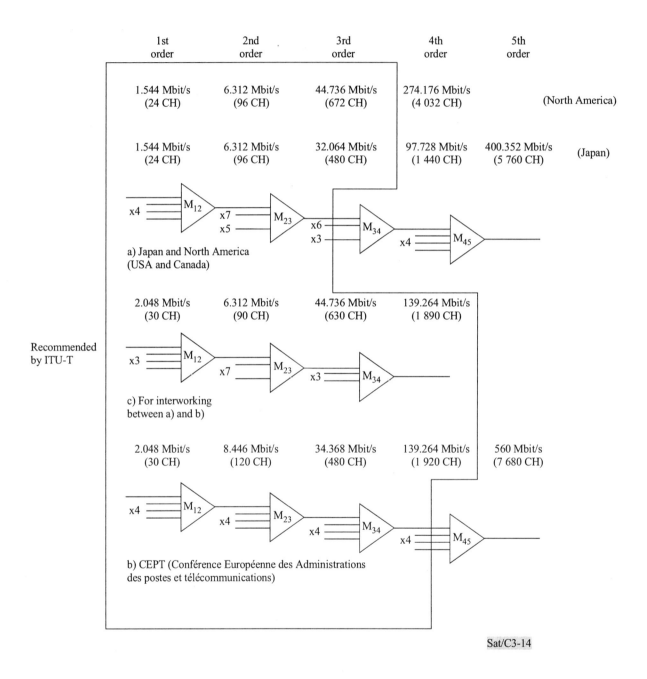

FIGURE 3.14

Digital hierarchy bit rates

Although capable of multiplexing and then transporting many different types of lower speed signals at higher rates, the PDH structure does not provide a means to access these original tributaries without completely demultiplexing the high speed frame. This may be illustrated by considering the provisioning of a 2 Mbit/s leased line: several multiplexing and demultiplexing operations have to be done to provide such a line from a high speed channel, say at 140 Mbit/s (colloquially named "Mux Mountains"). Figure 3.15 shows an add/drop process of low tributaries from/to a high bit-rate signal, where, for the sake of clarity, the scheme has been extremely simplified omitting all digital frames, accesses to the alarm and supervision collection system and engineering order wire connections required between remote sites and operation centres.

Not easy to modify, low performance adding and dropping of channels is a handicap for flexible connections or rapid provisioning of services. Control of the paths followed by circuits is also complicated within these (de)multiplexing operations and interconnection of equipment.

The inability to identify individual channels in a high-speed bit-stream, the absence of efficient means for monitoring the transmission quality (e.g. bit errors) and a frame structure with insufficient provision for the carrying of equipment and network management information are the main limitations of the PDH, which may be acceptable in telephony but are critical in a network that has to provide many other services.

In addition to the G.730, G.740 and G.750 Recommendations on PDH multiplexing equipment, a non-exhaustive list of ITU-T Recommendations on PDH includes:

G.702   Digital hierarchy bit rates

G.703   Physical/electric characteristics of hierarchical digital interfaces

G.704   Synchronous frame structures used at primary and secondary levels

G.706   Frame alignment and cyclic redundancy check procedures relating to basic frame structures defined in Recommendation G.704

G.797   Characteristics of a flexible multiplexer in a PDH environment

G.802   Interworking between networks based on different digital hierarchies and speech encoding laws

G.821   Error performance of an international digital connection forming part of an integrated services digital network

G.823   The control of jitter and wander within digital networks which are based on the 2 048 kbit/s hierarchy

G.824   The control of jitter and wander within digital networks which are based on the 1 544 kbit/s hierarchy

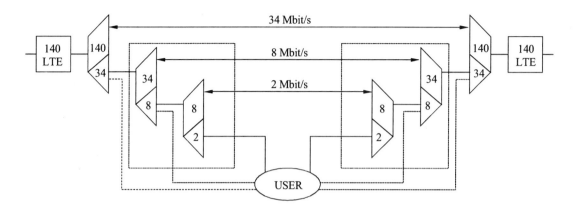

LTE: Line Terminator Equipment
----   If 8 Mbit/s or 34 Mbit/s to be dropped

Sat/C3-15

FIGURE 3.15

Add/Drop process of low tributaries from/to a high bit-rate signal

3.5.3   The synchronous digital hierarchy (SDH) [Ref. 3-1 and 3-7]

3.5.3.1   General considerations

Progress in electronics which makes available cheap processing capabilities to be included in equipment, together with the availability of transmission mediums with very wide bandwidth and high performance, the need to monitor and manage equipment and networks, and to build rapidly flexible, reliable and highly protected networks able to cope with the request of new, high quality and broadband services, among other factors, led to consider a new hierarchical structure overcoming the limitations of PDH systems.

The first initiatives in this area were taken in the United States under the title of Synchronous Optical Network (SONET). This title comes from the fact that this technique was thought to apply primarily to optical fibre links. The ITU-T standard version is called the Synchronous Digital Hierarchy. It is basically defined in ITU-T Recommendations G.707, G.708 and G.709. Through a proper coordination of ITU-T and ITU-R, there is a constant concern in ITU that Recommendations on SDH should be applicable, as far as possible, to all types of wideband transmission medium, including, of course, satellite links. Also for that purpose, new Recommendations on the applicability of ITU-T Recommendations to satellite communication are prepared by ITU-R SG 4.

The SDH is a hierarchical set of digital transport structures, standardized for the transport of suitably adapted payloads over physical transmission networks.

The SDH-based facilities are implemented in equipment that offers not only the conventional functions required for transmission networks but also new functions. This exploits the full potential of SDH and leads to new network structures.

SDH-based facilities offer significant advantages over earlier PDH systems, such as:

•   direct access to tributaries, that is to say, the ability to efficiently add/drop traffic without the need to multiplex and demultiplex the entire high-speed signal;

•   enhanced operation, administration and maintenance (OAM) capabilities;

•   easy growth to higher bit rates in step with evolution of transmission technology.

Network node interface (NNI) specifications necessary to enable interconnection of synchronous digital elements for transport payloads (including digital signals of the PDH) are defined in ITU-T Recommendation G.708 (see Figure 3.16).

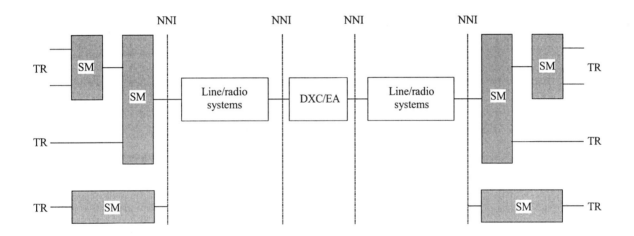

DXC: Digital cross-connect equipment
EA: External access equipment
SM: Synchronous multiplexer
TR: Tributaries

Sat/C3-16

FIGURE 3.16

Location of network node interfaces

3.5.3.2   Transmission rate levels

SDH bit rates are defined in ITU-T Recommendation G.707. Transmission rate levels are defined in such a way that the higher levels are obtained as integer multiples of the first level bit rate (155 520 kbit/s) and denoted by the corresponding multiplication factor.

Where there is a need to transport SDH signals on radio and satellite systems, which are not designed for the transmission of basic rate signals, relevant SDH elements may operate at a bit rate of 51 840 kbit/s across digital sections. However, this bit rate does not represent a level of the SDH nor a Network Node Interface bit rate.

3.5.3.3   Definitions of the frame structure and related elements

The information structure used to support section layer connections in the SDH is called synchronous transport module (STM). It consists of information payload and section overhead information fields organized in a block frame structure which repeats every 125 microseconds. The information is suitably conditioned for serial transmission on the selected media at a rate which is synchronized to the network.

The basic frame is the STM of level 1 (STM-1). Higher capacity STMs are such that the bit rate of a STM-N is exactly N times that of a STM-1 and its frame is derived by byte interleaving of N STM-1s. Already standardized STM-Ns are listed below, but higher capacity STMs are under consideration.

•   the STM-1 at 155 520 kbit/s,

•   the STM-4 at 622 080 kbit/s,

•   the STM-16 at 2 488 320 kbit/s.

Once the payload area of the STM-1 frame is formed (by signal and/or stuffing bytes), some more control information bytes are added to the frame to form the section overhead (SOH). Their purpose is to provide functions such as communication channels for OAM facilities, protection switching, section performance, frame alignment and operator specific applications.

The STM-1 frame is considered as a matrix structure of 270 byte columns and nine rows. The STM-N frame structure is shown in Figure 3.17. The three main areas of the STM-N frame are indicated:

•   section overhead,

•   administrative unit pointer(s),

•   information payload.

In order to accommodate signals generated by equipment from various PDH levels, SDH recommendations define methods of subdividing the payload area of an STM-1 frame in various ways so that it can carry different combinations of tributaries (see Appendix 3.5).

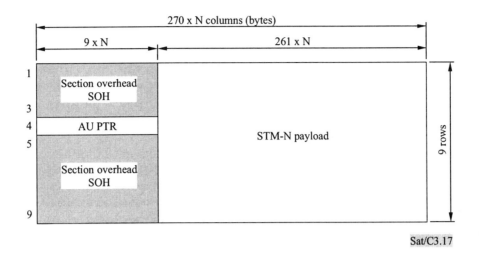

FIGURE 3.17

The STM-N frame structure

3.5.3.4   Summary of SDH features

The main characteristics of the SDH may be summarized as follows:

a) Flexibility of the frame structure. Synchronous multiplexing is used to map various tributaries within the STM-1 and to derive higher order levels (STM-4 and STM-16). This enables a direct visibility of the frame structure and the direct add and/or drop of lower order tributaries from high bit-rate line signals, facilitates the evolution towards higher bit rates and allows the realization of add/drop multiplexers and digital cross-connect systems. SDH allows a cost-effective handling of the traffic as well as the establishment of self-healing ring structures which provide service survivability, for instance in the event of cable failures.

b) Enhanced operation, administration, maintenance and provisioning capabilities, enabling the evolution of transmission networks towards truly managed networks. The basic frame structure includes functions for the operation and maintenance of transmission equipment and networks. Embedded data channels can transmit the operation and maintenance information data, gathered in each network element of the SDH subnetwork to a gateway management unit, and then, via a data communication network (DCN), for instance an

X.25 network, to a management centre. This capability allows the implementation of the ITU-T Telecommunication Management Network (TMN) concept.

c) Standardized interfaces for the most typical applications, such as intra-office, short-haul or long-haul inter-office (ITU-T Recommendation G.957). The detailed specification of all the parameters characterizing cables, transmitters and receivers allows inter-vendor (or transverse) compatibility (mid span meet) of equipment in the same optical section, i.e. the possibility to connect for instance transmitters and receivers from different manufacturers. Similarly, the electric interface for STM-1 is standardized (ITU-T Recommendation G.703) for interconnecting equipment in the same location. The simultaneous availability of SDH and PDH interfaces enables a gradual migration of the existing network to a fully SDH network.

Using essentially the same methods as in PDH technology, a synchronous network is able to significantly increase the bandwidth and reduce at the same time the amount of the equipment in the network. In addition, sophisticated network management, provided within the SDH, introduces more flexibility in the network.

The SDH defines a structure which enables PDH signals to be combined together and encapsulated within a SDH standard signal. Interworking between PDH and SDH equipment is straightforward. SDH equipment can therefore be deployed to suit particular needs, if constraints do not allow its fast introduction into the whole network.

SDH-based equipment exhibits a number of particularly attractive features. Its chief quality is its complete compatibility with plesiochronous networks, for which it requires only conventional plesiochronous interfaces. This allows a smooth introduction of SDH in PDH networks because, for example, it assures the transparent transport of PDH frames in SDH islands during the transition phase. Moreover, this equipment forms part of the worldwide standardization drive promoting a convergence of European, Japanese and North American hierarchies.

As already explained, although SDH was initially built and developed on optical fibre technology, microwave technology is being adapted to be compliant with SDH requirements, basically considering the STM-1 and lower bit rates (e.g. tributary) levels.

ITU-T Recommendations defining SDH principles, equipment and network include the following:

G.703   Physical/Electrical characteristics of hierarchical digital interfaces

G.707   Synchronous digital hierarchy bit rates

G.708   Network node interface for the SDH

G.709   Synchronous multiplexing structure

G.774   SDH management: Information model for the network element view

G.781   Structure of Recommendations of SDH multiplexing equipment

G.782   Types and general characteristics of multiplexing equipment for the SDH

G.783   Characteristics of SDH multiplexing equipment functional blocks

G.784   SDH management

G.803   The control of jitter and wander within digital networks which are based on the SDHT

G.831   Management capabilities of transport networks based on the SDH

G.957   Optical interfaces for equipment and systems relating to the SDH

G.958   Digital line systems based on the SDH for use on optical fibre cables

More details on SDH will be found in Appendix 3.5.

3.5.4   Asynchronous transfer mode [Ref. 3-1, 3-6]

3.5.4.1   Introduction and overview

The asynchronous transfer mode (ATM) was conceived to meet the sophisticated telecommunication requirements of the techniques and technologies of information transport.

ATM is a specific packet-oriented transfer mode which uses asynchronous time-division multiplexing techniques. It combines the flexibility of packet switching and the simplicity of circuit switching. The multiplexed information flow is organized into fixed length cells. A cell consists of an information field of 48 bytes and a header of 5 bytes (see Figure 3.18). The network uses the header to transport the data from source to destination. Transfer capacity is assigned by negotiation and is based on the source requirements and the available capacity. In general, signalling and user information are carried on separate ATM connections.

ATM is the solution selected for implementing broadband ISDN (B-ISDN) and so influences the standardization of digital hierarchies, multiplexing structures, switching and interfaces for broadband signals. It is a promising technology offering transmission at high bit rates, flexibility for multimedia, and capabilities for multipoint connections. Its introduction may result in a simpler and more flexible inner structure of networks. Furthermore, service control and operation functions can be enhanced by taking advantage of a high-speed and advanced signalling network. ATM is intended to accommodate the complete range of telecommunication services. Consequently, it is supported by very simple, highly efficient mechanisms.

ATM offers a single transfer technique, contrary, for example, to the approach of narrow-band ISDN (N-ISDN) that offers different variants of two basic transfer modes (the 64 kbit/s circuit and the D protocol packet handled at the core of the network by separate subnetworks).

The transfer mode is asynchronous because cells of a same connection are not restricted to have a relation in time, e.g. periodicity. The transmission rate of cells within a particular virtual channel can be variable and depends on the source rather than any clock reference within the network. Therefore, an ATM-based network is bandwidth transparent. This is a key element as both constant bit-rate and variable bit-rate services can be dynamically handled, new services can be introduced in a gradual manner and there is an insensitivity to inaccuracies in predictions of the mix in service demand. Moreover, exploiting the statistical properties of variable bit-rate services, statistical multiplexing of cells can be used, increasing the network efficiency.

Some of the specific, advantageous facilities of ATM-based B-ISDN are:

- high flexibility of network access due to the cell transport concept and specific cell transfer principles;

- dynamic bandwidth allocation on demand with a fine degree of granularity;

- flexible bearer capability allocation and easy provision of semi-permanent connections due to the virtual path concept;

- independence of the means of transport at the physical layer.

Questions relative to these ITU-R works on ATM and, more generally, to the problems of implementation of ATM satellite links and their interconnection with ATM networks are dealt with in some detail in Chapter 8, § 8.6.

More details on ATM are given in Appendix 3.4.

3.5.4.2   ATM entities and their relationships

In ATM the information is organized into fixed size cells which are the entities transported by the network. A cell consists of an information field which carries the user's data, and a header (Figure 3.18).

The header size (5 bytes or octets, of which one is for protection of the other four) and the information field size (48 octets) remain constant in all reference points, including the User Network Interface (UNI) and Network Node Interface (NNI), where the ATM technique is applied.

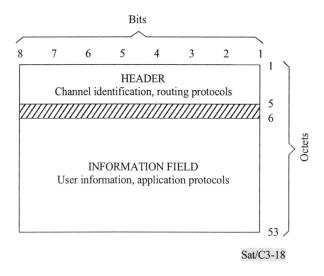

FIGURE 3.18

ATM cell structure

The ATM cell is both a multiplexing unit and a routing unit. Successive cells in a multiplex signal are identified and routed individually based on the information contained in the header, which includes two identifiers to allow a double routing involving the notions of channel (switching concept) and path (transmission concept).

The transport functions of the ATM layer are subdivided into two levels: The virtual channel (VC) level and the virtual path (VP) level. They are independent of the physical layer implementation.

VCs are aggregated in virtual path connections (VPCs) which may be routed as such through VP multiplexers/demultiplexers and VP switches and cross connects.

The functions of the cell header are defined by its own structure, shown in Figure AP3.4-3 (see Appendix 3.4). The virtual path identifier (VPI) and the virtual channel identifier (VCI) allow cell routing. Cell loss priority and payload type are identified. One bit is currently in reserve.

In the ATM cell format, at the UNI, the generic flow control (GFC) field contains 4 bits to carry the flow control information used to alleviate short-term overload conditions that may occur. The routing field (VPI/VCI) has 24 bits, 8 for the VPI and 16 for the VCI, for connection identification. The 3 bits available for the Payload Type (PT) identification provide an indication whether the cell payload contains user information (and service adaptation function information) or network information; from the eight possibilities, one is reserved. Depending on network conditions the bit for cell loss priority (CLP) is set (value 1, discard) or not. The remaining eight bits are the header error control (HEC) field, used for error management of the header.

3.5.4.3   Transmission systems for ATM cells

ATM cells can be transmitted in different ways. They can be inserted in the payload of the SDH, in the frames of the existing PDH or can be transmitted as a continuous sequence of cells. ITU-T Recommendation I.432 defines two mappings of the UNI interface: 1) SDH-based option. Cells are included in the SDH payload, and 2) Cell-based option. ATM cells may also be mapped in the payload of PDH signals.

An advantage of the PDH and SDH-based options resides in the possibility to exploit the PDH and SDH transmission systems and networks to transport ATM cells. This means that the transport network may be used for ATM services and the traditional applications, like the circuit switched ones.

Another important advantage is the possibility of using the same operation and maintenance (OAM) functions already defined in SDH, such as error monitoring, service channels, channels for TMN applications, which make use of the SDH overheads.

In order to map ATM cell streams in SDH virtual containers (VCs), the ATM cells are presented to the SDH VC by an adaptation function that inserts idle cells if the offered rate is not sufficient to fully load the SDH channel, and restrains the source when the offered rate becomes too high. The actual transmitted cell stream has exactly the same capacity as that of the SDH virtual container in which it is transported and that therefore limits the maximum information rate, although it is defined by the source.

The ATM mappings have been defined into SDH VC-4 and VC-4-4c (four VC-4 concatenated). Neither of these payloads carries an integer number of ATM cells. Cell boundaries will not be in the same relative position in successive SDH frames. This does not matter as cell streams carry their own cell delineation mechanism embedded in the Header Error Control.

The transfer capability at the user network interface (UNI) is 155 520 kbit/s, with a payload capacity of 149 760 kbit/s (due to the SDH header). With the ATM cell format of a 5 octet header and a 48 octet information field, the maximum rate available from the interface for all cell information fields is 135 631 kbit/s.

A second UNI interface is defined at 622 080 kbit/s with the service bit rate of approximately 600 Mbit/s.

3.5.4.4   ATM on satellite links

Significant studies and experiments have been conducted on the transmission of ATM over satellite links, in particular in Europe and by INTELSAT, including the possibility of on-board ATM switching. In fact, satellite links could prove to be a preferred medium for ATM-based transmission (e.g. for Internet links). This is why, again through a proper coordination of ITU-T and ITU-R, there is a constant concern in ITU that recommendations on ATM should be applicable to satellite links. New recommendations on the applicability of ATM to satellite communication are prepared by ITU-R SG 4.

3.5.5   Generalities on ISDN concepts – narrow and broadband [Ref. 3-1]

3.5.5.1   The ISDN concept and requirements

The integrated services digital network (ISDN) is conceived as a universally accessible digital network capable of providing a very wide range of telecommunications services using a limited number of powerful network capabilities.

The ISDN concept appears as a solution to provide all services in a more integrated manner. Each one of the current telecommunication networks provides a specific set of dedicated services (voice, circuit switched data, packet switched data, video). Many of these networks are supported by analogue and digital equipment for subscriber access, switching and inter-exchange transmission. Without an ISDN environment, a customer of a number of telecommunication services has separate access lines, user-network interface requirements and terminals for each service (see Figure 3.19).

The ISDN establishes bearer capabilities in the digital subscriber access to accommodate existing and new voice, data and video services in a full digital connection.

Basic technical requirements for ISDN implementation are:

1) Digital switching, transmission and user line;

2) Synchronization of exchanges, transmission lines and terminals to preserve transmitted digital information integrity;

3) ITU-T Common Channel Signalling System (CCSS) No. 7 between functional entities internal to the network;

4) Digital Subscriber Signalling System No. 1 (DSS 1/ISUP) involving message mode user-network signalling based on the D protocol.

Signalling messages are transported by the D channel; this allows a separate call establishment or release processing. The ITU-T CCSS No. 7 and the DSS 1 enable the introduction of new facilities.

Evolution towards ISDN implies that network digitization must cover: 1) subscriber exchanges with routing capabilities and their connection units (intra local area network); 2) transfer and distribution networks to allow the connection of ISDN customer equipment, for example private automatic branch exchange (PABX) and private installations; 3) all higher and lower rank trunk exchanges involved in the ISDN routing plan.

ITU-T has issued Recommendations covering the following ISDN aspects:

- General Introduction and Basic Principles (I.100-series)

- Services (I.200-series)

- Network Architecture and Capabilities (I.300-series)

- User-Network Interfaces (I.400-series)

- Interworking (I.500-series)

- Maintenance Principles (I.600-series).

ISDN implementation is expected to follow the sequence: 1) conversion of the public switched telephone network to digital switching and inter-exchange transmission (integrated digital network); 2) provision of digital transmission over existing copper pairs for the customer's connection to the network (integrated digital access); 3) end-to-end digital connections at 64 kbit/s (64 kbit/s ISDN); 4) upgrading of user access to broadband capability using optical fibres (broadband access); 5) end-to-end broadband switched services (broadband ISDN).

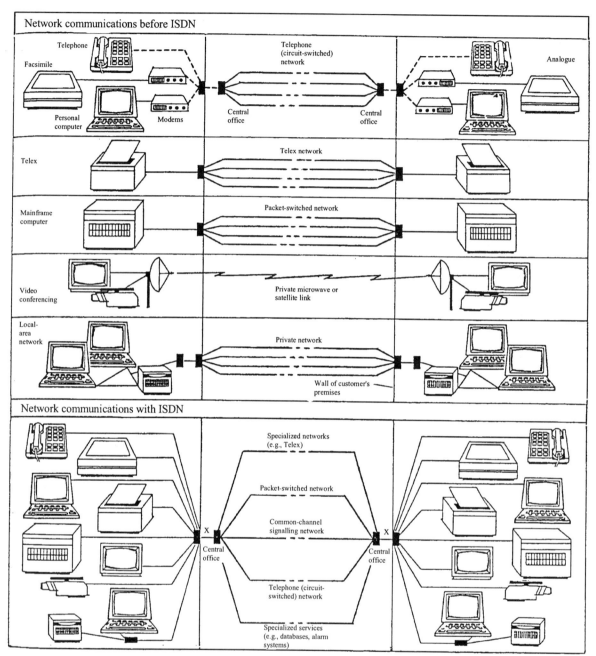

Sat/C3-19

FIGURE 3.19

Applications of conventional and ISDN networks

3.5.5.1.1 Architectural model

Network architectural models provide a framework for the definition of their component parts, interfaces between these parts and interfaces for user access. Interfaces are specified using layered protocols following the Open System Interconnection (OSI) reference model, which is the subject of the X.200-series Recommendations.

The architectural model for ISDN, shown in Figure 3.20, gives the partition of the network into its major blocks and the location of key reference points (S and T). These reference points are the locations of user/network interfaces. The S interface is for terminal connection and the T interface for connection of customer premises equipment (CPE), such as private automatic branch exchanges (PABX). Terminals may also be connected directly to public networks via a combined S/T interface.

Public networks comprise network terminations on customer premises together with the transport and signalling capabilities provided within the network. Transport capabilities are subdivided into 64 kbit/s and broadband, the latter seen as an overlay network. The signalling network serves both types of transport network and also provides user-to-user signalling as a special service.

Telecommunication "bearer services" are provided between user-network interfaces at S/T reference points. They are limited to the three lower layers of the OSI model, specifically the physical transmission layer of B and D channels, data link and network protocol layers of the D channel (see Figure 3.21 and also Appendices 8.1 "Introduction to signalling systems" and 8.2 "Introduction to the OSI model").

Teleservices include layers 2 and 3 of the B channel as well as the higher layer capabilities of layers 4 to 7 (transport, session, presentation, and application). These functions are implemented either in terminals or in special service facilities located at the edge of the network. This functional separation is consistent with the achievement of a high degree of transparency in the network leading to the flexibility needed to accommodate a wide range of services.

3.5.5.1.2 Standard channels

Bearer services are supported by network connection types which are characterized by attributes. A given connection type can be used to support all bearer services with compatible attribute sets. One of the key attributes is channel type/capacity. The following channel types have been standardized:

- B channels   at 64 kbit/s
- D channels   D(16) at 16 kbit/s

  D(64) at 64 kbit/s
- H channels   H0 at 384 kbit/s

  H11 at 1 536 kbit/s (North America and Japan)

  H12 at 1 920 kbit/s (Europe)

B channels, end-to-end circuit switched, can carry digitally encoded voice or user data. B channels can be used for signalling between network nodes but are not used for customer access signalling.

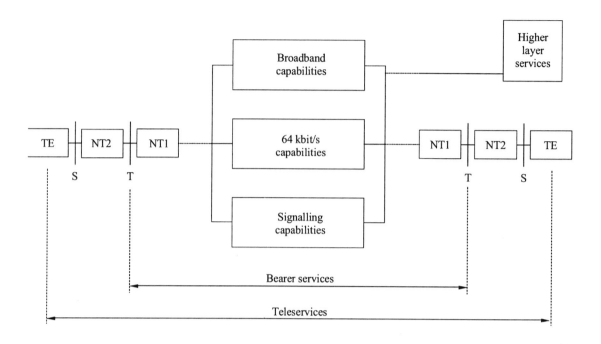

TE: Terminal equipment
NT2: Network termination 2
NT1: Network termination 1

Sat/C3-20

FIGURE 3.20

ISDN network architecture

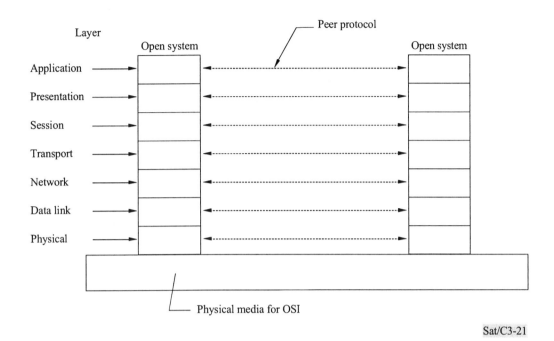

FIGURE 3.21

Seven layer reference model and peer protocols

D channels are access channels carrying control or signalling information and, optionally, packetized customer information, when used for access to X.25 packet switched network capabilities.

H channels can be used for reserved mode connections. Circuit-switched mode H channel connections are not directly supported by ISDN connection types. They are realized in the 64 kbit/s ISDN by multiple B channels, the terminal systems being responsible for the synchronization of the individual time slots.

As ISDN evolves, other H channels providing higher capacities may be available. H21 (34 Mbit/s), H22 (45 Mbit/s) and H4 (140 Mbit/s) channels have been proposed to be used, in circuit or fast packet mode, for applications such as LAN interconnection, high-speed data transmission, teleconferencing, imaging, etc.

3.5.5.1.3   Service provisioning

The ITU-T I.200-series of Recommendations covers ISDN service definitions. Teleservices and bearer services are defined by means of their attributes.

Bearer services are grouped according to the transfer mode. In the Blue Book Recommendations, 8 circuit and 3 packet mode bearer services have been defined as follows:

Circuit mode	Packet mode
• 64 kbit/s unrestricted • 64 kbit/s speech • 64 kbit/s 3.1 kHz • 2 · 64 kbit/s alternate speech/unrestricted • 384 kbit/s unrestricted • 1 536 kbit/s unrestricted • 1 920 kbit/s unrestricted	• Virtual call with permanent virtual circuit • Connectionless • User signalling

The variety of 64 kbit/s bearer services is required because of the need for special interworking treatment of speech or data carried on 3.1 kHz telephone type circuits using modems.

The ISDN, together with terminal functions, provides a range of teleservices. Several are standardized to allow public interoperability. Those included in the Blue Book Recommendations are: telephony; teletex; telefax (group 4 fax); mixed mode; videotex; telex.

All bearer service and teleservice specifications are limited to essential features but can be complemented by a series of supplementary services. Definitions have been produced for supplementary services in the following categories:

- Number identification

- Call offering

- Call completion

- Multiparty

- Community of interest

- Charging

- Added information.

ITU-T Recommendations dealing with basic aspects of ISDN include the following, in addition to the Recommendations already mentioned:

I.112   Vocabulary of terms for ISDN

I.411   ISDN user-network interface reference configurations

I.414   Overview of Recommendations on Layer 1 for ISDN and B-ISDN customer access

I.430   Basic user-network interface. Layer 1 specification

I.431   Primary rate user-network interface Layer 1 specification

I.500   General structure of the ISDN interworking Recommendations

I.501   Service interworking

I.510   Definitions and general principles for ISDN interworking

I.515   Parameter exchange for ISDN interworking

I.520   General arrangements for network interworking between ISDNs

I.525   Interworking between ISDN and networks which operate at bit rates of less than 64 kbit/s

I.530   Network interworking between ISDN and the Public Switched Telephone Network

I.570   Public/private ISDN interworking

I.580   General arrangements for interworking between ISDN and 64 kbit/s based ISDN

X.200   Reference model of Open Systems Interconnection for CCITT applications

3.5.5.2   Broadband aspects of ISDN

Broadband ISDN (B-ISDN) combines narrow-band and broadband services together into a single standard user/network interface. This network will require an important telecommunication infrastructure targeting high-speed data communications such as LAN interconnection, bulky data communications such as file transfer and cable television, but not limited to these services.

Residential or business subscriber communications could be at a bit rate from 64 kbit/s to 10 or 20 Mbit/s, of continuous type or sporadic, the connection established on demand (for example, a teleconference or the consultation of a database). Data transfer for business subscribers could be at a bit rate from 10 to 100 Mbit/s, sporadic, requiring permanent or semi-permanent connections or connections on demand. TV and broadcasting service communications could be up to 100 Mbit/s bit rates, continuous, point-to-point and point-to-multipoint.

Since B-ISDN is designed to carry constant bit rate (CBR) and variable bit rate (VBR) traffic types and handle different mixes of bandwidth demand and service types, flexibility appears as one of its fundamental features. Due to the transparency of line systems to the transported information, digital optical transmission networks, including the access part, meet this requirement. Other parts of the network may not have the required flexibility. For example, technology allows the conception and building of equipment, like terminals, multiplexers and switches, with a large amount of signal processing capability, which can pave the way to flexibility.

3.5.5.2.1   Basic principles

The B-ISDN supports switched, semi-permanent and permanent point-to-point and point-to-multipoint connections. It provides on demand reserved and permanent services.

Connections in B-ISDN support both circuit-mode and packet-mode services, of single medium or multimedia type, of a connection-oriented or connectionless nature, in a unidirectional or bidirectional configuration.

The B-ISDN architecture is detailed in functional terms. Therefore, it is independent of technology and implementation, which may be achieved in a variety of ways according to specific situations in each country.

The B-ISDN (and its network/terminal elements) will contain intelligent capabilities for the purpose of providing advanced service characteristics, supporting powerful operation and maintenance tools, network control and management.

Two major new aspects in the broadband era are the inclusion of distribution services (e.g. for cable TV) and the possibility of interactive videocommunications (e.g. videoconferencing). Services are classified in the following way:

Interactive services	Distribution services
• Conversational services • Messaging services • Retrieval services	• Broadcast services • Distribution services with user individual presentation control

Some examples of new broadband services are:

- High quality broadband videotelephony
- High quality broadband videoconference
- Existing quality and high definition TV distribution
- Broadband videotex.

3.5.5.2.2   Standardization aspects

The B-ISDN will be based on the asynchronous transfer mode (ATM), which is independent of the transport means at the physical layer.

ITU-T and other bodies, like ETSI, issued standards on B-ISDN and continue to further develop and complete them, covering:

- general B-ISDN aspects;
- specific service and network oriented issues;
- fundamental characteristics of the ATM procedure;
- relevant ATM parameters and their application at the UNI;
- impact on operation and maintenance of the B-ISDN access.

A non-exhaustive list of ITU-T Recommendations on B-ISDN includes:

I.113	Vocabulary of terms for broadband aspects of ISDN
I.121	Broadband aspects of ISDN
I.150	B-ISDN ATM functional characteristics
I.211	B-ISDN service aspects
I.311	B-ISDN general network aspects
I.321	B-ISDN Protocol Reference Model and its applications
I.327	B-ISDN functional architecture
I.361	B-ISDN ATM layer specification
I.362	B-ISDN ATM Adaptation Layer (AAL) functional description
I.363	B-ISDN ATM Adaptation Layer (AAL) specification
I.364	Support of broadband connectionless data service on B-ISDN
I.371	Traffic control and congestion control in B-ISDN
I.375	Network capabilities to support multimedia services
I.413	B-ISDN User Network Interface
I.432	B-ISDN UNI-Physical layer specification
I.610	OAM principles of the B-ISDN access

APPENDIX 3.1

Voice low bit-rate encoding
[Ref. 3-3, 3-4]

This Appendix provides some information on existing or planned speech coders standards and some technical aspects of two of the most recent techniques of speech coding, both based on the code excited linear prediction (CELP) method, namely:

- coding of speech at 16 kbit/s using low-delay code excited linear prediction (LD-CELP);

- coding of speech at 8 kbit/s using conjugate structure algebraic code excited linear prediction (CS-CELP).

These techniques have found their application in mobile communications: terrestrial radio mobile systems (such as proposed by CEPT GSM in Europe), and maritime radio mobile systems (such as proposed by Inmarsat).

AP3.1-1 Speech coders standardization aspects

Applications which have a major economic impact, such as mobile communications, provide a strong incentive to increase the efficiency of speech coders.

Speech coders are usually designed for a particular application. Speech coder attributes which can be optimized for that purpose are the following:

- bit rate;

- subjective speech quality;

- computational complexity and memory requirements;

- delay;

- channel-error sensitivity;

- signal bandwidth.

There are other attributes which may be important in particular speech-coding applications. These include the capability of a speech coder to transmit non-speech signals such as data, signalling and dial tones, and the capability to support speech and speaker recognition.

The following is a list of selected current and planned standards for speech coders:

Current ITU-T telephone bandwidth speech coders

G.711   64 kbit/s PCM   64 kbit/s

G.721	ADPCM	32 kbit/s
G.726	ADPCM	16, 24, 32, 40 kbit/s
G.727	ADPCM	16, 24, 32, 40 kbit/s
G.728	16 kbit/s LD-CELP	16 kbit/s

Planned ITU-T telephone bandwidth speech coders

ITU-T	8 kbit/s speech coder (G.729)
ITU-T	low bit-rate videophone coder
ITU-T	4 coder

ITU-T 7 kHz wideband speech coders

G.722	7 kHz audio coder
ITU-T	7 kHz speech coder

European digital cellular telephony speech coders

GSM	13 kbit/s RPE-LTP coder
GSM	half-rate speech coder

North American digital cellular speech coders

IS54	7.95 kbit/s VSELP
IS96	8.5 kbit/s QCELP and the EVRC

Japanese digital cellular telephony speech coders

JDC	Japanese digital cellular VSELP
JDC	half-rate 3.45 kbit/s PSI-CELP

Inmarsat

Inmarsat	4.15 kbit/s IMBE coder

Secure voice standards

FS1015	2.4 kbit/s LPC vocoder
FS1016	4.8 kbit/s CELP new 2.4 kbit/s secure coder

AP3.1-2 The code excited linear prediction (CELP) concept

AP3.1-2.1 Linear prediction coding (LPC) technique

Based on a parametric model of the voice signal, the LPC technique offers a low bit rate, typically in the range 8-16 kbit/s.

To achieve such low rates, the voice signal $S(x)$ is represented as the output of a time-varying autoregressive filter $H(x)$ driven by an idealized impulse-like source $E(x)$:

$$S(x) = H(x)E(x)$$

$S(x)$ is the x-transform of the signal time series:

$$S(x) = \sum_{k=0}^{\infty} s(k)x^{-k}$$

$H(x)$ is the x-transform of the impulse response:

$$H(x) = \sum_{k=0}^{\infty} h(k)x^{-k}$$

$E(x)$ is the x-transform of the excitation time series:

$$E(x) = \sum_{k=0}^{\infty} e(k)x^{-k}$$

$H(x)$ can now be expanded in terms of a pth-order non-recursive predictor $P(x)$:

$$P(x) = \sum_{n=1}^{q} a_n, x^{-n}$$

and:

$$H(x) - (1 - P(x))^{-1}$$

so that:

$$S(x) - H(x) E(x) - E(x)(1 - P(x))^{-1}$$

or:

$$S(x) = P(x) S(x) + E(x)$$

The signal $S(x)$ is represented by a prediction term, $P(x)S(x)$, plus an error term $E(x)$, which is the excitation function.

The predictor coefficients a_n best fit the speech production model when the expectation value for the mean square error $\langle E(k)^2 \rangle$ is minimized, tending to give a more idealized impulse source.

From the above equations,

$$\left\langle E(k)^2 \right\rangle = \sum_{k=1}^{\infty} (S(k) - \sum_{n-1}^{q} a_n s(k-n))^2$$

This function can be minimized by setting the partial derivatives to zero.

AP3.1-2.2 The CELP concept

The innovation brought up by the CELP coder is that it takes the replica of the decoder synthesis model in the analysis-by-synthesis process, for the selection of the excitation signal: i.e. the signal that is best approximating the signal S to be encoded according to a perceptual error criterion. Minimizing the CELP criterion is equivalent to vector quantization with a weighted error criterion.

The GSM codec is an example of codecs using linear prediction coding (LPC).

AP3.1-3 Coding of speech at 16 kbit/s using low-delay code excited linear prediction (LD-CELP) (ITU-T Recommendation G.728)

LD-CELP retains the analysis-by-synthesis approach to codebook search, which is the essence of CELP techniques; however it uses backward adaptation of predictors and gain to achieve an algorithmic delay of 0.625 ms. Only the index to the excitation codebook is transmitted. The predictor coefficients are updated through LPC analysis of previously quantized speech. The excitation gain is updated by using the gain information embedded in the previously quantized excitation. The block size for the excitation vector and gain adaptation is five samples only. A perceptual weighting filter is updated using LPC analysis of the unquantized speech.

AP3.1.3.1 LD-CELP encoder

After the conversion from A-law or μ-law PCM to uniform PCM, the input signal is partitioned into blocks of five consecutive input signal samples. For each input block, the encoder passes each of 1 024 candidate codebook vectors (stored in an excitation codebook) through a gain scaling unit and a synthesis filter. From the resulting 1 024 candidate quantized signal vectors, the encoder identifies the one that minimizes a frequency-weighted mean-squared error measure with respect to the input signal vector. The 10-bit codebook index of the corresponding best codebook vector (or "codevector"), which gives rise to that best candidate quantized signal vector, is transmitted to the decoder. The best codevector is then passed through the gain scaling unit and the synthesis filter to establish the correct filter memory in preparation for the encoding of the next signal vector. The synthesis filter coefficients and the gain are updated periodically in a backward adaptive manner based on the previously quantized signal and gain-scaled excitation (see Figure AP3.1-1 a).

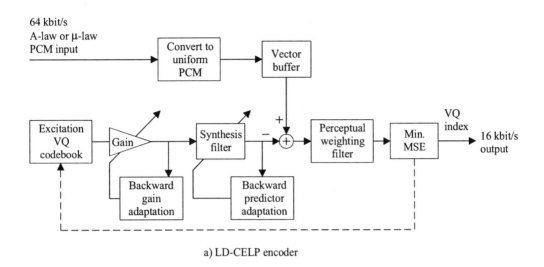

a) LD-CELP encoder

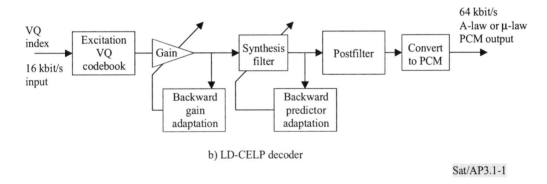

b) LD-CELP decoder

Sat/AP3.1-1

FIGURE AP3.1-1

G.728 Simplified block diagram of LD-CELP coder

AP3.1-3.2 LD-CELP decoder

The decoding operation is also performed on a block-by-block basis. Upon receiving each 10-bit index, the decoder performs a table look-up to extract the corresponding codevector from the excitation codebook. The extracted codevector is then passed through a gain scaling unit and a synthesis filter to produce the current decoded signal vector. The synthesis filter coefficients and the gain are then updated in the same way as in the encoder. The decoded signal vector is then passed through an adaptive postfilter to enhance the perceptual quality. The postfilter coefficients are updated periodically using the information available at the decoder. The five samples of the postfilter signal vector are next converted to five A-law or μ-law PCM output samples (see Figure AP3.1-1.b) LD-CELP encoder).

AP3.1-3.3 Speech performance

Appendix II to ITU-T Recommendation G.728 gives a broad outline of the speech performance of the 16 kbit/s LD-CELP algorithm when interacting with other parts of the network. Some general guidance is also offered on voice-like and non-voice signals.

i)   Single encoding

Under error-free transmission conditions the perceived quality of a 16 kbit/s LD-CELP codec is lower than that of a 64 kbit/s PCM codec, but equivalent to that of a codec conforming to 32 kbit/s ADPCM. The performance of the 16 kbit/S LD-CELP codec has been found to be substantially unaffected by noise (Gaussian distribution) with bit error ratio (BER) up to $1.0 \cdot 10^{-3}$, and has the equivalent performance to that of a codec conforming to 32 kbit/s ADPCM at a BER of $1.0 \cdot 10^{-2}$.

ii) Speech performance when interconnected with coding systems on an analysis basis

Multi-tandeming of 16 kbit/s LD-CELP codec (asynchronous tandeming)

When a 16 kbit/s LD-CELP codec is tandemed with multiple speech coding devices, its performance appears to be equivalent to that of 32 kbit/s ADPCM for up to three devices in tandem. Precise rules for tandeming 16 kbit/s LD-CELP codecs are found in ITU-T Recommendation G.113.

iii) Speech performance of the 16 kbit/s LD-CELP codec for synchronous tandeming

Subjective experiments have shown that the speech performance of synchronous tandeming (that is, interconnection of two or more LD-CELP codecs via 64 kbit/s PCM interfaces) is equivalent to the speech performance for asynchronous tandeming.

iv) Performance with encoding other than 16 kbit/s LD-CELP and 32 kbit/s ADPCM

In general, interconnection of 16 kbit/s LD-CELP with up to two other devices appears to be equivalent to interconnection of 32 kbit/s ADPCM with the same devices. However, the above conclusion may be altered when more results are available. Thus care must be taken in this scenario.

AP3.1-3.4 Performance with voiceband data

In general, the performance of the 16 kbit/s LD-CELP codec (even for a single encoding) has been found to be significantly inferior to that of 32 kbit/s ADPCM and 64 kbit/s PCM. However, it should be noted that 16 kbit/s LD-CELP is able to accommodate most (but not all) voiceband data modems operating at 2 400 bit/s or less, under realistic channel conditions, provided its perceptual weighting filter and postfilter are both disabled.

AP3.1-4 Coding of speech at 8 kbit/s using conjugate-structure algebraic-code-excited linear-prediction (CS-CELP) (ITU-T Recommendation G.729)

This coder is designed to operate with a digital signal obtained by first performing telephone bandwidth filtering (ITU-T Recommendation G.712) of the analogue input signal, then sampling it at 8 000 Hz, followed by conversion to 16-bit linear PCM for the input to the encoder; at the distant end the output of the decoder should be converted back to an analogue signal by similar means).

Other input/output characteristics, such as those specified by ITU-T Recommendation G.711 for 64 kbit/s PCM data, should be converted to 16-bit linear PCM before encoding, or from 16-bit linear PCM to the appropriate format after decoding.

AP3.1-4.1 General description of the coder

The CS-CELP coder is based on the Code-Excited Linear-Prediction (CELP) coding model. The coder operates on speech frames of 10 ms corresponding to 80 samples at a sampling rate of 8 000 samples per second. For every 10 ms frame, the speech signal is analysed to extract the parameters of the CELP model (linear-prediction filter coefficients, adaptive and fixed-codebook indices and gains). These parameters are encoded and transmitted. At the decoder, these parameters are used to retrieve the excitation and synthesis filter parameters. The speech is reconstructed by filtering this excitation through the short-term synthesis filter, as shown in Figure AP3.1-2. The short-term synthesis filter is based on a 10th order linear prediction (LP) filter. The long-term, or pitch synthesis, filter is implemented using the so-called adaptive-codebook approach. After computing the reconstructed speech, it is further enhanced by a postfilter.

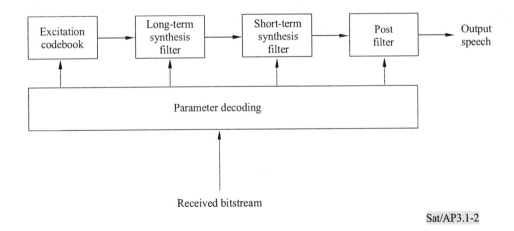

FIGURE AP3.1-2

G.729 Block diagram of conceptual CELP synthesis model

AP3.1-4.2 Encoder

The encoding principle is shown in Figure AP3.1-3. The input signal is high-pass filtered and scaled in the preprocessing block. The preprocessed signal serves as the input signal for all subsequent analysis. LP analysis is done once per 10 ms frame to compute the LP filter coefficients. These coefficients are converted to line spectrum pairs (LSP) and quantized using predictive two-stage vector quantization (VQ) with 18 bits. The excitation signal is chosen by using an analysis-by-synthesis search procedure in which the error between the original and reconstructed speech is minimized according to a perceptually weighted distortion measure. This is done by filtering the error signal with a perceptual weighting filter, whose coefficients are derived from the unquantized LP filter. The amount of perceptual weighting is made adaptive to improve the performance for input signals with a flat frequency-response.

The excitation parameters (fixed and adaptive-codebook parameters) are determined per subframe of 5 ms (40 samples) each. The quantized and unquantized LP filter coefficients are used for the second subframe, while in the first subframe interpolated LP filter coefficients are used (both quantized and unquantized). An open-loop pitch delay is estimated once per 10 ms frame based on the perceptually weighted speech signal. Then the following operations are repeated for each subframe. The target signal $x(n)$ is computed by filtering the LP residual through the weighted synthesis filter $W(z)/\hat{A}(z)$. The initial states of these filters are updated by filtering the error between LP residual and excitation. This is equivalent to the common approach of subtracting the zero-input response of the weighted synthesis filter from the weighted speech signal. The impulse response $h(n)$ of the weighted synthesis filter is computed. Closed-loop pitch analysis is then done (to find the adaptive-codebook delay and gain), using the target $x(n)$ and impulse response $h(n)$, by searching around the value of the open-loop pitch delay. A fractional pitch delay with 1/3 resolution is used. The pitch delay is encoded with 8 bits in the first subframe and differentially encoded with 5 bits in the second subframe. The target signal $x(n)$ is updated by subtracting the (filtered) adaptive-codebook contribution, and this new target, $x'(n)$, is used in the fixed-codebook search to find the optimum excitation. An algebraic codebook with 17 bits is used for the fixed-codebook excitation. The gains of the adaptive and fixed-codebook contributions are vector quantized with 7 bits (with moving average (MA) prediction applied to the fixed-codebook gain). Finally, the filter memories are updated using the determined excitation signal.

AP3.1-4.3 Decoder

The decoder principle is shown in Figure AP3.1-4. First, the parameter's indices are extracted from the received bitstream. These indices are decoded to obtain the coder parameters corresponding to a 10 ms speech frame. These parameters are the LSP coefficients, the two fractional pitch delays, the two fixed-codebook vectors, and the two sets of adaptive and fixed-codebook gains. The LSP coefficients are interpolated and converted to LP filter coefficients for each subframe. Then, for each 5 ms subframe the following steps are done:

• the excitation is constructed by adding the adaptive and fixed-codebook vectors scaled by their respective gains;

• the speech is reconstructed by filtering the excitation through the LP synthesis filter;

• the reconstructed speech signal is passed through a post-processing stage, which includes an adaptive postfilter based on the long-term and short-term synthesis filter, followed by a high-pass filter and scaling operation.

AP3.1-4.4 Delay

This coder encodes speech and other audio signals with 10 ms frames. In addition, there is a look-ahead of 5 ms, resulting in a total algorithmic delay of 15 ms. All additional delays in a practical implementation of this coder are due to:

• processing time needed for encoding and decoding operations;

• transmission time on the communication link;

• multiplexing delay when combining audio data with other data.

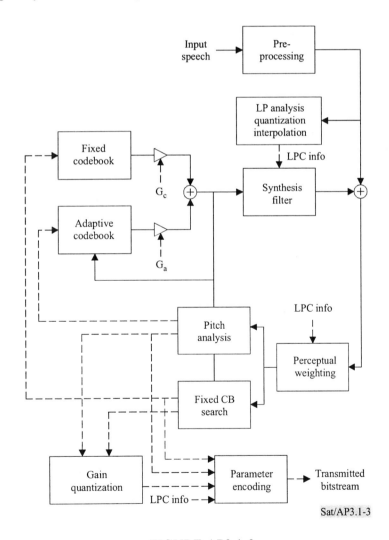

FIGURE AP3.1-3

G.729 Encoding principle of the CS-ACELP encoder

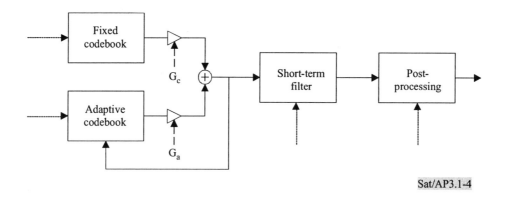

Sat/AP3.1-4

FIGURE AP3.1-4

G.729 Principle of the CS-ACELP decoder

APPENDIX 3.2

Error-correcting coding

AP3.2-1 Introduction

Error-correcting coding consists essentially of introducing a redundancy in the transmitted signal. Several types of redundancy can be used: in time, in spectrum, or in space.

For example, a time redundancy is introduced if the same message is transmitted at least twice. At the same rate, more time will be taken in order to transmit the same quantity of information. This strategy is in fact a repetition coding that can be seen as a very simple form of error control coding.

In (n,k) block coding, (n > k), n–k parity check symbols are added to the k information symbols to form a block of n symbols, called a codeword. The codewords from a block encoder are mutually independent.

In convolutional coding, each time a group of k information bits results in a group of n binary symbols. One difference with block coding is that each group of n binary symbols from a convolutional encoder is a function of not only the k input bits, but also of the K–1 precedent k-bit groups at input, where K is called the constraint length of the code. Another difference to block codes is that the encoding and decoding of a convolutional code are continuously performed.

In both cases, the ratio $r_c = k/n$ is called the code rate. The bandwidth W is inversely proportional to the symbol duration T_s in a digital transmission system. Without coding, if one symbol carries one bit, $T_s = T_b$ (with $T_b = 1/R_b$, R_b being the information transmission bit rate) and $W = W_b = T_s^{-1}$. With coding a shorter symbol duration $T_s = r_c T_b$ is needed, so a wider bandwidth $W = r_c^{-1} W_b$ must be occupied. A spectrum redundancy is then introduced, resulting in a bandwidth expansion.

The space redundancy is used for a better transmission scheme if an expanded signal constellation is used by the modulation system. For example this can be the case when an 8PSK modulation is associated with a 2/3 coding which provides the same bandwidth efficiency as the uncoded QPSK. If, in addition, the channel coding is associated with the digital modulation for a joint optimization of modulation and coding, more bandwidth and power efficiencies can be obtained. This case corresponds to the coded modulation schemes, treated in § 4.2.3, where it can be seen that the information transmission rate will be the same, and the bandwidth will be expanded slightly or not at all, but a performance improvement will be obtained, at the price of an increase of processing cost.

In the following, a classification of error control coding techniques will be performed. These techniques are still in evolution, some are obsolete, some others are becoming dominant in satellite communications. Only the widely used techniques will be presented in this appendix. Neither the obsolete techniques nor the very new but not mature ones will be included.

The basic FEC techniques used in satellite communication systems can be classified into two categories:

a) Convolutional codes, with trellis-coded modulation (TCM) as extensions, presented in Chapter 4, § 4.2.3.

The convolutional codes can be further divided into three classes:

- high-rate orthogonal codes (or majority decodable codes) associated to threshold decoding;

- short and medium memory codes and their punctured codes associated with Viterbi decoding using soft decision;

- long memory codes associated with sequential decoding.

The sequential decoding with long memory convolutional codes is a powerful coding technique, but it is not used in satellite communication systems, so will not be presented.

b) Binary and non-binary block codes associated with algebraic decoding. In this category, the most successful application is the Reed-Solomon code. The block coded modulations are another generalization of this second category, when they are jointly optimized with modulation schemes. The block codes can be divided into the following classes:

- BCH codes decodable by the algebraic decoding, including the Reed-Solomon codes as a non-binary subclass in BCH codes;

- Cyclic codes used for error detection;

- Simple codes such as the Hamming codes, Reed-Muller codes and certain particularly good ones: Golay, QR (Quadratic Residual), etc.

An excellent textbook for block codes is [Ref. 3-9]. The interesting subjects such as block codes, algebraic construction of codes, and Turbo codes will not be treated here. Some BCH codes have been used in TDMA systems, and self-orthogonal convolutional codes have been used in other early systems. These two techniques will not be studied in depth here; interested readers are referred to [Ref. 3-11] and [Ref. 3-12], respectively.

As important applications of FEC in satellite communications, the following two techniques will be presented and emphasized in this section:

i)   convolutional code associated with soft decision Viterbi decoding;

ii) concatenated coding using a Reed-Solomon outer code.

AP3.2-2 Convolutional codes with soft decision Viterbi decoding

Convolutional coding with soft decision Viterbi decoding is a standard technique for today's satellite communications.

AP3.2-2.1 Convolutional encoders

AP3.2-2.1.1 Convolutional encoder and generating matrix

For a standard and complete treatment of this subject, readers are referred to [Ref. 3-9]. This section is based on an easy and practical approach, in view of giving a quick introduction of convolutional codes.

Let us take the example of the most often used code (1/2, K=7) with the generating matrix G=[171, 133] that is widely used in satellite communication systems.

The generating matrix $G=[171, 133]=[g1(D), g2(D)]$ is represented in octal, the corresponding polynomials are $g1(D)=1+D+D2+D3+D6$ and $g2(D)=1+D2+D3+D5+D6$, with the convention as $171 = 001\ 111\ 001 = 1+D+D2+D3+D6$.

The non-systematic feedforward (1/2, K=7) encoder can be obtained at once from the generating matrix (Figure AP3.2-1), where $\oplus$ is the adder modulo 2, and "D" is a delay, a D-flip-flop, or a memory element.

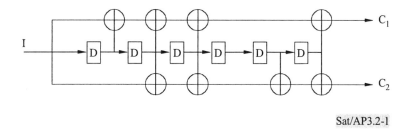

Sat/AP3.2-1

FIGURE AP3.2-1

A non-systematic feedforward encoder with matrix G = [171, 133]

More generally, a non-systematic feedforward (k/n, K) encoder can be constructed from its generating matrix $G(D)=[gij(D)]$, i=1, 2, …, k, j=1, 2, …, n. K is the constraint length of the code. K–1=m, is the number of the memory elements of the encoder.

The encoding is done in performing the D-domain matrix multiplication: $\underline{C(D)} = \underline{I(D)}*G(D)$, where $\underline{I(D)}$ is the information-carrying vector of dimension k and of semi-infinite length, representing the information sequences at the input of the encoder, $\underline{C(D)}$ the coded signal vector of dimension n and of semi-infinite length, representing the coded stream at the output of the encoder.

AP3.2-2.1.2 State diagram, tree, and trellis

Any convolutional encoder can be seen as a FSM (finite state machine). It is natural to use a state diagram to describe the code and study its properties. The code tree and the trellis are the two types of time developments of the state diagram, useful respectively in sequential decoding and Viterbi decoding.

Consider the code (1/2, K=3) with G=[7,5]=[1+D+D2, 1+D2] (Figure AP3.2-2).

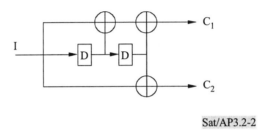

Sat/AP3.2-2

FIGURE AP3.2-2

A non-systematic feedforward encoder with matrix G = [7, 5]

If the contents of the memory elements are named as S1 and S2, the state of the coder at time t is (S_2^t, S_2^t). The space-state equations of the encoder can be written as:

$$S_1^t = S_1^{t-1}$$
$$S_1^t = I^t$$

(state transition equation)

$$c_1^t = I^t \oplus S_2^{t-1} \oplus S_1^{t-1}$$
$$c_2^t = I^t \oplus S_2^{t-1}$$

(output equation)

$S_1^{t-1}S_2^{t-1} \setminus I^t$	0	1
00	00/00	10/11
01	00/11	10/00
11	01/01	11/10
10	01/10	11/01

The corresponding state diagram, Figure AP3.2-3a), describes the relations between the possible states of the encoder.

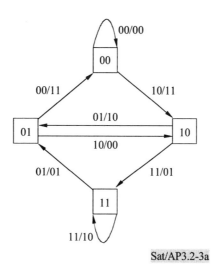

FIGURE AP3.2-3A)

State diagram of the non-systematic feedforward encoder with matrix G = [7, 5]

The code tree, Figure AP3.2-3b), is a semi-infinite time development of the state diagram. The root (or starting node) is the state "00". With an input "1", a lower branch is produced, and with an "0", an upper branch. After the root, each node represents a code sequence.

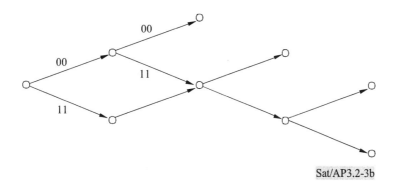

FIGURE AP3.2-3B)

Code tree of the non-systematic feedforward encoder with matrix G = [7, 5]

The trellis is also a graphic representation of the state diagram. Each node represents a state, and each branch is a state transition. A path is a succession of branches in the trellis, representing a code sequence (Figure AP3.2-3c)). What is different from the code tree is that the paths merge at all the states. This an important property very useful in understanding the Viterbi algorithm.

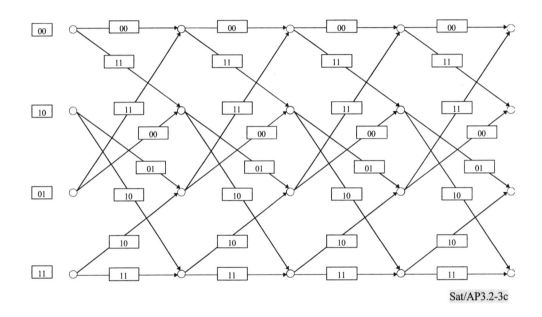

Sat/AP3.2-3c

FIGURE AP3.2-3C)

Trellis of the non-systematic feedforward encoder with matrix G = [7, 5]

The encoders, generating matrix, are logical representations of the code. The state diagram, the code tree and the trellis are graphic representations of the coded signal. A convolutional code is completely specified by its state diagram, from which its code tree and trellis can be defined easily. Inversely, from the code tree or the trellis, the state diagram can be determined.

AP3.2-2.2 Distance properties of convolutional codes

In a block code, only one distance is of interest, the minimum Hamming distance. In block codes the Hamming distance for two codewords is the number of the positions where the symbols differ.

For a convolutional code, several distance definitions, closely related, are useful, depending on the decoding techniques used. The minimum distance is related to the syndrome decoding, the free distance is meaningful for Viterbi decoding, and for sequential decoding, the distance profile is interesting.

- The minimum distance is defined as for a block code, but with truncated coded sequences that can be seen as a block code. It depends of the truncated length. If the truncated length increases, the minimum distance increases. The set of minimum distances indexed by increasing truncated length is called the distance profile.

- The free distance is defined as the distance between sequences with semi-infinite length. For a linear code, free is the distance from the code sequence generated by a "0" information sequence but beginning with a "1", and all the other code sequences. In a distance profile, the biggest term is the free distance. For sequential decoding, the codes with an optimum distance profile are the best choices. The optimum distance profile (ODP) codes have been found and listed in [Ref. 3-9]. As an example, refer to Figure AP3.2-3b). The distance profile can be found by inspection. The result is {2, 3, 4, 5, 5, ...}: the free distance is 5.

- The weight spectrum of a convolutional code is useful for determining its performance over the AWGN channel. It is defined as the sequence of the number of coded sequences $\{a_i\}$ of weight i, of the code, i=df, df+1, df+2, ... , df is the free distance.

In general, for computing these distances, special algorithms are necessary. The free distance can be obtained in the calculation of the distance profile. It can be also obtained using other procedures that are designed for an analysis of the weight spectrum of convolutional codes. The weight spectrum of a convolutional code determines fully the code performance over the AWGN channel. Much research effort has been spent on this topic, see § 3.3.5.5.

AP3.2-2.3 Soft decision decoding Viterbi algorithm

The convolutional coding has its main benefit from the use of soft decision, making full use of the channel information to improve the performance in the order of 2 dB with respect to the hard decision. In practice, finite wordlength has to be used in digital systems. The analogue output samples of the receive filter in a demodulator are usually quantized into 3 bits in the Viterbi decoders, as shown in Figure AP3.2-4:

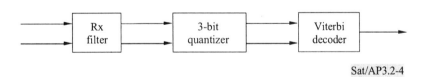

Sat/AP3.2-4

FIGURE AP3.2-4

The soft decision in Viterbi decoding

With the 3-bit quantized data from the channel, the Viterbi algorithm performs the optimum decoding of the convolutional code. The practical aspects of the Viterbi decoding can be found in [Ref. 3-10] and [Ref. 3-17]. To illustrate the principle, the trellis of the code is useful. Recall that the Viterbi algorithm is an application of the dynamic programming to decoding. The computation problem is to find the best path, or shortest path, in an oriented graph. To make sense the path length should be defined.

Each path on the trellis represents a code sequence $\underline{c}$, that corresponds to an information sequence $\underline{I}$ by the relation $\underline{c} = \underline{I}*G(D)$. Notice that this relation can be memorized implicitly in the trellis, and the encoding is not necessary in the Viterbi decoding. There is a "1-to-1" correspondence between $\underline{I}$, $\underline{c}$, and the trellis path.

The received sequence $\underline{r}$ is quantized and fed into the decoder. The optimum decoding is performed under the maximum likelihood criterion:

$$\text{To find } \underline{I}^* \text{ such that } \Pr\{\underline{r}/\underline{I}^*\} = \text{Max } \Pr\{\underline{r}/\underline{I}\}$$

$\Pr\{\underline{r}/\underline{I}\}$ 1 holds always. Over an AWGN channel, the positive quantity $-\log \Pr\{\underline{r}/\underline{I}\}$ is defined as the path length. Since the channel is memoryless, $-\log \Pr\{\underline{r}/\underline{I}\} = -\Sigma \log \Pr\{r_j/I_j\}$, the criterion has thus the following equivalent form:

$$\text{To find } \underline{I}^* \text{ such that } -\log \Pr\{\underline{r}/\underline{I}^*\} = -\text{Min } \Pr\{\underline{r}/\underline{I}\},$$

or

$$\text{To find } \underline{I}^* \text{ such that } -\log \Pr\{\underline{r}/\underline{I}^*\} = -\text{Min } \Sigma \log \Pr\{r_j/I_j\}$$

This is equivalent to:

$$\text{To find } \underline{c}^* \text{ such that } -\log \Pr\{\underline{r}/\underline{c}^*\} = -\text{Min } \Sigma \log \Pr\{r_j/c_j\}$$

From the received sequence $\underline{r}$ and each code sequence $\underline{c}$, a path metric can be further defined for an AWGN channel. Note that $\log \Pr\{r_j/c_j\} = \frac{1}{2} \log (2\pi\sigma) - \|r_j - c_j\|2$, the path length is now $\Sigma \|r_j - c_j\|2$, that is just the Euclidean distance between $\underline{r}$ and $\underline{c}$. This is why the maximum likelihood decoding is to find the code sequence, or the corresponding trellis path, that is the closest to the received sequence in Euclidean distance. The criterion is still expressed as:

$$\text{To find } \underline{c}^* \text{ such that } -\log \Pr\{\underline{r}/\underline{c}^*\} = \text{Min } \Sigma \|r_j - c_j\|2$$

The Viterbi algorithm solves this problem by applying the dynamic programming principle that assures the global optimality with local optimum decisions. The local decisions reduce the computation efforts on the trellis. In practice, much simpler integer metrics quantized to 3-bit are used, see [Ref. 3-10] and [Ref. 3-17].

Roughly speaking, to find the shortest path among the 2rL ones, r being the code rate, L the sequence length, 2rL path metrics are to be evaluated and compared, if a brute-force method is used. In the Viterbi algorithm (VA), much less calculation effort is necessary. From the first trellis section to the L-th one, the VA progresses section by section. At each node, it takes a local decision by eliminating one of the two paths merged to it by taking the one having the shorter path length. Each time a binary comparison is made, 1/2 paths have been eliminated in the future.

Once the L-th section is reached, the VA decoder begins to deliver a bit decision. At the L-th section, it gives the 1st bit decision. Then, it progresses on the trellis by one section, making the decision of the (t–L)-th bit at the instant t. In the decoder path memory, only L sections are stored. All the information stored for the sections preceding t–L sections are discarded. This is called truncated memory Viterbi decoding, that is implementable in practice.

The following example (Figure AP3.2-5) is used to show how the VA works on the code trellis.

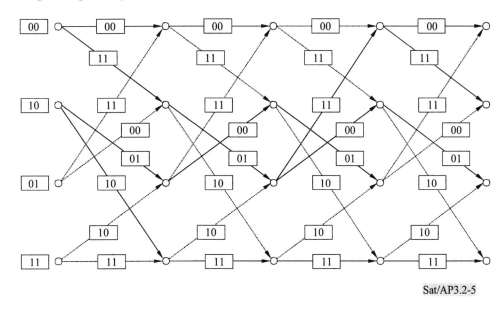

Sat/AP3.2-5

FIGURE AP3.2-5

Trellis for illustrating the Viterbi decoding

The solid branches constitute the survivor paths, the dashed ones are the eliminated ones by local decisions. The path "11 01 11 00 ... " is the optimum path. The paths "00 00 00 ...", "01 11 ...", and "10 00 10 ... " are all eliminated. If, for example, the path "10 10 10" were to have survived, all its descendants would have been evaluated. These computations, non-necessary for the global optimum decision, can be eliminated by the VA.

The hardware architecture of a Viterbi decoder is composed of four main parts. A branch metric calculator with storage, an "ACS" unit, a path memory with a decision logic, and a data bus with its controller. The "ACS" unit performs the so-called "accumulation-comparison-selection" operation. The path memory is organized in special data structure and is of a big volume, occupying usually a large part of the silicon surface in decoder chips. A Viterbi decoder ASIC is just a special purpose computer. The ACS unit acts as the CPU, the path memory is the data storage and bus, and the others perform as peripheries. The architecture may vary largely, depending on the bit rate in applications. For less than 1 Mbits/s, a DSP-based architecture may be suitable. An ASIC implementation is required if the bit rates are in 10 to 60 Mbits/s. See [Ref. 3-17] and recent technical literature for more details.

AP3.2-2.4 Punctured convolutional code and their decoding

A k/n convolutional code with k>1 is more complicated in trellis than a half-rate one. From each node, 2k paths depart and 2k paths arrive to it. For k=2, a quadruple local decision must be performed. But a quadruple decision is much more complex to implement than a binary one by electronic means. On the other hand, higher rate codes are necessary for some applications. The punctured codes have been found to meet these requirements. To construct a higher rate code, a part of code symbols are periodically punctured from the output of the encoder. For example, the 3/4 punctured code is constructed by puncturing periodically 2 symbols from 6 ones in each 3 sections in the trellis, so 3 bits, as shown in Figure AP3.2-6, where "x" is a puncturing position. The

puncturing matrix is thus defined as: $P = \begin{vmatrix} 1 & 1 & 0 \\ 1 & 0 & 1 \end{vmatrix}$

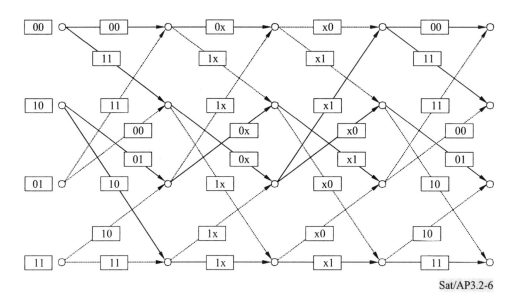

Sat/AP3.2-6

FIGURE AP3.2-6

Trellis for illustrating the 3/4 punctured code

In the Viterbi decoding (Figure AP3.2-7), for each punctured symbol, a null metric should be inserted to complete the branch metric calculation. This is obtained by inserting null samples in the sampled received signal. So, the punctured codes can be easily implemented by a VA decoder without any major hardware modification. The null metric insertion is transparent to the VA.

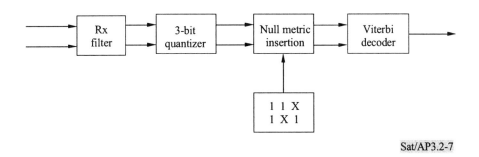

Sat/AP3.2-7

FIGURE AP3.2-7

The null symbol metric insertion in punctured code Viterbi decoding

The node synchronization of a VA decoding for a punctured code is more complicated, because the punctured positions must be identified correctly by the decoder. The following table provides an example of useful punctured codes, with the K=9 puncturing matrix and the free distance. More complete lists can be found in [Ref. 3-16].

TABLE AP3.2-1

Punctured codes, puncturing matrix and free distance

K=9	1/2	2/3	3/4	4/5	5/6	6/7	7/8
G1=561	1	11	111	1101	10110	110110	1101011
G2=753	1	10	100	1010	11001	101001	1010100
df	10	7	6	5	5	4	4

AP3.2-2.5 Association of a binary convolutional code with QPSK modulation

The most used configuration is shown in the following Figure AP3.2-8:

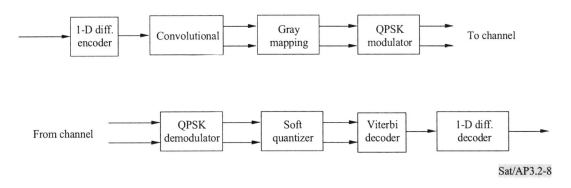

Sat/AP3.2-8

FIGURE AP3.2-8

Basic system of QPSK modulation and a half-rate convolutional coding

There are two mappings for a QPSK constellation, the natural one and the Gray mapping, as shown in Figure AP3.2-9.

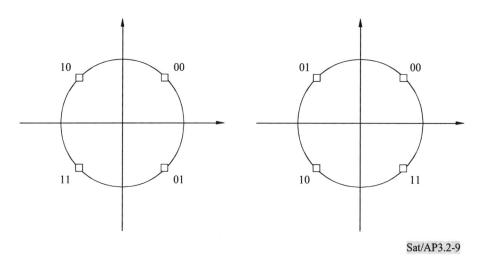

FIGURE AP3.2-9

The Gray (left) and the natural (right) mapping

A phase ambiguity is related to the carrier recovery of the QPSK demodulator. If the phase reference is established on the 3 points: 90°, 180° and –90°, the demodulation cannot work correctly with the decoder without a phase ambiguity removal device. It can deliver correct bits only if this phase reference is at the phase 0°.

The phase ambiguities can be partly removed by the use of the 1-dimensional differential coding. In the Gray mapping, if a phase ambiguity of 180° occurs, the 2 code symbols are all reversed, and the polarization of the QPSK symbols are reversed. If the code is transparent, the differential coding can solve this phase ambiguity. A convolutional code is transparent if any reversed coded sequence of a valid coded sequence is still valid, and the reversed coded sequence corresponds to the reversed information sequence of the one corresponding to the non-reversed coded sequence. This means that the decoder can decode a reversed QPSK symbol sequence, delivering a reversed information sequence. The reversion of the information sequence is corrected by the differential encoding/decoding put before convolutional encoder and after the Viterbi decoder.

The differential encoding/decoding can be put inside the convolutional codec. In this case, all the 3 phase ambiguities can be solved, as shown in Figure AP3.2-10. Two problems exist however in this configuration. The first one concerns the positions of the convolutional code and the Viterbi decoder into the system, because the input to the Viterbi decoder must be soft quantized data to which the differential decoding is hard to apply. The second one is the error doubling due to the 1-bit memory of the differential decoder.

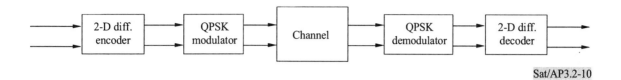

Sat/AP3.2-10

FIGURE AP3.2-10

QPSK modulation with 2-dimensional differential coding

In the system with a convolutional coding, the resolution of the ±90° phase ambiguities is achieved by the Viterbi decoder, depending on the path convergence property. The path metrics are non-decreasing functions of the time. In Viterbi decoders, a path metric normalization circuit is usually implemented to prevent these metrics from overflowing. Under normal conditions, this circuit takes action regularly. If one of the phase ambiguities of ±90° takes place, the Viterbi decoder accepts very noisy inputs, no path in the trellis is good, and all the path metrics increase rapidly. The path metric normalization circuit takes action very frequently. This is readily detected and an alarm is activated. The decoder chip then commands a 90° rotation of the input QPSK symbols, the phase ambiguity becomes either 0° or 180° and is removed. This method must however be statistically stable, a relatively long observation period must be used to prevent a false alarm due to the channel impulse noise. The number of path metric normalizations monitored over this observation period is tested for a threshold. It is the normalization rate that is meaningful.

The node synchronization in a Viterbi decoder can be also achieved with this same mechanism. Finally, a fully-transparent system with two encoder/decoder pairs is shown in Figure AP3.2-11.

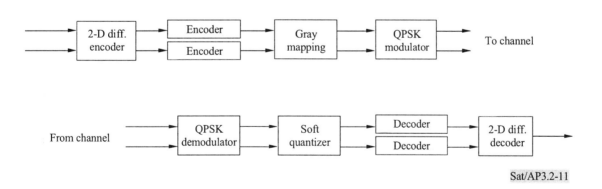

Sat/AP3.2-11

FIGURE AP3.2-11

Fully-transparent system of QPSK modulation with two 1-D transparent codes

The use of the punctured codes facilitates a trade-off between power and bandwidth. These codes can be decoded by the same Viterbi decoder chip as for the original half-rate with minor modification (Figure AP3.2-12).

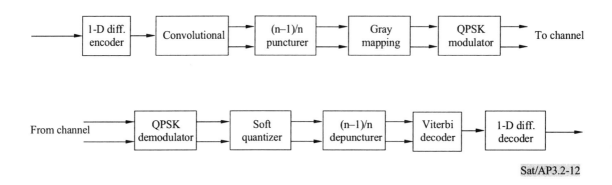

FIGURE AP3.2-12

Punctured codes and the Viterbi decoding with depuncturing

At the emission, the puncturer deletes periodically the symbols at the puncturing positions from the coded symbol stream output from the half-rate encoder. At the reception, the depuncturer inserts the null symbol into the punctured positions in the quantized data stream, used by the Viterbi decoder. In these operations, multiple clock could be necessary. The node synchronization of a punctured decoder is more complex, because the depuncturer must be synchronized with the received data. As mentioned above, both internal and external node synchronizers can be extended to perform this function.

AP3.2-2.6 Performance of convolutionally coded QPSK system

For specific convolutional codes, asymptotic bounds can be used for their error rate performance. These bounds are based on the union bounding technique combined with the transfer function of the code. Detailed theoretical treatments can be found in [Ref. 3-10]. These transfers are not easy to calculate for codes with long constraint length. Numeric methods have been developed to calculate partially these functions for almost all codes of practical interest, and complete lists of weight spectrum have been published [Ref. 3-16]. In fact, only the first terms of the expansion of the transfer function are enough for a quite accurate asymptotic error rate performance evaluation. The results meet very well the practical systems for large E_b/N_0 values, i.e. very low bit error rates. For low E_b/N_0 values, these bounds are not tight. However in this case, the error rate can be easily evaluated by simulation.

In the following, two figures are presented for the bit error ratio (BER) performance of some popular convolutional codes over the AWGN channel. Figure AP3.2-13 presents the punctured convolutional codes performance for K=7, Figure AP3.2-14 compares the half-rate codes with different constraint length.

Figure AP3.2-14 presents the AWGN performance of convolutional codes of different constraint length. It is shown that the long codes have better performance than the short ones. In the possibility of technology, according to the application, it is natural to choose a long memory code. In general, two basic architectures could be used for implementing a Viterbi decoder. For high bit rates in the range of 10-70 Mbit/s, or for the applications of a great quantity, an ASIC architecture is suitable that implements highly parallel computation. A DSP based architecture is usually adopted for lower bit rates below 10 Mbits/s for applications with a small quantity. It is noticed that the frontier at 10 Mbits/s cannot be very precise. With the DSP technology evolution, this boundary tends to be higher. At the same time, the upper bound of about 70 Mbits/s for the ASIC capacity will get higher also.

AP3.2-3 Concatenated coding system with Reed-Solomon outer code

The association of the soft decision decoded inner code and the rapid algebraic decoded Reed-Solomon outer code results in a powerful FEC scheme: the concatenated coding system. The inner code with Viterbi decoding can correct a large part of the random errors and the very short error bursts in the received signal. The residual errors at the outputs of Viterbi decoder, grouped in bursts, will not be easy to correct because the error correction capacity of the outer code has a limit. However, using a properly chosen interleaving that cuts the error bursts into short ones, a Reed-Solomon code can correct most of dispersed error bursts to achieve a very low bit error rate.

The Reed-Solomon code has a high rate and a very small bandwidth expansion is introduced. It will be shown that the design of a good concatenated coding system is achieved by a judicious trade-off between the choices of the codes, the interleaving (in function of the other performances and requirements such as the processing delay) the bit error ratio (BER), E_b/N_0 gain, and the bandwidth expansion, etc. Such systems are implemented for usually a very high performance, for example, a BER of the order of 10^{-10}. The high performance of concatenated coding meets very well the needs of digital TV and HDTV (high definition television) transmission by satellite. Several such concatenated coding schemes have been adopted by international standards for image and video transmission through satellite networks.

AP3.2-3.1 Basic configurations of the concatenated coding system

A general concatenated coding system for satellite transmission is presented in Figure AP3.2-15. It has been simplified by omitting all IF and RF parts. Link control, signalling and baseband processing methods such as pulse shaping and scrambling units are not included in the figure, but will be presented in other paragraphs.

The dashed blocks are optional. The solid ones are basic building blocks of a concatenated coding system.

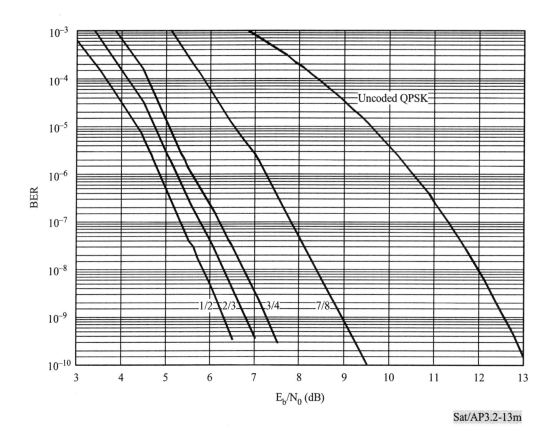

Sat/AP3.2-13m

FIGURE AP3.2-13

Performance of punctured convolutional codes over the AWGN channel

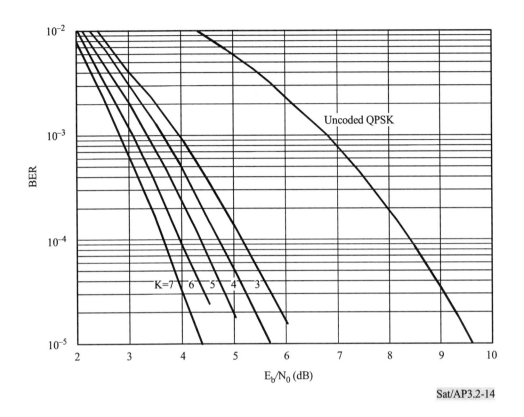

FIGURE AP3.2-14

Performance of half-rate convolutional codes over AWGN channel

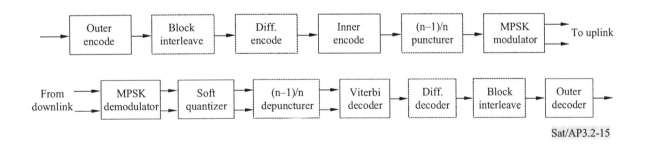

FIGURE AP3.2-15

A general concatenated coding system for satellite transmission

The outer encoder and decoder realize in general a Reed-Solomon encoding/decoding that performs finite field calculations. The block interleaver/deinterleaver are very useful functions that improve largely the whole system performance. The interleaving will be studied in a dedicated paragraph.

The differential encoding/decoding is supposed optional, depending on the applications. Its role is to remove automatically the phase ambiguities due to carrier recovery in the demodulator.

The inner encoder/decoder pair may be a convolutional/Viterbi decoder system corresponding to a QPSK modulation, it could be also a TCM (trellis-coded modulation) or BCM (block-coded modulation) scheme associated with an 8PSK or QAM-16 modulation. In the figure, an MPSK (M-ary Phase Shift Keying) modulation is supposed. Another important function not presented in the figure is scrambling; that will be covered by another dedicated paragraph later.

Compared to the classical coding system with a single code, another additional function is the word (interleaver) synchronization in the Reed-Solomon decoder, subject of a further dedicated paragraph.

AP3.2-3.2 Reed-Solomon codes and algebraic decoding

The inner code could be a convolutional one or a trellis code modulation. The mostly used outer code is in general a Reed-Solomon code based on a finite field, or a Galois field. For an introduction to the finite field calculation, and to the BCH and Reed-Solomon codes, see [Ref. 3-9], [Ref. 3-17 to 20] for a in-depth understanding.

The BCH codes are a class of block codes capable of correcting multiple symbol errors, named after their discoverers: Bose-Chaudhuri-Hocquenghem. These are cyclic codes in a symbol field, with their roots in another field called localization field. For example, the binary BCH codes of length 2m-1 are constructed over their symbol field GF(2), but for their algebraic decoding, operations are performed in their localization field $GF(2^m)$. When these two fields are the same, the Reed-Solomon codes result.

In general, the usual description of BCH codes is in cyclic code language, i.e., by using generating polynomials. To demonstrate their error correction capability, however, the block code language is more useful and matrix and determinant properties are used, see [Ref. 3-9].

The additional defining parameters of a BCH code, of length n and dimension k, designed to correct errors, are the symbol field KS and its localization field KL. The minimum Hamming distance is not given explicitly. The BCH bound holds always $d \geq 2t+1$, d being called the true distance. From the defining parameter t of the BCH codes, $d^*=2t+1$ is called the designed distance.

The Reed-Solomon codes are a special class of the BCH codes defined over the finite field $GF(q^m)$, q a prime. Here q is supposed to equal 2 to simplify the explanation. Its dimension is now given exactly as k=n–2t, and the Singleton bound holds with equality: d=n–k+1=2t+1. The designed distance is the true distance for the Reed-Solomon codes. This is an important property of these codes, because it has been shown that for given length and dimension, d=n–k+1 is the largest distance that can be obtained for any block code. The codes that meet the Singleton bound by equality are known as MDS (maximum distance separable) codes. The Reed-Solomon codes are MDS codes and are the best codes for concatenated coding systems. Usually the Reed-Solomon codes are constructed over $GF(2^m)$. In this case, since a symbol is equivalent to m bits, correcting t symbol errors implies that at most mt bits can be corrected. This shows the burst-error correction capability of the Reed-Solomon codes.

Encoding and decoding RS codes are presented by the following Figure AP3.2-16:

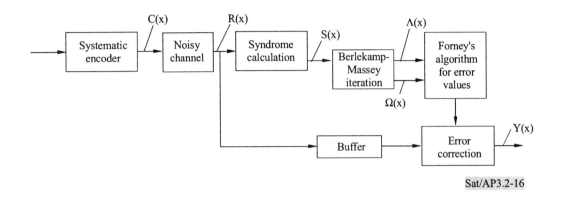

Sat/AP3.2-16

FIGURE AP3.2-16

Architecture of an encoding/decoding (codec) for BCH (RS) code

The encoding can be performed just as the cyclic codes. The decoding follows several steps:

- calculation of syndromes, [Ref. 3-17 to 19];
- formation of key equation by computing the error locator polynomial [Ref. 3-17 to 19];
- calculation of error locations by the Berlekamp-Massey iteration [Ref. 3-17 to 19];
- calculation of error values by the Forney formulae, [Ref. 3-17 to 19]; and
- error correction.

A detailed study of the algebraic decoding is beyond the scope of this section. For more details, see [Ref. 3-9], [Ref. 3-17 to 20]. The algebraic decoding has numerous variants and extensions. The extended Euclid iteration can be used in the place of the Berlekamp-Massey iteration.

Finally, for the bit-error performance of a concatenated coding system using a Reed-Solomon code, the following result is useful, but it assumes that the symbol error probability from the inner decoder Ps is known (by simulation in practice), and further that the symbol errors are uniformly distributed, and finally that the bits are in error with a probability of 1/2 in an erroneous symbol. Under these hypotheses, the bit error rate P_b can be estimated by:

$$P_b = \sum_{j=t+1}^{n} \frac{j}{2n} \binom{n}{j} P_s^j (1 - P_s)^{n-j}$$

In practice, if the interleaving is effective, these hypotheses are verified quite well and this formula is accurate enough.

In summary, to specify a Reed-Solomon code is to specify a polynomial p(x), usually a primitive polynomial of degree m, that defines the finite field $GF(2^m)$, and generates polynomial g(x). To implement the code, the first thing is to derive the methods to perform all operations in $GF(2^m)$, then encoding and decoding algorithms have to be implemented with suitable architectures. The finite field arithmetic, encoding/decoding algorithms, and Reed-Solomon codec architectures, are specific research topics and extensively studied for different practical applications.

AP3.2-3.3 Interleaving

The interleaving used in the concatenated coding system is a very important function (Figure AP3.2-17). The errors at the output of the Viterbi decoding are grouped into bursts. These bursts can be of different lengths, following an experimental distribution of exponential law. The interleaving breaks these long bursts into uniformly distributed symbol errors that are then corrected by the Reed-Solomon decoder. A block interleaver is usually used with a Reed-Solomon code. The interleaver is a data buffer organized in a matrix, each element being a Reed-Solomon code symbol, i.e., an element in $GF(2^m)$, or m bits.

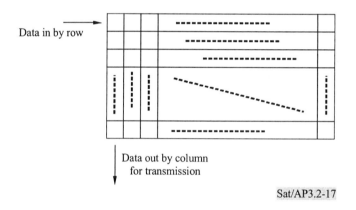

Sat/AP3.2-17

FIGURE AP3.2-17

Block interleaver and de-interleaver

At the emission, the m-bit symbols are written in row by row into the matrix. Once the matrix is full, the elements are read out column by column. The span of the interleaver, being the length of the row, is of length n, codeword length of the Reed-Solomon code. The number of rows per matrix is called the depth of the interleaver, denoted by M.

At the channel output, the burst errors will be distributed in columns. At the reception, another matrix that is the block de-interleaver, is written in again row by row and read out column by column. This means that a burst error is divided by M codewords. A burst error of length L will become M bursts of length L/M. Recall that the Reed-Solomon is capable of correcting t errors. If $L_{max}/M \leq t$ holds, this long error burst is corrected.

To design a concatenated coding system, it is interesting to evaluate the value of L_{max} or the burst error length distribution under nominal functioning of the system. For it is important to design the interleaver with a depth M large enough but not too large, because the interleaver introduces a delay and presents a hardware complexity at a certain cost. The time delay can be roughly evaluated by nMm/Rb (sec), where n is the span, M the depth of the interleaver matrix, m the number of bits of the elements or code symbols, Rb is the bit rate at the interleaver in (bits/sec).

In summary, the practical constraints for designing the interleaving in a concatenated coding system can be expressed as:

$$L_{max}/t \leq M \leq \tau_{max} R_b/nm,$$

τ_{max} being the maximum time delay that the system can tolerate. In high bit-rate applications, a large M could be taken. But for high bit rates, a very fast memory device would be necessary to implement the interleavers. It is limited to a small value if R_b is low and the interactive communication applications are concerned.

AP3.2-3.4 Word (interleaver) synchronization

For decoding the Reed-Solomon codewords in a concatenated coding system, the de-interleaver and the outer decoder must both be synchronized with the data stream to insure that the correct codewords are identified by the de-interleaver and decoder. This is the special problem of the word (interleaver) synchronization.

Several methods can provide the solution:

- insertion of a unique word;
- insertion of a unique word that replaces one or two Reed-Solomon code symbols in its parity-check part; or
- a frame synchronization device in the decoder.

The unique word detection performs in general a frame synchronization. If the unique word is put at the start or the end of a codeword, the codeword will be identified by the detection. It is possible to put repeatedly the unique word all M codewords, at the beginning of each interleaver matrix to perform the interleaver synchronization.

The insertion of a unique word may lead to an insertion loss, i.e. the unique word must be transmitted and a part of communication resource must be used. To avoid this loss, it is possible to replace one or two parity check symbols of a Reed-Solomon codeword by a unique word. At the decoding, these symbols are treated as erasures. This approach does not result in a noticeable performance loss, if the unique word does not appear very frequently in the data stream.

It can be seen that any frame synchronizer can be used to establish the word (interleaver) synchronization.

AP3.2-3.5 Block scrambling

In classical systems a sequential scrambler is used to disperse the data stream, in order to avoid long "0" or "1" runs in the data stream, and consequently to insure a good signal power spectrum.

Block scramblers are suitable for systems using concatenated coding with a Reed-Solomon outer code that requires a word synchronization as mentioned above. This is done by using a PN sequence generator of fairly long period and an addition to the bit stream of the generated binary sequence. The same sequence must be used at the both emission and reception sides in synchronization.

AP3.2-4 Conclusion

An engineering design of a FEC system must take into account several factors:

- the E_b/N_0 (C/N) value used in the link budget analysis;

- the bandwidth expansion;

- the bit error rate objective;

- the processing delay;

- the multiple access scheme, in particular the data format constraints and the multiple access interference analysis;

- the modulation and the demodulation schemes, in particular the synchronization threshold;

- the phase ambiguities resolution possibility with the code transparency;

- the availability of the decoder devices;

- the implementation complexity, the cost, and the development delay and economic efforts, etc.

Among the channel coding techniques, the two most important and largely used ones have been studied in detail: the convolutional coding with soft decision Viterbi decoding, and the concatenated coding systems using RS coding. The other techniques such as the self-orthogonal convolutional codes, the binary block codes such as Hamming, Golay and some high rate BCH codes, are omitted. Interested readers are referred to [Ref. 3-9], [Ref. 3-17 to 20] for an in-depth description.

Error control coding is a very active R&D domain and is of great importance for future satellite communication systems. As important recent events, the trellis-coded modulations and concatenated coding have been introduced into satellite transmission.

It is not easy to foresee the future evolution of the application of error control coding to satellite communication systems. Channel coding techniques will develop together with advanced systems in the future. For example, in on-board processing, the decoding function will be very advantageous for baseband regeneration, providing a significant saving in power efficiency of the payload. Other important domains that would demand very advanced channel coding techniques could be the CDMA satellite systems and other personal satellite communication systems based on low-earth-orbit satellite global constellation, and multimedia satellite systems with new applications, where new requirements for coding will be challenging and stimulating.

APPENDIX 3.3

Video signal compression techniques [Ref. 3-2, 3-5]

AP3.3-1 Principal bit-rate reduction methods

AP3.3-1.1 The first generation approach

AP3.3-1.1.1 Transform coding (spatial compression)

In transform coding (Figure AP3.3-1), a unit linear transformation is applied to an area of the picture, thus converting it into a new assembly possessing different statistical and psycho-visual properties. Unitary transformations preserve energy. However, advantage is taken of the fact that in the transformed domain, the energy is concentrated on a reduced number of coefficients. It should be noted that the transformation does not produce a bit-rate reduction but only a modification of the representation of the picture. The most commonly used transformation is the discrete cosine transform (DCT). The DCT is generally applied on rectangular or square blocks of the picture, the optimal size being around 8 x 8 or 16 x 16 pixels.

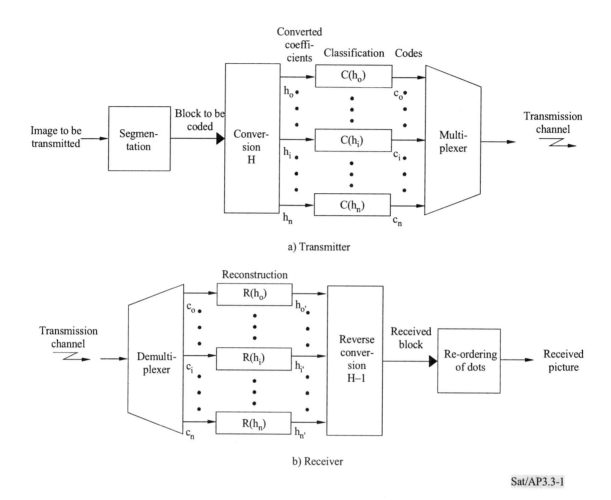

a) Transmitter

b) Receiver

Sat/AP3.3-1

FIGURE AP3.3-1

Block diagram of transform coding

The transformed elements of the blocks (generally called coefficients) can be coded in different ways. Two approaches can be cited:

- first, threshold coding consists of transmitting those coefficients that are above a predetermined threshold (this threshold can vary depending on the order of the coefficients). The block is scanned in a certain order (zig-zag scanning) until the amplitude of the last significant coefficient is transmitted. If a sub-threshold coefficient is encountered a special signal is transmitted to indicate this fact;

- secondly, zonal coding in which a dictionary of classes exists at the transmitter (coder) and at the receiver (decoder). A class is an indicator of which coefficients are transmitted. For the current block to be coded, the best class is chosen and its number is transmitted, as well as the amplitude of its coefficients.

AP3.3-1.1.2 Differential pulse code modulation (temporal compression)

In the case of differential PCM, in the same way as was explained in § 3.3.1, the transmission system uses the part of the picture already transmitted to develop a prediction $P(L(i,j))$ of the value $L(i,j)$ of the luminance signal sample to be transmitted. The difference $E(i,j)$ between the real and the predicted values is quantized and coded to be sent to the transmission channel. After decoding, the received value is added to the prediction effected by the receiver which makes it possible to evaluate $L(i,j)$.

Bit-rate compression is achieved by quantizing the prediction error less finely than the original signal. For a good reconstruction quality, the required bit rate is of the order of 4 to 5 bits per sample. The advantage of this system is its simplicity of implementation. Moreover, the separation into two functions – prediction and quantizing – permits a good adaptation of the system to the properties of the signal (prediction) and the observer (quantizing).

The quantizer makes use of the fact that the prediction error signal has more stable properties than the luminance signal, and that the visibility of faults depends on local contrast. The classifier works by partitioning the dynamic range of the error signal $E(i,j)$ into N zones, with a code corresponding to each zone. The reconstructor assigns to each code a coded signal reconstruction value which supplies the quantized value of the prediction error $Eq(i,j)$. The development of a quantizing law therefore consists in choosing the limits of N zones or thresholds of decisions and the reconstruction values. The quantizer applies a non-linear law so as to take advantage of the fact that the eye is less sensitive to restitution errors in the contrasted zones with large prediction errors than in quasi-uniform zones with slight prediction errors.

Whatever system is used, the symbols transmitted by the coder may be of fixed or variable length. In the latter case, a statistical coding is used which, for each elementary message to be transmitted, transmits a code, the length of which depends on the probability of this information. However, it must be borne in mind that this system, while reducing the mean bit rate, introduces all the problems bound up with the routing of a variable bit rate on constant bit-rate transmission channels.

The coding systems described above can be used to transmit a television signal in a 30 to 50 Mbit/s channel, but it is difficult to meet the quality objective at 30 Mbit/s. In practical terms, any increase in compression rate means that the coding system must be adaptable, which implies more complex processing.

AP3.3-1.1.3 Motion compensation

Motion compensation is the technique used to reduce the redundant information between frames, based on estimating the motion between them thanks to a limited number of motion parameters (vectors).

Only blocks (16 x 16 pixels usually) which differ from one image and its successor will need to be coded, leading to a significant reduction of the information to be transmitted. Coding of the blocks is performed using the techniques described above. Error is minimized by transmitting to the receiving end both the motion vector and the prediction error.

The recent development of hardware implementations in VLSI alleviated the drawbacks of the motion compensation technique (e.g. complexity, computation intensive process), and enabled its implementation in commercial video compression standards (for example the MPEG 2 standard, see below § AP3.3-2.3) and systems.

AP3.3-1.1.4 Hybrid coding

The principle of hybrid coding is to combine two or more coding techniques to gain more advantage from compression without sacrificing much in terms of quality.

The two types of approaches mentioned in § AP3.3-1.1.1 and AP3.3-1.1.2 which can be applied on intrafield blocks (spatial coding) as well as on temporal differences of blocks (interframe coding), can also be combined to form a hybrid coding scheme.

Another technique commonly applied is to combine the spatial DCT method with frame-to-frame (temporal) DPCM. To that aim temporal correlation is first reduced through prediction, and then the DCT is applied to the prediction. This method is central to the MPEG standard which is discussed in § AP3.3-2.3 below.

Hybrid coding opens up the possibility of offering a higher quality level (better resolution and motion performance) tailored to the customer's needs. While the basic service using DCT may be transmitted with maximum compression and minimum cost of transmission, enhancement through a second channel (adding the hybrid coding feature) would require more capacity together with some processing complexity.

AP3.3-1.2 The second generation approach

Image and video coding are basically carried out in the following two steps: firstly, image data are converted into a sequence of messages and, secondly, code words are assigned to the messages. As opposed to first generation methods which put the emphasis on the latter step, methods of the second generation put it on the first (object-oriented coding and recognition of the image), and use their outputs to implement the second step. Rather than in exploiting the spatial and temporal redundancy of the data, reduction is achieved by taking advantage of the human visual system specific features, i.e. in allocating more bits in visually important areas and fewer bits elsewhere.

Three schemes are representative of these second generation video coding techniques: 1) segmentation-based schemes, 2) model-based schemes, 3) fractal-based schemes. Applications foreseen are in telecommunications and information storage/retrieval.

In particular fractal compression is expected to provide a higher order of magnitude reduction as compared to the DCT, in terms of number of bits per second transmitted. Combination of statistical and fractal techniques is also a direction of investigation towards a higher degree of compression.

AP3.3-2 Selected standards for digital video and TV applications – bit-rate reduction techniques

AP3.3-2.1 Videoconferencing – ITU-T Recommendation H.261

ITU-T Recommendation H.261 provides a standard coding algorithm for the reduction of the bit rate of video signals.

The coding algorithm transforms the incoming video signal mainly by means of the discrete cosine transform (DCT) and a quantizer. The coding process is applied to blocks (sets of 8 x 8 pixels) of the video picture. There are two types of coding processes. The "Intra" process only handles the current picture or blocks. The "Inter" coding process, however, compares the current block with the previous one and transmits only the difference signal. If there is a movement in the picture, some pixels or blocks change their position. When the picture information does not change, a significant bit reduction can be achieved.

Following developments in digital video compression techniques, the availability of reduced bit-rate video codecs now makes it possible to implement videoconferencing for business applications on a cost-effective basis. Current bit rates used for videoconferencing are: 56/64 kbit/s, 112/128 kbit/s, 384 kbit/s, 738 kbit/s up to 1.544/2.048 Mbit/s, the most typical trade-off in picture quality vs. transmission cost often being at 384 kbit/s. Moreover, multi-rate codecs are available, permitting optimization of the channel bandwidth as a function of the quality requested for the immediate application. For example, a H.261 based videoconference codec provides a p x 64 kbit/s ($p = 1 \sim 32$) audiovisual service. A bit rate of 384 kbit/s ($p = 6$) is commonly used.

VSATs can actually offer a suitable communication medium for such applications. Of course in this case proper means for switching and managing the conference are to be provided in the VSAT network, especially in the case of N-way conferences.

AP3.3-2.2 The Joint Photographic Experts Group (JPEG) standard

The JPEG standard is described in ITU-T Recommendation T.81 ("Information technology – digital compression and coding of continuous-tone still images – requirements and guidelines"). This specification is composed of two parts: the first part sets out requirements and implementation guidelines for continuous-tone still image encoding and decoding processes, and for the coded representation of compressed image data for interchange between applications. These processes and representations are intended to be generic and to be applicable to a broad range of applications for color and grey scale still images within communications and computer systems (e.g. hard disk storage of photographs and CD-ROM media, desktop publishing, medical or scientific imaging).

The second part sets out tests for determining whether implementations comply with the requirements for the various encoding and decoding processes specified in the first part.

JPEG is based on the spatial DCT compression method, enabling the user to choose the degree of compression corresponding to its requirements ("lossy compression"). In this mode blocks of 8 x 8 pixels are processed at a time following the main steps: 1) the transformation of the digitized image using the DCT; 2) the quantization step (uniform for each of 64 DCT coefficients); 3) the coding of the coefficients using a variable length code ("entropy coding"). In addition a lossless compression using DPCM is also possible.

The JPEG can be implemented either in hardware or in software configurations. See Table AP3.3-1 below for a summary of the essential characteristics of the JPEG coding processes.

TABLE AP3.3-1

Essential characterization of coding processes (from ITU-T Recommendation T.81, 1992)

Baseline process (required for all DCT-based decoders)
• DCT-based process • Source image: 8-bit samples within each component • Sequential • Huffman coding: 2 AC and 2 DC tables • Decoders shall process scans with 1, 2, 3 and 4 components • Interleaved and non-interleaved scans

Extended DCT-based processes
• DTC-based process • Source image: 8-bit or 12-bit samples • Sequential or progressive • Huffman or arithmetic coding: 4 AC and 4 DC tables • Decoders shall process scans with 1, 2, 3, and 4 components • Interleaved and non-interleaved scans

Lossless processes
• Predictive process (not DCT-based) • Source image: P-bit samples ($2 \leq P \leq 16$) • Sequential • Huffman or arithmetic coding: 4 DC tables • Decoders shall process scans with 1, 2, 3 and 4 components • Interleaved and non-interleaved scans

Hierarchical processes
• Multiple frames (non-differential and differential) • Uses extended DCT-based or lossless processes • Decoders shall process scans with 1, 2, 3 and 4 components • Interleaved and non-interleaved scans

AP3.3-2.3 The Moving Picture Experts Group (MPEG) standards

MPEG is a series of standards for motion picture and video that supports a variety of picture formats while providing flexible encoding and transmission structures. Its algorithms can be implemented in hardware which decreases the delays associated with the coding and decoding phases. In addition the compression scheme used is compatible with a wide variety of transport and storage methods (see § AP3.3-2.4 concerning the digital video broadcasting (DVB) standard).

MPEG 1 has been standardized for digital storage media with minimal errors applications, like CD-ROM applications (storage/retrieval).

As opposed to MPEG 1 which was developed solely by the ISO/IEC, MPEG 2 is the outcome of a joint effort with ITU (ITU-T) with due regard to interlaced video signals (e.g. broadcast applications). MPEG 2 system and video specifications are dealt with in the following Recommendations:

- ITU-T Recommendation H.222 "Information technology – generic coding of moving pictures and associated audio information: systems";

- ITU-T Recommendation H.262 "Information technology – generic coding of moving pictures and associated audio information: video".

The MPEG standard may be used for transmission of video signals over VSAT systems.

AP3.3-2.4 The Digital Video Broadcasting (DVB) standard

The DVB originated in 1990 as a European initiative for governmental and industrial cooperation, that resulted in a Memorandum of Understanding which has today around 200 signatories, many of them from Europe and the United States.

Based on the MPEG 2 standard, the DVB standard can essentially be seen as a complete package for digital television and data broadcasting.

In particular the concept has been adapted to satellite digital television transmission (DVB-S). The DVB-S standard is intended to cover the scope of BSS as well as FSS satellite systems planned or operated (transponder bandwidths in the range of 26 MHz to 72 MHz), and describes a layered transmission architecture (see European Standard ETS 300421 "Digital broadcasting systems for television, sound and data services; Framing structure, channel coding and modulation for 11/12 GHz satellite services" for more information).

The following list illustrates possible DVB applications:

- Single programme generic TV:

  One MPEG 2 video signal (5 Mbit/s) and five stereo sound channels (256 kbit/s)

- Business TV with high data broadcasting channel:

  One MPEG 2 video signal (4 Mbit/s) and one stereo sound channel (256 kbit/s) and one 2 Mbit/s stream (G.703)

- Teleshopping:

  Two MPEG 2 video signals (3 Mbit/s) and two mono sound channels (128 kbit/s)

APPENDIX 3.4

Asynchronous transfer mode (ATM)
[Ref. 3-1]

As mentioned in § 3.5.4, ATM is a specific packet-oriented transfer mode which uses asynchronous time-division multiplexing techniques. It combines the flexibility of packet switching and the simplicity of circuit switching. The multiplexed information flow is organized into fixed length cells. A cell consists of an information field of 48 bytes and a header of 5 bytes (see Figure 3.18). The network uses the header to transport the data from source to destination. Transfer capacity is assigned by negotiation and is based on the source requirements and the available capacity. In general, signalling and user information are carried on separate ATM connections.

AP3.4.1 The ATM/B-ISDN protocol reference model

The ATM/B-ISDN protocol reference model is composed of a user plane, a control plane and a management plane (see Figure AP3.4-1; see also ITU-T Recommendations I.320 and I.321).

The user plane has a layered structure and provides for user information flow transfer along with associated controls. The control plane has also a layered structure and performs the call control and connection control functions; it deals with the signalling required to set up, supervize and release calls and connections. The management plane provides network supervision organized into two types of functions:

- plane management performs management functions related to a system as a whole and provides coordination between all planes;

- layer management performs management functions relating to resources and parameters residing in its protocol entities and handles the OAM information flows specific to the concerned layer.

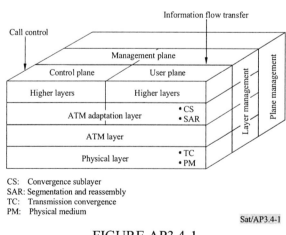

CS: Convergence sublayer
SAR: Segmentation and reassembly
TC: Transmission convergence
PM: Physical medium

Sat/AP3.4-1

FIGURE AP3.4-1

Protocol reference model for ATM

212

The layered structures of the user plane and of the control plane are:

- The ATM Adaptation Layer (AAL) which adapts service information to the ATM stream: for each connection the information is accommodated in the cell payload according to different rules depending on the type of service to be supported.

- The ATM Layer which mainly performs cell switching, routing and multiplexing: for each connection the proper header is added. In the case where many connections are transmitted on the same physical link, the cells from each connection are statistically multiplexed.

- The Physical Layer (PL) which primarily ensures reliable transport of ATM cells.

These layers are present in the equipment where appropriate. For example, in ATM Terminals or Terminal Adaptors, where cells are generated and terminated, AAL is present. In ATM switches and cross connects, AAL is not present because its information is not used, but the ATM layer is present in such equipment because the routing information contained in the cell header is used. The PL is present at every end of the transmission system.

An ATM network consists of interconnected ATM nodes. Each of these nodes would perform the functions of at least the two lower layers. The Higher Layers of the Protocol Reference Model are application dependent and reside in the customer premises equipment, or in specialized service provider nodes.

Above the Physical Layer, the ATM Layer provides cell transfer for all services and the AAL provides service dependent functions to the layer above the AAL.

The layers above the AAL in the control plane provide call control and connection control functions. The management plane provides supervision functions.

AP3.4-2 Main features of the ATM/B-ISDN protocol reference model layers

The Physical Layer consists of two sublayers: the Physical Medium (PM) sublayer and the Transmission Convergence (TC) sublayer.

The PM sublayer includes only physical medium dependent functions (like line coding and the reverse operation), and provides bit transmission capability including bit transfer and bit alignment.

The TC sublayer transforms a flow of cells into a flow of data units (bits) that can be transmitted or received over the PM. Cell rate decoupling includes insertion and suppression of idle cells in order to adapt the rate of valid ATM cells to the payload capacity of the transmission system actually used. The transmission frame adaptation function performs the actions necessary to structure the cell flow according to the payload of the transmission frame (transmit direction) and to extract this cell flow out of the frame (receive direction). The transmission frame may be a cell equivalent (no external envelope is added to the cell flow), an SDH envelope, a PDH envelope, etc.

To summarize, the TC functions are:

- Cell rate decoupling

- Header Error Control sequence generation/verification

- Cell delineation
- Transmission frame adaptation
- Transmission frame generation/recovery.

The characteristics of the ATM Layer are independent of the physical medium. Its functions are:

- Generic flow control
- Cell header generation/extraction
- Cell Virtual Path Identifier and Virtual Channel Identifier translation (mapping their values into new values if necessary)
- Cell multiplex/demultiplex.

The information field is transported transparently by the ATM layer, which does not perform any processing (like flow or error control) on it.

The ATM Adaptation Layer enhances the service provided by the ATM layer to support functions required by the next higher layer. The AAL performs functions required by the user, control and management planes and supports the mapping between the ATM layer and the next higher layer. The functions performed in the AAL depend upon the higher layer requirements. The AAL supports multiple protocols to fit the needs of the different AAL users. The AAL receives from and passes to the ATM layer the information in the form of a 48 octet ATM Service Data Unit. Different types of AAL are defined according to the services provided.

The AAL functions are organized into two logical sublayers: the Convergence Sublayer (CS) and the Segmentation And Reassembly (SAR) sublayer. The CS provides AAL service at the AAL service access point. The SAR functions are the segmentation of higher layer information into a size suitable for the information field of the ATM cell and the reassembly of the contents of ATM cell information fields into higher layer information.

A service classification (see Table AP3.4-1), intended for AAL and not general, is defined based on the following parameters:

- timing relation between source and destination (required or not required);
- bit rate (constant or variable);
- connection mode (connection- or connectionless-oriented).

Other parameters are considered as quality of service parameters and do not lead to different service classes of AAL:

- Cell loss ratio (CLR) is the ratio of the number of lost cells to the total number of expected cells (lost cells are cells which are discarded or misrouted). Due to the bursty error nature of a satellite link, corrupted headers will usually contain several errors which the HEC mechanism is not able to correct. The CLR will likely be the most critical parameter for satellite transmissions; the objective of ATM designers is to achieve a quality, in terms of cell loss, which remains excellent in all cases, defects due to these congestion phenomena remaining comparable with those generated by transmission systems ($1.0 \cdot 10^{-10}$ for example).
- Cell error ratio (CER) is the ratio of the number of cells having one or several errors in the payload to the total number of cells.

- Cell transfer delay: the propagation delay inherent to a geostationary satellite (one hop) is on the order of 240 ms and is the predominant contribution to the overall cell transfer delay. This delay must be must be accounted for in the higher data communication protocols.

- Cell delay variation (CDV) is mainly caused by queueing in network nodes. The impact of the satellite link is usually negligible except perhaps on some Constant Bit Rate (CBR) applications with very high clock frequency stability requirements: the residual movement of geostationary satellites around their nominal position causes a delay variation of about 1 ms within a one day period, and a Doppler shift on the clock frequency of about $1.0 \cdot 10^{-8}$.

TABLE AP3.4-1

Service classes for definition of AAL protocols

Service class	A	B	C	D
Timing relation between source and destination	Required	Required	Not required	Not required
Bit rate	Constant	Variable	Variable	Variable
Connection mode	Connection Oriented (CO)	CO	CO	Connectionless (CL)

Examples of services in the classes A, B, C and D are:

- Class A: Circuit emulation, voice, constant bit-rate video
- Class B: Variable bit-rate video and audio
- Class C: Connection-oriented VBR data transfer and signalling
- Class D: Messages, Connectionless VBR data transfer.

A newly defined application class, known as Class X, assumes that timing is not required, bit rate is variable and connection mode is either connection-oriented or connectionless (for some kinds of data transfer).

Corresponding to these classes, four types of AAL layers have been recommended by the ITU-T. It should be noted that the association of service classes and AAL types is somewhat artificial. In addition, other AAL protocols are under study (AAL type 5) or will be standardized. The boundary between the ATM layer and the AAL corresponds to the boundary between functions supported by the contents of the cell header (see Figure AP3.4-5) and functions supported by AAL specific information which is contained in the information field of the ATM cell.

AP3.4-3 ATM entities and their relationships (Figure AP3.4-2)

Channels are set up under the control of call and connection handling mechanisms, and paths are set up and managed on a semi-permanent basis, like transmission resources, under the control of network management mechanisms. However, these concepts evolve towards automatic access by users for the management of their paths (as in the case of leased lines).

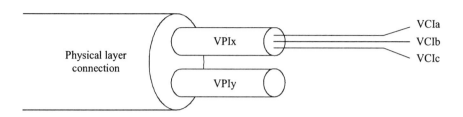

VCI: Virtual channel identifier
VPI: Virtual path identifier
VCIa, VCIb, VCIc: Possible values of VCI within the VP link with value VPIx
VPIx, VPIy :   Possible values of VPI within the physical layer connection

Sat/AP3.4-2

FIGURE AP3.4-2

ATM connection identifier

ATM is a connection-oriented technique. Connection identifiers are assigned to each link of a connection when required and released when no longer needed. ATM offers transfer capability common to all services, including connectionless services.

An ATM layer connection consists of the concatenation of ATM layer links in order to provide an end-to-end transfer capability to access points (Figure AP3.4-3).

The basic ATM routing entity for switched services is the virtual channel. It is handled in VC multiplexers/demultiplexers and switches (Figures AP3.4-3 and AP3.4-4).

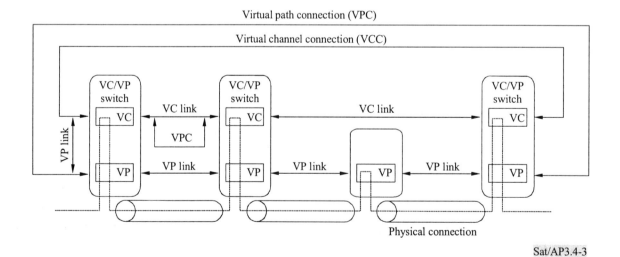

Sat/AP3.4-3

FIGURE AP3.4-3

ATM connections

A virtual channel connection (VCC) is a concatenation of VC links between two points where the ATM adaptation layer is accessed. VCCs can be provided on a switched or (semi)permanent basis.

The label of each ATM cell is used to explicitly identify the VC to which the cell belongs. The label consists of two parts: the virtual channel identifier (VCI) and the virtual path identifier (VPI). A specific VCI value is assigned each time a VC is switched in the network. A VPI identifies a group of VC links, at a given reference point, that share the same VPC. A specific value of VPI is assigned each time a VP is switched in the network. Thus, ATM channels are identified by their associated VCI/VPI.

A VC link is a means of unidirectional transport of ATM cells between a point where a VCI value is assigned and a point where that value is translated or removed.

A virtual path is a generic term for a bundle of VC links which have the same endpoints.

A VP link is a unidirectional capability for the transport of ATM cells between two consecutive ATM entities where the VPI value is translated. A VP link is originated or terminated by the assignment or removal of a VPI value.

At a given interface, in a given direction, the different virtual path links multiplexed at the ATM layer into the same physical layer connection are distinguished by the VPI. The different virtual channel links in a virtual path connection are distinguished by the VCI. Two different VCs belonging to two different VPs at a given interface may have the same VCI value. Therefore, a VC is only fully identified at a given interface by both VPI and VCI values.

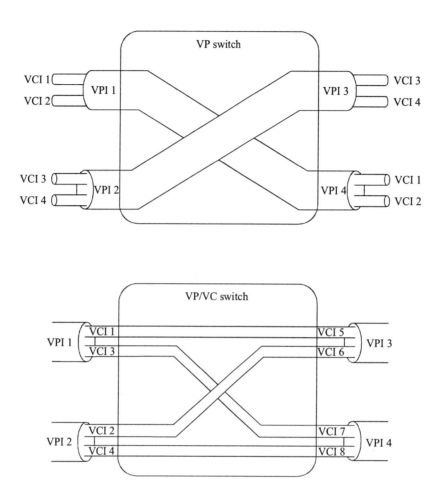

FIGURE AP3.4-4

VP and VC switching

A specific value of VCI has no end-to-end significance if the virtual channel connection is switched. VPIs may be changed wherever VP links are terminated (e.g. cross connects, concentrators and switches). VCIs may only be changed where VC links are terminated. As a consequence, VCI values are preserved within a VPC.

A VCC may be established or released at the user network interface (UNI) or at a network node interface (NNI). The NNI cell is identical to the UNI except that there is no GFC field and thus the VPI occupies 12 bits.

At the UNI, a specific value, independent of the service provided over the VC, is assigned to a VCI by one of the following: the network, the user, negotiation between the user and the network, or a standard procedure.

ATM network elements (ATM switches, cross connects and concentrators) process the ATM cell header (Figure AP3.4-5) and may provide VCI and/or VPI translation. Thus, whenever a VCC is established or released across the ATM network, VC links may need to be established or released at one or more NNIs. VC links are established or released between ATM network elements using inter- and intra-network signalling procedures; other methods are also possible.

A) Header structure at the UNI

8	7	6	5	4	3	2	1	bit / octet
GFC				VPI				1
VPI				VCI				2
VCI								3
VCI				PT			CLP	4
HEC								5

B) Header structure at NNI

8	7	6	5	4	3	2	1	bit / octet
VPI								1
VPI				VCI				2
VCI								3
VCI				PT			CLP	4
HEC								5

CLP   Cell Loss Priority
GFC   Generic Flow Control
HEP   Header Error Control
CT   Payload Type (8 possibilities, one reserved)
VPI   Virtual Path Identifier
VCI   Virtual Channel Identifier

Sat/AP3.4-5

FIGURE AP3.4-5

ATM cell header structure

A VPC may be established or released between VPC endpoints on a subscription basis or on demand (customer or network controlled).

VCCs and VPCs are provided with "quality of service" parameters such as cell loss ratio or cell delay variation. Cells with the CLP bit set are subject to discarding, depending on network condition.

Finally, Figure AP3.4-6 below summarizes the switching and cross-connections aspects of ATM networks.

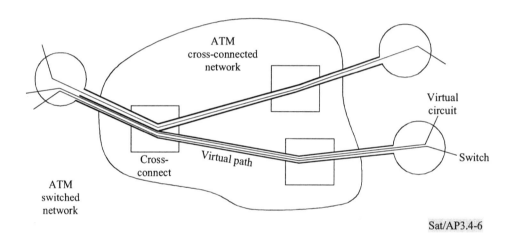

FIGURE AP3.4-6

ATM switched and cross-connected network

AP3.4-4 ATM resource control aspects

As opposed to switching networks where a fixed bit-rate resource is reserved throughout the connection and for the duration of each call, ATM is a queue-based packet switching technique which means that the assignment of a resource to a virtual channel is a purely logical operation, and that congestion phenomena are likely to happen, compromising the provision of the resource.

In addition, there is no absolute confidence that the user will not attempt to exceed the resource allocated to it.

Different techniques have been identified which provide an acceptable answer to this difficult problem of access control.

APPENDIX 3.5

The synchronous digital hierarchy (SDH)
[Ref. 3-1]

AP3.5-1 SDH multiplexing structure

In order to accommodate signals generated by equipment from various PDH levels, SDH recommendations define methods of subdividing the payload area of an STM-1 frame in various ways so that it can carry different combinations of tributaries (see § 3.5.3 for a definition of the STM-1 frame).

In SDH, a number of information structures, called "containers", are defined, each corresponding to an existing PDH rate signal, except for 8 Mbit/s. The signal to be carried on the synchronous digital network is mapped beforehand in an appropriate container of the synchronous frame structure. Each container has some control information known as the path overhead (POH) associated to it.

A path is a logical connection between the point at which a container for a signal at a given rate is assembled (source) and the point at which it is disassembled (sink), i.e. the originating and terminating points.

Path and section overheads associated at the appropriate multiplexing levels with individual SDH entities are intended to perform quality monitoring and end-to-end supervision as well as to offer communication channels and other facilities. For instance, path overhead (POH) functions include path identification, error rate detection, alarm transmission, payload composition, maintenance signalling and so on.

Together, a container of level n (C-n) and its associated POH form a Virtual Container of level n (VC-n), which is the entity managed by the SDH and the integer entity during its whole travel through the SDH network.

Table AP3.5-1 shows the containers defined for the respective basic PDH bit rates, where the digits represent the level of transmission hierarchy (Recommendation G.702) and, if the case arises, the lower (1) or higher (2) rate.

TABLE AP3.5-1

SDH containers adapted for PDH bit rates

Container	PDH bit rate
C-11	1 544 kbit/s
C-12	2 048 kbit/s
C-2	6 312 kbit/s
C-3	34 368 and 44 736 kbit/s
C-4	139 264 kbit/s

A Virtual Container is, therefore, the information structure used to support path layer connections in the SDH. It consists of an information payload (the container) and path overhead information fields organized in a block frame structure which repeats every 125 or 500 microseconds. Alignment information to identify the VC-n frame start is provided by the server network layer.

ITU-T Recommendation G.709 defines different combinations of VCs which can be used to fill up the payload area of an STM-1 frame. The process of loading containers and attaching overheads is repeated at several levels in the SDH, nesting of smaller VCs within larger ones, until the VC of largest size is filled and then loaded into the STM-1.

The assembly process enables virtual containers to be dropped or inserted. This gives, for instance, flexibility to allow a single wide bandwidth medium (e.g. one optical fibre) to be used for serving a number of physically separate locations in a ring structure. A related function is that of multiplexing containers into administrative units which can be handled as single entities by SDH digital cross-connect (DXC) equipment.

A pointer is an indicator whose value defines the frame offset of a virtual container with respect to the frame reference of the transport entity on which it is supported. Payload pointers are used to achieve phase alignment between SDH signals; they facilitate frame alignment and allow:

- to resolve jitter (i.e. difference between where a pulse is and where it should be) and wander effects, which, even though the network is synchronized, are due to accumulation of clock noise in the network and degrade transmission quality;

- to avoid long buffers in situations where VCs from different sources are multiplexed in the same STM-N, as is the case of cross connects.

Two types of virtual containers are identified:

- Lower order (LO) VC-n (n = 1, 2 or 3). This element comprises a single Container-n (n = 1 or 2) plus the lower order VC POH appropriate to that level;

- Higher order (HO) VC-n (n = 3 or 4). This element comprises a single Container-n (n = 3 or 4) or an assembly of tributary unit groups (TUG-2s or TUG-3s) together with VC POHs appropriate to that level.

The tributary unit (TU) is an information structure which provides adaptation between the lower order path layer and the higher order path layer. It consists of an information payload (the lower order VC) and a TU pointer which indicates the offset of the payload frame start relative to the higher order VC frame start.

A TU (n = 1, 2 or 3) consists of a VC-n together with a TU pointer.

In some applications, the VC-3 defined above is considered of lower order, because it can be transported in a VC-4. In this case, it is aligned to the TU-3.

The concept of tributary unit group (TUG) is used to define signals transported in a higher order VC, which can be formed in different ways to transport a variety of services at different rates. Then, mixed capacity payloads can be constructed to increase flexibility of the transport network.

A TUG-2 consists of a homogeneous assembly of identical TU-1s or a TU-2.

A TUG-3 consists of a homogeneous assembly of TUG-2s or a TU-3.

An administrative unit (AU) is the information structure which provides adaptation between the higher order path layer and the multiplex section layer. It consists of an information payload (the higher order VC) and an AU pointer which indicates the offset of the payload frame start relative to the multiplex section frame start.

Two AUs are defined: the AU-3 consisting of a VC-3 plus an AU pointer and the AU-4 consisting of a VC-4 plus an AU pointer.

An administrative unit group (AUG) is a homogeneous assembly of AUs, which may consist of three AU-3s or one AU-4. The STM-N payload can support N AUGs.

The following SDH procedures are defined:

- mapping, by which tributaries are adapted into virtual containers at the boundary of the SDH network;

- multiplexing, by which multiple lower order path layer signals are adapted into a higher order path, or higher order path layer signals into a multiplex section;

- aligning, by which the frame offset information is incorporated into the tributary unit or the administrative unit, when adapting to the frame reference of the supporting layer;

- concatenation, whereby a multiplicity of identical virtual containers is associated one with another, with the result that their combined capacity can be used as a single container across which bit sequence integrity is maintained.

ITU-T Recommendations G.708 and G.709 describe the relationships between the various multiplexing elements (shown in Figure AP3.5-1), the set of rules to construct STM-N frames, the arrangements for the international interconnection of STM-Ns, the functions, descriptions and usage of overheads, and the mapping of PDH signals.

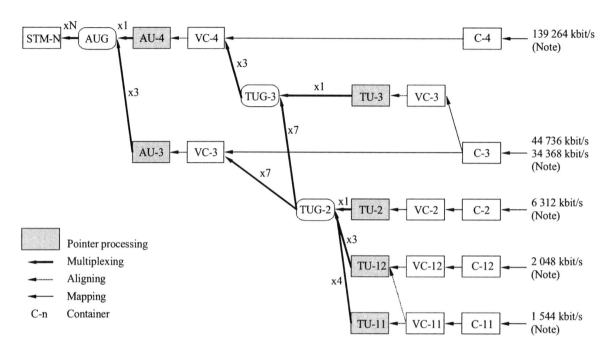

NOTE - G.207 tributaries associated with containers C-x are shown. Other signals, e.g. ATM, can also be accommodated.

Sat/AP3.5-1m

FIGURE AP3.5-1

SDH multiplexing structure

The European multiplexing scheme excludes the VC-11/TU-11/TUG-2/VC-3/AU-3/AUG and C-3/VC-3/AU-3 paths. On the other hand, the North American and Japanese schemes exclude the VC-11/TU-12, TUG-2/TUG-3/VC-4/AU-4/AUG, C-3/VC-3/TU-3/TUG-3 and C-4/VC-4 paths.

The European choice of the VC-4 as the virtual container of higher order corresponds to the present 140 Mbit/s and the North American and Japanese choice of the VC-3 to the level 45 Mbit/s. However, international interconnection between countries having different choices is based on the VC-4, requiring for that purpose equipment able to deal with that level on both sides.

The second European option is the use of the TU-12 for the transport of 1 544 kbit/s signals, simplifying their management by treating them in the same way as 2 048 kbit/s signals. However, at this level, the international interconnection is based on the use of TU-11, which requires the realization of the TU-11/TU-12 adaptation function on the European side in case of interconnection with countries using the TU-11.

When a transmission rate higher than 155 Mbit/s is required in a synchronous network, rates of 622 Mbit/s and 2.4 Gbit/s can be achieved by using a relatively straightforward byte interleaved multiplexing scheme. In this way, STM-4s and STM-16s may be considered as a combination of lower level STM signals. In fact, only the STM-1 signal can be directly accessed by low tributaries; an STM-N signal needs therefore to be previously demultiplexed into the STM-1 level, but this is a single process and thus economically carried out.

However, it should be noted that STM-4s and STM-16s terminating on a network node are disassembled to recover their section overheads and the VCs which they contain. Outgoing STM-Ns are reconstructed with new section overheads and new multiplexed path layer VC assemblies. regenerator section overheads (RSOH) and multiplexer section overheads (MSOH) are distinguished and used to "refresh" the signal overheads.

Therefore, in SDH, multiplexing may be regarded as taking place in two stages:

- a low level multiplexing of signals (of PDH origin in a AU-4) (AU-3 in North America);

- a high level multiplexing, from AU-4 (AU-3), to form the high speed signal.

AP3.5-2 Bandwidth provisioning

Some services may need transport at a rate that does not correspond to the defined frames; thus signals of varying capacity should be built.

The concatenation of M virtual containers of level n (noted VC-n-Mc), may be used for the transport of signals when their bit rates do not fit any defined C-n capacity. Concatenation mechanisms are defined in ITU-T Recommendation G.709.

The concatenation of a number of TUs within one STM-1 signal allows to provide a higher granularity to transmission rates. Concatenation of AU-4s is also defined. In addition, concatenation permits an increase in efficiency. To illustrate this, the concatenation of five TU-2 signals can carry a 32 Mbit/s video signal; in this manner, four video signals can be carried on a VC-4, while only three can be carried if mapped into C-3 containers.

With SDH, it will be possible to dynamically allocate on demand network capacity, or bandwidth. Then services requiring large bandwidth could be provided at very short notice, which increases user convenience and operator revenues.

The SDH offers an adaptable network solution for the future. It has been designed to support future services such as high definition television (HDTV), broadband ISDN, metropolitan area networks and personal communication networks.

AP3.5-3 Network equipment

Network functions required for SDH are essentially the same as those already performed in PDH networks, i.e. line transmission, cross-connection and multiplexing.

The specification method adopted for SDH, however, benefits from the principles of:

- higher levels and greater flexibility of functional integration (that is, various functional capabilities can be packaged in a single equipment); and

- independence between functions and their physical implementation in terms of hardware and software.

Network simplification is brought by SDH equipment, resulting in low operating costs, simplified maintenance, reduction of floor space and power consumption. The efficient drop and insert of channels offered by the SDH and the powerful network management capabilities facilitate provisioning of high bandwidth lines for new multimedia services as well as ubiquitous access to these services.

To ensure the compatibility of different vendors' equipment, interfaces between equipment and the management network are being specified. Currently multivendor compatibility is achieved at the optical line interface (cf. ITU-T Recommendation G.957).

A functional reference model has been made up, decomposing the network function of an equipment into a set of basic functions and describing each of them in rigorous detail in terms of its internal processes and the primitive information flows at the reference points at the input and output points of the basic elements.

Equipments are specified by the arrangement of the basic functions whose aggregation forms the required network function; an analogy is given by the selection of integrated circuits and their packaging to form a chip which performs the aggregated function.

Moreover this specification simplifies and systematizes the treatment of fault conditions and the subsequent actions to be undertaken.

Network equipment can be broadly classified into equipments intended for VC-4 signal transport and routing in the network and equipments for processing the content of the VC-4 (i.e. in order to insert/drop PDH signals into/from a VC-4 or to alter the content of a VC-4):

•   SDH line systems;

The simultaneous availability of synchronous and plesiochronous interfaces allows an immediate accommodation of the existing PDH networks and a smooth migration towards a fully synchronous network. Functionally speaking, the line systems permit the deployment of:

•   point-to-point links comparable to those implemented in the plesiochronous network;

•   line drop-and-insert links, using the tributary signal access facilities offered by the synchronous frame;

•   loop links for protecting transmissions from equipment and cable failures.

Interfaces have been standardized by the ITU-T.

•   SDH multiplexers (Recommendations G.781, G.782 and G.783).

A number of different SDH multiplexers have been defined:

•   Type I ("access multiplexer") is a simple terminal multiplexer which multiplexes a number of G.703 tributaries into an STM-N signal.

•   Type II takes a number of STM-N signals and multiplexes the constituent higher-order VCs (VC-3 or VC-4) into an STM-M aggregate signal.

•   Type III constitutes an add-drop multiplexer with STM-N aggregate ports and G.703 interfaces for the add/drop tributaries.

•   SDH cross-connects.

These equipments perform the cross-connection of virtual containers. Because of the frame synchronous feature, this operation, similar in principle to the distribution function of plesiochronous traffic, no longer requires the former mandatory multiplexing and demultiplexing stages.

AP3.5-4 Network management, availability and protection

An important part of SDH functionality is invested for management purposes:

- inclusion of parity checks in each layer of the network for performance monitoring and end-to-end supervision;

- existence of high bit-rate data communications channels in the section overheads for the transmission of management messages between network nodes; international and regional standards bodies have addressed protocol stacks, information models and object definitions that could be used across them.

Provision of network management channels within the SDH frame structure means that a synchronous network will be fully software controllable.

SDH network management follows the TMN standardization principles, in particular on object modelling in which management information is structured so as to allow:

- use of adapted OSI protocols in that application layer (CMISE and ROSE);

- handling of distributed management applications;

- eventual simplification of the difficult issues of interworking.

The telecommunication management network (TMN) transport network capabilities (Q interfaces) are offered through the use of control channels (ECCs) embedded in the SDH frame. At the level of the physical layer, the ECCs use the data communication channels formed by bytes D1 to D3 [resp. D4 to D12] accessible [resp. not accessible] in the repeater station (RSOH).

Network management systems perform the traditional functions as well as performance monitoring, and configuration and resource management.

With automatic management capabilities, the failure of links or nodes can be identified immediately. Using self-healing ring architectures, the network will be automatically reconfigured and traffic instantly rerouted until the faulty equipment has been repaired. Thus, failures in the network transport mechanism will be invisible in the end-to-end basis. Such failures will not disrupt services, allowing an extremely high availability of service and guaranteeing high levels of network performance.

Security and availability are among the most vital functional capabilities that need to be introduced in SDH networks. The network topology must be implemented in such a manner as to withstand link and node failures.

The upper level of the network can be meshed to provide reciprocal back up via rerouting, using digital cross-connect. In the lower part of the network, the adopted topology consists of physical loops generally using the SDH ring architecture.

BIBLIOGRAPHY

[1]   ITU Telecommunication Standardization Sector – 1993 – Introduction of new technologies in local networks.

[2]   Bruce, R. Elbert – 1997 – The Satellite Communication Applications Handbook, Ed. Artech House.

[3]   W.B. Kleijn and K.K. Paliwal – 1995 – Speech coding and synthesis, Ed. Elsevier.

[4]   Walter H.W. Tuttlebee – 1997 – Cordless Telecommunications Worldwide, Springer.

[5]   Joan L. Mitchell, William B. Pennebaker, Chad E. Fogg, Didier J. LeGall – 1997 – MPEG video compression standard, Chapman & Hall, International Thomson Publishing (ITP).

[6]   Wireless ATM – August 1996 – IEEE Personal Communications, Vol. 3, No. 4.

[7]   B. Cornaglia and M. Spini, SDH in digital systems, ICDSC-10 Proceedings, pp. 213-219.

[8]   C.E. Shannon – July 1948 – A mathematical theory of communications, Bell system Technical Journal, 27, pp. 379-423 (Part I).

[9]   S. Lin, D.J. Costello, Jr. – 1983 – Error Control Coding: Fundamentals and Applications, Prentice-Hall.

[10]   A.J. Viterbi, J.K. Omura – 1979 – Principles of Digital Communication and Coding, McGraw-Hill.

[11]   K. Feher – 1983 – Digital Communications: Satellite/Earth Station Engineering, Prentice-Hall Inc., Englewood Cliffs.

[12]   W.W. Wu – 1985 – Elements of digital satellite communications, Vol-II, Rockville, MD: Computer science.

[13]   Forney – November 1969 – Convolutional codes I: algebraic structures, IEEE Transactions on Information theory, Vol. IT-37, No. 11.

[14]   G.C. Clark, J.B. Cain, J. Geist – January 1979 – Punctured convolutional codes of rate n-1/n and simplified maximum likelihood decoding, IEEE Transactions on Information theory, Vol. IT-25.

[15]   Y. Yasuda et al. – October 1984 – High rate punctured convolutional codes for soft decision decoding, IEEE Trans. Comm., Vol. COM-32.

[16]   D. Haccoun, G. Begin – November 1989 – High-rate punctured convolutional codes for Viterbi and sequential decoding, IEEE Transactions on Communications, Vol. COM-37, No. 11.

[17]   G.C. Clark, J.B. Cain – 1981 – Error-correcting coding for digital communications, Plenum Press, New York.

[18]   E.R. Berlekamp – 1968 – Algebraic coding theory, McGraw-Hill, New York.

[19]   A. Michelson, A. Levesque – 1985 – Error-control techniques for digital communications, NY: Wiley.

[20]   R.E. Blahut – 1983 – Theory and practice of error control codes, Addison-Wesley, Reading, MA.

[21]   S.B. Wicker, V.K. Bhargava – 1994 – Reed-Solomon Codes and Their Applications, IEEE Press.

[22]   C. Berrou, A. Glavieux – October 1996 – Near optimum error correcting coding and decoding: Turbo-codes, IEEE Transaction on Communications, Vol.-44, No. 10.

[23]   S.B. Wicker – 1995 – Error control systems for digital communication and storage, Prentice-Hall.

CHAPTER 4
Carrier modulation techniques

The problem of effective spectrum use commonly arises from the lack of frequency spectrum and becomes more and more important in consequence of a trend to earth stations with antennas of small diameters and to more complicated multiple antenna beam satellites. In particular, interference problems arise from frequency reuse in multiple satellite antenna beams.

One of the techniques to solve this problem is the application of power effective and bandwidth effective modulation schemes.

In satellite communications as in the other communications systems we can see a transfer to digital transmission techniques in signal processing as well as in modulation. Analogue modulation is commonly used for television and FDM telephony transmission.

The most common methods of modulation used in the fixed-satellite service are FM for analogue signals and PSK for digital signals. Other modulation methods, such as combined amplitude and phase modulation are also discussed.

4.1 Analogue modulation

4.1.1 Frequency modulation (FM)

FM has been largely used in satellite communications. It is particularly convenient when a single carrier per transponder is used and where the constant envelope of the FM signal allows the power amplifiers to operate at saturation, thus making maximum use of the available power. FM is generally used in the two well-known FDM-FM and SCPC-FM systems.[*]

4.1.1.1 FDM-FM

The basic relationship between the carrier-to-noise and signal-to-noise ratios in a conventional FDM-FM system can be expressed by:

$$s/n = (c/n) \cdot \frac{3\,(f_{tt})^2}{f_2^3 - f_1^3} \cdot B_{RF} \cdot p \cdot w \tag{1}$$

---

[*] The SSB-AM transmission of an amplitude-modulated signal on a satellite link can be applied. This modulation technique makes it possible to get high transponder capacity by means of syllabical companding. However, high transponder e.i.r.p. with good linearity response, automatic frequency and gain adaptations devices, allowance of wider margins in the link budget and low interference conditions are required.

or:

$$s/n = (c/n) \cdot \frac{(f_{tt})^2}{(f_m)^2 + \dfrac{b^2}{12}} \cdot (B_{RF}/b) \cdot p \cdot w \tag{2}$$

When the value of f_m is more than about four times the value of b, the above relationship may be replaced with negligible error by the following approximate relationship:

$$s/n = (c/n) \cdot (f_{tt}/f_m)^2 \cdot (B_{RF}/b) \cdot p \cdot w \tag{3}$$

or in dB:

$$S/N = C/N + 20 \log (f_{tt}/f_m) + 10 \log (B_{RF}/b) + P + W \tag{3}$$

where:

S/N: the ratio of test-tone (i.e. 1 mW at the point of zero relative level) to the psophometrically weighted noise power in the highest telephone channel

C/N: the carrier-to-noise ratio in the radio-frequency bandwidth

N: $kT\, B_{RF}$

k: Boltzmann's constant ($k = 1.38 \cdot 10^{-23}$ J/K)

T: system noise temperature (K)

B_{RF}: the radio-frequency bandwidth (Hz) as defined in equation (4)

f_2: the upper frequency bound of the passband of the highest baseband channel (Hz)

f_1: the lower frequency bound of the passband of the highest baseband channel (Hz)

f_m: the mid-frequency (arithmetic mean) of the highest baseband channel (Hz), $f_m = (f_2 + f_1)/2$

b: the bandwidth of the telephone channel (Hz), $b = f_2 - f_1$

f_{tt}: the test-tone r.m.s. deviation per channel (Hz)

p: the pre-emphasis improvement factor

w: the psophometric weighting factor.

Typically the value of b is 3 100 Hz and p can be assumed to have the numerical values of 2.5 (4 dB) for the top baseband channel.

In the expression given above, only B_{RF} and f_{tt} are unknown. To solve for f_{tt} and hence B_{RF}, it is necessary to find an additional relationship between these two variables. It is assumed that the bandwidth B_{RF} is given by:

$$B_{RF} = 2\,(\Delta f + f_m) \tag{4}$$

i.e. the "Carson's rule bandwidth", where Δf is the multichannel peak frequency deviation.

To restrict the distortion noise due to over-deviation outside the available bandwidth ("truncation noise") to a tolerable level, it is necessary to define a suitable relationship between Δf and f_{tt}. A formula in common usage for frequency-division multiplex basebands is as follows:

$$\Delta f = f_{tt} \cdot g \cdot \ell \tag{5}$$

where:

g: "peak-to-r.m.s." factor as a numerical ratio. Thus, for a peak-to-r.m.s. value of:

$$13 \, dB, g = 10^{13/20} - 4.47$$
$$10 \, dB, g = 10^{10/20} - 3.16$$

ℓ: loading factor of the telephone multiplex (for n channels)

$$\ell = 10^{(-15 + 10 \log n)/20} \qquad \text{for } n \geq 240 \text{ channels} \tag{6}$$

$$\ell = 10^{(-1 + 4 \log n)/20} \qquad \text{for } n < 240 \text{ channels} \tag{7}$$

Earlier system designs used a peak-to-r.m.s. factor of 13 dB but later systems have used 10 dB with satisfactory results. In general, it can be expected that for carrier capacities below about 120 channels the 13 dB figure is more appropriate, while for those of higher capacity 10 dB is preferred.

Frequency modulation systems have a characteristic threshold of operation and for conventional demodulators the FM threshold occurs at a carrier-to-noise ratio of 10 dB. This threshold can be reduced by using such techniques as frequency feedback demodulators so that the threshold occurs about 3 dB lower than that for conventional demodulators.

The use of these so-called threshold extension demodulators in frequency modulation systems does not modify the basic relationship between signal-to-noise (S/N) and carrier-to-noise (C/N); it merely permits the use of lower C/N ratios than would otherwise be possible.

FDM-FM has been the most widely used multiple-access technique in the INTELSAT system. A table of the carriers used in the INTELSAT system is given in Annex II (Table AN2-1).

4.1.1.2   SCPC-FM

For one SCPC-FM demodulator the relationship between the carrier-to-noise and signal-to-noise ratios is also given by equation (1) where, now:

S/N: the ratio of test-tone to the psophometrically weighted noise power in the SCPC baseband channel

f_2: the upper frequency bound of the SCPC baseband channel (Hz)

f_1: the lower frequency bound of the SCPC baseband channel (Hz)

f_{tt}: the test-tone r.m.s. deviation (Hz)

The other parameters remain as defined before.

It should be noted that B_{RF} is now given by:

$$B_{RF} = 2 (\Delta f + f_2) \tag{8}$$

where Δf is the peak deviation in Hz. The parameter Δf is still given by $\Delta f = f_{tt} \cdot g \cdot \ell$, but now:

$$\ell = 10^{(m + 0.115\beta^2)/20} \tag{9}$$

where m and β are the mean (dBm0) and standard (dB) deviation of the statistical distribution of speech level in dBm0. Taking the usual values m = –16 dBm0 and β = 5.8 dB, this gives:

$$\ell = 0.248$$

The "peak-to-r.m.s." factor g depends on how clipping is defined and what is the accepted percentage of users suffering from clipping. Clipping is here defined as occurring when the r.m.s. frequency deviation produced by an active speech signal exceeds a level which is 15 dB below the peak frequency deviation. For 10% of the users suffering clipping the value of g is 8.4, while for 3% of the users suffering clipping, g is equal to 12.6.

4.1.1.3   System design (telephony)

The channel noise in the multiplexed telephone circuit is one of the most basic parameters in the design of satellite links. ITU-R recommends that the one-minute mean noise power in the hypothetical reference circuit for a telephone in the fixed-satellite service is required to be equal to or less than 10 000 pW0p for 80% of the time in the normal state of the circuit (Recommendation 353). Therefore, in the design of a satellite link, transmission parameters are usually determined in order to satisfy this criterion. The noise generated within the satellite link can be classified as follows:

- link noise (uplink thermal noise, downlink thermal noise, intermodulation noise, intra-system interference noise);

- earth station equipment noise (excluding thermal noise generated at the receiver)[*];

- interference noise from other systems (interference noise from terrestrial radio-relay systems and from other satellite systems).

In the INTELSAT satellite system, the total channel noise power of 10 000 pW0p is budgeted so as to allow 8 000 pW0p for link noise (including an allowance of 500 pW0p for unwanted emissions caused by multicarrier intermodulation from other earth stations in the system), 1 000 pW0p for earth station equipment noise and 1 000 pW0p for terrestrial interference noise. These allocations are typical of most existing systems. However, as demands on the satellite service increase with a consequent pressure to reduce satellite spacings, the current trend is to allocate larger allowances for interference which can be compensated for by the use of higher satellite powers and other new technologies.

---

[*]   This refers to noise due to amplitude and group delay distortion, etc.

4.1.1.4   TV-FM

Using equation (1) with $f_2 \gg f_1$ the S/N of TV-FM is given in dB by the following equations:

$$(S/N)_w = C/N_0 + 10 \log \left(\frac{3(r_1 \cdot \overline{\Delta F}_{p-p})^2}{f_m^3} \right) + P + W \qquad (10)$$

or:

$$(S/N)_w = C/N_0 + 20 \log \left(\frac{r_1 \cdot \overline{\Delta F}_{p-p}}{f_m} \right) - 10 \log \frac{f_m}{3} + P + W \qquad (11)$$

where:

$(S/N)_w$: ratio of the nominal amplitude (peak-to-peak) of the luminance signal to the r.m.s. amplitude of the noise measured after band limiting, emphasis and weighting with a specified network (dB)

C/N_0: carrier-to-noise density ratio (dB(Hz)) with

$N_0 = N/B_{RF}$

N: noise power in the radio-frequency bandwidth in (W)

r_1: ratio of the nominal amplitude (peak-to-peak) of the luminance signal to the peak-to-peak amplitude of a monochrome composite video signal (0.714 for 525/60, 0.7 for 625/50)

$\overline{\Delta F}_{p-p}$: peak-to-peak frequency deviation due to 1 V_{p-p} tone at pre-emphasis top cross-over frequency

f_m: top baseband frequency (Hz)

B_{RF}: radio-frequency bandwidth (Hz)

P: emphasis improvement (dB)

W: weighting factor (dB)

For measuring $(S/N)_w$, the bandwidth has been unified for all TV systems to 5 MHz. Therefore, equation (10) can be simplified as follows:

$$(S/N)_w = -196.2 + C/N_0 + 20 \log (r_1 \cdot \overline{\Delta F}_{p-p}) + P + W \qquad (12)$$

The weighting factor, which accounts for the average viewer's sensitivity to the various noise spectrum frequencies, has also been unified by ITU-R. Figure 4.1-1 shows the characteristics of this unified weighting network and Figure 4.1-2 the characteristics of the pre-emphasis networks.

Table 4.1-1 below gives P and W for the unified weighting curve and measurement bandwidth of 5.0 MHz. Information concerning other weighting curves and measurement bandwidth can be obtained from Table II, Report 637, CMTT, Vol. XII.

The accompanying audio programmes can be transmitted on an FM subcarrier (see Chapter 7, § 7.5.5.2).

TABLE 4.1-1

Improvement factors using unified weighting curves

No. of lines	Name of system	Emphasis improvement P (dB) [1]	Weighting factor W (dB) [2]	Emphasis improvement and weighting factor P + W (dB)
525	M	3.1	11.7	14.8
625	B, C, G, H, I, D, K, L	2.0	11.2	13.2

[1]   For triangular noise; the amount of improvement refers to pre-emphasis cross-over frequency.

[2]   For de-emphasized triangular noise.

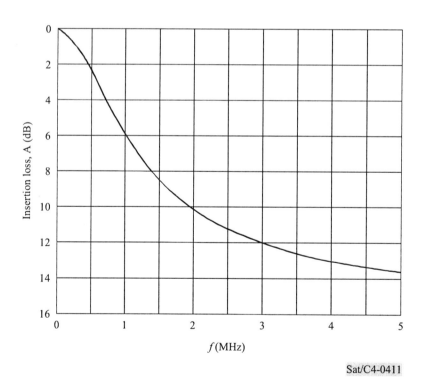

Sat/C4-0411

FIGURE 4.1-1

Unified weighting characteristic

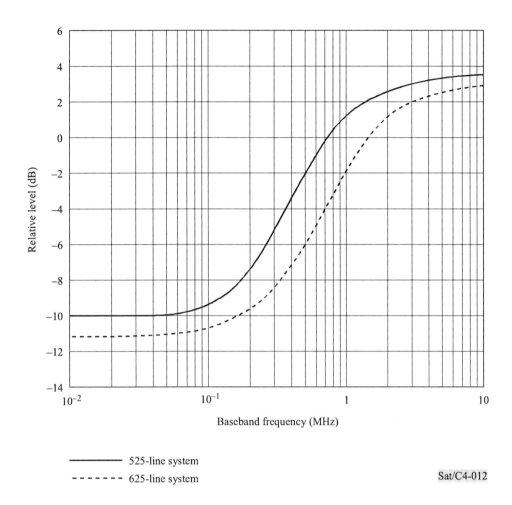

FIGURE 4.1-2

Pre-emphasis characteristics for 525/60 and 625/50 systems
(Recommendation ITU-R F.405)

4.1.2 Modulation

Satellite communications use larger frequency deviation than terrestrial microwave systems. Moreover, the modulator is usually designed to allow the adjustment of any modulation parameter which needs to be provided for in the communications system planning. In consequence, FM modulators (and also demodulators) must exhibit good linearity and group delay characteristics over a wide bandwidth (usually about 36 MHz).

Most FM modulators are Hartley (LC) oscillators using a variable capacitance (C) voltage controlled diode (varactor) with hyper-abrupt junction. Such a diode features an inverse square characteristic (capacitance versus voltage: $C = kV^{-2}$) which in turn gives a linear characteristic (frequency versus voltage) to the modulator.

4.1.3   Demodulation

The basic purpose of the demodulator is to recover the amplitude modulation of the baseband signal which is proportional to the instantaneous frequency deviation of the received carrier.

Conventional FM demodulators consist of a series-diode limiter followed by a triple-tuned-circuit discriminator.

However, when the carrier-to-noise ratio of the received carrier is too low, the use of a threshold extension demodulator (TED) becomes necessary.

For example, in the INTELSAT system, the (C/N) of the carriers with a baseband of $n \leq 252$ (n being the number of telephone channels) may be only a few dB over the threshold of a conventional demodulator. In consequence, INTELSAT recommends that, in this case, a TED be used in order to increase the operational margin and also to reduce adjacent carrier interference. The threshold phenomenon can be described as follows.

For each type of demodulator, a threshold value of the (C/N) can be defined: in the range 6 to 12 dB (10 to 12 dB for conventional demodulators).

Above threshold, the output signal-to-noise ratio (S/N) is proportional to (C/N), with a 1-to-1 correspondence (in dB) (see § 4.1.1.1): in this case, the output baseband noise is purely triangular, as given by FM theory.

Below threshold, the (S/N) departs from this 1-to-1 versus (C/N) relationship and deteriorates very rapidly as shown in Figure 4.1-3. This is due to the occurrence of additional impulsive noise. Figure 4.1-4 shows the classical representation of the addition of a noise vector to the signal vector. This noise vector is randomly variable both in amplitude and phase. If the r.m.s. noise voltage amplitude is sufficiently large (e.g. more than 1/3 of the signal voltage amplitude, i.e. C/N < 10 dB), the instantaneous noise voltage will occasionally take peak values larger than the signal voltage, as represented in Figure 4.1-4a) and $\pm 2\pi$ hops of the total (signal + noise) vector will occur, Figure 4.1-4. In which case, the output demodulated signal which is proportional to the instantaneous value of $d\varphi/dt$, takes the form of a pulse, so-called "click" with an area equal to 2π, Figure 4.1-4c).

The probability of such a noise peak is multiplied by about ten if the IF noise increases by 1 dB; for this reason, the degradation of the baseband signal-to-noise ratio is very rapid under the threshold.

As a normal FM noise spectrum is triangular and the impulsive noise spectrum is flat, the threshold effect is more important in the lower part of the baseband spectrum.

It should be noted that the occurrence of a threshold is a general phenomenon in all processes involving a so-called modulation gain (i.e. the positive difference (in dB) between S/N and C/N). Impulsive noise "clicks" can also be caused by:

- interference by adjacent carrier. In this case, the clicks occur when the combined vectors of noise and interference cause a rapid phase change of the total vector;

- band limitation by the band-pass filter. This is the case when the band-pass filter is relatively narrow. Then, impulsive noise occurs when the instantaneous frequency deviation is such that the carrier level is reduced by operation in the attenuated region of the filter.

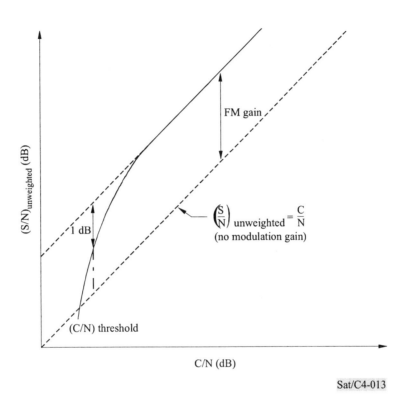

Sat/C4-013

FIGURE 4.1-3

Definition of the demodulator threshold

Figure 4.1-4*bis* gives the block diagram of the three main types of threshold extension demodulators used. For these three demodulators, threshold extension is obtained by decreasing the low-frequency components of the modulating signal, thus allowing a corresponding decrease in IF bandwidth or its equivalent, below the normal Carson bandwidth.

i)   Frequency modulation feedback (FMFB) TED: in the FMFB demodulator, the frequency deviation is decreased (at low frequencies) by the feedback loop, and the IF filter is tuned at a fixed frequency.

ii) Dynamic tracking filter (DTF) TED: in the DTF demodulator, the IF filter has only one pole and is varactor-controlled; it follows the low-frequency components of the modulation.

iii) Phase-locked loop (PLL) TED: in the PLL, the equivalent IF filter is implemented in baseband; its bandwidth is controlled by R-C components and by the loop gain.

More complex types of TEDs such as multiple (double) feedback demodulators may also be used.

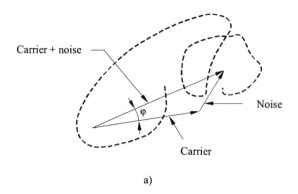

a)

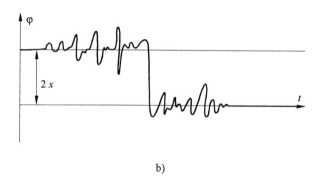

b)

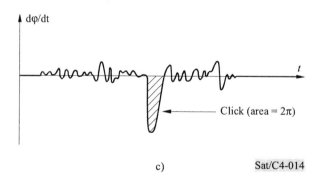

c)   Sat/C4-014

FIGURE 4.1-4

Generation of one "click"

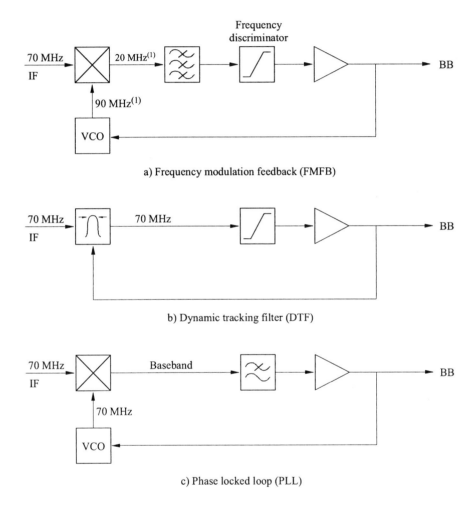

a) Frequency modulation feedback (FMFB)

b) Dynamic tracking filter (DTF)

c) Phase locked loop (PLL)

1) Typical value

Sat/C4-014bis

FIGURE 4.1-4 BIS

**The three main types of threshold extension
demodulators (TED)**

4.2 Digital modulation

4.2.1 Modulation schemes for satellite communications

In this section, digital modulation schemes for satellite communications are reviewed [Ref. 1]. First, the two most important performance measurements, power efficiency and bandwidth efficiency and also some other terms are defined. Second, an overall review of modulation schemes is given.

Individual modulation schemes are reviewed in detail in § 4.2.2 and 4.2.3.

Power efficiency and bandwidth efficiency

The power efficiency of a modulation scheme is defined straightforwardly as the required bit energy-to-noise spectral density ratio (E_b/N_0) for a certain bit error probability (P_b) of digital communication over an AWGN (additive white Gaussian noise) channel. $P_b = 10^{-5}$ is usually used as the reference bit error probability.

The determination of bandwidth efficiency is a bit more complex. The bandwidth efficiency is defined as number of bits-per-second that can be transmitted in one Hertz of system bandwidth. Obviously it depends on the definition of system bandwidth. Three different bandwidth efficiencies used in the literature are as follows.

Nyquist bandwidth efficiency – Assuming the system uses Nyquist (ideal rectangular) filtering at baseband [Ref. 2], which has the minimum bandwidth (0.5 R_s, symbol rate) required for inter-symbol interference-free transmission of digital signals, then the bandwidth $W = R_s$. Since $R_s = R_b/\log_2 M$, R_b = bit rate, for M-ary modulation, the bandwidth efficiency is:

$$R_b / W = \log_2 M$$

Even though practical filters always have wider bandwidths than those of Nyquist filters, which makes the real bandwidth efficiency less than $\log_2 M$, this definition still provides a good relative comparison between modulation schemes that employ rectangular baseband pulses, since all of them would suffer from the same degree of degradation in the baseband (hence in error performance) if a Nyquist filter were used. However, if baseband pulse shaping is used in the scheme, as in MSK, the assumption of Nyquist filtering is inappropriate and the null-to-null bandwidth is more appropriate for bandwidth efficiency comparison [Ref. 3].

Null-to-null bandwidth efficiency – For modulation schemes that exhibit power density spectral nulls, defining the bandwidth as the width of the main spectral lobe is a convenient way of bandwidth definition.

99 per cent bandwidth efficiency – If the spectrum of the modulated signal does not have nulls, as general continuous phase modulation (CPM), null-to-null bandwidth no longer exists. In this case, energy percentage bandwidth may be used [Ref. 4]. Usually 99 per cent is used, even though other percentages (e.g. 90 per cent, 95 per cent) are also used.

There are two methods for encoding the signals from the source:

- In direct encoding, the input signal is directly transmitted as information bits.

- In differential encoding, the information is transmitted in the form of the difference between two adjacent bits. Note that, in order to avoid a possible inversion of the information bits received, an ambiguity removal scheme may be needed (differential decoder).

There are two types of demodulation: coherent and differential:

- In coherent demodulation (e.g. coherent PSK, or CPSK), the received signal is demodulated by a locally generated reference signal provided by a carrier recovery scheme as explained below. Coherent demodulation is generally associated with direct encoding, but may also be associated with differential encoding (DEQPSK, see also below).

- In differential demodulation (or detection, e.g. differential QPSK or DQPSK), the received signal is multiplied by its copy delayed by one bit. Differential demodulation must be associated with a differentially encoded transmitted signal. Differential demodulation implementation is simpler than coherent demodulation. However the bit error ratio (BER) is degraded by a factor of about two for a given E_b/N_0 (since there is noise with the bits in the difference).

Carrier and clock recovery

- For carrier recovery which is needed with coherent demodulation, the reference signal is usually regenerated by a non-linear operation in the demodulator. This is because in the modulation process (e.g. PSK), the carrier frequency component is suppressed. For example, in QPSK, the received signal is quadrupled to remove the phase modulation and the result is divided by four (in BPSK, quadrupling is replaced by doubling). This entails a phase ambiguity (an integer multiple of $\pi/2$) in the recovered carrier, which must be removed by various processes. If differential encoding is employed, there is no need for removing the phase ambiguity, but this is also at the price of BER degradation.

- Clock recovery is needed with both coherent and differential demodulation, for restoring, through a synchronized local reference of the transmitted information bits (this is called also symbol timing recovery). Clock recovery also needs a non-linear operation which can be performed simultaneously with carrier recovery (in the coherent demodulation).

Digital modulation types

Baseband signals can be modulated onto a sinusoidal carrier by modulating one or more of its three basic parameters: amplitude, frequency, and phase. Accordingly, there are three basic modulation schemes in digital communications: amplitude shift keying (ASK), frequency shift keying (FSK) and phase shift keying (PSK). There are many variations and combinations of these techniques. Table 4.2-1 is a list of abbreviations and descriptive definitions of digital modulation schemes. Some of the schemes can be derived from more than one "parent" scheme.

TABLE 4.2-1

Abbreviation	Alternate abbreviation	Descriptive name
ASK FSK BFSK MFSK	 FSK	Amplitude shift keying Frequency shift keying (generic name) Binary frequency shift keying M-ary frequency shift keying
PSK BPSK DPSK QPSK DQPSK DEQPSK OQPSK $\pi/4$-QPSK $\pi/4$-DQPSK CTPSK MPSK	 2PSK 4PSK SQPSK 	Phase shift keying (generic name) Binary phase shift keying Differential BPSK Quadrature phase shift keying Differential QPSK (with differential demodulation) Differential QPSK (with coherent demodulation) Offset QPSK, staggered QPSK $\pi/4$ quadrature phase shift keying $\pi/4$ - differential QPSK $\pi/4$ - controlled transition PSK M-ary phase shift keying
CPM SHPM MHPM LREC CPFSK MSK DMSK GMSK SMSK TFM CORPSK	 FFSK 	Continuous phase modulation Single h (modulation index) phase modulation Multi-h phase modulation Rectangular pulse of length L Continuous phase frequency shift keying Minimum shift keying, fast frequency shift keying Differential MSK Gaussian MSK Serial MSK Tamed frequency shift keying Correlative PSK
QAM SQAM		Quadrature amplitude modulation Superposed QAM
Q^2PSK QPSK		Quadrature phase shift keying Differential Q^2PSK
IJF OQPSK SQORC		Inter-symbol-interference jitter free OQPSK Staggered quadrature-overlapped raised-cosine modulation

We can classify digital modulation schemes into two large categories: constant envelope and non-constant envelope. The constant envelope class is generally considered as the most suitable for satellite communications, because it minimizes the effects of non-linear amplification in the high power amplifiers. However, the generic FSK schemes in this class are inappropriate for satellite application since they have very low bandwidth efficiency in comparison with PSK schemes.

The PSK schemes have constant envelope but discontinuous phase transitions from symbol to symbol. Classic PSK techniques, i.e. BPSK and QPSK, are considered in § 4.2.2. More generally, modulation schemes with M-ary signals can be used. MPSK (M-ary PSK) have better power Nyquist efficiency than MFSK. And MFSK (M-ary FSK) have better power Nyquist efficiency but poorer bandwidth efficiency than classical modulation schemes.

The CPM schemes have not only constant envelope, but also continuous phase transitions from symbol to symbol. Thus they have less side-lobe energy in their spectra in comparison with the PSK schemes. Some CPM modulation types, i.e. OQPSK, MSK and $\pi/4$-QPSK are considered in § 4.2.3.1.

In CPM class, the MHPM is worth special attention since it has better error performance than single-h CPM by cyclically varying the modulation index h for successive symbol intervals. Binary multi-h CPFSK is a particular case of the MHPM applicable to a rectangular pulse of length L (LREC). It has good power and bandwidth efficiency.

By choosing different frequency pulses and varying the modulation index and size of symbol alphabet a great variety of CPM schemes can be obtained. Some of popular pulse shapes are raised cosine (LRC) (for example, used for TFM) and more smooth Gaussian pulse, used, for example, for Gaussian MSK (GMSK).

A special case of rectangular pulse of length L = 1 is 1REC which is usually referred to as CPFSK (see above). Further, if M = 2 and h = 1/2, 1REC becomes MSK (see § 4.2.1.2). Correlative phase shift keying (CORPSK) features correlation encoding of data bits before modulation which allows diminished spectral lobes.

As to the average power spectrum for some binary schemes, MSK has linear phase changes with sharp corners when data changes, GMSK and TFM have smooth phase changes, thus lower spectral side lobes. And they have better error probability when detected with maximum likelihood sequence detection which can be implemented by the Viterbi algorithms.

As mentioned above, demodulation may be coherent or non-coherent for modulation techniques shown.

Coherent demodulation performs well in the presence of Gaussian noise but is not very tolerant of other link disturbances, such as multipath fading, shadowing, Doppler shifts or phase noise. These effects have become more important in recent years due to the rapid growth of digital mobile communications (including satellite mobile communications), where these disturbances are more severe than Gaussian noise. The solution is to use differential demodulation, which avoids carrier recovery and achieves fast synchronization and resynchronization. Therefore it is more robust against these disturbances in mobile communications systems albeit with some loss in performance relative to coherent detection in Gaussian noise.

Among the modulation schemes described, MSK and $\pi/4$-QPSK can be non-coherently demodulated, and differential decoding can be added to BPSK, QPSK, MSK, Q^2PSK, and $\pi/4$-QPSK, to form DPSK, DQPSK, DMSK, DQ^2PSK, and $\pi/4$-DQPSK, respectively.

All of the above modulation techniques are constant envelope schemes.

As to non-constant envelope-modulation schemes (or amplitude pulse shaping modulation) the new trend is to exploit the pulse shape of transmitted symbols.

The generic non-constant envelope schemes, such as ASK and QAM are generally not suitable for satellite application. However some carefully designed non-constant envelope schemes, such as Q^2PSK, SQAM, etc. are applicable to satellite communications due to their excellent bandwidth efficiency. Even QAM, with a large signal constellation, could be considered for satellite applications where extremely high bandwidth efficiency is required. In this case, however, back-off in TWTA's input and output power must be provided to ensure the linearity of the power amplifier. Further information about new digital modulation schemes can be found in [Ref. 1].

Non-constant modulation-envelope schemes can be attractive if they have a compact spectrum, low spectral spreading caused by non-linear amplification, a good error performance, and simple hardware implementation.

There are some techniques which are suitable for non-linearly amplified satellite channels in a densely packed adjacent channel interference (ACI) environment.

This family of modulation schemes includes inter-symbol-interference jitter free (IJF) OQPSK, two-symbol-interval (TSI) hard limited OQPSK, superposed-QAM (SQAM), and cross-correlated QPSK (XPSK). The first three have one common feature, namely, the pulse shaping is always performed independently in the I and Q baseband channels and therefore there is no correlation between these baseband input signals. The XPSK has a controlled correlation between the I and Q baseband input signals, which reduces the envelope fluctuation of IJF-QOPSK and further reduces the spectral side lobes. The SQORC (the staggered quadrature-overlapped raised-cosine modulation) is actually a special case of the IJF-OQPSK according to [Ref. 1].

The modulators for this family of schemes are basically the same as that of QPSK, except that a special baseband encoder/processor is inserted in the I and Q channel prior to the carrier multiplier [Ref. 1].

It also should be pointed out that the BER performance of the new techniques are 2 to 3 dB inferior to QPSK or MSK in an ACI-free environment.

The most commonly used, classical, techniques, i.e. BPSK and QPSK, are reviewed in the next subsection (4.2.2). Then, some alternative techniques are discussed in § 4.2.3.

4.2.2   BPSK and QPSK

PSK modulation techniques include: the basic 2-phase modulation (2PSK) which represents a binary code by the dual phases 0, π; the 4-phase modulation (QPSK) which represents 2 binary codes by the quadratic phases 0, $\pi/2$, π, $3\pi/2$. In general, multi-phase modulation (MPSK) represents n binary codes by 2^n phases. However, since the allowances needed for noise or interference become larger with each increase in the degree of multi-phase, the higher order (greater than 4-phase) PSK systems require much more power than either 2- or 4-phase systems to achieve the same performance. Therefore, 2-phase and 4-phase modulation methods are widely used with present techniques. For most purposes, QPSK gives the best power bandwidth compromise. By suitable choice of modem filters, PSK signals can be transmitted with very small deterioration even in a non-linear satellite channel.

Transmission quality for the PSK system is evaluated by the bit error ratio. Bit errors are caused in the PSK system by thermal noise, inter-symbol interference, and phase jitter of recovered carrier and bit timing, etc. The BER caused by thermal noise is described first as it is the major cause of errors.

When Gaussian noise is added to 2-PSK signal, the bit error ratio P_e of the signal is given by:

$$P_e = \frac{1}{\sqrt{2\pi N}} \int_A^\infty \exp\left(-\frac{x^2}{2N}\right) dx = \frac{1}{2} \text{erfc} \sqrt{\frac{A^2}{2N}} \tag{13}$$

where A is the amplitude of the envelope of the PSK signal at the output of the receive filter at the decision timing, and:

$$\text{erfc}(x) = \frac{2}{\sqrt{\pi}} \int_x^\infty \exp(-t^2) dt \tag{14}$$

When a matched filter is used as the receive filter, $A^2/2N$ becomes maximum and equals E_b/N_0, where E_b is the energy per information bit of input PSK signal and N_0 is the noise power per Hertz at the input to the receive filter.

In the case of coherent quadriphase PSK (QPSK), the demodulation process is equivalent to the coherent detection of a 2-PSK signal with a level at 3 dB lower than that of the input QPSK signal, if the input PSK signal is coherently detected by a pair of reference carriers which are orthogonal with each other and are phase-offset by 45° relative to the input signal phase.

Then, the bit error ratio P_e of QPSK is given by:

$$P_e = \frac{1}{2} \text{erfc} \sqrt{\frac{A^2}{4N}} \tag{15}$$

If a matched filter is used as the receive filter, $A^2/4N$ equals $E/2N_0$ where E is the energy per symbol of a 2-PSK signal at the input to the receive filter, and $E = 2E_b$ because one symbol of a 4-PSK signal consists of a pair of bits. Consequently, the bit error ratio of quadriphase coherent PSK versus E_b/N_0 is identical to that of biphase coherent PSK. The bit error ratio versus E_b/N_0 for coherent detection of 2-PSK or QPSK is shown in Figure 4.2-1.

Note that comparing the required C/N of QPSK with that of 2-PSK for the same error ratio in coherent detection, the former requires a 3 dB higher C/N than the latter.

Note also that the bit error ratio of the differential phase modulation system is twice that of the non-differential phase modulation system.

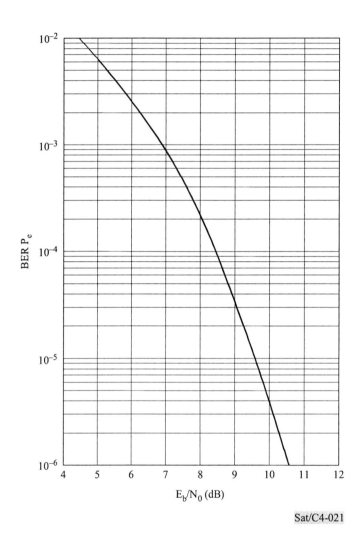

Sat/C4-021

FIGURE 4.2-1

Theoretical BER performance of coherent PSK

In computing the required C/N margins have to be added to the theoretical values to account for degradations other than thermal noise. As long as these other degradations are small compared with the thermal noise they can often be regarded as having Gaussian distributions, and can be added to the thermal noise on a power basis. However, the additional degradations will often not be small, in which case a more detailed analysis including computer simulations and measurements will be necessary.

Another important cause of errors is inter-symbol interference. This is caused by the characteristics of the IF filter of the PSK modem, the frequency response of the satellite's transponder, and especially by band limitations and TWT non-linearities. When the bandwidth of the filter is widened, inter-symbol interference decreases, but thermal noise increases. Consequently, a bandwidth of 1.05 to 1.2 times the symbol rate is usually used in which case the C/N degradation by inter-symbol interference is estimated to be about 1.5 to 2.0 dB.

Limiting the bandwidth of the PSK carrier results in the loss of the higher spectrum components and produces amplitude-modulation components. The amplitude-modulation components increase phase distortion of the PSK carrier by AM-PM conversion occurring in the satellite's transponder.

The signal normalized power spectral density of BPSK and QPSK signals is expressed by:

$$p(f) = \frac{7}{\beta R} \left(\frac{\beta \ln(\pi f / \beta R)}{\pi f / \beta R} \right)^2 \tag{16}$$

where R is the bit rate, f is the frequency and β has the value of 1 for BPSK and value of 1/2 for QPSK.

Assuming the in-phase and quadrature bit streams are independent, the bandwidth of power spectral density of QPSK is exactly half of that for BPSK for the same bit rate.

Therefore, in order to transmit data at R bit/s, the required bandwidths are R Hz for BPSK and R/2 Hz for QPSK in the ideal situations. However, taking into account the actual filter characteristics, the transmission bandwidths in the satellite communication are 1.2 times above the ideal values.

4.2.3 Other digital modulation techniques

4.2.3.1 Quadrature schemes

The most important modulations in satellite communications are quadrature schemes where the modulated signal is expressed in two channels, I and Q. Three schemes will be more particularly considered in this section, see [Ref. 5]: OQPSK, MSK and $\pi/4$-QPSK. A generic quadrature modulator is shown in Figure 4.2-2. This scheme can cover the QPSK, OQPSK, and MSK modulations.

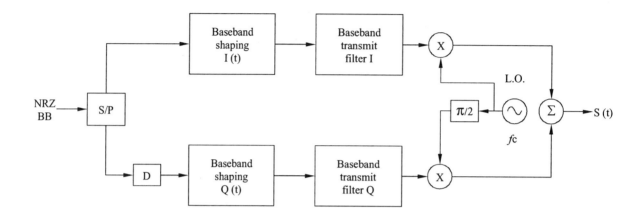

FIGURE 4.2-2

Generic quadrature modulator

The quadrature modulated signal can be expressed as:

$$S(t) = \sqrt{\frac{2E_s}{T_s}} \left[a(t)I(t)\cos\left(2\pi(f_c t)\right) - b(t-D)Q(t)\sin\left(2\pi(f_c t)\right) \right] \tag{17}$$

where E_s, T_s is the symbol energy and duration, $T_s = 2\,T_b$, T_b is the bit duration, fc is the carrier frequency, $a(t)$, $b(t) \in \{\pm 1\}$, are the information-carrying NRZ waveforms in two channels I and Q. The delay D is the time offset introduced at the channel Q. D = 0 for QPSK. For BPSK $I(t)$ or $Q(t)$ = equals zero. For OQPSK and MSK, $D = T_b = T_s/2$, and T_s is the symbol duration. The baseband shaping filters $I(t)$ and $Q(t)$ are rectangular functions for QPSK and OQPSK. For MSK, $I(t) = \cos\left(\pi t/2T_b\right)$, and $Q(t) = \sin\left(\pi t/2T_b\right)$.

The corresponding demodulator structure is shown in Figure 4.2-3.

The time offset is put on the I-channel, if it has been applied to the Q-channel at the transmission side.

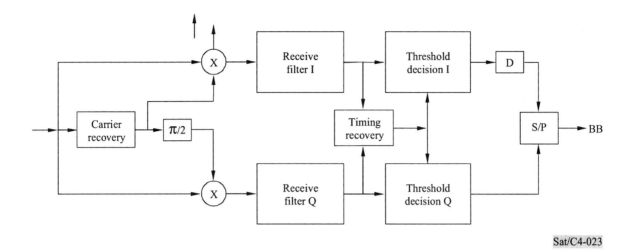

Sat/C4-023

FIGURE 4.2-3

Generic quadrature demodulator

Offset QPSK

Offset QPSK, abbreviated as OQPSK, or OKQPSK (Offset Keying QPSK), also known as SQPSK (staggered QPSK), is a modified form of QPSK (quadrature phase shift keying). The OQPSK baseband waveform is obtained in the same manner as the QPSK, except that a $T_s/2$ time delay is applied to the Q-channel, assuming that a(t) and b(t) are both information carrying baseband NRZ pulse streams:

$$S_{OQPSK}(t) = \sqrt{\frac{2E_S}{T_S}}[a(t)\cos(2\pi(f_c t)) - b(t - \frac{T_s}{2})\sin(2\pi(f_c t))] \qquad (18)$$

where T_s is the symbol duration. The OQPSK system has the same error rate performance and so the same bandwidth efficiency. The OQPSK's phase has no π transitions. A phase transition of π occurs if, and only if, the I- and Q-signals change their value at the same time instant, if a Gray mapping is assumed. (In Gray mapping, each point of the constellation in the complex I, Q plane is associated with the bits of the I, Q channels with only one bit different from its adjacent points.) In the OQPSK scheme, the Q-channel baseband signal is delayed by $T_s/2$, the value changing of the I- and Q-channels cannot occur at the same instant. This property leads to a reduction of its non-linear distortion when passed through a non-linear element. Compared to the QPSK signal, a better performance can be expected with OQPSK over a non-linear satellite channel.

MSK

MSK: minimum shift keying, is a special form of coherent frequency shift keying modulation with the modulation index 1/2. It is also known as FFSK (fast FSK). It can also be seen as a continuous phase modulation (CPM): 1REC (h = 1/2). Numerous references exist on the MSK system and its applications, due to its special role in digital modulation theory. The MSK signal is attractive for satellite transmissions because of its special properties. In this description, however, MSK will be introduced as a member of the quadrature modulation family. In fact, in the OQPSK scheme, if baseband pulse shapings with the weightings "cos" and "sin" are applied to the I- and Q-channels, the MSK signal is produced:

$$
\begin{aligned}
S_{MSK}(t) &= \sqrt{\frac{2E_s}{T_s}} \left[a(t)\cos\left(\frac{\pi t}{T_s}\right) \cos\left(2\pi(f_c t)\right) - b\left(t - \frac{T_s}{2}\right) \sin\left(\frac{\pi t}{T_s}\right) \sin\left(2\pi(f_c t)\right) \right] \\
&= \sqrt{\frac{E_s}{T_b}} \left[a(t)\cos\left(\frac{\pi t}{2T_b}\right) \cos\left(2\pi(f_c t)\right) - b\left(t - T_b\right) \sin\left(\frac{\pi t}{2T_b}\right) \sin\left(2\pi(f_c t)\right) \right]
\end{aligned}
\tag{19}
$$

where T_s is the symbol duration, and $T_s = 2T_b$, T_b is the bit duration. The MSK signal has a continuous phase that changes linearly with slopes $\pm\pi/2T_b$ due to the baseband pulse shaping. There is no abrupt phase transition. Consequently, the MSK signal has less non-linear distortion over a non-linear satellite channel. Since $a(t)$, $b(t) \in \{\pm 1\}$, and $\cos^2 x + \sin^2 x = 1$, the MSK signal has a constant envelope.

The MSK system can be implemented in serial, instead of the parallel structure in Figures 4.2-1 and 4.2-2. The serial MSK (SMSK) can be more advantageous, for example, no I-Q balancing is needed [Ref. 6], [Ref. 7], [Ref. 8], [Ref. 5].

GMSK

An MSK signal has all of the desirable properties, except for a compact power density spectrum. This can be alleviated by filtering the modulating signal with a low-pass filter prior to modulation. The impulse response of a low-pass filter has Gaussian shape:

$$
h(t) = \sqrt{(2\pi / \ln 2)}\, B \exp\left\{ (2\pi^2 B^2 / \ln 2)\, t^2 \right\}
$$

where B is the 3 dB bandwidth of the filter. Under decreasing B the shape of the frequency pulse becomes smoother and the spectrum occupancy of the modulated signal decreases.

π/4-QPSK

The π/4-QPSK is a modulation scheme particularly well adapted to mobile communication channels, because of its low complexity implementation using differential detection. The carrier recovery is avoided and a fast signal acquisition can be achieved. It has no π phase transition, so that its reduced envelope fluctuation results in a smaller non-linear distortion. These properties make it attractive not only for the mobile radio links, also for the TDMA burst transmissions [Ref. 9].

The baseband symbols are expressed as: $S_k = \exp\{j\theta_k\}$, $k = 0, 1, \ldots$, where the channel symbol phase $\theta_k = \psi_k + \theta_{k-1} \pmod{2\pi}$, $\theta_0 = 0$, ψ_k is the phase transition. Therefore, $\theta_k \in \{0, \pi/4, \pi/2, 3\pi/4, \pi, -3\pi/4, -\pi/2, -\pi/4\}$, and S_k take the 8 signal points on the constellation of Figure 4.2-4. The data to phase transition mapping rule is shown in the following table:

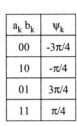

$a_k b_k$	ψ_k
00	$-3\pi/4$
10	$-\pi/4$
01	$3\pi/4$
11	$\pi/4$

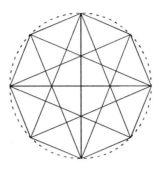

Sat/C4-024

FIGURE 4.2-4

$\pi/4$-QPSK constellation and phase transition trajectories

The $\pi/4$-QPSK has the same power spectrum density as the QPSK.

However, experimental measurements in research laboratories indicate that in fully saturated, non-linearly amplified systems, $\pi/4$-QPSK still has a significant spectral restoration. Further reduction of the spectral restoration has been achieved by sinusoidally shaping and offsetting the pulse transitions in the I and Q channels. This is called $\pi/4$ - controlled transition PSK (CTPSK).

Either coherent or non-coherent demodulation can be used. When non-coherently demodulated with the advantage of avoiding the carrier recovery, 3 dB is lost with respect to the coherent demodulation.

Relation with CPM signals

The CPM schemes have been extensively studied [Ref. 6]. MSK is a CPM scheme. When memory is introduced into the CPM signal phase, more compact spectrum shapes are obtained. The optimum coherent demodulation can be performed by a maximum likelihood sequence estimation (MLSE) receiver using the phase trellis, if the modulation index is rational. Theoretically, better power and bandwidth efficiencies can be achieved with CPM.

It is evident that a CPM signal can be represented as a quadrature modulation signal. It can be shown that a quadrature modulation signal s(t) is a continuous phase signal, if I(t) and Q(t) are continuous time functions, and continuous at the time interval boundaries $t = nT_b$, for n = 0, 1, ...

Power spectrum density

PSDs of some members in the quadrature modulation family

i)   QPSK, OQPSK and π/4-QPSK: $S_{QPSK}(f) = 2T_b\left(\dfrac{\sin 2\pi fT_b}{2\pi fT_b}\right)^2$ (20)

ii)   MSK: $S_{MSK}(f) = T_b\left(\dfrac{16}{\pi^2}\right)\left(\dfrac{\cos 2\pi fT_b}{1-16f^2T_b^2}\right)^2$ (21)

T_b is the bit duration, and f can be seen as $f - f_c$, f_c the carrier frequency.

Figure 4.2-5 presents the PSD curves of the QPSK, OQPSK, π/4-QPSK and MSK.

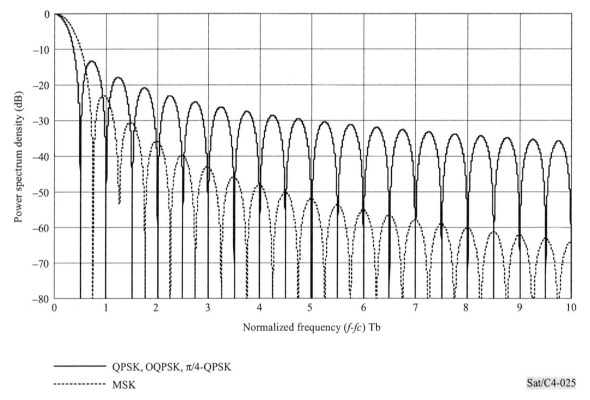

——————— QPSK, OQPSK, π/4-QPSK

------------ MSK   Sat/C4-025

FIGURE 4.2-5

Power spectrum densities of QPSK, OQPSK,
π/4-QPSK and MSK

The PSD of QPSK, OQPSK and π/4-QPSK, have the first null at $(f - f_c)\,T_b = 0.5$, while MSK's first null is at $(f - f_c)\,T_b = 0.75$. The bandwidth efficiency of MSK can be considered as 1.333 bit/s/Hz, if special baseband pulse shaping technique is applied. It has been shown that this bandwidth can be efficiently improved to 2 bit/s/Hz, see [Ref. 10] for more details. The envelope of the MSK's PSD decreases as f^{-4}, compared to f^{-2} for those of QPSK, OQPSK and $\pi/4$-QPSK. The MSK has also a much lower side lobe.

Bit error probability

Bit error probability expression over an AWGN channel [Ref. 10], [Ref. 11]

The bit error rate expressions of some quadrature modulations are collected here for reference. These results are analytical, over the ideal AWGN channel. For practical channels with band limiting filtering, non-linear elements and channel coding, the performance can be evaluated by computer simulations.

i) QPSK, OQPSK and MSK: $P_b = Q\left(\sqrt{\dfrac{2E_b}{N_0}}\right)$, and $Q(x) = \dfrac{1}{\sqrt{2\pi}}\displaystyle\int_x^\infty e^{-\frac{y^2}{2}}\, dt$

where E_b is the bit energy, N_0 the noise power spectral density. Another frequently used expression is:

$$P_b = \frac{1}{2}\mathrm{erfc}(\sqrt{\frac{E_b}{N_0}}), \text{ with } \mathrm{erfc}(x) = \frac{2}{\sqrt{\pi}}\int_x^\infty e^{-z^2}\, dz = 2Q(\sqrt{2}x) \tag{22}$$

ii) The differentially encoded coherent BPSK and QPSK with the Gray bit mapping, the following result holds approximately:

$$P_b = \mathrm{erfc}(\sqrt{\frac{E_b}{N_0}}) = 2Q\left(\sqrt{\frac{2E_b}{N_0}}\right) \tag{23}$$

iii) The differentially phase modulated BPSK, denoted DPSK or DBPSK, with non-coherent 2-symbol differential detection, the bit error rate is:

$$P_b = \frac{1}{2}e^{-\frac{E_b}{N_0}} \tag{24}$$

For the differentially encoded QPSK (DQPSK) [Ref. 11],

$$P_b = Q(a,b) - \frac{1}{2}I_0(ab)\exp(-\frac{1}{2}(a^2 + b^2))$$

where $Q(x,y)$ is the Marcum's Q-function [Ref. 12], with,

$$a = \sqrt{2(1-\frac{1}{\sqrt{2}})\frac{E_b}{N_0}}, \quad b = \sqrt{2(1+\frac{1}{\sqrt{2}})\frac{E_b}{N_0}},$$

and $I_0(x)$ the modified Bessel function of zero-order.

This expression can be written also as:

$$p_b = e^{-2\gamma_b} [\sum_{k=0}^{\infty} (\sqrt{2}-1)^k I_k(\sqrt{2}\gamma_b) - \frac{1}{2}I_o(\sqrt{2}\gamma_b)] \tag{25}$$

where $I_k(x)$ is the first modified Bessel function of the k-th order, and $\gamma_b = E_b/N_0$.

iv)   For coherent detection of $\pi/4$-QPSK, the performance is the same as the coherent QPSK. The differential detection of $\pi/4$-QPSK has the same error performance as the DQPSK that loses 2.3 dB with respect to the coherent QPSK.

Figure 4.2-6 shows the error-rate performance of the coherent QPSK, the differentially encoded coherent QPSK, and the non-coherently detected differentially phase modulated BPSK.

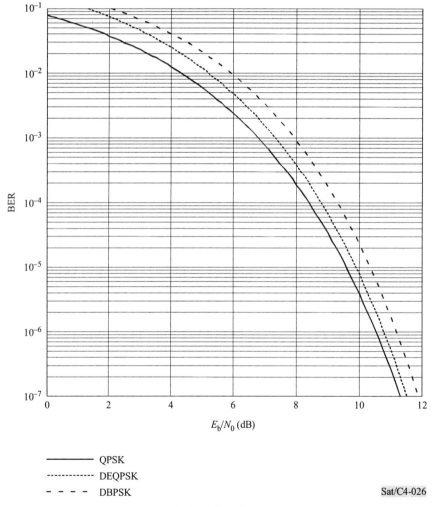

Sat/C4-026

FIGURE 4.2-6

**Bit error rate performance of QPSK,
DEQPSK and DBPSK**

The digital modulation techniques presented here are far from complete, in-depth treatments of the subject of digital modulations can be found in the references [Ref. 10], [Ref. 11], [Ref. 13], [Ref. 6].

In summary, some members in the quadrature modulation family are compared in Table 4.2-2.

TABLE 4.2-2

Comparison between the quadrature modulation family

Scheme	Bandwidth efficiency (Bits/s/Hz)	Power efficiency (E_b/N_0 for 10^{-5} over AWGN)	Complexity	Phase transitions	Envelope fluctuation dynamic	Other observations
QPSK	2	9.6 dB (C)	Moderate	$0°, \pm\pi/2, \pi$	Big	C
OQPSK	2	9.6 dB (C)	Moderate	$0°, \pm\pi/2$	Smaller	C
MSK	NB	9.6 dB (C)	High	$0°$ (CPM)	Few	C and D
DQPSK and $\pi/4$-QPSK	2	12.6 (D)	Low	$0°, \pm\pi/2$	Smaller	C and D

"NB" depends on implementation method, see [Ref. 10].
"C" means the coherent demodulation.
"D" means differential demodulation.
"C and D" means that the modulation scheme can be demodulated either coherently or differentially.

4.2.3.2   Trellis coded modulation

In trellis coded modulation (TCM) error correction coding and modulation are combined. Because this is a relatively new technique – at least as concerns its application to satellite communications – this subsection will provide a rather detailed introduction to the subject.

Terminology

First, some terminologies are recalled in the following:

Signal constellation is an ensemble of signal points from which the baseband modulator selects a signal point according the binary inputs. A **2D constellation** is composed of two-dimensional signal points, for example, the QPSK, MPSK and QAM. An **MD constellation** is a multi-dimensional constellation that can be usually a concatenation of 2D ones. It can be also an orthogonal signal set, a bi-orthogonal one, etc.

Set partitioning is a partition of a signal constellation into subsets according to a certain rule.

Free distance of a convolutional code is the minimum Hamming distance between all the coded sequence pairs of infinite length generated by the encoder. It determines roughly the performance of the code when decoded by the Viterbi algorithm over an AWGN channel.

Systematic feedback encoder, for a convolutional code, is one of its encoder forms, in opposition to feed forward non-systematic encoder, with the input information bits unchanged, and present at the output of the encoder. Part of its output is fed back to the input of the encoder. It is specified by parity check polynomials. The feed forward non-systematic encoder, defined by generator polynomials, has no feedback. These two encoders are equivalent in performance. For details see [Ref. 8].

Free squared Euclidean distance of a trellis code is the minimum squared Euclidean distance between all the coded signal constellation point sequence pairs of infinite length generated by the trellis encoder/mapper. It determines roughly the performance of the trellis code when decoded by the Viterbi algorithm over an AWGN channel.

Intra-subset distance is the minimum squared Euclidean distance in a subset, i.e., the distance between the two closest points of all possible point pairs in the subset.

Inter-subset distance, for two subsets, means the minimum distance between all the signal point pairs, with one point from a subset and another from the other subset. It can be defined for several subsets.

Mapping for a trellis code is a rule that assigns a group of binary symbols to the signal constellation point, used in the transmitter. **Demapping** is the inverse operation of mapping, used in the receiver.

For other terminologies related to convolutional and trellis coding, see [Ref. 14], [Ref. 8] and [Ref. 15].

TCM principle

The trellis coded modulation (TCM) is a technique for joint optimization of modulation and channel coding. With a channel capacity argument, the following results have been shown in [Ref. 16]:

- For efficiently transmitting information over a power-limited and band-limited channel, the modulation and the channel coding must be optimized jointly.

- For transmitting m bits per two-dimensional symbol over the AWGN channel, an expanded signal constellation of 2^{m+1} points, and a coding rate of m/(m + 1) are sufficient.

For example, a signal coded with a code rate R = 2/3, then modulated in 8-PSK, has the same spectrum occupancy as a 4-PSK modulated signal, but it will have a better performance owing to the use of the Viterbi algorithm for decoding (see Appendix 3.2).

Also using an information theory approach, these two conclusions have been confirmed by using the cut-off rate criterion, that is, a practical limitation on the information rate, in the case of the non-linear satellite channel [Ref. 17]. In recent years, it has been generally proved that a TCM using an expanded signal constellation can achieve a considerable coding gain either slightly or without expanding the bandwidth.

A TCM scheme is characterized by four functional parts:

- a signal constellation of two or multiple dimensions, expanded from 2^m points to 2^{m+1} points;

- a convolutional encoder of rate k/k + 1, with k ≤ m, and v memories;

- a set partitioning that partitions the original constellation into 2^{k+1} subsets;

- a mapping rule that defines the relation between the m binary, the encoded and the uncoded, symbols, and the constellation signal points, that are two- or multiple-dimensional.

A differential encoder may be used on the bits affected by the phase rotations before the convolutional encoder. Among the m incoming binary source bits, not all may be encoded. In fact, if k bits, k ≤ m, are encoded, then the other m – k bits uncoded, the total rate is always m/m + 1. The k + 1 encoded binary symbols are used to select the subset, the remaining m – k symbols then select the signal point inside the selected subset, according to the mapping rule. If k = m, the subsets become single signal points, the above statement remains true.

A general TCM encoder/modulator is shown as follows (Figure 4.2-7).

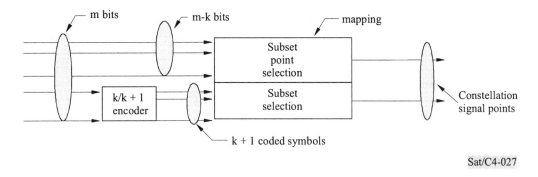

Sat/C4-027

FIGURE 4.2-7

A general TCM encoder/modulator

Figure 4.2-8 presents a set partitioning example with 8-PSK. Figure 4.2-8 a) presents an 8-PSK mapping from $(b_2b_1b_0)$ to the two-dimensional signal points. Figure 4.2-8 b) shows the signal point selection process. For example, if $b_0 = 0$, the subset {000, 100, 010, 110} is selected. Then, $b_1 = 1$ selects {010, 110}, and $b_1 = 0$ selects {010}. This means that $(b_2b_1b_0) = (010)$ is mapped into the signal point {010}, that is the point $5\pi/8$.

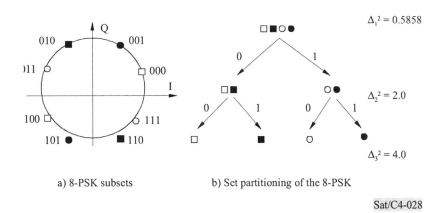

a) 8-PSK subsets   b) Set partitioning of the 8-PSK

Sat/C4-028

FIGURE 4.2-8

8-PSK subsets and set partitioning

The intra-subset distances at the first, the second, and the third level are: $\Delta_1^2 = 0.5858$, $\Delta_2^2 = 2.0$ and $\Delta_3^2 = 4.0$. Suppose that the first level subsets are indexed by z_0, the second level by z_1, and so on.

To understand how the TCM designs generally improve the coded modulation system performance over an AWGN channel, the following two observations are important:

- the set partitioning is performed such that the intra-subset Euclidean distance is enlarged;

- the inter-subset Euclidean distances are enlarged in the sequence sense by the coding.

The combined effect is an increase of the sequence Euclidean distance in the ensemble of the generated signal sequences, resulting in a performance improvement. TCM can be seen as an error control coding technique using the space redundancy instead of the time redundancy. A trellis code can be seen as a generalization of a convolutional code in that the coded binary symbols in the convolutional codes are replaced by the coded subsets in a TCM system. The code design is, however, different than the convolutional codes based on the Hamming distance. In the TCM schemes, the inter-subset Euclidean distances at the k-th level may not be the same, the Hamming distance optimum codes are not optimum for the TCM design. Therefore, the code must be optimized jointly with the subsets. Ungerboeck has obtained a number of optimal 2D schemes by exhaustive research by computer [Ref. 16], [Ref. 18]. For example, with k = m, for a spectral efficiency of 2 bits/s/Hz, instead of an uncoded QPSK, a trellis coded modulation can be constructed with an 8-PSK and a 2/3 convolutional code. For 3 bits/s/Hz, a 16-QAM associated with a 3/4 code can be used to construct a TCM that is better than an uncoded 8-PSK. Theoretically, up to 6 dB can be obtained as coding gain for the case of 8-PSK and QAM. Practically, 4 to 5 dB can be obtained, without expanding the bandwidth.

The convolutional code is expressed using the systematic feedback encoder, the equivalent feed forward minimal non-systematic encoder can be used also, for details see [Ref. 8]. The encoder structure is deduced from the parity check relation $\underline{z}H^T = \underline{0}$. As an example, for k = 2, v = 5, $H = [h_0, h_1, h_2] = [45, 10, 20] = [1 + D^2 + D^5, D^3, D^4]$, the encoder is shown in Figure 4.2-9.

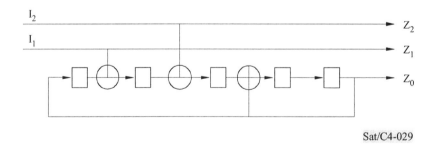

Sat/C4-029

FIGURE 4.2-9

Example of 2D 8-PSK optimum trellis code with v = 5

(I_1, I_2) are the two input information bits ($k = 2$), (z_2, z_1, z_0) the three encoded binary symbols, the mapping is defined as shown in Figure 4.2-8 a).

Demodulation and decoding of TCM systems

An optimum MLSE (maximum likelihood sequence estimation) receiver based on the Viterbi algorithm [Ref. 15] will realize the optimum error rate performance. Compared to the convolutional decoding, a TCM receiver contains a demapping unit that follows the Viterbi decoder. In other words, the Viterbi decoder makes the decision on the optimally decoded subset, that is the closest one to the received signal point, the signal point or the decoded bits will be determined by the demapping unit as the closest signal point to the received signal point. A differential decoder could be put at the output bits affected by the phase rotations.

Figure 4.2-10 shows a general architecture of a trellis decoder, the eventual differential decoder is omitted.

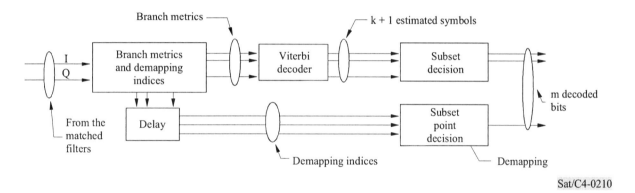

Sat/C4-0210

FIGURE 4.2-10

A general TCM demodulator/decoder

TCM with octal phase modulation (OPSK) or 8-PSK

The octal phase modulation (OPSK), also known as 8-PSK, is a constant amplitude scheme with a higher bandwidth efficiency of 3 bits/s/Hz. The first TCM applications to satellite transmissions occurred with the 8-PSK trellis codes. Using a 72 MHz transponder bandwidth, transmissions at up to 155.52 Mbits/s have been realized.

An important feature of high bit-rate satellite transmission is the demand for high performance, because the high bit-rate applications are related to the images and TV, HDTV services that require a BER in the order of 10^{-9} to 10^{-11}. A concatenated coding system must be used. The high rate 8-PSK trellis codes such as 5/6 and 8/9 are interesting for the inner codes in an RS/TCM concatenated coding system, most suitable to the high bit-rate satellite transmissions.

The optimum trellis codes for 2-D 8-PSK have been found by computer searches [Ref. 16], [Ref. 19] and are listed in Table 4.2-3.

TABLE. 4.2-3

2/3 trellis coded 2-D 1 x 8-PSK codes

$v2_v$ = # state	k	H_2	h_1	h_0	phase inv	d^2_{free}	N_{free}	G_{asym} (dB)
1	1		1	3	180°	2.586	2	1.12
2	1		2	5	180°	4.0	1	3.01
3	2	04	02	11	360°	4.586	2	3.60
4	2	14	06	23	180°	5.172	4	4.13
4	2	16	04	23	360°	5.172	2.25	4.13
5	2	14	26	53	180°	5.172	0.25	4.13
5	2	20	10	45	360°	5.757	2	4.59
6	2	074	012	147	180°	6.343	3.25	5.01
6	2	066	030	103	360°	6.343		5.01
7	2	146	052	225	180°	6.343	0.125	5.01
7	2	122	054	277	360°	6.586	0.5	5.18
8	2	146	210	573	180°	7.515	3.35	5.75
8	2	130	072	435	360°	7.515	1.5	5.75

In the first row of the above table, v denotes the memory amount of the encoder, so k = v + 1 is its constraint length, k is the number of the bits being encoded among the m information bits. The item "Phase Inv" means the maximal degree to which the code is transparent. For example, "180°" means that the code is transparent up to 180°, but not to 45° and 90°, the code has a partial transparency, while 360° means that the code is not transparent at all. The encoders are given in systematic feedback form, defined by the parity check polynomials, see the example shown in Figure 4.2-9. The term "d^2_{free}" is the square Euclidean free distance of the code, "N_{free}" is the first error coefficient, "G_{asym}" is the asymptotic coding gain, the practical coding gain is lower bounded by this quantity.

Pragmatic trellis codes with 8-PSK modulation

The drawbacks of the optimum trellis codes are:

- relatively small and dense Euclidean distance spectrum with large error coefficients;

- highly dense trellis for k > 1, and dedicated Viterbi decoder with a high complexity.

The first one concerns the performance, the second is related to the complexity issue. These problems limit the application of the optimum trellis codes.

A class of trellis codes using the same convolutional code optimum to QPSK, associated with an 8-PSK constellation has been found, called "pragmatic trellis codes" (PTCM) [Ref. 20]. The PTCM codes avoid the above drawbacks by using the standard code (k = 7, r = 1/2) with G = (171, 133), and its widely used decoder chips. In addition, the main advantage of this design is the possibility of a fast development of almost optimum TCM systems at a low cost, without developing dedicated devices, especially the decoder.

The PTCM 8-PSK subsets, set partitioning and mapping are shown in Figure 4.2-11.

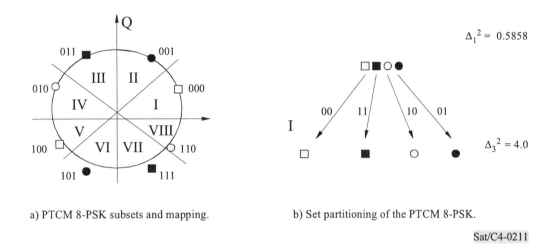

a) PTCM 8-PSK subsets and mapping. b) Set partitioning of the PTCM 8-PSK.

Sat/C4-0211

FIGURE 4.2-11

PTCM 8-PSK subsets, set partitioning and mapping

Inside each subset, the intra-subset square Euclidean distance is 4.0. The inter-subsets errors are protected against the channel noise by the (S, K = 7) coding. The set partitioning used is special because it is 4-ary at a single level. The two coded symbols select one of these four subsets, the uncoded bit selects one signal point inside each subset.

The PTCM 8-PSK system is shown in Figure 4.2-12.

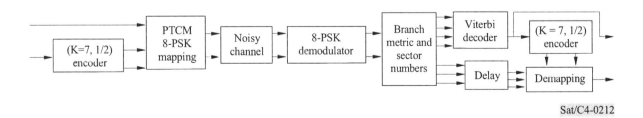

Sat/C4-0212

FIGURE 4.2-12

PTCM 8-PSK encoder/decoder scheme

Denote the input bit streams B_0 and B_1. B_1 is encoded by the (K = 7, 1/2) encoder to deliver the two encoded symbols, C_0 and C_1, that select one of the four subsets shown in Figure 4.2-11 b). Each subset, being itself a BPSK constellation with an enlarged distance 4.0, contains two signal points, and B_0 selects one of these two points. The mapping given in Figure 4.2-11 a) defines the 8-PSK signal (I, Q) to be sent to the noisy channel. The mapping rule is shown by the following table:

$B_0\,C_0\,C_1$	000	001	011	010	100	101	111	110
φ	$\pi/8$	$3\pi/8$	$5\pi/8$	$7\pi/8$	$9\pi/8$	$11\pi/8$	$13\pi/8$	$15\pi/8$

B_0 and B_1 specify three binary symbols that correspond to an 8-PSK symbol, the coding rate is 2/3. The PTCM 8-PSK has a bandwidth efficiency of 2 bits/s/Hz.

At the output of an 8-PSK demodulator, the (I, Q) signals are quantized into n bits each. In practice, n could be 6 to 8. The branch metric calculator performs two operations: i) the branch metrics, defined as the angle distance from the received signal point to the four subsets, are evaluated and sent to the Viterbi decoder that makes the decision on B_0; ii) the geometric location of the received signal is determined and denoted as sector number, see Figure 4.2-11. The sector number is delayed and fed into the demapping unit to assist in the decision on B_0. The demapping works by the following principle: the optimum subset is determined by the optimum decided bit B_1, and, inside the optimum subset, the signal point determined by the value of B_1 is decided using the sector number. The signal point that is the closest to the received one is taken for the optimal decision. The optimum subset is selected effectively by the two encoded symbols from the encoder, this is why an additional encoder must be used to regenerate these two encoded symbols for the demapping. The delay is put on the sector number signal to compensate the Viterbi decoder's decision delay.

The PTCM 8-PSK system has a square Euclidean free distance of 4.0, and an asymptotic gain of 3 dB over the uncoded QPSK [Ref. 20].

A differential encoder could be applied to the input of the PTCM encoder, and a differential decoder to the output of the PTCM decoder. The PTCM 8-PSK system is partially transparent to $\pm\pi/2$, and π, but not to the remaining four phases: $\pm\pi/4$, $\pm 3\pi/4$. A special two-dimensional differential encoder/decoder pair can be used to remove the $\pm\pi/2$ and π ambiguities, then the remaining ones are removed by the Viterbi decoder's path metric monitoring circuit.

Punctured pragmatic TCM (PPTCM)

Another advantage of the pragmatic approach to TCM is the use of the punctured codes to construct schemes with higher rates. We present briefly the 5/6, 8/9 schemes, constructed using the 3/4 and 5/6 punctured codes. A general PPTCM scheme is shown in Figure 4.2-13.

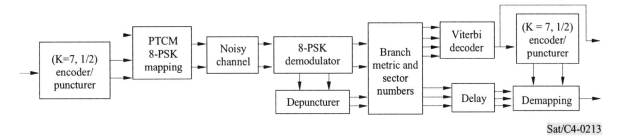

Sat/C4-0213

FIGURE 4.2-13

PPTCM 8-PSK encoder/decoder scheme

For a 5/6 PPTCM, the 3/4 punctured encoder is used to generate four coded symbols from 3 B_1 bits. These four symbols, combined with 2 bits from the stream B_0, (each two coded symbols combined with one uncoded bit), generate two 8-PSK symbols, resulting in a coding rate of 5/6. It has a bandwidth efficiency of 2.5 bits/s/Hz.

In a similar way, to construct a 8/9 PPTCM, the 5/6 punctured encoder is used. From the stream B_1, five information bits generate six coded symbols. Combined with 3 bits from the stream B_0, these six symbols generate three 8-PSK symbols, resulting in a coding rate of 8/9. It has a bandwidth efficiency of 2.667 bits/s/Hz.

About TCM applications to satellite communications

The PTCM 8-PSK codes are interesting for satellite transmissions. For example, in the INTELSAT IDR service, a new coding system using the PTCM 2/3 8-PSK code concatenated with an Reed-Solomon code is expected to improve the capacities up to 25%, based on the Standard A earth stations and INTELSAT VII space segment conditions. With the concatenation with an outer Reed-Solomon code, the performance will be also largely improved over the QPSK with 3/4 punctured coding.

A typical application is the satellite transmission of 139.236/155.52 Mbits/s PDH/SDH satellite interface equipment over 72 MHz transponder bandwidth. A 8-PSK trellis code of six dimension of rate 8/9 can be used as the inner code in a concatenated coding system with a Reed-Solomon outer code RS (255,239). Several such systems have been studied and tested, cf. [Ref. 21], [Ref. 22] and their cited references.

Concatenated RS/PPTCM 8-PSK systems have been used also in high and variable bit-rate transmissions of HDTV signals by satellite channel [Ref. 23].

In general, for a spectral efficiency less than 3 and more than 2 bits/s/Hz, 8-PSK TCM schemes can be used. For more than 3 bits/s/Hz, trellis coded modulation schemes with QAM constellations should be considered. With increasing demand for higher bandwidth efficiency, the QAM trellis codes could find potential applications in HDTV satellite transmissions in the future. The difficulties of applying the trellis codes using QAM constellations to satellite communications is their non-constant amplitude. More non-linear distortion occurs when passed through a non-linear amplifier such as an HPA. A precise amplitude control (AGC: automatic gain control) must be used at the receiver. A deep input back-off must be put in the HPA. However, the QAM trellis codes seem to be the only solution in the case where the bandwidth is the major system limitation for a very high bit-rate transmission.

Variable bit-rate applications

Variable bit-rate transmission is a new feature of today's satellite communications in the frequency bands higher than 10 GHz, for example, the Ku- (12/14 GHz) and the Ka- (20/30 GHz) bands, where rain fading is severe [Ref. 23].

To compensate for this fading, flexible channel coding with variable bit rates is an efficient approach. The coding rate is changed to adapt to the channel variation, keeping the bandwidth occupation unchanged. Thanks to the advanced variable bit-rate data compression techniques, the source bit-rate is adaptively changed to cope with the variable channel coding rate [Ref. 23]. Under heavy rain conditions, more powerful, lower rate channel coding with a high compression ratio source coding is used to compensate the propagation attenuation, while a higher rate channel coding is associated with a higher source bit-rate source coding for a higher quality when the channel is normal. All the RF units and the space segment remain unchanged.

To implement a variable bit-rate channel coding, punctured convolutional codes QPSK is suitable. In addition, the pragmatic trellis codes with 8-PSK can offer several coding rates to support different bandwidth requirements by using punctured codes, as we have seen in this section. The same Viterbi decoder can be implemented for several constellations and coding rates.

Shortened Reed-Solomon codes, the optimum trellis codes can all be used to implement variable bit-rate coding in different ways.

4.2.4   Modems for digital communications

The role and the functions of a modem are briefly described in this paragraph, see [Ref. 24].

4.2.4.1   Digital transmissions

As shown in Figure 4.2-14, a significant number of processing steps must be properly carried out to build a complete *digital* transmission. It must clear at this point that the term *digital* is applied exclusively to the *kind* of transmission used. The transmission can be said to be *digital* whereas the modem remains *analogue*: this means simply that the transmitted signal carries *symbols* chosen among a given alphabet, these being processed by analogue circuits and components.

Only digital transmission will be treated in this section. However, both analogue and digital modems will be detailed, even if digital techniques have to be distinctively emphasized to reflect the state-of-the-art and the main evolutions foreseen in the digital communications domain.

Around the modem itself, a few blocks of Figure 4.2-14 require a short explanation:

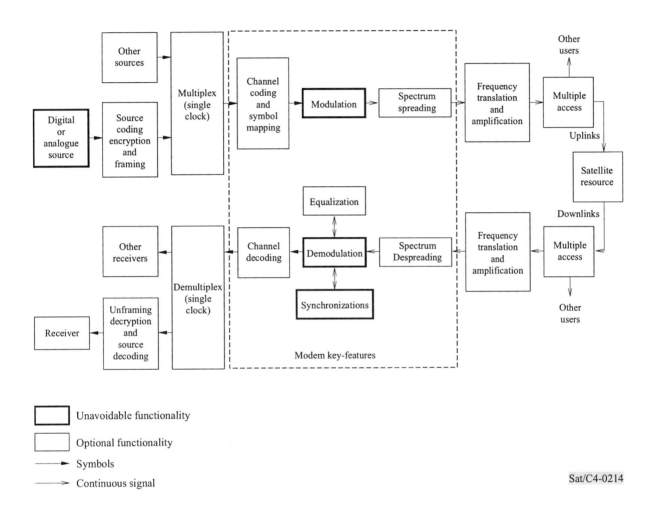

FIGURE 4.2-14

Overall digital transmission scheme

• Source coding/decoding

This operation takes as input an analogue signal carrying the information of interest and processes it so as to output a stream of bits. An efficient source coding is expected to output the *minimal* number of bits required to transmit properly a given message. The source decoding is a synthesis processing, aiming at reconstructing as accurately as possible the original signal.

- Encryption/decryption

The bits are "encrypted" or else coded by means of a secret sequence – the "clue" – known exclusively by a reduced number of users. A systematic method for breaking a 200-bits clue would typically require 10^{20} years with today's most powerful processors!

- Framing/unframing

Under this generic title different functions are brought together:

- specific encoding and decoding, along with the important scrambling and unscrambling functions that greatly help transmitting messages on severely fading channels;

- insertion of preambles/postambles in the burst modems: this is of great interest in the case of time division multiplexing (TDM) schemes (see § 4.2.5.2 and 4.2.5.4 about "burst modems" in the sequel for further details). In the case of time division multiple access (TDMA) schemes, the preamble/unique word recovery is necessary for locating the beginning of every burst, and therefore being able to identify the bursts of interest.

- Multiplexing/demultiplexing and the multiple access

Take for instance a user wishing to transmit four 72 kbit/s channels: a solution is to multiplex those four channels into one single 288 kbit/s signal and to transmit it on the appropriate sub-band of the satellite. Within this sub-band, the allocation is merely static and completely controlled by the user. Synchronization matters must be solved by the user himself and do not appear at the system level. No confusion should be made between this multiplexing/demultiplexing function (*the reference is not found*) and the *multiple access* function which allows different users to establish dynamically a communication through the same satellite. Such a multiple access provides the users with a complete and efficient scheme, thus avoiding interference among the communications.

- Frequency translations and amplifications

These operations are necessary to transmit the signal through the allocated and predefined frequency bandwidth.

4.2.4.2   Going through the key modules of the demodulator

The main demodulation modules are listed below and will be described in detail later.

- Synchronization

- The clock recovery (timing correction): as a matter of fact, the information is carried here by discrete symbols, transmitted at regularly spaced time intervals. The role of the receiver consists in coming to a decision about which symbol is currently transmitted. But those decisions cannot be taken at any time: of course, they have to respect the symbol rate but must occur as well as the short period of time where a given symbol is not perturbed by the neighbouring symbols. This makes a timing synchronizer mandatory.

- Frequency recovery: there always remain uncertainties about oscillator frequencies, involving deviations in the multiple frequency translation stages of a demodulator. This is usually taken into account at the end by applying an adequate frequency estimator.

- Phase estimator: whenever the information is carried by the phase of the transmitted signal, it may be useful to achieve an efficient phase estimation, which can be considered as a fine extension of the frequency estimation. In this case, the demodulation will be said to be *coherent*.

- Filtering

  The goal is twofold:

- Let the overall transmission channel respect the Nyquist criterion: through a proper *matched* filter, the inter-symbol interference (ISI) is nearly cancelled. See the filtering § 4.2.6 for details about ISI cancellation and filtering techniques.

- Reject efficiently possible perturbators, like neighbouring channels or jammers.

4.2.5   Conventional modem architectures

4.2.5.1   Description of conventional modems

i)   Example: QPSK modems

The description given here is restricted to QPSK modems, with coherent demodulation, because this type of modulation is the most widely used for satellite communications. Figure 4.2-15 shows the basic principles of QPSK modulation and demodulation: modulation is performed by combining two 2-PSKs in quadrature. Demodulation is performed in the same way, using the recovered carrier for coherently demodulating the two received bit streams by two demodulators in quadrature.

- The input binary data stream is split into two data streams A and B (for instance, even bits in A and odd bits in B). The baud (or symbol) rate, i.e. the modulation rate, is equal to the bit rate of A and B, that is a half of the actual (input) data bit rate. In the demodulator, the same data streams are recovered and are combined in order to build the output data.

- If the input data bit rate is R, the symbol rate is R/2, the symbol period is $T_s = 2/R$ and the occupied bandwidth is approximately $B = 1.2/T_s = 0.6\ R$. For example, in the INTELSAT/ EUTELSAT TDMA-QPSK system, the nominal bit rate is R = 120.832 Mbit/s, the symbol rate is 60.416 Mbaud and the bandwidth is approximately 72 MHz.

- Filtering can be done equally at IF using band-pass filters, or at baseband, using low-pass filters; in the latter case, filtering and modulation/demodulation are closely related functions as explained in § 4.2.6.

As a matter of fact, the demodulator filters introduced after the sampling stage are in charge of two different functions. They must first demodulate the signal, which corresponds to a low-pass filter behaviour: this cancels the remaining negative frequency signal component. But at the same time, they have to carry out an efficient "matched" filtering (see § 4.2.6) through which adjacent symbols are shaped so as not to interfere with each other.

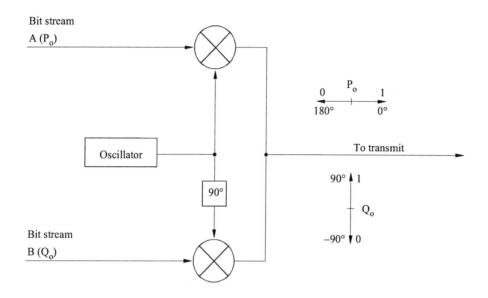

a) QPSK modulation

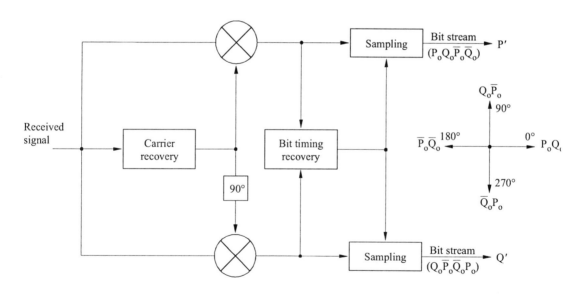

b) QPSK coherent demodulation and pulse regeneration

Sat/C4-0215

FIGURE 4.2-15

Basic principles of QPSK modulation and demodulation

ii)   Removal of the phase ambiguity

In a QPSK modem applying a coherent demodulation (such as the one described above), it is mandatory to recover the phase of the transmitted signal, since this phase itself carries the information. The phase estimation first removes the modulation by multiplying the signal phase by four in the case of a QPSK and by two for a BPSK (26):

$$4 \cdot \text{Phase} (a_k) = 4 \cdot (n_k \cdot \pi/2 + \Phi) \, [2\pi] = (2n_k \cdot \pi + 4\Phi) \, [2\pi] = 4\Phi \, [2\pi] \qquad (26)$$

where Φ denotes the unknown transmit phase, a_k the current transmit symbol, n_k an integer, and $[2\pi]$ the 2π-modulus.

This result is divided by four to obtain the desired estimated phase:

$$\text{Phase} (a_k) = \Phi \, [\pi/2]$$

It appears clearly that a $\pi/2$-ambiguity is unavoidable.

Due to the $\pi/2$ phase ambiguity of the recovered carrier of the demodulator, the output data of the demodulator/regenerator I' and Q' are not always identical to the input data, I and Q, of the modulator.

The four possible configurations are:

$$I' = I \quad Q' = Q$$
$$I' = Q \quad Q' = \bar{I}$$
$$I' = \bar{I} \quad Q' = \overline{Q}$$
$$I' = \overline{Q} \quad Q' = I$$

The same configuration will continue for a long period (minutes/hours), until a $\pi/2$ hop of the recovered carrier, due to noise, leads to another configuration.

One simple way to overcome this difficulty is to use a differential encoder/decoder scheme. In this case, a single encoded symbol does not carry the phase of the corresponding transmitted symbol, but *the difference between the phases of two consecutive encoded symbols* is equal to the phase of the transmitted symbol. The differential decoder only has to implement a phase subtraction to recover the information.

Thus, the ambiguity is removed but with the following penalty: one transmission error now corresponds to two consecutive errors in the decided phase variation, which results (with the convenient modulation algorithm, e.g. Gray coding) in two bit errors and therefore a greater BER for a given C/N.

It is interesting to note that a burst of N errors due to a $\pi/2$-hop does not result in 2N errors but only in two errors, one at the beginning of the burst and one at the end. In other words, two consecutive errors do not give rise to an error if they are "coherent", i.e. phase-shifted by the same value. However, such a case of two coherent consecutive errors has a low probability of occurrence. The global error probability is thus approximately doubled by a differential encoding/decoding scheme.

iii)   Carrier recovery

Two examples of carrier recovery circuits are shown in Figure 4.2-16:

a)   The first circuit uses a frequency multiplication by four. At the multiplier output, an unmodulated carrier at four times the nominal IF is generated, since modulated phases which are an integral multiple of $\pi/2$ are changed into multiples of 2π. This carrier is then filtered and the frequency of the filtered signal is divided again by four (by logic circuits). Filtering of the line may be done either by a passive LC filter or by a phase-locked loop (PLL).

PLL is preferred for continuous mode carriers, because there is no phase shift of the output signal due to frequency drifts. Passive filters are often preferred for burst modulation (see § 4.2.5.2).

b)   The second circuit called a Costas loop (which is mainly used for continuous mode carriers) uses a voltage controlled oscillator (VCO) at the nominal IF frequency and an appropriate control algorithm.

If θ is a phase rotation applied at the demodulator input (or, in an equivalent way, between the VCO and the demodulator), it can easily be shown that the open-loop output of a circuit having the following algorithm is periodical with respect to θ, with a $\pi/2$ period: $X \cdot \mathrm{sgn}\,(Y) - Y \cdot \mathrm{sgn}\,(X)$ (sgn is the algebraic sign) which makes it a convenient algorithm for use with QPSK. Another algorithm: $\mathrm{sgn}\,(X) \cdot \mathrm{sgn}\,(Y) \cdot \mathrm{sgn}\,(X + Y) \cdot \mathrm{sgn}\,(X - Y)$ is often preferred because it can be implemented using only logic circuits.

A filtering bandwidth of 2/100 of the baud rate is sufficiently low to avoid any noticeable degradation of the bit error ratio due to jitter of the recovered carrier. When cycle-skipping of the recovered carrier is critical (ambiguity is removed by using known words as explained above), the relative bandwidth of filtering of the recovered carrier must be kept as low as 0.5/100.

Automatic frequency tracking should be provided in the carrier recovery circuits in order to compensate for transmitting and receiving earth station equipment frequency drifts (especially due to local oscillators) and also satellite transponder frequency drifts (including Doppler effect).

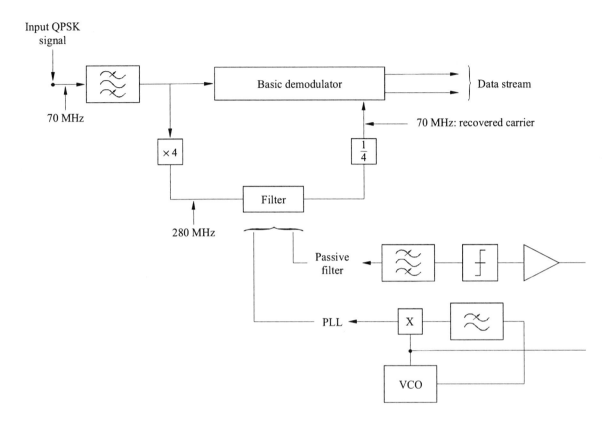

a) Multiplication by four

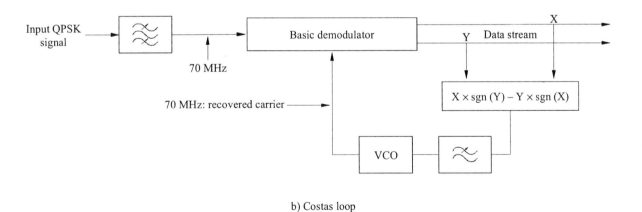

b) Costas loop

Sat/C4-0216

FIGURE 4.2-16

Carrier recovery in QPSK

iv)   Clock recovery

A frequency (spectral line) corresponding to the clock frequency can be obtained by any non-linear process, either from the input signal of the demodulator (IF) or the output signal (baseband). This frequency is filtered by a passive or a PLL filter.

The order of magnitude of the necessary bandwidth is the same as that for carrier recovery.

The problem due to frequency drifts is less stringent than for carrier recovery since carrier drifts do not affect the bit rate.

4.2.5.2   Specific requirements of burst modems

In the case of non-continuous digital communications, i.e. when the signal is transmitted only during a limited period of time (burst mode transmission), the carrier and clock frequency must be recovered very rapidly, using dedicated bits which are included in the preamble at the beginning of each burst. Of course, since the preamble does not carry actual information data bits, it must be kept as short as possible to keep good transmission efficiency.

In this section, two special features of burst modems will be discussed, with special attention to TDMA-QPSK transmission:

i)   modulator carrier on/off circuit;

ii) and iii)   demodulator circuits for rapid carrier and clock recovery.

i)   Modulator carrier on/off circuit

This circuit consists of a simple IF electronic switch, with 40-50 dB isolation, e.g. connected to the output of the crystal oscillator of the modulator. IF filtering is carried out by low-pass filters (i.e. before the basic modulation process). A fixed delay, compensating for the group delay of the filters, must be inserted in the switch control input.

ii)   Demodulator circuits (carrier recovery)

Most TDMA demodulators use the frequency multiplication (x4) technique for carrier recovery as described in § 4.2.5.1 ii). A passive filter is usually preferred to a phase-locked loop (PLL) for recovered carrier filtering as explained below.

If a PLL is used for filtering the recovered carrier, the acquisition time is highly dependent on the initial phase shift between the VCO and the input frequency to the PLL, at the beginning of the burst.

As the different transmitting earth stations have uncorrelated local oscillators, this phase shift is purely random.

The most unfavourable case occurs at point A in Figure 4.2-17 because the error voltage allowing the loop to move towards either the lower stable operating point, or the higher one, is very low. In this case, the acquisition time goes to infinity.

The more straightforward solution, in TDMA, consists of using a passive filter, the acquisition time of which is finite.

A typical block diagram of the carrier recovery and demodulator circuits is given in Figure 4.2-18. This particular implementation refers to the INTELSAT/EUTELSAT 120 Mbit/s TDMA-QPSK system. The IF is at 140 MHz.

- The main filtering is carried out by a passive filter after frequency quadrupling and down-conversion to 20 MHz. The bandwidth of this filter is 0.5/100 times the baud rate (60 Mbaud) = 300 kHz.

- Such a small bandwidth cannot be implemented at 560 MHz. For this reason, the signal is down-converted.

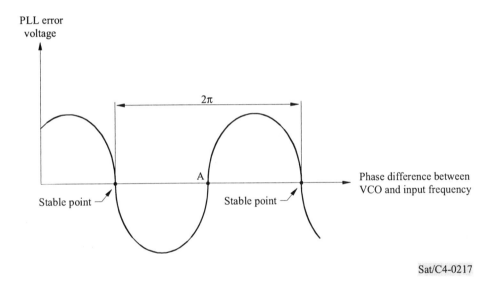

Sat/C4-0217

FIGURE 4.2-17

Acquisition of a PLL

CHAPTER 4 Carrier modulation techniques

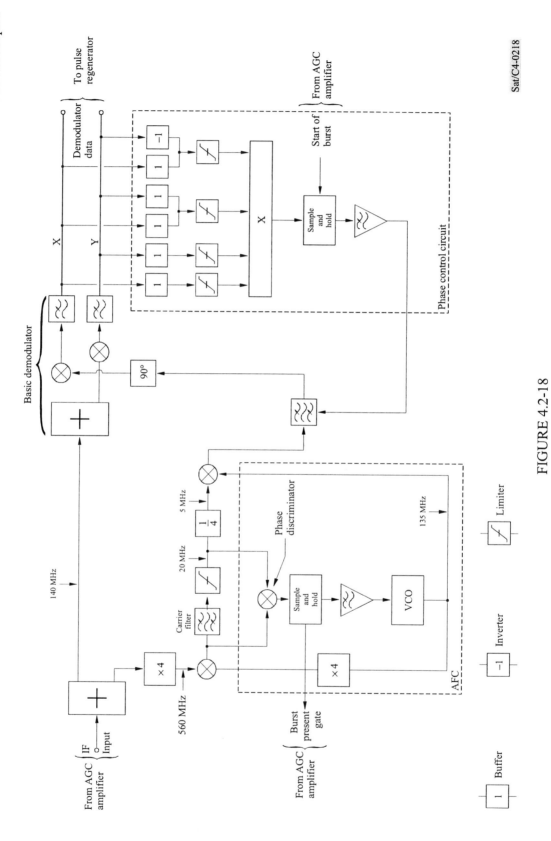

FIGURE 4.2-18

Carrier recovery (120 Mbit/s TDMA modem)

Sat/C4-0218

In order to avoid the phase shift given by the filter and slow frequency drifts (due to ageing, temperature and Doppler), the down-converter local oscillator is controlled by the output of a phase discriminator. The circuit described above is an AFC, which keeps the input frequency exactly at the central frequency of the filter. The time constant of the AFC is large ($\approx$1 ms).

In order to compensate for fast phase shifts, i.e. from one burst to another, a "phase control circuit" (see the right side of Figure 4.2-18) is implemented. This circuit uses the algorithm: sgn (X) $\cdot$ sgn (Y) $\cdot$ sgn (X + Y) $\cdot$ sgn (X − Y) to deliver the sign of sin 4φ used for controlling the phase shifter (φ being the instantaneous phase error between the reference and the received signal). The time constant of this phase control is small ($\approx$ 1 μs).

As can be seen in Figure 4.2-18, the AFC and the phase control circuit are only activated during bursts.

NOTE – An automatic gain control (AGC) with sample-and-hold circuits should also be provided in the demodulator in order to compensate for possible variations:

*   of the average power of all the bursts in the TDMA frame (this is particularly useful in the case of very low duty cycles and avoids noise amplifications between bursts);

*   of burst-to-burst amplitude variations (the AGC reference being taken during the preamble of each burst).

iii)   Clock recovery

For the same reason as for carrier recovery, a passive filter is generally used.

At 60 Mbaud, the bandwidth of this filter is about 300 kHz (0.5/100).

In order to compensate for drifts in L/C components or filters, an AFC can be used, as shown in Figure 4.2-19, which is similar to the AFC used for carrier recovery.

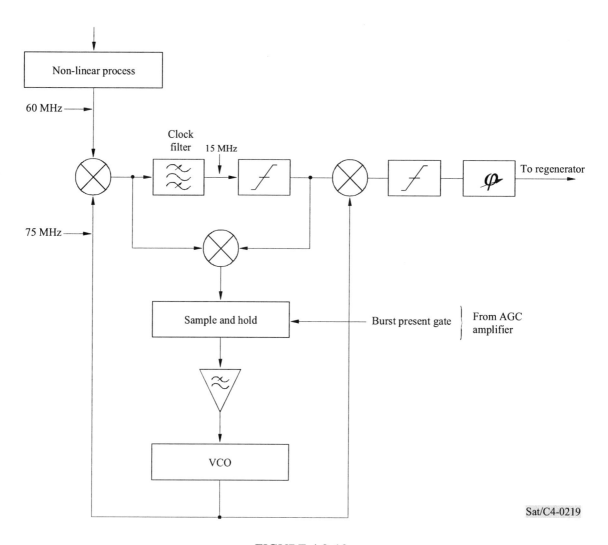

FIGURE 4.2-19

Clock recovery (120 Mbit/s TDMA modem)

4.2.5.3   Sampling considerations

Conventional (i.e. commercial) demodulators often implement architectures where the sampling process, which is a part of analogue to digital converter (ADC) functions, is driven through a voltage controlled oscillator (VCO), using both the symbol rate and the timing estimate as inputs. These inputs allow the A/D conversion to be precisely carried out at time intervals corresponding to the optimal aperture of the demodulated eye.

The scheme of a QPSK modem has been given in § 4.2.5.1 i). In this architecture, the matched filter is often realized in IF, before the baseband conversion achieved by the double balanced mixers and the low-pass filters. But it is becoming a common practice to carry out the pulse shaping (i.e. the matched filter) by means of digital filters, placed after the AD converter. Such a demodulator appears then as described in Figure 4.2-20.

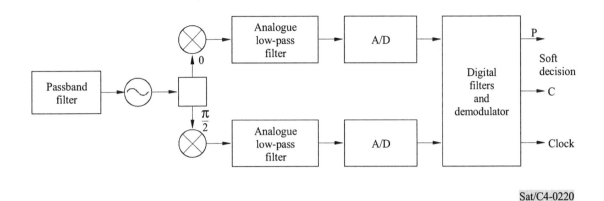

Sat/C4-0220

FIGURE 4.2-20

Typical block diagram of a digital demodulator

4.2.5.4   New techniques for TDMA mode transmission

Specific and severe constraints characterize burst mode transmissions. Indeed, since bursts are transmitted at different instants from different stations, an uncertainty exists on the knowledge of the parameters which characterize the transmitter, the receiver and thus the received waveform (the accurate location of the beginning of the slot, the phase shift and the frequency offset between transmitting and receiving oscillators). These parameters can only be extracted from the received symbols of the current burst, even for very short bursts. This operation has to be performed again for each burst. Moreover, the use of the satellite channel implies very low signal-to-noise ratio operating points.

With such very particular constraints, digital feed forward structures are well suited to this kind of receiver. Figure 4.2-21 shows an example for the implementation of a TDMA receiver synchronization structure.

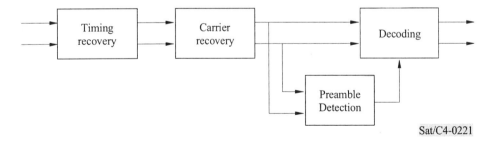

Sat/C4-0221

FIGURE 4.2-21

Implementation example for a TDMA modem

In such a modem designed to operate at a low signal-to-noise ratio (about $(E_b/N_0)_{min} = 2$ dB) in burst mode, the maximum normalized frequency offset $(\Delta f T_s)_{max}$ is one of the main parameters, where Δf is the frequency offset itself and T_s designates the symbol period. Indeed, the minimum data rate (512 kbit/s for instance) and the maximum frequency offset (2 kHz for instance) between transmitting and receiving oscillators determine $(\Delta f T_s)_{max}$. This input specification allows an optimal configuration to be determined for the carrier recovery unit. For such a synchronization structure using feed forward algorithms, an example for burst length is from 200 to 10 000 symbols. The minimum value is defined by the carrier recovery unit performance in extracting the symbol phase and carrier information, from a reduced number of symbols.

4.2.6   Filtering techniques

4.2.6.1   Simplified theory

When a Dirac pulse (very small duty-cycle pulse), as shown in Figure 4.2-22 is applied to a low-pass filter, the output signal at the sampling times of the following and preceding pulses is called inter-symbol distortion and acts as perturbative noise.

From Figure 4.2-22, it appears that a given filter will not introduce inter-symbol interference (ISI) under the following condition:

$$h\,(kT_s) = 0 \text{ if } k \neq 0$$

where h(t) represents the filter impulse response, $1/T_s$ the symbol rate and k an integer.

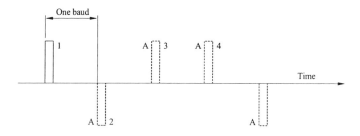

a) Low-pass filter input

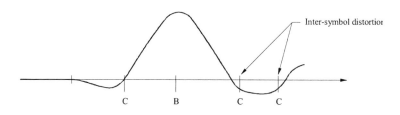

b) Low-pass filter output for impulse No.1

A: Following pulses

B: Sampling time for impulse No.1

C: Sampling time for following and preceding pulses

Sat/C4-022

FIGURE 4.2-22

An impulse response example

Thus, to obtain an ISI-free demodulated signal, the impulse response of the transmission channel must respect the condition stated above. Under the name "transmission channel" all the processing stages that modify the information carrying signal are brought together: the digital and analogue filters, the D/A and A/D converters impulse responses and so on.

The ISI-free condition mentioned above can be stated under the following equivalent form: the overall frequency channel response must be constant. It means simply that the channel becomes transparent, or else that the information is unaffected by the transmission. If, as usual, the overall attenuation is sufficiently high for frequencies above $1/T_s$, an explicit condition on the filter frequency response can be derived, which is currently referred to as the "first Nyquist criterion": the frequency response of the overall transmission channel must have a hermitic symmetry around the point $(1/2T_s, 1/2)$ (see Figure 4.2-23).

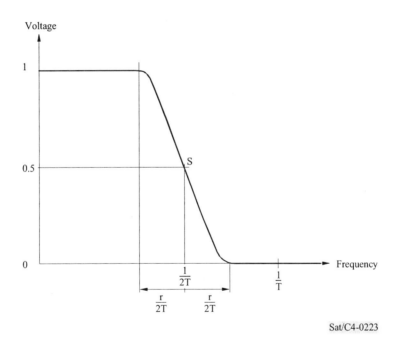

Sat/C4-0223

FIGURE 4.2-23

Inter-symbol distorsion and roll-off filter characteristics

A common practice is to use a nominal filter characteristic as follows:

- constant response for: $0 < f < (1 - r)/2T$

- sinusoidal response for: $(1 - r)/2T < f < (1 + r)/2T$

- null response for $f > (1 + r)/2T$

The coefficient r $(0 < r < 1)$ is called "roll-off".

In satellite transmissions, the nominal value of the roll-off coefficient is between r = 0 and 0.5.

These considerations lead to the design of the global channel filter; there still remains to be described the repartition of the filtering. Since at least one filter is needed on the modulator side, to provide the transmitted signal with an appropriate shape, and since another filter is mandatory in the demodulator to reject noise and perturbation interference, the global filter is split in two parts. Furthermore, it can be shown that the optimal repartition consists of two equal filters, if the channel simply adds a white noise to the signal. Thus, the transmitter and receiver filter frequency response is equal to the square root of the global filter frequency response.

The same theory applies for band-pass filters which are perfectly symmetrical to the central IF frequency. If imperfect symmetrical band-pass filters are used, another type of inter-symbol distortion appears, from one carrier to the other (from I to Q and Q to I, see Figure 4.2-15).

4.2.6.2   Basic presentation of FIR filters

A finite impulse response (FIR) filter operates on a digital signal.

When this signal is sampled the sampled-data signal is called a discrete-time signal. As a result of the Shannon sampling theorem (1949), it is possible to consider a discrete-time signal as a realistic image of the original continuous-time signal.[*] When the discrete-time signal has a finite number of different values for its amplitude, it is called a digital signal. This process is called quantization: each sample becomes a number represented by a group (word) of ones and zeros (bits) and can be processed by digital computation methods (see Figure 4.2-24).

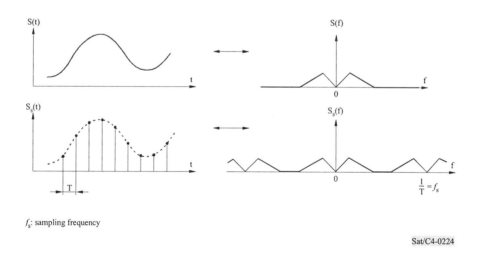

f_s: sampling frequency

Sat/C4-0224

FIGURE 4.2-24

Spectral relationship between continuous and sampled signal

---

[*]   Any frequency-bounded continuous-time signal can be represented without loss of information as a series of samples of the original signal when the sampling frequency is greater than the Nyquist rate, i.e. twice the maximum frequency of the continuous-time signal.

When the sampling theorem is satisfied, the spectrum of the discrete-time signal is a periodic replica of the continuous-time signal.

A low-pass filter (often called anti-aliasing filter) is needed to prevent aliasing caused by frequencies above half the sampling frequency.

The definition of a FIR filter can be easily understood, if not strictly defined, by the use of an analogue filtering formula:

$$y (u) = \int_{-\infty}^{\infty} h (t) \cdot x (t-u) \cdot dt \tag{27}$$

In its discrete form, this can be stated as follows:

$$y (t) = \sum \chi (k \cdot T_e) \cdot h (t-k \cdot T_e) \tag{28}$$

where T_e corresponds to the sampling period. Since the filter impulse response can be neglected outside a given time segment $[-M.T_e, M.T_e]$, this convolution becomes:

$$y (t) = \sum_{k=-M}^{M} \chi (k \cdot T_e) \cdot h (t-k \cdot T_e) \tag{29}$$

This is exactly the form of FIR convolution of length 2M + 1 (Figure 4.2-25).

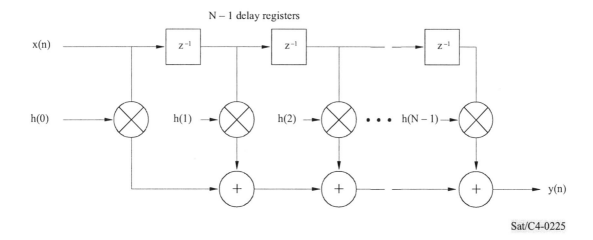

Sat/C4-0225

FIGURE 4.2-25

Typical FIR filter structure

REFERENCES

[1] Fuqin, Xiong, [August, 1994] "Modem techniques in satellite communications", IEEE Communications Magazine, p. 84-98.

[2] Sklar, B., "Digital Communications: Fundamentals and Applications".

[3] Haykin, S., [1988] "G. D. Digital Communications", (Wiley).

[4] Sundberg, C.-E., [April, 1988] "Continuous Phase Modulation: A Class of Jointly Power and Bandwidth Efficient Digital Modulation Schemes with Constant Amplitude", IEEE Commun. Mag., Vol. 24, No. 4, pp. 25-38.

[5] Juing, Fang, [March, 1996] "The other digital modulation techniques". ITU-R, Doc. 4HB/14, 25-28 March 1996.

[6] Anderson, J.B., Aulin, T., Sundberg, C.-E., [1986] "Digital Phase Modulations", Plunum Press.

[7] Zimer, R.E., Ryan, C.R., [1983] "Minimum-shift-keyed Modem Implementation", IEEE Comm. Mag. Vol. 21, No. 7, 1983.

[8] Forney JR, G.D., [November, 1970] "Convolutional codes-I: algebraic structure", IEEE Trans. on IT-16, No. 6.

[9] Lju, C.L., Feher, K., [March, 1991] "π/4-QPSK Modems for Satellite Sound/Data Broadcast Systems", IEEE Trans. on Broadcasting.

[10] Feher, K., [1983] "Digital Communications: Satellite/Earth Station Engineering", Prentice-Hall Inc., Englewood Cliffs.

[11] Proakis, J.G., [1983] "Digital Communications", McGraw-Hill, New York.

[12] Marcum, J., [1950] "Tables of Q-functions", RAND Corp., Rep. M-339, Jan.

[13] Bic, J.C., Dupontel, D., Imbeaux, J.C., [1986] "Elements of digital communications I, II", Dunod, Paris.

[14] Biglieri, E., Divsalar, D. *et al.* [1991] "Introduction to Trellis-Coded Modulations", New York: MacMillan.

[15] Viterbi, A.J., Omura, J.K., [1979] "Principles of Digital Communication and Coding", McGraw-Hill.

[16] Ungerboeck, G., [January, 1982] "Channel coding with multilevel/phase signals", IEEE Trans. on IT-28.

[17] Bigliery, E., [May, 1984] "High-level modulation and coding for non-linear satellite channels", IEEE Trans. on Communications, Vol. COM-32.

[18] Ungerboeck, G., [February, 1987] "Trellis-Coded Modulation with Redundant Signal Sets, Part-I: Introduction, Part-II: State of the Art", IEEE Comm. Mag. Vol. 25, No. 2.

[19] Pietrobon, S.S. *et al.* [January, 1990] "Trellis-coded multidimensional phase modulation", IEEE Trans. on IT-36, No. 1.

[20] Viterbi, A.J., Wolf, J.K. *et al.* [July, 1989] "A pragmatic approach to trellis-coded modulation", IEEE Comm. Mag.

[21] Pietrobon, S.S., Kasparian, *et al.* [1994] "A multi-D trellis decoder for a 155 Mb/s concatenated codec", International Journal of Satellite Communications", Vol. 12.

[22] Delarulle, D., [1995] "A pragmatic coding scheme for transmission of 155 Mb/s SDH and 140 Mb/s PDH signals over 72 MHz transponders", Proc. ICDSC-10, p. 319-324.

[23] [Flash-TV] "Flash-TV: Flexible and Advanced Satellite Systems for High Quality Television, with Interconnection with IBCNs", Race-II Project R2064.

[24] Bertrand, D. [1996], "Modems for Digital Communications". Document 4HB/21.

CHAPTER 5

Multiple access, assignment and network architectures

5.1 Introduction

Multiple access is the ability for several earth stations to transmit their respective carriers simultaneously through the same satellite transponder. This feature allows any earth station located in the corresponding coverage area to receive carriers originating from several earth stations. Conversely, a carrier transmitted by one earth station through a given transponder can be received by any earth station located in the corresponding coverage area. This enables a transmitting earth station to group several signals into a single multi-destination carrier.

A communication satellite acts as a relay station and as a nodal point in the circuits connecting the earth stations involved. It contains one or more transponder chains, each of which has the capability of frequency translating, amplifying and retransmitting the signals received from earth stations in the system. Some remote controlled switching capability may exist between some of the transponder chains. Moreover, in some advanced systems, on-board (switching and) processing (OBP) is provided in order to perform one or more of the following functions: switching (in frequency, time or space, i.e. between antenna spot-beams), signal regeneration and processing (notably baseband processing). In fact, the available information transmission capability (traffic capacity) of a transponder chain is generally greater than that needed by a particular transmitting earth station. Therefore, in order to optimize the utilization of the transponder capacity, more than one earth station is allowed to access a transponder chain with either single- or multi-destination transmissions, which is precisely the function of multiple access.

Three main modes of multiple access can be implemented:

- Frequency division multiple access (FDMA), where each concerned earth station is assigned its own carrier frequency, inside the transponder bandwidth (§ 5.2).

- Time division multiple access (TDMA), where all concerned earth stations use the same carrier frequency and bandwidth with time division (i.e. they do not simultaneously transmit their signals) (§ 5.3).

- Code division multiple access (CDMA), where all concerned earth stations simultaneously share the same bandwidth and recognize the signals by various processes such as code identification (§ 5.4).

Note that:

- FDMA can be associated either with analogue or digital modulation. On the contrary, TDMA and CDMA need practically to be associated with digital modulation.

- Contrarily to FDMA or CDMA in which continuous signals are transmitted, in TDMA the signals are time shared, i.e. are carried in the form of non-continuous bursts.

- The three categories of multiple access may be combined. For example, low or medium bit rate TDMA carriers can share, in FDMA, the same transponder (see below § 5.3.3) and the same is true with CDMA carriers.

There are also two different modes for assigning the communication channels in the transmitted carriers:

- Pre-assignment multiple access (PAMA), in which the channels required between two earth stations are assigned permanently for their exclusive use.

- Demand-assignment multiple access (DAMA), in which the channels' allocation is changed in accordance with the originating call (see below § 5.5). The channel is automatically selected and is connected for transmission only while the call is continued. This approach substantially increases the efficiency of the satellite transponder utilization and, more generally of the whole communication system in comparison with PAMA.

Note that DAMA can be associated either with FDMA or TDMA (and with analogue or digital modulation).

5.2 Frequency Division Multiple Access

FDMA was the first multiple access technique employed in satellite communications. Because of its simplicity and flexibility, it remains very commonly used. In FDMA, a different frequency is allocated in a transponder to each carrier (possibly multi-destination) to be transmitted by an earth station, and then a given bandwidth, in proportion with the carrier capacity, is also allocated. Therefore, the satellite resource is used in common (Figure 5.1).

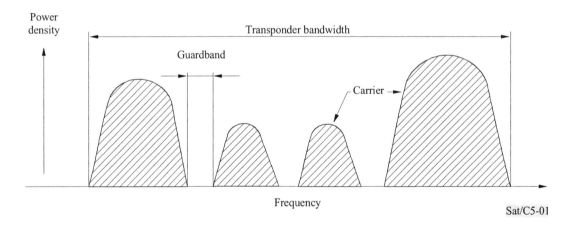

FIGURE 5.1

Frequency plan of a satellite transponder for FDMA transmission

A detrimental effect of this type of multiple access is that, due to non-linearities in the transponder chain and notably in the power amplifier, simultaneously transmitting several carriers in the same transponder causes intermodulation between these carriers, resulting in unwanted emissions (intermodulation products).

In order to reduce the level of such interference, it is necessary to keep the transmitted power significantly lower than the maximum available output power (saturation). This is called "back-off". Furthermore, the power transmitted by each earth station needs to be controlled.

Intermodulation effects are dealt with in detail in Chapter 2 (§ 2.1.5), Appendix 5.2 and, as concerns these effects in earth stations' power amplifiers, in Chapter 7 (§ 7.4.5.2).

FDMA may be implemented with various modulation-multiplexing methods, the most common being:

- FDM-FM (analogue), in which carriers are frequency-modulated by a frequency division multiplexed baseband signal.

- TDM-PSK (digital), in which carriers are PSK modulated by a time division multiplexed baseband signal.

- SCPC (for earth stations with small traffic), in which each individual telephone (or data) channel modulates the carrier, either by FM (analogue) or PSK (digital).

5.2.1 Multiplexed FDMA

In multiplexed (FDM or TDM) FDMA, each carrier is assigned a separate, non-overlapping frequency channel as shown in Figure 5.1. Power amplifier intermodulation products are controlled to acceptable levels by appropriate frequency selection and/or reduction of input power levels to permit sufficiently linear operation. Although such a control may also be needed in medium and high traffic earth stations[1], this is particularly critical in the satellite transponder where the output power is a very important and costly parameter. Output back-offs up to 3 dB (50% reduction of the output power) could typically be needed.

Signal distortions and adjacent channel interference are also to be accounted for.

The extent of the guardbands depends in part on the residual sidebands in each transmitted signal. They must also take into account the frequency drifts of the satellite and of earth stations' local oscillators. Doppler shifts of the satellite can also be significant for very low data rate transmissions.

Calculations of the effects of intermodulation products created by satellite and earth station amplifiers must also take into account changes in the relative signal strength received at the satellite, due to possible variations in the earth stations e.i.r.p., rain losses, antenna pointing deviations, etc.

---

[1] In earth stations equipped for transmitting several carriers through the same power amplifier, there is also a need for reducing the intermodulation products by operating the amplifier with some back-off.

FDM-FM multiplexing and modulation are dealt with in Chapter 3 (§ 3.4), Chapter 4 (§ 4.1.1.1), Chapter 7 (§ 7.5.3) and also, as concerns multiplex equipment and baseband formats, in Chapter 8 (§ 8.1.3).

In pace with the current trend towards a general digitalization of telecommunications and also towards ISDN networking, FDM-FM-FDMA, the first and once dominant technique to be implemented in satellite communications (especially in international communications), is now frequently replaced by TDM-PSK-FDMA.

In fact, the use of digital techniques enables significant increases in capacity. TDM-PSK-FDM can be extremely efficient for point-to-point and point-to-multipoint links and also gives the possibility to further increase this traffic capacity by implementing low bit rate encoding (LRE) telephony and digital processing techniques such as digital speech interpolation (DSI) and even digital circuit multiplication equipment (DCME) (see Chapter 3, § 3.3.7).

TDM-PSK multiplexing[2] and modulation are dealt with in Chapter 3 (§ 3.5), Chapter 4 (§ 4.2), Chapter 7 (§ 7.6.2.2) and, as concerns multiplex equipment, DCME, etc. in Chapter 8 (§§ 8.1.3, 8.1.4, etc.).

In order to optimize the link budget and the power-bandwidth efficiency, forward-error correction (FEC) techniques are very generally applied (Chapter 3, § 3.3.5 and Appendix 3.2). Since various coding methods and code rates are available, the use of FEC makes it possible to adapt the link budget design in a flexible way thanks to a trade-off between the quality (BER) and the occupied bandwidth. FEC encoding can be used either to relax the link budget parameters (e.g. smaller earth stations) or to improve the BER of a given link.

In conclusion, the possibilities of TDM-PSK-FDMA are intermediate between analogue FDMA and TDMA. As a result of digital processing (LRE, DSI, etc.), transponder capacities can be significantly increased (e.g. by a factor 2) compared to FDM-FM-FDMA. Although the potential capacity of full-transponder TDMA (see below § 5.2) is even greater, TDM-PSK-FDMA allows an easier transition to digitalization in the equipment of existing earth stations (it may be sufficient to change the FDM-FM modems and, of course the terrestrial interfacing, including multiplex, equipment).

5.2.2   Single Channel Per Carrier (SCPC)

In an SCPC system, each carrier is modulated by only one voice (or low to medium bit rate data) channel. In the case of voice (telephony), the channel can be processed in a number of ways.

Some – older – analogue systems use companded frequency modulation (CFM, see Chapter 3, § 3.2.1), but most systems are digital (PSK modulated). Although some of them remain using

---

[2] Because in TDM-PSK-FDMA, contrarily to SCPC, several communication channels can be multiplexed on a single carrier (possibly towards several destinations). This type of transmission is often called MCPC (multiple channels per carrier).

conventional 64 kbit/s pulse code modulation (PCM) with 4-PSK (QPSK) carrier modulation[3], 32 kbit/s ADPCM and low bit rate encoding (LRE), combined with powerful FEC schemes and with 4-PSK (QPSK) or 2-PSK (BPSK) are nowadays generally preferred, especially for domestic and very small aperture terminal (VSAT) systems.

In voice transmissions, the carrier is voice activated and this permits up to 60% power saving in the satellite transponder (the carriers are active only 40% of the time on average).

SCPC-CFM systems use bandwidths of 45, 30 or 22.5 kHz per carrier, while bandwidths for digital SCPC (SCPC-PSK) can extend also from 45 kHz (64 kbit/s) down to 22.5 kHz or even less (LRE). A 36 MHz transponder can therefore accommodate from 800 up to 1 600 simultaneous SCPC channels (see Figure 5.2). It is also possible, in FDMA, to share a transponder between a part used for SCPC carriers and another part used for TDM carriers.

Much more details on digital SCPC operation and earth station equipment will be found in Chapter 7 (§ 7.6.2.1).

The assignment of transponder channels to earth stations may be fixed (PAMA) or variable. In the latter case (DAMA), the channel slots of the transponder are assigned to different earth stations according to their instantaneous needs (see below § 5.5).

SCPC systems are cost effective for networks consisting of a significant number of earth stations, each needing to be equipped with a small number of channels (thin route, e.g. rural, telephony). In the case of relatively heavy traffic, TDM-PSK-FDMA systems tend to be more economical.

---

[3] Historically, this was the first SCPC system used in international communications (INTELSAT IESS-303 specifications).

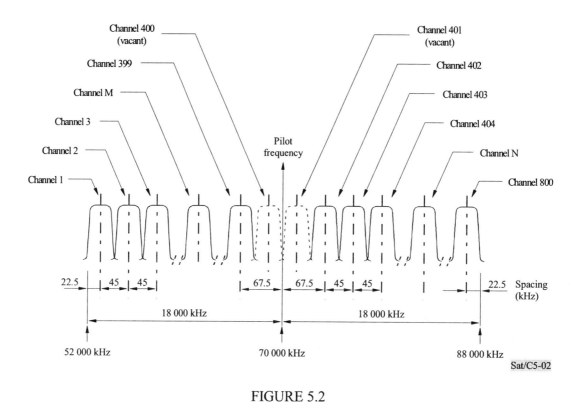

FIGURE 5.2

Typical frequency plan of a satellite transponder for 45 kHz SCPC channels

5.3 Time Division Multiple Access

Time division multiple access (TDMA) is a digital multiple access technique that permits individual earth station transmissions to be received by the satellite in separate, non-overlapping time slots, called bursts, in which information (e.g. digital telephony) is buffered. Each earth station must determine the satellite system time and range so that the transmitted signal bursts, typically quadriphase PSK (QPSK) modulated, are timed to arrive at the satellite in the proper time slots as shown by Figure 5.3. Signal timing and details of signal formats are discussed below. It should be noted that the bit rate of the transmitted bursts is generally many times higher than the bit rate of the continuous bit streams at the input of the earth station terminal.

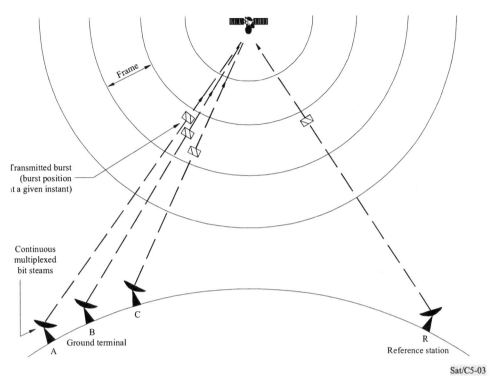

Transmitted burst
(burst position
at a given instant)

Continuous
multiplexed
bit steams

Frame

A
B
C
Ground terminal

R
Reference station

Sat/C5-03

FIGURE 5.3
Configuration of a TDMA system

Compared with FDMA, TDMA offers the following features:

i) The satellite transponder carries only one radio-frequency carrier (comprising the traffic from several earth stations). In consequence, there is no intermodulation caused by non-linearity and the satellite transponder can be driven nearly at saturation[4], giving a more efficient use of the satellite power.

ii) In TDMA the capacity does not decrease steeply with an increase in the number of accessing stations. Typically an 80 MHz transponder (INTELSAT satellite) driven by a single TDMA carrier can support about 1 600 telephone channels at 64 kbit/s channel bit rate. Note that the application of DSI and DCME techniques permits an increase of the transponder capacity by a factor from two up to five, but this is not exclusive to TDMA and, as seen above (§ 5.2.1), this can be implemented with any TDM carriers.

iii) The introduction of new traffic requirements and changes is easily accommodated by altering burst length and burst position.

---

[4] However, due to filtering effects on practical digital signals, non-linearity may impose a small back-off in the amplifiers (satellite and earth stations). See Chapter 2, § 2.1.5.3.

As explained in § 5.1, TDMA can be combined with FDMA by sharing a transponder between one TDMA carrier and others, e.g. multiplexed carrier(s). This mixed FDMA-TDMA process is discussed below (§ 5.3.3). However, it is only when TDMA is used alone in a satellite transponder ("full transponder TDMA") that the full efficiency quoted above in i) is obtained.

5.3.1   System configuration

As shown on Figure 5.4, in each TDMA earth station parallel input signals are entered, through Terrestrial Interface Modules (TIMs), under the form of digital bit streams (or of analogue streams which are first digitized) and are addressed to different receiving earth stations (multiple access function). The bit streams are assembled in the TIMs to form groups of bits called sub-bursts. Each sub-burst contains information originating from a given TIM and destined for a given receiving earth station (or a given set of receiving earth stations). The TDMA buffers convert these serial bit streams to a burst of data bits. In this burst, the sub-bursts are allocated separate portions of the transmit TDMA burst, following a group of bits referred to as the preamble. The burst is then converted to IF signals by the modulator (generally QPSK).

In a receiving earth station (Figure 5.5), the TDMA receiver demodulates each of the bursts from separate transmit earth station TDMA terminals, then extracts only the required portions (sub-bursts) to recover, through TIMs, separate serial bit streams.

As shown in Figure 5.6, the input signal to a satellite transponder carrying TDMA consists of a set of bursts originating from a number of earth stations. This set of bursts is referred to as a TDMA frame. The figure shows that a small time gap, called guard time, is left between bursts to ensure that they do not overlap with one another.

The first burst in the frame contains no traffic and it is used for synchronization and network control purposes. This burst is called the reference burst and is generally sent by a special reference earth station which is in charge of providing synchronization, monitoring and managing the traffic for the whole system.

Each burst is periodically transmitted at the TDMA frame rate which is generally a multiple of 125 microseconds (corresponding to the conventional PCM sampling frequency of 8 kHz). In each traffic burst, the preamble is used to synchronize the burst and assist in controlling the network. The reference burst contains only the preamble. In general, the burst preamble contains the following parts:

*   a carrier and bit timing recovery sequence, which has the function of providing the carrier reference and the bit timing clock required for demodulation in the receive terminals;

*   a bit pattern called the unique word which is used to identify the starting position of the burst in the frame for each receiving traffic earth station and the position of the bits in the burst, and to resolve bit ambiguity of the received PSK carrier.

Note that the transmit frame timing is generally obtained from the received frame timing by adding a delay (Dn) which can be provided by the reference station, via the reference burst:

*   order wire bits, carrying telephone and other service information for inter-station communications;

- control bits, carrying network management information. This section of the preamble is referred to as the control channel.

A supplementary feature can be implemented in TDMA systems: this consists in extending TDMA operation to several satellite transponders.

In this case (which is actually used in the INTELSAT TDMA standard), a single TDMA frame can correspond to different carrier frequencies and/or polarization, thanks to "transponder hopping" (i.e. by rapid switching, in earth station TDMA terminals, between the relevant transponders' frequencies and/or polarizations).

5.3.2 Synchronization in TDMA systems

Several different synchronization problems are involved in operating the TDMA system.

To demodulate the input burst mode PSK carriers, it is necessary to recover the carrier and bit timing within the recovery sequence at the start of each burst. Thus the TDMA demodulator usually has very high speed circuits for carrier and clock recovery.

Another crucial synchronization problem arises in the burst transmission timing at each accessing station in order to prevent overlapping of bursts at the satellite transponder. This control is called burst synchronization. The burst synchronization is performed so that each burst maintains a predetermined timing difference referenced to the position of the reference burst (received from the reference station) at the satellite transponder. To achieve the burst synchronization, the following methods have been studied:

- "global beam" synchronization – the transmit timing error is detected at each transmitting station by examining the received signal sequence which includes its own transmission as well as that from a reference station;

- feedback synchronization – the detection of the timing error is performed at the receive station or the reference station and the information on the burst position error is sent back to the transmitting station via the control channel;

- open loop synchronization – transmit timing is determined by knowing (calculating or measuring) the range from each station to the satellite.

It should be noted that in the first two methods, synchronization is based on a closed loop method. The final method requires that the station is able to receive its own transmitted bursts from the satellite.

In the case of satellite switched TDMA (SS-TDMA) where the downlink transmission uses several non-global coverages (spot beams), an additional type of synchronization is needed. Since it is necessary to route the bursts through the different spot beams according to their destination, by means of an on-board satellite switching matrix, the TDMA frame has to be organized with a synchronous switching sequence in which one or several time slots of the overall frame are assigned to each coverage area.

5.3.3   A typical example – The INTELSAT/EUTELSAT TDMA system

Table 5.1 shows the main transmission parameters of the most important TDMA systems under current utilization, i.e. those standardized for and implemented on the INTELSAT and EUTELSAT satellites.

These systems are designed to operate with 72 MHz or 80 MHz transponders at 6/4 GHz and 14/11 GHz and also to be compatible with satellite switched transponders (SS-TDMA). Terrestrial interfaces can be equipped for operating with direct digital interfaces (DDI, at 2 Mbit/s) or with digital speech interpolation (DSI) modules.

The systems make use of reference stations to achieve acquisition and synchronization of the traffic bursts, as explained above. Each reference station is equipped with sufficient redundancy to provide a high degree of reliability and, in order to satisfy INTELSAT/EUTELSAT continuity of service requirement, each reference station operating in a given coverage area has a separate secondary reference station. In SS-TDMA, however, only two reference stations are required per system to achieve the same continuity of service. The reference stations are also in charge of the function of monitoring the TDMA system.

Note that, thanks to the utilization of modern technology (VLSI circuits, advanced digital modems and FEC codecs, PC-based monitoring and control, etc.), new compact TDMA terminals are now available. More information on TDMA systems and earth station terminals is given in Chapter 7 (§ 7.6.3).

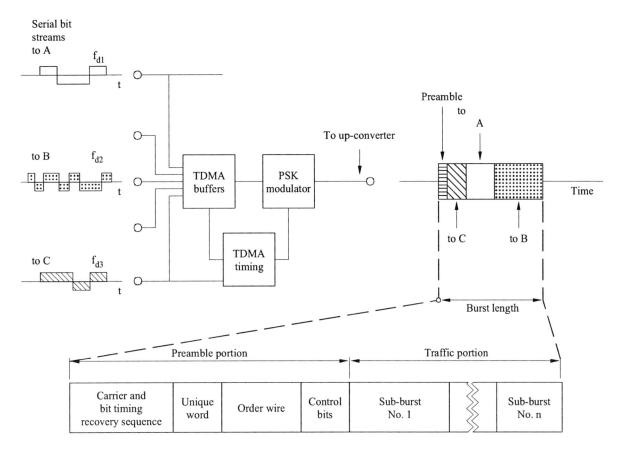

Sat/C5-04m

FIGURE 5.4

Simplified TDMA terminal transmit
operation and data format

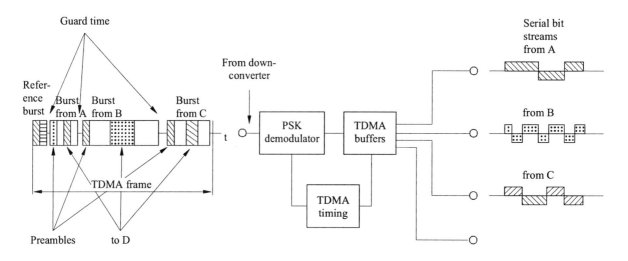

Sat/C5-05

FIGURE 5.5

Simplified TDMA terminal receive operation and data format

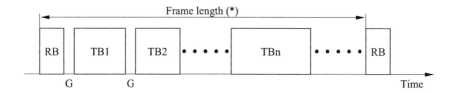

RB: Reference burst
(N.B: There can be a secondary reference burst in the frame e.g. in the INTELSAT/EUTELSAT TDMA)
TB: Traffic burst
G: Guard time
(*): e.g. 2 ms or 120 832 symbols (QPSK, in the INTELSAT/EUTELSAT TDMA)

Sat/C5-06

FIGURE 5.6

Simplified TDMA frame stucture

TABLE 5.1

Main transmission parameters of the INTELSAT/EUTELSAT TDMA system

Transmission bit rate	120.832 Mbit/s
Radio frequency band	Transmit: 6 GHz (INTELSAT) or 14 GHz (EUTELSAT)
	Receive: 4 GHz or 11 GHz
Modulation	4-PSK (QPSK)
Symbol bit rate	60.431 MSymbols/s
Demodulation	Coherent
Encoding	Absolute (not differential)
Error correction coding (FEC)	Rate 7/8 BCH (optional)
Nominal IF/RF bandwidth	72 MHz
Maximum earth station e.i.r.p.	81.5 dBW (6 GHz, INTELSAT)
	NOTE – E.i.r.p. for the INTELSAT VII satellite, minimum elevation angle, i.e. 10° and beam centre (correcting terms to be added for other angles and locations).
	83 dBW (14 GHz, EUTELSAT)
	NOTE – E.i.r.p. for 14.5 GHz and worst-case location (correcting terms to be added for other frequencies and locations).
Bit Error Ratio (BER)	In compliance with ITU Recommendation (ISDN)
Terrestrial interfaces	Digital Speech Interpolation (DSI) or Digital Non-Interpolated (DNI)

5.3.4   Frequency and Time Division Multiple Access (FDMA-TDMA)

As explained above, TDMA can be combined with FDMA by transmitting a particular TDMA carrier in a fraction of the satellite transponder, i.e. by sharing, by FDMA, the transponder between this TDMA carrier and other carriers (TDMA or continuous). Of course, while full-transponder TDMA transmissions normally use high bit rates (e.g. 60 to 120 Mbit/s) and occupy the full bandwidth of the transponder, such FDMA-TDMA transmissions use lower bit rates (e.g. 30 to 10 Mbit/s or even less).

This type of multiple access method requires a significant back-off in the power amplifiers (satellite and possibly earth stations), so that the transponder efficiency is lower than with pure TDMA.

However, this technique may be well adapted to some traffic configurations, by keeping some advantages of pure TDMA, especially for data transmissions[5]. The possibility of implementing demand assignment (DA) of the channels within the TDMA frame(s) may also be beneficial. In demand assignment TDMA (DA-TDMA) an earth station is assigned time slots on demand by a central or control earth station.

5.3.5   Packet transmissions

In packet data communications, the information is transmitted by grouping the data into packets. Each packet usually includes a header and a payload. While the latter contains the information proper, the former contains all service data necessary for identifying and routing the packet. Packet-oriented data communications are often implemented in VSAT networks, with a star architecture and the central earth station (called the "Hub") acting as a packet switch.

A particular case of packet communications directly transmitted through satellites is dealt with below (§ 5.3.6).

More details on packet transmission and switching in VSAT networks will be found in Supplement No. 3 of the Handbook on Satellite Communications, "VSAT Systems and Earth Stations".

The most modern form of packet-oriented communications is the asynchronous transfer mode (ATM) standardized by the ITU-T, especially for the Broadband ISDN (B-ISDN). In ATM, the packets are small, fixed size, blocks of data called cells (53-byte long, including a 5-byte header).

However, it must be noted that, in ATM communications (e.g. for B-ISDN), a satellite link, when implemented in the network, is generally used simply as a transmission medium where a well-adapted transmission mode could be RA-TDMA. It is only in the case of satellites with on-board processing (OBP), that the satellite can be used as an ATM Switch. For a more detailed description of ATM and its compatibility with satellite communications, see Chapter 3, § 3.5.4 and Appendix 3.4.

5.3.6   Random Access TDMA (RA-TDMA)

Random Access TDMA, the access and assignment mode used in the so-called Aloha systems, is a particular type of low bit rate TDMA, where a large number of users share asynchronously the same satellite transponder capacity by randomly transmitting short bursts or packets.

There are several types of Aloha systems, the main ones being pure Aloha and slotted Aloha.

In the pure Aloha method, the timing for packet transmission is purely random. Each user terminal (local earth station) will transmit a packet whenever there arises the need to transmit one. The packets so transmitted from different user terminals, however, may overlap at the satellite

---

[5] Low bit rate TDMA can be of advantage, for example, in very small earth station (VSAT) networks for corporate communication networks.

transponder. Such overlapping, or collisions, are monitored, either by the terminals themselves or through a control station and an acknowledgment process. Unsuccessful packets are retransmitted until correct reception by the end user.

When the traffic volume is small, the probability of successful transmission in one single satellite hop, i.e. within a short delay, is high. However, the probability of collision increases with the traffic volume up to a limit where the system becomes unstable, which means that a control mechanism should be implemented.

In the slotted Aloha method, a synchronization mechanism is introduced in the TDMA: a control station transmits to the terminals timing information which defines time slots the duration of which is approximately equal to a packet length. Each user terminal is allowed to transmit a packet in the period of each time slot. Two packets, or more, transmitted in the same time slot will result in a collision in the satellite transponder and will need retransmission, but as there will be no partial collisions, the overall probability of packet collision is lower.

In supplement to their rather high average transmission delay (due to retransmissions), the most significant operational parameter of Aloha systems is their throughput, i.e. the ratio between the actually transmitted traffic and the full capacity of the TDMA carrier. It can be shown that the maximum throughput of pure Aloha is 18.4% and is increased to 34.8% for the slotted Aloha.

Due to their implementation simplicity, satellite Aloha systems are largely used for low bit rate transmissions (where delay is not a major problem), especially in business networks (VSATs). Another very usual application is to common signalling channels, e.g. for DAMA reservation systems.

Again, more details on RA-TDMA will be found in Supplement No. 3, on "VSAT Systems and Earth Stations", to Edition 2 of the "Handbook on Satellite Communications (FSS)" (ITU-R, Geneva 1995).

5.4 Code Division Multiple Access (CDMA)

5.4.1 CDMA and Spread Spectrum basic concepts

There is a third multiple access category called code division multiple access (CDMA). CDMA systems were originally designed for military purposes, but they are now used also for commercial systems.

With FDMA, the signals from the various users are amplified by the satellite transponder in a given allocated bandwidth at the same time, but at different frequencies. In TDMA, they are amplified at different times but at the same nominal frequency (being spread by the modulation in a given bandwidth). In the third category of multiple access, called code division multiple access (CDMA), the signals operate simultaneously at the same nominal frequency, but are spread in the given (allocated) bandwidth by a specific encoding process. The bandwidth may extend to the entire transponder bandwidth, but is often restricted to a part of the transponder (in fact, CDMA can possibly be combined, if needed, with FDMA and/or TDMA).

As concerns the specific encoding process, each user is actually assigned a "signature sequence", i.e. with its own characteristic code, chosen in a set of codes assigned individually to the various users of the system. This code is mixed, as a supplementary modulation, with the useful information signal. On reception, from all the signals that are received, a given user earth station is able to select and recognize, by its own code, the signal which is intended for it and then to extract the useful information. Note that the other received signals can be those intended for other users, but they can also originate from unwanted emissions, which gives CDMA certain anti-jamming capability. For this operation, where it is necessary to identify one signal among several others sharing the same band at the same time, correlation techniques are generally employed.

The two most common CDMA techniques are based on:

- Direct sequence (DS), also called pseudo-noise (PN) modulation, which is the dominant technique;

- Frequency hopping (FH) modulation.

Although FH systems do find application in satellite communications and other techniques have been proposed, such as time hopping, this section will mostly deal with DS systems.

As a result of these techniques, the transmitted bandwidth (the given allocated bandwidth referred to above) is much larger (e.g. by 10^3 or more) than the baseband bandwidth of the information signal. This is why these processes are also called spread spectrum or spread spectrum multiple access (SSMA) techniques.

Some of the features of CDMA systems are summarized below:

- Unlike FDMA and TDMA, only minimum dynamic (frequency or time) coordination is needed between the various transmitters.

- The system intrinsically accommodates multiple users (each with their own code in the set) and new users can easily be introduced. In principle, no channel assignment control is needed[6]. Only the transmission quality (signal-to-noise) is subject to a gradual degradation when the satellite transponder loading increases. This is because, in a given receiving earth station, each other user signal (not destined to this station) is received as a supplementary unwanted, noise-like, signal (i.e. the other users transmit their signal in the same extended bandwidth).

In fact, the capacity of the system is limited by the quality of transmission which is acceptable in the presence of this "self-noise" or "self-interference" (also called MAI: multiple access interference) caused by the other users of the system:

- The power flux-density (pfd) of the CDMA signals, as received in the service area, is automatically limited, with no need for any other energy dispersal process.

- As already explained, CDMA brings out significant anti-jamming capability.

- It also provides a low probability of intercept by other users and some kind of privacy, due to individual characteristic codes.

---

[6] In CDMA, there is, in principle, no need for frequency or time assignment to the users, as is the case in DAMA controlled systems (e.g. instantaneous frequency assignment in SCPC with DAMA).

As a consequence of these features, CDMA allows a good flexibility in the management of the traffic and of the orbit/spectrum resources.

At present, these advantages prove to be particularly effective in new systems devoted to communications towards very small (possibly handheld) terminals, either for MSS applications[7] (e.g. the Globalstar system) or for FSS (e.g. the Skybridge system).

5.4.2   Direct sequence systems

Figure 5.7 summarizes the typical operation of a DS system (of course, there can be variations in the block-diagram, e.g. the spreading can be performed after modulation).

The PN code generator generates a pseudo-random binary sequence of length N at a rate R_c, with $R_c = N \cdot R_b$, R_b being the information bit rate. This sequence is combined, i.e. modulo-2 added with the information signal: this means that each information signal bit is cut in N small "chips" (whence the name "chip rate" for R_c), thus spreading the combined signal in a much larger bandwidth $W_{ss} \sim R_c$. This signal modulates, usually by PSK (BPSK or QPSK[8]), the carrier before transmission. At receive, a replica of the PN sequence is generated and combined, by a correlation process, in synchronism with the transmitted sequence. This results in "de-spreading" of the received signal which is restored, by coherent or non-coherent (see below § 5.4.5.2) demodulation.

5.4.3   Frequency Hopping systems

Figure 5.8 summarizes the typical operation of a FH system (again, there can be variations in the block-diagram). The system works similarly to the DS system, since a correlation process ("de-hopping") is also performed at receive. The difference is that here the pseudo-random sequence is used to control a frequency synthesizer, which results in the transmission of each information bit in the form of (N) multiple pulses at different frequencies in an extended bandwidth $W_{ss} = N \cdot \_f$ (_f being the frequency synthesizer step). Note that:

- coherent demodulation is difficult to implement in FH receivers because it is difficult to maintain phase relation between the frequency steps;

- due to the relatively slow operation of frequency synthesizers, DS systems permit higher code rates than FH systems. In fact, combined FH/DS systems have been proposed.

5.4.4   CDMA parameters and performance

In supplement to the spreading factor $M = R_c/R_s$ (where R_s is the symbol rate), the two basic parameters defining CDMA system performance are the processing gain and the jamming margin:

---

[7] Note that, in MSS CDMA systems, CDMA is generally also implemented for connecting gateway earth stations to the satellites. These feeder links (also called connection links) operate in the FSS.

[8] Note that, in the case of QPSK, two independent spreading sequences may be used for the two quadrature components (I and Q).

- the so-called processing gain g_p (G_p when expressed in dB) is the ratio of the transmitted (expanded) bandwidth to the information bandwidth (or bit rate)[9]:

$$g_p = W_{ss}/R_b \sim R_c/R_b$$

or, in dB:

$$G_p = 10 \log (W_{ss}/R_b) \sim 10 \log (R_c/R_b).$$

This expresses the ratio of the signal-to-noise (or e_b/n_0) of the output signal (after de-spreading) to the signal-to-noise of the signal at the receiver input[10].

Processing gains G_p can typically extend from about 20 dB up to 60 dB.

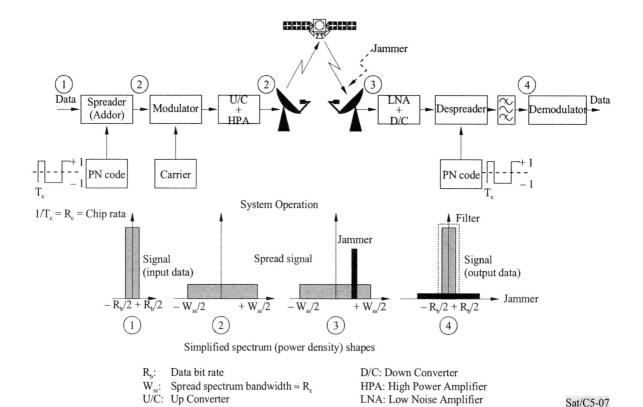

R_b: Data bit rate   D/C: Down Converter
W_{ss}:   Spread spectrum bandwidth $\approx R_c$ HPA: High Power Amplifier
U/C: Up Converter   LNA: Low Noise Amplifier

Sat/C5-07

FIGURE 5.7

Direct sequence (PN: pseudo-noise) CDMA

---

[9] $g_p = M$ for BPSK modulation and $g_p = M/2$ for QPSK ($R_s = 1/2\ R_b$).

[10]   In fact, in the absence of interference, G_p is not really a system gain, since signal de-spreading at reception only compensates spreading at transmission. As shown in Figure 5.7, it is against in-band jammers that G_p actually gives a gain.

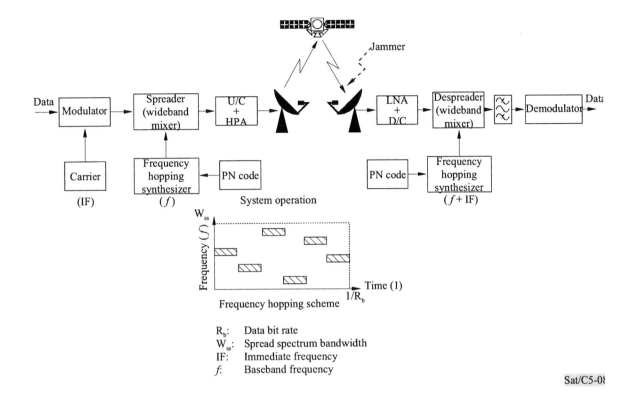

FIGURE 5.8

Frequency hopping (FH) CDMA

- the jamming margin m_j (M_j when expressed in dB) is the maximum tolerable interference-to-signal power ratios, i.e. the degree of interference which a CDMA system can accept for a given specified transmission quality (E_b/N_0, whence BER). By interference, this is meant any type of undesired signal, either "self-noise" due to the other users of the system and/or an external jamming signal (in the same extended bandwidth).

If j is the undesired signal power, its spectral density is j/W_{ss} and the total noise density is $n_0 = n_{0Th} + j/W_{ss}$. Assuming that j/W_{ss} is dominant against the thermal noise n_{0Th} and $s = e_b \cdot R_b$ being the desired signal power, then the theoretical jamming margin is:

$$m_j = j/s = (W_{ss} \cdot n_0)/(e_b \cdot R_b) = g_p/(e_b/n_0)$$

or, in dB:

$$M_j = G_p - (E_b/N_0) - L$$

In the last expression, the term L is introduced as an implementation margin (about 1 to 3 dB) representing various losses (spreading and de-spreading, demodulation, etc.).

This expression can also be seen as a rule-of-thumb value of the system capacity, i.e. of the number of users that can be accommodated (assuming that each user is received at the same level and that there is no external jammer). For example, if the specified E_b/N_0 is 8 dB and L = 2 dB, the capacity is about one tenth of the processing gain.

5.4.5   Asynchronous and Synchronous CDMA[11]

5.4.5.1   General

The description, in § 5.4.2, of direct sequence (DS) systems, actually applies to asynchronous CDMA (A-CDMA) links. Such links are particularly suited when synchronism cannot be achieved between the signals generated by the various users. This is specifically the case of the return links from small earth stations to base stations (e.g. in mobile-satellite systems)[12].

Whenever, on the contrary, all user signals are generated or may be coordinated within a single station, it can be more appropriate to synchronize the various signals. This is called synchronous CDMA (S-CDMA), although such a process should be considered more relevant to multiplexing than to multiple access.

The advantage of S-CDMA is that multiple access interference (MAI) can be avoided, therefore enhancing the system capacity[13]. This is particularly appropriate to satellite links where multiple paths do not destroy signal orthogonality (which could be the case in mobile terrestrial systems).

In consequence, in current CDMA mobile systems projects (satellite and terrestrial cellular), S-CDMA has been adopted for the forward links and A-CDMA for the return links.

5.4.5.2   A-CDMA

A-CDMA can actually be considered as the "true" CDMA. A-CDMA signals can be received by coherent detection or by non-coherent detection.

A-CDMA with coherent detection:

As usual in digital communications, coherent detection provides the best BER vs. E_b/N_0 performance. Furthermore, this performance is usually improved, just as in conventional PSK modulation, by combining forward error correction (FEC) coding with the spreading signature code.

---

[11] In the following sections (§§ 5.4.5 to 5.4.7), [Ref. 4] has been largely used. This reference also includes an important bibliography on the subject.

[12] In such systems, the forward link connects, through the satellite, a base (or gateway) station to the user terminals, while return links connect the user terminal to the base (gateway) station. Note that this is to be compared to the terminology used in VSAT networks, where the traffic from the base ("Hub") station to small remote earth stations (VSATs) is carried by "outroute" (or "outbound") channels, while, in the reverse way, "inroute" (or "inbound") channels are used (refer to "VSAT Systems And Earth Stations", Supplement 3 to Edition 2 of the Handbook).

[13] However, the capacity of S-CDMA is limited by the number of possible orthogonal codes in a given sequence and demand assignment could be again required if the number of possible users is greater than this capacity.

For example, convolutional codes are used or even trellis-coded modulation (TCM, see Chapter 4, § 4.2.3.2) can be used, depending on the system characteristics. It should be noted that using a rate r code does not mean that the transmitted bandwidth should be further increased by a factor 1/r.

In fact, FEC encoding can be considered as a first stage of the bandwidth extension that can be integrated with the CDMA proper spreading, thus keeping, in principle, M and G_p at their specified value[14].

The BER obtained by coherent detection of PSK modulated A-CDMA signals is the same as in conventional PSK, except that a term relevant to the MAI noise caused by the other users of the system must be added to the thermal noise in the formulas (Chapter 4, § 4.2.1, formulas (13) and (14)).

A-CDMA with non-coherent detection:

Notwithstanding its performance advantages, A-CDMA with coherent detection is seldom practicable for low or medium bit rate communication systems using small (even portable or handheld), low-cost terminals. This is especially the case in the higher frequency bands (14 and 11-12, 30 and 20 GHz), where significant phase noise can occur in the up- and down-converters.

Systems implementing robust, while efficient, demodulators with differential or non-coherent detection have been proposed and are actually implemented in various satellite and terrestrial mobile systems. In a particular system, FEC convolutional encoding is combined with an inner WH encoding and with a very long outer PN sequence. The period of this sequence is much longer than the chip rate and is common to all users (each user being assigned a different specific start epoch). The design of the demodulator and decoder used (including an estimator of the most likely transmitted WH function) gives to this system near to optimum performance, especially in the case of satellite links (where multiple-path and deep fading are much softer than in terrestrial systems).

5.4.5.3 S-CDMA

The most common codes for S-CDMA are the so-called orthogonal codes like the Walsh-Hadamard (WH) binary functions or the quasi-orthogonal codes (QO-CDMA), e.g. "preferentially phased" gold codes. As explained before, the utilization of such codes permits a very good isolation of the multiplexed channels, while avoiding MAI.

This is particularly true in satellite applications, where chip rates up to about 5 Mbit/s can be used without significant deterioration of the channels orthogonality (due to multipaths, the situation is much less favourable in terrestrial systems).

---

[14] Note that, just as in conventional FEC techniques, the implementation of "soft-detection" sampling (e.g. with two or three bits per sample) can improve by about 2 dB the performance (E_b/N_0 for a given BER), compared to the "hard-detection" with single-bit samples.

Carrier and chip clock synchronization is usually achieved in S-CDMA links by the transmission of a reference pilot carried in the (spread) S-CDMA multiplex and which is common to all user terminals' coherent demodulators.

Again, various implementation processes and demodulation schemes have been proposed and are actually implemented in satellite and terrestrial mobile systems. For example, just as for A-CDMA, a system, based on the combination of an inner WH encoding with a long outer PN sequence can be used. Quasi-synchronous CDMA systems, with slightly lower performance, have also been proposed.

5.4.6   Power control techniques in CDMA systems

As stated before, CDMA systems are particularly subject to mutual interference caused by the multiple users of the system (MAI). Furthermore, this effect is highly dependent on the relative signal levels received from the intended (wanted) user and from the various (unwanted) interferers. Basic system capacity calculations assume that each user is received at the same level. But actual situations can be very different and A-CDMA return links can be subject to so-called "near-far" effects. This term means that the intended (wanted) signal level may be more distant (i.e. possibly weaker) than an interfering ("nearer") signal[15].

In the case of satellite links, the unbalance between the received signals depends mainly on the path loss variation due to the user location with respect to the centre of the satellite antenna beam.

In the case of non-geostationary-satellite systems (NGSO, often using low-earth orbiting (LEO) satellite constellations), this location, and thus the path loss, is time-dependent and interfering signals may also arise from other satellites in the constellation. In addition, in the case of mobile systems, the path can be randomly obstructed by vegetation, buildings, etc. and also, due to the low directivity of the earth station (user terminal) antenna, multiple path effects can occur. In consequence, power-control techniques are generally needed in such systems[16].

Power control can be implemented by a coarse open-loop technique (generally based on the correction of the uplink power, through estimation of a pilot signal level) and/or by closed-loop technique (e.g. by a signal reception at the other end of the link and transmission of correction data via return signalling packets).

---

[15] Multiuser detectors (MUD) have been proposed as a means for counteracting interference effects, in particular MAI. A MUD is designed to receive not only the designated user signal, but also the other signals (their timing and codes being assumed to be known) and to cancel them from the useful signal. Various cancellation processes are possible but are generally complex.

[16] Of course the situation is much more favourable in satellite systems than in (cellular) terrestrial systems. While, in the latter case, multiple paths cause intense signal fluctuations which can result in power dynamic range of the received signals up to about 80 dB, in the former case this range should be of the order of 15 dB, at least for outdoor users.

5.4.7 Other considerations on CDMA systems

5.4.7.1 Satellite transponder operation

Non-linear effects in digital transmission have been discussed in Chapter 2 (§ 2.1.5.3). In fact, even in the case of a single CDMA carrier, satellite transponder non-linearities can modify cross-correlation properties of the code sequences, thus increasing the BER (for a given E_b/N_0).

This is particularly significant in S-CDMA systems.

In order to cope with these problems, the satellite transponder must be operated in a sufficiently linear region, i.e. with a given back-off (BO) below saturation.

But, in turn, this reduces the signal power emitted by the transponder. Therefore, as usual, an optimum operating BO must be found.

5.4.7.2 Diversity implementation

In CDMA, time diversity can be implemented to improve signal reception.

If the same user signal is received via multiple paths, i.e. with different delays (time offset being greater than one chip), the receiver can be designed to demodulate separately and to optimally recombine these "echoes" (this is called a RAKE receiver). This is a rather common practice in CDMA terrestrial (cellular) mobile systems (e.g. US IS-95), due to the multiple path environment found in urban areas. Moreover, space diversity can be implemented in such systems by receiving the same user signal from the neighbouring cells.

In the case of satellites, multiple paths are generally not significant within a single satellite spot beam. However, an important application of diversity operation in CDMA can be found in LEO multi-satellite systems. Here, each satellite moving over a given service area can reuse the same (spread) carrier frequency (with the same code or even with different codes), thus permitting "soft hand-over" during the changing of the satellite currently communicating with a given user. Moreover, full time-diversity effect can be used if a RAKE receiver is implemented on the user terminal.

5.4.7.3 Capacity of CDMA systems

The estimation of the traffic capacity of CDMA systems (or of its bandwidth efficiency, in bit/s/Hz) proves to be a very difficult and controversial task. Intrinsically, it should be about the same (in the case of S-CDMA) or generally lower (in A-CDMA, due to MAI) than the capacity of other access methods (FDMA/SCPC, TDMA). However, the following factors could improve this situation:

- Activity factor: As explained in Chapter 3, in telephony, if the RF signal is actually activated and transmitted only during the effective speaking periods, i.e. (statistically) 40% of the time, a 2.5 dB (activity factor) capacity advantage is obtained.

  In FDMA and TDMA, it is difficult to take into account this activity factor, except by using special techniques (voice detector in SCPC, digital speech interpolation – DSI – in digital multiplexed transmission). In CDMA, the activity factor is automatically accounted for (at least in A-CDMA), since transponder loading and MAI depends on the number of active channels.

- Diversity implementation, as discussed above.

- Frequency reuse: In satellites with spotbeams, the same frequency band can be used in the different spotbeams, with the same code set, or different code sets. Polarization discrimination frequency reuse could also be envisaged, although the cross-polarization isolation of the earth station (terminal) antennas could be insufficient in some applications.

5.4.7.4   Technological issues

As stated before, the main marketing "niche" for satellite CDMA systems currently appears to be communications towards very small (possibly handheld) terminals, either for MSS or for FSS applications. The success of such systems relies on the development of very low-cost user terminals. The design of such terminals, and, in particular of the modems, should be made possible by an intensive utilization of digital processing (DSP) techniques. Due to the progress in microprocessors and application-specific integrated circuits (ASIC), the feasibility of integrating a complete modem on a single VLSI chip can be envisaged.

5.5 Demand Assignment (DAMA)

5.5.1   General definitions

In most satellite networks, several stations have to share one transponder, which thus becomes a multiple access transmission resource. Apart from random access systems (Aloha type for example), the methods used for sharing this resource consist of assigning to each station (or channel):

- an available frequency band and power (for frequency division multiple access); or

- a time slot in a frame (for time-division multiple access); or

- a spreader code (for coded multiple access).

The allocation of an operating frequency, time or code for transmission and reception of a signal by a given earth station, obviously establishes the characteristics of each link in the network, as regards capacity and destination.

This allocation may be defined once only, and remain fixed throughout the operational period of the network: this is known as a pre-assigned, multiple-access system (PAMA).

Conversely, this allocation may be established instantaneously for the duration of each transmission or session: this is known as a demand-assignment multiple-access (DAMA) scheme.

The second category of system is more complicated, but in many cases it has the following advantages:

- Economical use of transmission resources

   Unused links do not actually consume resources in the satellite in DAMA mode. The reserve capacity on the satellite is not, therefore a function of the number of stations and channels, as in

PAMA mode, but rather, is a function of the overall volume of traffic in the network. Whenever the circuits to be linked are carrying little traffic, there is a gain (see § 5.5.7).

- Economies in the sizing of stations

This is especially true in the case of networks using SCPC access, where each channel corresponds to a separate modem in the earth station. When the DAMA is further developed, it is possible to assign an SCPC channel equipment to a circuit only when the link is established. If the circuits to be linked are not very busy, there is a real saving in the number of SCPC channels needed from a particular size of station (see § 5.5.7).

- Improvement in network connectivity

Unless all stations were to be over-engineered, it is rare for a network operating in PAMA mode to be fully meshed (i.e. for direct links to be provided between all stations in the network). Some communications are therefore only possible with a double link. The DAMA mode, on the other hand, allows direct links to be set up on demand between all stations, once radio dimensioning allows.

5.5.2   Functional description of a DAMA system

The following functions are implemented in a DAMA system:

- Signalling control

This function involves detecting a communication request (telephone call, or request for data transmission connection), recording the destination number and the additional information needed to set up the link (quality needed, for example), and then detecting a clear signal to release the link. This function is normally performed by the traffic terminal itself, running suitable software.

- Communication routing

This function involves determining the destination access of the communication, according to the signalling information recorded. This operation relies on a dialling plan, or a list of all the numbers handled by the system, very similar to the operating method of a switching centre. It may be performed by each traffic station or at a central site.

- Resources management

This function involves determining the transmission resources to be used for a new communication, on the satellite segment (frequencies or time slots) and on the ground segment (choice of modems). This operation has a varying degree of complexity, depending on the level of resource optimization required. If the system allows, a decision may be made during this operation to rearrange the resources used by other links (often the case in TDMA/DAMA systems).

- Resource assignment

This function involves configuring various traffic stations, and supervising the establishment and release of the link after configuration.

Figure 5.9 illustrates the relationships between these main functional parts in the context of a link establishment.

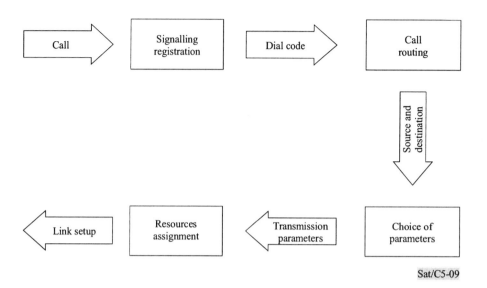

FIGURE 5.9

Main functional blocks of a DAMA system (link establishment)

5.5.3   Services available in DAMA mode

Most services operating in non-permanent circuit mode can be connected to a demand assignment system. Those most frequently met are as follows:

- telephone subscriber service ("long-line" DAMA);
- telephone links between exchanges (public or private);
- data transmission in connected mode;
- videoconferencing.

For data transmission many systems offer telephone modem emulation (usually with Hayes type signal processing). There are also DAMA systems using X.25 protocol (links are set up when the first virtual circuit to the caller is opened).

5.5.4   SCPC/DAMA systems

Single channel per carrier (SCPC) systems are especially well suited to DAMA operation for telephone applications. Each transmission or reception channel for each station on an SCPC network may actually be configured independently of the others. It is thus possible to modify frequency allocations for each new incoming communication to the network.

5.5.4.1   Operating principle for a telephone SCPC/DAMA system

When a call is detected, and after the signalling has been recorded and the destination determined, the system then chooses a transmission and reception channel (modem and terrestrial interface) in the two stations involved, as well as a frequency band for the two directions of the communication. If all these resources are available, and if the power required for the new bidirectional link does not exceed the maximum power allocated for the service in the transponder, the system proceeds to assign the chosen frequencies to the transmission and reception channels concerned, then monitors the setting up of the two new links in the network. These links will remain in place until the system detects a clear signal (for example, if one of the callers hangs up), which then stops the transmissions on the two frequency bands used, and returns the resources concerned (frequencies, available repeater power, and transmission and reception channels in the stations) to the "available" state.

5.5.4.2   The SPADE system

For international telecommunications, the INTELSAT SPADE system was the first example of a global network operating on the principle described above.

From 1973 onwards, the INTELSAT organization was operating a group of SCPC channels with 4-PSK modulation in DAMA mode, intended to provide operators of standard A stations with an economical means of communication with multiple destinations, if the traffic for each route did not justify an FDM-FM carrier. The network thus developed was known as SPADE (for "SCPC PCM multiple Access Demand assignment Equipment").

This system was characterized by distributed control of frequency allocations, unlike most other DAMA systems operating within domestic networks, which were based on centralized control. In the SPADE system, stations were thus able to configure their own channels, independently of each other. For the sake of system control, all network stations were able to transmit on a 128 kbit/s common channel, with TDMA (one time slot per station).

SPADE system stations consisted of a group of SCPC systems, controlled by a control and switching subassembly. The latter continually updated a table of active frequencies in the system. If a local communication was detected, it chose a new frequency pair from the list of available frequencies, and immediately notified the destination station of this assignment over the common signalling channel. After this, a free SCPC channel could be chosen, and the terrestrial link switched to this channel to set up the communication. Communication release, and setting up a link to a remote communication, were carried out in the same way.

Today, this system is abandoned, and is soon to be replaced by the TRDS system (thin route DAMA System) broadly inspired by the SCPC/DAMA systems for domestic networks, whose precise parameters are presently being finalized.

5.5.4.3   SCPC/DAMA systems for small-scale networks

Two important factors have considerably accelerated the development of low-cost SCPC/DAMA systems: development of digital modulation SCPC channel equipment, and the significant increase in the number of VSAT data type stations which can be used as signalling channels in DAMA systems.

Figure 5.10 shows an example of a small, star network SCPC/DAMA system for telephone applications. All the traffic stations are equipped with SCPC channels, and a signalling channel for receiving signalling information broadcast on a TDM carrier from the central signalling terminal (CST) located in the central station. They then send data packet signalling on a TDMA common carrier to the CST. This two-way signalling channel is used for demand assignment functions, and also for remote management of traffic stations from the network control centre (NCC) of the central station.

When a communication is detected in a traffic station, the SCPC terminal records the call number, then sends the request to the NCC via the signalling channel and the CST (which acts as a multiplexer for the signalling). The NCC analyses the request and chooses new working frequencies to set up the communication. This allocation is then sent to the traffic stations for action (see Figure 5.11, as an example, which shows the exchanges needed to set up a telephone call using ETSI QSIG signalling).

This is a centralized type system, since the signalling analysis and all allocation decisions are made at a central site (by the CST/NCC pair, in our example). This choice of architecture allows the greatest simplification of the traffic stations, essential in small network systems, and gives optimum use of satellite resources. Conversely, the signalling traffic is greater, and the central site may be a rather weak nodal point.

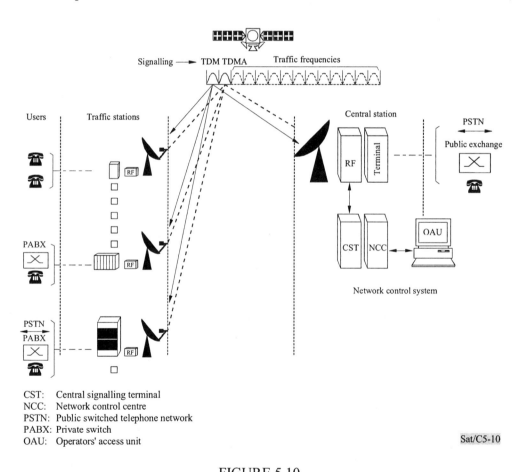

CST: Central signalling terminal
NCC: Network control centre
PSTN: Public switched telephone network
PABX: Private switch
OAU: Operators' access unit

Sat/C5-10

FIGURE 5.10

Example of an SCPC/DAMA for a small,
star-configuration telephone system

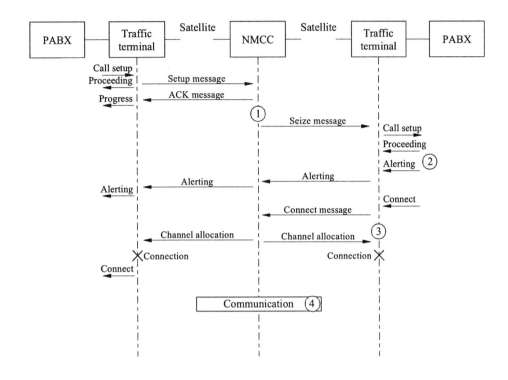

Sat/C5-11

FIGURE 5.11

Example of internal signalling messages for a
centralized DAMA system (QSIG signalling)

5.5.5 TDMA/DAMA systems

Medium rate TDMA terminals provide great flexibility in traffic reconfiguration. They even allow
time-division resources to be completely reorganized without any break in the associated traffic (this
is not generally possible with a traditional SCPC terminal). For this reason, the medium rate TDMA
systems are very often associated with DAMA mode for multiple services applications (telephony,
video, data transmission).

Some manufacturers produce centralized architecture systems, like the example of the SCPC
telephone network described in the previous chapter, where the terminal registers the signalling and
carries out the resource allocation instructions, whilst allocation analysis and decisions are made at a
central site. It should be noted that the internal signalling channels are transmitted and received in

this case by the traffic terminals themselves, unlike most SCPC systems, where a specific modem is added for signalling transmitted in TDM/TDMA mode.

There are also distributed architecture systems, in which each traffic station has a certain default transmission time slot, which it can organize independently of other stations, depending on its own traffic requirements. In this case, each station sends a header packet at the start of its reserved time slot providing callers with real time information on the destination of each subsequent transmitted burst.

5.5.6   MCPC/DAMA systems ("Bandwidth on Demand")

Multiplexed channel systems (MCPC), of which the most well-known examples are those which comply with INTELSAT IDR and IBS standards, have been defined and are generally used in PAMA mode. These systems do not actually provide the same degree of reconfiguration flexibility as the SCPC or TDMA terminals. In particular, it is not possible to modify the allocation of a single channel without affecting transmission on the other channels of the same multiplexer.

There are, however, more and more MCPC type networks, where a TDM/TDMA signalling channel is added to each traffic station, allowing all terminals to be controlled from a central site, using the main principle as for the SCPC/DAMA centralized architecture systems (see § 5.5.5.3). The operators may reconfigure the remote stations using a control computer linked to the signalling channels, to set up or remove the links. The computer is responsible for establishing frequencies, rates and transmission levels, according to operator demand. In some systems, an operator terminal is placed in each traffic station to gather user demands and transmit them to the central site via the signalling channels. It is therefore possible to request a link be set up, removed or modified, directly from the user site.

This type of operation is well suited to occasional data transmissions, or videoconferencing sessions, when more sophisticated signalling (ISDN for example) is unsuitable or unavailable. It allows a less efficient sharing of satellite resources than for SCPC/DAMA, but results in extremely simple, low-cost traffic stations. This type of operation has often been called "BOD: bandwidth on demand"[17].

5.5.7   Dimensioning DAMA systems

A DAMA system automatically assigns satellite circuits, e.g. SCPC channels, between any earth station according to the actual traffic demand, as described in § 5.5.1. In the following, it is shown how the required number of DAMA channel units is found for a given number of user channels and a given blocking factor.

As described in § 5.5.4.1, each channel unit in the earth station is able to transmit and receive at different frequencies due to frequency agile local oscillators (synthesizers). A DAMA control signal is provided for the channel frequency selection by the local oscillators.

---

[17] This term has also been used to indicate a TDMA/DAMA system.

The allocation of the satellite capacity to the user demand is dynamic. Therefore, the DAMA system achieves an enhanced use of the satellite capacity compared to PAMA mode.

The main system parameters of DAMA are the satellite transponder capacity (frequency slots), the number of circuits and earth stations, and the number of channel units.

When a domestic satellite network is established, the operation may start with a few channels in the pre-assigned (PAMA) mode. At a later stage, when the traffic increases, DAMA operation can be implemented as discussed below.

For the connection of local telephone lines to a satellite local (remote) earth station, two methods are possible (see Figure 5.12):

Method a) (Figure 5.12 a)):

Each subscriber accesses the satellite network directly through its local earth station. Each terrestrial telephone channel is connected to its corresponding satellite channel. Therefore, each satellite channel can be considered as a "long line" connected to the telephone exchange (switch) of its correspondent (in fact, in the case of a star network, this is usually the exchange located in the central earth station). These long lines, which are usually very lightly loaded (e.g. 0.1 Erlang) are well adapted to the implementation of DAMA and the number of channel units in the central earth station is calculated by using Erlang Tables (see Appendix 5.3).

Method b) (Figure 5.12 b)):

A local exchange is connected between the subscribers and the satellite earth station. It reduces the number of access lines to the earth stations and the satellite channels by switching the lines. Only the active access channels are switched to the earth station. The required number of concentrated satellite channels (i.e. the number of SCPC channels) can be calculated or taken from the Erlang Tables (see Appendix 5.3). It is obvious that the satellite channels are heavily loaded in this case.

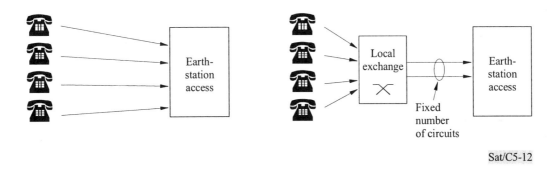

Sat/C5-12

FIGURE 5.12 A)   FIGURE 5.12 B)

Direct access to the satellite   Access to the satellite via local exchange

Figure 5.13 shows the two corresponding blocking levels in the satellite network.

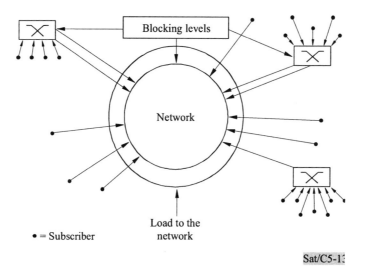

FIGURE 5.13

Two possible blocking levels in a network

An illustrative example is given in the case of method b):

- Assumptions:

    - Average number of subscribers connected to their local exchange: 24.

    - Traffic load from/towards the satellite (no distinction between incoming/outcoming circuits).

    - Blocking factor (i.e. probability of incoming circuits overload): 5% (0.05).

- Result: 24 subscriber circuits with 0.1 Erlang produces a total local average traffic of 2.4 Erlang. Referring to the Erlang Table (Appendix 5.3), six SCPC channels can support 2.96 Erlang and are therefore sufficient. The traffic load per channel is about 40%.

Under these conditions, the two network configurations are to be considered. Note that the definition of these configurations (Star and Meshed) will be discussed in more detail in § 5.6:

- **Configuration 1: Star network with a central earth station ("hub") with fully pre-assigned SCPC channels (PAMA):**

The traffic is switched in the hub as an exchange node. No DAMA is needed. Other advantages of this star architecture, especially as concerns the possible utilization of small remote earth stations will be explained below (§ 5.6). In fact, this simple configuration should be considered for use whenever most of the traffic is exchanged between the remote stations and the central earth station, which is often the case in rural networks. However, an obvious disadvantage is that the remote-to-remote traffic requires a double hop with its inherent double delay.

• Configuration 2: Meshed network with DAMA:

If a significant proportion of the traffic is exchanged between remote earth stations, then a DAMA controlled network is the best solution. However, at the beginning of operation, a pre-assigned – DAMA compatible – configuration is generally sufficient.

Later on, the implementation of a DAMA control should permit the traffic growth to be coped with, while keeping about the same transponder occupancy and the same number of channel units. This increase in network efficiency due to DAMA operation is illustrated by the example given below:

This example shows how, in a fully SCPC meshed network, the number of satellite RF carriers (i.e. the number of SCPC half-circuits) and the number of channel units equipping each earth station can be calculated. In this example, the network includes five earth stations (A, B, C, D, E) with an assumed traffic given by Table 5.2. Using a blocking factor of 2% (grade of service: 1 in 50), the network equipment required, calculated by the Erlang Table, is given by Table 5.3. The sum of the matrix elements is 98 (98 half-circuits). Therefore 49 pre-assigned SCPC channels should be installed for a PAMA network.

If a DAMA system is implemented at a later stage, the Erlang Table shows that the traffic capacity of these 49 two-directional channels will be 39.3 Erlang, whereas the PAMA network enables a traffic of 15 Erlang (sum of the elements in Table 5.2). Therefore the subsequent implementation of DAMA will:

• increase the traffic capacity by a factor more than two, with the same equipment;

• give full flexibility for the distribution of the traffic.

TABLE 5.2

Example of traffic matrix

Station	To/from stations				
	A	B	C	D	E
A		3.5	2	1.5	1
B			1.5	1	1
C				1.5	1.5
D					0.5

TABLE 5.3

**Equipment of the network
(no DAMA implemented)**

Station	Required number of SCPC channel units					Total
	Pre-assigned to the traffic to/from:					
	A	B	C	D	E	
A		8	6	5	4	23
B	8		5	4	4	21
C	6	5		5	5	21
D	5	4	5		3	17
E	4	4	5	3		16

- Total of pre-assigned RF carriers: 98
- Total of pre-assigned SCPC circuits: 49

Table 5.4 is another typical example[18] of a PAMA vs. DAMA comparison. The saving factor in channel unit (c.u.) equipment as a function of the number of destinations and of the level of traffic per station for a given blocking factor (e.g. 5%) is shown when the DAMA switching function is additionally implemented.

TABLE 5.4

**Typical example of Pre-Assignment (PAMA) vs. Demand Assignment (DAMA) comparison
(c.u.: number of channel units; S: saving factor in channel units)**

Number of Destinations	Traffic density								
	0.1 Erlang/station			0.5 Erlang/station			1 Erlang/station		
	PAMA c.u.	DAMA c.u.	S	PAMA c.u.	DAMA c.u.	S	PAMA c.u.	DAMA c.u.	S
1	1	1	1	2	2	1	3	3	1
2	2	2	1	4	3	1.3	6	5	1.2
4	4	2	2	8	5	1.6	12	7	1.7
8	8	3	2.7	16	7	2.3	24	12	2
10	10	3	3.3	22	9	2.4	30	14	2.1
20	20	5	4	40	14	2.9	60	25	2.4
40	40	7	5.7	80	25	3.2	120	45	2.7

---

[18] This example is taken from: "The INTELSAT DAMA System", Forcina, G., Oei, S., Simha, S., Conference publication No. 403 of IEE, Vol. II, IDCSC-10, 1995.

5.6 Network architectures

This section classifies satellite communication networks according to the arrangement of the links between the various earth stations of the network. This is called the architecture, the configuration or the topology of the network. The selection of multiple access/multiplexing/modulation as a function of the network architecture is dealt with in Appendix 5.1, § AP5.1-2.

NOTE – Referring to the Open System Interconnection (OSI) Model (see Appendix 8.2), this subsection describes architectures only at the layer 1 – or physical layer (as opposed to possible interconnections at layer 3 – or network layer).

5.6.1   One-way networks

In one-way networks, information is transmitted in only one direction, generally by using the satellite to relay this information for distribution (distribution networks) or collection (collection networks) from/towards a central earth station towards/from a number of receive-only/transmit-only remote earth stations. Consequently, these networks provide the first example of a star architecture (see Figure 5.14 a)).

i)   Distribution networks

In distribution networks, the central earth station, sometimes called the "hub" is, in principle, a transmit-only station (although receive facilities are usually provided for control and network supervision) and the remote earth stations are receive-only (RO) stations (often called terminals). The distribution can be selective, i.e. addressed to a group of terminals (or even to a single one) or it can be received by any terminal in the coverage. It can be emphasized that the satellite is really a unique transmission medium for such services of the broadcast type. In particular, terrestrial packet switched networks cannot, at present, be configured for data broadcast using standardized protocols.

There are many applications of satellite information distribution to RO terminals. For example, the operation of the global positioning system (GPS) is based on the reception, by small, handheld, terminals, of reference signals broadcasted by several orbiting satellites. Another example is offered by paging systems where dedicated signals are addressed to the users.

Another important application is data distribution towards RO microstations or VSATs (very small aperture terminals). This is generally done in the framework of business communication networks and the distribution can be either general or restricted to closed users groups (subnetworks)[19]. But the main and most popular application of satellite distribution is the broadcasting of television programmes (or radio programmes) towards small and inexpensive TVRO stations, either for individual (DTH: direct-to-home) or for collective reception. There are now tens of millions of such TVROs throughout the world and the importance of this application is currently and very rapidly

---

[19] The case for two-way (transmit and receive) VSATs is discussed below (§ 5.6.2). More generally, VSATs are the subject of Supplement No. 3 ("VSAT Systems and Earth Stations", Geneva, 1995).

growing, due to the advent of TV "bouquets", i.e. the simultaneous digital transmission of multiple programmes (e.g. eight per satellite transponder, see Chapter 7, § 7.10).

The design of TV broadcasting systems offers a first and very simple example of service cost optimization. On the one hand, the cost of the central component of the system, i.e. the investment cost of the central earth station and of the signal processing/broadcasting facilities and the recurrent operational cost (space segment leasing, personnel, spares, etc.) are shared between a multiplicity of users.

On the other hand, by implementing a satellite with a high e.i.r.p. (spot beams, full transponder operated near saturation), the TVRO stations can use small antennas and can be mass-produced at very low cost. For example, 0.6 m to 0.8 m diameter antennas are commonly used to receive digital TV bouquets broadcast at 11 GHz from a high power FSS satellite transponder (e.g. 49 dBW e.i.r.p.). Note the reduction of the antenna diameter may be limited by the need to avoid interference from nearby satellites. This should not be the case in the BSS where the satellites are significantly spaced apart.

In the case of data distribution to VSATs, the cost and the limited availability of the space segment often remain important factors; the system design is usually based on the use of a portion of a transponder and the optimization is a compromise between VSAT cost and space segment occupancy.

ii)   Collection networks

Collection networks operate in a reverse manner to distribution networks, i.e. they are used to transfer collected data one way from remote earth stations towards one central earth station. Since most such networks are currently operated outside the FSS, especially for remote sensing from unattended stations, they do not appear to have found many applications, up to now, in the framework of the FSS.

5.6.2   Two-way networks

Two-way networks provide the most general applications of satellite communications. The following types of architecture will be examined: point-to-point and point-to-multipoint links, mesh, star and mixed configurations.

i)   Point-to-point and point-to-multipoint links (Figure 5.14 b))

Point-to-point links are elementary and are just a two-way connection between two earth stations. The term point-to-multipoint link describes a network connecting reciprocally a limited number of earth stations (say, up to five), which are preferably all of the same type. Point-to-point and point-to-multipoint networks are the most common and conventional types of satellite networks. Their most usual applications are the interconnection of main communication centres for heavy route telephony, high bit rate data transfer, television programme exchange, etc.

ii)   Mesh networks (Figure 5.14 c))

The term mesh network describes a network capable of fully interconnecting a significant number of earth stations, usually all of the same type (a point-to-multipoint link can be considered as a simplified form of mesh network). Mesh networks allow interconnection between non-hierarchical centres. When locally connected to terrestrial networks, the earth stations are operating as nodes of a complete communication system and there is no need for a transit exchange inside the satellite mesh network. However, if the network is operated in TDMA, the TDMA reference station may include traffic management facilities (including demand assignment) which plays the role of a transit exchange. Also, in advanced rural satellite systems, mesh interconnections can be implemented to allow direct (single hop) communications between local (remote) earth stations, usually under the control of a centralized DAMA control.

iii)   Star networks (Figure 5.14 a))

Star architecture, which has already been considered in the case of one-way networks, now finds many applications for two-way networking between hierarchical centres, i.e. one main centre in which the central earth station, e.g. the "hub" is installed and a number of remote sites, with relatively low traffic demand, at which small stations or microstations ("VSATs") are installed. The two most common applications of this architecture are:

- rural telecommunications, i.e. (according to a CCITT (ITU-T) definition) communications for scattered dwellings, villages or small towns in areas having one or more of the following characteristics: difficulty in obtaining electricity, lack of technically skilled local personnel, isolated area, severe environmental conditions, cost restrictions and where thin-route telephony is generally the main service required;

- corporate communications between a central data processing facility, e.g. the "host" computer and remote DTEs (digital terminating equipment). The host computer is connected to the hub (central earth station) and the DTEs are directly connected to the VSATs.

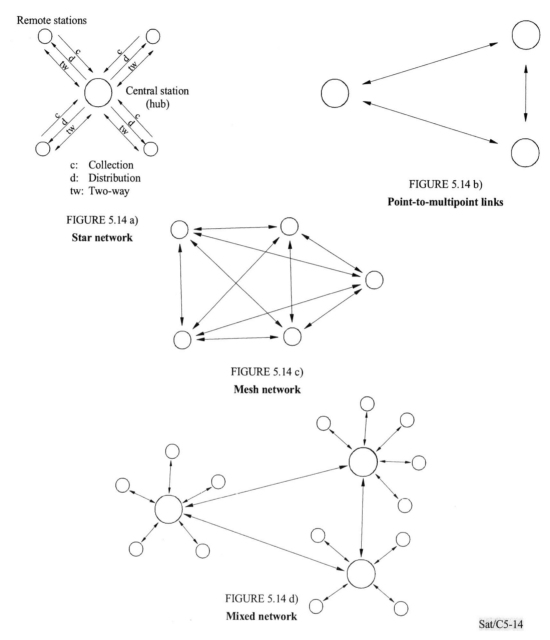

c: Collection
d: Distribution
tw: Two-way

FIGURE 5.14 a)
Star network

FIGURE 5.14 b)
Point-to-multipoint links

FIGURE 5.14 c)
Mesh network

FIGURE 5.14 d)
Mixed network

Sat/C5-14

NOTE - In all of the above figures, the satellite is not represented and the earth stations which are operating as communication nodes are schematized by a circle.

Two-way satellite star networks implemented with a large central station and small remote stations have special features which can be summarized as follows:

• as in the case of one-way star network, the overall system cost is minimized because the high cost of the central facility, which is necessarily implemented with a relatively large antenna and with costly radio and control equipment, is shared between multiple users;

- the high receive figure of merit (G/T) of the central station permits the remote users to be equipped with low-cost stations because only a small antenna gain and a low power transmit amplifier are needed;

- the outroute carriers (from the central station towards the remote stations) require a high RF power from the satellite transponder. However, the inroute carriers (from the remote stations towards the central station) require much less power[20]. In this case, the total power required from the satellite transponder is primarily controlled by the outroute carriers;

- a maximum limit to the remote stations' amplifier power (and a minimum limit to their antenna diameter) can be imposed by the maximum permissible off-axis e.i.r.p. (uplink interference towards adjacent satellites). For example, assuming a 1 m antenna for a remote station in the 6/4 GHz band, the maximum transmit power should be about 0.5 W/4 kHz (-3 dB(W/4 kHz)) if Recommendation ITU-R S.524 is to be respected. The consequences are that a sufficiently large hub antenna should be used and that a significant amount of energy dispersal may also be needed;

- in principle, communications between two remote stations cannot be established directly and communications necessitate a double hop through the central station (this is a difference from mesh networks). Generally, this does not cause problems in the case of data networks, but it can be a drawback in the case of telephony (e.g. rural telephony). This is why INTELSAT, in the case of the VISTA service which uses star networks, provides for the possibility of some direct links between small (standard D1) remote stations. This, however, involves supplementary costs as the space segment charge is higher and the e.i.r.p. from the small earth stations must be increased for the channels transmitted towards another standard D1 station. This may exclude the utilization of solid-state amplifiers;

- in the case of rural telecommunications, the central station, e.g. the hub, is usually provided with switching equipment and often acts directly as a "remote" telephone (or data) exchange for the remote stations. If provided for in the network, it can also manage the connection/disconnection process of the direct links between the remote stations: this provides for the possible DAMA operation of the central station. Note that, in this case, the network uses a mesh architecture;

- if double-hop delay is acceptable[21], the star architecture can be used for providing fully interconnected links between small, low-cost remote stations (equipped with small antennas and low-power, fully solid-state amplifiers). Of course this involves the use of a high satellite e.i.r.p. for each outroute carrier. In such a configuration, the central station, e.g. the hub, operates as a relay between the remote stations. Moreover, it can be used to perform many more functions than simple demodulation/remodulation and it can be used to optimize separately the outroute

---

[20] As a first approximation, the outroute to inroute transponder power ratio is equal to the central to remote antenna aperture ratio.

[21] Double-hop telephony should, in general, be avoided. However, it could be accepted in certain situations, in particular, in rural telephony networks, for remote-to-remote station communications in the case where the cost of the implementation of direct (single hop) communications should be too high and should not be justified by sufficient traffic (see above: mesh networks, § 5.6.2 ii)). In this case, high-quality echo cancellers need to be used.

and the inroute links. Some signal processing operations which could be implemented in the hub are: demultiplexing the inroute carriers, switching the signals towards their destination and multiplexing them in a different arrangement, if necessary. In the simplest case of SCPC inroute and outroute carriers, only switching is required (this is a simplified DAMA function). In the case of fully digital networks, other possible hub operations are: signal regeneration, framing and formatting, inserting/dropping of auxiliary signals (such as control and service channels), error correction implementation, etc. In fact, these operations in the hub are very similar to those of an active satellite, i.e. a satellite with a signal processing/switching payload.

iv)   Mixed networks

An example of a mixed network is given in Figure 5.14 d). Here, three high-traffic stations, located in provincial capitals, are in a multipoint configuration and each of them constitutes the central station of a star subnetwork.

Another example is show in Figure 5.14 e), where the hub is equipped with two antennas looking at two satellites, (D) and (I). (D) is supposed to be a domestic satellite with a high e.i.r.p. and (I) to be an international satellite. Here, the hub serves as a relay allowing VSATs in Country A to get access to a huge "host" data processing facility (e.g. a big database) located in a Country B. A simplified scheme would consist of a single satellite used both for linking the VSATs to the hub (e.g. through a high e.i.r.p. transponder leased on an international satellite) and for linking the hub to the host. In this case, the hub would be equipped with a single antenna. Note that such configurations are applicable only in the cases where double-hop delay does not cause problems.

These are only typical examples among many possible mixed configurations.

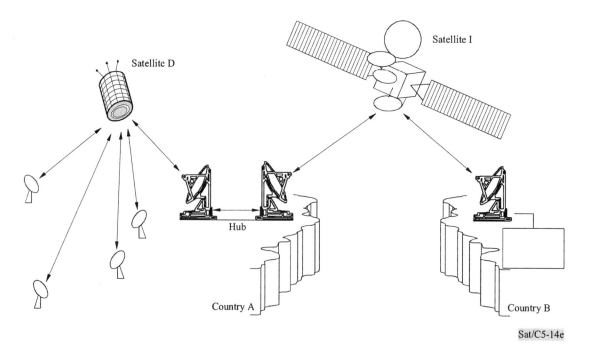

FIGURE 5.14 E)

Mixed data network and international relaying

5.6.3   LEO satellites network architectures

Non-geostationary-satellite systems (non-GSO) are currently under development and construction and some of them should begin operation before the end of the century, both for mobile services (MSS) and fixed services (FSS). Many (but not all) of these systems are based on constellations of low-earth orbit (LEO) satellites (e.g. at an altitude around 1 000 km).

The architectures of these systems are generally derived from the concepts developed for terrestrial mobile cellular systems, i.e. are based on "cells". This means that a given user terminal (a remote earth station) within a cell communicates with the satellite, the instantaneous location of which is closest to the terminal. From this satellite the communication is transferred, either directly, through inter-satellite (ISL) links, or through a gateway earth station, to its destination, which can be either another user terminal or a terrestrial communication node. Handover procedures are provided to avoid interruption of the communication when a satellite is about to go out of view, with the change occurring to the next available satellite of the constellation (i.e. passing over the cell). The process of communication transfer is managed and routed through service access points and switches located in control centres and/or in the gateways.

More details on FSS LEO systems are given in Chapter 6.

APPENDIX 5.1

Additional considerations on coding, modulation and multiple access

This appendix gives a few additional considerations to coding, modulation and multiple access. These considerations apply to systems based on geostationary satellites with transparent transponders.

AP5.1-1 Bandwidth limited versus power limited operation

A significant comparison is worth emphasizing between bandwidth limited and power limited operation of a satellite system. This is illustrated by Figure AP5.1-1 which shows, in relative terms, transponder traffic capacities, i.e. the number of channels available through the transponder as a function of two parameters, the transponder e.i.r.p. and the earth station G/T. Of course, these two parameters can be strictly added (in decibels) only when the downlink carrier-to-thermal noise ratio is largely limiting the overall link budget. However, similar, and more realistic, graphs can be drawn if the calculations take into account the other noise sources, i.e. uplink thermal noise, intermodulation noise and interference noise and also the transponder operating points, i.e. the actual input and output back-offs.

For each type of modulation (BPSK or QPSK) and for each type of possible error-correcting scheme (forward error correction (FEC) with rate ½ or ¾), the figure shows two regions corresponding to two types of operation:

- the bandwidth limited operation region (horizontal line) – in this part of the graph, the traffic capacity is maximized and the sum of the channel bandwidths, including guardbands, is equal to the available transponder bandwidth. The border of the region gives the minimum value of e.i.r.p. + (G/T) for this type of operation, i.e. the minimum earth station G/T for a given satellite e.i.r.p. (or conversely, the minimum satellite e.i.r.p. required for operating with a given earth station G/T);

- the power limited operation region – this part of the graph corresponds to a lower traffic capacity, due to the fact that the available transponder bandwidth is not fully occupied by the signals. This is the consequence of implementing the system with earth stations with reduced (G/Ts), i.e. equipped with smaller antennas (and/or with more noisy LNAs).

It should be noted that Figure AP5.1-1 gives only typical cases of digital transmission. In fact, such a figure should be drawn for each actual application case, with due account taken of the following factors:

• type of multiple access – the optimum type is full transponder TDMA, because this allows the transponder to be operated at saturation (or very near saturation). In all other types of operation (FDMA), the traffic capacity of the transponder is reduced because of the additionally required back-off and intermodulation products noise;

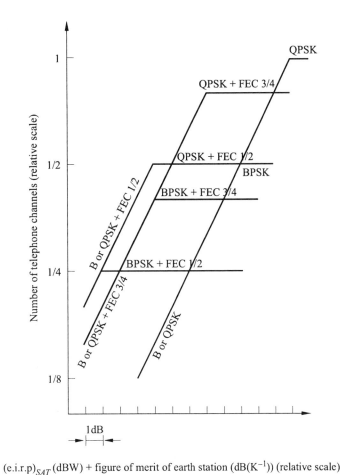

Sat/C5-AP5-11

FIGURE AP5.1-1

Relative traffic capacities (digital transmission)

- type of modulation – for a given signal, the occupied bandwidth and the required power are directly dependent on the type of modulation. For example, bandwidth occupancy is reduced to R/N when using M-ary digital modulation (M-PSK) (with R = information bit rate in bit/s and N = log₂ M. This means that the maximum traffic capacity of the bandwidth-limited region is increased by a factor N (compared to 2-PSK, i.e. BPSK), but simultaneously, for M > 4, the power-limited region is extended, because the required power per transmitted symbol is increased. Ex-CCIR Report 708 (Geneva, 1982) gives typical examples of transponder traffic capacities with M-PSK (M = 2, 4, 8 and 16, as indicated by the number of phases Φ). In practice, only BPSK (M = 2) and QPSK (M = 4) types of modulation are commonly used in current FSS systems;

- error-correction encoding/decoding – as explained in Chapter 3, Appendix 3.2 and Chapter 4, the utilization of error-correcting codes is becoming more important and popular in modern digital satellite transmission. This is because FEC codecs, often with highly efficient soft decoders, are now available as low-cost LSI "chips". For example, if a rate ½ FEC codec with Viterbi decoding is implemented in the earth stations of a power-limited system for a given earth station characteristic (antenna diameter and G/T, etc.) the traffic capacity of the system will be increased by a factor of 3 (due to the 5 dB decoding gain). Of course, this simultaneously reduces the maximum traffic capacity of the bandwidth-limited region by a factor of 2 (because the bandwidth required for transmitting the information is double). This means that error-correction should not be used when operating with very high e.i.r.p. satellites unless needed by an interference environment. In actual applications, error correction code rates and schemes must be carefully balanced with bandwidth limitations. In fact, a combination of FEC coding and a higher M-ary modulation (e.g. M = 8) can be a good trade-off between power and bandwidth limitations;

- baseband processing – in Figure AP5.1-1, the various possible baseband processing schemes have not been considered and the graphs should be interpreted as giving the basic traffic capacity for a given type of channel, e.g. 64 kbit/s. However, various baseband processing schemes, in particular data compression, should be mentioned as having a direct impact on the actual traffic capacity of a transponder. As concerns telephony, voice activation, which is particularly important in SCPC links, and digital speech interpolation (DSI) can both multiply by a factor of about 2.5 the basic traffic capacity. When combined with reduced bit rate telephony, the multiplication factor can be increased, e.g. to 5 with 32 kbit/s ADPCM (and more with very low bit rate (LRE) codecs, e.g. 4.8 kbit/s which are now rather common, at least for mobile telephony). Note the combination of DSI with reduced bit rate telephony can be implemented on most types of TDM and TDMA links by using a digital circuit multiplication equipment (DCME, see Chapter 3, § 3.3.7).

AP5.1-2 Selection of multiple access, multiplexing and modulation

i)   Point-to-point, point-to-multipoint and mesh digital networks

Table AP5.1-1 compares various possible multiple access methods for application to point-to-point, point-to-multipoint and mesh digital FSS satellite communication networks. The associated multiplexing methods are also indicated. PSK (QPSK or BPSK modulation is generally used, in combination with forward error correction (FEC by convolutional encoding with soft-decision

Viterbi decoding plus, possibly, concatenation of an outer Reed-Solomon (RS) code (see Chapter 7, § 7.6.2.2). Note that this table summarizes typical examples and should not be by any means considered as fully comprehensive. New concepts have recently appeared, especially for non-GSO systems.

Tentative conclusions are as follows:

- FDMA remains the simplest multiple access for low or medium traffic capacity links with a few destinations. Although not mentioned in the table, analogue multiplexing (FDM) and Modulation (FM), were, until a recent past, the most common processes. In the general trend towards full digitization, they are now very generally and very advantageously replaced by TDM/PSK multiplexers/modems (e.g. INTELSAT IDR and IBS standards);

- for high traffic capacity links with multiple destinations, full transponder TDMA should be considered as the more effective mode. This is also shown by Figure AP5.1-2. This figure illustrates the fact that the equipment cost in FDMA operation increases rapidly with the number of modems (i.e. with the number of destinations). However, with TDMA, the equipment cost is nearly independent of the number of destinations. In this figure, the cross-over point can vary typically from five for medium bit rate, simplified TDMA equipment to ten for high bit rate TDMA equipment;

- it is for the same reasons that, even for smaller traffic capacity, TDMA, in the form of FDMA-TDMA (or shared transponder TDMA) remains attractive for mesh networks, with a medium or large number of nodes (e.g. more than five) for possible use with integrated services business communications networks.

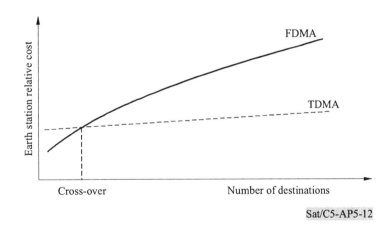

Sat/C5-AP5-12

FIGURE AP5.1-2

TDMA versus FDMA - Cost comparison

TABLE AP5.1-1

**Point-to-point, point-to-multipoint and mesh digital networks –
Comparison between various multiple access methods**

Case No.	Multiple access	Multiplexing	Advantages	Disadvantages	Preferred applications
1	FDMA	No multiplexing (SCPC) + (possibly) DAMA	• Earth station (E/S) HPA power proportional to the traffic. • Simple implementation: no problems of synchronization, etc.	• Back-off in satellite transponder and E/S HPA. • In the E/S, as many modems (channel units) as separate point-to-point links (without DAMA) or as required by the traffic capacity (Erlangs) (with DAMA).	• E/Ss with small traffic and/or multiple destinations ("thin route", rural telephony, etc.).
2	FDMA	Multiple Channels Per Carrier (MCPC) by TDM + (possibly) DSI and DCME	• Earth station (E/S) HPA power proportional to the traffic. • Simple implementation: no problems of synchronization, etc. • Low cost for a few destinations.	• Back-off in satellite transponder and E/S HPA. • Complex modem and mux/demux equipment for multiple destinations. • Reduced flexibility in traffic configuration.	• Point-to-point and point-to-multipoint traffic with a few destinations: probably the most attractive solution in such cases.
3	TDMA (full transponder, typically 40 to 120 Mbit/s)	TDM + (possibly) DSI and DCME	• Very well adapted to many destinations and high traffic capacities. • Only small back-off in satellite transponder and in E/S HPA. • Easy traffic reconfiguration and demand assignment (DAMA).	• Network synchronization and acquisition by a reference E/S needed. • Relatively complex operational problems. • Same HPA power in all E/Ss, independently of their traffic.	• High traffic capacity point-to-multipoint and mesh networks.
4	FDMA – TDMA (shared transponder TDMA, typically <40 Mbit/s)	TDM + (possibly) DSI and DCME	• Very well adapted to many destinations and medium or low traffic capacities. • Easy traffic reconfiguration and demand assignment (DAMA).	• Network synchronization and acquisition by a reference E/S needed. • Back-off in satellite transponder and E/S HPA.	• Medium traffic capacity mesh networks.

ii)   Star digital networks

Table AP5.1-2 compares various possible multiple access and multiplexing methods for application to star digital FSS satellite communication networks. Just as in i) above, PSK (QPSK or BPSK) modulation is generally used, in combination with forward error correction. Again, this table summarizes only typical examples and should not be considered as complete. However, as it is, this table summarizes most types of star networks which are in current operation for two main applications: thin route (rural) telephony and data networks implemented with microstations (VSATs) for business applications.

TABLE AP5.1-2

Star networks – Comparison between various multiple access methods

Case No.	Multiple Access/Multiplexing		Advantages	Disadvantages	Preferred applications
	Outbound link	Inbound links			
1	FDMA No multiplex (SCPC) or TDM (MCPC)	FDMA (SCPC) + (possibly) DAMA	• Wide range of applications. • See also same as in case 1 of Table AP5.1-1.	• In the E/S, as many modems (channel units) as separate point-to-point links (without DAMA) or as required by the traffic capacity (Erlangs) (with DAMA). • See also same as in case 1 of Table AP5.1-1.	• Rural telephony. • Data networks for business communications (VSAT systems).
2	FDMA-CDMA (e.g. S-CDMA)	FDMA-CDMA (e.g. A-CDMA with Random Assignment (RA))	• Very simple implementation • Activity factor directly accounted for. • Intrinsic energy dispersal can allow transmission by very small antennas.	• Inroute traffic limited by Multiple Access Interference (MAI). • Back-off needed in satellite transponder for sufficient linearity.	• Data networks for business communications (VSAT systems).
3	FDMA-TDM	FDMA-TDM/RA (Aloha or slotted Aloha or Demand Assignment)	• Wide range of possibilities in messages bit rates. • Flexibility in traffic rearrangement. • High bandwidth and power transponder efficiencies. • Intrinsic compatibility with many simultaneous inbound carriers.	• Relatively complex implementation.	

The following comments apply:

• Until a recent past, most rural communication networks were operating with SCPC/CFM links. However, digital SCPC links based on lower bit rate voice encoding at 32 kbit/s, 16 kbit/s ADPCM or less (e.g. 4.8 kbit/s) are becoming of common utilization, thanks to the availability of LRE codecs using VLSI technology[22], making it possible to build very compact, low-cost rural telephony microstations.

• Case 2 of Table AP5.1-2 refers to CDMA. It is the availability of signal processing VLSI circuits which also allowed to build economical systems based on CDMA technique. CDMA can offer advantages, especially in the case of potential interference. Their main drawback is their rather low transponder utilization efficiency, due to their "self-noise" (Multiple Access Interference – MAI, see above § 5.4).

• Case 3 of Table AP5.1-2 refers to the TDM/TDMA technique often used in business communication networks (VSAT networks). The central earth station of this star network is called the hub and, according to this technique:

• the hub station broadcasts to the remote stations (VSATs) an outbound continuous-wave carrier (e.g. at 512 kbit/s) on which the various data messages are time division multiplexed;

• each remote station (VSAT) transmits towards the hub its own (inroute) message in time division multiple access (low bit rate TDMA, e.g. at 64 kbit/s), i.e. by sending a signal "burst" only during the duration of the properly buffered message.

A typical example of system characteristics is given below:

• the TDM outroute carrier is modulated in BPSK with an overall information bit rate of 256 kbit/s (including framing). After encoding by an error-correcting code (rate ½ FEC), the actual transmitted bit rate will, in this case, be 512 kbit/s;

• N inroute TDMA carriers are shared between the remote VSATs and transmitted by them to the hub at 64 kbit/s. Each carrier accesses the satellite in random assignment mode (RA/TDMA), also called "Aloha access mode", which means that each remote VSAT sends independently its instantaneous data packet as a burst in a TDMA frame (in case of collision with a burst sent by another remote station, the message is automatically repeated until acknowledgement is received. See § 3.4.2.5). In practice, the frame efficiency of this process is about 12%, thus giving a total information bit rate of about 8 kbit/s per carrier. Taking into account the error-correction encoding (rate ½ FEC), the transmitted bit rate is 128 kbit/s in this example.

The TDMA access mode can offer, as a specific advantage, very flexible frame rearrangement capabilities. For example, switching from the random access mode (RA/TDMA) to a demand assigned mode (DA/TDMA) is possible. In the DA/TDMA mode, long data files can be quickly transferred at high bit rates (up to 64 kbit/s with full occupancy of one TDMA carrier). Telephony channels (e.g. service channels) can also be optionally implemented in some systems. It should be noted that the management of this assignment, as well as the general control of the network, has to be completely carried out from the hub station.

For more information on VSAT systems, refer to Supplement No. 3, "VSAT Systems and Earth Stations" (ITU-R, Geneva, 1995) to the Handbook on Satellite Communications.

---

[22] These LRE codecs are often derived from those developed and mass-produced for terrestrial mobile cellular networks.

APPENDIX 5.2

Intermodulation interference in travelling wave tubes for multiple access communication satellites

AP5.2-1 Introduction

Travelling wave tube type amplifiers are currently used in communication satellite transponders and exhibit two kinds of non-linear characteristics:

- amplitude: non-linearity of output amplitude vs. input amplitude of signal (AM-AM conversion);

- phase: non-constant phase shift between input and output signals, as a function of input power (AM-PM conversion).

With frequency division multiple access (FDMA) several carriers are simultaneously amplified by the same travelling wave tube (TWT). Non-linear amplification produces distortions and intermodulation interference. The amplified signal contains intermodulation products at frequencies of:

$$f_x = k_1 \, f_1 + k_2 \, f_2 + ... + k_N \, f_N \tag{40}$$

where:

$f_1, f_2 ..... f_N = $ input carrier frequencies;

$k_1, k_2 ..... k_N = $ integer numbers.

The order of intermodulation product x is defined as:

$$|k_1| + |k_2| + ... + |k_N| \tag{41}$$

For example, $2f_1 - f_2$ is one of the third-order products. When the centre frequency of the transponder is large compared to the bandwidth of the transponder, odd-order intermodulation products are the only ones falling in the useful frequency band. This is the case in satellite communications.

Intermodulation product power decreases with input power and with the order of the product. So, numerical calculations are generally restricted to third-order (or sometimes fifth-order) intermodulation products.

The number of intermodulation products increases very quickly with the number N of input carriers (see Table AP5.2-1). Numerical calculations are normally feasible only when N is small. When N is too large, computation time becomes prohibitive and approximate formulas are used to estimate intermodulation interference (several hundreds of carriers).

TABLE AP5.2-1

Number of intermodulation products

Type of product	Order	Number of products of the type	N–5	N–10
$2f_1 - f_2$ $f_1 + f_2 - f_3$	3	$N(N-1)$ $1/2N(N-1)(N-2)$	20 30	90 360
$3f_1 - 2f_2$ $2f_1 + f_2 - 2f_3$ $3f_1 - f_2 - f_3$ $2f_1 + f_2 - f_3 - f_4$ $f_1 + f_2 + f_3 - 2f_4$ $f_1 + f_2 + f_3 - f_4 - f_5$	5	$N(N-1)$ $N(N-1)(N-2)$ $1/2N(N-1)(N-2)$ $1/2N(N-1)(N-2)(N-3)$ $1/2N(N-1)(N-2)(N-3)$ $1/12N(N-1)(N-2)(N-3)(N-4)$	20 60 30 60 60 10	90 720 360 2 520 2 520 2 520
		Total	290	9 180

AP5.2-2 Model representing non-linear satellite transponder

The model most frequently used to represent non-linear TWT amplifiers is the so-called envelope model.

Representing the unmodulated single carrier input signal as:

$$S_i(t) = A \cos (2\pi f_0 t + \varphi) \qquad (42)$$

with A, f_0 and φ being its amplitude, frequency and phase, respectively, the output signal in the first zone (around frequency f_0) can be expressed as:

$$S_0(t) = g(A) \cos [2\pi f_0 + \varphi + f(A)] \qquad (43)$$

In the above equation, g(A) and f(A) are the amplitude and phase characteristics of the amplifier, commonly referred to as the AM/AM and AM/PM characteristics. They are single carrier characteristics and can be easily measured.

Assuming that a TWT amplifier is a memoryless device, g(A) and f(A) can be considered as not being frequency dependent. This condition is practically satisfied when the ratio of the transponder bandwidth and its centre frequency is much smaller than unity. Equations (42) and (43) will apply also for the case when the envelope and phase of the input signal are time functions. So, one can write:

$$S_i(t) = A(t) \cos [2\pi f_0 t + \varphi(t)] \qquad (44)$$

and

$$S_0(t) = g[A(t)] \cos \{2\pi f_0 t + \varphi(t) + f[A(t)]\} \qquad (45)$$

Both g[A(t)] and f[A(t)] are the same single carrier characteristics as those in (43). As they are functions of the envelope of the input signal, the model is therefore called an envelope model.

AP5.2-3 Intermodulation products level

When the input signal consists of N modulated carriers,

$$S_i(t) = \text{Re} \left\{ \sum_{\ell-1}^{N} A_\ell(t) \exp\left[j2\pi(f_o + f_\ell)t + j\varphi_\ell(t) \right] \right\} \qquad (46)$$

and $A_\ell, f_o + f_\ell$ and $\varphi_\ell(t)$ represent the amplitude, frequency and phase modulation of the ℓth carrier.

The measured transponder characteristics g(A) and f(A) can be conveniently approximated by a Bessel function series expansion as:

$$g(A) \exp\left[jf(A) \right] \approx \sum_{p=1}^{P} b_p \, J_1(\alpha pA) \qquad (47)$$

In the above approximation, J_1 denotes the first-order Bessel function of the first kind, b_p the complex coefficients and α a real number. These coefficients and α are determined to give the b least squares fit to the measured characteristics.

Using the above approximation, the output signal of the transponder, assuming the input signal is as given by (46), can be written as:

$$s_o(f) = R_o \left\{ \sum_{k_1 = -\infty}^{\infty} \sum_{k_2 = -\infty}^{\infty} \sum_{k_N = -\infty}^{\infty} M(k_1, k_2, k_3..k_N) \cdot \exp \sum_{\ell=1}^{N} \left[j2\pi(f_o + k_\ell f_\ell)t + jk_\ell \varphi_\ell(t) \right] \right\} \quad (48)$$

$$(k_1 + k_2 + .. \, k_N = 1)$$

with complex amplitude $M(k_1, k_2 ... k_N)$ given as:

$$M(k_1, k_2..k_N) = \sum_{p=1}^{P} b_p \prod_{\ell=1}^{N} j_{k\ell}(\alpha pA_\ell) \qquad (49)$$

The output signal in (48) contains N fundamental components and intermodulation products whose frequencies are:

$$f_o + k_1 f_1 + k_2 f_2 .... k_N f_N = f_o + \sum_{\ell=1}^{N} k_\ell f_\ell \qquad (50)$$

Since only the components around f_o are of interest, there is a constraint that:

$$\sum_{\ell=1}^{N} k_\ell = 1 \qquad (51)$$

The set $\left(k_\ell\right)$ defines the components in the output signal. So, the set $k_1 = 1$; $k_2, k_3...k_N = 0$ gives the fundamental component at frequency $f_o + f_1$; the set $k_2 = 1$; $k_1, k_3, k_4...k_N = 0$ the fundamental component at frequency $f_o + f_2$.

The sum:

$$O = |k_1| + |k_2| + ... |k_N| = \sum_{\ell=1}^{N} |k_\ell| \qquad (52)$$

defines the order of the intermodulation product in the output signal. For the third order products, $O = 3$ and, for example, the set $k_1 = 2$; $k_2 = -1$; $k_3, k_4 ... k_N = 0$ gives the intermodulation product at frequency $f_o = 2f_1 - f_2$. The fifth order intermodulation products are obtained for $O = 5$. As an example, the set $k_1 = 3$; $k_2 = -2$, $k_3, k_4...k_N = 0$ defines the product at frequency $f_o + 3f_1 - 2f_2$.

Complex amplitude of these products can be obtained by using (49).

Practical computations have shown that ten terms ($P = 10$) in the Bessel function series (47) provide a very good approximation of the measured transponder characteristics. A set of coefficients for a typical TWT is given below:

Coefficients b_p

	Real part	Imaginary part
b_1	1.465	1.593
b_2	0.8983	0.1681
b_3	0.07757	−0.4212
b_4	0.466	0.6159
b_5	−0.4006	−0.9712
b_6	0.4711	0.9114
b_7	−0.2911	−1.011
b_8	0.04761	0.7479
b_9	0.07479	−0.4785
b_{10}	−0.079	0.1657

$\alpha = 0.8$

Figure AP5.2-1 gives an example of the output power per carrier vs. total input power for different numbers of identical unmodulated input carriers ($N = 1, 2, 3, 5, 10$ and 20) for a typical TWT.

The saturation phenomenon is obvious on these curves and the saturation point for one carrier is used as the reference for input and output powers (0 dB point). Powers or back-offs are in decibels relative to this saturation point.

Figure AP5.2-2 shows the output power of each carrier and the third-order intermodulation products for three identical unmodulated input carriers. A general rule is that $2f_1 - f_2$ type products are 6 dB lower than the $f_1 + f_2 - f_3$ type product when the input power is not too large.

Figure AP5.2-3 shows the same results for third- and fifth-order intermodulation products for five identical unmodulated carriers. As mentioned in the introduction, fifth-order intermodulation products are normally less powerful than third-order intermodulation products, at low values of input power. Fifth-order components may be neglected in an approximate calculation.

With modulated carriers, intermodulation products have a spectrum that depends on the modulation of the input carriers. The derivation of this type of spectrum is easy only in some particular cases.

A general formula for angle modulation (phase or frequency modulation) is:

$$\gamma_x(f) = \gamma_1, k_1(f) * .... * \gamma_N, k_N(f) \tag{53}$$

where:

$* =$ symbol of convolution;

$\gamma_x(f) =$ power spectrum density of intermodulated product x with centre frequency:

$$f_x = k_1 f_1 + ... + k_N f_N \tag{54}$$

and

$\gamma_i k_i(f) =$ the power spectrum density of an angle modulated carrier with $k_i \cdot \varphi_i(t)$ as a modulating signal (where $\varphi_i(t)$ is the modulating signal of the ith input carrier).

If $k_i = 1$, γ_i, $1(f)$ is the spectrum of the ith modulated input carrier. If $k_i \geq 2$ calculation of γ_i, $k_i(f)$ is not straightforward, except with gaussian approximation.

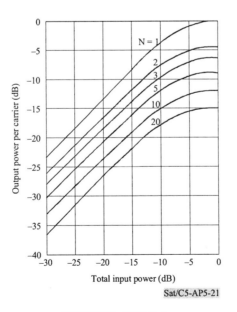

Sat/C5-AP5-21

FIGURE AP5.2-1

Output/input power

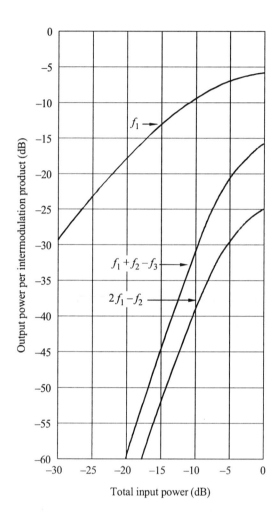

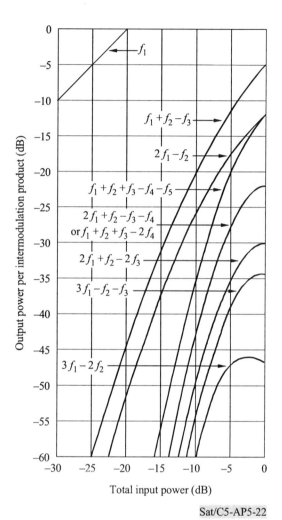

Sat/C5-AP5-22

FIGURE AP5.2-2

Intermodulation – Three carriers

FIGURE AP5.2-3

Intermodulation – Five carriers

For frequency modulation with large index and a gaussian type modulating signal (that is a good approximation of a multiplex telephone signal) the spectrum of ith modulated carrier is gaussian with a variance σ_i^2. In this case the spectrum $\gamma_x(f)$ is also gaussian, with variance:

$$\sigma_x^2 = \sum_{i=1}^{N}\left(k_i \cdot \alpha_i\right)^2 \tag{55}$$

When the power and spectrum of each intermodulation product are calculated separately, the addition of all the elementary spectrums gives the overall output spectrum. The complete sum of the modulated intermodulation products is equivalent to a noise product which is called intermodulation noise.

Figure AP5.2-4 gives the intermodulation noise spectrum for a typical TWT with ten carriers. Arrows show the frequency and relative level of each input carrier. The six central carriers have small levels (24 telephone channels), two other carriers have average levels (60 telephone channels) and two side carriers have large levels (132 telephone channels). Input back-off is 8 dB and output back-off is 3.45 dB. Intermodulation noise power spectrum in dB(W/kHz) is shown in Figure AP5.2-4. From this spectrum it is convenient to define a power spectral density N_o (function of frequency) and an intermodulation noise temperature T, using the relationship $N_o = k\,T$ (k: Boltzmann's constant).

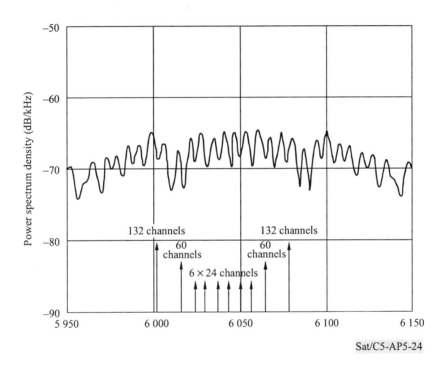

Sat/C5-AP5-24

FIGURE AP5.2-4

Intermodulation noise

The output power of each carrier divided by the intermodulation noise temperature in the frequency bandwidth of the carrier is the C/T intermodulation ratio that is used in link budget calculations.

Figure AP5.2-5 gives C/T intermodulation ratio with the same frequency arrangement as in Figure AP5.2-4. One solid curve is for the 132 channel carriers and the other one for 24 channel carriers. Dashed curves are the same carriers, but without taking into account AM-PM conversion. Comparison between the solid and dashed curves shows that the additive degradation produced by AM-PM conversion is 3 to 6 dB (value increasing at low back-off) compared to the results with amplitude non-linearity alone.

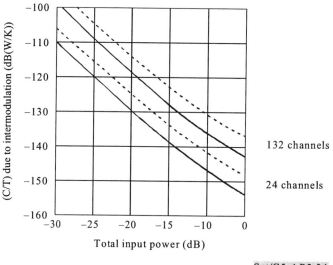

Sat/C5-AP5-24

FIGURE AP5.2-5

Intermodulation – Ten carriers
(with the same frequency arrangement as in Fig. AP5.2-3)

AP5.2-4 Frequency plan selection

In order that intermodulation products do not fall on useful carriers, a method based on different triangle sets may be used [Ref. 6]. The drawback of this method is that all carriers are supposed to be identical and that it does not take into account the spectrum of the intermodulation products, but only their frequency. Furthermore, the method gives a smaller number of usable carriers compared to the total number of frequency slots in the repeater.

The number of usable carriers can be increased by using a plan that accepts some intermodulation interference, but which is optimized so as to reduce it to a minimum [Ref. 7]. However, the drawbacks of this method are the same as before (identical carriers, no spectrum consideration).

General rules for obtaining a good frequency assignment plan are to avoid a constant frequency spacing between carriers (because with constant frequency spacing intermodulation products always fall on carriers) and to put large carriers on the sides of the frequency band of the repeater (so that large intermodulation products fall outside the useful band).

AP5.2-5 Formulas for SCPC

With single-channel-per-carrier (SCPC) multiple access systems the number of simultaneous carriers in the same repeater is very large (several hundreds). Extrapolation of results obtained by computer calculation with up to one hundred carriers and a typical TWT leads to the following formulas:

$$BO = 0.82 \ (BI - 4.5) \tag{56}$$

where:

> $BI =$ input back-off (dB);
>
> $BO =$ output back-off (dB);

$$C/T = -150 - 10 \log n + 2 \ BO \tag{57}$$

where:

> $n =$ number of active carriers;
>
> $C/T =$ carrier power to intermodulation noise temperature (per carrier).

The exact value of C/T depends slightly on the frequency assignment and may vary within ±1.5 dB compared to the approximate value. Frequency bandwidth of carriers has little effect on the results.

APPENDIX 5.3

Erlang Tables

The Erlang formulas and tables, published for the first time in 1917 by A.K. Erlang, enable the calculation of the probability that a telephone call is lost because the telephony network is overloaded.

Several users may offer traffic to a transmission system with n traffic paths or lines. To calculate the blocking probability of the incoming calls or offered traffic, the Erlang B equation is applied. The Erlang B equation is based on the assumption that the arrival and duration of new calls are independent of the number of current calls ("infinite source" condition) and that a new call, which is blocked, is considered as lost.

The Erlang B equation is:

$$p = \left(y^n / n!\right) / \left(\sum_{x=o}^{x=n} \left(y^x / x!\right)\right)$$

where:

        p = probability of loss or blocking;

        n = number of traffic paths (or circuits, devices, etc.);

        y = traffic offered (Erlangs).

The "traffic quantity" is a measure which is equal to the total of occupancy durations of a traffic path or circuit. This measure is also applicable not only to traffic paths, but also to group of paths, group of circuits and other equipment or device carrying communications.

Another measure is the "traffic intensity". It is defined as the ratio of the traffic quantity to the time duration or traffic quantity per unit time. The unit of the traffic intensity is the Erlang or Traffic Unit (TU). Sometimes it is given for the "busy hours" of the network. One Erlang means that the considered traffic path is continuously occupied. Erlang is a dimensionless quantity; it is also adopted by ITU.

The "offered traffic" is the product of the number of calls per unit time and the holding time of the calls expressed in units of time. The "offered traffic" is measured in Erlangs (dimensionless).

The "grade of service" is the ratio of calls lost at the first attempt to the total number of attempts of incoming calls. It is also named "blocking probability".

The "grade of service" depends on the distribution in time of the incoming calls (e.g. Poisson process) and the duration of calls, the number of traffic sources, conditions under which calls are lost or blocked.

When no channel can be assigned to an offered call and the call is placed in a queue which has an infinite length, the Erlang C equation has to be applied. The call in the queue is then served in the order of its arrival.

Example – According to Table AP5.3-1, a transmission system comprising $n = 50$ lines has a blocking probability of $p = 0.05$ (5%) for a given (offered) traffic of 44.53 Erlangs.

This should allow, for example, to reliably connect, through a switch, more than 200 subscribers, each one with an average traffic of 0.2 Erlang.

TABLE AP5.3-1

Erlang Tables

Blocking probability for a given number of trunk lines and the offered traffic

n	P					
	0.001	0.002	0.005	0.01	0.02	0.05
1	0.001	0.002	0.005	0.01	0.02	0.05
2	0.05	0.07	0.11	0.15	0.22	0.38
3	0.19	0.25	0.35	0.46	0.60	0.90
4	0.44	0.53	0.70	0.87	1.09	1.52
5	0.76	0.90	1.13	1.36	1.66	2.22
6	1.15	1.33	1.62	1.91	2.28	2.96
7	1.58	1.80	2.16	2.50	2.94	3.74
8	2.05	2.31	2.73	3.13	3.63	4.54
9	2.56	2.85	3.33	3.78	4.34	5.37
10	3.09	3.43	3.96	4.46	5.08	6.22
11	3.65	4.02	4.61	5.16	5.84	7.08
12	4.23	4.64	5.28	5.88	6.62	7.95
13	4.83	5.27	5.96	6.61	7.41	8.83
14	5.45	5.92	6.66	7.35	8.20	9.73
15	6.08	6.58	7.38	8.11	9.01	10.63
16	6.72	7.26	8.10	8.87	9.83	11.54
17	7.38	7.95	8.83	9.65	10.66	12.46
18	8.05	8.64	9.58	10.44	11.49	13.38
19	8.72	9.35	10.33	11.23	12.33	14.31
20	9.41	10.07	11.09	12.03	13.18	15.25
21	10.11	10.79	11.86	12.84	14.04	16.19
22	10.81	11.53	12.63	13.65	14.90	17.13
23	11.52	12.27	13.42	14.47	15.76	18.08
24	12.24	13.01	14.20	15.29	16.63	19.03
25	12.97	13.76	15.00	16.12	17.50	19.99
26	13.70	14.52	15.80	16.96	18.38	20.94
27	14.44	15.28	16.60	17.80	19.26	21.90
28	15.18	16.05	17.41	18.64	20.15	22.87
29	15.93	16.83	18.22	19.49	21.04	23.83
30	16.68	17.61	19.03	20.34	21.93	24.80
31	17.44	18.39	19.85	21.19	22.83	25.77
32	18.20	19.18	20.68	22.05	23.73	26.75
33	18.97	19.97	21.51	22.91	24.63	27.72
34	19.74	20.76	22.34	23.77	25.53	28.70
35	20.52	21.56	23.17	24.64	26.43	29.68
36	21.30	22.36	24.01	25.51	27.34	30.66
37	22.03	23.17	24.85	26.38	28.25	31.64
38	22.86	23.97	25.69	27.25	29.17	32.63
39	23.65	24.78	26.53	28.13	30.08	33.61
40	24.44	25.60	27.38	29.01	31.00	34.60
41	25.24	26.42	28.23	29.89	31.92	35.59
42	26.04	27.24	29.08	30.77	32.84	36.58
43	26.84	28.06	29.94	31.66	33.76	37.57
44	27.64	28.88	30.80	32.54	34.68	38.56
45	28.45	29.71	31.66	33.43	35.61	39.55
46	29.26	30.54	32.52	34.32	36.53	40.54
47	30.07	31.37	33.38	35.21	37.46	41.54
48	30.88	32.20	34.25	36.11	38.39	42.54
49	31.69	33.04	35.11	37.00	39.32	43.54
50	32.51	33.88	35.98	37.90	40.25	44.53

REFERENCES

[1] Bedford, R., Chaudhry, K., Smith, S. – Improvement and application of the INTELSAT (SS) TDMA system. Conference publication No. 403 of IEE, Vol. II, IDCSC-10, 1995.

[2] Lunsdorf, J., *et al.* – INTELSAT second-generation TDMA terminal. Conference publication No. 403 of IEE, Vol. II, IDCSC-10, 1995.

[3] Forcina, G., Oei, S., Simha, S. – The INTELSAT DAMA system. Conference publication No. 403 of IEE, Vol. II, ICDSC-10, 1995.

[4] De Gaudenzi, R., Giannetti, F., Luise, M.L. – Advances in satellite CDMA transmission for mobile and personal communications. Proceedings of the IEEE, January 1996, p. 18.

[5] Rappaport, Th.S. – Wireless communications. Prentice Hall, New Jersey, 1996.

[6] Fang, R.J.F. and Sandrin, W.A. [Spring, 1977] – Carrier frequency assignment for non-linear repeaters. COMSAT Tech. Rev., Vol. 7, 1, 227-244.

[7] Hirata, Y. [Spring, 1978] – A bound on the relationship between intermodulation noise and carrier frequency assignment. COMSAT Tech. Rev., Vol. 8, 1, 141-154.

BIBLIOGRAPHY

Fuenzalida, J.C., Shimbo, O. and Cook, W.L. [Spring, 1973] – Time domain analysis of intermodulation effects caused by non-linear amplifiers. COMSAT Tech. Rev., Vol. 3, 1, 89-141.

Shimbo, O. [February, 1971] – Effects of intermodulation, AM/PM conversion and additive noise in multicarrier TWT systems. Proc. IEEE, Vol. 59.

Shimbo, O., Nguyen, L. and Albuquerque, J.P. [April, 1986] – Modulation transfer effects among FM and digital signals in memoryless non-linear devices. Proc. IEEE, Vol. 74.

CHAPTER 6

Space segment

6.1 Satellite orbits

6.1.1 Introduction

This section describes the mechanics of satellite orbits and their significance with regard to communication satellites. The reader is introduced to the fundamental laws governing satellite orbits and the principal parameters that describe the motion of artificial satellites of the Earth. The types of orbits are also classified and compared from a communication system viewpoint in terms of Earth coverage performance and environmental and link constraints.

During the last two decades, commercial communication satellite networks have utilized geostationary satellites extensively to the point where portions of the geostationary orbit have become crowded and coordination between satellites is becoming constrained. Recently, however, non-geostationary satellite systems have grown in importance, both because of their orbit characteristics and Earth coverage capabilities in high latitudes. Some of their special features are described in this section.

6.1.2 Background

The problem of determining the position and path of a satellite in space as a function of time has occupied scientists and philosophers for thousands of years. It finally evolved on Kepler in the 17th century to discover the properties of planetary motion from observations of our Sun and its planets. Kepler's Laws state that:

1) Each planet moves around the Sun in an ellipse, with the Sun at the focus (motion lies in a plane).

2) The line from the Sun to planet, or radius vector (r), sweeps out equal areas in equal intervals of time.

3) The ratio of the square of the period (T) to the cube of the semi-major axis (a) is the same for all planets in our solar system.

It remained for Newton's discovery of the Universal Law of Gravitation to identify the forces associated with Kepler's Laws. Specifically, every mass (M_1) attracts another mass (M_2) with a force directed along the line joining the two masses and having a magnitude (F) of $G M_1 M_2/r^2$ where G is the Universal Gravitational Constant.

6.1.3   Orbital dynamics

6.1.3.1   Keplerian orbits

The "Keplerian assumption" which underlies Keplerian orbits is a first order approximation of the forces that apply to any satellite orbiting around a central body (e.g. an artificial communication satellite around the Earth, or a planet around the Sun). This is very useful and efficient to describe the fundamental laws of satellite orbits. Under this assumption, the mutual interaction between the attracting bodies is restricted to a two-body problem (e.g. the satellite and the Earth, no moon or sun attraction), where the only acting force is the Newtonian force (or central force) μ/r^2.

When the mass m of the satellite can be neglected with respect to the mass M of the central body:

* $\mu = GM$ where G is the gravitational constant ($\mu \approx 398\ 600\ \text{km}^3/\text{s}^2$ for the Earth).

* r = distance between the centre of the Earth and the satellite.

In the case of an Earth satellite, the first law of Kepler states that the trajectory of any satellite which experiences such a central force μ/r^2 is a conic section whose focus coincides with the centre of the Earth. In the following, we shall only deal with the ellipse since other conic sections (parabola and hyperbola) are not of interest for Earth satellite communications.

A simplified diagram of an elliptical orbit around the Earth, in the plane of the orbit, is shown in Figure 6.1.

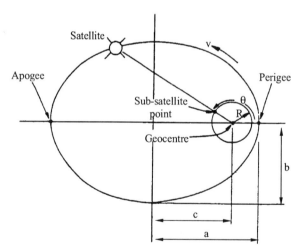

a: semi-major axis
b: semi-minor axis

e: eccentricity = $(1 - (b/a)^2)^{0.5}$; $0 \le e < 1$; (circle: e = 0)
c: distance between centre of ellipse and focal point = ae
R: mean radius of Earth
r, θ polar coordinates of satellite; θ (the true anomaly) is measured
   from perigee

Sat/C6-01

FIGURE 6.1

Geometry of an elliptical orbit

Considering an inertial reference frame (i.e. fixed with respect to stars), an ellipse can be defined by the following five parameters (see Figure 6.2):

* semi-major axis of the ellipse a: (a > R = equatorial Earth radius ≈ 6 378 km);

* eccentricity of the ellipse e: ($0 \leq e < 1$; e = 0 circular orbit);

* inclination of the orbital plane i: ($0 \leq i \leq 180°$);

* right ascension of the node Ω: ($0 \leq \Omega \leq 360°$);

* argument of perigee ω: ($0 \leq \omega \leq 360°$).

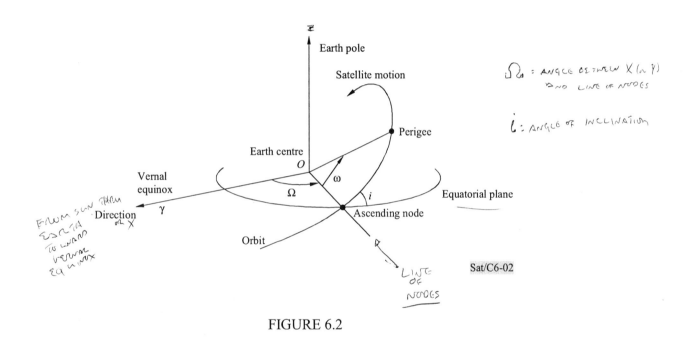

FIGURE 6.2

Orbit parameters

The position of the satellite along the ellipse is geometrically given by the true anomaly v or the eccentric anomaly E (see Figure 6.3) but it is generally replaced by the mean anomaly A_M which is proportional to time and mathematically more useful:

$$\operatorname{tg} \frac{v}{2} = \sqrt{\frac{1+e}{1-e}} \operatorname{tg} \frac{E}{2}$$

$$A_M = E - e \ \sin E$$

NOTE – In the particular case of circular orbits, $A_M = E = v$.

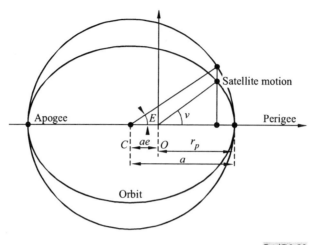

Sat/C6-03

FIGURE 6.3

Ellipse geometry

The magnitude r of the position vector of the satellite is:

$$r = a(1 - e^2)/(1 + e \cos v)$$

The apogee radius r_{ap} and perigee radius r_p can then be straightforwardly derived and can usefully define the shape of the ellipse through the use of both perigee height h_p and apogee height h_{ap}:

• $h_p = a (1 - e) - R$;

• $h_{ap} = a (1 + e) - R$.

The second law of Kepler states that the area that is swept by the radius vector ρ_r by unit of time is constant, which explains why a satellite is slowing at its orbit apogee and why it is accelerating at its orbit perigee:

$$r^2 dv/dt = \text{constant} = n_k a^2 \sqrt{1 - e^2}$$

with n_k = Keplerian mean motion of the satellite = 2π/orbital period (rad/s).

Another useful relationship is expressing the conservation of the total energy and allows the calculation of the satellite absolute velocity V all along its trajectory:

$$V^2 = \mu (2/r - 1/a)$$

Finally, the very important third law of Kepler relates the Keplerian mean motion n of the satellite (or its Keplerian orbital period $T = 2\pi/n_k$) to the semi-major axis of the ellipse a, in a very simple and remarkable manner:

$$n_k^2 a^3 = 4\pi^2 a^3/T^2 = \text{constant} = \mu$$

6.1.3.2   Orbit perturbations

Although it provides an excellent reference, the Keplerian approximation is only valid at the first order, when very small disturbing forces are neglected; the most important perturbations are due to dissymetries and inhomogeneities of the Earth gravity field, with amplitudes thousands of times lower than the central Newtonian force.

Because of these small perturbations, the trajectory of the artificial satellites is not a closed ellipse with a frozen orientation in space, but an open curve that continuously evolves with time, both in shape and orientation.

There are two types of orbit perturbations:

- the perturbations of gravitational origin which derive from a potential and for which total energy is conserved (e.g. non-central terms of the Earth's gravity field when expanded into spherical harmonics, third body attraction by Moon, Sun etc.);

- the non-gravitational perturbations that do not derive from a potential and that are dissipating energy (e.g. atmospheric drag, solar radiation pressure...); they are sometimes called "surface forces" since they depend on satellite surface characteristics (e.g. area to mass ratio, aerodynamic coefficient, reflectivity properties...).

The relative amplitudes of these forces is mainly governed by the altitude of the satellite and the effect on the trajectory can become very important, even for very small amplitudes, due to cumulative effects or resonance effects.

The most important disturbing force of an artificial satellite is due to the first zonal J_2 term of the Earth gravity field. This J_2 term ($J_2 \sim 10^{-3}$) takes into account the dynamical influence of the non-spherical shape of the Earth. The main perturbations are secular drifts (or precession rates) of the ascending node and of the perigee, together with a slight modification of the Keplerian mean motion n_k of the satellite:

- $\Omega = d\Omega/dt = -3/2\ n_k J_2\ (R/a)^2\ (1-e^2)^{-2}\ \cos i$;

- $\omega = d\omega/dt = -3/4\ n_k J_2\ (R/a)^2\ (1-e^2)^{-2}\ (1-5\cos^2 i)$;

- $M = dM/dt = n_k\ [1 + 3/4\ J_2\ (R/a)^2\ (1-e^2)^{-3.2}\ (3\cos^2 i - 1)]$.

It can be noted that the precessing rate of the ascending node can be either positive (i < 90°), negative (i > 90°), frozen (i = 90°) or adjusted in inclination in order to exactly compensate for the orbital motion of the Earth around the Sun for the so-called sun-synchronous orbits ($d\Omega/dt \approx 0.986°/\text{day}$). The order of magnitude ranges from a few degrees per day in low-Earth orbits to very small values at high altitudes.

It can also be noted that the motion of the perigee is of the same order of magnitude as Ω and can be frozen for two values of inclination (i = 63.4° and 116.6°) called critical inclinations; this property is very useful when it is desirable that the apogee be geographically frozen: it has been put into practice in several satellite communication systems such as the Russian Molnya network.

More details about geostationary perturbations and their consequences in terms of orbit and attitude control are given in section 6.2.3.

6.1.3.3   Orbit determination

For mission and control purposes, it is necessary to know at each instant both position and velocity of the satellite in a given reference frame or, which is mathematically equivalent, the six orbital parameters defined in section 6.1.3.1. This knowledge is the output of a complex orbit determination process where a trade-off has to be made between the number and accuracy of orbit measurements and the level of accuracy of disturbing force modelling, which depends upon the level of orbit accuracy that is required.

Due to recent and remarkable achievements in space geodesy, it can be considered that gravitational forces can be modelled as accurately as desired for communication purposes, whatever the orbit and provided that enough computation resources are allocated, which is not a problem with ground facilities, but may be a problem on board satellites. However, the modelling of surface forces described in section 6.1.3.2 and the prediction of their influence in terms of precise orbit accuracy are still a challenge, especially for low-Earth orbits below about 700 km altitude. Regular orbit measurements are then needed to compensate for that uncertainty.

Finally, it can be noted that this orbit determination process will occur more and more in real time aboard the satellite so that the satellite will be able to navigate autonomously. This is a general trend that is clearly illustrated by the increasing use of new systems such as the United States Global Positioning Satellite System (GPS) aboard spacecraft, for both orbit and attitude determination.

6.1.3.4   The geostationary orbit (GSO)

Satellites launched in the equatorial plane (zero degrees of inclination) with a circular orbit at 35 800 km of altitude appear to an observer on the surface of the Earth to be fixed in space. In reality, it is circling the Earth approximately every 24 hours in synchronism with the Earth's rotation about its axis. This orbit has a number of attractive features for communication satellites such as: Earth station antennas can be pointed at the satellite without the need for tracking mechanisms; Doppler shift is essentially non-existent; and approximately one-third of the Earth's surface is visible from one satellite (direct line of sight). These features have resulted in the implementation and operation of a large number of GSO satellites, especially for broadcasting services. Figure 6.4 shows the coverage provided by a GSO satellite to the visible Earth's horizon. The useful coverage area is limited to elevation angles of 5° or greater depending on the transmission frequencies and ground obstacles.

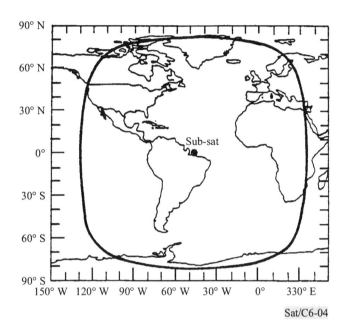

Sat/C6-04

FIGURE 6.4

Coverage of a GSO satellite at 47 degrees West longitude

6.1.4 Orbit design

6.1.4.1 Environment constraints

When selecting the orbit of a space communication system, there are mainly two kinds of space environment constraints that are both highly dependent on the altitude (the atmosphere will not be considered here since it has already been addressed as a disturbing force in section 6.1.3.2):

- the Van Allen radiation belts where energetic particles such as protons and electrons are confined by the Earth magnetic field and can cause dramatic damage to electronic and electrical components of the satellite;

- space debris belts and graveyard orbits where old spacecrafts are injected at the end of their lifetime; this is currently a concern of growing importance with the advent of large satellite constellations.

The electromagnetic constraint of the Van Allen belts is currently far more important to take into account than the space debris. It is characterized by two zones of high energy (inner and outer zones) with a climax around 3 500 km and 18 000 km. In practice, low orbits below 1 600 km and geosynchronous orbits experience significantly lower levels of radiation than medium-Earth orbits for which more favourable orbits are around 9 000 km and worst orbits are around 18 000 km (see Figure 6.5).

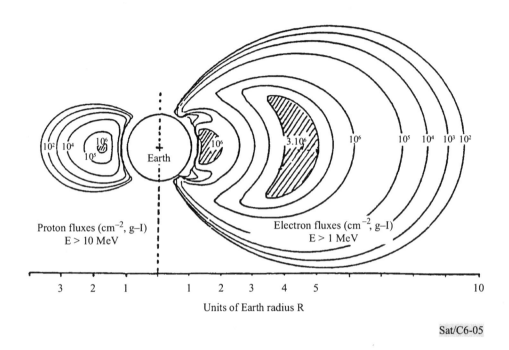

FIGURE 6.5

The Van Allen radiation belts

Other space environment constraints such as solar illumination, temperature, vacuum, etc. are not envisaged here since they are usually dealt with when the spacecraft is designed and cannot be considered as primary constraints for satellite communication orbits.

6.1.4.2   Communication constraints

Although geostationary satellites are currently operating with an overall roundtrip propagation of about 250 or more milliseconds, the utilization of non-geostationary satellites with shorter delays may offer significant advantages, in particular when a high interactivity is required (e.g. voice transmission, interactive video, etc.).

At first order, this time delay depends on the geometric length of the space link and is more favourable for low-Earth orbits than for geostationary orbits. Typical two-way propagation delays range from a few milliseconds at low altitudes, to several tens of milliseconds at medium altitudes and up to 250 milliseconds or more at geosynchronous altitudes.

However, additional delays have to be added in a practical communication system, taking into account processing, buffering, switching delays, etc. that lead to increases over the geometric propagation delay by several tens of milliseconds. And when intersatellite links are used in satellite systems, an additional delay of the same order of magnitude must be taken into account.

Another communication constraint is to assure that the space link can be established and maintained when considering:

- natural obstacles around the user terminal (geometrical and radioectrical masks, multipath...);
- satellite antenna pointing (spacecraft attitude, masks);
- propagation attenuation (rain, troposphere...);
- link budget parameters (BERP, distance...).

In general, a minimum elevation threshold above the local user horizon is defined, under which the link suffers significant impairments. Typical thresholds range from 5° to 10° for 6/4 GHz systems, 10° to 15° for 14/10-12 GHz systems and >15° for 30/20 GHz systems. As a first approximation these are generally considered as azimuth independent. Higher thresholds around 30° or 40° are sometimes considered for specific applications, as for example mobile communications (urban areas) or wideband multimedia (attenuation).

Maintaining a communication link as long as possible is highly desirable and depends on the visible duration of the satellite by the user, and on the number of satellites, if it is not a geostationary satellite. This problem is addressed at a global system level in sections 6.1.4.3 and 6.1.5.1.

6.1.4.3   Satellite ground track and coverage

The satellite ground track is the trace of the points of the Earth overflown by the satellite. This concept is important to understand which part of the Earth is seen by the satellite when it moves along its trajectory. The ground track is obtained by the vector combination of the absolute satellite velocity projected onto the subsatellite point and the Earth velocity of this subsatellite point. From kinematic considerations, it can be shown that, in an Earth-fixed reference frame, a satellite that accomplishes $N + m/q$ revolutions per day (N, m, q are integers) has a ground track that repeats itself after a time period (Prep) having completed $(qN + m)$ revolutions, such that:

$$\text{Prep} = (qN + m)\,\text{Porb} = (N + m/q)\,\text{Pnod}$$

where:

Porb	=	satellite orbital period between two successive ascending passes at the equator	=	$2\pi(\omega + M)$ (s)
Pnod	=	nodal (or draconitic) period of the ascending node	=	$2\pi(\theta - \Omega)$ (s)
θ	=	sideral rotation rate of the Earth	$\approx$	$2\pi/86\ 164$ (rad/s)

It should be noticed that the above formulas are very general and are not restricted to the Keplerian assumption due to the Earth dynamic flattening J_2 mentioned in section 6.1.3.2, ω, and Ω are different from zero, and M is different from the Keplerian mean motion n_k. In particular, the orbital period Porb is slightly different from the Keplerian orbital period T given by the 3rd law of Kepler in section 6.1.3.1.

The ground track after one orbital period is called a loop. The number of orbital loops in 24 hours is roughly given by N and the total number of different orbital loops is qN + m.

The shape of the loop is dependent on the choice of both eccentricity e and argument of perigee ω, but it is remarkable that the inclination i is equal to the maximum latitude. In the particular case of circular orbits (e = 0), there is a North/South symmetry of the ground track.

Typical values for N range from N = 12 to 15 rev/day for low-Earth orbits, to N = 3 to 4 rev/day for medium-Earth orbits and down to N = 2 for 12-hour orbits (e.g. Molnya, GPS) and N = 1 for geosynchronous orbits (N = 0 corresponds to supersynchronous orbits with orbital periods greater than 24 hours). Figure 6.6 shows the relationship of orbit altitude versus orbital periods for satellites in low-altitude circular orbits.

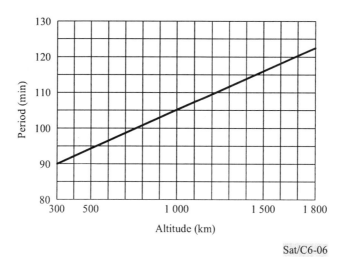

Sat/C6-06

FIGURE 6.6

Orbital periods vs. altitude for circular orbits

The ground track of a geostationary satellite (N = 1; 1 revolution per day; a = 42 164 km; i = 0; e = 0; Porb = 23 h 56 mn) is very particular since it is a point on the equator, characterized by its constant geographical longitude.

The ground coverage of the communication satellite antenna can be modelled in a first approximation by the intersection of a cone and a sphere, which is a circle of constant elevation whose value is the minimum elevation threshold defined in section 6.1.3.2. The size of this instantaneous field of view is defined by the Earth-centred half-angle θ and is governed by the altitude h of the satellite and the elevation ε:

$$\theta = Arc\ sin\ [(R\ /(R + h))\ cos\ \varepsilon].$$

The users can also see the satellite during time intervals which are all the more important as the satellite is at high altitude (see the 2nd law of Kepler recalled in section 6.1.3.1). A drawback is that the line-of-sight geometry is rapidly changing for low-Earth orbits which creates Doppler effects and specific pointing constraints. Finally, it can be said that from a pure coverage point of view, high- and medium-altitude orbits appear to be more effective than low-Earth orbits.

6.1.4.4 Orbit comparison for communication applications

Table 6.1 summarizes the main coverage performances and link constraints that have been detailed in previous sections 6.1.4.1, 6.1.4.2 and 6.1.4.3.

At this level of comparison, only orbits are compared and cannot lead to definite answers in terms of system design without a careful review of mission requirements and satellite constellation design considerations.

TABLE 6.1

Orbit comparison for satellite communications applications

ORBITS	LEO	MEO	HEO	GEO
Environment constraints	Currently low (space debris: growing concern)	Low/medium	Medium/high (Van Allen belts: 4 crossings/day)	Low
Orbital period	1.5-2 h	5-10 h	12 h	24 h
Altitude range	500-1 500 km	8 000-18 000 km	40 000 km apogee (perigee below 1 000 km)	40 000 km (i = 0)
Visibility duration	15-20 mn/pass	2-8 h/pass	8-11 h/pass (apogee)	Permanent
Elevation	Rapid variations; high and low angles	Slow variations; high angles	No variations (apogee); high angles	No variation; low angles at high latitudes
Propagation delay	Several milliseconds	Tens of milliseconds	Hundreds of milliseconds (apogee)	> 250 milliseconds
Link budget (distance)	Favourable; compatible with small satellites and handheld user terminals	Less favourable	Not favourable for handheld or small terminals	Not favourable for handheld or small terminals
Instantaneous ground coverage (diameter at 10° elevation)	≈ 6 000 km	≈ 12 000-15 000 km	16 000 km (apogee)	16 000 km
Examples of systems	IRIDIUM, GLOBALSTAR TELEDESIC, SKYBRIDGE, ORBCOMM...	ODYSSEY, INMARSAT P21...	MOLNYA, ARCHIMEDES...	INTELSAT, INTERSPOUTNIK, INMARSAT...

LEO   : low-Earth orbits
MEO   : medium-Earth orbits
HEO   : highly-eccentric orbits
GEO   : geostationary orbits

6.1.5   Constellation design

6.1.5.1   Orbit and constellation trade-off

For non-geostationary orbits, it is not possible to insure the continuous (or permanent) coverage of any region with a single satellite. Since the visibility duration of a satellite by the users is finite, a "constellation" of several satellites is then needed. The constellation design process consists in optimizing the number and the relative placement of the satellites that best satisfies space and time coverage requirements (and constraints) at the lowest cost.

This is a complex trade-off that must take into account satellite and launch costs and service availability among other factors. This optimization process does not automatically lead to simply minimizing the number of satellites in the constellation. For instance, it may be cheaper to launch three dozen small satellites in six launches into low-Earth orbits than 12 big satellites in six launches into medium-Earth orbits.

The continuous coverage of the Earth (or regions of the Earth) by one or more satellites has been analytically studied by several authors from a purely geometrical coverage point of view. Their results are useful to provide a first estimate of the number of satellites and are summarized in section 6.1.5.2. The problem of non-permanent coverage of the Earth is more complicated and cannot be solved without numerical simulations. Some insights are given in section 6.1.5.3.

6.1.5.2   Continuous single and multiple satellite coverage

It is well known that three geostationary satellites cannot permanently cover 100% of the Earth since polar regions are not accessible. Recent studies have shown that four highly elliptical satellites is the minimum number of satellites required to insure continuous coverage of the entire Earth, while it was generally thought previously that five circular satellites in medium-Earth orbit was the absolute minimum.

The minimum number of satellites N_1 that are required to insure permanent coverage of the Earth by a single satellite depends on the instantaneous field of view of the satellite communication antenna (defined by the radius θ of the instantaneous field of view (see section 6.1.4.3) and is given by the following simple approximation formula:

$$N_1 \approx 4/(1 - \cos \theta)$$

Figure 6.7 presents the above relation as a function of both satellite altitude h and minimum elevation ε, and for optimal constellations in low-Earth sun-synchronous orbits:

The continuous single satellite coverage above latitude φ requires roughly $N_1 \cos \varphi$ satellites.

The continuous coverage of the Earth by more than one satellite is sometimes required for specific applications (e.g. localization) or specific constraints (e.g. frequency coordination with other satellite communication systems).

An estimation of the number N_2 (resp N_3) of satellites that are required to insure a continuous double (resp. triple) satellite coverage of the Earth is given in a first approximation by:

$$N_2 \approx 7.5/(1 - \cos \theta); \; N_3 \approx 11/(1 - \cos \theta)$$

It can be outlined that these coverage-optimized constellations generally follow very symmetrical and well-known design schemes (e.g. Walker constellations, Rider constellations...), but it should also be emphasized that additional considerations (e.g. service continuity, tolerance to satellite loss) generally lead to non-standard constellations at the end of the optimization process.

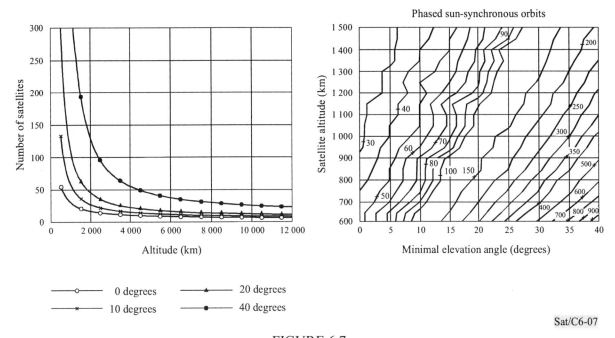

FIGURE 6.7

Continuous single satellite coverage of the Earth

Owing to their strong symmetry properties, a small set of parameters is generally sufficient to fully describe the relative geometry of these symmetrical constellations, but it should be noted that, in this case, all the satellites have the same orbit altitude and inclination. For instance, a classical Walker pattern is entirely described by the following three integer parameters T/P/F, where:

T =   total number of satellites of the constellation;

P =   number of orbital planes which are supposed to be evenly distributed in the equatorial plane of the planet (i.e. there is a constant spacing of 360/P degrees in ascending node between adjacent planes).

NOTE – P must exactly divide T since there are T/P satellites per orbital planes that are evenly distributed.

F =   in-plane phasing parameter, expressed in units of 360/T degrees, between two satellites in adjacent orbital planes.

As an example, the first Globalstar (a mobile LEO satellite system) constellation is a 48/8/3 Walker pattern. This means that there are 48 satellites in eight orbital planes that are separated by 45° in ascending node, with six satellites separated by 60° in each orbital plane and with a 22.5° phasing angle between two consecutive planes.

6.1.5.3 Non-permanent coverage

Satellite constellations with intermittent coverage are generally useful for specific communication applications such as store-and-forward messagery or short voice transmission.

Non-permanent coverage occurs only for non-geostationary orbits and when there are less satellites than the minimum number required for continuous coverage (see section 6.1.5.2), or when their number is large enough but when their relative placement is not optimized. A worst case is achieved with a constellation reduced to only one satellite. Depending on the altitude and the minimum elevation angle, typical average gaps in low-Earth orbits range from about 10 hours at 5° elevation and mid-latitudes with one satellite, to less than one hour with four satellites and about one half hour with six satellites.

More quantified results require precise definitions of the zones of interest and of the constraints, together with specific numerical simulations. Generally, the constellations that are optimal for permanent coverage are not optimal for non-permanent coverage, but they generally provide fairly good initial conditions for the optimization process.

6.2 General description of communication satellites

6.2.1 Introduction

All satellite systems, regardless of their primary functions, rely on radiocommunications to complete their objectives. Communication satellites, i.e. satellites whose primary functions are telecommunication services, have been allocated by ITU-R frequency bands which are listed in Chapter 9, Appendix 9.1. The most commonly used of these bands are the 6/4 GHz band (i.e. the 6 GHz band for the uplink and the 4 GHz band for the downlink), often called the "C-Band", 14/10-12 GHz band ("Ku-Band") and, more recently, the 30/20 GHz band ("Ka-Band").

The two general Earth orbital configurations used for communication satellites have been the geostationary orbit (GSO) and the non-geostationary orbits (non-GSO) associated with low-, medium- and high-altitude Earth orbits. A description of the different orbit characteristics was provided in the previous section 6.1. GSO satellites have dominated the national, regional and international communication satellite networks during the last two decades. During the last few years, however, non-GSO satellites have become the focus of new developments for the FSS.

The general subsystems of a communication satellite include the space platform or enclosure, the power generating system, the environmental and orbital control components and the communication payload.

In the case of GSO satellite communication systems, the main constraints that the mission imposes on the spacecraft are:

- high degree of station-keeping accuracy and attitude control;
- high antenna pointing accuracy;
- long lifetime in the nominal orbit position (10-15 years);
- reliable electric power supply;
- effective thermal control of electrical and other components;
- operation during solar eclipse (in the Earth's shadow);
- large launch vehicle capable of insertion into GSO orbit.

In the case of non-GSO systems, the same mission constraints apply with the exception that a) station-keeping is not usually required since the satellite follows its normal Keplerian trajectory after it is launched into orbit, although some systems with synchronized orbit spacing requirements may need station-keeping; b) lifetimes are usually shorter because low orbital altitudes introduce drag forces which accelerate orbit decay; and c) launch vehicle requirements are considerably less for low-altitude orbits.

The main subsystems of the spacecraft to support the payload which are described in this section include:

- the enclosure or platform structure;
- the thermal control system;
- the power supply system;
- the attitude and orbit control subsystem;
- the telemetry, command and ranging system;
- the apogee motor.

6.2.2 Satellite structural designs

The structure of a spacecraft is designed to house, support and protect the various subsystems and components of a satellite during its lifetime. The greatest inertial and dynamic stresses on the structure and accompanying satellite equipment occurs during the launch phase. During this cycle, the satellite is subjected to mechanical and acoustical shocks and vibrations from the launch vehicle engines and the thermal stresses associated with the booster rocket plumes and air friction. The spacecraft is protected during this "aerodynamic" phase with an enclosure or "nose cone". Thereafter, the nose cone is jettisoned and the spacecraft must survive the inertial and thermal stresses of the additional propulsion stages until it is finally inserted into its proper orbit. During the orbital or operational phase, the structure must protect the satellite equipment from the environment of space (weightlessness, vacuum, thermal, etc.) and other relatively small dynamic forces produced by the station-keeping and attitude control engines or inertial/momentum devices. The deployment of structures, such as solar panels and large antennas, requires special techniques in the vacuum of space because of the lack of a damping medium, such as air.

Extensive testing of the structure, with actual or simulated components and environments, is an essential function for the successful design of a satellite. Test facilities, such as vibration machines and "cold" vacuum chambers, were introduced to the industry during the mid-1950s and 1960s.

There are two principal structural types that have been used extensively in the design of satellites. They are the spin stabilized spacecraft and the three-axis body stabilized spacecraft.

Spin stabilized structures are usually shaped like drums, in which part of the drum rotates (50-100 rpm), and part is despun so that an antenna mounted in this part is always facing the Earth. The spinning part is covered with solar cells and its spin axis is oriented perpendicular to the Sun (North-South axis for GSO satellites) so that the solar rays are providing maximum energy to the solar cells. The despun part, containing the antennas and some Earth sensors, rotates once with every circling of the Earth. Slip rings are required to conduct electrical power from the spinning solar cell assembly to the despun communications system and antennas. An example of this type of structure is shown in Figure 6.8.

Body stabilized structures are usually shaped like a "box" with square or rectangular sides, and with external attachments to accommodate the subsystems and components that need to function outside of the enclosure. The box rotates once for every circling of the Earth so that the side with externally mounted antennas will always face the Earth. For GSO satellites, the spacecraft appears fixed in space to an observer on the surface of the Earth. This structure utilizes a deployed set of solar panels with solar cells mounted on one side of the panel surfaces. To maintain the cells in a normal position relative to the Sun, the panels, which are oriented with a North-South axis, rotate once per orbit revolution. An example of this type of structure is shown in Figure 6.9. An exploded view showing the various structural components of this type of satellite is shown in Figure 6.10.

Since low mass is an extremely important feature of the spacecraft structure, the main frame is comprised of strong lightweight metals, such as aluminum or magnesium alloys, and composite structures made of special plastic or fibre materials. Auxiliary structures for solar panels, antennas and other relatively large equipment rely on a number of advanced structural designs and materials, such as carbon fibres and epoxy resins. New developments in strong lightweight structures and materials (such as carbon nanotube filaments) are likely to provide significant advances in the future.

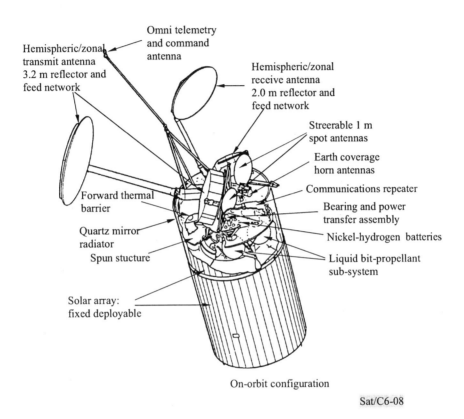

Omni telemetry
and command
antenna

Hemispheric/zonal
transmit antenna
3.2 m reflector and
feed network

Hemispheric/zonal
receive antenna
2.0 m reflector and
feed network

Streerable 1 m
spot antennas

Earth coverage
horn antennas

Communications repeater

Bearing and power
transfer assembly

Forward thermal
barrier

Nickel-hydrogen batteries

Quartz mirror
radiator

Spun stucture

Liquid bit-propellant
sub-system

Solar array:
fixed deployable

On-orbit configuration

Sat/C6-08

FIGURE 6.8

Example of spin stabilized satellite

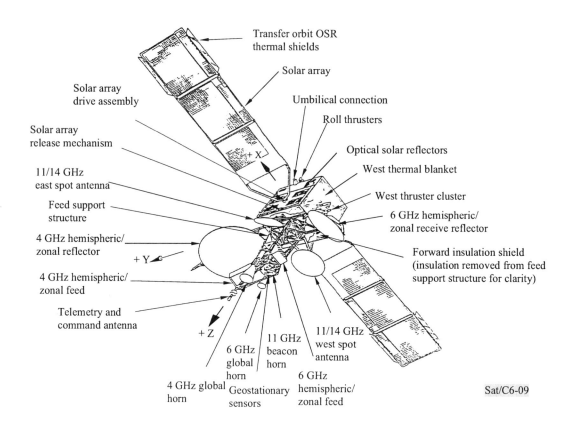

FIGURE 6.9

Example of three-axis stabilized satellite

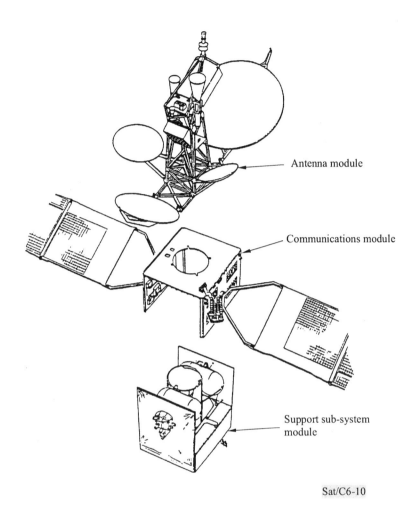

Sat/C6-10

FIGURE 6.10

Exploded view of early three-axis stabilized satellite

6.2.3   Thermal control system

The obvious objective of the thermal control system (TCS) is to assure that the equipment in and about the spacecraft structure is maintained within temperatures that will provide successful operations. There are many factors that need to be considered in the design of the TCS since satellites in space are subjected to a thermal environment that is very different from that on the Earth where gravity and a fluid medium (air) exists and convection, conduction and radiation are the principal mechanisms for heat transfer.

The "extreme" vacuum in space limits all heat transfer mechanisms to and from the spacecraft and its external environment to that of radiation. The second law of thermodynamics states that the direction of heat flow is invariably from a "hot" to a "cold" body. Thus, a spacecraft will receive thermal energy from the Sun (the Sun's corona is approximately 6 000° K), reflected solar energy from the Earth and Moon, and, depending on the satellite surface temperatures, it will receive or transmit thermal energy with the Earth and Moon. The planets and stars are relatively insignificant in determining the heat balance of a spacecraft.

With regard to the other aspects of the thermal environment, the spacecraft is surrounded by cold space which acts as an infinite thermal "sink" (the estimated temperature of deep space is equivalent to about 4° K).

There are several techniques for controlling thermal radiation exchange of the spacecraft. The geometry or projected area of surfaces directly facing the Sun, the Earth and deep space are important factors. Often, movable shutters or louvers are used to provide some means of dynamic control at the critical surfaces. Also, the spectral absorptivity (alpha) and emissivity (epsilon) properties of material surfaces or coatings must be carefully selected since there is a significant difference between the alpha/epsilon values of coatings between solar temperatures (short wavelengths) and infrared temperatures (long wavelengths) associated with the Earth and deep space. Many coatings have been discovered or developed that have unique alpha/epsilon ratios that selectively either absorb or reflect solar radiation but have opposite properties for infrared radiation.

The main structure of the satellite, which contains the communications and support equipment and "housekeeping" components, is constructed with a combination of thermal insulation and conductive materials so that temperatures within the structure are held within satisfactory limits. Most equipment can tolerate a relatively wide range of temperatures in order to function, but considerations of life expectancy and reliability often impose the need for limited temperature ranges. For example, batteries usually require narrow limits and liquid propellants need a high level of thermal protection.

"Super" insulations, with orders of magnitude improved performance over earthly equivalents, are feasible in the vacuum of space. These insulations are made of multiple layers of plastic sheets coated with highly reflective materials, such as aluminized mylar. They are often used on tanks to reduce the rate of evaporation of cryogenic fuels as heat is absorbed from the Sun or Earth or from heat generating equipment.

Heat generated from power amplifiers, motors, heaters, etc. must be conducted or transported to the outside surfaces where it can be dissipated. High power TWTs may require "heat pipes" to convey large amounts of heat to external radiators. Care must be exercised in the design of "joints" since the vacuum of space can negatively affect conduction across metallic interfaces.

External structures, such as solar arrays and antennas, require special considerations since they are subjected to the full radiation environment of an earth satellite. Some of these are evident in Figures 6.8 and 6.9. Solar arrays, which are continuously facing the Sun, need special coatings (optical reflectors, etc.) to limit the absorption of the Sun's rays and radiate this energy to space. Antennas, usually facing the Earth, are subject to continuously varying solar exposures. The antenna structure and thermal design must be such that temperature variations over the surface do not distort the antenna shape beyond its operational design limits. This requires good conductivity properties and special alpha/epsilon coatings.

Finally, it must be recognized that the thermal environment in space for an Earth satellite is constantly changing. Also, each side of a spacecraft may be exposed to its unique environment. The variations likely to be experienced by a spacecraft are:

1) Varying solar exposure as the satellite circles the Earth.

2) Seasonal variations relative to the Earth and Sun.

3) Eclipse periods (72 mins maximum) when the satellite is in the Earth's shadow.

4) Varying reflections of solar energy from the Earth due to cloud cover (changing albedo's).

6.2.4 Satellite attitude and orbit (station-keeping) control systems

6.2.4.1 Background

The focus in the past for the in-orbit control of satellite communication systems has been on equipping spacecraft with the necessary sensors and functional devices so that monitoring control centres located on the Earth can maintain attitude and orbit location of the spacecraft within specified limits throughout its operational life. GSO satellites, for example, are usually controlled on an individual basis and are maintained in a particular orbit location within ±.05° of longitude and latitude. Antenna beam pointing requirements are often restricted to ±.10°. Non-GSO satellites, on the other hand, may or may not require station-keeping, depending on the coverage characteristics of the satellite constellation. The overall design trends for orbit and attitude systems is to automate the control functions so that small groups of controllers can perform the ground operations. Following is a brief discussion of the essential requirements and functions of existing and planned systems.

6.2.4.2 Attitude control

The objective of the attitude control subsystem is to maintain the antenna RF beam pointed at the intended areas on Earth. The axis of the antenna-bearing platform carrying the antennas is made to point towards the Earth's centre and the antennas are mounted in relation to this platform so as to be directed towards the area required.

The attitude control procedure involves:

- measuring the attitude of the satellite by sensors;

- comparing the results of these measurements with the required values;

- calculating the corrections to be made to reduce errors;

- introducing these corrections by operating the appropriate torque units.

During normal operation, the satellite experiences only smooth cyclic drift disturbances of the order of 10^{-5} Newton metre (N · m). However, when station-keeping is performed, torques of the order of 1 N · m are applied to the spacecraft (typically for 30 s to 2 min per week, or 05 to 2 x 10^{-4} of total time).

The sensors normally used for attitude measurements are of the infrared type and measure the difference in the infrared signature between space and the Earth's disc in the CO_2 emission/absorption band: (15 ± 1 micron). Since the background noise is of the order of 0.02° (in a

1 Hz bandwidth) and there are seasonal cyclic variations of the zero and infrequent transient phenomena known as "cold clouds", the typical measurement accuracy that can be obtained is 0.05°. In addition to Earth sensors, Sun sensors are also used to determine body orientation. They can be used to directly measure yaw angle (over much of the daily orbit), as well as to provide additional roll-pitch data.

A higher degree of beam pointing control is used in many domestic and new generation international systems (INTELSAT-VI). This attitude control mode uses a ground-based pilot beam (beacon) which is sensed on board the spacecraft to directly obtain the antenna orientation. When more than one antenna is mounted on the same platform, this approach permits independent control of the orientation of each antenna, in response to the error signals generated by the respective sensor, tracking the same beacon or separate beacons. Motor driven gimbals are required in this case. Such a control system can correct the effects of relative misalignment between the various antennas, due to mechanical errors and thermal variations. This control method can improve the net beam pointing accuracy by a factor of 2 or 3 compared with body orientation. In addition, if pilot beacons from two well separated earth stations are used, direct sensing of beam rotation (yaw) error can also be obtained.

Currently all types of attitude stabilization systems have relied on the conservation of angular momentum in a spinning element. Stabilization systems can be classified in two categories:

- spin stabilization;

- three-axis stabilization.

The principles of attitude control systems can be illustrated by the evolution of the INTELSAT family of satellites:

- INTELSAT-I and II were solid spinners (implying a toroidal antenna directing only some 4% of RF energy to Earth);

- INTELSAT-III, IV and IVA included a despun element ("dual spinners"):

  - an offset parabolic reflector on INTELSAT-III,

  - the entire RF payload in INTELSAT-IV and IVA (while the "platform" remains spun and feeds power to the payload through a slip ring rotary mechanism);

- in INTELSAT-V and VA, the spinning element is reduced to a fast rotating wheel internal to the spacecraft (which allows the use of a solar array pointed permanently to the Sun): this configuration is called "three-axis". In all cases, the angular momentum of the rotating body is nominally normal to the orbit plane.

Whichever stabilization system is used, the drift rate of the platform orientation (due primarily to solar radiation pressure) is of the order of 0.02° per hour. When the drift shows up as an error in the North-South pointing (usually called "roll" error), it is corrected through the use of small jets, magnetic torquers, or altering the position of "solar sails". The errors around the spacecraft/Earth line (undetectable by Earth pointing) are allowed to build up, as such errors have only very small effects on communication performance, and show up within the next six hours as North-South pointing errors.

The pointing errors in the East-West direction (usually called "pitch" errors) are corrected by accelerating/decelerating the relative rotation between the despun and spun elements.

The latter control loop is on board the spacecraft and is always automatic. The correction of the orientation of the angular momentum (precession) can be either automatic on board (usually with the so-called "three-axis configurations") or by ground command (usually with "dual spinners"), or a mixture of the two.

In every case the zero reference can be adjusted by ground command:

• INTELSAT-VI, launched starting from 1989, is a "dual spinner", but it also uses automatic on-board precession control.

The types of stabilization used on spacecraft are as follows:

i) Spin stabilization (see Figure 6.8)

The satellite is rapidly spun around one of its principal axes of inertia. In the absence of any perturbing torque, the satellite attains an angular momentum in a fixed direction in an absolute frame of reference. For a geostationary satellite, therefore, the spin axis (pitch) must be parallel to the axis of the Earth's rotation.

The perturbing torques produce two effects: they reduce the spin speed of the satellite; and they affect the orientation of the spin axis.

Satellites using this type of stabilization generally have a despun platform, one direction of which is servo-controlled so as to point towards the centre of the Earth. This platform usually carries the antennas and the payload. It should be noted that spin stabilization is normally used during the phase between injection into the transfer orbit and arrival in the geostationary orbit, even when the satellite is of the three-axis design. This arrangement neutralizes the effects of the perturbing torques caused by the distance between the direction of thrust of the apogee motor and the centre of mass of the satellite.

The earliest spin stabilized satellites were spun about their axis of maximum moment of inertia, and were thus inherently stable. However, as antenna patterns became more complex and power requirements grew, it became necessary to enlarge both the antenna and solar array drums while still meeting the geometric constraints of the launch vehicles. The resultant spacecraft designs could no longer be spun about the maximum inertia axis, and hence were nutationally unstable. In response, two types of nutation control have been developed: pulsed thruster firings (quantized control) and antenna platform de-spin motor control (linear). The use of these has allowed spin stabilized spacecraft to become more powerful (currently 2-3 kW) and to carry large and complex antenna subsystems.

ii) Three-axis stabilization (see Figure 6.9)

The antenna is part of the satellite, which maintains a fixed orientation with respect to the Earth, except for the solar arrays which are rotated to point towards the Sun.

The simplest method uses a momentum wheel, which simultaneously acts as a gyroscope, as in spin stabilization, and as a drive. Certain perturbing torques can be resisted by changing its spin speed and, consequently, the resulting angular momentum of the satellite. Nutation control systems are also used in three-axis stabilized satellites.

New innovations to reduce fuel requirements of thrusters for attitude corrections include a solar sailing feature (INMARSAT and TELECOM II), a three-axis stabilization system of momentum wheels with magnetic bearings (ETS-VI) and propulsion fuel heaters [1].

6.2.4.3   Orbit (station-keeping) control

On-board propulsion requirements for both geostationary and non-geostationary (LEO, MEO and HEO) satellites account for a significant part of the total mass of a spacecraft system, especially if the operational life extends to ten years and beyond. The importance of these systems to the mission of space systems has resulted in the establishment of several research and development programmes in government (ESA, NASA, NASDA, etc.) and industry to improve performance. A few recent innovations include arcjet thrusters, an advanced solar electric propulsion system, pulsed plasma thrusters (stationary and anode layer thrusters), ion propulsion systems, iridium coated rhenium chambers for chemical propellants, etc. Weight savings by the use of efficient propulsion systems (for final orbit insertion, station-keeping and relocation in orbit) provide such benefits as increased payloads (transponders, etc.), increased life, lower launch requirements, etc. Following is a brief description of the environmental and other factors affecting orbit control [2].

A geostationary satellite is subject to disturbances which tend to change its position in orbit. They lead to spurious orbit plane rotation and semi-major axis and eccentricity errors. As viewed by an observer on the Earth, the satellite displays an oscillatory movement with a periodicity of 24 hours. This motion is characterized by a North-South component due to orbit inclination (the so-called "figure of eight"), and an in-plane component. In turn, this component is made up of a longitudinal drift, due to the semi-major axis variation, and of a daily in-plane oscillation (altitude and longitude) due to the eccentricity error. The most apparent and relevant component of the in-plane oscillation is that in East-West direction.

The objective of orbit control is to maintain the spacecraft inside the allocated position "box" in latitude/longitude (current Radio Regulations only limit the longitudinal variations to ±0.1°, for satellites using frequencies allocated to the fixed-satellite or broadcasting-satellite service).

The sources of disturbance are:

• the lunisolar attraction. This disturbance tends to rotate the orbit plane about an axis which, on the average is perpendicular to the astronomical direction of the Aries constellation. The corrections must be performed at six or 18 hours sidereal time (which gains one day per year or approximately four minutes a day with respect to solar time) with the direction of thrust along the North (or South) axis. The magnitude of the corrective impulse is around 45 m/s per year, slightly modulated by an 18.7 year cycle of Moon orbit motion.

The disturbance shows up as a sinusoidal daily motion in elevation (latitude), growing in half amplitude by nearly 0.02° per week. It should be noted that there exists an orbit inclined at 8° to the equator on which there is no disturbance at all. (This can only be used with steerable ground antennas or a beamwidth wider than ±8°):

- the longitudinal component of the acceleration of gravity due to the ellipticity of the Earth's equatorial section known as "Earth triaxility". It is fixed for a given longitude. Its maximum value is around 2 m/s per year. Corrections must be performed with the direction of thrust along the orbit. Its effect shows up as a uniformly accelerated drift in longitude;

- the effect of the solar radiation pressure which builds up eccentricity in the orbit (by decelerating the spacecraft in the morning and accelerating it in the afternoon). It depends upon the area-to-mass ratio of the spacecraft, mostly dictated by the power-to-mass ratio of the payload. It shows up as a daily oscillation in longitude and altitude, the amplitude of which grows at a rate that is proportional to the area-to-mass ratio of the spacecraft. The correction must be performed with thrust along the orbit.

The corrections for the last two disturbances are conveniently combined, resulting in thrust periods at either six hours solar time or 18 hours solar time (depending upon station longitude), and in a minimum expenditure of propellant.

The duration of the thrusting period (which implies extremely high disturbance torques) is of the order of:

- 30 to 150 s per week since the previous North/South correction, for North-South station-keeping;

- 2 to 20 s per week since the previous East/West correction, for East-West station-keeping.

For most current satellites, the maximum time between corrections to keep within ±0.1° is:

- around two months in North/South;

- around two to three weeks in East/West.

The role of the orbit control subsystem is to reduce the amplitude of this undesired movement. Small thrusters are fired at appropriate points in the orbit to provide the required corrective velocity increments. Early systems used monopropellant thrusters in which hydrazine was decomposed at about 1 300 K in a catalyst bed in the rocket chamber. More recently, other propulsion systems have been developed, to either supplement or replace the basic hydrazine system. Examples of the latter are electrically augmented hydrazine thrusters in which the hydrazine is electrically heated (up to 2 500 K or above) after decomposition to increase its enthalpy, and ion engines in which a gas (mercury or zenon) is ionized and then accelerated in an electric field. Both of these have been designed to increase the fuel usage efficiency (specific impulse) during the North-South manoeuvres.

The most common replacement for hydrazine monopropellant reaction control systems is a bipropellant (fuel/oxidizer) system. This type of propulsion system is not only more fuel-efficient than the monopropellant type, but can also be integrated with a liquid apogee motor (see § 6.2.7). Whichever system is used, the thrusters provide attitude control as well as orbit control.

The propellant mass consumed is equivalent to slightly below 2.5% spacecraft mass per year, where the catalytic decomposition of hydrazine is employed, and approximately 1.8% of spacecraft mass per year, with either electrically augmented hydrazine or bipropellant thrusters.

LEO, MEO and HEO satellites that rely on constellation configurations in which the precise maintenance of orbits is required, employ orbit control systems and equipment similar to those for GSO satellites. This would apply to cases of circular orbits in which the inertial forces are small and orbits are designed to be repeatable. In addition to assuring coverage over preferred areas on the Earth, another reason for this type of constellation would be to limit or avoid coverage areas where potential radio interference to and from other radiocommunication systems can occur.

There are some planned non-GSO systems which do not require precise orbit maintenance, thus reducing the housekeeping weight and complexity of the spacecraft compared to those with accurately synchronized orbits. Real time analysis of the orbit parameters would be required to effectively predict communication coverage functions. The initial orbit injection conditions (errors in inclination and altitude) of the satellites are important since they remain with the system throughout its life. The orbit altitude, inclination and orbit plane spacing can affect the drift rates (ascending node and orbit plane spacings) of the satellites and must be carefully selected based on the ultimate accuracy required in predicting the average coverage characteristics and deviations of each satellite in the system [3].

6.2.5   Power supply

Electrical power requirements for communication satellites have increased considerably during the last 20 years as launch systems have become more powerful and are able to insert payloads into orbit in the order of 2 000 kg to 5 000 kg. A large number of transponders can now be accommodated on a single spacecraft, thus multiplying its communication capacity several-fold compared to early systems. For GSO systems, the introduction of "hybrid" satellites, in which both 6/4 GHz and 14/10-12 GHz are utilized for communication services, have increased the electrical requirements from approximately 1 kW to 15 kW. The number of transponders per spacecraft have more than doubled in this period of time (40 to 50 transponders is commonplace at present) and power amplifiers for each transponder have increased from 10-20 Watts to as much as 60-100 Watts of transmission power. The emerging 30/20 GHz satellite systems that intend to use very small earth station antennas will place increasing burdens on satellite power supplies.

The power requirements of non-GSO satellite systems are generally lower per satellite due to shorter transmission distances, often with less bandwidth availability and fewer transponders, among other factors. However, recent reports on the characteristics of these systems have indicated that power requirements can vary from 0.5 kW ("little" LEOs) to as much as 3 kW ("big" LEOs) per spacecraft. Large satellite "constellations" could demand over 1 megawatt of electrical power [4].

The Sun provides the primary source of power for communication satellites. The choice of systems for converting solar power to electrical power was explored by engineers and scientists during the early years of space systems development when the efficiency of the process was an important consideration. Early experimenters tried to develop heat engine (Carnot) power plants using parabolic solar collectors. Atomic cells or batteries were also considered. All of these approaches were overtaken by solar cells, made of silicon and gallium arsenide wafer-like or coating materials which convert a portion of the Sun's radiation equivalent to approximately 1 kW per square metre of projected area normal to the Sun's rays directly into electrical power. In the ensuing decades, efficiencies of solar cells were increased from a few per cent to slightly over 18%. Increasing the efficiency of these devices is a continuing process of experimentation and development [5].

The principal components of the power supply system of a communication satellite include 1) the power generators, usually solar cell arrays located on the spinning body of a spinning satellite, or on "paddles" for a three-axis stabilized satellite (see Figures 6.2-1 and 6.2-2); 2) electrical storage devices such as batteries for operation during solar eclipses; 3) the electrical harness for conducting electricity to all of the equipment demanding power; 4) the converters and regulators delivering regulated voltages and currents to the equipment; and 5) the electrical control and protection subsystem which is associated with the telecommand and telemetry subsystem.

The equipment requiring electrical power includes 1) the communication system in which transponders demand as much as 70% to 80% of the total power; 2) battery charging, about 5% to 10%; 3) thermal control, about 7% to 12%; 4) tracking, telemetry and control, about 2% to 4%; 5) attitude control and station-keeping, about 3% to 5%; and 6) the remainder, about 2% to 4%.

The satellite power supply components are subject to the same dynamic and environmental constraints as those described in the previous sections. Low mass, long life and reliability are of major importance since the system must function satisfactorily in space for 10 to 15 years without being serviced.

a)   Solar array

A necessary consideration for the design of the solar arrays is to assure that sufficient surplus capacity exists initially so that end-of-life performance will meet the communication mission requirements. The electric power supplied depends on the conversion efficiency of the solar cells and the size of the solar array. Although steady progress is being made in improving the performance of solar cells, most existing operational satellites have solar cell systems with overall efficiencies of about 12%.

These are generally made of P type silicon single crystal wafers on which a thin N type layer is created by doping to form a diode. Each cell is covered by a window of molten silica to reduce the effect of radiations in space (30% efficiency loss in five years). An individual solar cell supplies about 50 mW of power (under 0.5 V) and an array is made up of a large number of cells connected in a series/parallel arrangement. For an adjustable array, the end-of-life specific powers are about 21 to 23 W/kg and 60 to 67 W/m^2.

b)   Secondary sources

Since most telecommunication satellite equipment (attitude and orbit control, telecommand, payload, etc.) has to be in permanent operation, energy has to be stored for use during eclipse periods.

Electrochemical generators are best for this purpose. Nearly all telecommunication satellites are equipped with nickel-cadmium batteries, despite their low power-to-weight ratio (35 W/kg), because they are hermetically sealed and have a very long life. During the next decade, they are likely to be replaced by nickel-hydrogen batteries.

The mass of the battery depends, *inter alia* on the utilization factors on the spacecraft – depth of discharge, temperature. The battery life depends on the depth of discharge, i.e. the ratio between the capacity discharged during an eclipse period and the normal capacity. To obtain a lifetime of over five years, the value of the ratio should not be less than 60-70% for the maximum duration of an eclipse period, which is 72 min. The lifetime also depends on the battery temperature, the best results being obtained between +5°C and +15°C.

c)   Eclipse periods

When the satellite moves into the shadow of the Sun's rays caused by the Earth, a thermal shock is caused to the satellite components and the solar cells which supply the primary energy are no longer illuminated. This eclipse period is maximum when the Sun is in the direction of the intersection between the ecliptic plane and the equatorial plane, viz. twice a year at the equinoxes (about 21 March and 22 September). On these days, the shaded part of the orbit extends over 17.4° (the angle of the Earth, as seen from a geostationary satellite) and the eclipse lasts 72 min. This maximum duration decreases and the daily eclipses cease when the Sun's inclination from the equatorial plane becomes equal to 8.7° (17.4°/2), i.e. about 21 days before and after each equinox. The middle of the eclipse duration occurs at midnight, satellite longitude time. In consequence, if a regional or national (domestic) satellite can be located west of its service area, the eclipse will occur after midnight local time, and the on-board secondary power source could be reduced (or, even, possibly, omitted) if the traffic at that time does not justify the full communication capacity.

d)   Regulators and converters

A battery which supplies power during an eclipse period has to be recharged during the sunlight period. Certain precautions have to be taken while recharging the battery.

Two main procedures are used:

•   the battery is connected directly in parallel with the solar array and establishes its potential (unregulated line);

- the solar array is kept at a fixed voltage and a charge control circuit is connected in series with the battery (regulated line).

The advantages of these two types of regulation are as follows:

Regulated line	Unregulated line
Better adaptation of power sources to equipment requirements.	Simplicity of system (although d.c.-d.c. converters are more complex).
Good compatibility with modular and standardization concepts.	Configuration adapted to very highly pulsed load operation (e.g. TDMA).
Possibility, subject to electromagnetic compatibility, of supplying equipment directly from the regulated line.	

As a general rule, power should be supplied to equipment in the form of regulated d.c. voltages. High-efficiency chopping converters are at present used to supply these various voltages.

6.2.6 Telemetry, command and ranging (TCR)

The status and condition of the active components and resources of a satellite system are collected from special sensors distributed throughout the spacecraft. This data is then transmitted periodically or upon command to the ground control centre (GCC) by the telemetry subsystem. Command signals are transmitted from the GCC to the satellite to satisfy operational mission requirements or to respond to emergency conditions. The satellite location is tracked by sending a ranging signal to (command link) and from (telemetry link) the satellite. This signal is then processed to obtain the level of precision required for station-keeping or relocating the satellite.

The usual sequence of TCR functions can be described as:

- reception and demodulation (sometimes decoding) of command signals intended to keep the satellite operational and to adapt the payload to mission requirements;

- collection, shaping and emission of telemetry signals for the permanent control of the whole satellite;

- transfer, after demodulation and remodulation, of signals used for ranging.

The TCR system must function reliably through all phases of a satellite's life and requires redundancy for some of its equipment. An omnidirectional antenna is used during transfer-orbit operations or in cases of disorientation of the satellite. During normal on-station operations, the high gain communication antennas are used for the transmission and receiving of TCR signals. The satellite manufacturer usually designs the TCR system from existing standards which they have developed. The frequency bands used for the TCR signals are generally from the edges of the operational FSS bands (6/4 GHz, 14/10-12 GHz, etc.), although the omnidirectional antenna system may use a lower frequency band (2 GHz) to minimize link power requirements.

6.2.7   The apogee motor

Most non-GSO satellite systems are likely to be launched directly into their final orbit either as a single satellite or a number of satellites in the same orbit plane (see section 6.9.3.2). Apogee motors are not required in these circumstances.

For GSO satellites, however, there is the option of either a direct insertion into orbit from a single launch system or the use of a transfer orbit for the first phase, after which an apogee motor is used to place the satellite into its final circular GSO orbit (see section 6.9.3.1). The latter case is more cost effective since the launcher equipment bay (guidance system, power supply, etc. weighing approximately 200 to 300 kgs) can be jettisoned before the final-stage rocket firing.

The transfer orbit employed during the last two decades has a perigee of about 200 km, an apogee at the geostationary altitude (about 36 000 km) and an inclination which is close to the latitude of the launching site. The apogee motor is fired at a precise time to produce a circular orbit and to eliminate the inclination of the orbit. Consequently, the apogee of the transfer orbit should be situated at the geosynchronous altitude and in the equatorial plane. The further the launching site is from the equator, the more power is required. Compared to a launching site on the equator, the mass that can be put into geostationary orbit using the same apogee motor is some 20% less if the launching site is at a latitude of 30° and approximately 35% less at a latitude of 45°.

With current propulsion technologies, the net mass ratio (transfer orbit mass over spacecraft on station minus apogee propulsion tanks, empty) is between 1.7 (ARIANE from Kourou) and 1.9 to 2 (launches from Cape Canaveral).

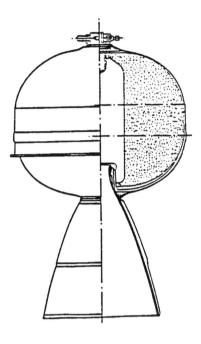

European "MAGE" powder engine

Sat/C6-11

FIGURE 6.11

Example of solid-propellant apogee motor

The apogee motor should be able to deliver a velocity increment between 1 500 m/s (ARIANE from Kourou) and 1 850 m/s (launches from Cape Canaveral). The velocity increment must be delivered at apogee (5.25 hours after perigee plus an arbitrary number of transfer orbit periods of 10.5 hours). The control is effected through the spacecraft TTC and attitude control system, usually in a spun mode (arrays and antennas in stowed position, as thrusts are orders of magnitude higher than in station-keeping).

Like all anaerobic rocket engines, the apogee motor is characterized by a specific impulse and by a ratio of structural mass to propellant mass.

There are two types of apogee motor:

• traditional technology has made use of "solid propellant apogee motors" (mixtures of ammonium perchlorate and aluminium powder in a matrix of elastomeric plastics). This type of apogee motor, (see Figure 6.11) is very efficient and has been used, up to now, for most satellites. It requires, on firing, the spin stabilization of the combined apogee motor and satellite;

- current trends are towards "bi-liquid propulsion" (nitrogen tetroxyde and methylated hydrazine) usually with common tankage both for the apogee motor (Figure 6.12) and for the separate thrusters of the on-station propulsion system.

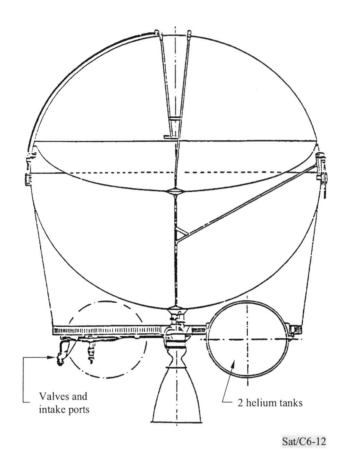

Valves and
intake ports

2 helium tanks

Sat/C6-12

FIGURE 6.12

Example of bi-liquid propellant apogee motor with central membrane tank

6.2.8   Characteristics of the payload of some communication satellite

Table 6.2 contains data on the BOL (beginning-of-life) mass, the EOL (end-of-life) primary power, the RF (radio frequency) power, the eclipse capacity and the design life of some communication satellites. The examples given in this table are concerned with the space segment payload for systems of the FSS such as:

- international systems: INTELSAT-IVA, V, VI;
- regional systems: ARABSAT, EUTELSAT I, TELECOM I;
- national systems: ANIK-C, SATCOM-V, SBS.

As a comparison, one of the examples (TVSAT-A5) gives details of the payload of a BSS (television) satellite.

TABLE 6.2

Characteristics of the payload of some communication satellite

Satellite	Satellite mass, BOL* (kg)	Primary power (EOL)** (W)	RF power (W)	Eclipse operation	Design life (years)
SBS(F-3)	563	900	200	yes	8
ANIK-C	567	800	240	yes	10
SATECOM-V	598	1 000	204 ***	yes	10
ARABSAT	695	1 440	312.5	partial	7
TELECOM-1	690	1 150	194	yes	7
EUTELSAT I (ECS)	680	950	180	yes	7
EUTELSAT II	1 700	3 000	800	yes	7-10
INTELSAT-0IVA	790	600	120	yes	7
INTELSAT-V	1 200	1 300	240	yes	7
TVSAT-A5	1 600	4 200	1 250	no	10
INTELSAT-VI	2 250	2 260	460	yes	10

* Beginning-of-life
** End-of-life
*** 212.5 W during eclipse

6.3 The communication payload

6.3.1 Introduction

The satellite, or space station, is the focal point for the receipt and transmission of radio signals to a network of earth stations. Because of their high altitude or elevation, satellites can view large surfaces of the Earth by direct line-of-sight. For example, the much used geostationary orbit provides coverage of at least one-third of the Earth's surface (see Figure 6.4). This characteristic allows the use of higher frequencies (above 2 GHz) for radiocommunications than were practised in the early days of radio. The frequency bands in most common use for commercial satellite systems are the 6/4 GHz, 14/11-13 GHz and, more recently, 30/20 GHz. The bandwidth requirements that have been established for most systems are 500 MHz for each direction (space-to-Earth and Earth-to-space). Larger bandwidths are expected for the frequencies above 20 GHz. The specific frequencies allocated to the fixed-satellite service (FSS) by the ITU Radio Regulations are shown in the table in Appendix 9.1.

The subsystem known as the "payload" is comprised of the communication transponders, antennas and associated equipment involved directly in the receipt and transmission of radio signals from and to the earth station network. Most transponders in operational use at the present time are "transparent", i.e. they simply translate the frequency of received signals, amplify them and route them to the appropriate transmitting antennas. More sophisticated transponders have been put into operation recently (INTELSAT-VI and VII among others) that contain switching elements that rapidly transfer signals between multiple satellite beams. Also, more advanced transponders have been under development in several experimental programmes that feature demodulation, baseband processing and regeneration of the signals. At present, there are several programmes under way in which 30/20 GHz communication systems are being designed with multiple spot beams and elaborate on-board processing and switching mechanisms.

The ability of new launch systems to insert large masses into orbit has allowed designers to increase the reliability of satellite equipment by increasing redundancy, electromagnetic compatibility and such support functions as the communication payloads have increased in size and capacity. Some of these improvements with regard to support subsystems were reviewed in section 6.2. A brief description of the components and functions of payload subsystems is provided in the following sections.

6.3.2   Antenna subsystem

6.3.2.1   General

The satellite antenna design depends upon the objectives of the communication services and the type of spacecraft to be employed.

Important considerations include the orbit, attitude and station-keeping control method, radio frequencies and bandwidth, service area or "footprint", spacecraft structure, launch vehicle capabilities and user terminal characteristics. As the industry developed, the need for efficient utilization of the orbit (GSO) and radio spectrum resulted in antennas with cross-polarization and beam discrimination properties (multiple small beams) to achieve "frequency reuse". This introduces additional problems for antenna designers since the size of the antenna is an inverse function of the size of the beam. By exhibiting considerable innovative techniques, such as the use of multiple feeds for a single parabolic antenna, beams have been "shaped" to cover many geometrically complex service areas.

In conjunction with the solar panels, the design of high gain antennas for the principal communication services usually dictates the spacecraft design. Since each frequency band requires a receive and a transmit antenna, satellites using multiple frequency bands increase the structural and logistic requirements of the spacecraft. If the uplink and downlink frequencies are relatively close (for example, 6/4 GHz or 14/12 GHz), the same antenna reflector can be used for both receive and transmit functions. Diplexers are used to separate the signals. In some designs in which linear cross polarization is used, superimposed reflectors provide a compact antenna system that occupies the space of a single reflector.

During the launch phase and spacecraft orientation procedures, special omnidirectional antennas are needed for telemetry, command and ranging operations (see section 6.2.6). Bi-cone antennas with annular shaped beams are usually used during the spacecraft spin cycle, when stability is required for the spacecraft prior to the perigee or apogee motor firings. Once on station, the telemetry or command functions are transferred to the high-gain communication antennas. The bi-cone antennas then become back-up units in the event the spacecraft is respun or disoriented.

6.3.2.2 Structures

Early versions of low altitude satellites used monopole antennas (whip elements and dipoles) with passive magnetic stabilization for attitude control. These grew in sophistication and capability as larger antennas were designed which had to be folded during the launch phase, and deployed after insertion into orbit. New LEO satellite systems currently being developed plan to use a combination of phased arrays and reflector antennas. A glimpse of new sophisticated satellite antenna designs planned for LEO satellites is provided in section 6.5.

Spin stabilized satellites require the antenna to be mounted on the despun portion of the spacecraft. Many systems use off-set fed paraboloid antennas which continuously face the Earth, as depicted in Figure 6.8. Early satellites restricted the size of the antennas to fit within the nose fairing of the launch vehicle without folding. Later, as service demands increased and designs improved, the antennas grew in size and had to be stowed during the launch phase and deployed in orbit.

Three-axis stabilized spacecraft increased the design options for satellite antennas, both in size and complexity. The side of the satellite that always faces the Earth is the logical platform for the location of centre or near-centre fed antennas. However, the adjacent sides also provide excellent platforms for off-axis fed antennas. Some of the structural options which have been used for GSO satellites include:

- a tower structure (see Figures 6.9 and 6.10) in which antenna reflectors are mounted on the side facing the Earth. Although this arrangement provides for a relatively simple antenna design, the disadvantage is the long length of the waveguides connecting the transponder power amplifiers with the antenna feed;

- mounting the antennas on the "East" or "West" side of the satellite equipment housing or box as shown in Figure 6.13. The feed waveguides are shorter than in the previous example. This arrangement requires folding or stowing of the antennas against the spacecraft body during launch and deployment of the reflector upon insertion into orbit as shown in Figure 6.13. Flexibility is provided for those satellites employing more than one frequency band and associated antennas. Many satellite systems operate with both 6/4 GHz and 14/10-12 GHz frequency bands;

- using direct radiating arrays which can be mounted on the side facing the Earth as shown in Figure 6.14. This type of antenna does not require a reflector, and can be installed with a relatively simple structure.

6.3.2.3 Performance characteristics

The main characteristics of a satellite antenna are:

- coverage contour (beam configuration);

- pattern shape and side-lobe level;

- polarization purity;

- power handling;

- RF sensing capability.

a)   Coverage

The coverage area as seen from the satellite is defined by the iso-gain (or iso-e.i.r.p.) contours.

The first satellite antennas were formed by a reflector fed by circular horns. This structure radiates only circular beams.

However, for a given e.i.r.p. the RF power provided by the transponders and therefore the electrical power consumed can be reduced if the transmitting antennas concentrate their radiation on the regions to be served (coverage area).

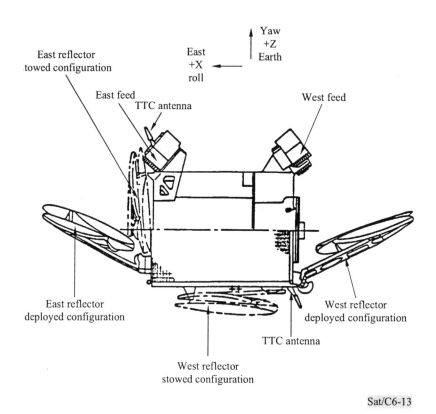

Sat/C6-13

FIGURE 6.13

EUTELSAT II spacecraft configuration

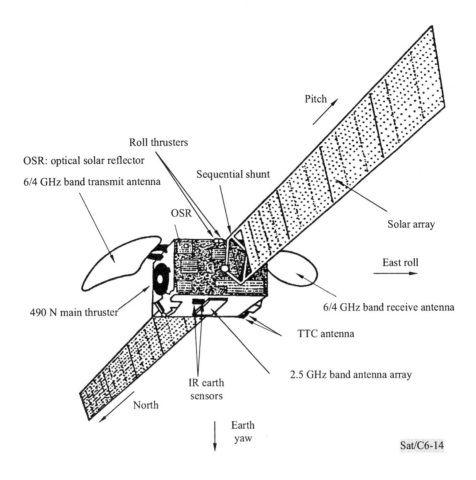

Roll thrusters

OSR: optical solar reflector

6/4 GHz band transmit antenna

Sequential shunt

OSR

Pitch

Solar array

East roll

6/4 GHz band receive antenna

490 N main thruster

TTC antenna

2.5 GHz band antenna array

IR earth sensors

North

Earth yaw

Sat/C6-14

FIGURE 6.14

ARABSAT satellite

Consequently, recent satellites use shaped beam antennas which radiate within the service area contours and avoid over-spilling.

Although the problem is less critical on reception, this type of antenna often has to be used here too in order to reduce the earth station uplink RF power requirement, and hence the cost of earth stations.

It can be demonstrated theoretically that the geometry of the satellite footprint is a good approximation of the antenna feed geometry (Figure 6.15 shows the feed geometry and the corresponding coverage contours of a satellite antenna located at 319° W).

A better approximation of the desired service area contours would require a larger number of feeds and an increase in the reflector dimensions. Table 6.3 illustrates the impact of improved shapings on the antenna dimensions.

TABLE 6.3

INTELSAT antenna dimensions

INTELSAT spacecraft generation	IVA	V	VI
Antenna system weight (kg)	51	69	313
Reflector maximum dimension (m)	1.34	2.44	3.2
Maximum number of horns per feed	37	90	146

b)   Pattern shape and side-lobe level

Only in the case of high power direct broadcasting satellites, are the pattern shape and side-lobe levels specified by § 3.13.3, Annex 5 to Appendix 30 of the Radio Regulations and Recommendation ITU-R BO.652-1 (see Figure 6.16 for Regions 1 and 3). The extension of the use of this type of mask to all communication satellites would increase the feed complexity and the reflector dimension of future spacecraft.

c)   Polarization purity and frequency reuse

•   Polarization discrimination

The limitations on available frequencies and congestion of the geostationary orbit result in an increasing need for frequency reuse by means of polarization discrimination.

Either circular or linear polarization can be used. For circular polarization, the feed horns are generally circular or hexagonal. However, in the case of linear polarization, the horns can be rectangular and the number of horns necessary to generate shaped beams can be reduced.

For linear polarization, the use of a dual-gridded dish is the best way to achieve good polarization discrimination, the advantages over a solid reflector with polarization discrimination in the feed are less complex feed, no frequency sensitivity and separation of foci which allows implementation of two different feeds for frequency reuse.

The gridded dish is composed of a dielectric (transparent) dish, supporting a metallic (reflecting) grid. Two dishes of this type may be superposed, e.g. the first with "horizontal" wires and the second with "vertical" wires (see Figure 6.17) to form a "dual-gridded dish".

The polarization isolation necessary for frequency reuse is typically 27 dB. The EUTELSAT II gridded reflector achieves an isolation better than 36 dB.

When orthogonal polarizations are used, the axial ratio (AR) of the polarization indicates the contribution of the cross-polarization level (X polar). For instance, an AR of 0.7 dB gives an X polar level of –27.9 dB.

•   Beam discrimination

When the service areas can be covered by well isolated beams like the different "hemi" or "zone" beams of the INTELSAT-VI satellites (see Figure 6.18) two separated beams can use the same frequency bands (INTELSAT-VI reused six times the same frequency band, two times by polarization discrimination and four times by beam discrimination).

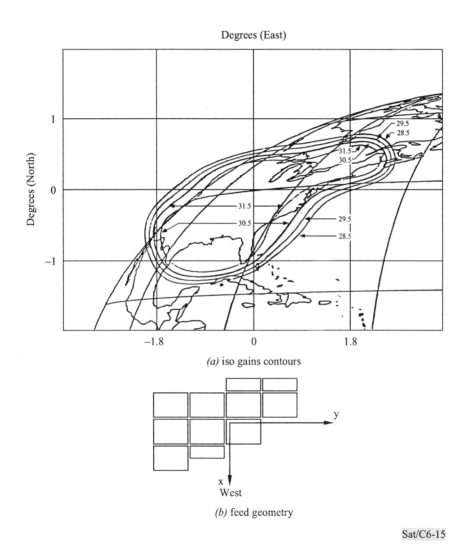

FIGURE 6.15

Example of antenna feed geometry and corresponding coverage contours

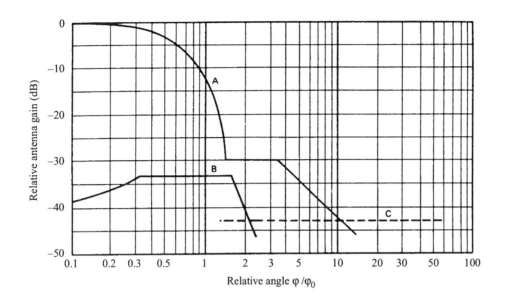

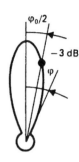

Sat/C6-16

FIGURE 6.16

**Reference patterns for co-polar and cross-polar components
for satellite transmitting antennas in Regions 1 and 3**

Curve A: Co-polar component (dB relative to main beam gain)

$$-12\left(\frac{\varphi}{\varphi_0}\right)^2 \qquad\qquad \text{for } 0 \le \varphi \le 1.58\ \varphi_0$$

$$-30 \qquad\qquad \text{for } 1.58\ \varphi_0 < \varphi \le 3.16\ \varphi_0$$

$$-\left[17.5+25\log\left(\frac{\varphi}{\varphi_2}\right)\right] \qquad\qquad \text{for } \varphi > 3.16\ \varphi_0$$

after intersection with curve C: as curve C.

Curve B: Cross-polar component (dB relative to main beam gain)

$$-\left(40+40\log\left|\frac{\varphi}{\varphi}-1\right|\right) \qquad\qquad \text{for } 0 \le \varphi \le 0.33\ \varphi_0$$

$$-33 \qquad\qquad \text{for } 0.33\ \varphi_0 < \varphi \le 1.67\ \varphi_0$$

$$-\left(40+40\log\left|\frac{\varphi}{\varphi}-1\right|\right) \qquad\qquad \text{for } \varphi > 1.67\ \varphi_0$$

after intersection with curve C: as curve C.

Curve C: Minus the on-axis gain (Curve C in this figure illustrates the particular case of an antenna with an on-axis gain of 43 dBi).

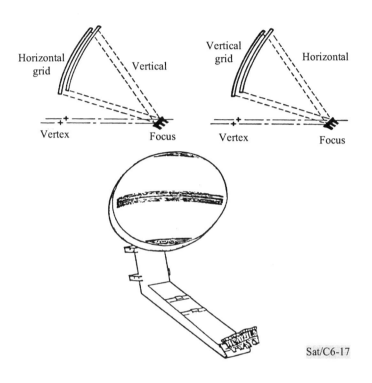

FIGURE 6.17

Dual-gridded reflector antenna system with superposed grids

d)   Power handling

Each generation of satellites radiates higher e.i.r.p.

Typical values of maximum RF power per antenna port are 600 W at 1 1/12 GHz band for EUTELSAT II and 1 kW for DBS satellites like TDF 1 or TVSAT.

This evolution requires an increasing effort in the fields of primary feed thermal control and intermodulation products.

e)   RF sensing capability

When the beamwidth is small (less than 2°), an RF sensing system is used, which automatically corrects any deviation in beam direction.

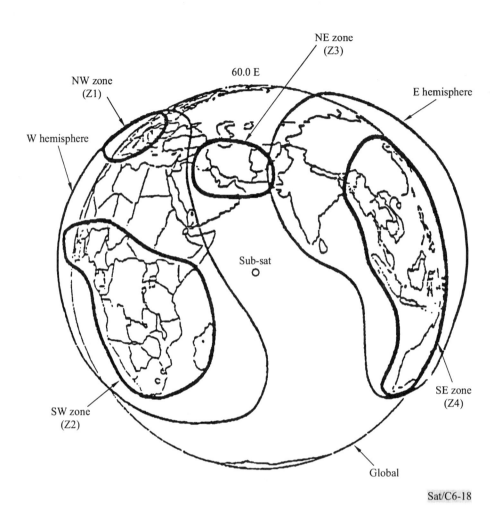

FIGURE 6.18

INTELSAT-VI typical antenna pattern coverage (10R, 60° E)

The RF sensing works with a beacon on ground. The satellite antenna radiates a difference pattern and the null direction is locked to the ground beacon.

Different solutions are used to obtain the difference pattern. It is possible to use four horns of a multifeed antenna to generate the sum and difference patterns or the higher modes of a single corrugated horn.

6.3.2.4   Antenna research and development trends

For commercial satellite systems, antenna research has focused on a) techniques for accurately shaping radiation patterns through the use of computer software programs; b) improvements in cross or orthogonal polarization isolation and main beam-to-side lobe isolation; c) modifying or reconfiguring radiation patterns in orbit; and d) developing rapidly scanned multiple spot beams, phased arrays and hybrid antenna systems. The development of new materials and devices, such as high temperature superconductors and carbon fibres, GaAs monolithic microwave integrated circuits (MMICs) and opto-electronic devices, have accelerated improvements in satellite antenna performance.

Phased arrays, as an alternative to large reflectors, deploy an array of small antennas over a large area, connected in such a way that their received or transmitted signals are in the correct electrical amplitude and phase with each other. The frontview of a typical example of a phased array antenna is shown in Figure 6.20 [Ref. 1]. The individual antennas are called elements; groups of elements are referred to as sub-arrays, which, in turn, can be combined into large arrays. The circuitry which performs the phase and amplitude distribution (or even multiple beam generation) is referred to as a beam forming network (BFN). Phase shifters can shape the size and direction (i.e. performing beam scanning or switching) without the need of physically moving the antenna, as is the case for conventional (single-horn fed) reflector antennas. The BFN may be controlled by electrical or even by electronic means. In the latter case, beam scanning or switching can be performed very rapidly (for example, in satellite switching TDMA (SS-TDMA)).

However, arrays can also can be applied to reflector antennas by employing BFN feeds which can reconfigure a satellite beam by ground command. A typical example of a complex multi-horn feed system with a BFN constructed in waveguide technology is represented in Figure 6.20. The use of multi-horn-fed reflector antenna technology predominates the commercial communication satellite industry at present. As noted above, this design is used to shape antenna beams and focus radio energy onto desired service areas. INTELSAT has used three-layer BFNs to provide three region coverage with the same feed-horn-reflector system.

Intersatellite link (ISL) antennas may require large scanning capabilities, especially for LEO satellites. Direct radiating phased arrays, phased array feeds for reflectors and mechanically scanned sub reflectors are promising antenna techniques for ISL applications. Europe's data relay satellite (DRS) has scanning antennas at 27.5 GHz capable of scanning 20° within 120 seconds. Japan's communications and broadcasting engineering test satellite (COMETS) will employ a number of antennas for satellite broadcasting, feeder links for mobile satellite and ISL applications. The latter will use a 2 m reflector operating at 47/44 GHz (see Figure 6.9). The USA/NASA tracking and data relay satellite system (TDRSS) has been a pioneer (1970-80s) in the development of ISS systems (2 GHz and 15/13 GHz).

Experiments and developments of communication systems in the 30/20 GHz frequency bands have been in progress for over a decade. Most employ array-fed reflector antenna designs to achieve multiple narrow beams in the service areas. A few of the research and development organizations active in this area include ESA/Europe (ARTEMIS), Italy (ITALSAT), Japan (COMETS and ETS-VI), NASA/USA (ACTS), Russia (KOMETA), among others.

Much of the advance developments in phased arrays and large deployable antenna structures is being driven by mobile satellite (MSS) communication initiatives (8-30 m mesh reflector antennas, folding petal antennas, direct transmitting radiating array and focal array fed receiving reflector, etc.) This technology will provide significant improvements to the FSS as well.

6.3.3   Transparent (repeater) transponders

6.3.3.1   General

As was noted above, satellite transponders receive signals from earth stations or relay satellites and transform them into appropriate signals for return transmissions. They may simply be repeaters that amplify and frequency shift the signals or be much more complex, performing additional functions such as signal detection, demodulation, demultiplexing, remodulation and message routing (see section 6.3.4). This section describes the features of transparent transponders.

However, many considerations relative to the radio-frequency (RF) and intermediate frequency (IF) input stages as well as to the output power amplifiers may also apply to the more complex transponders described in section 6.3.4.

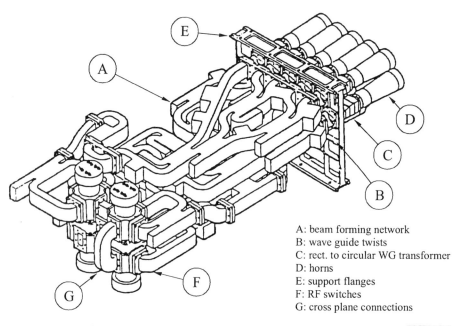

A: beam forming network
B: wave guide twists
C: rect. to circular WG transformer
D: horns
E: support flanges
F: RF switches
G: cross plane connections

Sat/C6-19

FIGURE 6.19

A multiple horn feed system and the accompanying beam forming network (from [Ref. 1])

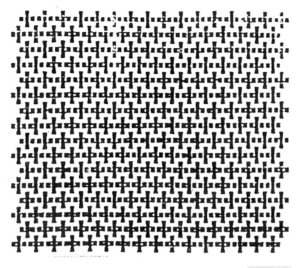

Sat/C6-20

FIGURE 6.20

A phased array antenna (from [Ref. 1])

A key element in any spacecraft transponder is the high power amplifier (HPA). For a single carrier, the HPA is usually operated at or near the maximum output power level, or saturation, to attain a high overall efficiency in the conversion of dc energy into RF energy. The HPA or transmitter is also required to amplify the signals without distortions or other impairments. Two types of HPAs are in common use, electron beam devices or travelling wave tube amplifiers (TWTAs), and solid state power amplifiers (SSPAs).

The signals reaching a satellite receiving antenna are extremely weak and the transponder has to amplify them and send them in a translated band to the retransmitting antennas. Amplification is of the order of 100-110 dB (see Figure 6.21) and can even exceed 120 dB in broadcasting satellites.

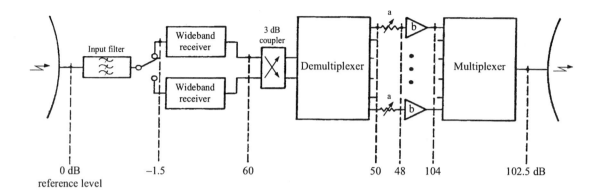

a: switchable attenuator
b: high power transmit amplifier

Sat/C6-21

FIGURE 6.21

Typical diagram of the relative levels in a transponder

In the transponder, the signals are usually degraded to some extent. These impairments, which must be kept within tolerable limits, are caused by many factors including the following main ones:

- amplifier non-linearity which gives rise to intermodulation (several carriers in one amplifier) cross-modulation and inter-symbol interference in the case of digital transmission;

- interference between signals transmitted in neighbouring frequency bands;

- multiple paths taken by a signal between input and output;

- amplitude and phase variations in the passband, caused essentially by the filters which give rise to signal distortion, resulting in baseband noise or errors.

A single amplifier is generally not sufficient to deliver the total power required by all the RF channels. In addition, non-linearity of the amplifiers means that operations must be conducted with amplifier back-off which increases with the number of signals to be amplified and limits the output level. Amplification must therefore be effected in two stages:

a) common low-level amplification of all the desired signals in the total band supplied by the receiving antenna;

b) signal amplification by sub-bands or frequency channels (fractions of the total band), up to the desired output levels. Filter and coupler systems are used for splitting the initial band into sub-bands after the common low-level amplification section and at the input to the amplifying chains and recombining them at the output before they reach the transmitting antennas.

The positioning of the transponder components within the spacecraft body is often a delicate matter in view of the constraints involved, particularly the following:

- the need to remove heat given off by the heat dissipating sub-assemblies (particularly the power amplifiers), which entails the use of sufficiently large panels radiating directly into space;

- the layout of the components and the fitting of the protective devices needed for meeting electromagnetic compatibility requirements, avoiding coupling and reducing the effects of thermal variations;

- reduction of antenna feeder losses, particularly by limiting the length of the waveguide or cable connections.

6.3.3.2   The wideband-receiver subsystem

This provides both the first amplifying stage and the translation from the receive to the transmit band in case of a single conversion system. For dual conversion systems, the wideband receivers provide amplification and translation to the intermediate frequency. Its noise factor must be sufficiently low so as to have as little effect as possible on the C/N ratio of the uplink.

Intermodulation products limit the output level of the receiver depending on the transistors used. Typically the receiver system provides an amplification (gain) of roughly 50-60 dB (see Figure 6.22). Allowance must be made for losses in the frequency converter and in the filters and couplers. The gain distributions must ensure an acceptable level of intermodulation products and a maintenance of a satisfactory noise factor.

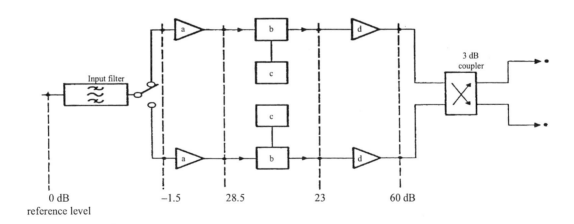

a: preamplifier   d: amplifier
b: mixer   e: channelized portion
c: local oscillator

Sat/C6-22

FIGURE 6.22

**Example of wideband receiver (with redundancy)
with typical diagram of relative levels**

Wideband receivers are fully solid state. They must be particularly reliable since any fault would affect all the signals transmitted in the total band. They must therefore be redundant.

Preamplifiers may incorporate tunnel diodes, transistor or parametric stages. Advances in gallium arsenide (GaAs) field effect transistors (FET) technology have superseded the tunnel diode and the parametric amplifiers. Preamplifiers must be preceded by a bandpass filter to eliminate image frequencies and all signals outside the receiving band.

Amplifiers may incorporate either bipolar transistor (4 GHz) or GaAs FETs (4 GHz and above). For 30/20 GHz band operation High Electron Mobility Transistors (HEMT) are promising devices for low noise performance.

Frequency converters, generally with diodes, are designed not so much to minimize losses but to suppress the transmission of local oscillator frequencies, their harmonics and higher order products. Single ended or symmetric double circuits are frequently used. The local oscillator, which has to produce a power of about 10 dBm, may be a conventional crystal controlled oscillator followed by multiplier stages, a phase-locked loop or a dielectric cavity oscillator stabilized by a crystal reference. The oscillator must be highly stable in terms of temperature and time and its phase noise must be reduced to a minimum.

6.3.3.3   The channelized subsystem

The second amplifying stage is effected in the transmit band by a group of amplifying chains, each of which amplifies a fraction of the total band subdivided into frequency channels.

The output from the wideband receiver is channelized by a set of circulators and bandpass equalized filters known as the input demultiplexer. After high power amplification, the channels are recombined at the output of the transponders by one or more sets of filters known as output multiplexers (Figure 6.23).

The channelized subsystem of the transponder therefore consists of an input demultiplexer, an amplifying chain and an output multiplexer (Figures 6.23 and 6.24).

The channelization plan varies from system to system. In a system such as INTELSAT where multiple frequency reuse is carried out by polarization and spatial beam isolation (see section 6.3.3.4), the channelization plan is identical for each beam so that connectivity between beams can be maintained by simple RF switching at the intermediate frequencies within the payload. For domestic systems, most designs opt for an interstitial (interleaved) channelization plan for the two orthogonally polarized receive and transmit beams. This has the advantage of reducing intra-system interference by having the high energy density portions of the spectrum in one polarization placed against the guardbands of the orthogonal polarization.

As for the channel bandwidth, there is no unified standard. In the 6/4 GHz band, 40 MHz spacing is common but not universal, and in the 14/11-12 GHz band, various bandwidths are being used, e.g. 27 MHz, 36 MHz, 45 MHz, 54 MHz, 72 MHz, etc.

The input demultiplexer, which is fed by the receiver, divides the total transmission bandwidth into frequency channels corresponding to the amplification chains (see Figure 6.23). The selective filters in the demultiplexer must have sufficiently steep slopes to avoid multiple paths through adjacent amplifying chains and a sufficiently flat response curve in the passband to keep distortions to tolerable levels. The filters will generally have to be amplitude- and phase-equalized. The losses in the demultiplexer do not have any significant effect and must simply be taken into account in the overall gain budget. The filters can thus be linked by circulators in tandem, either in a single group or more frequently in two groups (fed by the two leads from the receiver coupler), one for even channels and the other for odd channels.

Several types of filter may be envisaged, but only the least bulky and lightest are actually used taking into account the effects of the whole temperature range inside the spacecraft both in orbit and during the launch phase.

Use is often made of dual mode cavities, through which a single signal passes twice and therefore halves the number of cavities required in the filter. The technology used must be such as to make them extremely reliable, highly stable, easy to assemble and light. In the 6/4 GHz band carbon fibre (which makes for the lightest equipment) or thin invar technology is used. At 14/11 GHz thin invar is generally employed. Since these filters are passive components, they do not require any redundancy.

Each amplifier chain corresponds to a specific band selected by the input demultiplexer filters. The main component of a chain is the power amplifier which is connected to the corresponding filter in the output multiplexer, generally through an isolator. The amplifier may be a TWT or a solid-state power amplifier (SSPA). For the 4 GHz band, the output power of TWTAs is usually between 5 W and 10 W although TWTAs with output powers of up to 40 W are sometimes used. Solid-state power amplifiers working at 4 GHz with power outputs of up to 10 W are also now commercially available and efforts are being made to develop units producing up to 30 W.

At the 11-12 GHz band, low power TWTAs of between 10 and 20 W and medium power TWTAs of between 40 and 65 W are available. In addition, high power TWTAs with power levels up to 250 W have been space qualified for broadcast-satellite applications. Solid-state power amplifiers are not yet commercially available for the 11-12 GHz band due to their poor DC to RF conversion efficiency. However, development efforts are taking place to produce suitable units. In addition, the amplifier chain may contain the following options:

- a fixed attenuator to compensate for channel loss deviations in the input demultiplexer and gain variations of the TWTA. At its output there will be the same nominal level in each chain;

- an attenuator operated by ground command, according to the overall required gain, the type of signal transmitted, equipment degradation, etc.;

- sometimes an amplifier with automatic gain control to compensate for fluctuations due to atmospheric conditions in the uplink (in the case where a single signal is transmitted in the channel);

- a driver amplifier when the power amplifier gain is not large enough to provide the total required amplification under all conditions;

- a linearizer to compensate for the non-linearities of the TWTA or SSPA;

- a limiter circuit in certain modes of operation.

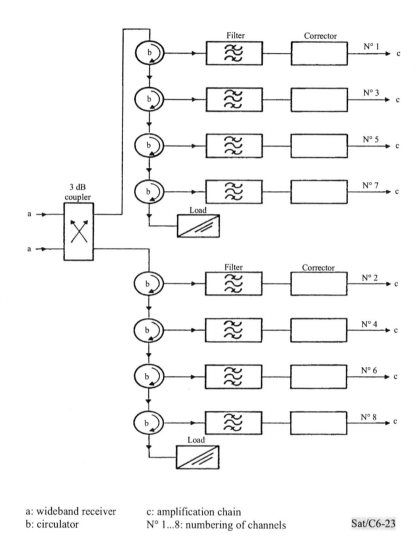

a: wideband receiver   c: amplification chain
b: circulator   N° 1...8: numbering of channels   Sat/C6-23

FIGURE 6.23

Example of an input demultiplexer (simplified diagram)

Linearization (pre-distortion) can be used to enable operation with less TWTA or SSPA back-off. Owing to reliability requirements, the amplifying chains and in any case the power amplifiers must be redundant. Different types of redundancy can be employed depending upon the available volume and mass, e.g. one standby amplifier for one active amplifier, or one common standby amplifier for two active amplifiers, or several standby amplifiers for all the active amplifiers. The corresponding switching configurations (at the input and output) can be more or less complex. However they should not cause a significant increase in the losses between amplifier outputs and filter inputs in the multiplexer or impair the quality of the transmission.

Arrangement of the power amplifiers may sometimes pose problems, since these amplifiers must be mounted on panels for radiating the heat towards space. The cabling to the corresponding output multiplexer filters must be short and operation of redundant equipment must not impair working conditions (increase in losses, change in the thermal equilibrium).

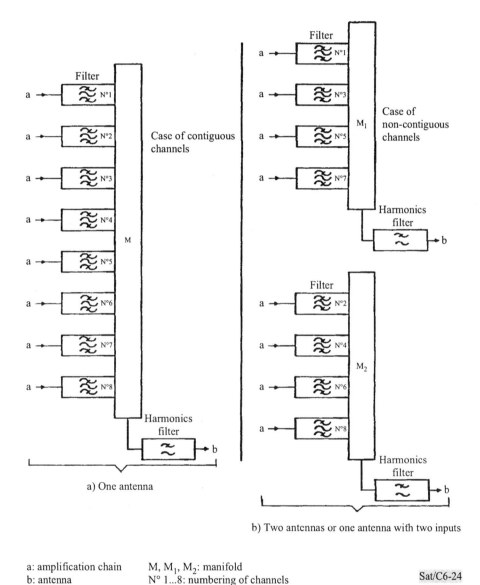

a: amplification chain   M, M_1, M_2: manifold
b: antenna   N° 1...8: numbering of channels   Sat/C6-24

FIGURE 6.24

Example of output multiplexers (simplified diagrams)

Losses in output multiplexers must be kept low since any loss between the final amplifier outputs and the antennas has a direct effect on the e.i.r.p. For the same reason filters are directly coupled to a single waveguide (manifold) feeding the transmit antenna (see Figure 6.24). The resulting installation is rigid and sometimes quite bulky. For the output multiplexer, equalizers are not needed in the satellite since they can be installed in the earth stations. The technology used for input demultiplexer and output multiplexers is basically the same, however, output filters and manifolds dissipate significant heat which must be removed.

It is easier to connect filters of non-adjacent even or odd channels to a single manifold, and so two manifolds with their outputs linked to a twin-port antenna are employed. The construction of the output multiplexers is thereby simplified, but the antenna is more complex and less efficient.

Adjacent channel filters can also be coupled to a single manifold, i.e. contiguous multiplexer, but then the filter slopes must be made steeper, losses are increased and the manifold becomes more complicated to build; on the other hand, the single-port antenna is simpler and more efficient. It is thus necessary to strike an optimum balance between the type of transmission antenna and output multiplexer used.

6.3.3.4   Transponders without on-board processing

Basically, satellite transponders receive, amplify, frequency translate and retransmit various types of communication signals.

Wideband as well as channelized transponder configurations can be envisaged. Most of the actual satellite repeater configurations, in the fixed-satellite service, are based on wideband receivers followed by channelized transmitters.

As far as connectivity among different RF channels is concerned, there are two main cases depending on whether the transponders are connected to a single transmit beam or to several beams.

Consider the simple case where the received signal is routed to only one transmit beam. The signals in the receiving band are amplified and translated into the transmit band. Two types of frequency translation may be used:

*   a single conversion system which translates the frequency directly from the receive band to the transmit band;

*   a dual conversion system which first translates the received signals to an intermediate frequency for part of the amplification process, and then translates them again to the final transmit frequency bands.

The second type, i.e. dual conversion, is sometimes preferred because it offers the advantages of:

*   eliminating potential instabilities due to feedback within high gain amplifying chains;

*   avoiding cross modulation products or harmonics of the received signals and the local oscillator of the frequency translator falling within the desired signal bands; and

*   providing a convenient intermediate frequency for switching and cross-strapping between payloads that work with different receive and transmit frequency bands. The disadvantage is that two local oscillators and two frequency translators are required.

The transponder provides an amplification of roughly 100-110 dB in two stages: low-level amplification in a wideband receiver, followed by high-level amplification by high power amplifiers in the channelized subsystem. Attention must be paid to electromagnetic compatibility (EMC) owing to the high gains and large bandwidths involved.

Systems operating with several beams are currently in service, e g. INTELSAT-IVA and V. These beams can provide independent transmissions. However, it is frequently necessary to connect users covered by different beams. In such cases, one beam's uplink channel will be connected to the corresponding channel of another beam's downlink.

This does not invalidate the descriptions given in section 6.3.3.3. There is still one receiver subsystem per beam or polarization, but some of the outputs of the demultiplexer filters are connected to amplifying chains belonging to different transponders. The interconnections between an uplink beam and 3 downlink beams may use one of the following methods:

- interconnections whose configuration can be changed from time to time by means of electromechanical switches operated by ground command (Figure 6.25). This type of interconnection is used in INTELSAT-V;

- dynamic interconnections in TDMA with satellite switching between bursts (SS-TDMA). This is carried out by means of an on-board diode or transistor switching matrix which connects each burst of an up-beam frequency channel to the amplifying chain of the destination down beam. This switching is sequentially operated by a command unit according to the traffic organization; the sequential order can be modified by ground command. The matrix introduces insertion losses which must be offset by additional amplifiers (Figure 6.26). Such a system is found in INTELSAT-VI.

The switches, command units, matrix and all the interconnections must be set up in such a way that the level of the signals at the transponder inputs remains relatively constant irrespective of the switching and paths traversed. Electromagnetic compatibility (EMC) must also be ensured and spurious coupling and induction caused by switch command pulses must be avoided.

The characteristics of the satellite transponders are mainly influenced by the performance of the TWTAs. Some of the factors involved are considered below:

i)   Hard-limiting transponders

Hard-limiting transponders include a limiting device which clips the incoming signal. This permits the travelling wave tube amplifier (TWTA) to be operated at saturation with output power almost independent of the repeater input power. However, multicarrier operation is restricted because of intermodulation noise.

ii)   Quasi-linear transponders

A TWT will behave in a more linear fashion if the output power is backed off from saturation. Typically a 4 to 6 dB back-off is necessary for FDMA operation. A smaller back-off is usually sufficient to reduce the spectrum spreading and in-band distortion with single carrier PSK transmission.

iii) Linear transponders

An ideal linear transponder would have a linear transfer characteristic up to the point where saturation occurs. Methods for improving the linear portion of the transponder transfer characteristic over that achievable with TWTA transponders are being approached in two different ways.

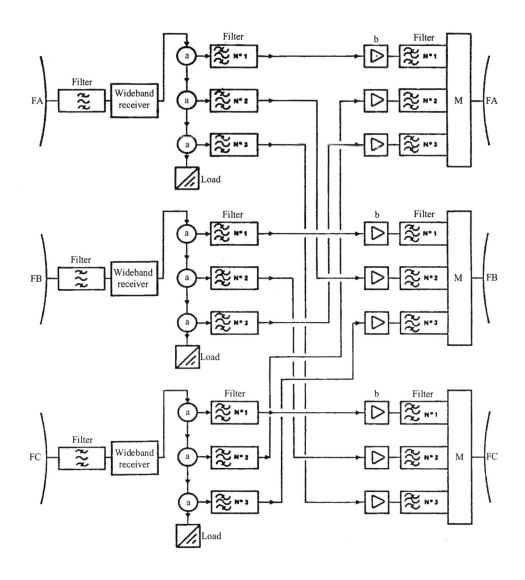

FA, FB, FC: antenna beam
a: circulator
b: amplification chain

M: manifold
N° 1, 2, 3: numbering of channels

Sat/C6-25

FIGURE 6.25

Example of an interconnection diagram between antenna beams

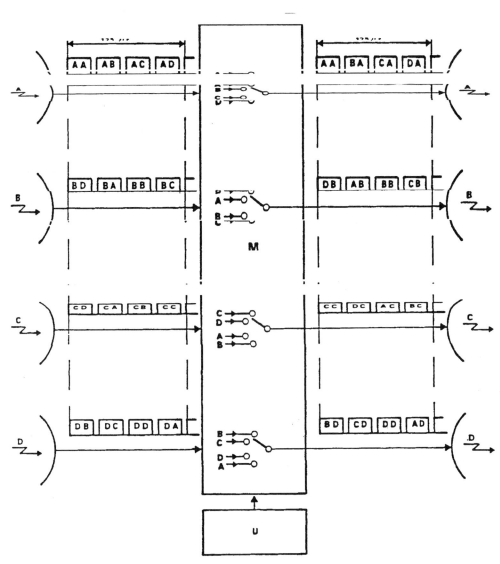

Sat/C6-26

TDMA in a network including:

- a satellite with a multibeam antenna and only one frequency (transmission and reception)
- synchronized earth stations
- a synchronized switching matrix on board

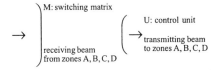

M: switching matrix

receiving beam
from zones A, B, C, D

U: control unit

transmitting beam
to zones A, B, C, D

FIGURE 6.26

Example of TDMA using a switching matrix on the satellite
(satellite switched TDMA (SS-TDMA)

One approach is to improve the linearity of TWTA transponders through the use of various linearizer configurations which have been developed using feed forward, Butler matrices and pre-distortion techniques.

The other approach is to replace a TWTA with a solid-state amplifier based upon FET technology, which has an intrinsically more linear transfer characteristic when near saturation than a TWTA. Such amplifiers are being developed and show promise of improved linearity (and reliability) while avoiding the on-board mass and power penalties associated with the previously mentioned TWTA linearization techniques.

6.3.3.5   Transponder research and development trends

The current major manufacturers of space qualified TWTAs include Thomson-CSF Tubes Electronics in Europe and Hughes in the United States. All have made significant improvements in the size, weight, power and efficiencies of these units during the last decade. Other groups involved in making space related TWTAs include NEC and Toshiba in Japan. Table 6.4 (A and B) shows a summary of recent TWT developments at Thomson-CSF and NEC/Toshiba. Efficiencies have progressed steadily to the extent that 60% are now achievable for 12 GHz bands and additional improvements of from 5-7% are expected in future versions. Parallel with the development of TWTAs are electronic power conditioners (EPCs) to meet the multi-kilovolt ranges required by TWTAs. EPC efficiencies of 90-93% (DC input to high voltage output) have been achieved for high power (200-260 W) TWTs at 12 GHz.

Solid state power amplifiers (SSPAs) have been used extensively for lower power (10-20 W) and lower frequency (4 GHz) applications and for phased arrays where power gain is distributed over a large number of elements. The status of SSPA development in Europe and Japan is shown in Table 6.5 (A and B). At the present time, these devices have inferior efficiencies and power capabilities compared to TWTAs. However, their inherent high reliability and convenient physical characteristics for space applications (arrays, etc.) have made them the focus of intense development. It has been estimated by manufacturers of these devices that within a few years 20 GHz SSPAs will be available capable of 30 W of RF power. Higher frequency applications are also forecasted for active phase array systems along with advance MMIC technology advancements.

TABLE 6.4 (A)
TWTA developments by Thomson-CSF Tubes Electronics

Frequency band (GHz)	RF output power (W)	Broadband efficiency (%)	Typical gain (dB)	Non-linear phase (degrees)
1.5	150	49	34	40
2.3	215	54	35	45
4	17-130	54-59	55	35-45
8	12-160	56-63	63	45
12.0	25-240	60-64	55	35-45
20	15-140	52-60	60	40-55
32	Broadband operating models			
64				
80	Technology available			
NOTE – An Electronic power conditioner (EPC) efficiency of 93% is assumed.				

TABLE 6.4 (B)

TWTA developments in Japan

Frequency (GHz)	Saturated power (W)	Efficiency (%)	Voltage (kV)	Weight (kg)	Manufacturer	Comments
12	120	59	6.7	3.3	NEC	4 stage MDC, M-Type cathode, conductive cooling, PPM focusing
12	200	59	6.9	3.5	NEC and Toshiba	Integral pole piece construction
20	10	33	4.9	1.7	NEC	Used on ETS-VI
22	200	50.7	12.3	5	Toshiba and NEC	Coupled cavity design used on COMETS
44	20	32	10	6.2	NEC	Used on COMETS

TABLE 6.5 (A)

SSPA developments reported in Europe

Frequency (GHz)	Output power (W)	Gain (dB)	Efficiency (%)	Manufacturer	Comments/applications
2.5	10 W module up to 40 W combined		40	Alcatel	Up to 40 W combined
	-	55	>37	ANT	Used on HERMES
	20			Matra-Marconi	Used with Internet III
	40	47	>40	MBB	
4	10	65	31	ANT	Used on DFH 2 and 3 (China)
	12		30	Alcatel	Used on Telecom 2
	30	55 to 75	29	ANT	
18	10	60	17	ANT	MMIC; 1-2 W modules
20	0.5			ANT	Used for ACTS
	5			Thomson-CSF	5 W combined

TABLE 6.5 (B)

SSPA developments reported in Japan

Frequency (GHz)	Output power (W)	Efficiency (%)	Manufacturer	Comments/ applications
2.5	2		MELCO	Used on ETS-VI
2.5	110	42	Fujitsu	5 beams, 8 amplifiers
4	30	38	NEC	Used on INTELSAT-VII
12	30		MELCO	MMIC, GaAs, FET
20	3	18	MELCO	MMIC, 2 modules
21	20	10	MELCO	MMIC, 1W-2W modules
38	~0.8	~5	NEC	Used on ETS-VI, 4 devices in parallel
38	0.5	~3	Fujitsu	Used on ETS-VI, MMICs, chip power combining

6.3.4   Transponders with on-board processing

In digital satellite systems, performance and efficiency can be improved by using on-board processing transponders which are capable of performing switching, regeneration or baseband processing. These techniques have not been widely used up to now in commercial satellites, but will find applications in the future.

6.3.4.1   RF switching

Three types of RF switching operations can be envisaged at microwave frequencies:

a)   switching information bursts from one transmission RF channel to another;

b)   switching information bursts from one fixed spot beam to another;

c)   switching a scanning spot beam from one earth station to another.

Operation a) can be performed by an on-board frequency-agile frequency converter. It can provide interconnectivity among earth stations operating at different carrier frequencies.

Operation b) provides interconnectivity among earth stations accessing different spot beams by cyclically interconnecting the TDMA signals via an on-board time-domain switch matrix (SS-TDMA). The microwave switch matrix is controlled by a programmable distribution control unit clocked by a stable time reference. Each earth station in the network synchronizes the transmission of its information bursts to the programmed on-board switch sequence, in order to communicate with the earth stations which access other beams. Several configurations of this type have been analysed and proposed.

Operation c) requires the implementation of a steerable beam. A satellite with a high gain steerable spot beam can give total coverage to the entire service area, while still providing the advantages of high antenna gain and a good efficiency of TDMA. In these systems the earth stations in the network share a single RF channel in TDMA mode, while the steerable beam scans the service area in synchronism with the time division format.

6.3.4.2 Regenerative transponders

If the transmitted digital information is recovered on board by regeneration, the uplink will be separated from the downlink and the following main advantages are obtained:

- the overall error ratio of the communications system is the sum of the error ratios of the up and downlinks instead of being determined by the total carrier-to-noise ratio. Earth station and satellite e.i.r.p. can be thereby reduced;

- the effects of non-linear distortions of the two links do not accumulate;

- any on-board multipath interference arising from RF coupling among different channels is eliminated;

- the on-board modulators allow the on-board RF amplifier to be fed by signals which are virtually free from amplitude modulation components; therefore the RF amplifiers can be operated at saturation level without significant performance degradation;

- in SS-TDMA systems, on-board switching matrices can be realized by baseband circuits; these offer substantial advantages (in terms of mass, size, power consumption) compared to microwave switching matrices which must be used with non-regenerative repeaters; the baseband switch matrix can consist of simple logic gates. Custom-built components could allow the realization of high-speed, large-dimension matrices with low power consumption and size;

- more flexible on-board redundant schemes can be conceived;

- a system master clock can be placed on board, thus simplifying clock recovery and frame synchronization in SS-TDMA systems;

- the downlink can be made to operate in a continuous carrier mode (even in the case of TDMA) by the use of on-board demodulation.

6.3.4.3 Bit stream processing

With the availability of the digital information on board the satellite, a number of advanced techniques can be used. One of the most attractive features of on-board processing is that of bit rate conversion. In SS-TDMA systems with a large number of participating earth stations with very different traffic requirements, the low-traffic stations usually have a low efficiency, since they have to be designed for a high transmission rate but can exploit their resources for only a low percentage of the time. In these cases substantial advantages will be achieved if the satellite is equipped with regenerative repeaters operating at different bit rates. Each participating station will operate at the most suitable bit rate for its own traffic requirements.

Interconnectivity among stations, operating at different bit rates, will be achieved by on-board bit rate converters and switch matrices.

Additional digital processing options can be envisaged, such as on-board error correcting decoding and the implementation of store and forward concepts by appropriate buffers. Reliability would be the major problem with these applications.

Finally it would be possible to adopt on the downlink a modulation scheme different to that of the uplink.

6.3.4.4   SS-TDMA with on-board TST switching stage

Referring to the terminology commonly used in terrestrial switching techniques, an SS-TDMA satellite system can be regarded as a time-space-time (TST) switching centre in which the time stages are located in the earth stations, and the space stage in the satellite.

As the technology of digital circuits progresses, it is logical to envisage going one step further and entrusting the satellite with the complete TST switching function. By transferring the time stages from all earth stations to the satellite, one can expect to reduce significantly the complexity of the earth segment. This system concept has a number of advantages and useful features because:

*   since it is no longer necessary for the earth stations to send no more than one burst per frame, regardless of the final destination of the traffic, a higher frame efficiency can be achieved;

*   it becomes possible to incorporate the bit stream processing (see section 6.3.4.3 above) to enable the system to cope more efficiently with different traffic requirements;

*   variable traffic patterns can be more easily accommodated by means of a demand-assignment technique, so that the utilization efficiency of the system is increased;

*   the system lends itself to incorporation in an integrated services digital network (ISDN), being able to route different kinds of signals (voice, data, videoconference, etc.) simultaneously.

6.3.4.5   ISDN applications

The establishment of a broadband ISDN will provide accommodation for most forms of digital transmission on a circuit or packet switched basis, and satellites are likely to achieve a deeper penetration into the network.

The optimization of the switching functions may require a distribution of these functions at suitable nodes.

User-to-user digital switching consists of time domain stages (T), which reorganize the frame by changing the time allocations, and of space domain stages (S), which switch the frame without reframing.

In particular, the cascade connection of a T-stage, an S-stage and a T-stage provides full switching flexibility. A first way of distributing these functions in a network including a satellite link is by placing the S-stage on board the satellite and the remaining stages of the T-type at the transmit and receive side of suitable earth stations located in connection with ISDN nodes. The on-board S-stage would consist of an on-board switching matrix, which operates either at RF level in a non-regenerative satellite or at baseband level in a regenerative satellite. With these devices the input digital channels are interconnected in a real time to the output digital channels and the input/output interconnections are dynamically reconfigured, so as to allow the appropriate routing of each unit of digital information. Figure 6.27 provides illustrations of the distribution of the switching functions over the satellite link (end-to-end). The example on the left side of the figure indicates that the user traffic within each coverage beam can be concentrated by earth stations providing time domain switching. The access to the satellite is by beam discrimination for stations located in different spot beams or by frequency discrimination for stations located within a common coverage beam.

A further development of the distributed switching exchange could be obtained by locating the whole time-space-time (TST) switching facility on board the satellite. Actually satellite TST stages are being actively considered at an advanced research and development level.

With this architecture the satellite would act as a real "switchboard in the sky" and the switching facility would be made available to all the terrestrial parts of the ISDN which is covered by the satellite antenna beams; access from ISDN nodes would likely be achieved by very small dish terminals of reduced complexity.

This architecture is shown on the right side of Figure 6.27 and it can be adapted both to purely TDMA systems and to mixed systems. For example, a mixed system has been proposed which combines SCPC carriers on the uplink and TDM (MCPC) carriers on the downlink.

It should be noted that all the architecture shown in Figure 6.27 can also find application in a context other than ISDN.

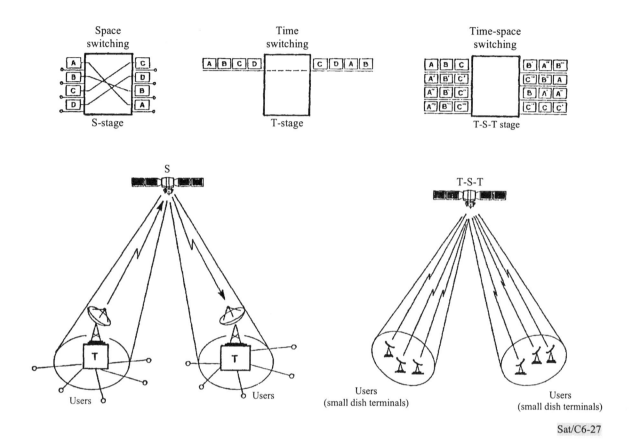

FIGURE 6.27

Examples of configurations of the switching functions in a satellite network

6.3.4.6   Digital transponders: an example

The present example applies to new digital systems featuring the characteristics listed below. Although they currently apply to mobile satellite communication systems, they are not specific to those systems and could also find applications in the framework of FSS systems implementing small earth terminals:

- A large number of high gain narrow beams pointed towards the service area. The use of active antennas allows flexibility in the selection of service area elements.

- Improved efficiency in the use of satellite power resources. The implementation of small earth terminals (e.g. VSATs) imposes requirements on the satellite transponder e.i.r.p. Hence, high gain antennas are required on the spacecraft.

- Operation with frequency reuse in order to save frequency resources. This induces a constraint on the space isolation to be provided between the beams operating at the same frequency.

Several payload configurations can be candidates. The first alternative is whether the payload is regenerative or not. Of course regenerative processes provide significant improvements on the system performance. However, one important reason why regenerative payloads are currently not frequently used is their lack of flexibility, i.e. the difficulty in adapting themselves to various types of terminals, modulation and coding. Moreover, in these regenerative payloads, various functions that are provided at the individual channel level (e.g. gain control or routing towards the various beams), require demultiplexing at the smallest level of the individual channel.

The architecture involved in the payload described by Figure 6.28 below is actually transparent at the level of the individual channels. Those are demultiplexed by frequency filtering. In consequence, the process is independent of the channel modulation technique. This process, which uses a multi-beam antenna, is as follows:

- Demultiplexing: it separates the individual channels contained in the uplink frequency multiplex.

- Beam forming: it provides steering of the various antenna beams towards the coverage area elements. This is implemented by adequately weighting (in amplitude and phase) the different channels of a beam forming network (BFN). The weighting is performed by using complex coefficients in digital registers.

- Multiplexing: it recombines the output signals to form an adequate output frequency multiplex that is further routed towards the antenna radiating elements, to be finally transmitted to the service area element of the destination user.

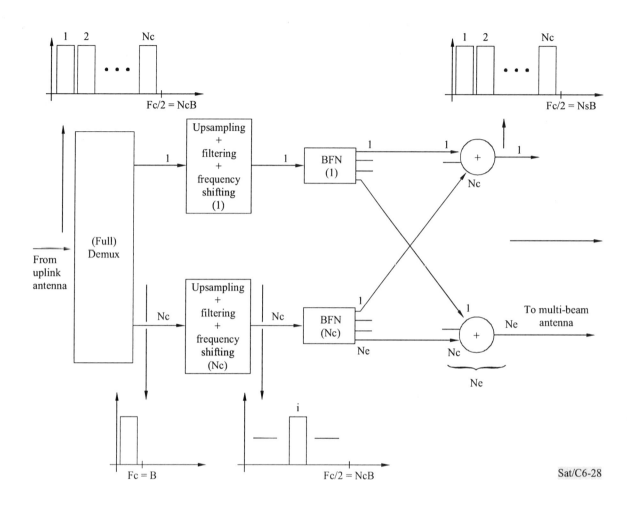

FIGURE 6.28

Examples of a payload architecture implementing digital demultiplexing/multiplexing of individual channels and multi-beam antenna with a beam forming network (BFN)

6.3.4.7   Recent developments in OBP technology

A summary of the recent status of satellite systems using on-board processing (OBP) technologies is shown in Table 6.6. They depict a wide range of capabilities from RF switching to selected downlink transmitters to complex baseband switches with either ground control or autonomous satellite control (ASC) features. On-board RF/IF switching can be reconfigured on a near-real-time basis by ground control and is the preferred method at present because of its higher reliability. However, a great deal of research is being focused on the technologies required for ASC systems.

In addition to the organizations mentioned above, there are from twenty to thirty groups involved in state-of-the-art OBP developments, many of them in Europe, Japan and North America. In addition, among many others throughout the world, there are about a dozen companies in the United States that plan to implement 30/20 GHz satellite communication systems with OBP capabilities in the next decade.

TABLE 6.6

Satellites with on-board processing technologies

SATELLITES WITH ON-BOARD PROCESSING TECHNOLOGIES					
SATELLITE TECHNOLOGY	**ACTS**	**COMETS**	**ETS-6**	**ITALSAT F1-F2**	**ARTEMIS**
BUILDER/PRIME CONT.	GE	TOSHIBA-NASDA/CRL	TOSHIBA, NASDA & NTT	ALENIA SPAZIO	ALENIA SPAZIO
MULTI-BM. ANTENNAS	Ka 2 AGILE. 4 SPOT	2 BEAM	13 Ka - 5 S BAND	6 SPOT BEAMS	3 OR 4 BEAMS MMICS FOR BFN
BB PROC. ENC/DECODE		BPSK DEMOD COM BNC & VITERBI DECODE			
REGENERATION	REGENERATION	8 CHAN PER BEAM (2)		REGENERATION	
RF/IF SWITCHING	RF, 4 X 4 IF & 4 X 4 SS	2 X 2 IF FILTERS FOR BEAM INTERCONNECT	12 X 12 IF FOR ON-BOARD ROUTING OR SS CONT.		
BB SWITCHING	ROUTING & CKT SW GND CONT.	BEAM CONNECT AND BASEBAND SWITCHING		6 X 6 BASEBAND SWITCH MATRIX	
DIGITAL FILTERING		POLYPHASE FFT			
MULTIPLE ACCESS	FDMA SS-TDMA			SS-TDMA	CDMA-FDMA
INTER-SAT LINKS	NO	INTER-ORBIT	S-BAND		S-BAND INTERORBITAL & OPTICAL INTERSAT
KU - KA TECH.	Ka 900 MHz BW	Ka: 20/23 GHz Ku: 11/14 GHz	Ka: 30/20 GHz (8 BAND FOR MOBILE)	Ka	Ka
OB NET PROTOCOLS	@ MASTER GND STATION	@ MASTER GND STATION	NO	@ MASTER GND STATION	NO
OB SW. MICROPROC	NO	NO		DUAL MICRO-PROC. FOR BB SW. MATRIX	
ON-BOARD MEMORY	64 kbps Support				
LAUNCH DATE	1993	1997	1993	1991-93	1994

More complete information on OBP will be found in "VSAT systems and Earth Stations" (Supplement No. 3 to the Handbook, ITU-R, Geneva, 1995, often called: "VSAT implementation of satellite with multiple spot beams and on-board processing"). In the section, some details, more specifically oriented towards VSAT applications, are given on such subjects as: typical examples of OBP, satellite switched, transponder architectures (FDMA, TDMA, FDMA/TDM), multicarrier demodulators and demultiplexers, etc.

6.4 Intersatellite links (ISL)

6.4.1   Introduction

Intersatellite links are employed to provide connections between earth stations in the service area of one satellite to earth stations in the service area of another satellite when neither of the satellites covers both sets of earth stations. Figure 6.29 shows a simplified sketch of this concept.

If there is an overlap between the two service areas, an alternative to employing an ISL is to link an earth station in the non-overlapping part of one area to an earth station in the non-overlapping part of the other area by a double satellite hop via an intermediate earth station in the overlap region. However, for satellites at geostationary height the "round trip" time delay of the double hop (more than a second) is unacceptable for PSTN-quality voice circuits. For networks using geostationary satellites which are separated by relatively small geocentric angles ISLs can be employed to reduce the "round-trip" delay to an acceptable value.

If the satellites are separated by angles large enough to make parts of the Earth's surface visible to only one or other of the two, then an ISL may be used to extend the service area available to users of the network. However, in the case of geostationary satellites this involves fairly long ISLs, so the problem of excessive time delay is not fully solved.

Constellations of satellites in low-Earth orbits sometimes incorporate a pattern of ISLs to enable long overland routes to be covered without recourse to multiple satellite hops via intermediate earth stations.

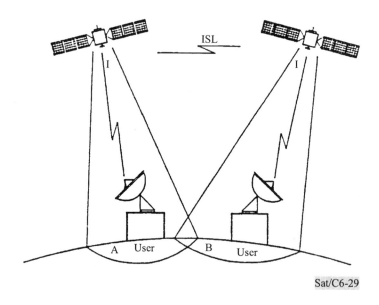

Sat/C6-29

FIGURE 6.29

Interlinking of satellite networks via intersatellite link (ISL)

For ISLs the connection between satellites is fully at space level and the interfacing of the network formats and protocols have to be performed on board the satellites. It is likely that the introduction of ISLs will coincide with the introduction of on-board regeneration and processing techniques.

Actually the concept of a "gateway" is entirely transferred to the space level and the Earth segment requires no modifications, each terrestrial node needing only one earth station for complete network connectivity.

It is likely that the introduction of both gateway stations and ISLs to interlink satellite networks at various hierarchical levels may lead to an optimized architecture at worldwide level allowing for flexible routing of traffic.

6.4.2   ISL benefits

i)   Increased coverage

A single geostationary satellite can view 40% of the Earth's surface. An ISL connecting two satellites 60° apart (a limit set by path delay considerations) could cover 57% of the Earth's surface. This would make the installation of earth stations at the customers premises more attractive, because satellites should be able to carry the bulk of their traffic over their conventional (non-lSL) up and downlinks, leaving the land masses to the East and West which would be visible from the satellite to make full use of the available (i.e. 60°) satellite separation. These considerations would not apply to links between regional or domestic systems.

ii)   Flexibility in arc positioning

An ISL can enhance the attractiveness of trans-oceanic satellite service by allowing the satellites to be positioned closer to the land masses, resulting in higher earth station elevation angles than otherwise possible. Even if an earth station can be seen from a satellite, there are cases where the earth station elevation angle may be too low to allow a practical link to the satellite. This is particularly true if the earth station is in an urban area where there may be interference from buildings and from RF sources and where off-axis emission of the transmit earth station could be a major concern. Since use of small earth stations close to, or at, the customer premises represents one of the fastest growing markets for satellite services, this consideration will become increasingly important. The increased use of 14/11-12 GHz band services, with its greater rain attenuation and depolarization, also makes the use of improved elevation angles extremely important. These considerations become even more important for potential 30/20 GHz band satellite services.

The improved elevation angles made possible by the ISL allow a more attractive domestic service to be offered from the same satellite that is serving an international communications function. The flexibility in positioning a satellite made possible by an ISL can, in some cases, improve the orbital arc utilization. For example, in some regions there are very few orbital slots that can provide single hop links between major markets. An ISL can increase the number of attractive orbital locations as well as provide improved service through higher elevation angles.

Flexibility in orbital arc positioning made possible by the use of ISLs can minimize potential interference between international business services and direct broadcasting-satellite services (DBS). The interference arises because the business frequency bands in some areas overlap with DBS frequency allocation for other areas. An ISL reduces this interference problem by allowing customers to point their antennas only at the segment of the orbital arc occupied by the ISL fed satellite.

iii) In-orbit connectivity

An ISL can provide inter- and intra-connectivity between domestic/regional/international systems, increasing the coverage obtained from a single earth station. It can also make satellite clusters* more attractive especially if the link allows connectivity between an all 6/4 GHz band satellite and an all 14/11-12 GHz band satellite.

6.4.3   Basic technologies

There are two basic technologies for implementing ISL: microwave or optical transmission.

Microwave ISLs

ITU has allocated the bands 22.55 to 23.55 GHz, 32.0 to 33.0 GHz, 54.23 to 58.2 GHz and 59 to 64 GHz to ISLs. Because of the large available ISL bandwidths, expansion can be used to reduce the required e.i.r.p. Depending on the system frequency requirements, direct heterodyne or RF FM remodulation techniques can be used. The first technique simply consists of upconverting the signals destined for ISL transmission without any bandwidth expansion. The capacity and range of this transmission technique is limited since the (C/N) of the ISL adds directly to the uplink and downlink (C/N)s. The second technique involves first a down conversion of the signals from IF to a conveniently low frequency which is then used to FM modulate the carrier at the ISL transmit frequency, i.e. a double modulation has occurred with this technique. This technique provides good immunity to interference, and allows the ISL capacity or range to be extended at the expense of using up more of the allocated ISL RF bandwidth. Figure 6.30 b) shows the basic block diagrams of an ISL using the FM remodulation technique.

---

*   "Satellite cluster" is a group of different satellites closely spaced in geostationary orbit so that they can operate like one single larger satellite.

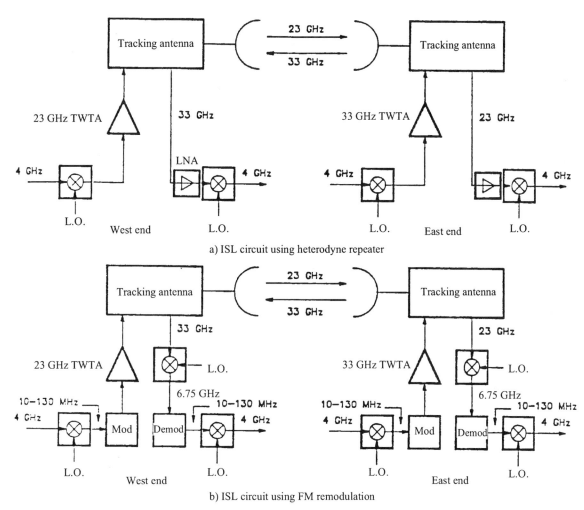

a) ISL circuit using heterodyne repeater

b) ISL circuit using FM remodulation

Sat/C6-30

FIGURE 6.30

ISL baseline systems

It has been experimentally shown that acquisition and tracking with the microwave ISL antennas is readily achievable. Acquisition could potentially be achieved via ground commands if the positions and the attitudes of the two satellites are known to a fair degree of accuracy.

Microwave ISLs can also be designed to provide acceptable services through solar conjunctions. A larger link margin would be required to maintain service continuity through solar conjunctions. However, if a few minutes (highly predictable) of outage a year was considered acceptable, a substantial saving in ISL mass and power could be realized.

Optical ISLs

Microwave ISLs cannot economically provide a high capacity link over a significant orbital arc separation. For example, a 240 Mbit/s microwave ISL operating over a 60° angular arc separation would require 500 W of RF power or 1 000 W of d.c. power assuming the antenna diameter is 2 m (which is about the maximum feasible size). In contrast, an optical ISL using a 30 cm antenna (lens) would require only about 40 mW of optical power. The reduction in power is a result of the greater antenna gain, as seen below.

Antenna gain is proportional to $(D/\lambda)^4$ where D is the lens diameter and λ the wavelength. The wavelength of a 30 GHz microwave signal is 1×10^{-2} m while for a typical laser it is 1×10^{-6} m. Thus, the antenna gain is 1×10^{16} or greater for an optical ISL than for a microwave ISL. Even if the optical antenna is one order of magnitude smaller than a microwave antenna, its gain is 1×10^{12} greater.

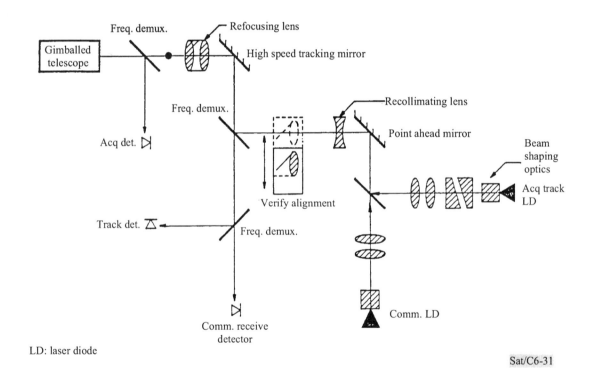

FIGURE 6.31

Optics of a laser ISL

The very large antenna gain implies a very narrow beamwidth of the order of 5 μrad for a 30 cm antenna. For an orbital arc separation of 60°, the beamwidth is about 200 m.

The narrowness of the beam will require a complex acquisition sequence and a very accurate antenna pointing subsystem. The pointing subsystem must be sufficiently fast (of the order of 1 kHz) to cope with spacecraft disturbances such as thruster valves opening and closing, vibration from momentum wheels, solar array motion, etc. This can be accomplished using a simple quadrature detection system as shown in Figure 6.31. One result of the high speed tracking subsystem is that four sigma pointing errors will occur once every 30 s. A BER calculation must include the effects of such mispointing.

Most of the efforts to develop optical ISLs have used on-off key (OOK) amplitude modulation. This is accomplished by driving the laser diode with a d.c. bias added to the non-return to zero (NRZ) signal. The use of OOK requires on-board demodulation to recover the bit stream, limiting its use to satellites with on-board regeneration. Another disadvantage of OOK modulation is that it is difficult to operate the link during solar conjunction when the Sun will be shining directly on the detector. While this occurs only for 2 min per year per spacecraft (4 min per year for the link), a longer outage may occur under near solar conjunctions depending upon system margin and screening from extraneous light.

An optical heterodyne link would, in principle, have higher capacity and be better able to cope with solar conjunction. The use of an analogue link, which would not require on-board regeneration, imposes a much smaller penalty on a heterodyne link than on an amplitude modulation link. While much work is being done on both modulation formats, modulation and detection are far more difficult for heterodyne systems than for OOK systems. Most commercial interest is centred on OOK, although in-orbit tests using both kinds of modulation formats are planned.

The optical power required for a high capacity link is surprisingly small, typically under 1 W. Most of the d.c. power and mass requirements of an optical ISL are in the tracking subsystem. As a result, the mass and d.c. power are only secondarily dependent upon the capacity of the link. Thus, a high capacity link would be less costly per bit than a low capacity link.

6.5 Special features of non-GSO (NGSO) satellite systems

6.5.1 Introduction

There are many non-GSO satellite systems which have been operational for over three decades, performing special functions and services such as weather observations, remote earth exploration, radionavigation, communications, surveillance, etc. One of the unique features of these satellites is the ability to view the entire Earth's surface periodically from a single satellite. If simultaneous viewing of the Earth is required, a number of satellites can be employed depending on their orbit altitude. These systems include 1) low-Earth orbit (LEO) satellites such as several weather satellite systems; 2) medium-Earth orbit (MEO) satellites; and 3) high-Earth orbit (HEO) satellites such as the navigational satellites GPS and GLONASS. The MOLNIYA satellite system uses a highly elliptical orbit to provide coverage in latitudes of 60-70°. A description of the characteristics of a range of non-GSO orbits is provided in section 6.1.

This section and Appendix 6.1 provide technical descriptions of some newly conceived and developed non-GSO fixed-satellite service communication systems which have created a great deal of interest in industry circles and the ITU Radiocommunication Sector.

6.5.2   Types of LEO networks

The evolution from geostationary to low-Earth orbit (LEO) satellites has resulted in a number of proposed global satellite systems, which can be grouped into three distinct types. These LEO systems can best be distinguished by reference to their terrestrial counterparts: paging, cellular, and fibre.

System Type	Little LEO	Big LEO	Broadband LEO
Examples	ORBCOMM	Iridium, Globalstar	Teledesic, Skybridge
Main applications	Low bit rate data	Mobile telephony	High bit rate data
Terrestrial counterpart	Paging	Cellular telephony	Fibre

On the ground, cellular and fibre services are complementary, not competitive, because they offer fundamentally different kinds of services. Similarly, systems shown as Big-LEO and Broadband LEO are expected to be complementary rather than competitive because they propose to provide distinctly different services to different markets. The Big LEOs, for example, plan to provide narrow-band mobile voice service, whereas Teledesic and Skybrige plan to provide primarily fixed, broadband connections comparable to urban wireline service.

6.5.3   Benefits of broadband LEO networks

Access to information is becoming increasingly essential to the conduct of business, education, health care and public services. Yet, most people and places in the world do not now have access even to basic telephone service. Even those who do have access to basic phone service get it through 100-year-old technology – analogue copper wire networks – that for the overwhelming part will not likely be upgraded to an advanced digital capability.

While many places in the world are connected by fibre – and the number of places is growing – it is used primarily to connect countries and telephone company central offices. Even in the most developed countries, high cost will prevent the deployment of a significant amount of fibre for local access to individual offices and homes. In most parts of the world, fibre deployment in the local access network is not likely to happen.

This lack of broadband local access is a major problem for all of the world's societies. If the powerful new communication technologies are available only in advanced urban areas, people will be compelled to migrate to those areas in search of economic opportunity and to fulfil other needs and desires. The one-way information dissemination made possible through broadcast technologies has created a means for nearly all of the world to view the benefits of advanced technology. People living outside of advanced urban areas will undoubtedly desire access to these benefits. Universal two-way network links are needed to allow people to participate economically and culturally with the world at large without requiring that they move to places with modern telecommunication infrastructures.

Low-Earth orbit (LEO) satellite systems can help by providing global access to the telecommunications infrastructure currently available only in advanced urban areas of the developed world. Just as networks on the ground have evolved from centralized systems built around a single mainframe computer to distributed networks of interconnected PCs, space-based networks are evolving from centralized networks relying on geostationary satellites to distributed networks of interconnected low-Earth orbit satellites. Because their low altitude eliminates the delay associated with traditional geostationary satellites, these networks can provide communications that are compatible with existing, terrestrial, fibre-based standards.

These LEO systems have themselves evolved from small systems that will provide the satellite equivalent of paging to the "Big LEOs" that will provide the satellite equivalent of cellular service. The next generation of "Broadband LEOs", like the Teledesic network, hold the potential to emulate and extend the Internet – providing many users access to broadband communications while adding real-time capability to all parts of the globe.

Because LEO satellites move in relation to the Earth, they have another characteristic with profound implications: continuous coverage of any point on Earth requires, in effect, global coverage. In order to provide service to the advanced markets, the same quality and quantity of capacity has to be provided to the developing markets, including those areas to which no one would provide that kind of capacity for its own sake. In this sense, LEO satellite systems represent an inherently egalitarian technology that promises to radically transform the economics of telecommunications infrastructure to enable universal access to the Information Age.

Many LEO systems are currently proposed, including some to be operated in very high RF bands (30/20 GHz, 50/40 GHz, etc.). Some of these systems are planning or even under development and construction. Others are under preliminary study at the ITU-R.

Table 6.7 summarizes some salient characteristics of some of these systems in the FSS.

Two systems, now at an advanced stage of development and to be operated in the FSS, are described in Appendix 6.1 as examples of the possibilities now offered by telecommunication satellites. The first one, TELEDESIC (also known as LEO SAT-1 in ITU-R documents), uses ISLs between satellites, whereas the second, SKYBRIDGE (also known as FSAT-MULTI1 in ITU-R documents), relies on Earth gateways for linking the satellites.

TABLE 6.7

Main characteristics of some planned non-GSO FSS systems (from Documents 4/19, 5 November 1998 and 4A/213, 2 February 1999)

System name	FSAT-MULTI 1-B (SkyBridge, See NOTE 1)	FSAT-MULTI 1-A	LEO N	LEO SAT 1 (Teledesic, See NOTE 2)	LEO SAT 2	MEO J	MEO K	MEO V	NGSO-KX
Frequency band (GHz) -up -down	18 10-12	27.5-30 17.3-20.2	12.75-13.25 10.7-10.95	28.6-29.1 18.8-19.3	28.6-29.1/29.5-30 18.8-19.3/19.7-20.2	29.5-30 19.7-20.2	27.6-28.6 17.8-18.8	29.5-29.9 19.8-20.2	28.6-29.1 18.8-19.3
Type of orbit	circular	circular	circular	circular	Circular	circular	circular	circular	circular
Number of orbit planes	20	161	7	21	7	9	9	4	4
Number of satellites per orbit plane	4	1	13	40	9	1	1	6	5
Inclination angle (degrees)	53	87.1133	82	98.2	48	75	75	82.5	55
Number of antenna beams	24 max. per satellite		37	<49	432 (up) 260 (down)	256	256	625	20
Altitude of orbit (km)	1 469.3	1 675	700	700	1 400	13 900	13 900	10 360	10 352
Required spectrum in each direction (MHz)	1 650		200	500	1 000	500	1 000	200	500
Minimum operating angle (degrees)	5; 10	20 at the equator	10	40	16		10	40	30
Earth station switching strategy	Track the best elevation satellite within the operating range		by programme	track nearest satellite				by programme	interference avoidance/ highest elevation angle
Number of earth stations	up to 20 million			up to 20 million		unlimited			unlimited

NOTE 1 – Updated data. See description of the SkyBridge system in Appendix 6.1 (§ AP 6.1.2).

NOTE 2 – See description of the Teledesic system in Appendix 6.1 (§ AP 6.1.1).

6.6 Launching, positioning and station-keeping

6.6.1 Launching

Although the processes involved in launching depend on:

- the type of launcher;
- the geographical position of the launching site;
- constraints associated with the payload,

the most economical conventional method based on using the Hohmann transfer orbit consists of the following (see Figure 6.32 (A and B)):

i) placing the satellite in a low circular parking orbit with an altitude of approximately 200 km;

ii) at an equatorial crossing, impart a velocity increment to change the parking orbit into an elliptical transfer orbit with the apogee at 36 000 km;

iii) producing a circular equatorial orbit when the satellite is passing through the apogee of the transfer orbit by igniting the apogee motor.

Because the apogee motor is considered as a part of the satellite equipment, the launching with an expendable launch vehicle normally involves the first two operations, i) and ii). A reusable launcher, such as the space transportation system (STS) only carries out the first phase i) of the launch and an additional perigee stage is needed for the second operation ii).

In order to successfully insert the satellite into the transfer orbit, the launcher must operate with great precision with regard to the magnitude and the orientation of the velocity vector.

The degree of precision generally required of the launcher is approximately ±200 km for the height of the apogee.

The launching operations necessitate, either at the launching base or at the stations distributed along the trajectory, ranging, telemetry and possibly command facilities.

The use of a launcher imposes a number of constraints on the satellite. These fall into two categories:

- technical constraints concerning dimensions, mechanical and electrical interfaces and the environment, which must be taken into account from the outset in the design of the satellite;
- operational constraints associated with the preparation of the launcher and the length of the launch window.

6.6.2 Positioning on geostationary orbit

Starting with the highly elliptical transfer orbit into which the launcher has inserted the satellite, the positioning phase consists of carrying out the manoeuvres necessary to circularize and equatorialize this orbit, and then gradually directing the satellite to its stationary position.

This short phase, lasting one to two weeks, calls for different methods from those used for the operational phase. The short duration of this phase permits the use of the same means for the various programmes that are required.

a)   Control stations

In transfer orbit, the control of the satellite as it travels around the Earth requires tracking stations located on different parts of the globe. This network of stations can operate within the same frequency bands as the stations responsible for station-keeping, i.e. the telecommunication frequency bands (6/4 GHz or 14/11 GHz), since it is only used during a short period of time. However, if a number of satellites operating at different frequencies are planned, it may be better to establish a network of stations in a specific frequency band (for example, the space-operation service band between 2 and 2.3 GHz) and to use that band to position all the different satellites in turn. These stations, like those responsible for station-keeping, must be equipped with location facilities (ranging and/or angle measurements).

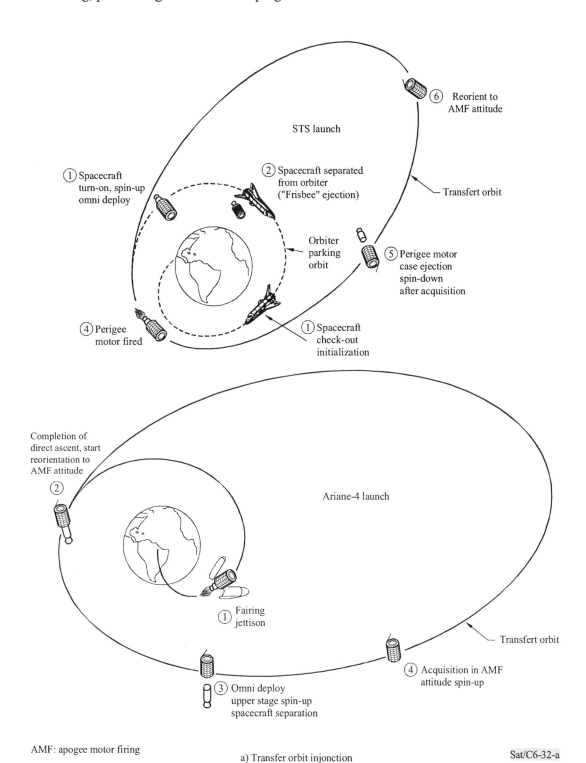

STS launch

① Spacecraft turn-on, spin-up omni deploy

② Spacecraft separated from orbiter ("Frisbee" ejection)

⑥ Reorient to AMF attitude

Transfert orbit

Orbiter parking orbit

⑤ Perigee motor case ejection spin-down after acquisition

① Spacecraft check-out initialization

④ Perigee motor fired

Completion of direct ascent, start reorientation to AMF attitude

②

Ariane-4 launch

Transfert orbit

① Fairing jettison

④ Acquisition in AMF attitude spin-up

③ Omni deploy upper stage spin-up spacecraft separation

AMF: apogee motor firing

a) Transfer orbit injonction

Sat/C6-32-a

FIGURE 6-32-A

Mission summary

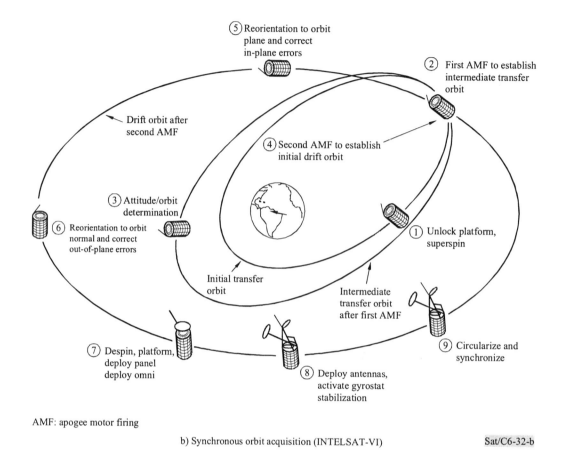

AMF: apogee motor firing

b) Synchronous orbit acquisition (INTELSAT-VI)   Sat/C6-32-b

FIGURE 6-32-B

Mission summary

b)   Control centre

As in the station-keeping phase, the control centre, using the same facilities, controls the satellite from the technical point of view. In addition, it needs to be temporarily connected to more powerful computer facilities which enable the manoeuvres specifically involved in this positioning phase to be prepared and carried out. These manoeuvres consist of the following:

•   determining the satellite's attitude;

•   calculating the optimum parameters for the orbit-circularization manoeuvre, otherwise known as the apogee manoeuvre and for subsequent manoeuvres;

•   determining the characteristics of the correction thrusts which orient the satellite in the precise direction required by the apogee manoeuvre;

•   monitoring and controlling the manoeuvre during which the satellite assumes the final attitude required for the service planned.

6.6.3 Station-keeping and payload operation monitoring

Tracking, telemetry, telecommand and monitoring of telecommunication satellites are performed by special TTCM earth stations which are operated by the authority responsible for the satellite.

The main tasks to be carried out during the station-keeping phase are as follows:

- the satellite must be kept on station within the narrow window formed by the longitude and latitude assigned to it;

- its attitude must be monitored and corrected to ensure that the antennas are always pointed in the right direction towards the Earth.

This is a very long phase (seven to ten or more years) and the tasks involved are sufficiently long term to warrant the use of specialized control facilities for the satellite or family of satellites.

a) Station-keeping facilities

The above-mentioned functions are carried out on the basis of telemetry information transmitted by the satellite, ranging measurements of various kinds and command orders sent from the ground to the satellite.

Radio contact with the satellite is provided by one or more tracking, telemetry and telecommand stations, while the data processing and formulation of telecommand orders are carried out in a Control Centre.

Since the satellite appears to be stationary in relation to the Earth's surface, the network of stations can be relatively simple. A single well-sited station equipped with an antenna with a very limited coverage, or even in certain simple cases a fixed antenna, is sometimes sufficient. However, it is usually necessary to plan for a number of stations to provide redundancy. It is often economically advantageous to use the same station both to monitor the satellite and to handle traffic, especially since the telemetry and telecommand signals use the same frequency bands as the user service.

b) Maintenance in orbital position

The positioning window of the satellite is very narrow, i.e. approximately ±0.1° (or less) in longitude, given the current longitudinal station-keeping regulations and the technical constraints imposed by the use of the satellite. Although the Radio Regulations do not prescribe any limit for the latitude station-keeping, it is technically feasible to achieve the same ±0.1° window as for the longitude station-keeping.

Various types of orbit perturbations, such as the Earth's gravitational irregularities or the attraction exerted by the Sun or Moon and solar pressure, tend to shift the satellite from the required position. It is therefore necessary to activate the satellite's thrusters in order to bring it back into position.

This means that the orbit must be determined very precisely. From the parameters of the orbit which have been determined and from the knowledge of the effects of the orbit perturbations, predictions of the future satellite positions are made. This permits planning appropriate station-keeping manoeuvres to ensure that the satellite never leaves its window. These manoeuvres imply the firing of thrusters in the North-South direction (normally single firings) and in the East-West direction (single firings or occasionally double firings spaced at an interval of 12 hours). The frequency of the orbit correction depends on the width of the window and the satellite's position over the equator. It may prove necessary to carry out one correction per week for a window of 0.05°.

Satellite ranging is carried out by measuring the distance from two tracking, telemetering and telecommand stations (if these stations are at a sufficient distance from each other) or by combining the measurement of the distance to a station with an angular measurement of the direction of the satellite as seen from that station. The accuracy of these measurements must be of the order of 50 to 100 m in distance and an order of magnitude smaller than the width of the window for the angular measurement. Earlier methods of orbit determination are based on batch processing of tracking/ranging sets of data collected over an extended period of time (up to two days) and on the usage of estimation techniques such as least squares. More recent methods are based on the real-time processing of data by mean or recursive estimation techniques (Kalman filtering).

c)   Technical control of the satellite

It is necessary to monitor, by means of telemetry, the correct operation of the various satellite systems and to transmit command signals in order to correct any anomalies, to operate the redundancy equipment in the event of a malfunction and to operate the payload.

The main points of interest for monitoring are as follows:

- behaviour of the satellite thermal control system;

- behaviour of the electrical power supply system, particularly during eclipse periods;

- behaviour of the on-board TTC subsystem;

- the operation of the attitude and orbit control system;

- the operation of the payload;

- the location of operational malfunctions by means of routine or spot checks;

- immediate action in emergencies to ensure adequate protection of the services provided by the satellite of the on-board equipment and of the satellite as a whole.

As a part of normal procedure, the values of relevant satellite parameters are checked with a computer which warns the controllers when any value exceeds a pre-set limit.

According to the different situations encountered, the controllers, who are fully informed of the operation and behaviour of the satellite, decide which command signals to send and then analyse the received telemetry signals in order to check that the commands have been carried out correctly.

d)   Attitude control

The satellite is equipped with Earth or Sun sensors, from which telemetry signals are derived, which enable determination of the satellite attitude. The attitude of the satellite is corrected, when necessary, by thrusters operated by telecommand or by automatic on-board control. These thrusters are similar to those used for orbit control.

6.7 Reliability and availability considerations

For general considerations on quality and availability, see Chapter 2, section 2.2.

6.7.1 Satellite reliability

An important aspect of satellite reliability is the fact that satellites must operate in space without benefit of repairs or maintenance. Under these circumstances, it is imperative that all parts, components and subsystems of the satellite system must be exposed to extensive testing in a simulated space environment. Since the dawn of the space age, approximately 50 years ago, life testing of electronic parts and equipment designed for space use has been conducted on a massive scale by many government and industrial organizations. The test results of parts and equipment must provide a very high probability of success before they can be selected for satellite applications. Items associated with critical functions, such as launch operations, station-keeping, attitude control, telemetry and command and communications are often designed with redundancy to assure successful performance. The operational lifetime of communication satellite systems have increased from a few years in the 1960s to approximately 15 years with current designs. One of the principal limitations for the operational lifetime of a satellite is the amount of propellant fuel that can be carried at launch in order to maintain in-orbit station-keeping and attitude control. Also, subsystems that experience substantial wear and decay are batteries, solar cells and high power amplifiers.

There are several key phases associated with spacecraft reliability during space operations. The first is the launch phase in which a primary launch vehicle is used (see section 6.6) for inserting the spacecraft into a low or transient orbit, followed by staged boosters and, finally, by an apogee motor that inserts the satellite into the desired orbit. During the 1970s and early 1980s, most launch systems that were used for inserting geostationary satellites into orbit were government type boosters, examples of which were the Ariane, Atlas/Centaur, Delta, Proton and Space Shuttle. Since these were mature systems in which many development problems had been solved earlier, the success rate was nearly 90%. In the last several years, however, many new or modified launch systems have been introduced that exhibit embryonic design deficiencies that have resulted in reduced reliability expectations. The impact on insurance costs for the launch phase of a satellite system has increased considerably during the last few years, to the extent that some countries prefer to insure themselves when using unproven launch systems. However, these are expected to improve as additional experience is achieved. Section 6.9 provides some brief descriptions of some of the more prominent launch systems that are currently available or are planned for the near future.

After insertion into orbit, the reliability of spacecraft can be divided into three phases with different mechanisms being primarily responsible for possible failures. The first phase may experience failures due to non-deployment of such subsystems as solar panels, antennas and other previously stored or folded components. The failure of the initiation of household functions, such as attitude controls, station-keeping or power generation, may manifest themselves in this phase if design deficiencies exist, especially in new or unproven equipment. The second phase is characterized by random failures of components in the satellite. The third phase includes the depletion of propellant which renders the satellite unusable due to lack of station-keeping and attitude control, even though the electronics may still be operational. However, geostationary satellites have been able to extend their operational life by several years by confining their station-keeping to the East-West plane and allowing the spacecraft to drift in the North-South direction (see section 6.2.4.3).

Maintaining the reliability of a spacecraft in orbit for periods of many years has been the concern of the satellite communication industry since its inception. Doubling the operational life of a satellite provides many cost advantages, since fewer expensive launch systems need to be employed. However, this does increase the overall complexity of the spacecraft design and construction costs. Ordinarily, the manufacturer provides a reliability estimate of the spacecraft lifetime in orbit, with special attention to the performance of the communication subsystem. As an example, the manufacturer of the INTELSAT-V satellite estimated the probability of the satellite to be operational after ten years in orbit to be approximately 60%. As expected, shorter periods had improved estimates, as shown in Figure 6.33. After seven years, the probability estimate was quoted as 76%, with at least 65% availability of the channels in any coverage area. A review of the in-orbit performance of the INTELSAT-IVA and V satellites is summarized in Table 6.8.

The failures in these satellites have mainly been related to receiver and TWTA problems. A review of INTELSAT satellite performance in recent years, encompassing IVA, V and VI models, shows increasingly better reliability results with the expectation of full performance of the satellite systems over periods exceeding seven years. Although the current trend implies that new satellite systems have increasingly better reliability performance, there have been some recent spectacular failures of spacecraft in orbit without clear explanations of the cause or causes. Environmental factors, such as solar storms or interstellar explosions have been alluded to as possible causes, but it is much too early to evaluate these sources as being significant. The more likely cause is some unexpected design failure.

In order to calculate the probability for the joint event of the successful launch of a spacecraft and its successful operation in orbit after a given time, the individual probabilities of these two events must be multiplied.

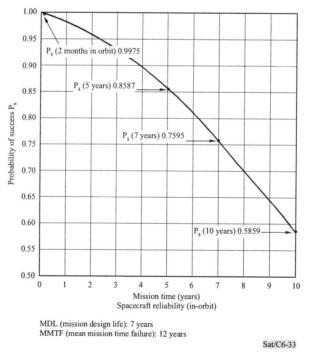

MDL (mission design life): 7 years
MMTF (mean mission time failure): 12 years

Sat/C6-33

FIGURE 6.33

Typical prediction of spacecraft in-orbit reliability

NOTE – Number of parts considered in the evaluation: 54 878. Apogee motor and deployment success probabilities are not included.

TABLE 6.8

INTELSAT satellite lifetime estimates

(July 1987)

Satellite	Launch date	Design life	End of estimated full performance life[1]
INTELSAT-IVA (F-1)	September 1975	September 1982	March 1984
INTELSAT-IVA (F-2)	January 1976	January 1983	March 1984
INTELSAT-IVA (F-3)	January 1978	January 1985	January 1986
INTELSAT-IVA (F-4)	May 1977	May 1984	June 1985
INTELSAT-IVA (F-6)	March 1978	March 1985	December 1985
INTELSAT-V (F-1)	May 1981	May 1988	August 1988
INTELSAT-V (F-2)	December 1980	December 1987	April 1989
INTELSAT-V (F-3)	December 1981	December 1988	October 1989
INTELSAT-V (F-4)	March 1982	March 1989	August 1989
INTELSAT-V (F-5)	September 1982	September 1989	June 1990
INTELSAT-V (F-6)	May 1983	May 1990	April 1991
INTELSAT-V (F-7)	October 1983	October 1990	September 1990
INTELSAT-V (F-8)	March 1984	March 1991	September 1990
INTELSAT-V (F-9)	March 1985	March 1992	May 1993
INTELSAT-V (F-11)	June 1985	June 1992	September 1993
INTELSAT-V (F-12)	September 1985	September 1992	January 1994

[1] These dates have been adjusted not to exceed the manoeuvre life.

6.7.2 Satellite availability

Communication satellites are usually linked to earth stations that form networks. They may range from a group of stations linked together into a private network or, in the case of wideband digital services, linked to the public switched network. In addition to equipment failures, one of the main sources of satellite link interruptions is rain attenuation, especially in the higher carrier frequencies above 10 GHz. To compensate for this problem, satellite and earth station transmitters employ higher powers than that required merely to compensate for natural noise. This additional power is referred to as "margin" if it is continuously applied, or, in some cases, applied on an as needed basis by the application of "power control". Satellite availability of 99.5% was achievable in early satellite communication systems, although 12/14 GHz systems suffered lower availability performance in rainy areas. The trend during the last twenty years has been to increase the transmission power of satellites substantially, especially for wideband digital systems. In these cases it is very important to have a high level of satellite availability, since many interface protocols require nearly continuous connections of very high quality (low bit error rates). Within a communication circuit that includes other modes of transmission (terrestrial, etc.), special specifications are imposed on the satellite to assure the reliable functioning of the overall system. In many cases, the availability requirement of a satellite forming part of a long-distance circuit can be as high as 99.99%. To achieve this degree of performance the satellite needs to have high quality components with sufficient link margins and redundancy. Satellite availability is calculated from the duration of an outage in which the particular circuit is interrupted, compared to the period of time that is scheduled for the user or client. The definition of availability becomes:

$$\frac{\text{scheduled operating hours} - \text{hours of outage}}{\text{scheduled operating hours}} \times 100$$

In order to compete with fibre optic or other terrestrial systems, some satellite systems have been designed with extremely high availability capabilities through the use of reliable equipment and innovative redundancy schemes, including earth station diversity and power control.

6.8 Management of satellite communication network operations

Efficient management of satellite communication network operations is essential for the proper utilization of the high investment costs involved in satellite systems and for maintenance of the vital long-distance communications provided by these systems. In international and regional satellite systems such as INTELSAT, INTERSPUTNIK, ARABSAT and EUTELSAT, the space segment is generally owned and operated by the international or regional organization established by the member countries and the ground segment is managed by various entities in the respective countries. The use of the space segment is subject to certain entry conditions and its tariffs are decided by the international organization. In the case of national satellite systems, the management structure depends on government policy, the existing set up of telecommunication entities and user agencies, and varies considerably depending on whether these entities are government or privately owned. As typical examples, the management of an international system (INTELSAT) is described in the following sections.

6.8.1 Management of international satellite communication network (INTELSAT) operations

In the INTELSAT system the space segment is owned, operated and maintained by the INTELSAT organization, while earth stations are owned and operated by telecommunication entities in the countries where they are located. Satellite control and coordination of communication services are effected through the INTELSAT headquarters located in Washington DC, United States. Detailed information on INTELSAT operational management, coordination and control, and line-up procedures is given in the satellite system operations guide (SSOG). Figure 6.8.1 illustrates the management organization for the INTELSAT system. The INTELSAT operations centre (IOC), located within INTELSAT headquarters, is manned 24 hours a day and is responsible for the continuous control of the space segment.

6.8.1.1   Technical and operational control centre (TOCC)

A technical and operational control centre is established for each region and for other major work centres. TOCCs are located within INTELSAT headquarters. Each TOCC is primarily a regional coordinating office and, through the INTELSAT operations centre, is responsible for planning and coordinating the following:

- those activities, including monitoring, which are necessary to ensure the proper utilization and performance of the space segment facilities;

- verification tests of new earth stations, initial line up, implementation of new satellite links, alterations to existing satellite link parameters (i.e. centre frequency, bandwidth and channel capacity), verification of existing technical parameters, and for ensuring that the earth stations concerned and the IOC are advised of the results;

- those maintenance activities of earth stations which affect the space segment;

- ensuring the availability of communications for operational management activities;

- control of space segment utilization for the implementation of contingency plans, transitional plans and emergencies.

6.8.1.2   Spacecraft control centre (SCC)

The spacecraft control centre located in INTELSAT headquarters is responsible for spacecraft control and operational functions, other than use of the communication transponders. These functions include:

- satellite positioning including command and firing of the apogee motor following a satellite launch;

- maintenance of satellite operating parameters;

- analysis of telemetry data received from the tracking, telemetry and telecommand (TTC) stations to access satellite performance;

- monitoring satellite orbits;

- provision of pointing data for transmission to earth stations.

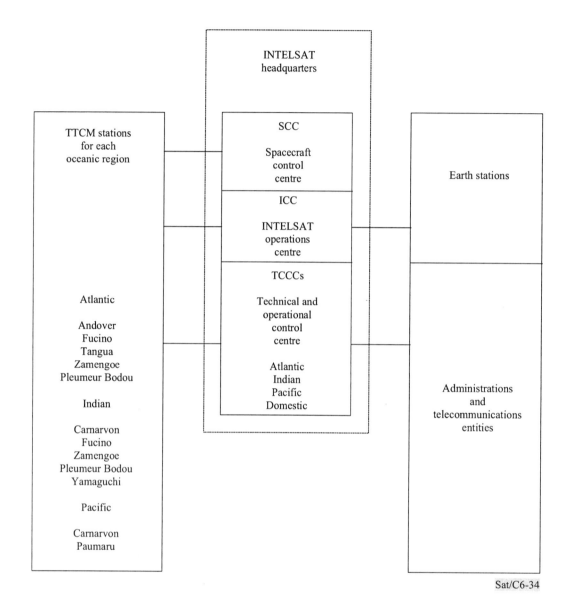

Sat/C6-34

FIGURE 6.34

INTELSAT operational organization

6.8.1.3   Tracking, telemetry, telecommand and monitoring (TTCM)

The tracking, telemetry, telecommand and monitoring stations are located for each region as indicated in Figure 6.34. The tracking, telemetry and telecommand (TTC) functions are associated with the SCC and the monitoring function is associated with the TOCC. Apart from monitor equipment, TTC stations are equipped to measure angle tracking data and ranging data, receive satellite telemetry from the satellite and transmit commands to the satellite. They are also equipped to accumulate and to format selected spacecraft telemetry data and earth station angle tracking data, range data, beacon signal strength, beacon identification and beacon frequency for transmission to the SCC.

6.8.1.4 Communication system monitoring (CSM)

Communication system monitoring stations are usually co-located with the tracking, telemetry and telecommand facilities, and form a part of the TTCM stations. Each monitor facility is closely associated with the TOCCs and the principal responsibilities of a monitoring station are:

- to assist earth stations and INTELSAT in the verification of certain mandatory performance characteristics of earth stations;

- to measure the satellite e.i.r.p., deviation and the centre frequency of carriers. Additionally, the out-of-band noise and energy dispersal frequency (EDF) of the carriers may also be measured;

- to monitor the frequency spectrum of operating satellites;

- to perform measurements as directed by the IOC or TOCC when such assistance has been requested by earth station operators and to perform other measurements as directed by the IOC or TOCC.

6.8.1.5 PCM-PSK network reference and monitor stations (NRMS)

For each PCM-PSK network, one of the earth stations functions as a network reference and monitoring station for that network.

6.8.1.6 Operation of engineering service circuits (ESCs)

Engineering service circuits (ESCs) provide the primary communication requirements for the management, operation and maintenance of the INTELSAT system. Two 4 kHz channels located in the 4-12 kHz portion of the baseband spectrum are assigned for engineering service circuit use on each satellite carrier. Each of these 4 kHz channels is arranged to provide one speech channel and up to five telegraph channels. Both speech and telegraph channels carry signalling information. Each earth station has been assigned a specific address digit or code which enables selective dialling among earth stations and the regional TOCC. ESCs are also available on digital links. Separate SCPC channels are available for ESCs for the SCPC/SPADE network.

6.8.1.7 TDMA-DSI network operations

The INTELSAT TDMA-DSI network is controlled by TDMA reference stations under the command of the INTELSAT operations centre TDMA facility (IOCTF). Each INTELSAT TDMA-DSI network consists of two pairs of reference stations which supply reference bursts to each 80 MHz transponder of the satellite for burst synchronization and also perform acquisition of traffic terminals and TDMA system monitoring. Some of the transponders will be connected in an East-West and West-East configuration, thus requiring one pair of reference stations in the East zone beam and another in the West zone beam. Each reference station is connected to the IOCTF by high speed data lines, together with voice and telegraph circuits. The transponders receive two reference bursts per frame, one from each of a pair of reference stations forming an automatic back-up facility. Both open-loop acquisition and closed-loop frame synchronization are used by terminals in the system. Within each network, the functions of corrective feedback of timing information and overall control TDMA system are performed by the reference stations. Therefore, these reference stations operate as controlling stations. Normally one reference station acts as the primary and the other as the secondary station, but the roles are interchangeable.

In a simple East/West, West/East configuration, the traffic terminal in the East coverage area can receive bursts only from the pair of reference stations in the West. These controlling West zone reference stations measure the traffic terminal's frame position against the East zone primary reference burst. For establishing transmit and receive side frame timing, computer-based logical procedures are employed at the reference and traffic terminals. This forms the central control technique for the TDMA system.

In the TDMA/SSOG testing operation, the station is verified to the maximum extent possible prior to the transmission to the satellite. When the station has been fully evaluated locally, transmission to the satellite follows under the close supervision of the controlling reference stations and the IOCTF. This phase includes tests of the various transmit and receive protocols. Further tests and adjustments follow and finally traffic checks are performed. Thus, by using SSOG tests and the SSOG protocol tests at the reference station, uniform testing standards are applied.

The TDMA system monitor (TSM) at the reference station cyclically monitors each traffic burst and reference burst to provide a record of system performance and bring any deviations to the notice of operators at the reference station and the INTELSAT operations centre (IOC). TSM measurements include relative burst power, burst position error, 4-PSK carrier frequency, pseudo bit error ratio and satellite transponder operating point.

6.8.2   Management of domestic satellite communications network operations

The management of domestic satellite communication networks is relatively less complex in comparison to management of an international network due to a smaller service area, and since the network including earth stations is generally owned and operated by one or a few entities within the country, it is possible to centralize operation, monitoring and control functions. An earth station associated with a network operations and control centre could be co-located with a TTCM earth station for spacecraft operations or with one of the main traffic earth stations. Proper coordination between the agency responsible for spacecraft operations, earth station operators and customers like common carriers and broadcasting agencies is essential for efficient working of the system. Network management structure in different countries may vary considerably depending on the organizational set up of the agencies, which may be totally government owned with a single telecommunication administration or, at the other extreme, may involve several private agencies. In any case, an operational group responsible for implementation of operational plans, coordination and control is required to be set up and the responsibilities of individual representatives from concerned agencies identified.

Initial earth station check-out procedures, line-up procedures, operational procedures, inter-agency interfaces and procedures for periodic review of operations, reports and coordination meetings have to be laid down. Monthly, weekly and daily schedules of operations generated by customer agencies and ratified by the spacecraft manager from the viewpoint of spacecraft operational constraints should be the primary mechanism for inter-agency coordination. An operations directory should provide full information for inter-agency communication/correspondence. Log books should be maintained for all subsystems of space segment and ground segment on a shift by shift basis for keeping track of operational status.

Transmissions from all communication transponders are commonly monitored and network discipline is maintained by a network operation and control centre (NOCC). Functions and monitoring facilities of a typical NOCC are described in section 6.8.2.1. Apart from monitoring and control of carrier transmission, in some networks, particularly those having remote isolated regions and unattended earth stations, network management may involve the centralized supervision, fault monitoring and control of remote stations.

6.8.2.1 Network operation and control centre

The main functions of the network operation and control centre (NOCC) are:

- verification testing of earth stations including performance tests of antenna radiation patterns and initial line ups for new services;

- routine periodical scanned measurements of all satellite transmissions and repetitive measurements and recording of selected carrier parameters for resolving anomalous performance;

- coordination of maintenance activities for effective utilization of space segment and ground segment, and implementation of transmission plans, transition and contingency plans, and emergency measures;

- control of uplink transmission from earth stations to maintain network transmission plans.

Network operation and control centre (NOCC) measurements: the following typical parameters on different carriers may be monitored by the NOCC:

- e.i.r.p.;

- e.i.r.p. stability;

- centre frequency;

- centre frequency stability;

- multichannel r.m.s. deviation;

- energy dispersal peak deviation;

- energy dispersal frequency;

- baseband pilot level;

- out-of-band baseband noise.

Measurements may also include the following satellite transponder parameters:

- transponder saturated e.i.r.p.;

- illumination flux-density for transponder saturation;

- spacecraft receive G/T;

- transponder small signal gain.

Certain parameters such as e.i.r.p. and frequency may be subjected to a limit check with provision for alarms when these parameters exceed pre-determined limits.

The network operation and control centre can also be provided with optional equipment for manual, semi-automatic and automatic operation, and facilities for storage and recall of information on selected transmission measurements and, on request, the display on a CRT of any measurement, printout or chart recording of any or all of the measurements. The type of parameters to be monitored, degree of automation and facilities to be provided at the NOCC may vary depending upon satellite system requirements, its size and technical and economic considerations. A large satellite system such as INTELSAT has a highly sophisticated and elaborate automated network operation and control system. On the other hand, a small domestic satellite network may limit NOCC functions essentially to the monitoring of e.i.r.p. and centre frequency on a semi-automatic or manual basis only.

Two-way voice and teleprinter links between the NOCC and the TTCM station which monitors and controls spacecraft operations and also between the NOCC and other earth stations, are provided to achieve coordination and control of transmission by the NOCC. These links may be established through satellite-derived engineering service channels and/or dedicated terrestrial connections for operational use with a back-up mode to provide redundancy.

6.8.2.2   Management of very small aperture earth terminals

A recent trend is to use very small aperture terminals (VSATs) to provide low bit rate data or a message service using a packet mode for business communications to a very large number of remote locations. These VSATs are linked through a master station in a star network configuration and are generally unattended. The master station, apart from handling traffic to and from VSATs, provides automatic monitoring and fault analysis, statistical information on traffic, accounting and billing. VSAT parameters monitored periodically include transmit frequency offset, transmit gain and power output, receive gain, temperatures of upconverter and low noise converter assemblies and power supply voltages, etc. Some of the parameters like transmit frequency offset and data packet formats can also be set on command from the master station. If any VSAT goes off-line, the master station automatically tries to bring it on-line through repeated acquisition attempts. If these attempts fail, the system prompts the operator at the network operation and control centre with audio and visual alarms. The operator can then initiate simple step by step diagnostic procedures to locate and take measures to rectify the problem. In case the terminal parameters deviate beyond specified limits, it is possible to switch off the RF power of the terminal from the master station. A network processor at the master station maintains records of number of packets received and sent, number of retransmissions and bit errors.

6.9 Satellite launch systems

6.9.1   Background

The first launch systems to place satellites into orbits around the Earth were developed by government agencies in the 1950s to insert satellite communication and observation systems into low-Earth orbits (150-200 km altitude). Most of these launchers were modelled after the intercontinental ballistic missiles of the period. In the 1960s era, space exploration programmes associated with flights to the Moon and planets resulted in the development of powerful rockets that were capable of inserting satellites into the geostationary orbit, commonly referred to as the "GSO" (35 786 km altitude). The era of the extensive use of GSO communication satellites started in the 1970s and has continued without interruption to the present time.

Recently, considerable interest has been shown for the development of new non-GSO communication satellites which have very different launch requirements from GSO satellites. The technology, however, is well developed since many non-GSO satellites with a variety of service missions (weather, earth mapping, navigation, etc.) have been launched during the last several decades. Also, many launch systems with GSO capabilities are able to insert several LEO satellites into low- or medium-Earth orbits with one launch operation.

6.9.2   Launcher considerations

The basic requirements for the selection of a launch system are 1) lift capability to the desired orbit; 2) availability after the satellite construction and test phase has been completed; and 3) cost of equipment and services. Until recently, the choice has been limited and negotiations have normally been with government agencies. Now, a new era has evolved in which a range of launch vehicles are being offered internationally on a commercial basis by competing private companies and government organizations. The launch industry is expanding rapidly and new performance capabilities and services are constantly being featured. Thus, this section should only be regarded as a guide to what may be available. Direct contact with the suppliers will be necessary in order to obtain all the necessary details associated with contracting for a launch system.

6.9.3   Types of launch systems

6.9.3.1   Geostationary orbit (GSO)

The predominant launch systems for GSO satellites have **expendable** boosters which employ several steps for inserting a satellite into its final orbit. The first step usually involves a few rocket firing phases which place the satellite and its attached apogee rocket motor (ARM) into a transfer orbit with a perigee of approximately 200 km in altitude and an apogee at the GSO altitude. At apogee, the ARM is fired to circularize the orbit into a geosynchronous mode. Some available launch systems with these characteristics include the ARIANE, ATLAS, DELTA, H-Series, LLV, LONG MARCH, M-Series, PROTON, TITAN, ZENIT, among others. A brief description of the capabilities of these systems is provided in the following sections.

There has been interest in developing reusable launchers in which the launch vehicle is returned to Earth intact and then readied for the next launch. An example is NASA's space transportation system (Space Shuttle), which places satellites into low-Earth orbit from which an intermediate rocket inserts the satellite into a GSO transfer orbit. Then the ARM can be fired to achieve the final orbit. Since the Space Shuttle carries a human crew, its costs are too high to be practical for the many commercial communication satellites that need to be placed into orbit. It is reserved for launching special payloads or performing special operations that require human intervention. New initiatives have been reported about the development of small reusable launch vehicles (Kistler Co.) for operations in the next decade.

6.9.3.2 Non-geostationary orbits (non-GSO)

Launch systems for low-Earth orbit (LEO) satellites usually require much lower booster capabilities than for GSO systems and have shown greater flexibility in their designs. For example, some LEO launch systems have been carried aloft in aircraft to improve their payload delivery capabilities. Others are designed to launch several satellites in a particular orbit or constellation, thus reducing the number of launches and the overall costs.

The basic design or vehicle of non-GSO launch systems are similar to that for the GSO satellites when multiple satellites or large payloads need to be inserted in non-GSO orbits. Rocket stages may be added or deleted depending on the payload and orbit requirements.

Non-GSO launch systems have enjoyed a long period of operations reaching back to the first earth satellite (Sputnik) in 1957. New developments to increase the reliability and reduce the cost of these systems has continued so that, at present, there are several new or modified systems available to the communication satellite industry. A few examples of LEO type launch systems include Atlas I (United States), Aussroc (Australia), Capricornio (Spain), Delta Lite (United States), ESA/CNES Series (Europe), J-Series (Japan), Kosmos (Russia), Lockheed Astria (United States), Long March CZ-1 (China), PacAstro (United States), Pegasus (United States), Sea Launch (United States/International), Shavit (Israel), SLV Series (India), Soyuz/Vostok (Russia), and VLS Series (Brazil), among others.

6.9.4 Launcher selection

A preliminary review for the selection of a launch system would entail equating the performance capabilities against such requirements as satellite system weight to be injected into a specific orbit, the volume available in the nose cone or housing of the rocket, the injection accuracy for transfer orbits or final orbit insertion, and other technical factors. An equally important set of considerations is the reliability and costs of the launch system, including launch services. In addition, transportation costs to the launch site need to be assessed as well as related insurance fees.

The recent commercialization of the launch industry has introduced a high level of competition among suppliers, both for governments and private organizations. The latest information should be obtained in this highly dynamic environment before any commitment is made for a particular launch system. New data services, such as the "Internet", and technical journals, such as NASA's "Transportation Systems Data Book", provide general information about the status of many launch systems and their manufacturers or distributors. For up-to-date technical details and costs, suppliers should be contacted directly.

6.9.5 Current and future launch systems

This section provides some preliminary information on some of the recently employed satellite launch systems and some of the modified systems reported in trade journals and reports. It is not an exhaustive summary of all the launch systems that have emerged during the last decade, but a brief view of some examples of launch systems that have operational experience.

a)   Ariane Series

At present, Ariane 4 is the most prominent launch system in the international commercial satellite communication industry. This system was developed by the European Space Agency (ESA) and CNES, the French Space Agency, and operations are conducted by Arianespace. The 4 Series, which has a reliability of over 90%, is capable of inserting from 1 900 to 4 200 kg into a geostationary transfer orbit (GTO). The Ariane vehicles are launched from Kourou, French Guyana where the latitude is approximately 5° N. A larger version, Ariane 5, has recently completed development and has become available for commercial launches. A schematic of Ariane 5 is shown in Figure 6.35.

b)   Atlas Series

These systems represent the larger of the commercial launchers in the United States of America. Developed in the 1960s, the Atlas is currently operated for commercial services by the Lockheed Martin Company and some Russian aerospace companies. The Atlas 1 and 2 versions, which had a reliability close to 90%, are capable of inserting from 2 250 to 3 490 kgs into a GTO.

Figure 6.36 shows a schematic of the Atlas launch system and lists the technical characteristics of its subsystems and components.

c)   Delta Series

These systems have a long history of reliable operations (98%) in the United States. Currently, they are manufactured and marketed by the McDonnell Douglas Company. The Delta II version is capable of inserting from 950 to 1 820 kgs into a GTO. A Delta III version is currently under development to more than double the lift capability of its predecessors. Figure 6.37 shows a summary of the Delta's growth from its start in 1960 until the present. Also shown are the present models intended for LEO applications.

d)   H-Series

The H-2 launch system, which was developed by Japan from their earlier N-Series of vehicles, has been successful in their initial flights in inserting heavy payloads into the GSO and space. Lift capability to the GTO is 4 000 kgs. Figure 6.38 shows a history of the development of the H-Series of launchers by Japan.

e)   LLV Series

The Lockheed Launch Vehicle is another flexible system that utilizes small solid rocket boosters to increase its lifting capacity. The LLV-2 and -3 versions are able to place 1 305 and 2 500 kgs into a GTO. This system is another good candidate for launching LEO satellite systems into orbit.

f)   Long March Series

This system, developed by China, includes a range of vehicles from the small CZ-1D to the large CZ-2E. These launchers are available for commercial satellite services. The range of lift capabilities to the GTO vary from 200 to 3 370 kgs. This series of vehicles has plans to launch LEO satellites within a few years, where the lift capability will be greater by a factor of 2 or 3. Figure 6.41 below summarizes the characteristics of Long March launchers.

g)   M-Series

The M vehicles, which have all solid propellants, were also developed by Japan, but for smaller payloads. The M-V model has the ability to insert 1 215 kgs into the GTO. This series is planned for a variety of space missions.

h)   Proton Series

These systems, which were designed to lift very heavy payloads into space, have a long history of operations in the former USSR. Presently, it is being marketed by International Launch Services, a joint venture between Krunichev of Russia and Lockheed Martin of the United States This vehicle has a lift capability of 5 500 kgs into a GTO. Figure 6.39 shows the major hardware components of the Proton D-1 launch system.

i)   Titan Series

This launch system was developed in the United States several decades ago as a ballistic vehicle and was subsequently revised to insert heavy satellites into orbit. The Titan IV version is capable of inserting from 6 350 to 8 620 kgs into a GTO. This system is primarily employed for United States government operations.

j)   Zenit Series

This launch system was modified from earlier USSR large lift vehicles and is presently manufactured in the Ukraine by NPO Yuznoye. It is capable of lifting 4 300 kgs into a GTO. In a joint venture with Boeing (United States) and Kvaerner (Norway), the Sea Launch Company plans to increase this capability to 5 400 kgs by launching the Zenit from a modified ocean oil platform located on the equator. A schematic of the Zenit system in shown in Figure 6.40 with a sketch of the ocean platform under development by the Sea Launch Company's joint venture programme.

With regard to other systems mentioned in this section, the reader is advised to seek out information from the organizations mentioned above for up-to-date information.

Figure 6.35   -   Schematic of Ariane 5 launch system

Figure 6.36   -   Atlas launch system design

Figure 6.37   -   Growth of Delta family of launch systems with schematics of "Med-Lite" launchers

Figure 6.38   -   History of Japanese launch systems

Figure 6.39   -   Expanded view of Proton D-1 launch system showing major hardware components

Figure 6.40   -   Schematic of Sea launch platform and Zenit 3SL launcher

Figure 6.41   -   Long March Series of launch vehicles

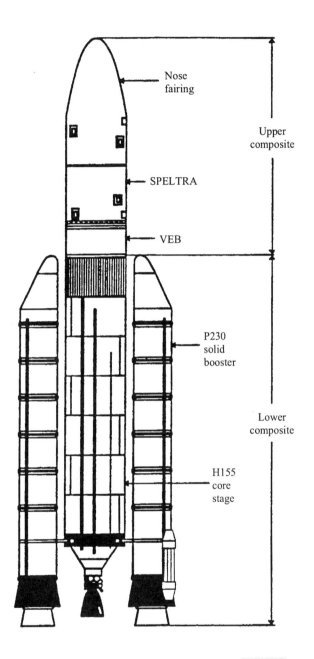

Sat/C6-35

FIGURE 6.35

Schematic of Ariane 5 launch system

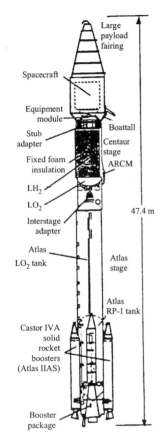

PAYLOAD FAIRING

Features:	large (Shown)	Medium
Diameter:	4.2 m	3.3 m
Length:	12.2 m	10.4 m
Mass:	2 087 kg	1 409 kg
Subsystems:		
Fairing:	aluminium skin stringer and frame	
	Clamshell	
Boattail/Split:	aluminium skin stringer and frame	
Barrel:	Clamshell	
Separation:	Pyro bolts & spring thrusters	

CENTAUR UPPER STAGE

Features:	
Size:	3.05 m dia x 10.06 m length
Inert mass:	2 200 kg
Propellant:	16 780 LH$_2$ & LO$_2$
Guidance:	Inertial
Subsystems:	pressure stabilized stainless steel tanks
Structure:	Separated by common ellipsoidal bulkhead
Propulsion:	Two Pratt & Whitney restartable engines
Option:	RL10A-4
Thrust:	181.5 kN RL10A-4-1
Isp:	449.0 s 198.4 kN
	Two electromechanically actuated 51 cm
	Colombium extendible nozzles
	Twelve 27-N hydrazine thrusters
Pneumatics:	Helium & hydrogen autogenous
	(tank pressurization)
Hydraulics:	Fluid-provides two separate systems
	for gimballing main engines
Avionics:	Guidance, navigation & control, vehicle
	sequencing, computer-controlled vent &
	pressurization, telemetry, C-band
	tracking, range safety command,
	electrical power
Insulation:	Polyvinyl chloride foam (1.6 cm thick),
	modified adhesive bonding

INTERSTAGE ADAPTER (ISA)

Features:	
Size:	3.05 m dia x 3.96 m length
Mass:	545 kg
Subsystems:	
Structure:	aluminium skin stringer and frame
Separation:	Flexible linear-shaped charge
Control:	Roll control module installation

ATLAS STAGE

Features:	
Size:	3.05 m dia x 24.90 m length
Propellant:	156 400 kg LO$_2$ & RP-1
Guidance:	From upper stage
Subsystems:	
Structure:	pressure stabilized stainless steel tanks
	integrally machined thrust structure
Separation:	Booster package - 10 pneumatically
	actuated separation latches
	Sustainer section - 8 retro rockets
Propulsion:	Rocketdyne MA-5A
	Booster engine (2 chambers)
	SL thrust = 1 854 kN, Isp = 262.1 s
	VAC thrust = 2 065 kN, Isp = 293.4 s
	sustainer engine
	SL thrust = 266.0 kN, Isp = 216.1 s
	VAC thrust = 380.6 kN, Isp = 311.0 s
Pneumatics:	Helium (pressure tanks), hydraulics and
	lubrication system
Hydraulics:	Fluid - provides two separate systems
	for main engine gimbal
Avoinics:	Flight control, flight termination
	Telemetry, computer-controlled Atlas,
	pressurization system, rate gyro unit

SOLID ROCKET BOOSTERS (SRB)

Four Thiokol castor IVA
Size:	102 cm dia x 13.6 m length
Mass:	11 567 kg (each fuelled)
Thrust:	433.7 kN (each)
Isp:	237.8 s

Ground-lit motors 11° cant nozzles
Air-lit motors 7° cant nozzles
Inadvertent separation destruct system (ISDS)

Sat/C6-36

FIGURE 6.36

Atlas launch system design

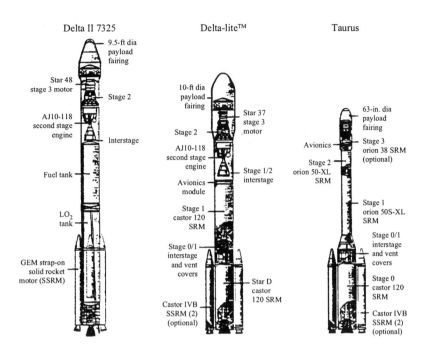

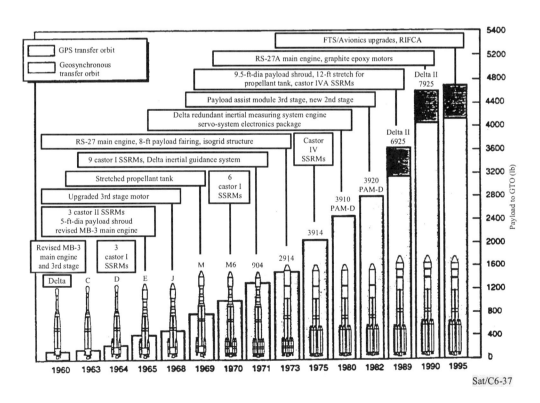

FIGURE 6.37

Growth of Delta family of launch systems with schematics of "Med-lite" launchers

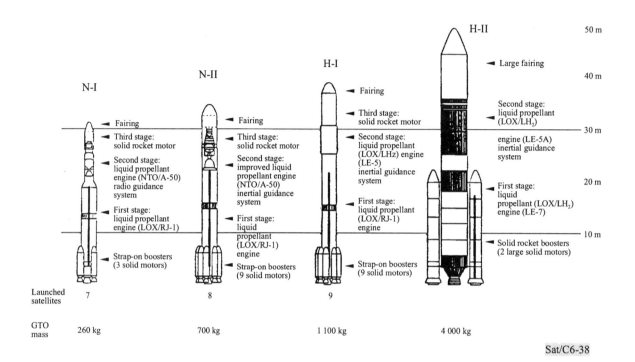

Sat/C6-38

FIGURE 6.38

History of Japanese launch systems

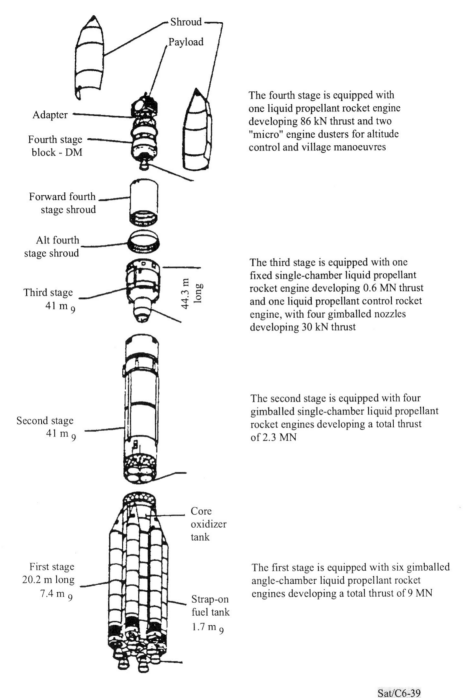

The fourth stage is equipped with one liquid propellant rocket engine developing 86 kN thrust and two "micro" engine dusters for altitude control and village manoeuvres

The third stage is equipped with one fixed single-chamber liquid propellant rocket engine developing 0.6 MN thrust and one liquid propellant control rocket engine, with four gimballed nozzles developing 30 kN thrust

The second stage is equipped with four gimballed single-chamber liquid propellant rocket engines developing a total thrust of 2.3 MN

The first stage is equipped with six gimballed angle-chamber liquid propellant rocket engines developing a total thrust of 9 MN

Sat/C6-39

FIGURE 6.39

Expanded view of proton D-1 launch system showing major hardware components

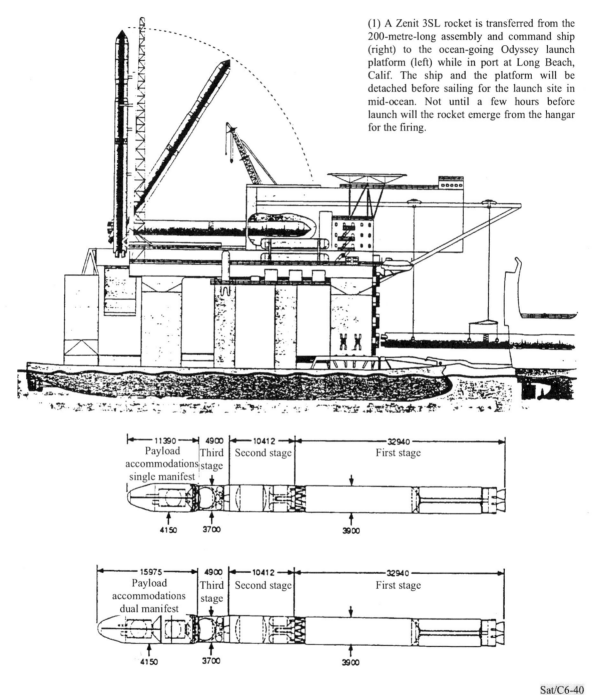

(1) A Zenit 3SL rocket is transferred from the 200-metre-long assembly and command ship (right) to the ocean-going Odyssey launch platform (left) while in port at Long Beach, Calif. The ship and the platform will be detached before sailing for the launch site in mid-ocean. Not until a few hours before launch will the rocket emerge from the hangar for the firing.

Sat/C6-40

FIGURE 6.40

Schematic of Sea launch platform and Zenit 3SL launcher

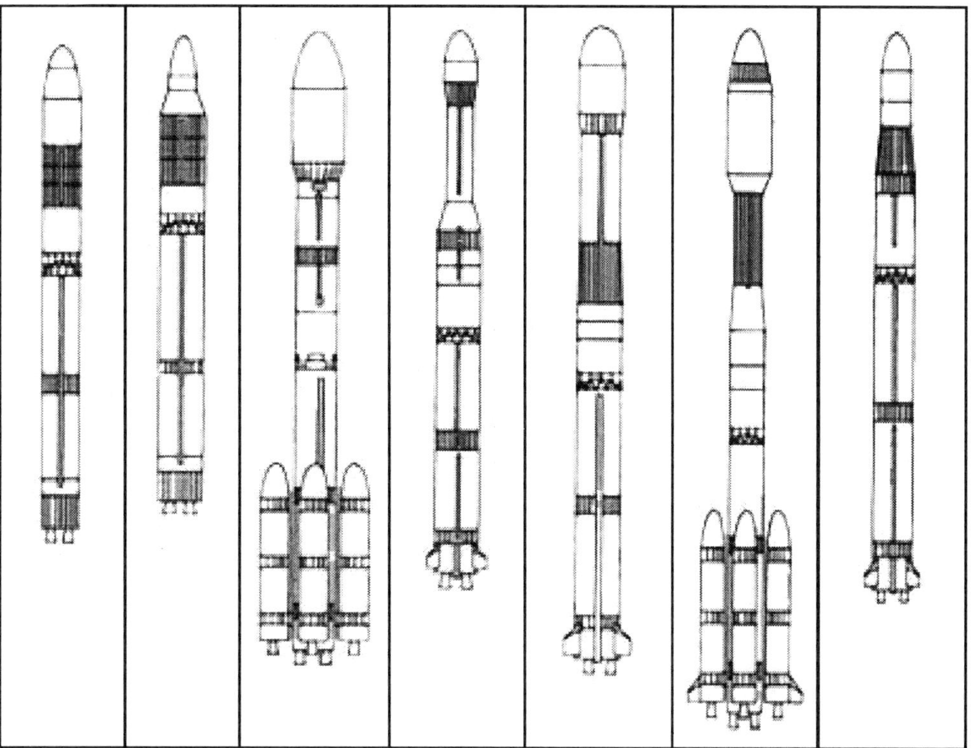

Launch Vehicle	LM -2C	LM -2D	LM -2E	LM -3	LM -3A	LM -3B	LM -4
Overall Length (m)	40.0	37.7	49.7	44.6	52.5	54.8	45.8
Lift -off Mass (ton)	213	233	460	204	241	425.8	249.2
Lift -off Thrust (kN)	2 962	2 962	5 923	2 962	2 962	5 923	2 962
Fairing Di ameter (m)	2.60/3.35	2.90	4.20	2.60/3.00	3.35	4.00/4.20	2.90/3.35
Stages 1 and 2	Propellant: UDMH/N_2O_4/Diameter: 3.35 m						
Stage 3	N/C	N/C	N/C*	LOX/LH_2	LOX/LH_2	LOX/LH_2	UDMH/N_2O_4
Primary Mission	LEO	LEO	LEO/GTO	GTO	GTO	GTO	SSO
Payload Capacity (kg)	2 800	3 500	9 500/3 500	1 500	2 600	5 000	1 650
Launch Site **	JSLC TSLC	JSLC	JSLC/XSLC	XSLC	XSLC	XSLC	JSLC TSLC

* Long March 2E, with a solid Perigee Kick Motor (EPKM), can perform GTO launch missions with launch capacity up to 3   500 kg.

** XSLC/JSLC/TSLC re   fer to three Satellite Launch Centres of Xichang, Jiuquan and Taiyuan in China.

Sat/C6-41

FIGURE 6.41

Long March Series of launch vehicles

REFERENCES

[1]   Ref: NASA/NFS Panel Report on Satellite Communications Systems and Technology, July 1993, Volume 1. Chapter 2, "Review Assessment of Satellite Communication Technologies".

[2]   Ref: AIAA-96-1142-CP paper, On-Board Propulsion for Communications Satellites, by L.W. Callahan, F.M. Curran and T.J. Wickenheiser.

[3]   Ref: AIAA-96-1082-CP paper "Design and Operation of a Non-Station Kept LEO Constellation", by R.J. Cenker (Aerospace Consulting Group, Inc., Mark Halverson and Robert Nelson (GE American Com Inc.).

[4]   (Ref 1): "Solar Arrays Meeting the Power Requirements of Communication Satellites" by P.A. LLes, F.F. Ho and E.B. Linder, AIAA -96-1024-CP paper.

[5]   (Ref 2): SCARLET: "A High-Payoff, Near-Term Concentrator Solar Array", by P. Alan Jones, D.M. Murphy, T.J. Harvey, D.M. Allen, L.H. Caveny and M.F. Piszczor, AIAA-96-1021-CP paper.

[6]   (Ref 1): "Satellites in the NII/GII", by N.R. Helm and B.I. Edelson, AIAA-96-0994-CP paper.

[7]   Ref: Chapter 2, NASA/NSF Panel Report on "Satellite Communications Systems and Technology", Volume 1, July 1993.

[8]   Ref: Chapter 2, NASA/NSF Panel Report on "Satellite Communications Systems and Technology", Volume 1, July 1993 on Transponders, TWTAs and SSPAs.

APPENDIX 6.1

Examples of non-geostationary satellite systems

AP6.1-1 Teledesic satellite system

AP6.1-1.1 Introduction

The proposed network uses a constellation of 840 operational interlinked low-Earth orbit satellites and up to four operational spares per orbital plane to provide global access to a broad range of voice, data and video communication capabilities in the FSS. The network will provide switched digital connections between users of the network through its standard terminals with transmission rates from 16 kbps to 2.048 Mbps and through its gigalink terminals at 155.52 Mbps, and multiples of this rate up to 1.24416 Gbps, depending upon the instantaneous capacity requirement.

Terminals at gateway and user sites communicate directly with the satellite-based network and through gateway switches, to terminals on other networks. Figure AP6.1-1 is an overview of the network. Each satellite in the constellation is a node of a fast packet switch network, and has intersatellite communication links ("ISLs") with eight adjacent satellites. Each satellite is normally linked with four satellites within the same plane (two in front and two behind) and with one in each of the two adjacent planes on both sides.

AP6.1-1.2 Constellation description

The satellite constellation is organized into 21 circular orbit planes that are staggered in altitude between 695 and 705 km. Each plane contains a minimum of 40 operational satellites plus up to four on-orbit spares spaced evenly around the orbit. The orbit planes are at a sun-synchronous inclination (approximately 98.2°), which keeps them at a constant angle relative to the Sun. The ascending nodes of adjacent orbit planes are spaced at 9.5° around the equator. Satellites in adjacent planes travel in the same direction except at the constellation "seams", where ascending and descending portions of the orbits overlap. There is no fixed phase relation between satellites in adjacent planes: the position of a satellite in one orbit is decoupled from those in other orbits.

As satellite communication links operating in the Ka-band frequencies are subject to high rain attenuation and terrain shadowing, the network is designed to operate with high elevation service angles. The attenuation and shadowing problems diminish as the elevation angle of the signal path is increased. The satellite constellation is designed to ensure that there is always at least one satellite visible above a 40° elevation angle over the entire coverage area. Coverage is provided 24 hours-a-day between 72° North and South latitudes, with partial day coverage to higher latitudes.

This system uses a nominal 700 km altitude to meet the network requirement for low end-to-end delay. The altitudes of satellites in different orbit planes are staggered to eliminate the possibility of collision between satellites in crossing orbits. The nominal 700 km altitude and 40° elevation mask angle yield a satellite footprint approximately 1 400 km in diameter.

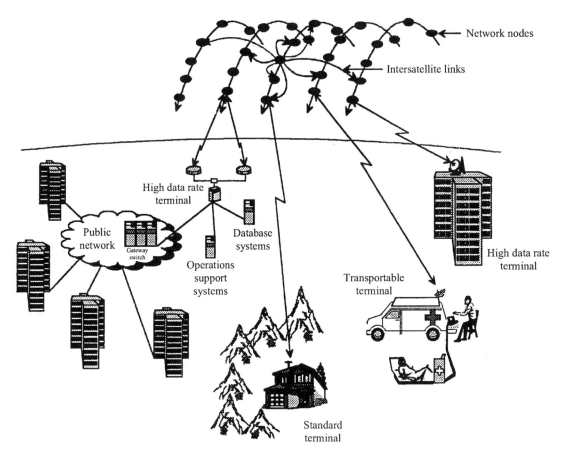

Sat/AP6.1-1

FIGURE AP6.1-1

The communication network

AP6.1-1.3   Communications links

All of the communications links transport data and voice as fixed-length (512) bit packets. The basic unit of channel capacity is the "basic channel", which supports a 16 kbps payload data rate and an associated 2 kbps for signalling and control. Basic channels can be aggregated to support higher data rates. Standard terminals will include both fixed-site and transportable configurations. Within its service area, each satellite can support a combination of standard terminals with a total throughput equivalent to over 125 000 simultaneous basic channels, plus as many as 16 Gigalink Terminals.

The uplinks use dynamic power control of the RF transmitters so that the minimum amount of power is used to carry out the desired communication. Minimum transmitter power is used for clear sky conditions. The transmitter power is increased to compensate for rain. Power spectral density produced by the standard terminals is -38 dBW/Hz (clear sky) and -21 dBW/Hz (rain). Power spectral density produced by a gigalink terminal is -40 dBW/Hz (clear air) and -23 dBW/Hz (rain).

The uplink terminals can use antennas with diameters from 16 cm to 1.8 m as determined by the terminal's maximum transmit channel rate, climatic region and availability requirements. The average transmit power for standard terminals varies from less than 0.01 W to 4.7 W depending on antenna diameter, transmit channel rate and climatic conditions. All data rates, up to 2.048 Mbps, can be supported with an average transmit power of 0.3 W by suitable choice of antenna size.

AP6.1-1.4 Network structure

One benefit of a small satellite footprint is that each satellite can serve its entire coverage area with a number of high-gain scanning beams, each illuminating a single small cell at a time. Small cells allow efficient reuse of spectrum, high channel density and low transmitter power. However, if this small cell pattern swept the Earth's surface at the velocity of the satellite (approximately 25 000 km per hour), a terminal would be served by the same cell for only a few seconds before a channel reassignment or "hand-off" to the next cell would be necessary. However, frequent hand-offs result in inefficient channel utilization, high processing costs and lower system capacity. The network uses an Earth-fixed cell design to minimize the hand-off problem.

The system maps the Earth's surface into a fixed grid of approximately 20 000 "super-cells", each consisting of nine cells (see Figure AP6.1-2). Each super-cell is a square 160 km on each side. Super-cells are arranged in bands parallel to the equator. There are approximately 250 super-cells in the band at the equator, and the number per band decreases with increasing latitude.

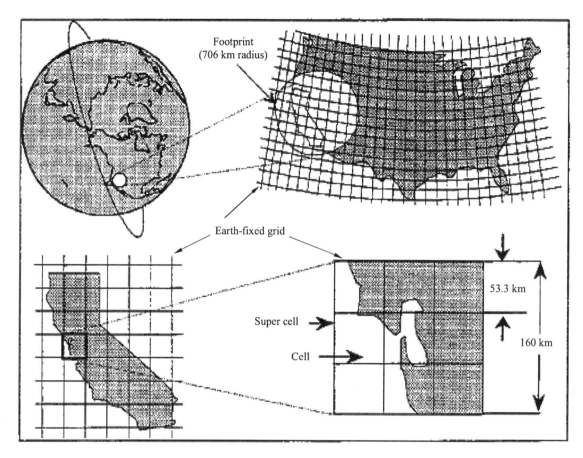

Sat/AP6.1-2

FIGURE AP6.1-2

Earth-fixed cells

A satellite footprint encompasses a maximum of 64 super-cells, or 576 cells. The actual number of super-cells for which a satellite is responsible at any given moment varies by satellite with its orbital position and its distance from adjacent satellites, but does not exceed 49. In general, the satellite closest to the centre of a super-cell has coverage responsibility. As a satellite passes over, it steers its antenna beams to the fixed cell locations within its footprint. This beam steering compensates for the satellite's motion as well as the Earth's rotation. This concept is illustrated in Figure AP6.1-3.

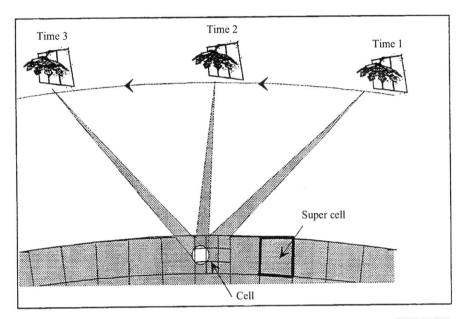

Sat/AP6.1-3

FIGURE AP6.1-3

Illustration of beam steering to an Earth-fixed cell

Channel resources (frequencies and time slots) are associated with each cell and are managed by the current "serving" satellite. As long as a terminal remains within the same Earth-fixed cell, it maintains the same channel assignment for the duration of a call, regardless of how many satellites and beams are involved.

A database contained in each satellite defines the type of service allowed within each Earth-fixed cell. Small fixed cells allow the system to avoid interference to or from specific geographic areas and to contour service areas to national boundaries. This would be difficult to accomplish with large cells or cells that move with the satellite.

AP6.1-1.5 Multiple access method

The network uses a combination of multiple access methods to ensure efficient use of the spectrum (see Figure AP6.1-4). Each cell within a super-cell is assigned to one of nine equal time slots. All communication takes place between the satellite and the terminals in that cell during its assigned time slot. Within each cell's time slot, the full frequency allocation is available to support communication channels. The cells are scanned in a regular cycle by the satellite's transmit and receive beams, resulting in time division multiple access ("TDMA") among the cells in a super-cell. Since propagation delay varies with path length, satellite transmissions are timed to ensure that cell N (N=1, 2, 3, ...9) of all super-cells receives transmissions at the same time. Terminal transmissions to a satellite are also timed to ensure that transmissions from the same numbered cell in all super-cells in its coverage area reach that satellite at the same time. Physical separation (space division multiple access or "SDMA") and a checkerboard pattern of left and right circular polarization eliminate interference between cells scanned at the same time in adjacent super-cells. Guard time intervals eliminate overlap between signals received from time-consecutive cells.

Within each cell's time slot, terminals use frequency division multiple access ("FDMA") on the uplink and asynchronous time division multiple access ("ATDMA") on the downlink. On the uplink, each active terminal is assigned one or more frequency slots for the call's duration and can send one packet per slot each scan period (23.111 msec). The number of slots assigned to a terminal determines its maximum available transmission rate. One slot corresponds to a standard terminal's 16 kbps basic channel with its associated 2 kbps signalling and control channel. A total of 1 800 slots per cell scan interval are available for standard terminals.

The terminal downlink uses the packet's header rather than a fixed assignment of time slots to address terminals. During each cell's scan interval the satellite transmits a series of packets addressed to terminals within that cell. Packets are delimited by a unique bit pattern and a terminal selects those addressed to it by examining each packet's address field. A standard terminal operating at 16 kbps requires one packet per scan interval. The downlink capacity is 1 800 packets per cell per scan interval for standard terminals. Each satellite beam transmits only as long as it takes to send the packets queued for a cell.

The combination of Earth-fixed cells and multiple access methods results in very efficient use of spectrum. The system can reuse its requested spectrum over 20 000 times across the Earth's surface.

Cell scan pattern

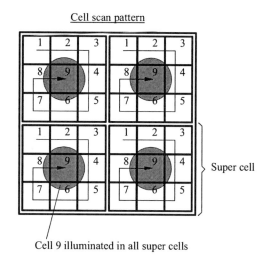

Super cell

Cell 9 illuminated in all super cells

Cell scan cycle

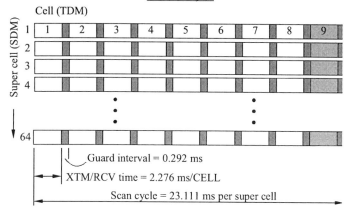

Guard interval = 0.292 ms

XTM/RCV time = 2.276 ms/CELL

Scan cycle = 23.111 ms per super cell

Channel multiplexing in a cell

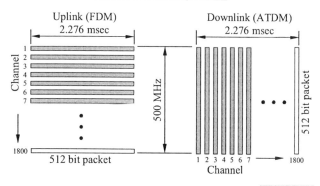

Sat/AP6.1-4

FIGURE AP6.1-4

Standard terminal multiple access method

AP6.1-1.6 Technical characteristics

For each link, the operating frequency bands, specific operating frequencies and the total bandwidth are shown in Table AP6.1-1 and the polarizations are shown in Table AP6.1-2.

TABLE AP6.1-1

Frequency bands, requested frequencies and total bandwidth

	Frequency band (GHz)	Requested frequencies (GHz)	Total requested bandwidth (MHz)
Standard terminal uplink	27.5-30.0	28.6-29.1	500
Standard terminal downlink	17.7-20.2	18.8-19.3	500
High data rate uplink	27.5-30.0	27.6-28.4	800
High data rate downlink	17.7-20.2	17.8-18.6	800
ISL	59-64	59.5-60.5 and 62.5-63.5	2 000

TABLE AP6.1-2

Polarizations

Standard terminal uplink	RHC and LHC
Standard terminal downlink	RHC and LHC
High data rate uplink	RHC and LHC
High data rate downlink	RHC and LHC
ISL	RHC and LHC
Command uplink	Vertical
Telemetry downlink	Vertical

The satellites use three different antenna gains to partially compensate for the variation in free space loss to different positions in the satellite footprint. The satellite transmitter antenna gains, output powers and maximum e.i.r.p.s for each antenna beam are shown in Table AP6.1-3.

TABLE AP6.1-3

Satellite transmitter output power and maximum e.i.r.p.

	Transmit antenna peak gain (dB)	Transmit power (W)	Maximum e.i.r.p. (dBW)
TSL			
Centre circle	29.8	93.75	49.52
Middle ring	30.9	93.75	50.62
Edge ring	32.0	93.75	51.72
GSL	41	4.6	47.6
ISL	48	5.5	55.4
Telemetry	0	2	3.0

The satellites are processing satellites. They demodulate and decode all received packets. The decoded packets are routed by a digital switch, encoded, modulated and retransmitted. The satellites do not contain conventional bent-pipe transponders. Thus the concepts of saturation flux-density and transponder gain have no meaning in the proposed system. Received packets can be routed to any of the antenna beams by the digital fast packet switch.

Gain contours for the satellite transmit and the receive antenna beams are shown in Figure AP6.1-5.

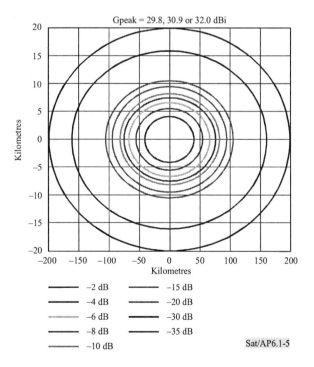

FIGURE AP6.1-5

Satellite scanning beam transmit and receive gain contours

AP6.1-1.7 Summary of teledesic network description

Teledesic uses small, "Earth-fixed" cells both for efficient spectrum utilization and to respect nations' territorial boundaries. Within a 53 by 53 km cell, the Network can accommodate a sustained load of over 1 800 simultaneous 16 kbps voice channels, 18 simultaneous T-1 (1.544 Mbps) channels, or any comparable combination of channel bandwidths. This represents a significant system capacity, equivalent to a sustained capacity of 20 000 simultaneous T-1 lines worldwide, with the potential for growth to higher capacities. The Network offers high-capacity, "bandwidth-on-demand" through standard user terminals. Channel bandwidths are assigned dynamically and asymmetrically and range from a minimum of 16 kbps up to 2 Mbps on the uplink and up to 28.8 Mbps on the downlink. Teledesic will provide a smaller number of high-rate channels at 155 Mbps to 1.2 Gbps for gateway connections and users with special needs. The low orbit and high frequency (30 GHz uplink/20 GHz downlink) allow the use of small, low-power terminals and antennas, with a relatively low cost.

AP6.1-2 The SkyBridge satellite system

AP6.1-2.1 System overview

The SkyBridge satellite system is a broadband access system for multimedia services using a constellation of non-geostationary (non-GSO), low-Earth orbit (LEO) satellites and operating in the 10-12/18 GHz frequency bands (see below section AP6.1-2.2). From the dawn of the 2000 decade, SkyBridge will provide directly to end users a large variety of high quality broadband and narrow-band interactive services, including access to Internet and on-line services, remote access for business and also entertainment services.

Through its constellation of 80 LEO satellites, SkyBridge is designed for worldwide coverage between 72° N and 72° S. The system will allow:

- asymmetric interactive services (e.g. high speed Internet applications);
- symmetric services (e.g. for conversational applications, like videoconferencing).

Each satellite provides a 3 000 km radius coverage divided into fixed cells of typically 350 km radius (as shown by Figure AP6.1-6). In this coverage, a satellite can share its capacity (4.7 Gbps, forward and return) between the cells, i.e. between a maximum of 24 spot-beams. This leads to a total system average throughput of 200 Gbps (sum of busy hour traffic) over land areas.

The SkyBridge system will reuse the frequency bands allocated to FSS, BSS and terrestrial networks in its operating frequency band, i.e. 1.65 GHz bandwidth for up- and downlinks in the 10-12/18 GHz bands. However, thanks to the careful design of its constellation and to an effective technique of ceasing relevant transmissions during potentially interfering periods, the system will ensure continuous operation while protecting other systems (see below sections AP6.1-2.2 and AP6.1-2.3).

The system provides on-demand dynamic bandwidth allocation (up to 60 Mbps forward link and 20 Mbps return link[1]). Its intrinsically low path delay (20 ms typical) makes it compatible with current terrestrial protocols such as Internet Protocols, thus allowing seamless integration into terrestrial networks.

In supplement to the service links, i.e. the communication links directly available to end users, the SkyBridge system has the capability to provide a few gateway-to-gateway communication links, allowing a gateway isolated from adequate terrestrial infrastructure to access by a second hop to terrestrial networks or servers, and relay links allowing a gateway to serve user terminals located in another cell which would not be served otherwise.

The SkyBridge system is composed of:

- the space segment comprising a constellation of 80 LEO transparent satellites. It is controlled by the satellite operations control centre (SOCC) linked to tracking, telemetry and control (TT&C) earth stations;

- the earth segment, divided into:

  - the user terminal earth segment comprising very small earth stations (0.3 m to 1 m), each one being connected, through the satellites, to the closest gateway;

  - the gateway earth segment which allows the connection with the terrestrial broadband infrastructure. Each gateway is responsible for all users within a fixed cell of typically 350 km radius, called a gateway cell.

The complete SkyBridge system operation is managed by the mission management centre (MMC).

It should be noted that, although the SkyBridge space segment will be owned and operated by the system's owner on a global basis, the gateways will be owned and operated by local operators, under their own regional or national regulations, in each of the territories serviced by the system. Also, local service providers will offer services to users on a regional or national basis.

AP6.1-2.2 Space segment

The SkyBridge is based on a constellation of 80 transparent LEO satellites (see Figure AP6.1-7). Each satellite is in a circular orbit at an altitude of 1 469 km above the Earth.

The constellation is a Walker sub-constellation 80/20/15 (i.e. 80 satellites distributed in 20 orbital planes with a phase shift of 67.5° between the first satellites of adjacent planes). The main parameters of the constellation are given in Table AP6.1-4.

---

[1] A forward link (also called outroute or outbound in VSAT terminology) is a two-way (up and down) link between a gateway and one of its user terminals via the satellite and a return link (also called inroute or inbound in VSAT terminology) is a two-way link between a user terminal and its gateway.

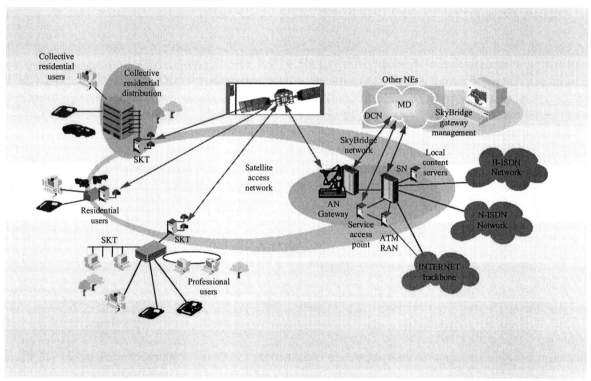

Sat/AP6.1-6

FIGURE AP6.1-6

SkyBridge system architecture

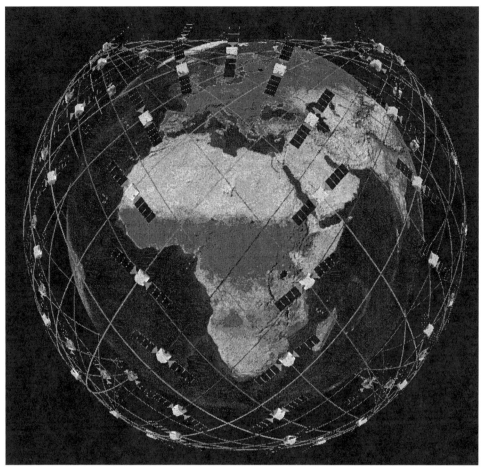

Mean anomaly (º)   80/20/15 Constellation

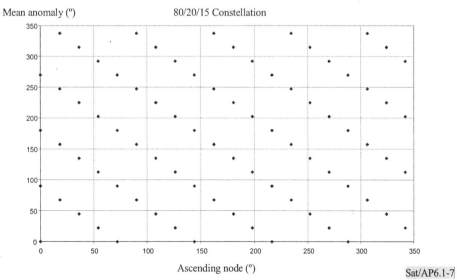

Ascending node (º)

Sat/AP6.1-7

FIGURE AP6.1-7

SkyBridge constellation

Table AP6.1-4
Sub-constellation parameters

– Number of planes	20
– Satellites per plane	4
– Inclination	53°
– Altitude	1 469 km
– Argument of perigee	90°
– Eccentricity	0
– Plane spacing (equator)	18°
– Phasing between satellites	67.5°
– Orbit period	115 min.

It can be noted that the design of the constellation has been chosen to allow a beginning of service prior to the launch of the 80 satellites. For example a 40/10/15 Walker intermediate constellation can be used to open service in temperate latitudes.

This unique constellation design allows:

• to ensure that, at all times and at any point within the ±72° latitude Service Region, at least one satellite is available for communication from the SkyBridge earth stations above a 10° elevation angle and outside of a non-operating zone around the GSO (as explained below). In fact, the design of the constellation permits each satellite and earth station to avoid an interference condition, while still ensuring continuous real-time service to all users, by having available another satellite to transfer the traffic to. Moreover, at latitudes between 30° and 57°, two satellites are always available and three between 35° and 40°;

• to optimize the protection of the GSO FSS networks, due notably to the possibility of shutting down a spot-beam of a satellite when it crosses a ±10° non-operating zone around the geostationary orbit, while keeping the traffic in operation through another satellite;

• to minimize the number of required satellites to offer a continuous coverage, while taking into account the above defined non-operating zone.

As already shown in Figure AP6.1-2, each satellite has a 3 000 km radius coverage divided into ground fixed cells of typically 350 km radius. This is obtained by the utilization of on-board antennas using advanced phased-array technology, which creates a maximum of 24 steerable spot-beams. The spot-beams achieve high antenna gain and permit maximum frequency reuse. Two (circular) polarizations and the entire authorized frequency band can be used within each cell.

Some characteristics of the satellites are listed below:

• Mass ≤1 250 kg

• Two solar panels: 3.5 kW (end of life)

• Full operation during eclipses

• Consumption ~ 3 kW

• Operational lifetime: 8 years

• Launching: By groups of one to ten satellites. Various possible launchers (Atlas, Proton, Delta, Ariane, etc.)

• Communication payload: Phased array transmit and receive antennas (with miniaturized filters and MMIC technology for the LNAs and TWTAs). Twenty-four forward and return transparent transponders (modems). Transparent RF cross-connections are implemented to fulfil the relay links needs.

The main satellite transmit and receive parameters are summarized in Tables AP6.1-5 and AP6.1-6 below.

TABLE AP6.1-5

SkyBridge satellite transmission parameters

Transmit (downlink)	e.i.r.p. (dBW)	Power per carrier (W)
• Service link, forward (satellite to user)	21.4 dBW/22.6 MHz	6.4 dBW
• Service link, return (satellite to gateway)	7.9 dBW/2.93 MHz	−7.1 dBW
• Infrastructure link (satellite to gateway)	21.4 dBW/22.6 MHz	6.4 dBW

NOTE – The values depend on the earth station antenna dimensions and G/T (see below, section AP6.1-2.6).

TABLE AP6.1-6

SkyBridge satellite reception parameters

Receive (uplink)	System noise temp. (K)	G/T (dB.K^{-1})
• Service link, forward (user to satellite)	455	−8.4
• Service link, return (Gateway to satellite)	455	−14.5
• Infrastructure link (Gateway to satellite)	455	−14.5

AP6.1-2.3 Frequency band and polarization – Coordination process

The service links (both forward and return) will use at least a 1.65 GHz continuous frequency band between 12.75 GHz and 18.1 GHz for the uplinks and a 1.65 GHz continuous frequency band between 10.7 GHz and 12.75 GHz for the downlinks. The T&C links will be also implemented in the same bands.

Circular polarization (RHCP and LHCP) is employed in order to have simple terminals.

The precise frequency to be operated in a given country or region will be chosen in the above quoted bands according to ITU Radio Regulations and allocations[2] and to local regulations and allocations.

Full protection of other services, and, in particular, of the GSO satellite services is ensured as follows (see also above in § AP6.1-2.2).

When a SkyBridge satellite with a spot-beam on a given gateway cell enters the cell non-operating zone (i.e. within a ±10° zone around the GSO, when an interference condition potentially exists), the spot-beam is shut down and the traffic is handed over to another satellite in the constellation not then in the non-operating zone for this cell. The implementation of this interference avoiding by proper frequency sharing is managed by the MMC.

Moreover, severe conditions are imposed to the SkyBridge emissions (power flux-density limitations, out-of-band emissions, etc.) in order to protect other services and, in particular, terrestrial communication services.

AP6.1-2.4 Spot-beam steering and handover operation

The satellite spot-beams are steered in order to compensate the satellite motion and to generate fixed cells on the Earth (the Earth surface has been divided in approximately 500 cells).

According to geometrical criteria and resource allocation optimization (including the shutting down process in non-operating zones), a traffic hand-over, which is transparent to the user, occurs between satellites from time to time. Following the instructions received from the MMC, the satellite creates a spot-beam above that gateway for the duration of the assignment. After the release of the spot-beam, the corresponding resources are made available to another gateway. The process is illustrated by Figure AP6.1-8.

---

[2] As detailed in Chapter 9 (§ 9.3) of this Handbook, three general categories of allocations are relevant in the ITU frequency plans: "Planned FSS" (Appendix 30B of RR), "Unplanned FSS" (Articles 11 and 13), or "Planned BSS" (Appendices 30 and 30A).

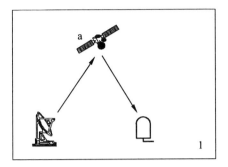

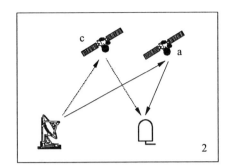

a: outgoing satellite
c: incoming satellite

Phase 1: Nominal traffic is routed through
satellite a.
Phase 2: Terminals are synchronized first on
the incoming satellite while transmitting and
receiving from satellite a. Then synchronized
terminals receive resource allocation on satellite c
and hand-off their traffic to satellite c.
Phase 3: Nominal traffic is routed through satellite c.

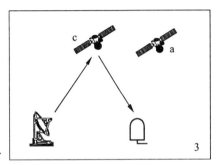

Sat/AP6.1-8

FIGURE AP6.1-8

Handover process

For handing over, a user terminal automatically follows instructions received from a signal broadcasted by its gateway, via the satellite providing coverage at the time. On the user terminal, there are two antenna beams and modems available, which permit a soft transfer from one satellite to another.

AP6.1-2.5 Transmission characteristics

Spread Spectrum is used on all communication links.

CDMA is combined with TDMA and FDMA for separating forward and return links. In the case of service links with very small, medium traffic capacity earth stations (residential users), the forward and return links may be shared by TDMA (time duplexing). FDMA (frequency duplexing) is used for higher capacity earth stations equipped with larger earth stations.

The information bit stream is encoded with FEC codes and then PSK modulated. The chosen codes lead to a reference $E_b/N_0 = 3.5$ dB. For frequency spreading, orthogonal and spreading overlay codes are employed. These transmission characteristics are summarized in Table AP6.1-7.

TABLE AP6.1-7

SkyBridge transmission characteristics

	Residential	**Pro**	**Gateways**
Modulation	QPSK/BPSK		
Coding ratio	½		
Peak forward data rate	20.48 Mbps	11 x 20.48	11 x 20.48
Peak return data rate	2.56 Mbps	11 x 2.56	Infrastructure 11 x 20.48
Uplink bandwidth	1.6 GHz		Service links 11 x 2.56
Downlink bandwidth	1.6 GHz		
Reference E_b/N_0	3.5 dB		

AP6.1-2.6 Earth segment

The earth segment of the SkyBridge system includes the gateways and the user terminals.

AP6.1-2.6.1 Gateways

As shown on Figure AP6.1-9, each gateway is composed of:

i)   A gateway access subsystem, i.e. the gateway earth station proper which provides the physical links with the user terminals. It includes the RF equipment (antenna, transmit and receive equipment) and an access system in charge of hand-over and radio resource management.

ii)   A gateway switching subsystem which includes switches for routing the traffic. It provides access to local servers and permits interfacing with broadband and narrow-band (or Internet) networks. The gateway switching subsystem is ATM based, the ATM technique being only used for transport.

iii) A service access subsystem which provides a user service access point for managing subscriber registration, user service profile, user identification, etc.

The gateways are designed for easy extension in order to accommodate traffic increases. The number of antennas implemented in a gateway may range from 2 to 6.

Typical RF characteristics of a gateway are:

•   Antenna size: 2.5 m to 4.5 m

•   G/T ~ 25 dB/K

•   e.i.r.p.: 60 to 70 dBW (for Infrastructure links ~ 51 dBW + 10 log N, N being the number of modems).

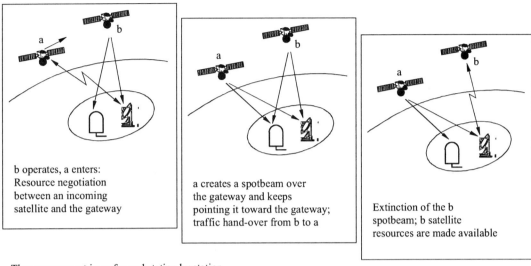

b operates, a enters:
Resource negotiation
between an incoming
satellite and the gateway

a creates a spotbeam over
the gateway and keeps
pointing it toward the gateway;
traffic hand-over from b to a

Extinction of the b
spotbeam; b satellite
resources are made available

The management is performed station by station

NGSO FSS system operating mode   Sat/AP6.1-9

FIGURE AP6.1-9

SkyBridge operation and gateway architecture

AP6.1-2.6.2 User terminals

The user terminals are very small earth stations composed of:

- An outdoor unit (OU) which provides the physical link with the dedicated gateway. It includes the antenna, the transmit and receive equipment, the modem and a control unit.

- An indoor unit (IU) for delivering and collecting the user traffic via adaptation functions and interfaces (NT: network termination) with terminals, e.g. PCs, local area networks (LAN), TV set top boxes, network computers, PBX, etc.

Different types of user terminals will be available, depending on the applications:

- Residential user terminals offering low cost and easy installation and operation for individual users.

- Professional user terminals with higher capacity.

The design of the user terminal antenna is simplified by the fact that the satellite constellation is stable with monotonously repetitive ground tracks. The position of the satellites in view is thus fully predetermined.

Table AP6.1-8 gives the main characteristics of the user terminals.

TABLE AP6.1-8

User terminal characteristics

		Residential	Professional
•	Overall size of outdoor unit (OU)	≈ 50 cm	≈ 80 cm
•	Transmitted bit rate	2.56 Mbit/s	n x 2.56 Mbit/s
•	Received bit rate	20.48 Mbit/s	m x 20.48 Mbit/s
•	Duplexing technique	Frequency	Frequency
•	e.i.r.p. (dBW)	≈ 34	≈ 44
•	G/T (dB.K^{-1})	≈ 8	≈ 14

CHAPTER 7

Earth segment

7.1 Configuration and general characteristics of earth stations

7.1.1 Configuration, block diagrams and main functions

The earth station is the transmission and reception terminal of a telecommunication link via satellite. The general configuration of an earth station is not substantially different from that of a radio-relay terminal, but the very large free-space attenuation (about 200 dB) undergone by the carrier radio waves on their path between the station and satellite (approximately 36 000 km) usually requires the main subsystems of an earth station to have a much higher performance level than those of a radio-relay terminal.

The general operational diagram of an earth station is given in Figure 7.1, from which it will be seen that the station consists of the following main subsystems:

- the antenna system;

- the receiver amplifiers (low-noise);

- the transmitter amplifiers (power);

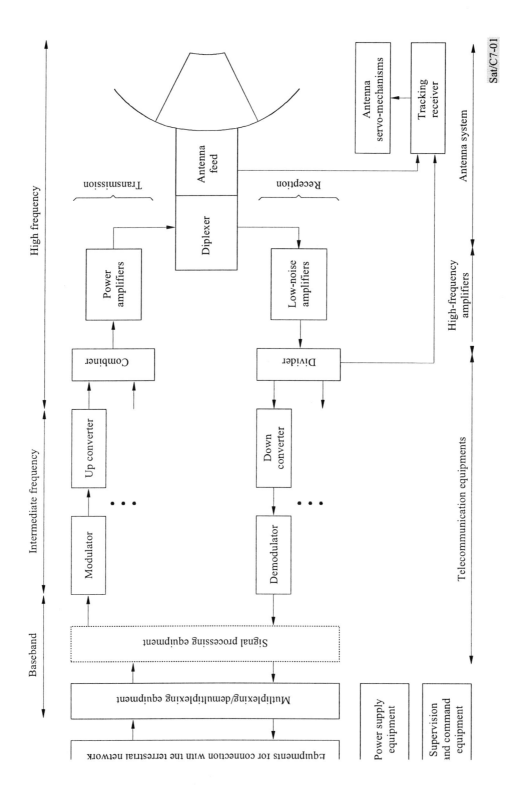

FIGURE 7.1

General operational diagram of an earth station

- the telecommunication equipment (frequency converters and modems);
- the multiplexing/demultiplexing equipment;
- the equipment for connection with the terrestrial network;
- the auxiliary equipment;
- the power-supply equipment;
- the general infrastructure.

To achieve the required availability, the practice of providing equipment redundancy is widely used. In fact, an earth station with multiple access links comprises two categories of subsystems, i.e. those which are common to all RF links and those which are specific to a particular link.

The first category includes primarily the antenna system, the receive low noise amplifiers and the transmitter power amplifiers. Low noise amplifiers and power amplifiers are usually backed up by hot, automatically switched, stand-by units. On the contrary, the antenna system (feed, tracking device, mechanical structure, drive and servo-controlled mechanism) cannot generally be provided with redundant units and great care must be applied to its design and construction in order to guarantee a very long MTTF (mean time to failures). Note also that power supply equipment is often designed for uninterruptible operation.

Some configurations of redundancy for high power amplifiers are shown in § 7.4.6.

The second category may include telecommunication equipment, multiplex units, etc. when the configuration of the station, the link distribution and also the MTTF of the units allows it, those subsystems may possibly be provided with one-for-n type redundancy, i.e. with a single stand-by unit (also possibly automatically switched) common to n (identical) active units.

The subsystems are summarized below and described in greater detail in subsequent sections.

7.1.1.1 The antenna system

The antenna, with a diameter which may vary between about 33 m and 3 m or even smaller, is the most conspicuous and often the most impressive subsystem of an earth station.

The antennas of earth stations are common to transmission and reception and must have the following performance features:

- high gain for transmission and reception, requiring reflectors which are large in relation to the wavelength, and have high efficiency;
- low level of interference (for transmission) and of sensitivity to interference (for reception), calling for radiation diagrams with low levels outside the main lobe (small side lobes);
- radiation with high polarization purity;
- for reception, low sensitivity to thermal noise due to ground radiation and various losses.

From the mechanical point of view, the radio performance calls for great structural accuracy: the surface accuracy of the main reflector must be about 1/50th of the wavelength, for example, 1 mm for the 6 GHz band, or a relative accuracy of 1/15000 in the case of the large INTELSAT

Standard A antennas of about 16 m. This accuracy must be maintained under all environmental operational conditions (e.g. with wind gusts to 20 m/s). Moreover, the antenna beam must remain pointed towards the satellite under all environmental operational conditions and irrespective of the residual movements of the satellite: for example, in the case of the INTELSAT Standard A 16 m antennas, the angular accuracy must be about 0.015° which requires an automatic tracking device to control the drive mechanism of the antenna.

The antenna system generally consists of the following[5]:

- the mechanical system, comprising the main reflector, the pedestal, the driving gear and servo-system;

- the primary source, comprising the illuminating horn, the associated mirrors (auxiliary mirrors of Cassegrain-type antennas and sometimes periscopic mirrors) and "non-radiating" components (tracking coupler, polarizers, duplexers, etc.);

- the receiver of the automatic tracking device.

Types of earth station for 6/4 GHz and 14/11-12 GHz bands are often classified only according to the size of their antennas:

- large stations: antennas dimension more than 15 m;

- medium-sized stations: antennas of approximately 15 m to 7 m;

- small stations: antennas of 7 m to 3 m or less;

- microstations for VSAT (very small aperture terminal): 4 m to 0.7 m.

It should be noted that this is a broad classification which also covers a classification in respect of the complexity of the other subsystems. The general design of an earth station should be such that the performance levels, and hence the costs of its component subsystems, should be consistent with each other.

7.1.1.2 Low noise amplifiers (LNAs)

To receive the very weak signals from a satellite, the earth-station antenna must be connected to a highly sensitive receiver, i.e. one with very low inherent thermal noise. The thermal noise of a receiver is characterized by its "noise figure" but for very low-noise receivers it is preferable to apply the concept of "noise temperature", measured in Kelvin (see Chapter 2, § 2.1.4). The basic parameter that characterizes the sensitivity of the earth station for reception is the G/T, or the ratio of the antenna gain (G) to the total noise temperature (T). The noise temperature itself is the sum of the equivalent noise temperature of the antenna (T_A) and the noise temperature of the receiver (T_R).

A low-noise amplifier (LNA) is thus always used as a microwave preamplifier in the reception chain of the earth station. It should be placed as close as possible to the diplexer of the antenna feed, to avoid the additional noise caused by losses in the waveguides. The low-noise amplifier is usually wideband: a single amplifier simultaneously amplifies all the carriers emerging from the receiver port of the antenna diplexer.

---

[5] The low noise amplifier (§ 7.1.1.2) of the receiving chain is sometimes also included in the antenna system.

A stand-by amplifier is usually provided (1 + 1 redundancy).

Until 1972, nearly all INTELSAT Standard A stations in the 4 GHz band were equipped with parametric preamplifiers cooled to 20 K by a helium gas cryogenic device operating in a closed circuit. Since then, progress in microwave circuits and components (parametric diodes, very high frequency "pumps", circulators, etc.) resulted in the development of parametric amplifiers which, even uncooled, showed performance nearly equivalent to the previous cooled ones.

However, current advances in field effect transistors using gallium arsenide (GaAs) have led to the introduction of LNAs based on simple transistor amplifiers achieving low noise temperature performance at lower cost. Such LNAs are now widely used in modern earth stations.

HEMTs (high electron mobility transistors), a type of GaAs FET with improved high frequency performance, are widely used especially for LNAs to achieve good noise temperature performance.

NOTE – The performance of low-noise amplifiers declines (i.e. the T_R increases) in the high frequency bands (e.g. 11-12 GHz and even more at 20 GHz and above). It should be borne in mind, however, that in these bands a higher equivalent antenna noise temperature (T_A) must be expected, due to the influence of atmospheric precipitation, and that this reduces the relative effect of an increase in T_R.

7.1.1.3   Power amplifiers (PAs or HPAs: high power amplifiers)

The order of magnitude of the power required at the transmitter output is 1 W or less per telephone channel and 1 kW per television carrier.

The two main types of microwave tube used in earth-station power amplifiers are travelling wave tubes (TWT) and klystrons. Furthermore, in the case of small stations, solid-state power amplifiers are more and more used.

a)   Travelling wave tube (TWT) amplifiers

The travelling wave tube is intrinsically a wideband amplifier covering the entire usable band of the satellite (500 MHz or more) with the necessary uniformity of gain and group delay. Because of these features, the TWT appears to be the ideal power amplifier for earth stations, since it allows several telephone carriers to be transmitted simultaneously with a single tube, irrespective of the repeaters and the frequencies allocated to these carriers.

It should be noted, however, that the simultaneous transmission of several carriers in the same tube produces intermodulation noise components which increase as the operating point of the tube approaches saturation. Since the maximum level of intermodulation components is subject to specification, the degree of linearity of the tube parameter at the operating point is stipulated in each configuration. This leads to a back-off of the operating point in relation to saturation and hence to a loss of available power. This loss during multi-carrier operation can now be partly offset by using new equalizing devices known as linearizers.

b)   Klystron amplifiers

Klystrons are essentially narrow passband devices: about 40 MHz for 6 GHz klystrons, and 80 MHz for 14 GHz klystrons and sometimes more. These bandwidths suffice for traditional frequency modulated (FDMA mode) carriers, but may be inadequate for carriers with phase modulation and digital coding (TDMA mode). In any case, the choice of the klystron as the type of tube to be used generally entails the use of an amplifier for each of the carriers transmitted except for SCPC (and even possibly for SCPC and TV in certain domestic systems). Klystrons may be fitted with a mechanical (remote control) tuning device whereby they can be adjusted to the operational centre frequency (or to the centre frequency of the repeater) so that this frequency can easily be changed.

Despite the drawbacks of a narrow bandwidth, klystron power amplifiers are generally more economical than TWT amplifiers and have the following advantages:

- high efficiency (e.g. 39%);

- very simple power supply (heating circuit and anode circuit, focusing by a permanent magnet);

- great sturdiness and long working life (30 000 to 40 000 h);

- possibility of operating with reduced power consumption (for reduced HF power).

Klystron power amplifiers will generally be used in preference to TWT amplifiers when a station has to transmit only a small number of FDMA carriers. They are particularly well adapted to the transmission of television carriers. It must also be noted that klystron tubes are not available for the lower power ranges (e.g. lower than 700 W).

In the case of large stations, it may be necessary to use several power amplifiers, either klystrons or TWTs, for the following reasons:

- in the case of klystron amplifiers, there should generally be as many amplifiers (n) as transmitted carriers;

- in the case of TWT amplifiers, the output power of a single amplifier may be insufficient;

- one or more (m) stand-by amplifiers are usually provided to ensure the necessary availability. Redundancy is often of the 1 + 1 type in the case of TWT amplifiers and of the n + m type in the case of klystron amplifiers.

These multiple amplifiers are connected to the transmit port of the antenna diplexer through a system comprising switches (for redundancy) and an output combiner.

c)   Solid-state amplifiers

In the case of small low-capacity stations, a solid-state amplifier, normally with field effect transistors, may suffice.

Because of current advances in field effect transistors using gallium arsenide (GaAs) and advanced circuit technology, recent power level of solid-state amplifiers reached 100 W in 6 GHz band and 20 W in 14 GHz band.

Solid-state power amplifiers are very reliable and economical and may provide an ideal solution for small earth stations.

7.1.1.4 Telecommunication equipment

The term telecommunication equipment usually refers to the equipment which modulates the very high-frequency carrier with low-frequency signals (baseband) for emission and extracts (demodulates) these low-frequency signals on reception.

The low-frequency signals may be analogue telephone signals (usually multiplexed), digital signals, audio and video (television) signals, etc.

Telecommunication equipment comprises frequency converter equipment, modulating and demodulating equipment and signal-processing equipment.

7.1.1.4.1 Frequency converter equipment

Up converters change intermediate frequency (IF) signals (e.g. IF of 70 MHz, 140 MHz, 1 GHz, etc.) from the modulator into radio-frequency signals (e.g. in the 6 GHz or 14 GHz band). These signals are then amplified by the power amplifier before being transmitted through the antenna.

Down converters change the radio-frequency signals (e.g. 4 GHz or 11 GHz) received by the antenna and pre-amplified by the low-noise amplifier into intermediate frequency signals. These signals are then translated to the baseband in the demodulator.

7.1.1.4.2 Modulating and demodulating equipment (MODEM)

This equipment superimposes the audio-frequency signals on the IF carrier (modulators) or extracts them from the IF carrier (demodulators). In analogue transmission, frequency modulation is the normal process, whereas for digital transmission phase modulation (PSK) is used, frequently four-phase (4-PSK). Modulation with two phase conditions, called bi-phase modulation (2-PSK), or with eight phase conditions or more is also used. Recently the trellis coded modulation (TCM), a technique of jointly optimizing modulation and coding is also used.

A transmission chain (i.e. a modulator and converter) is required for each carrier. Redundancy is frequently of the 1 + 1 type (one stand-by for each chain in service). When the high-frequency carriers (produced by the converters) are amplified in a common power amplifier, they are added in the input combiner of the power amplifier subsystem.

Similarly, each carrier received has its own reception chain (i.e. a converter and a demodulator). For frequency division multiple access (FDMA) a station generally receives more telephone carriers than it transmits, since each carrier emitted is intended for several stations (multi-destination carriers). For reception, each of the carriers transmitted by the corresponding stations must be received in order to extract (at baseband) the signals intended for the station considered. The reception chain redundancy of the type n + m (m stand-by chains for n chains in service), must be calculated according to the desired availability. A divider located at the output of the common low-noise amplifier distributes the received high-frequency carriers among the n chains.

7.1.1.4.3   Signal-processing equipment

For digital transmissions using time-division multiplex (TDM), and in particular time-division multiple access (TDMA), signal-processing equipment is required, as shown in Figure 7.1. The functions of this equipment include:

i)   Formatting of the digital data on transmission, this equipment adapts the continuous input stream of digital data bits for transmission via satellite through the modulator. This usually means that the data is inserted into a frame and in TDMA, converted (by means of buffer stores) into a very fast stream of bits in short bursts included in the frame. The station may thus transmit multi-destination bursts in the same way as a multi-destination carrier is transmitted in FDMA.

Changing the format of the data requires the insertion of additional bits in the frame and in the bursts for synchronization, addressing, etc.

On reception the reverse process recreates the continuous data streams at the output baseband interfaces. In TDMA the signal-processing equipment receives the bit streams from the demodulator in the form of bursts from the various corresponding stations.

ii)   Synchronization on transmission and reception, positioning of bursts in the frame (on transmission), recovery of bursts (on reception).

iii) Encoding/decoding operations may be necessary to modify the bit stream for transmission via satellite, e.g. for low bit rate encoding (LRE) and/or forward error correction (FEC).

iv)   Various additional data-processing operations to improve transmission and make it more reliable.

For TDMA transmissions the modulator and demodulator units (MODEM) are frequently physically incorporated in the signal-processing equipment, and constitute part of the common equipment of the "TDMA terminal".

For digital video transmission, video compression/decompression operations using MPEG standards are also used.

7.1.1.5   Multiplex/demultiplex equipment

a)   Analogue transmissions (telephony)

Even when all the transmissions are analogue, i.e. where the transmission via satellite is analogue (FDMA-FDM) and the interface with the terrestrial network is also analogue, it is still almost always necessary to modify the distribution of the telephone channels (groups, in particular) within the baseband multiplexers. Prior to transmission, the multiplexed telephone signals from the interface with the terrestrial network are rearranged to form the baseband(s) which are to modulate the multi-destination carrier(s) according to the standard adopted for transmission via satellite. On reception, the signals from the various demodulators (which come from the carriers transmitted by the corresponding stations) are filtered in order to extract only those telephone channels (or more generally the groups and supergroups) intended for the station involved and which are then combined according to the standard multiplex arrangements for terrestrial transmission.

It can be seen that satellite telecommunications are characterized by an asymmetric structure between the transmission and reception multiplex divisions. Provision therefore has to be made in the earth station for special multiplex/demultiplex equipment, which constitutes the interface between satellite transmission and connection to the terrestrial network.

It should be noted that this equipment also carries out the separate function of multiplexing/demultiplexing the service channels (which are normally transmitted in the 4-12 kHz sub-basebands).

b)   Analogue transmissions (television)

For television, multiplex/demultiplex units are commonly used to insert and extract the sound (audio) channel(s) which may be transmitted on a subcarrier by frequency modulation at the same time as the video signal.

c)   Digital transmission

For digital transmission via satellite, the telephone channels to be transmitted, or more frequently the standard multiplexes from the terrestrial transmissions (e.g. 30 channel PCM), are re-combined and rearranged to form the bit streams to be transmitted by the station (in particular, after grouping in bursts, for TDMA transmission). On reception the reverse process is used to extract the bit streams intended for the station (from the bursts from the corresponding stations in the case of TDMA transmission).

NOTE 1 – In digital transmission the borderline between the functions performed by the multiplex/demultiplex equipment, and the signal-processing equipment described above (§ 7.1.1.4), is not always clear. In particular:

•   in the case of continuous transmissions, some of the signal-processing functions listed in § 7.1.1.4.3 may be assigned to the multiplex/demultiplex equipment. This applies in particular to i) formatting and ii) synchronization, both of which are considerably simplified for continuous mode transmissions;

•   for burst mode transmissions, i.e. in TDMA, the interface between the signal-processing equipment and the multiplex/demultiplex equipment may be in either burst mode or continuous mode.

NOTE 2 – The multiplex/demultiplex equipment may perform an important additional function, namely, digital speech interpolation (DSI). This function ensures that the inactive periods in both directions of a duplex telephone call may be used profitably to combine the bits from the telephone channels of a multiplex and thereby increase the capacity of the transmission channel (in a ratio of 2 to 2.5).

When combined with LRE, the traffic capacity can be even further increased: such DSI/LRE equipment is called digital circuit multiplication equipment (DCME).

NOTE 3 – For TDMA transmissions another important additional function may be performed, i.e. rapid rearrangement of the bursts within the frame. This is a demand assignment function which makes it possible at any moment to adapt the transmission channels to real-time requirements.

7.1.1.6   Equipment for connection to the terrestrial network

For telephony, the earth station is usually connected to the terrestrial network through a switching centre. This may be a transit centre in the case of international stations and large or medium-sized stations in national systems, or possibly a subscriber exchange in the case of small local stations within national networks.

The specific equipment usually required for such a connection is:

* the terrestrial link between the earth station and the switching centre. It may use a coaxial cable, although more often than not the geographical conditions make it necessary to use a radio-relay link.

    NOTE – For small stations in national networks the station and the switching centre may be located at the same site;

* echo cancellers and various peripheral signalling equipment.

For television the earth station is connected:

* to the programme studios for the transmission function;

* to local broadcasting transmitters for the reception function.

The connection is usually done by radio-relay. For small stations the (receiver) station is often connected directly to a local television distribution network.

7.1.1.7   Auxiliary equipment

The auxiliary equipment of an earth station includes:

7.1.1.7.1   Supervisory and command equipment

The supervisory and command equipment comprises:

* alarm signals from the station subsystems;

* controls, sometimes automatic, for switching (normal/stand-by) of spare equipment;

* controls for operation of the subsystems;

* analogue information for supervising the operation of the subsystems; and

* sometimes, equipment for storing and/or recording the station's most important operating parameters.

The alarm signals and the controls are frequently grouped on a display panel showing the main functions of the station (mimic diagram).

The supervisory and monitoring equipment has often, in the past, been installed in operator consoles. The current trend is to replace these bulky consoles by interactive microprocessor based terminals with keyboards and visual display screens. It should be noted that for this type of supervision and monitoring each item of the station equipment must be associated with a command/data capture interface, which may be incorporated in it.

7.1.1.7.2 Measuring equipment

Provision is usually made in the station equipment for a set of specific measuring apparatus (generally commercially available) which is used to adjust certain equipment items, to check performance, and for maintenance and certain repair operations. Some of the test equipment may be operated automatically to ensure systematic routine maintenance. It may also be incorporated in the supervisory and control equipment referred to in § 7.1.1.7.1 above, in which case it should have a standard command/data interface (which usually complies with standard GP/IB-IEEE 488).

7.1.1.7.3 Service channel equipment

In most telecommunication networks in the fixed-satellite service, service telephone and telegraph (telex) channels are associated with each telephone channel multiplex transmitted by an earth station. After multiplexing/demultiplexing (see § 7.1.1.5) the service channels are used directly to establish connections between the various stations, between the stations and the system control centre and between a given station and the switching centre associated with it. To facilitate the establishment of such connections, specialized automatic telephone and telegraph switching equipment is normally provided in each station. This equipment is generally designed to provide modern switching functions (simultaneous and conference calls, transfers, parking, etc.).

7.1.1.8 Power supply equipment

The satisfactory operation and service continuity of an earth station depends on the correct design of its electric power supply. There are two main sources of power:

- the main power supply, with stand-by capability;

- the uninterrupted power supply (UPS).

In addition, an auxiliary low voltage (24 V or 48 V) d.c. source may be required to supply certain automatic equipment. The main supply distribution network is via the station transformer unit. It is backed up by an independent generating set (or better still by two sets with 1 + 1 redundancy) driven by a fast-start (5 to 10 s) diesel engine. This generator, which for large stations would have a power of usually about 250 kVA, supplies the whole station, including the antenna motors, lighting and air-conditioning. Maintaining the stand-by generator and keeping a stock of diesel fuel is one of the basic tasks in the management of the station.

The purpose of the uninterrupted power supply, which receives its primary energy from the main power supply, is to provide a constant high-quality power supply (stable voltage and frequency with no significant transients), while the stand-by sets are starting up following a power cut in the distribution network. This source supplies all the electronic equipment. For a large station the power would be roughly from 50 to 100 kVA, most of which (80% to 90%) is required for the high-power amplifiers.

The three most common uninterrupted power supply systems are:

- alternator motors with an inertia flywheel. The flywheel provides reserve mechanical energy which continues to drive the alternator while the diesel motors are starting up;

- alternator converter systems. In this case the reserve energy is provided by accumulator batteries. The batteries, which are kept charged by the main power supply (via rectifiers), feed an alternator which constitutes the uninterrupted source;

- static converter systems in which the alternator mentioned above is replaced by a solid-state a.c. generating unit using thyristor bridges. This is the most commonly used system at present.

The converter systems described above require a large set of electric batteries. The size of the unit depends on the installed power of the uninterrupted source but also, and most importantly, on the duration of autonomy required (permissible duration of interruption of the general source). This duration usually ranges from a few minutes to half an hour.

Clearly, if there is no adequate and reliable local electricity network, the power supply equipment may be designed quite differently. It might, for instance, be based entirely on diesel generating sets with built-in redundancy and switching systems guaranteeing continuous operation.

It is often the case for small stations that no electricity supply is available and that the station must operate without technical staff. Here the engineer must endeavour to design low consumption equipment, if possible completely solid state, which can operate without ancillaries (air conditioners, etc.) in all local environmental conditions. Power supply systems should be provided that require the least possible maintenance and replenishment (fuel, etc.). Solar power units are particularly suitable and can generally be used for power consumption not exceeding 500 W.

7.1.1.9   General infrastructure

The general infrastructure of an earth station includes all premises, buildings and civil engineering works. Its size obviously depends on the type of station. For "large stations" (see § 7.1.1.1) and particularly the INTELSAT Standard A stations, there are two possible types of construction:

- a station with a single antenna. Here all the equipment may be installed in one building located under the antenna (apart from the power supply equipment which for sound-proofing reasons is usually housed in its own building). In this way the general infrastructure is particularly compact and economical;

- a station with several antennas. Each antenna is erected on a separate building which houses the equipment directly associated with it (low-noise amplifiers, tracking receiver, power amplifier and sometimes frequency converters). A central operations building, common to all the antennas, contains the actual telecommunication and operating equipment, some of which may be shared (e.g. supervisory and command equipment) or multi-purpose (e.g. the transmission and reception chains may be assigned to the various antennas according to requirements). Microwave (waveguide) or intermediate frequency (coaxial cable) links, called inter-facility links, are used to connect the equipment in the antenna buildings to the equipment in the main building. The general infrastructure of a large multi-antenna station can account for a high proportion of the total cost of the centre: between 20% and 50%, or even more if additional installations and premises are necessary, as is frequently the case (conference rooms, staff accommodation, etc.). Construction costs should also cover the general preparation of the site and access (roads, etc.), supply networks (electricity, air conditioning, fluids, etc.) and miscellaneous fittings (grounding, internal communications, fire prevention, drainage, etc.).

For medium-sized stations, which require less equipment and consume less power than large stations, the infrastructure can be designed more simply and economically. It may be advantageous for ready-wired prefabricated buildings to be delivered to the site.

Finally, small stations may be designed as compact units with all the equipment assembled and installed at the factory in a single shelter or box which can be transported, installed and brought into service directly after minimum preparation of the site (antenna foundations in particular).

7.1.2 Main earth stations

7.1.2.1 Stations for international operation

Although INTELSAT is no longer the only global international operator, INTELSAT standards are described below as typical examples of international earth stations.

To introduce a new earth station into the "INTELSAT global system", i.e. for international[2] traffic, the concerned administration should refer to an INTELSAT general document entitled "Procedures governing application, approval, verification and operation of earth stations in the INTELSAT system". Seven standard types of earth stations are normally admitted for operation in the "INTELSAT global system", though other types ("non-standard") may be taken into consideration (at least for provisional operation) on a case-by-case basis. INTELSAT's earth stations technical specifications (IESSs) define these seven types of stations as Standards A, B, C, D, E, F, and G.

- Standard A stations are the most commonly used. They operate in the 6/4 GHz band and are equipped with a big antenna (larger than 15 m diameter), with very low noise receiving amplifiers and with high power transmit amplifiers. They can handle any kind of traffic (multiplexed telephony, data, TV programmes, etc.) and can be easily adapted to any increase or modification in the traffic configuration. The main features of the INTELSAT Standard A stations are summarized in Table AP7.1-1 (see Appendix 7.1).

- Standard B stations also operate in the 6/4 GHz band. They are equipped with medium sized antennas (about 11 m) and with rather simple receive and transmit communication chains. Due to constraints put on their transmission modes (SCPC or companded FDM-FM telephony), to their restricted TV receive and transmit capabilities and to higher space segment charges, they are usually only cost-effective when limited to small or medium traffic capacities (e.g. less than 60 telephone circuits). The main features of INTELSAT Standard B stations are summarized in Table AP7.1-1 (see Appendix 7.1).

- Standard C stations operate in the 14/11 GHz band with antennas of about 11 m diameter and are specially intended for high capacity message transmission[3] . Their main features are summarized in Table AP7.1-3 (see Appendix 7.1).

---

[2] Under INTELSAT's regulations, certain national communication links may be included in the global system. This is the case for territories which are separated from the mainland by geographical obstacles such as oceans, mountains, etc.

[3] Through cross-strapped connections in the satellite transponders, communications can be established between Standard C and Standard A stations.

- Standard D stations operate in the 6/4 GHz band and are specifically designed for use with INTELSAT's VISTA service to provide basic satellite service to rural and remote communities. The Standard D-1 is a low-cost, small (5 m) antenna providing one to four voice grade channels. In the interest of reduced cost the Standard D-1 performance requirements have been eased to allow polarization axial ratios of up to 1.3 dB. The Standard D-2 earth station is similar in performance to the Standard B earth station. Their main features are summarized in Table AP7.1-1 (see Appendix 7.1).

- Standard E stations operate in the 14/11 GHz or 14/12 GHz bands and range in size from 3.5 m to 10 m and are specifically designed for use with the totally digital INTELSAT Business Services (IBS), to provide integrated service networks for international and domestic business service applications. Their main features are summarized in Table AP7.1-3 (see Appendix 7.1).

- Standard F stations operate in the 6/4 GHz band and range in size from 5 to 10 m and are designed for use with the totally digital IBS to provide integrated service networks for international and domestic business service applications. Their main features are summarized in Table AP7.1-3 (see Appendix 7.1).

- Standard G stations operate in the 6/4, 14/11 or 14/12 GHz bands and include a wide range of international earth-station antenna sizes with minimum constraints on the earth-station owner. The performance characteristics do not include the following parameters:

  - maximum e.i.r.p. per carrier;

  - modulation method;

  - G/T;

  - transmit gain; and

  - channel quality.

The earth-station owner/user has great flexibility and freedom in deciding upon the best transmission method for his requirements. He then submits his transmission plan to INTELSAT for review and approval, to ensure that the operation of his station will not cause harmful interference to other users on the INTELSAT system, or to other satellite systems.

NOTE – There are some other operational international networks besides INTELSAT.

In 1988 PANAMSAT started its international satellite network operation as the first international satellite network operated by a private company. After PANAMSAT, some private companies also started their international satellite networks. At present COLUMBIASAT, ORIONSAT, PANAMSAT and RIMSAT are operational international networks besides INTELSAT.

7.1.2.2   Stations for regional or domestic systems

A number of earth-station types are available for regional or domestic applications. The selection of a specific type depends on the general system operation and on satellite communications payload performance characteristics. These stations which use medium-size antennas may be categorized according to the following guidelines:

i)   Stations operating in connection with leased 6/4 GHz space segment (transponders) on INTELSAT satellites. These stations are generally in compliance with Standard B, with the following characteristics:

- the antenna diameter may vary typically from about 5 m to 15 m;

- the communication modes (modulation and multiplexing methods) may be different and are usually chosen in order to optimize the overall system operation. In particular, telephony, which was previously transmitted in analogue modulation (SCPC-FM or FDM-FM, generally with companding), is now generally implemented in digital modulation (e.g. in SCPC-PSK or TDM-PSK, often with low bit rate encoding).

Specific optimization of the transmission parameters and of the link budget makes it possible to implement medium size, economic earth stations while transmitting rather high traffic capacity.

The earth-station owner/user has great flexibility and freedom in deciding upon the best transmission design method for his requirements. He then submits his transmission plan to INTELSAT for review and approval to ensure that the operation of his station will not cause harmful interference to other users of the INTELSAT system or to other satellite systems.

ii)   Stations operating at 6/4 GHz in the framework of dedicated satellite systems – such as Indonesia's PALAPA, ARABSAT, etc.: these stations are again often similar to the INTELSAT Standard B stations. This is due to the fact that the limited required earth coverage allows for high transponder e.i.r.p.s through directive satellite antennas. Therefore high traffic capacities can be accommodated with rather simple earth stations, equipped with medium-sized antennas.

iii)   14/11 GHz stations: the 14/11 GHz (and the 14/12 GHz) bands are increasingly used in regional and domestic satellite systems.

The EUTELSAT system described in Annex 3 is an example of a regional system exclusively based on these bands.

There are different categories of earth stations depending on the applications:

- EUTELSAT standard TDMA/TV earth stations used for high capacity telephony and for high quality TV programme exchange.

- EUTELSAT SCPC/SMS earth stations, used for Business Services with SCPC access mode to the satellite. Three different earth station standards are defined with dish diameters ranging from about 2.4 to 5.5 m. The technical characteristics of these stations are given in Table AP7.1-4 of Appendix 7.1.

- EUTELSAT TDMA/SMS earth stations, used for Business Services with TDMA access mode to the satellite. These stations generally have dishes of about 3.5 m diameter.

- Non-standard earth stations used for other types of applications, established by administrations that are members of the EUTELSAT organization, over leased transponders. The earth station requirements for these applications are determined by the providers of the services within a set of EUTELSAT guidelines that protect other users.

7.1.2.3   Hub earth stations for VSAT (very small aperture terminal) systems

VSAT networks are generally designed according to a star architecture in which a central earth station, usually called the "hub", is linked to a large number of geographically dispersed remote VSATs (see below § 7.1.3.2). In most applications the hub is connected, possibly through terrestrial lines, to a host computer.

The general design of the hub station is very similar to that of a classical earth station as regards the RF/IF equipment, the main differences being in the digital processing and baseband equipment. In fact, the RF equipment of an existing suitable earth station (including its antenna) may be used in order to limit investment costs.

The block diagram of the hub station is shown in Figure 7.2.

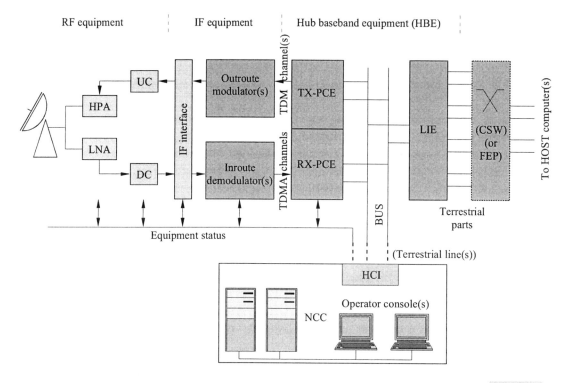

Sat/C7-02m

HPA:	High power amplifier	LIE:	Line interface equipment
LNA:	Low noise amplifier	BUS:	Traffic and utility bus
UC:	Up-converter	HCI:	Hub control interface
DC:	Down-converter	NCC:	Network control centre
PCE:	Processing and control equipment	(FEP:	Front end processor)
	(TX: transmit, RX: receive)		

FIGURE 7.2

Hub simplified block diagram

The main functional elements of the hub station are as follows:

- the RF equipment, note that, in the case of shared hub networks this subsystem is common to the various subnetworks;

- the IF equipment, including the transmit outroute modulator(s) and the receive reroute demodulators;

- the baseband equipment, which may comprise:

  - the transmit and receive processing and control equipment (TX and RX PCEs);

  - the terrestrial line interface equipment (LIE);

  - the hub traffic and utility bus;

  - the network control centre (NCC) and its associated operator console(s).

    Note that this configuration may differ depending on each particular system or manufacturer. Also, redundancy equipment is not represented in the figure.

Data to be transmitted to the remote VSATs incoming via terrestrial lines (e.g. from host computer(s)) enter the hub through the LIE and are routed to the TX PCE, then to the modulator.

In the reverse direction, data received from the VSATs pass through the demodulators and the RX PCE before being sent to the users applications (e.g. host) through the LIE. The complete VSAT network operation is controlled and monitored by the operator console(s) associated with the NCC.

For more details on VSAT systems, refer to "VSAT Systems and Earth Stations (Supplement 3 to the Handbook)".

7.1.2.4   Other main earth stations

7.1.2.4.1   Broadcasting-satellite service (BSS) feeder-link stations

The frequency band allocated for the feeder links of the BSS is either 14.0 to 14.5 GHz or 17.3 to 18.1 GHz except in a few countries where both bands are allocated. Countries using the 14.0 GHz band are likely to use stations for BSS feeder links that are similar to those used for the FSS in the same frequency band. A typical BSS feeder-link station might have the following parameters:

- frequency range: 14.0 to 14.5 GHz. In addition the station will have facilities for monitoring the broadcast downlinks in the 11.7 to 12.3 (or 12.5) GHz band;

- antenna diameter: 5-8 m;

- transmit amplifier: 1-2 kW klystron HPA for each broadcast carrier.

An important feature of BSS feeder-link stations is the provision of a facility for uplink power control (UPC). This is used to maintain a constant power flux-density at the satellite so that the re-transmitted picture quality is not affected during periods of precipitation. Alternatively a back-up station located at a suitable distance from the main earth station could be used to provide space diversity.

7.1.2.4.2   Mobile-satellite service (MSS) earth stations

The satellites of the INMARSAT system currently provide a range of communications services (voice, telex, fax and data) to different users using a variety of terminals and applications (aeronautical, maritime or land). There are two fundamental types of user earth stations in the INMARSAT system that carry traffic, namely, the land earth stations (LES) – sometimes also referred to as coast earth stations (CES) – operating in the 6/4 GHz band, and the mobile earth station (MES) operating in the 1.6/1.5 GHz band.

The FSS feeder-link gateway to the mobile earth stations is via the INMARSAT land earth stations. As of November 1998, there were about 40 LESs distributed around the globe, with at least one in every continent. A land earth station need not necessarily be located on a "coast" but it does need to be located within the coverage beam of one or more INMARSAT satellites. The INMARSAT antenna beams are designed to cover the three major ocean regions.

Land earth stations are owned independently by telecommunications operators. An LES operator is often, but not always, the signatory (the organization nominated by its government to invest in and work with INMARSAT) of the country in which the LES is located.

The parameters of typical INMARSAT earth stations are given below:

Receive system performance

Parameter	C-band (3.6 GHz)	L-band (1.5 GHz)
Receive gain	49.2 dBi	$\leq$29 dBi
System noise temperature	71 K	$\leq$501 K
G/T at 5° elevation	30.7 dB(K^{-1})	$\leq$2 dB(K^{-1})

Transmit system performance
(for one telephone voice channel and Mini-M antenna)

Parameter	C-band (6.4 GHz)	L-band (1.6 GHz)
Total e.i.r.p. at edge of coverage	59.5 dBW	17 dBW
Antenna transmit gain	54 dBi	8 dBi
Transmitter carrier power	5.5 dBW	9 dBW
Transmitter harmonics	<60 dBc/4 kHz	<60 dBc/4 kHz

7.1.3   Small earth stations

7.1.3.1   General

As previously explained (§ 7.1.1.1), earth stations are often classified by the dimension of their antenna. However, the term "small earth station" should be taken in a broader sense and should include a wide variety of earth stations implemented in the FSS (at 6/4 GHz, 14/10-12 GHz, 30/20 GHz) in the framework of GSO or non-GSO satellite systems.

In fact, this sector of the earth segment is, at the moment, in very rapid evolution in terms of technical progress and market evolution.

Small earth stations will be characterized here by their applications and by their technical features.

i)   Applications

In terms of applications and services, small earth stations can be characterized by their proximity to the users. In consequence, they are generally used as access communication systems, by permitting connection to public or private communication networks or to shared information or computer means. The connection can be through a local distribution or switching system (PBX, local area network, etc.) or can be direct to the user. In the latter case, the market trends are towards very small and low-cost stations ("microstations"), often simply called "terminals".

The utilization of small stations generally fall under two main categories:

*   remote area communications systems (rural telecommunications and services to isolated sites such as off-shore platforms, pipelines, mines, etc.);

*   business communications (corporate networks), generally in the framework of closed users communities. The so-called VSATs (very small aperture terminals), described below (§ 7.1.3.2) enter in that category.

Some important applications are listed below:

*   telephony;

*   fax;

*   data transmission;

*   TV reception for local distribution (e.g. cable TV) or re-broadcasting;

*   access to communication networks (Internet, Intranet), including electronic mail (Email), data distribution and access to data banks, etc.;

*   computer load sharing, distributed data processing, etc.;

*   teleconference and videoconference.

ii)   Technical features

Small earth stations have in common some or all of the following technical characteristics:

*   space segment: the utilization of small earth stations involves a high level of power received from the satellite. In the case of a GSO satellite, this can be achieved thanks to a high e.i.r.p. GSO transponder.

*   antenna diameter: indicatively, small earth station antenna diameters are less than 7 m (and generally less than 5 m), 2.5 m and 1 m respectively in the 6/4 GHz, 14/10-12 GHz and 30/20 GHz frequency bands.

*   antenna pointing: in most cases no tracking is required for GSO systems but can be needed in the case of non-GSO systems.

- transmit amplifier: about 500 W (RF tube) to 1 W (solid state).

- multiple access and modulation. The most common modes are:

  - analogue FDMA-SCPC (companded FM, see § 7.5.4). However this mode is practically disappearing in current offerings, being replaced by digital modulation and coding.

  - FDMA-SCPC-PSK (see § 7.6.2.1).

  - FDMA-TDM-PSK (MCPC, see § 7.6.2.2), with bit rates indicatively from 64 kbit/s to 45 Mbit/s.

  - TDMA-TDM-PSK (FDMA-TDMA) with medium (e.g. 25 Mbit/s) or even low (e.g. 256 kbit/s) bit rates (see § 5.3.4 and § 7.6.3.3).

  - Random Access TDMA (RA-TDMA, see § 5.3.5), often used in VSAT systems (see below § 7.1.3.2) for low bit rate return ("outroute") carriers.

  - CDMA-PSK, now proposed, especially for non-GSO systems, either in the MSS or in the FSS (see § 5.4 and Appendix 2.1, § 2).

- voice coding: digital transmission with low bit rate encoding (LRE) is now commonly implemented in small station systems for telephony (see § 3.3.2 and Appendix 3.1).

- demand assignment: demand assignment (see § 5.5) is often implemented in small earth station systems, especially in the case of thin route (e.g. rural) telephony.

- television: small earth stations are often equipped for the reception of analogue or digital television programmes.

So as to make better use of the orbit-spectrum resource and to protect other systems, attention needs to be paid to the reduction of side-lobe radiation of small station antennas. Recommendations have already been issued by the ITU-R (case of VSATs, see below § 7.1.3.2) and studies remain in course on the subject.

As concerns the earth station construction, small earth stations are characterized:

- by a compact equipment assembly, all equipment being generally contained, either in a shelter near the antenna, or even, as explained below, in two "boxes", the first one (the "outdoor unit" or ODU comprising the main parts of the radio equipment) being located in the antenna system and the other one (the "indoor unit" or IDU comprising the signal processing and the interfacing equipment) being inside the user premises. Ultimately, in the case of very small terminals, the complete equipment could form a single unit;

- by an unmanned operation (permitted by remote control and monitoring from central sites);

- by a modular architecture for allowing flexibility in operation and for facilitating the maintenance.

Note that these modern features, as well as the decrease in cost, are already, and will be more and more, permitted by the utilization of advanced technologies (ASIC, etc.) and of an intensive utilization of digital signal processing.

7.1.3.2 VSATs (very small aperture terminals)

The term "very small aperture terminal", or simply "VSAT" was introduced in the eighties to designate small earth stations generally implemented in the framework of "VSAT systems" (or "VSAT networks") used for private corporate communications. In fact, if the term remains widely used, although it does not cover any precise definition, this is because it has marked the success of a really new market with the introduction of low cost direct connections, by satellite, of multiple users to central communication or computing facilities and of the production, by quantities (tens of thousands), of the small earth stations needed for these utilizations.

The response of the ITU to this generalization of such earth stations and systems has been two-fold:

- Recommendations were specially issued for the earth stations, covering generalities (ITU-R S.725), spurious emissions (ITU-R S.726), cross-polarization isolation (ITU-R S.727), off-axis e.i.r.p. (ITU-R S.728), control and monitoring function (ITU-R S.729), interconnection with public switched data networks (ITU-T X.361) and with public ISDN (ITU-T I.571)[4];

- a special Handbook was published on the subject (Supplement 3 to the main Handbook: "VSAT systems and earth stations", Geneva, 1995).

    Note that only summary information is given below on VSATs and that the reader is referred to the above-mentioned Supplement for detailed technical descriptions on VSATs;

- VSAT earth stations are usually implemented to form closed networks for dedicated applications either for information broadcasting (receive only VSATs) or for information exchange (transmit/receive VSATs);

- VSAT (remote) earth stations are generally directly installed, on users' premises and unattended. Their location density may be high;

- VSAT earth stations are often part of a network which has a "star" topology, consisting of a relatively large central station (hub station: see above § 7.1.2.3), and many VSAT (remote) earth stations. However, some networks operate in a point-to-point- or "mesh"- configuration, without a hub;

- VSAT earth stations usually employ digital transmission with a low or medium bit rate (≤ 2 Mbit/s); and

- VSAT (remote) earth stations are equipped with small antennas: the antenna diameters are normally limited to 2.4 m, however, in some circumstances, large diameter up to 5 m may be required.

Note that:

1) the most commonly used FSS bands are 14/11-12 GHz and 6/4 GHz.

2) the coding modulation and access techniques could be very diverse, corresponding to the most effective technologies for a given application.

3) TVRO earth stations are not classified as VSATs. However, the reception of video signals by a VSAT is often implemented.

---

[4] Specifications on VSAT operation were also issued in the USA, in Europe, in Japan, etc.

7.1.3.2 a)   General construction and installation of VSAT

The VSAT earth station is functionally divided into three major elements: the antenna, outdoor unit (ODU) and indoor unit (IDU). The typical configuration of a VSAT is shown in Figure 7.3. All three components are compact and designed for low cost mass production. This section mainly describes the configuration of VSATs for 14/11-12 GHz band operation. The configuration for 6/4 GHz band operation is similar except for the antenna and RF circuit.

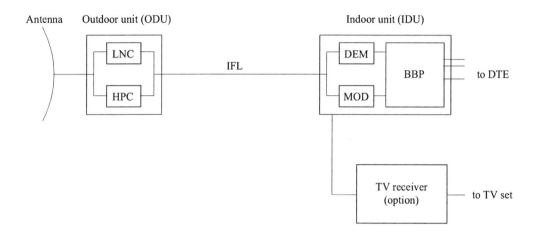

LNC: Low noise amplifier and converter
HPC: High power amplifier and converter
IFL: Inter-facility link
DEM: Demodulator
MOD: Modulator
BBP: Baseband processor
DTE: Data terminal equipment

Sat/C7-03

FIGURE 7.3

Typical configuration of a VSAT

Small offset parabolic antennas with typically 1 to 2 metre diameters are widely used. The ODU typically contains RF electronics such as a low noise converter (a low noise amplifier with a down converter) and a high power converter (a high power amplifier with an up converter) in a compact weatherproof housing with an integrated antenna feed horn, and is installed behind the antenna focal point. The antenna with the ODU can be installed very easily on the roof top, on the wall, or in the car park of the user's office buildings, where the user data terminals are located. The IDU typically contains an IF circuit, a modem and a baseband signal processor. Sometimes the modulator circuits are contained in the ODU instead of in the IDU.

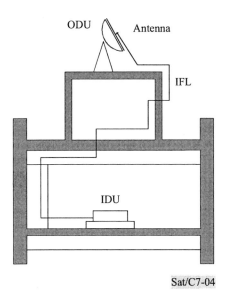

FIGURE 7.4

Typical installation of a VSAT

The IDU is usually installed near the user data terminals and connected to them directly through the standard data communication interface. Usually an optional TV receiver can be connected to the IDU to receive TV signals transmitted by another transponder of the same satellite. The ODU and the IDU are connected by inter-facility link (IFL) cable(s). The IDU can be separated from the ODU by as much as 100 m to 300 m.

A typical installation layout of the VSAT is shown in Figure 7.4. Consideration should be given to safety aspects to protect operational personnel and the public, such as tolerance to high winds, avoidance of electrical shock, lightning protection, and protection from radio-frequency radiation hazards.

Again, for more details on VSAT systems, refer to "VSAT Systems and Earth Stations (Supplement 3 to the Handbook)".

7.1.3.3   Other very small terminals

Following the success of VSATs, new types of very low cost terminals (i.e. at the price, say, of a PC) are appearing on the market for a very wide range of applications and with the hope of sales by mass quantities.

These terminals (sometimes called USATs, for ultra small aperture terminals) will operate with new satellite systems which are currently proposed all over the world and which are based on powerful GSO satellites or on non-GSO satellites (e.g. TELEDESIC, SKYBRIDGE[5]). The reduction in cost

---

[5] See in Chapter 6, § 6.5 and Appendix 6.1.

of the terminals should be permitted by mass production and should be based on the utilization of advanced technologies (ASIC, etc.) and of digital signal processing, most of the functions (including possibly demodulation) being implemented by software on a small number of VLSI components[6].

The proposed systems generally fall in two categories with different applications:

- implementation of telecommunications means in isolated sites (villages, islands, etc.), at acceptable economical conditions, especially for serving developing countries;

- wideband direct access from user terminals to any type of communication networks on a global basis.

7.1.4   Transportable and portable earth station

7.1.4.1   Transportable earth station

Transportable earth stations are used in the following applications:

- satellite news gathering (SNG) (see § 7.9);

- emergency communications in cases of natural disasters;

- temporary communications;

- additional communications capacity for special events.

These earth stations are transportable via a van, truck or aircraft and provide voice, data and video transmission through an international, regional and domestic satellite systems. Various transportable earth stations are used in the 6/4 GHz, 14/12 GHz and 30/20 GHz frequency bands. Examples of transportable earth station realizations are described below (refer to Recommendation ITU-R S.1001).

Various types of small earth station equipment have been developed for the use of new satellite communication systems in the 14/12 GHz band. To implement small earth stations, efforts have been made to decrease the size and to improve transportability so as to ease their use for general applications. This allows the occasional or temporary use of these earth stations for relief operations within a country or even worldwide. Such temporary earth stations are installed either in a vehicle or use portable containers with a small antenna. It is thus possible to use them in an emergency.

The vehicle equipped earth station in which all the necessary equipment is installed in the vehicle, e.g. a four-wheel drive van, permits operation within 10 min of arrival including all necessary actions such as antenna direction adjustments.

A transportable earth station is disassembled prior to transportation and reassembled at the site within approximately 15 to 30 min. The size and weight of the equipment generally allow it to be carried by hand by one or two persons, and the containers are within the limit of the IATA checked

---

[6] Reduction of the size and the cost of the terminals can go even further in the case of mobile earth stations (MSS) for S-PCN applications.

luggage regulations. Total weight of this type of earth station including power generator and antenna assembly is reported to be as low as 150 kg, but 200 kg is more usual. It is also possible to carry the equipment by helicopters.

Some characteristics of small transportable earth stations in the 14/12 GHz band are shown in Table 7.1.

TABLE 7.1

Characteristics of transportable earth stations for the 14/12 GHz band

Type of transportation	Air transportable
Antenna diameter (m)	1.2~21.8
e.i.r.p. (dBW)	62.5~72
RF bandwidth (MHz)	20~30
Total weight	200~275 kg
Package:	
• Total dimensions (m)	<2
• Total number	8~13
• Max. weight (kg)	20~45
Capacity of engine generator (kvA)	0.9~93
Required number of persons	1~3

7.1.4.2 Portable earth stations

The INMARSAT-M system provides direct-dial telephone, facsimile or 2.4 kbit/s data transmissions in the maritime or land-based services.

The physical size of Standard-M terminals are the size of a small suitcase (11 kg).

Furthermore the INMARSAT-M system has evolved into a mini-M system using INMARSAT 3rd-generation satellites. The transponders on the satellites, when used with spot beams, are powerful enough to allow small A4 notebook-size terminals (2.6 kg) to operate with a land earth station (LES) and thus be connected to the international public telecommunications network.

The terminal uses the standard INMARSAT transmit frequency band of 1 626.5 to 1 660.5 MHz and receive band of 1 525.0 to 1 559.0 MHz. Voice channels are available and encoded at 4.8 kbit/s. Facsimile is transmitted and received using the G3 standard at 2.4 kbit/s and asynchronous data transmission is available up to 2.4 kbit/s. A built-in rechargeable lithium ion battery is used to power the equipment. When fully charged it will power the terminal for 4.5 hours in receive mode and for about 1.2 hours in transmit mode.

In the framework of the new projected satellite systems for mobile-satellite communications (MSS) or personal satellite communications (PCS), such as IRIDIUM, GLOBALSTAR and others, very compact handheld earth stations (often called simply "terminal") are currently under development. These terminals should be designed under the same technological principles as those used in the cellular mobile terrestrial systems.

7.2 Antenna system

7.2.1 Main antenna parameters

The following basic parameters of an antenna are defined in § 2.1.2: gain, effective aperture, radiation diagrams and beamwidths, side lobes, polarization and noise temperature. The figure of merit of a receiving station is defined in § 2.1.4.

Some of the more fundamental parameters of an earth station are considered in more detail from a specific "earth segment" point of view, in the following sections.

7.2.1.1 Antenna side lobes

The side-lobe characteristic of earth station antennas is one of the main factors in determining the minimum spacing between satellites and therefore the orbit/spectrum utilization efficiency.

Recommendation ITU-R S.465-5 gives a reference radiation diagram for use in coordination and interference assessment. This reference diagram is defined as follows:

$$G = 32\text{--}25 \log \varphi \text{ dBi} \qquad \text{for } \varphi_{min} \leq \varphi < 48°$$

$$-10 \text{ dBi} \qquad \text{for } 48° \leq \varphi \leq 180°$$

For antenna $D/\lambda \leq 100$ in network coordinated prior to 1993

$$G = 52\text{--}10 \log (D/\lambda) \text{--}25 \log \varphi \text{ dBi} \qquad \text{for } \varphi_{min} \leq \varphi < 48°$$

$$10\text{--}10 \log (D/\lambda) \text{ dBi} \qquad \text{for } 48° \leq \varphi \leq 180°$$

where

φ: off-axis angle referred to the main-lobe axis

φ_{min}: 1° or 100 λ/D degrees, whichever is the greater

G: gain relative to an isotropic antenna.

Recommendation ITU-R S.580-5 stipulates that new antennas of an earth station operating with a geostationary satellite should have a design objective such that the gain, G, of at least 90% of side-lobes peaks does not exceed:

For antenna $D/\lambda > 50$

$$G = 29\text{--}25 \log \varphi \text{ dBi} \quad \text{for } \varphi \text{ min} \leq \varphi \leq 20°$$

These requirements apply to any off-axis direction which is within 3° of the geostationary-satellite orbit (see Recommendation ITU-R S.580-5).

Figure 7.5 shows various examples of good quality radiation patterns.

For VSAT antennas, Recommendation ITU-R S.728 should also be met. The recommendation specifies the off-axis power density which is determined not only by antenna side-lobe performance but also by the transmit power and the bandwidth.

Other recommendations for non-GSO systems are being prepared.

7.2.1.2   Antenna noise temperature

Antenna noise temperature has been defined in § 2.1.4.

Antenna noise temperature (or "antenna temperature") must be kept as low as possible by proper design in order to obtain a high figure of merit (G/T) (see § 7.2.1.3).

The main sources of antenna noise are listed again below (with reference to Figure 7.6 which shows the antenna radiation pattern in polar coordinates):

*   atmospheric attenuation noise: this noise decreases rapidly with the elevation angle of the antenna, since the higher the elevation, the shorter the path length in the atmosphere;

*   cosmic noise (a few K);

*   "ground noise" due to emission of noise energy by the soil (the soil is absorbent for RF waves and thus acts as a "grey body"). The lower the side-lobe level in the direction of the ground, the lower is this noise contribution. Good earth station antennas are called "low noise antennas" or "cold antennas" when this side-lobe level is sufficiently low;

*   miscellaneous losses (cross-polarization leakages, ohmic losses, etc.).

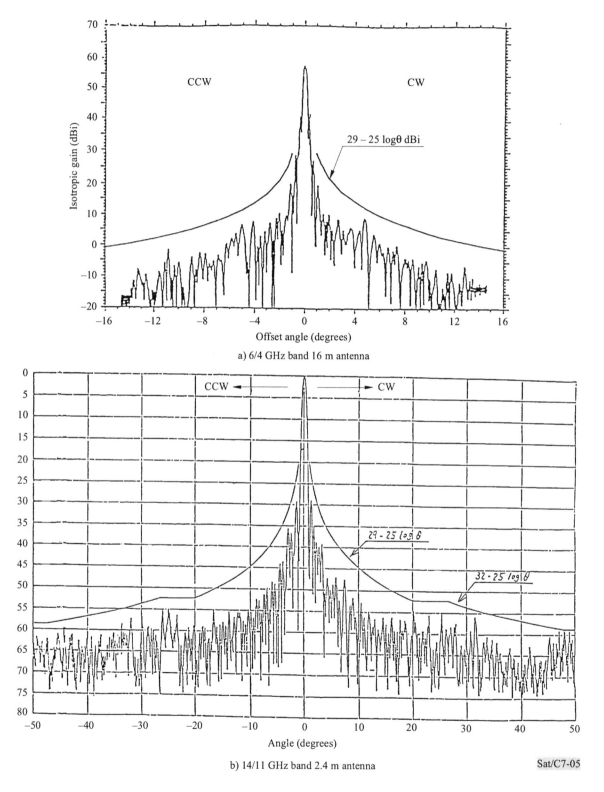

a) 6/4 GHz band 16 m antenna

b) 14/11 GHz band 2.4 m antenna

Sat/C7-05

FIGURE 7.5

Measured radiation patterns of various antennas

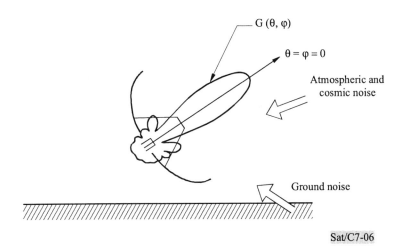

FIGURE 7.6

Contributions to antenna noise temperature

Typical values of noise temperature (T_{A0} versus elevation angle) for a well-designed earth station are given by Figure 7.7 a). These values do not include miscellaneous losses (e.g. ohmic losses) in the antenna feed nor possible additional losses in the atmosphere (e.g. due to rain).

The following equation covers the case of feeder losses and additional atmospheric losses:

$$T_A = \frac{1}{l_F}T_{A0} + (1 - \frac{1}{l_F})T_0 + \frac{1}{l_F}(1 - \frac{1}{l_{atm}})(T_{atm} - T_c) \tag{1}$$

where:

l_F: feeder losses (including "passive" components of the antenna system such as polarizer, duplexer, etc.) ($l_F > 1$, or: $L_F = 10 \log l_F$ in dB);

l_{atm}: additional atmospheric attenuation ($l_{atm} > 1$, or: $L_{atm} = 10 \log l_{atm}$ in dB);

T_c: noise temperature due to sky (in the absence of atmospheric attenuation);

T_{A0}: "clear sky" antenna noise temperature (Figure 7.7 a))

$$T_{A0} = \frac{1}{4\pi} \iint g(\Omega) \cdot T_b(\Omega) \, d\Omega$$

$g(\Omega)$: antenna radiation pattern;

$T_b(\Omega)$: brightness temperature including ground noise (Figure 7.7 b) or § 2.1.4.3, Figure 2.5);

T_{atm}: physical temperature of the surrounding atmosphere ($T_{atm} = 270$ K);

T_0: physical reference temperature = 290 K.

The two first terms of Equation (1) give the total clear sky antenna noise temperature (including feeder losses).

It should be noted that:

- feeder losses L_F include antenna reflector system leakage and losses due to primary source (horn), duplexer, mode transducer, polarizer and other circuit losses (ohmic and dielectric losses). These losses must be kept as low as possible. In addition the (waveguide) link loss between the antenna system output receive port and the low noise amplifier (LNA) input port is included in L_F. Therefore, this link must be very short (the LNA is usually located directly at the antenna receive port). The overall L_F loss can be kept to about 0.15 dB (in the 4 GHz band) by good design;

- for an approximate calculation of the total clear sky antenna noise temperature, add 7 K per 0.1 dB feeder loss to T_{A0}, obtained from Figure 7.7 a).

The third term of Equation (1) gives the additional noise temperature (ΔT_A) due to atmospheric losses. This term is negligible in the 4 GHz band but may be quite significant at higher frequencies.

The following table gives typical examples of additional noise temperature (for $L_F = 0.1$ dB and $T_c = 15$ K).

Atmospheric attenuation (L$_{atm}$ dB)	Additional antenna noise temperature (ΔT_A)
0.5 dB	27 K
1 dB	51 K
2 dB	92 K
3 dB	124 K
5 dB	170 K

Effect of solar interference

Cosmic noise, i.e. antenna noise temperature due to extra terrestrial radio sources, may increase if a strong discrete radio source falls into the antenna beam (or possibly, in the side lobes). The main sources causing an increase in antenna noise temperature are first the sun and secondly the moon.

In practice, an earth station is subject to interference and even to outage when the sun is in conjunction with the satellite, i.e., near to the earth station-to-satellite direction.

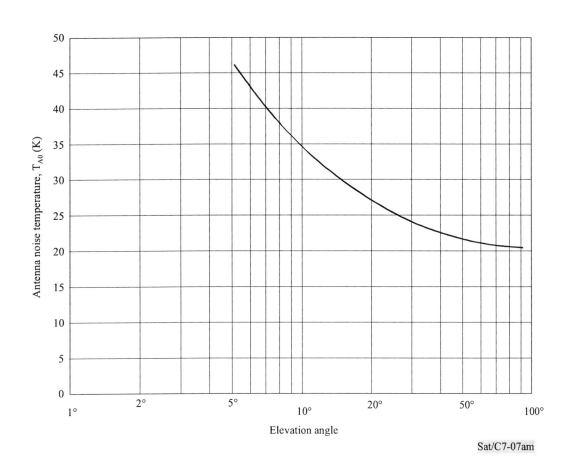

Sat/C7-07am

FIGURE 7.7 A)

Typical "clear sky" earth station antenna noise temperature

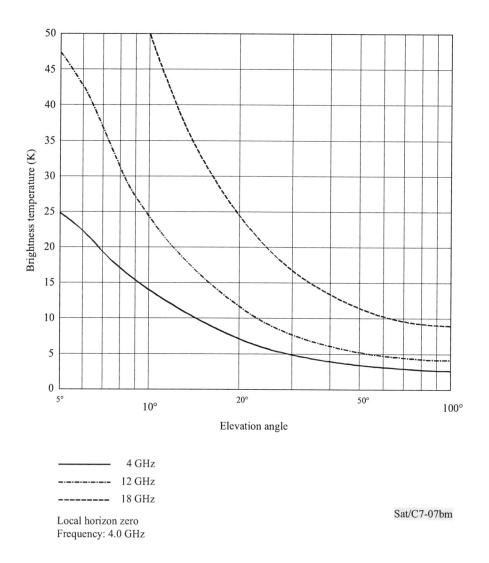

Local horizon zero
Frequency: 4.0 GHz

Sat/C7-07bm

FIGURE 7.7 B)

Typical brightness temperature

An approximate formula for the direction of the sun is:

$$i = 23\sin\frac{2\pi t}{365}$$

where i, in degrees, is the sun's inclination to the equatorial plane and t, in days, being the time referred, for example, to the autumnal equinox (about 22 September).

Conjunction occurs when i', the earth station-to-satellite direction inclination to the equatorial plane is equal to i (0 < i' < 8.7°, 8.7° being the half-angle of the Earth as seen from a geostationary satellite).

It is easy to calculate that, depending on the earth station location, conjunction will occur daily during two periods in the year (between 0 and 21 days before or after the equinoxes) for about $5 \cdot \theta$ day (θ being the earth station beamwidth, in degrees).

This happens:

- before the vernal equinox and after the autumnal equinox if the earth station is in the northern hemisphere;

- after the vernal equinox and before the autumnal equinox if the station is in the southern hemisphere;

- each daily conjunction lasts the time it takes for the sun to pass through the antenna beam, i.e. approximately $480 \cdot \theta$ s.

For example, with an 11 m antenna at 4 GHz, the half-power antenna beamwidth is $\theta_0 = 0.44°$ but if the full beam and the first side lobes are taken into account, interference may be caused to the earth station operation for about five days twice a year with a daily duration of about 7 min.

During the conjunction, the antenna noise temperature (T_{A0}) rises abruptly to very high values.

If the antenna beamwidth θ is smaller than the sun's angular width ($\alpha = 0.5°$), T_{A0} is raised approximately up to the sun's noise temperature (e.g. 20 000 K or even more at 4 GHz). If $\theta > \alpha$, this maximum T_A is reduced by a ratio equal to about $(\theta/\alpha)^2$.

More details on interference caused by the sun and extraterrestrial sources are given in Annex II to ex-CCIR Report 390.

7.2.1.3   Figure of merit (G/T)

7.2.1.3 a) General

The figure of merit, which is the most fundamental performance characteristic of an earth station in the receive mode, has already been defined in § 2.1.4 as the ratio between the gain G of the antenna (at the receive frequency and in the direction of the satellite) and the system noise temperature T (referred to the receiver input). Though it is not solely a characteristic of the antenna, some additional data on the G/T are worth mentioning in this section, since good antenna design involves optimization of the overall G/T.

The G/T is usually expressed in decibels per Kelvin ($dB(K^{-1})$), i.e.:

$$G/T = G - 10 \log T \quad dB(K^{-1}) \tag{2}$$

Figure 7.5 gives T (clear sky) and G/T (at 4 GHz) as a function of the elevation angle.

In systems operating at frequencies above 10 GHz, the specifications of the earth stations, in particular the figure of merit, must take account of G/T losses due to atmospheric effects and precipitation. These losses are generally specified for a percentage of time determined by the quality desired for the system.

The specification of the G/T must take account of these losses:

- directly, since they lead to an increase in the required G/T;

- indirectly, since they entail an increase in the noise temperature T.

The general formula used to specify the G/T of earth station antennas at frequencies above 10 GHz is usually written as follows:

$$G/T_i > K + 20 \log(f/f_0) + L_i \quad dB(K^{-1}) \tag{3}$$

for at least $(100 - P_i)\%$ of the time.

L_i, expressed in dB, is the additional loss on the downlink caused by precipitation (L_{atm} in Equation (1)). It depends on the local meteorological conditions and on the required availability (specified by P_i).

T_i: receive system noise temperature taking L_i into account

K: clear sky G/T

f: actual frequency

f_0: lowest frequency of the receive band

Examples of such specifications are given in Table AP7.1-1.

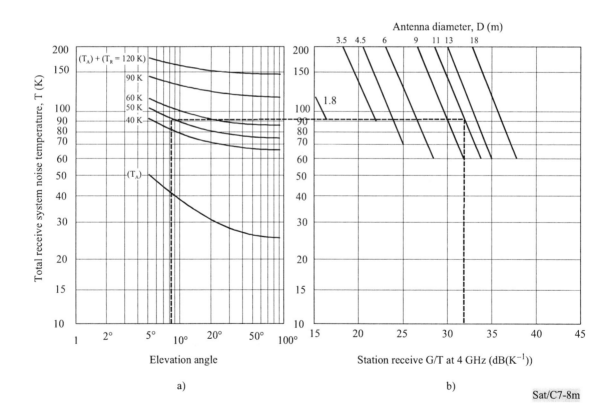

FIGURE 7.8

Figure 7.8 a): total receive system noise temperature, T, as a function of antenna elevation angle with antenna noise temperature, T_A[7], and receiver noise temperature, T_R, as parameters.

Figure 7.8 b): earth station receive G/T[8] at 4 GHz as a function of total receive system noise temperature (given by Figure 7.8 a)) with antenna diameter as a parameter.

An example: G/T = 31.7 dB(K^{-1}) for 11 m antenna, 50 K receiver and 8.3° elevation.

7.2.1.3 b) Measurement of the G/T ratio

There are two measurement methods of the G/T ratio other than the calculation of G/T from G and T measured separately.

One method is performed with the aid of radio stars and the other method utilizes a signal from a geostationary satellite instead of the emission of a radio star.

i)   Measurement of the G/T ratio with the aid of radio stars

This method utilizes the radiation of a radio star. However this method has the following disadvantages:

*   accuracy is not very good for smaller earth stations;

*   the radio star method is not suitable for stations in the southern hemisphere where the stars are visible only at low elevation angles or not at all;

*   tracking of the radio stars is required which may not be possible for stations with limited steerability.

Method of measurement

By measuring the ratio, r, of the noise powers at the receiver output, the G/T ratio can be determined using the formula:

$$\frac{G}{T} = \frac{8\pi k(r-1)}{\lambda^2 \Phi(f)}$$

where

        k: Boltzmann's constant

        L: wavelength (m)

     $\Phi(f)$: radiation flux-density or the radio star at frequency (f) at measurement ($Wm^{-2} Hz^{-1}$)

        r: (Pn + Pst) / Pn

      Pn: noise power corresponding to the system noise temperature T

---

[7] T_A includes miscellaneous feed losses.

[8] Antenna efficiency is 70% for large antennas, 60% for small antennas.

Pst: additional noise power when the antenna is in exact alignment with the radio star

G (antenna gain) and T (system noise temperature) are referred to the receiver input.

In this equation, account is taken of the fact that the radiation of the star is generally randomly polarized and only a portion corresponding to the received polarization is received. The radiation flux-density Φ(f) is obtained by radio astronomical measurements.

This method has a basic advantage when compared with the calculation of GN from G and T measured separately as only one relative measurement is necessary to determine the ratio, instead of two absolute measurements.

ii)   Measurement of the G/T ratio with a signal from a geostationary satellite

In this method, a satellite signal substituted for the signal emanating from the radio star. Instead of measuring the ratio of the radio star signal plus noise to the noise, the ratio of the total signal coming from the satellite plus noise to the noise power is measured. Since there is noise emanating from the satellite, due to factors such as the noise figure of the spacecraft receiver, this additional noise must be taken into account. Further, a reference earth station with known receive gain with respect to the satellite being used for the measurement, must be available to make a measurement of the satellite output power simultaneously with the measuring earth station.

Method of measurement

By measuring the ratio r, of the satellite signal power plus noise power to the noise power, the ratio G/T can be determined using the formula:

$$G/T=[(kBLA) / A / E]\cdot[(r{-}1) - (Tsat / T)]$$

where:

k: Boltzmann's constant

B: noise bandwidth of the earth station receiver (Hz)

L: free-space loss

A: satellite antenna aspect correction factor

E: satellite beam centre e.i.r.p. (W)

Tsat: noise temperature of the earth station originating from the satellite (K)

T: earth station system noise temperature (K)

r: (C + kTsatB + kTB)/(kTB)

C: satellite carrier power at the receiving earth station (W)

7.2.1.4   Wideband performance

High antenna gain, low side-lobe levels, good polarization purity, low antenna noise temperature and also good impedance matching (i.e. low feed circuit VSWR) must be maintained over both the whole transmit and receive bandwidths. This can be a difficult engineering problem; for example, in the INTELSAT-VI 6/4 GHz band, the total antenna bandwidth extends from 3.625 to 6.425 GHz (more precisely 3.625 to 4.2 GHz plus 5.850 to 6.425 GHz) and several subsystems of the antenna system must operate correctly over this very wide band.

Operation over the total bandwidth of the new extended frequency bands (see Table 1.2) is feasible, but in many cases, it may be necessary to retrofit the earth station equipment.

Multi-band antennas, covering simultaneously the 6/4 GHz and 14/11 GHz bands (for example) are feasible and may be useful for some applications.

7.2.2 Radioelectrical design

7.2.2.1 Main types of antennas

The most frequency used antennas are reflector antennas. The most common types are classified in Table 7.2 according to their configuration:

- axisymmetric or offset;
- single reflector or dual reflector.

The following types are discussed in more details below:

a) Cassegrain antennas;

b) Gregorian antennas;

c) Cassegrain antennas with beam waveguide (BWG) feed;

d) offset antennas.

Array antennas have also been recently introduced for small earth stations. See below § 7.2.5.

7.2.2.1.1 Cassegrain antennas

While the conventional parabolic antenna makes use of a single paraboloid reflector fed by a primary radiator (usually a conical horn) located at the focus of the paraboloid ("front-fed antenna"), the Cassegrain antenna makes use of a dual reflector system fed by a primary radiator located at the focus of the system.

The reflector system consists of a main reflector (which is nominally a paraboloid) and a secondary reflector, also called "sub-reflector" (which is nominally a hyperboloid).

The focus of this system, i.e. the focus of the virtual paraboloid, which is optically equivalent to the system, is in the vicinity of the main reflector apex. Therefore, the Cassegrain antenna is a "rear-fed antenna" which provides a very convenient location for the complete feed system (primary radiator, duplexer, junctions, couplers, polarizers, etc.).

Another major advantage of the Cassegrain configuration is that antenna efficiency can be improved by applying the so-called "reflector shaping" (or "modified reflectors") technique. This technique is illustrated in Figure 7.9 a): the principle is to modify the shape of the sub-reflector in order to improve the distribution of the energy that is reflected towards the main reflector. For example, an elementary beam of FA rays radiated from the primary horn towards the edge of the sub-reflector is reflected along AB towards the edge of the main reflector. As shown in Figure 7.9 b), the effect of the shaping is to increase the energy concentration on the outer part of the main reflector and also to reduce the energy loss outside the main reflector (the so-called "spill-over" loss). On the other hand, the effect of the shaping in the inner part of the reflectors is to decrease the energy concentration on the main reflector. The overall effect on antenna performance is twofold: first the spill-over loss is reduced and second (and more important) the distribution of the energy on the antenna radiating aperture is made more uniform (as shown in the right-hand part of the figure). This means that the radiation efficiency of the main reflector (aperture efficiency), which is the dominant factor in the overall antenna efficiency, is improved.

The shape of the main reflector must also be modified in order to maintain a constant total path length (FA + AB + BD) of the rays all over the antenna (see Figure 7.9 b)), in accordance with the conventional laws of optics. In terms of diffraction calculations, this is the condition for uniform phase distribution over the antenna aperture.

NOTE 1 – The performance of Cassegrain antennas may be more or less impaired by the partial blockage and scattering caused by the sub-reflector and also by its supporting struts.

The main effects are:

*   efficiency loss

    This loss can be reduced to some tenths of a decibel if the sub-reflector diameter is not greater than about 1/10 of the main reflector diameter. The blocking effect can be reduced by application of the so-called "shaping technique". This technique allows a reduction of the energy directed from the sub-reflector to the central part of the main reflector (and also to the primary feed). The sub-reflector supporting struts also have the blocking effect, although the loss is lower than that of sub-reflector blockage.

*   side lobe effects

    Direct radiation of the primary feed outside the sub-reflector diameter (so-called "sub-reflector spill-over" radiation) increases the side lobes of the antenna and must, therefore, be kept as small as possible. The sub-reflector edge diffraction must also be combined with the spill-over radiation.

    The sub-reflector struts are located in the field of two radiated waves. One is the radiated power from the main reflector (plane wave) and the other is the radiated power from the sub-reflector (spherical wave). The both waves are scattered by the struts and degrade the side lobe.

NOTE 2 – Cassegrain and Gregorian antennas exhibit a lower noise temperature than front-fed antennas.

This is due to the fact that the direct spill-over rays from the primary radiator are directed towards the sky, while they are directed more or less towards the ground in the case of front-fed antennas.

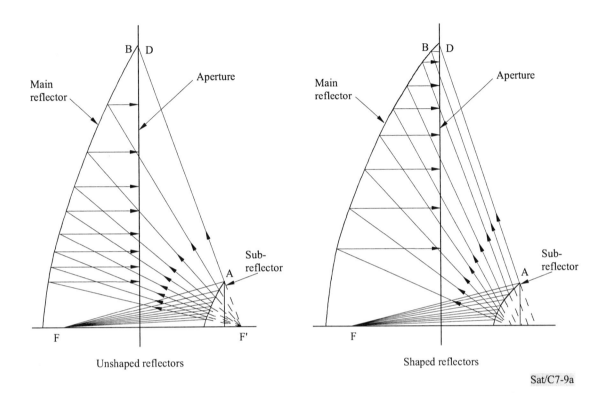

FIGURE 7.9 a)

Reflector design concept

(this figure shows half part of the reflector)

TABLE 7.2

Classification of antennas (radioelectrical design)

Antenna type	Axisymmetric type			
	Single reflector type	**Dual reflector type**		
Example	**Parabolic antenna**	**Cassegrain antenna**	**Cassegrain antenna fed by 4-reflector beam waveguide**	**Gregory antenna**
Schematic views				
Features	Simple configuration Low aperture efficiency because reflector-shaping cannot be applied High noise temperature due to large spill-over power from main reflector Bad accessibility for large diameter antenna because feed and LNA should be adjoined to primary radiator (horn)	Sub-reflector convex (Cassegrain) High efficiency and low noise temperature because reflector-shaping can be applied Feed and LNA can be installed in the equipment room behind the main reflector, so rather good accessibility can be obtained Frequency range is narrower than that of 4-reflector beam waveguide feed type Fairly good radiation pattern (Fig. 7.9 a))	High efficiency and low noise temperature over super-broadband Good accessibility because feed and LNA can be installed in the room free from El – and Az – rotations No waveguide run and rotary joint is required, so transmit power can be about 2 dB higher at 6 GHz band for 30 m diameter antenna Good radiation pattern (Figure 7.9 b))	Sub-reflector concave (Gregory) High efficiency and low noise temperature because reflector-shaping can be applied Feed and LNA can be installed in the equipment room behind the main reflector, so rather good accessibility can be obtained Frequency range is narrower than that of 4-reflector beam waveguide feed type Fairly good radiation pattern (Fig. 7.9 a))
Applications	Small size earth station antenna	Medium station earth antenna	Large size earth station antenna (D/λ about 500)	Medium earth station antenna

TABLE 7.2

(continued)

Antenna type	Offset type			
	Single reflector type		Dual reflector type	
Example	Parabolic antenna	Torus antenna	Cassegrain antenna	Gregory antenna
Schematic views	 Sat/C7-T725	 Sat/C7-T726	 Sat/C7-T727	 Sat/C7-T728
Features	Excellent radiation pattern and low noise temperature because of no blocking Excellent VSWR	Tracks a quasi-stationary satellite without moving its main reflector at all Beam steering can be made by moving only the primary radiator Poor aperture efficiency Rather poor radiation pattern Multi-beam capability with multiple primary radiators	Excellent radiation pattern because of no blocking High efficiency and low noise temperature because of no blocking and reflector-shaping Excellent VSWR Small windload if limited steerable amount is selected Good accessibility because feed and LNA can be installed in the room free from El – and Az – rotations	Excellent radiation pattern because of no blocking High efficiency and low noise temperature because of no blocking and reflector-shaping Excellent VSWR Small windload if limited steerable amount is selected Good accessibility because feed and LNA can be installed in the room free from El – and Az – rotations
Applications	Small size earth station antenna (e.g. TVRO)	TVRO Antenna for multiple satellites reception	Small and medium size earth station	Small and medium size earth station

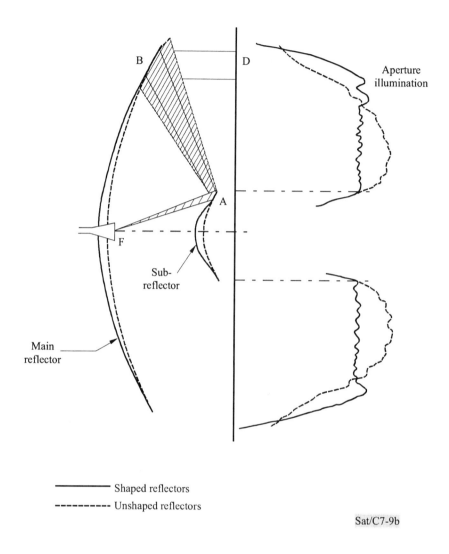

Sat/C7-9b

FIGURE 7.9 b)

Field distribution of Cassegrain antenna

7.2.2.1.2   Gregorian antennas

When Gregorian antennas are used, the concave and not the convex sub-reflector side is facing the feed. The rays from the feed hitting the sub-reflector intersect after reflection before they impinge on the main reflector. Since the waves reflected from the sub-reflector edge should not collide with the opposite one, the sub-reflector of a Gregorian antenna has to be placed before the main reflector aperture (F/D ≥ 0.3). For this reason, the structure of a Gregorian antenna cannot be as compact as that of a Cassegrain type (F/D = 0.25).

For a concave surface, the reflection points on the sub-reflector are more evenly spaced from the feed. The curvature in the vertex of the sub-reflector can be kept smaller because, due to the radiation spread, less power has to be transported from the centre to the periphery. The feed and the sub-reflector edge (after reflection at the main reflector) are subjected to less radiation, and consequently less interference is caused.

Wherever the feed can be installed close to the sub-reflector, the use of a Gregorian antenna is of advantage because of its good spill-over characteristics and its favourable side-lobe behaviour. This applies in particular to small and medium-size antennas required to meet Recommendation ITU-R S.580.

7.2.2.1.3   Cassegrain antennas with beam waveguide feeds

Prior to 1975, nearly all the INTELSAT Standard A earth stations were equipped with Cassegrain antennas. A cabin, located behind the main reflector, was used to house the components which required short low-loss RF links to the antenna feed ports (i.e., low noise amplifiers, tracking receiver and, if possible, high power amplifiers). This not very practical arrangement is avoided by associating the feed system with a beam waveguide (BWG).

In this arrangement (see the figure in Table 7.1), the primary radiator (horn) and its associated cumbersome RF circuits are very conveniently located at the antenna base. The primary radiation from the horn is guided up to the sub-reflector by the BWG which is a four-mirror system usually consisting of two plane mirrors and of two elliptical (or parabolic) ray focusing mirrors; each mirror causes 90° reflection of the rays. Two mirrors are coaxial to the elevation antenna axis and two are coaxial to the azimuth antenna axis. They are equivalent to rotary joints and avoid the problems associated with antenna rotation.

However this type of antenna has a rather complex structure as shown in Table 7.2.

Because of complexity, this type of antenna is less used now for Standard A but could remain useful for special cases where low transmit loss of the BWG is necessary.

NOTE – BWG systems with only two mirrors have also been proposed but they necessitate an offset azimuth axis mechanical arrangement.

7.2.2.1.4   Offset antennas

Most antenna reflector systems, either parabolic with front feed or the Cassegrain or Gregorian types, are axisymmetrical. However, offset antennas, i.e. antennas using a non-symmetrical reflector system (as shown in Table 7.1) can be used when superior performance is desired. This is because offset antennas do not suffer from blockage effects. Specific feed designs have been proposed to solve field distribution problems due to the asymmetry (especially polarization purity problems).

7.2.2.2   Feed systems

7.2.2.2.1 Feed system composition

The feed system of an antenna is composed of the primary radiator (horn) and of its associated RF circuits. A few typical functional block diagrams of complex feed systems (including dual polarization frequency reuse implementation) are shown in Figure 7.10. Of course, where there is no need for frequency reuse, the feed systems are much simpler.

It consists of:

i)   a primary radiator (horn);

ii)   a tracking mode coupler (TMC),which is used only in the case of monopulse type tracking (see § 7.2.4.2 b));

iii) an orthomode junction system (OMJ) which separates the dual polarized transmit path (e.g. at 6 GHz) from the dual polarized receive path (e.g. at 4 GHz);

iv)   a polarizer system: in the case of dual circular polarization, the two orthogonal linearly polarized high power signals are transmitted through the two input ports of an OMT (orthomode transducer) and then converted into the two LHCP and RHCP transmit signals by a polarizer system. Conversely, the two LHCP and RHCP received signals are converted by a polarizer system into two orthogonal linearly polarized signals and then delivered to the receivers (low noise amplifiers (LNAs)) through the two output ports of an OMT.

A second polarizer system may be installed for (rain) polarization compensation.

Polarizer operation is described in § 7.2.2.2.3;

v)   transmit and receive orthomode transducers (OMTs) which separate the orthogonally polarized signals.

NOTE – This feed system provides independent polarizers for the transmit and receive frequency bands.

Other types use a wideband polarizer system in the common transmit/receive path (this system should be located between the horn (or the TMC, if included) and the OMJ (or the OMT) (see Figure 7.10).

7.2.2.2.2 Primary radiators

Various types of horn antenna have been developed for the primary radiator of an earth station antenna. They can be classified as shown in Table 7.3 with respect to their features. Of the horns listed, the corrugated conical horn is the most widely used. The corrugated horn (and also the dual mode and the multi-flare horn) is a so-called "scalar feed" because it provides a good axial symmetric radiation pattern whatever the polarization.

TABLE 7.3

Classification of primary horn

Horn type	Clear aperture horn			Rectangular aperture horn		
Items	Conventional conical horn	Dual mode horn (step type)	Corrugated conical horn	Pyramidal	Diagonal horn	Multi-flare horn
Shape						
Aperture field shape	TE°_{11}	$TE^\circ_{11} + TM^\circ_{11} =$ hybrid mode	EH (hybrid mode)	TE°_{10}	TE°_{10}	$TE^\circ_{10} + TM^\circ_{12} =$ hybrid mode
Frequency characteristics	Wide	Less than 5%	Approximately 1 octave	Wide	Wide	Approximately 20%
Beam axisymmetricity	Poor	Good	Excellent	Poor	Fair	Good
Side lobe	Poor	Good	Excellent	Poor	Fair	Good
Cross-polarization level	Poor (−18 to 20 dB)	Good (less than −25 dB)	Excellent (less than −30 dB)	-	-	Good
Power in beam	Poor	Good	Excellent	Poor	Poor	Good
Remarks				Only linear polarization use	Only linear polarization use	

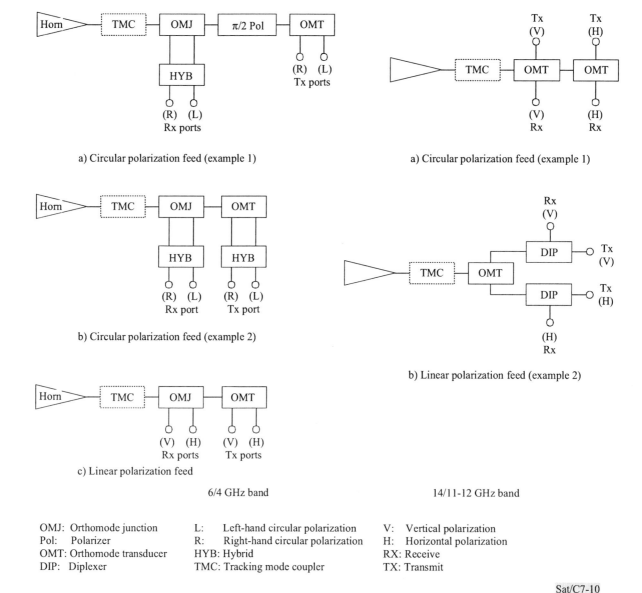

a) Circular polarization feed (example 1)

a) Circular polarization feed (example 1)

b) Circular polarization feed (example 2)

b) Linear polarization feed (example 2)

c) Linear polarization feed

6/4 GHz band   14/11-12 GHz band

OMJ: Orthomode junction	L:	Left-hand circular polarization	V:	Vertical polarization
Pol:	Polarizer	R:	Right-hand circular polarization	H:	Horizontal polarization
OMT: Orthomode transducer	HYB: Hybrid	RX: Receive
DIP: Diplexer	TMC: Tracking mode coupler	TX: Transmit

Sat/C7-10

FIGURE 7.10

Typical feed system block diagram

As shown in Figure 7.10, feed systems in the 14/11-12 GHz band are generally simple because of linear polarization implementation and of a relatively narrower bandwidth.

7.2.2.2.3 Polarizer operation

i)   Circular polarization (CP)

As explained in § 2.1.2, any polarization can be considered as the vector sum of two orthogonal, linearly polarized components (LPC). More precisely, circular polarization can be considered as the sum of two LPC with equal amplitude and a $\pi/2$ phase difference. RF polarizer operation is based on this principle: a dielectric plate (or a metal fin array, etc.) called a quarter-wave ($\lambda/4$) plate, inserted in a circular waveguide, causes a $\pi/2$ phase difference between the RF components parallel to, and perpendicular to, the quarter-wave plate. This is due to the lower phase velocity of the waves when propagating with the electric field (E) vector parallel to the plate.

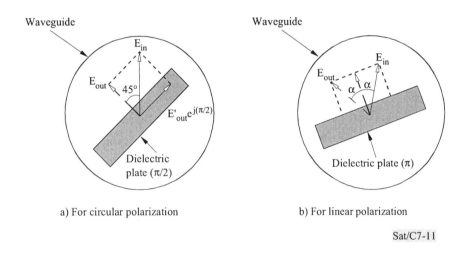

a) For circular polarization   b) For linear polarization

Sat/C7-11

FIGURE 7.11

Polarizers: principle of operation

Referring to Figure 7.11 a), a quarter-wave plate is set into the waveguide inclined at 45° to the incoming linearly polarized wave E_{in}, so that the component of E_{in} perpendicular to the plate propagates through the plate without phase or amplitude alteration (E_{out}), while the component parallel to the plate is delayed by $\pi/2$ and gives an output component $E'_{out} \exp (j \, \pi/2)$. The loss in the dielectric plate is usually negligible ($E'_{out} = E_{out}$), so that the output wave is circularly polarized (e.g. RHCP). Should the input LP have been horizontal, the output would have been an LHCP wave. Polarization conversion from CP to LP is also performed by the same process.

Other devices can be used for converting LP to CP (or the reverse). For example, a conventional waveguide "3 dB coupler" which divides an RF field into two equal components with a $\pi/2$ phase difference can be used to form a CP wave in a circular waveguide.

In addition to the conventional devices mentioned above, another device, now used, is called the septum polarizer which combines the two functions: polarizer and OMT (orthomode transducer).

As represented by the top part of Figure 7.12, this device looks like a three-port coupler with two adjacent rectangular waveguides at one end and a square waveguide (or a circular waveguide) at the other.

The separation, in the square waveguide, between the two rectangular waveguides is made by a septum, i.e. a metallic plate, with a tapered or stepped transition.

The operating principle (represented at reception) is shown in the lower parts of the figure.

Taking the example of left-hand circular polarization, an input LHCP fields is represented in the square waveguide (at the left-hand part of the figure). This circularly polarized field can be split in two horizontal and vertical ($\pi/2$ phase-shifted) components, as illustrated separately on the two superposed sets of pictures, which schematize the field distortions during the propagation in the transition. Finally, as represented at the right-hand part of the figure, due to the superposition of in-phase and out-of-phase fields, the output field is concentrated in the right-hand rectangular waveguide.

Of course, the same schematic representation holds also at transmission, i.e. the propagation from the right-hand to the left-hand parts of the figure, and also for RHCP (bottom set of pictures), which corresponds to the left-hand rectangular waveguide.

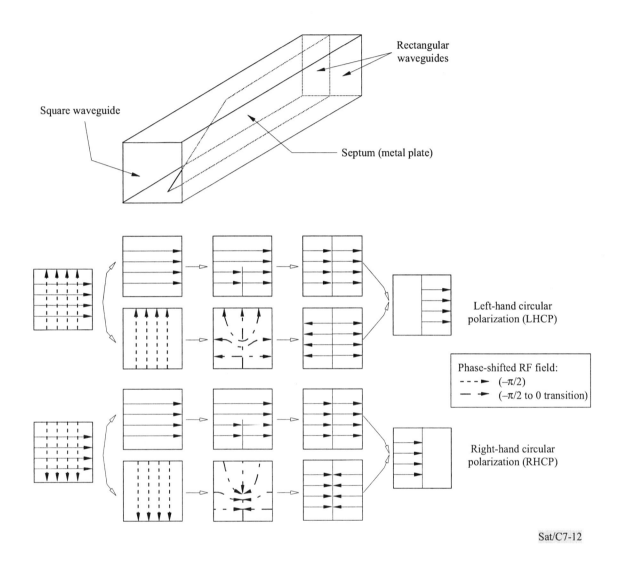

Sat/C7-12

FIGURE 7.12

Septum polarizer

ii)   **Linear polarization (LP)**

In principle, no polarizer is needed when the antenna is operated in LP. However, the feed must be positioned in order to transmit (and receive) the E vectors with the correct orientation. If this orientation changes due to the attitude of the satellite, rotating joints must be provided. Another solution is to use a half-wave ($\lambda/2$) polarizer.

The operating principle is shown in Figure 7.11 b), where a dielectric plate (or a metallic fin array, etc.) called a half-wave plate is inserted in a circular waveguide, causing a π phase shift of the parallel component of the input (E_{in}) electric field. Therefore, if α is the inclination of the plate, the output (E_{out}) electric field is rotated by 2α. For example a vertical electric field is converted to horizontal if $\alpha = 45°$.

iii)   Elliptical polarization, polarization compensation

Elliptical polarization is not normally used but imperfections (e.g. of the satellite or earth station antenna) may cause a nominally circular (or linear) polarized wave to be transmitted or received as an elliptical wave (e.g. nearly circular or linear).

However, the most common effect which causes unwanted conversion of CP (or LP) into elliptical polarization is rain (and other types of atmospheric precipitation). This is due to the oblate shape of the rain drops, which induces differential phase shift and differential attenuation (in a similar way to the dielectric plate of a polarizer).

If an elliptically polarized (e.g. nearly LHCP) signal is received by a dual polarized (frequency reuse) antenna, the wanted, co-polarized component will be received at the nominal (e.g. LHCP) output port, but an unwanted cross-polarized component will be also received at the other (e.g. RHCP) output port and will interfere with the (nominally) RHCP wanted signal which is normally received at this port.

In the 6/4 GHz bands, only differential phase shift needs to be considered, but at higher frequencies, differential attenuation is not negligible. In the former case, polarization compensation, i.e. conversion of an elliptically polarized wave received into an LP output (or the inverse process at transmit for "pre-compensation" on the uplink), can be effected by using a supplementary half-wave polarizer (as represented by a broken line on Figure 7.8). The operation is as follows (e.g. at receive): the first quarter-wave polarizer, when properly oriented, converts the input elliptically polarized wave into LP. The second half-wave polarizer, when properly oriented, changes the LP orientation to adapt it to the polarization of the output port.

NOTE 1 – It can be shown that following another procedure, a second quarter-wave polarizer could be used.

NOTE 2 – Compensation of differential attenuation is much more difficult and would require the use of variable couplers either at RF or IF.

NOTE 3 – The implementation of automatic polarization compensation would involve an adaptive process: for example, a satellite beacon signal could be used to minimize the received cross-polarized component by controlling the orientation of the rotatable polarizers.

Uplink compensation is more complicated: a method has been proposed that controls the orientation of the transmit path polarizers by processing the receive path data. This is possible only if differential phase shifts at the receive (e.g. 4 GHz) and transmit (e.g. 6 GHz) frequencies are sufficiently correlated.

NOTE 4 – The compensation methods described above are based on wideband RF circuits (polarizers).

Compensation with IF circuits has also been proposed and could be useful, especially when rain effects prove to be frequency sensitive.

7.2.3 Mechanical antenna design

7.2.3.1 General

The mechanical construction of an earth station antenna is composed of:

- the electrical assembly (as described in § 7.2.2) which consists of the reflector(s) system and the feed system;

- the antenna mount system (or pedestal) which supports the electrical assembly (usually on two orthogonal movable axes);

- the drive subsystem which permits the electrical assembly to be steered in any possible orientation around the mechanics axes of the antenna mount.

NOTE – Some antenna types do not need any movable axis. This is the case when the antenna beam can be positioned by moving the feed system only, while keeping the reflector system fixed (e.g. Torus antenna in Table 7.1).

It should be noted that some parts of this section apply primarily to large and medium size antennas. This is the case, in particular, for § 7.2.3.3 and 7.2.3.4.

7.2.3.2 Antenna mechanical accuracy

In order to achieve the desired antenna gain pattern, it is necessary that the antenna reflectors, both main- and sub-reflectors, have high surface accuracy and that the antenna continuously points towards the satellite with the required accuracy under all environmental conditions.

The gain variation (degradation) ΔG_r due to reflector surface errors can be expressed approximately as follows:

$$\begin{aligned} \Delta G_r &= 10\log(\exp\!-(4\pi\varepsilon/\lambda)^2) \\ &\approx -686(\varepsilon/\lambda)^2 \end{aligned} \quad \text{dB} \qquad (4)$$

where:

ε: r.m.s. manufacturing tolerance

λ: wavelength in free space

Thus, if the gain degradation is to be kept better than, for example, 0.2 dB, the reflectors must be fabricated with a dimensional accuracy of less than 0.017λ, i.e. < 1 mm for a 6/4 GHz antenna.

Other antenna deformations due, in particular, to weight, wind and thermal effects, can also degrade the gain and the radiation pattern (and especially the side-lobe level).

The gain variation (degradation) ΔG_p due to the antenna beam positioning error $\Delta\theta$ (in degrees) is given approximately as follows:

$$\Delta G_p \approx -12(\Delta\theta/\theta_0)^2 \quad \text{dB} \qquad (5)$$

where θ_0 is the half-power beamwidth and is given approximately by the following equation:

$$\theta_0 \approx 65(\lambda/D) \quad \text{degrees} \tag{6}$$

where D is the antenna diameter (m). Therefore:

$$\Delta G_p \approx -0.003(D/\lambda)^2 \cdot (\Delta\theta)^2 \quad \text{dB} \tag{7}$$

(for examples, see § 7.2.4.3).

7.2.3.3   Antenna steering and mount systems

The earth station antenna should at least be capable of being steered over the expected movement range of the satellite position. It is desirable that the antenna should be steerable over as wide an angular range as possible so as to enable it to be maintained and to be used for tests, etc. An antenna with full sky coverage, is called a "fully steerable" antenna. On the other hand, a "limited steerable" antenna can steer the beam only over a limited portion of the celestial sphere.

Antenna mounts can be classified into three types as shown in Figure 7.13.

The Az-El mount system is the type most widely used for satellite communication earth station antennas.

In this design, one of the axes (Az-axis) is set vertical to the ground while the other (El-axis) is parallel to the ground. This mount has an advantage in that only the El-rotation causes deformation of the antenna due to its weight, the mount configuration is simple and the antenna weight is the lightest.

Therefore, most antennas which require a high accuracy in the reflector surface and antenna pointing use an Az-El mount.

One possible drawback of the Az-El mount may occur when the earth station is installed near the sub-satellite point. This is due to the fact that the Az-El mount has a mechanical pole in the direction of the zenith, which means that it is difficult to track a satellite passing near the zenith because of the high azimuth velocity which would be required.

Three types of Az-El mounts are used for earth station antennas, i.e. a yoke-and-tower type (or "king post" type), the wheel-and-track type and jackscrew drive type.

The X-Y mount system has X and Y axes which are orthogonal to each other. In this system, both axes have to be positioned at rather a high level above the ground to realize a fully steerable antenna, so that the antenna weight is increased. Therefore this type of mount is mainly used in limited-steerable antenna systems. It has the advantage that it does not have a mechanical pole in the operating directions. Another advantage is that it can be driven by simple mechanisms using jackscrew actuators.

The polar mount system has a right ascension axis (RA axis: calibrated in hours) parallel to the Earth's polar axis and the declination axis (Dec axis) perpendicular to the RA axis. This mount is commonly used for radio telescopes because it enables the antenna to track a radio star with rotation of the RA axis only.

Polar mounts are rarely used in earth station antennas due to the asymmetry of the weight and wind loading effects. However, they can be a good choice for medium and small antennas, when no declination tracking is needed as a consequence of accurate satellite North-South station-keeping.

7.2.3.4 Drive and servo subsystem

Antenna drive systems based on gear drive mechanisms and on jackscrew drive mechanisms are widely used. The use of the latter is restricted to antennas with limited steering because of its configuration. For generating the drive torque, either d.c. or a.c. motor drives are used. The former is mainly used for fully-steerable antennas and the latter is used for limited-steerable, medium or small antennas, especially when combined with the step-track system.

A typical block diagram of the antenna drive and servo subsystem is shown in Figure 7.14.

The steering accuracy of the antenna beam must be kept within about one tenth of the half-power beamwidth of the antenna in order to maintain stable communication. It follows that the required accuracy is of the order of $0.02°$ for an INTELSAT Standard A antenna. Since the accuracy of the drive and servo subsystem is determined by that of the rotating axes and angle pick-off (read-out) mechanism, careful attention should be paid to the design and installation procedures to minimize mechanical deflection, backlash of the gear trains and angle pick-off mechanisms, and mechanical deflecting of the antenna reflector system.

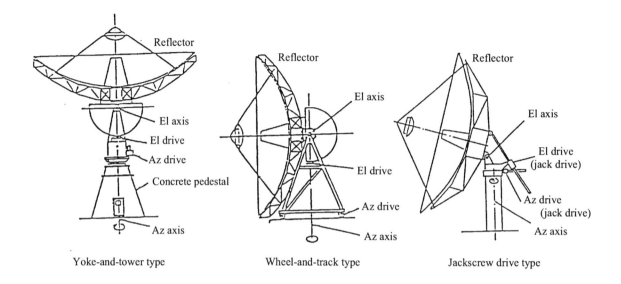

a) Az-El mount

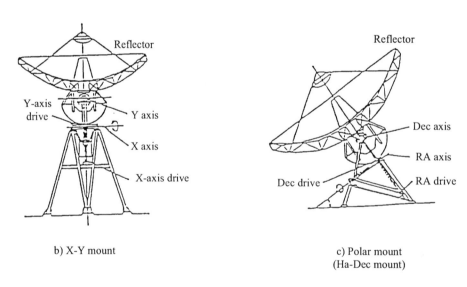

b) X-Y mount

c) Polar mount
(Ha-Dec mount)

Sat/C7-13

FIGURE 7.13

Antenna mount types

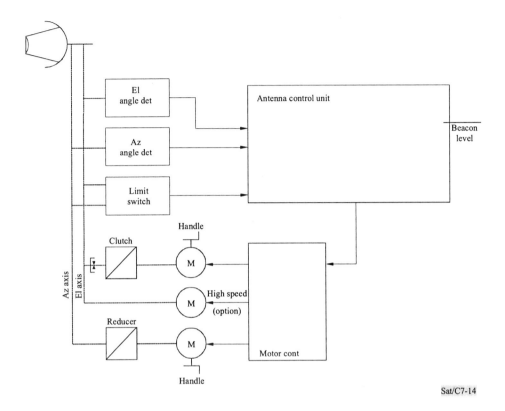

Sat/C7-14

FIGURE 7.14

Block diagram of antenna drive and servo-system

7.2.4 Antenna beam positioning – tracking systems

7.2.4.1 General

Table 7.4 gives a summary of the tracking systems used for keeping the antenna beam axis (direction of maximum gain) pointed towards the satellite, in spite of:

- satellite North-South and East-West residual movements;

- mechanical loads on the antenna reflector and structure (weight, wind);

- variations of propagation conditions (specially high frequencies, i.e. in the 14/11-12 GHz and higher bands).

Type No. 1 where no tracking system is needed is the simplest and most economical. In the case of small stations (e.g. for rural communications, business systems or VSAT), system planners should strive to design stations with non-tracking antennas.

Type No. 2 is also rather simple, because no beacon receiver nor special feed is needed.

In Types Nos. 3 and 4, a special beacon receiver and a complete servo-controlled drive system are needed, both for azimuth and elevation.

TABLE 7.4

Tracking systems

Type No.	Tracking system	Conditions and/or description	Satellite beacon needed	Example	Remarks
1	No tracking (fixed pointing)	Possible for small antennas $D/\lambda \le k/(SSK+APE)$ SSK: satellite station-keeping(d°) APE: antenna pointing error (See NOTES)	No	SSK=0.05°(N/S) and 0.05°(E/W) APE=0.05° $D/\lambda \le 150$ e.g. $D \le 3.1m$ for 14. 5 GHz operation e.g. $D \le 7.5m$ for 6 GHz operation	Capability for manual orientation of the antenna must however be provided
2	Programme tracking	Permanent calculation of satellite orbital position at the station or transmission of local antenna pointing data by a control station	No	These methods can be advantageous with medium size antenna in domestic system	
3	Step tracking	– Servo system with automatic permanent searching for maximum received signal – May be impaired by strong wind gusts and by propagation fluctuation	Yes	– All medium sized antennas (INTELSAT Standard B) domestic and regional systems – Large antennas (INTELSAT Standard A) in good environmental conditions	Step tracking is currently considered as the most cost-efficient system at 6/4 GHz
4	Monopulse tracking	– Servo system with instantaneous detection of pointing errors – High precision system – Tracking coupler needed in the antenna feed	Yes	Large antennas $(D/\lambda \ge 400)$: INTELSAT Standards A and C, EUTELSAT ECS antennas	Monopulse is the most precise and reliable system for large antennas especially at high frequencies
5	Tracking with beamscanning	– High precision system – Beam scanning needed	Yes	Small to large size antennas	

NOTES relevant to Type Nos. 1 and 2.

NOTE 1 – Satellite station-keeping (SSK) angle must account both for North-South and East-West satellite movement (e.g. if N/S and E/W SSK are both ±0.1° total SSK = ±0.14°).

NOTE 2 – Antenna pointing error: see § 7.2.4.3 a).

NOTE 3 – $K = 18° \sqrt{n}$ if acceptable antenna gain loss $\le n$ dB (n = 1 in table examples).

7.2.4.2   Tracking systems

Apart from the programme tracking (described below in d)), the two main methods for automatically tracking the satellite are Type No. 3 (Step-tracking, described below in a)) and Type No. 4 (Monopulse, described below in b)). Another method, Type No. 5, using scanning of the antenna beam (described below in c)) is less frequently used.

a)   Step-track

The step-track method uses a so-called "climbing the hill" servo-system. The antenna beam is steered step-by-step so as to obtain a stronger signal from the satellite than that received in the previous position as shown in Figure 7.14. If the step-steering of the antenna beam has decreased the receive signal level, the step-track processor will command the antenna to be steered in the opposite direction.

The receive signal is usually derived from the carrier of a satellite beacon. In the step-track system, no special tracking feed is needed and only a simple beacon receiver and step-track processor are required. However, a disadvantage of the system is that the tracking accuracy is directly affected by rapid variations of the incoming signal due to such effects as atmospheric scintillation, rain absorption and beacon instability.

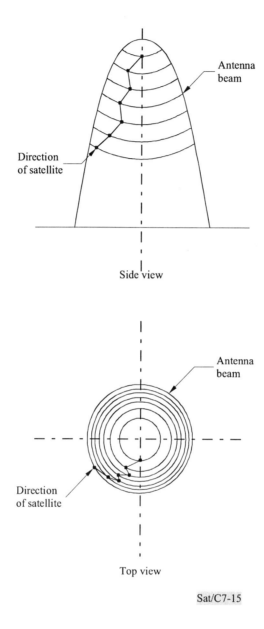

FIGURE 7.15

Step-track system

This disadvantage can be partially overcome by combining the step track with a program or memory which is activated in the case of rapid or excessive level variations or loss of the beacon signal.

The memory allows the tracking process to continue until normal operational conditions are restored, using tracking information acquired in the previous 24 hours.

b)   Monopulse

The monopulse method derives its name from radar technology. In this method, signals caused by azimuth and elevation misalignments are generated instantaneously in the antenna feed system and available at special "error" or "difference" antenna feed output ports.

The first monopulse systems (multi-horn monopulse systems) made use of four primary horns symmetrically located around the focus. These horns provide beams slightly offset from the antenna boresight axis. Tracking signals are obtained by comparing the amplitude of the received signals between these beams. A one-dimensional explanation of the principle of this system is given in Figure 7.16 a). Two primary horns, denoted as A and B, respectively, are symmetrically positioned on both sides of the focus of the antenna. Their radiation patterns are shown in Figure 7.16 b). The angle difference between the antenna boresight axis and the satellite direction is obtained by coherently detecting the error signal (Δ signal) with reference to the sum signal (Σ signal), as shown in Figure 7.16 c), the two tracking signals being made available at the output parts of a hybrid circuit in the feed system.

The disadvantage of these systems is that they require complicated and cumbersome feeds and that good radiation patterns cannot be obtained when these feeds are used.

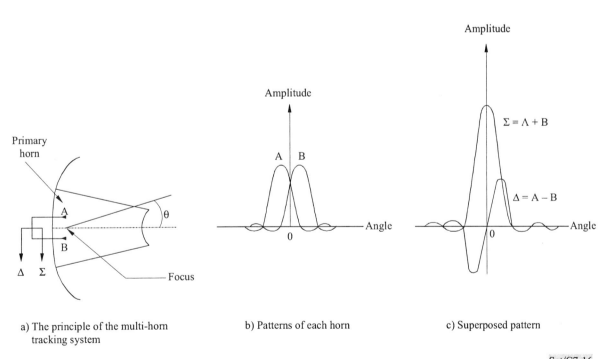

a) The principle of the multi-horn   b) Patterns of each horn   c) Superposed pattern
   tracking system

Sat/C7-16

FIGURE 7.16

Multi-horn monopulse tracking system

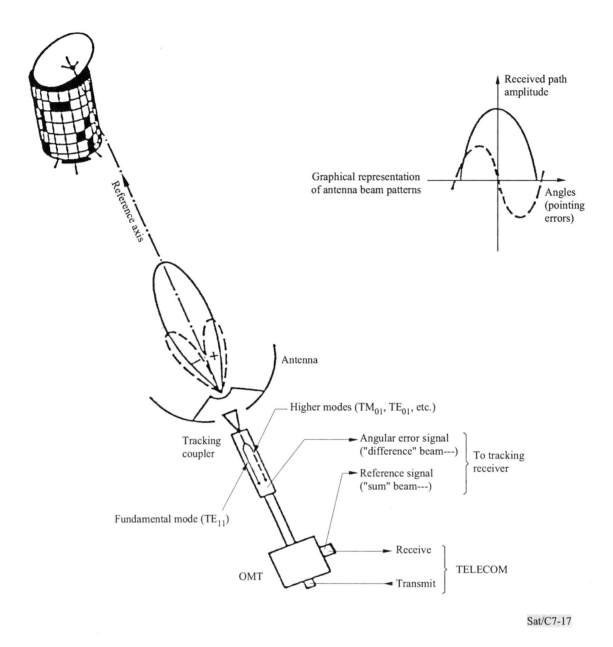

FIGURE 7.17

Multi-mode monopulse tracking system

Modern monopulse systems (multi-mode monopulse systems) make use of a special microwave coupler (monopulse tracking coupler) inserted in the antenna feed as shown on Figure 7.17. This coupler picks up the higher mode signals which are excited in the feed horn when the antenna beam axis is offset from the satellite direction. Such higher mode signals correspond to odd mode radiation patterns with a null in the beam axis direction. After coherent detection by a reference signal (which is the normal fundamental mode signal), bipolar discrimination error voltages are obtained and are directly fed to the servo-system. An example of a tracking receiver is shown in Figure 7.18.

When operating in circular polarization (which is the case for the INTELSAT 6/4 GHz system), only one higher odd mode – usually the TM_{01} mode of circular waveguides – is required. This is because the phase and the amplitude of the TM_{01} component, when compared with the fundamental TE_{11} mode component (used as a reference), contain the angular error information.

By in-phase and in-quadrature coherent detection, azimuth and elevation errors can be obtained (as shown in Figure 7.18).

When operating in linear polarization, the TM_{01} detection only delivers one error signal (e.g. azimuth) and a second higher odd mode is required for elevation. This second mode can be the TE_{01} mode.

Combination of other modes is also possible (e.g. TE_{21} with proper orientation or $TM_{01} + TE_{21}$, etc.). In this case, the tracking receiver of Figure 7.18 must be modified to accept two input error signals.

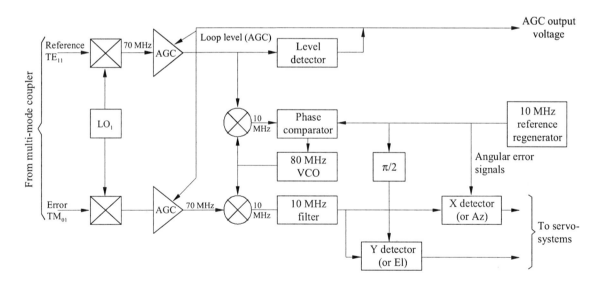

LO: Local oscillator
AGC: Automatic gain control
VCO: Voltage controlled oscillator

Sat/C7-18m

FIGURE 7.18

Typical block diagram of monopulse tracking receiver (processing unit)

c)   Tracking with beam scanning

In this tracking system, the main beam of the antenna is shifted in two orthogonal planes, at two symmetrical positions in each plane. (See Figure 7.19.)

Measurement of the signal level provides the amplitude of the antenna misalignment, and the direction of the misalignment referred to the antenna coordinates system.

Scanning of the main beam creates an amplitude modulation of the received signal. Signal processing gives two error signals at the output of the tracking receiver.

The beam shift is achieved by adding, in the feedhorn, a higher mode (example TM 01 or TE 21 in conical horn), whose level and phase relative to the fundamental mode, creates a phase gradient on the main dish aperture. The slope of the gradient is proportional to the relative level of the higher mode. The relative phase permits management of the plane of the beam shift.

This device requires a beacon receiver with a signal processing board and control of the amplitude and phase of the higher mode.

The system is tuned at the beacon frequency with a narrow-band RF filter so that the beacon shift is cancelled at every other frequency.

The tracking accuracy is similar to the tracking accuracy of the monopulse mode.

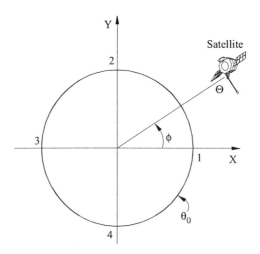

θ_0: Beam scanning amplitude

Θ: Satellite coordinates

ϕ: Satellite coordinates

Levels at points 1 and 3 give the misalignment along X
Levels at points 2 and 4 give the misalignment along Y

Sat/C7-19

FIGURE 7.19

d)   Programme tracking

Programme tracking uses a computer program to calculate the satellite position.

There are three categories of program tracking.

1)   The first method is to calculate the antenna pointing position using the orbital position of the satellite from the control station, or using the satellite position parameters.

2)   The second method is to use stored previous tracking data (for example during 24 hours) of satellite coordinates and to estimate the present position.

3)   The third method is a more intelligent or more accurate method than the second method and predicts the satellite position from recent data during a few days. The prediction algorithm is more sophisticated and includes the daily satellite drift correction.

7.2.4.3   Antenna beam positioning errors

The accuracy of antenna positioning is usually specified by two parameters:

a)   The pointing accuracy which is the angle between the antenna beam axis and the wanted direction (the former is often indicated by angular read-out devices). The pointing accuracy has to be taken into account:

*   in Types Nos. 1, and 2 of Table 7.4;

*   in Types Nos. 3, 4 and 5 during antenna positioning for initial satellite acquisition.

The pointing accuracy is limited by:

*   mechanical axis and feed alignment errors;

*   angular read-out errors;

*   errors due to deformation of antenna structures (and foundations), wind, gravity and thermal effects;

*   servo-system errors (wind and gravity torques, dynamic lag) (Type Nos. 2, 3, 4 and 5 only).

These partial errors can be added quadratically if they are uncorrelated. A typical overall figure for the pointing accuracy of an INTELSAT Standard A 16 m antenna system is 0.06° for a wind of 13 m/s with 20 m/s gusts.

Pointing accuracy has to be taken into account when the antenna beam axis is pointed at a designated angular position (e.g. the satellite position, as given by its ephemeris). Other pointing methods can be devised (e.g. by searching for a maximum of the signals received from the satellite).

b)   The tracking accuracy which is the residual error angle between the antenna beam axis and the direction of arrival of the satellite received signal under auto-tracking operation (Type Nos. 3 or 4 of Table 7.4).

The tracking accuracy is limited by:

*   shift of the difference beam axis (Type No. 4 of Table 7.3);

*   errors due to step size and to beacon signal level measurement (Type No. 3);

- errors due to propagation variations and to satellite beacon instability (Type No. 3);

- receiver thermal noise (Type Nos. 3 and 4);

- errors due to the effect of wind gusts on a servo drive system (Type Nos. 3 and 4);

- tracking mechanism and servo drive system basic errors (back-lash – if not compensated – dynamic lag, etc.) (Type Nos. 3 and 4).

These residual pointing errors can be added quadratically if they are uncorrelated.

The errors due to the effect of wind gusts are often specified for two wind conditions: full antenna performance must be obtained under condition 1 (e.g. 13 m/s wind with 20 m/s gusts), and a degraded but sufficient performance for operation under condition 2 (e.g. 20 m/s wind with 27 m/s gusts).

A condition 3 is often specified for antenna survival (e.g. at 55 m/s wind, the antenna system must not be damaged).

The receiver thermal noise introduces an r.m.s error which is given (for Type No. 4: monopulse) by the following formula:

$$(\Delta\theta)_{\text{noise}} = \frac{1}{S}\sqrt{kTB/\text{Pr}} \qquad (8)$$

where:

> S: slope of the difference pattern, i.e. sensitivity of the discrimination signal delivered by the tracking mode coupler (e.g. 3 degrees^{-1} is typical for a Standard A antenna)

It can be shown that $S = \eta_{TC} \cdot D/\lambda$ (in radian^{-1}) or $S = (1/57)\,\eta_{TC} \cdot D/\lambda$ (in degree^{-1}), where η_{TC} is the mode coupling efficiency. Typically, $\eta_{TC} = 0.4$;

> kTB: noise power (k = $1.38 \cdot 10^{-23}$ J/K), T being the total noise temperature in the error channel of the tracking receiver and B being the actual error integration bandwidth, i.e. the servo bandwidth (1 Hz)

> Pr: power received from the satellite beacon in the reference channel. Note that the formula is valid for a noise-free reference signal (i.e. noise in the reference channel much lower than in the error channel: this is the normal case when the reference signal is picked up at the output of the normal telecommunication channel low noise amplifier)

A typical overall figure for the tracking accuracy of an INTELSAT Standard A 16 m antenna under wind condition 1 is 0.02° (r.m.s.) with monopulse tracking (Type No. 4) and 0.03° (r.m.s.) with step tracking (Type No. 3).

NOTE – The antenna gain loss due to the total beam positioning error (either a pointing error or a tracking error) is given by the approximate formula given in § 7.2.3.2.

The following Table gives typical examples for the Standard A 16 m antenna errors quoted above:

	(degrees)	Receive loss (4 GHz)	Transmit loss (6 GHz)
Pointing error	0.05	0.34 dB	0.77 dB
Tracking error (monopulse) step trac(k)	0.02 0.03	0.05 dB 0.12 dB	0.12 dB 0.28 dB

7.2.5 Array antennas

7.2.5.1 General

In satellite earth stations, reflector antennas such as paraboloid antennas or Cassegrain antennas have been widely used. More recently, array antennas have appeared as an attractive solution for small earth stations or portable terminals.

In the reflector antenna, the radiated power of the primary radiator or the power from the subreflector is reflected and directed in a specific direction. On the other hand, in the array antenna the electromagnetic wave is radiated by each element. The array antenna, in general, consists of the radiation elements and feed network (Figure 7.20). If the feed network is a simple power divider, and the phases of power to the elements are the same, the radiation from each element is summed up in one direction.

The array antenna is capable of incorporating various functions which cannot be attained by antenna with a single radiation element. For example, by furnishing variable phase shifters in the feed network to change the phase of the supplied wave of each element, the antenna beam can be steered as shown in Figure 7.20.

7.2.5.2 Types of array antennas

Array antennas used for satellite earth stations are categorized as follows.

1) Categorization with respect to radiation elements

   There are many type of antennas with respect to the radiation elements.

   Typical radiation elements used for satellite earth stations are as follows.

   a) Printed array (Figure 7.21);

   b) Waveguide slot array (Figure 7.22) etc.

2) Array antennas are categorized as to whether the phases of the element excitation are fixed or variable

   a) array antenna with fixed phase excitation;

   b) array antenna with variable phase excitation (Figure 7.23).

      Array antennas categorized as b) are called "phased array antennas" and have a capability of beam scanning.

3)   Array antennas are categorized as to whether its elements include an active device or not.

a) array antenna which has only passive devices (phase shifters, etc.);

b) array antenna which has active devices (transmit amplifiers or receive amplifiers).

Array antennas categorized as b) are called "active phased array antennas" (Figure 7.24).

4) Antennas that process the input signal by combination of radiation elements and circuit devices are called signal processing antenna. One typical signal processing antenna is the "adaptive antenna".

An example of an adaptive antenna is the adaptive beam forming antenna, which directs the main beam to the incoming wave. Another example is the adaptive null steering antenna, which forms a null in the direction of the unwanted wave.

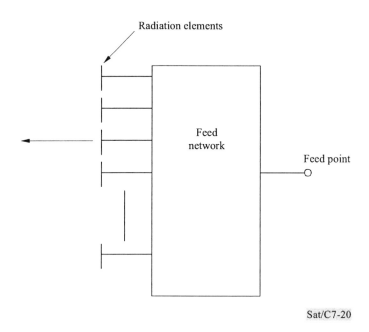

FIGURE 7.20

Array antenna

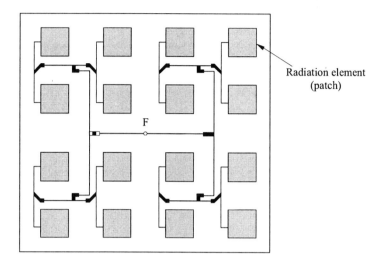

F: Feed point

Sat/C7-21

FIGURE 7.21

Printed array antenna (an example)

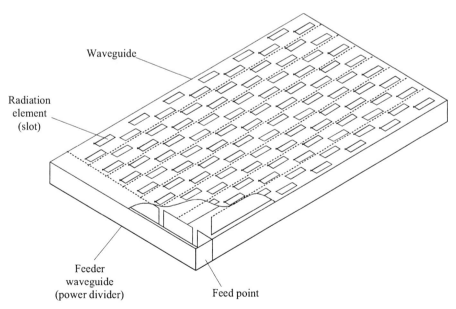

Sat/C7-22

FIGURE 7.22

Waveguide slot array

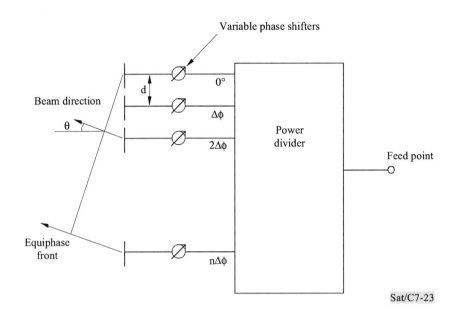

FIGURE 7.23

Phased array antenna

The beam direction is steered by θ if sφ = d sinθ in Figure 7.23.

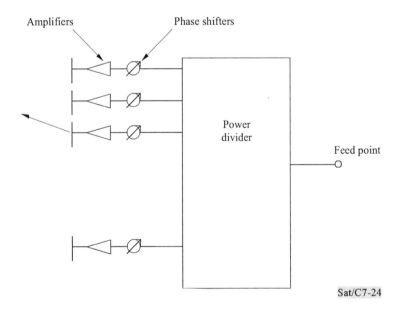

FIGURE 7.24

Active phased array antenna

7.2.6   Installation

The civil works associated with the antenna are described below.

Two types of antennas are considered in this section: large antennas (>15 m) used in high traffic capacity earth stations and medium-sized antennas (15 m to 10 m) used in domestic network stations.

As concerns the installation of small earth stations and, in particular of VSATs, refer to Supplement 3 to the Handbook ("VSAT Systems and Earth Stations", § 5.3).

The pointing accuracies of the antennas require that the antenna be built on ground with stable soil conditions. The soil should be analysed in depth to determine the actual condition on which the foundations are to be laid. These analyses are usually conducted by the antenna supplier/erector and the type of foundation is determined after a study of the soil report.

The foundation must be designed to meet the operational and survival loads imposed by the antenna, the support structure and the pedestal. These include wind loads, gravitational loads and seismic loads.

7.2.6.1   Large antennas

Most large antennas are mounted on an azimuth-over-elevation (Az-El) mechanical assembly of the wheel and track type. In this design, a large diameter circular rail supports the mechanical assembly and allows its rotation in azimuth, while the horizontal stresses are transmitted to a central bearing.

The foundation pedestal can consist of eight columns of reinforced concrete supporting the azimuth rail tied to four columns supporting the central bearing (pintle), connected by radial beams and a top slab.

The pedestal can also consist of a continuous polygonal concrete shell with a self-bearing dome-shaped roofing embedded in the circular beam, the pintle being included in this roofing. In this latter case the foundation is an integral part of the antenna building.

As the diameter of the azimuth rail ring is about 15 m to 18 m for large (30 m) antennas, the area under the top slab or the roofing is an ideal place to locate RF equipment. The internal construction and finish of this area should follow the general guidelines previously given for the main building.

The antenna foundation and pedestal should have a good drainage system which is capable of gathering and dispersing stormwater captured by the antenna and support structure.

7.2.6.2   Medium-sized antennas

Medium-sized antennas (10-15 m) are generally mounted on limited motion assemblies (Az-El or X-Y type) attached to a fixed support structure secured directly to the foundation by means of embedded anchor bolts.

The foundation should be designed after completion of a soil analysis. In general, it consists of footings or of a raft with piles tied in the upper part by groundsels, the whole being constructed in reinforced concrete. The foundation should have a sufficient mass, conveniently distributed to compensate for the wind, gravitational and seismic loads imposed by the antenna. A less stringent soil condition can thus be tolerated.

Small footings should also be provided for supporting an electronic equipment shelter situated, if possible, between the rear legs of the antenna support structure.

7.3 Low noise amplifiers

7.3.1 General

As explained in § 7.1.1.2, the receive subsystem of an earth station always includes a preamplifier with low intrinsic noise, usually referred to as the LNA (low noise amplifier). Keeping in mind that each 0.1 dB feeder loss adds about 7 K to the total receive noise temperature, this low-noise amplifier must be connected to the antenna feed receive port by a very low loss RF link: in practice, the LNA is either directly attached to the feed port (in the case of small or medium antennas) or connected through a low loss waveguide.

FET (field effect transistor) amplifiers[9] are currently used in earth-station receive subsystems. HEMTS (high electron mobility transistors), a type of GaAs FET with improved high frequency performance, are widely used especially for LNAs to achieve good noise temperature performance. The amplifiers can be operated near ambient temperature but superior noise performance is obtained – if necessary – by lowering the physical temperature of the most sensitive components of the LNA (by refrigeration), thus reducing their ohmic losses.

In calculating the total noise temperature of the receive subsystem, it is necessary to take account of the noise temperature (T_i), the gain (g_i) of the successive amplifying stages of the LNA and also the noise temperature (T_n) of the receiver proper (down converter, etc.). The total noise temperature is given by the following equation:

$$TR = T1 + \frac{T2}{g1} + \frac{T3}{g1 \cdot g2} + \cdots + \frac{Tn}{g1 \cdot g2 \cdots gn-1} \qquad (9)$$

This equation shows that the noise temperature of stage i is attenuated by the total gain of the preceding stages. Therefore, the noise performance of the first (input) stage is the dominant factor and this is why, in modern LNAs amplification with refrigeration is used only in the first (and possibly the second) stage(s). Three stages are commonly used with a total gain ($g1 \cdot g2 \cdot g3$) of 40 to 60 dB, thus "masking" (i.e. making negligible) the contribution of the noise (Tn) of the receiver proper.

---

[9] Maser LNAs were employed at the beginning of the satellite communications era. Parametric amplifiers using an inversely biased semiconductor diode were commonly employed until recently. Tunnel diodes and low-noise travelling wave tubes have also been proposed for LNAs. However, these LNA types are no longer in practical use and will not be described.

The physical temperature of the LNA can be controlled by various means.

a)   Thermoelectric cooling

Thermoelectric cooling by Peltier effect diodes provides a simple and efficient solution for refrigerating the LNA sensitive components to about −50°C.

Such cooling is sufficient to obtain LNA noise temperatures as low as $T_e = 28$ K at 4 GHz and 60 K at 11-12 GHz. Since the thermoelectric cooling devices do not include any moving parts (unlike cryogenic cooling) and since they can be fitted directly to the LNA (in a sealed pressurized enclosure), this solution combines high performance with sturdiness and good maintainability.

Cryogenic cooling using gaseous helium is now being abandoned in modern stations, because of its complexity (especially for maintenance), bulkiness and high cost.

b)   Temperature compensation

When moderate noise performance is acceptable (e.g. $T_e = 30$-50 K at 4 GHz and 70-150 K at 11-12 GHz) thermoelectric cooling can be avoided and replaced by simpler control systems allowing operation with stable performance in the normal range of ambient temperatures. One important advantage of such systems is that problems related to gas leaks (dew point, etc.) are simplified, which improves reliability.

7.3.2   FET low-noise amplifiers

Bipolar transistors exhibit complex noise (shot noise, etc.) mechanisms and give relatively poor noise temperature performance at high frequencies (i.e. above 1 GHz).

By contrast, noise in field effect transistors is mostly thermal and can be reduced by proper choice of semi-conductor material and by using micron and submicron fabrication technologies for semi-conductor layer processing and for electrode etching.

In recent years, substantial progress has been made in gallium arsenide (GaAs) field effect transistors (FET) for microwave low noise applications. Noise temperatures which could previously be obtained only with parametric amplifiers is now available with GaAs FETs at considerably reduced overall size and cost.

In addition, GaAs FETs offer other advantages when compared with parametric amplifiers. They are transmission-type amplifiers (in contrast with negative resistance, reflection-type parametric amplifiers) and therefore they are more stable and less critical of circuit impedance and can be optimized for wider bandwidths. They are also powered by d.c. (in contrast with the very high frequency pump power oscillator required for parametric amplifiers).

A typical GaAs FET is made of a 0.15 μm thick n-type GaAs epitaxial layer grown on a semi-insulating GaAs substrate and uses three planar electrodes: the source, the gate and the drain. Current flow between the source and the drain is controlled by the voltage on the Schottky-barrier gate.

At present the high-electron mobility transistor (HEMT), a kind of FET, is commonly used because of its lower noise performance. HEMTs are based on a heterojunction such as GaAs/n-AlGaAs (Figure 7.25) and InGaAs/n-AlGaAs.

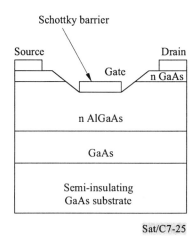

FIGURE 7.25

Basic configuration of HEMT

The mostly thermal character of FET noise makes it possible to reduce the amplifier noise temperature by lowering its physical temperature. With thermoelectric cooling, noise temperatures of about 28 K at 4 GHz and 60 K at 11-12 GHz are achievable. Uncooled current performance is about 30-50 K at 4 GHz, 70-150 K at 11-12 GHz and 180-250 K at 20 GHz.

7.3.3   Summary of the present characteristics of LNAs

Table 7.5 and Figure 7.26 give typical examples of the performance of LNAs in the 4 GHz band. Table 7.6 gives typical noise temperature of LNAs currently available in various frequency bands.

TABLE 7.5

Typical characteristics of LNAs (4 GHz band)

	Item	Thermoelectrically cooled FET amplifier	Uncooled FET amplifier
1	Noise temperature (K)	28	30-50
2	Frequency bandwidth (MHz)	500-800	500-800
3	Gain (dB)	60 (6 stages)	60 (6 stages)
4	Saturation output power (dBm)	+10	+10

TABLE 7.6

Typical noise temperature of current FET LNAs

Frequency band (GHz)	Type	Typical noise temperature (K)
3.7-4.2	Thermoelectric cooling	28
	Uncooled	30-50
11.7-12.2	Thermelectric cooling	60-70
	Uncooled	70-150
17.7-19.5	Uncooled	180-250

7.3.4 Low noise converters

Recent low noise amplifiers and down converters are usually integrated into low noise converters. The low noise converter is compactly housed in a single package. This permits simple installation with antennas.

A Ku band low noise converter unit consisting of a 11/12 GHz band GaAs FET low noise amplifier and a block down converter is covered in this section.

The low noise converter (LNC) converts the satellite signal from the 11/12 GHz band to the 1 GHz band and is normally mounted on the antenna.

The technical features of the Ku band LNC and redundancy configuration are described below.

• Low noise performance

The 11/12 GHz band low noise amplifier consists of the low noise GaAs FETs.

The noise temperature of the LNC unit is 80-180 K depending on the earth station G/T requirement and antenna size.

• Hybrid integration technology

A microwave integrated circuit (MIC) device composed of a semi-conductor chip is employed for 11/12 GHz band GaAs FET low noise amplifier, 1 GHz band IF amplifier, down converter, and phase comparator, to realize a compact equipment configuration, wideband coverage and high reliability.

- Simple installation

The low noise converter is compactly housed in a single package. This permits simple installation with antennas.

As shown in Figure 7.26, the low noise converter consists of a 11/12 GHz band low noise GaAs FET amplifier, a 10 GHz band local oscillator, for high frequency stability, a down-converter, a 1 GHz band IF amplifier, and power supply circuits.

The low noise converter is housed in a single compact package.

The gain of the 11/12 GHz low noise amplifier is about 32 dB, and that of the 1 GHz band IF amplifier is about 32 dB. The conversion loss of the down converter unit is about 6 dB.

Therefore, the overall conversion gain of the low noise converter is in excess of 55 dB.

The 11/12 GHz band low noise amplifier consists of a three-stage GaAs FET amplifier.

The 3rd order intermodulation intercept point is more than +15 dBm. The receive signal, amplified by the low noise amplifier, is converted into a 1 GHz band intermediate frequency signal by a down converter using a Shottky barrier diode mixer. The signal is then amplified by the 1 GHz band IF amplifier.

The 3rd order intermodulation intercept point is about +20 dBm.

The reference signal is supplied from a 25 MHz crystal oscillator externally, which is mounted in the LNC power supply unit or 1 GHz down converter unit. The crystal oscillator is selected to meet required stability for the communications system.

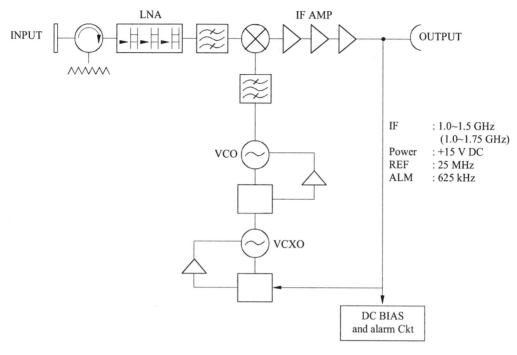

IF   : 1.0~1.5 GHz
          (1.0~1.75 GHz)
Power   : +15 V DC
REF   : 25 MHz
ALM   : 625 kHz

ALM:	Alarm	Ckt:	Circuit
IF:	Intermediate frequency	REF:	Reference
IF AMP:	IF amplifier	VCO:	Voltage controlled oscillator
LNA:	Low noise amplifier	VCXO:	Voltage controlled X'tal oscillator

Sat/C7-26m

FIGURE 7.26

Block diagram of LNC unit

7.4 Power amplifiers

7.4.1   General

As shown in the general block diagram of an earth station (see § 7.1.1), the various input chains are connected to up converters (U/Cs), the outputs of which pass through a combiner (input combiner) which feeds the final power amplifiers (usually called HPAs: high power amplifiers). The output power ($P\tau$) of an HPA must be sufficient to deliver the equivalent isotropically radiated power (e.i.r.p.) which is required for each transmit carrier, taking into account the antenna gain gτ (e.i.r.p. = pτ · gτ). The simplest stations include a single active HPA, which is usually associated with a stand-by HPA, through an input and an output switch (1 + 1 redundant arrangement). Medium and high traffic capacity earth stations often include more than one active HPA. In this case, the various active and stand-by HPAs are connected to the antenna diplexer through a system comprising switches (for redundancy) and an output power combiner (see § 7.4.8).

The reader is referred to § 7.1.1.3 for general information about power amplifiers and for a comparison between TWT (travelling wave tube) amplifiers and klystron amplifiers. The present section gives more details on microwave tubes, HPA and output combiner technologies and also on various problems such as intermodulation, AM/PM conversion, etc.

7.4.2   Microwave tubes

7.4.2.1   Microwave tubes general operation

Earth station HPAs commonly use klystron or TWT amplifier tubes. Both klystrons and TWTS are classified as microwave linear beam tubes. Other types of microwave tubes have been developed, such as backward-wave oscillators and crossed-field tubes (magnetron, carcinotron, etc.) but they are not in current use for earth stations HPAs.

Microwave linear beam tubes are composed of:

- an electron gun which uses a cathode and a beam focusing electrode to generate the electron beam;

- a magnetic focusing system which confines the electrons in a long, narrow, cylindrical beam (in spite of the space charge effect, i.e. mutual electron repulsion). This magnetic focusing system can use either electromagnetic coils or tubular permanent magnets (often of the periodic permanent magnet (PPM) type in modern tubes);

- a slow-wave RF structure. In this structure, the electrons are grouped into bunches under the influence of velocity modulation. These electron bunches act as an amplitude modulated current which induces a progressively increasing RF voltage in several microwave cavities or in a microwave periodic structure (such as helix). This mechanism, in which the microwave energy is obtained from the electron kinetic energy, requires proper phasing between the electron bunches in the beam and the RF waves in the slow-wave structure;

- a collector electrode which closes the d.c. circuit by gathering the beam electrons which impinge upon it.

7.4.2.2   Klystron tubes

Klystrons used in earth station HPAs are of the multi-cavity amplifier type. The slow-wave RF structure is composed of a series of cavities, i.e. microwave resonant circuits (e.g. 5 cavities) as represented in Figure 7.27 a). The low level RF input signal excites the input cavity and alternately accelerates and decelerates the electrons passing in the interaction gap, thus creating electron bunching. This induces an RF voltage in the second cavity. An additional modulation of the beam velocity is produced by this resonant voltage which increases the modulation of the beam current.

This amplification process is repeated in each intermediate cavity, thus generating a high power RF voltage which is extracted from the last cavity by the output circuit.

The salient features of klystron tubes and HPAs are given in § 7.1.1.3 b). Other important characteristics are listed below:

i) Instantaneous bandwidth: the higher the output saturation power, the wider the possible instantaneous bandwidth. In the 6 GHz band, 40 MHz bandwidth is common for output power greater than 300 W (3 kW klystrons with 80 MHz bandwidth are now available). In the 14 GHz band, 80 to 100 MHz bandwidth klystrons are currently available with output power up to 3 kW.

ii) Tuning range: the operating central frequency of a klystron can be changed by mechanically tuning the cavities. 500 to 600 MHz tuning ranges are common and are usually provided with remote control capability, e.g. with 6 to 24 switchable pre-set frequencies for channel selection.

iii) Focusing system: permanent magnet focusing is much more simple to operate and maintain than electromagnetic coil focusing. Klystrons with permanent magnet focusing are available up to about 3 kW power.

iv) Cooling system: in the same way, cooling by forced air is much preferable to cooling by liquid circulation systems. Again, air cooled klystrons are available up to about 3 kW.

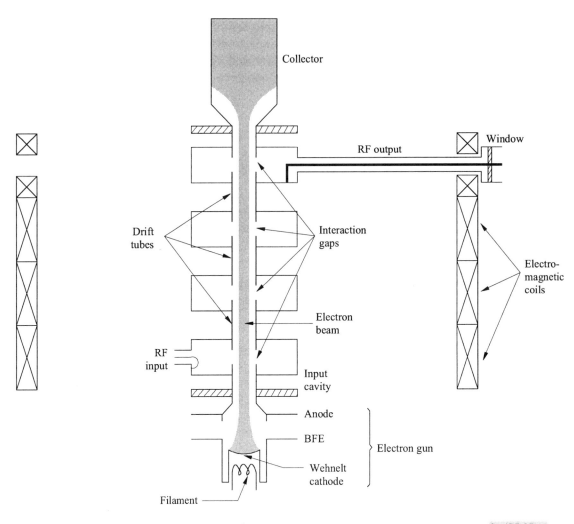

Sat/C7-27am

FIGURE 7.27 A)

Schematic cross-section of a klystron BFE: beam focusing electrode

Table 7.7 gives basic data on typical klystron tubes currently available for earth station HPAs in the 6 GHz and 14 GHz bands. Other available klystrons include:

- very high power klystrons at 14 GHz (up to 10 kW);

- klystrons in the 17/18 GHz band (1.5 kW);

- klystron in the 30 GHz band (2 kW).

TABLE 7.7

Klystrons in the 6 GHz, 14 GHz, and 30 GHz bands

Tuning range (GHz)	Output power (W)	Gain (dB)	Instantaneous bandwidth (MHz)
5.925-6.425[1]	150 400 750 1 000 1 500 2 000 3 000 3 400 14 000[3, 4]	50 36 40 35 37 38 35-42[2] 40 52	23 42 40 40 40 40 40-80[2] 45 70
14-14.5	1 500 2 000 3 000	40 40 42	100 100 85
27.5-30.5[2]	350 450	40 40	100 120

[1]   Klystrons are also available in the new increased band: 5.850-6.425 GHz.

[2]   Depending on type and manufacturer.

[3]   Focusing by electromagnetic coils (all other tubes are focused by permanent magnets).

[4]   Water cooling (all other tubes are air-cooled).

7.4.2.3   Travelling wave tube

As explained in § 7.1.1.3, the major advantage of the TWT is its intrinsically wide bandwidth. This broadband capability features high operational flexibility:

- any carrier frequency change can be effected, in the whole operational bandwidth (e.g. 500 MHz or more), without any tuning or modification of the HPA system (including the output combiner);

- several carriers at different frequencies can be simultaneously transmitted in the same HPA and, therefore, traffic expansions of an earth station are possible without increasing the number of tubes. Of course, this multiple carrier capability is limited by the overall TWT output power and, more precisely, by the output power available for sufficiently linear operation of the TWT (see § 7.4).

In a TWT, the electron beam interacts with a forward RF wave propagating in the periodic, non-resonant slow-wave structure. Velocity modulation, created along the beam, induces an RF current that excites waves in the structure in both directions. However, the synchronization conditions are such that only forward waves add in phase and are amplified.

Most TWTs use a helicoidal RF transmission line (helix) as a slow wave structure (see Figure 7.27). Modern technologies have improved the heat dissipation along the structure, thus allowing helix TWTs to be used for up to about 3 kW output power at 6 GHz, 1 kW at 14 GHz.

However, at higher frequencies, due to its reduced cross-section dimensions, the helix structure is no longer usable for high power. High power TWTs at 14 GHz (and higher frequencies) use coupled-cavity slow-wave structures.

In TWTs, the electrons when leaving the interaction space retain considerable energy (unlike the case of the klystron). This is why, in many modern tubes, these electrons are collected at a potential below the slow-wave structure potential. This technique, which is called depressed collector operation, reduces the collector heat dissipation and increases the tube efficiency. Further improvement can be obtained with multiple collector stages.

A periodic permanent magnet structure is commonly used for focusing beam to minimize size and weight. A similar cooling technique as used for klystrons can apply to TWTs.

Table 7.8 gives basic data on current typical TWTs used for earth station HPAs in the 6 GHz, 14 GHz and 30 GHz bands. Figure 7.28 gives the power consumption of the various types of HPAs.

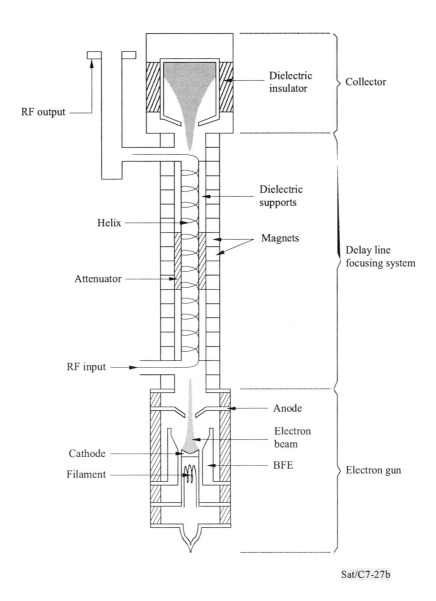

Sat/C7-27b

FIGURE 7.27 B)

Schematic cross-section of a travelling wave tube

TABLE 7.8

TWTs in the 6 GHz, 14 GHz and 30 GHz bands

Frequency band (GHz)	Output power (W)	Gain (dB)	Focusing[3]	Cooling[4]
5.925-6.245[1]	40	39	PPM	CD
	75	35-41[2]	PPM	CD
	150	40-44[2]	PPM	CD
	400-600[2]	35-50[2]	PPM	FA
	700-750[2]	43	PPM	FA
	1 200-1 500[2]	36	EM	FA
	3 000-3 200[2]	33-50[2]	PPM or EM[2]	FA or L[2]
	8 000-8 500[2]	35	EM	L/FA
	13 000	35	EM	L/FA
14-14.5	16-25[2]	40-55[2]	PPM	CD
	50	50	PPM	CD
	130-160[2]	45-50[2]	PPM	CD or FA[2]
	250-300[2]	42-50[2]	PPM	FA
	500-750[2]	50	PPM	FA
	1 000	45	PPM	FA
	2 000-2 500[2]	30-45[2]	PPM or EM[2]	FA or L/FA[2]
	3 000	50	EM	FA
27.5-30.5	25	40	PPM	CD
	100-150[2]	37-45[2]	PPM	CD or FA[2]
	200	30	PPM	FA
	600-1 000[2]	33-45[2]	PPM or EM[2]	FA or L[2]

[1]   TWTs are also available in the new increased band: 5.850-6.425 GHz.

[2]   Depending on type and manufacturer.

[3]   PPM: periodic permanent magnet; EM: electromagnetic coils.

[4]   CD: conduction; FA: forced air; L: liquid; L/FA: partially liquid and forced air.

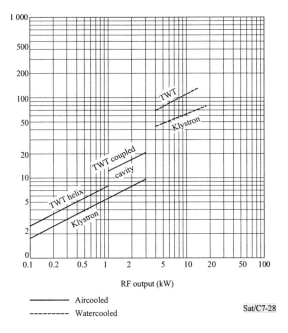

FIGURE 7.28
Typical power consumption of HPAs

7.4.3 Microwave tube amplifier design

As explained before there are two main types of high power amplifier (HPA) which are used in earth terminals for satellite communications:

• those using a broadband travelling wave tube (TWT) with a bandwidth of 500 MHz or more;

• those, much simpler, using a klystron with a bandwidth of 40 or 80 MHz, tunable in the 500 or 600 MHz frequency band occupied by the satellite transponders.

The following sections give the essential features to be taken into account for the design of HPAs. Figures 7.29 and 7.30 represent the block diagrams of HPAs using respectively a TWT and a klystron.

7.4.3.1 TWT power supply

In addition to the heater supply, three different high voltages are necessary for the anode, the helix (or the coupled-cavity) grounded structure and the collector. In some medium power TWTs the use of a separate collector is optional (operation with either a depressed or a grounded collector) but the high power TWTs are all operated with a depressed collector.

The switching on of a helix TWT requires either a fast rise time of the helix voltage or a means of preventing beam current until the voltages are at operating values. One method is to use a switch which allows the anode to be clamped to the cathode when the collector and helix voltages are applied ("off" position) and then afterwards the anode voltage, derived from the helix supply, is applied ("on" position). Another means is to use a separate power supply for the anode providing a voltage which is delayed with respect to the helix voltage.

554

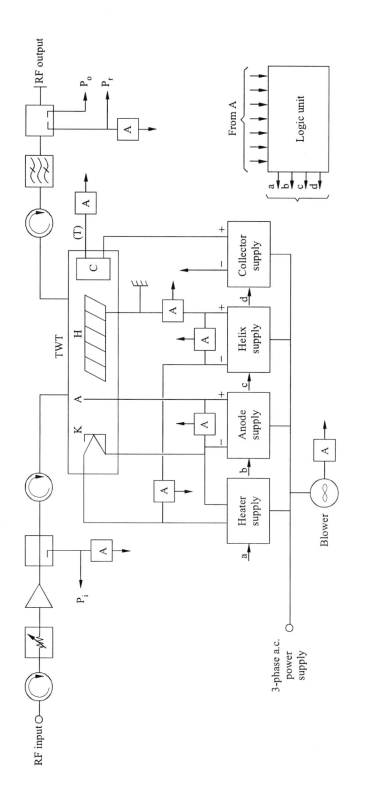

Sat/C7-29

FIGURE 7.29

Block diagram of a TWT HPA (500 to 700 W)

P_i: input power measurement

P_o: output power measurement

P_r: reflected power measurement

A : alarm

K: cathode

(T): temperature protection

C: collector

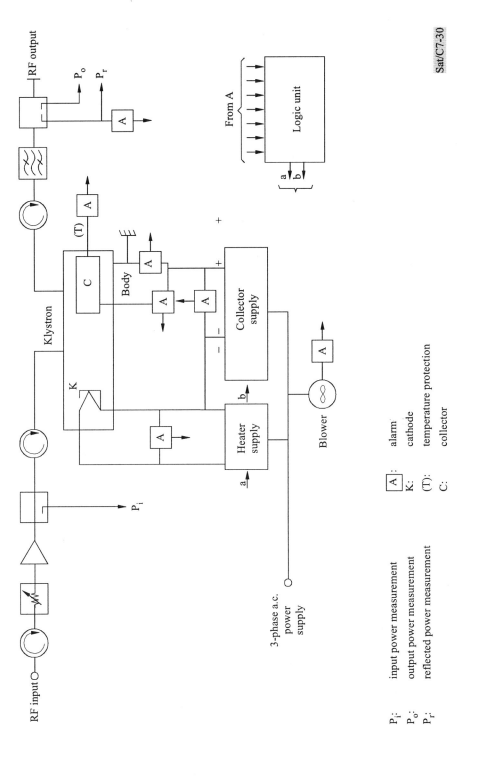

FIGURE 7.30

Block diagram of a klystron HPA (1 to 2 kW)

Sat/C7-30

P$_i$: input power measurement
P$_o$: output power measurement
P$_r$: reflected power measurement

A : alarm
K: cathode
(T): temperature protection
C: collector

The heater voltage can be either a.c. or d.c. (d.c. is recommended).

Some typical values for a 500 W TWT are as follows:

- helix voltage: 10 kV. Helix current to be limited to 15 mA. The helix voltage should be well regulated (up to 10^{-3} of the specified value) and well filtered (up to 10^{-6} of the specified value);

- anode voltage: 6 kV (regulated up to 10^{-3}), current: 1 mA;

- collector voltage: 5.5 kV not regulated, current: 400 mA;

- heater voltage should be regulated within 0.1 V.

7.4.3.2 Klystron power supply

In addition to the heater supply, only a single high voltage for the beam is required between the cathode and the collector (or body). Unlike the helix TWT, the beam voltage can be run up slowly. The klystron amplifier is sensitive to the beam power supply variations. A typical value of RF output power sensitivity is 0.01 dB/V. In order to ensure high gain stability, it is necessary either to regulate the beam voltage or to use an automatic level control (ALC) which reduces the gain variations caused by fluctuations of the beam voltage.

Typical values for a 3 kW klystron are: 8.5 kV for the beam voltage and 1A for the collector current.

7.4.3.3 Type of high voltage supply

Each of the above-mentioned high voltages can be provided in two different ways:

- the usual is to obtain the high voltage directly from the a.c. main line through a conventional power supply chain of transformer, rectifier and filtering components;

- another way is to obtain a low d.c. voltage (e.g. 300 to 600 V) from a chain similar to the previous one, with only rough filtering and then to use a d.c.-d.c. converter to provide the required high voltage. This converter employs a solid-state switching device working at a frequency of one to several tens of kHz. The advantages are the reduced size and weight of the transformer and filtering components and also the reduced energy stored in the filtering capacitors. On the other hand, it is necessary to prevent noise effects at the switching frequency and its harmonies.

7.4.3.4 Tube protection

Helix current protection is essential for high power TWTs and is usually provided in the form of a relay which removes helix voltage in the event of high helix current. Body current protection, in the form of a relay, should also be provided for the klystron.

All high power air-cooled TWTs and klystrons are protected from loss of sufficient cooling by an integral thermal switch.

Where liquid cooling is employed care must be taken to ensure the purity of the coolant.

Arc detectors are recommended at power levels of 1 kW and above. The time response of the detection circuit should be about 15 ms.

The heater must be at its operating conditions prior to applying beam voltage (up to 5 min delay may be necessary).

7.4.3.5 Driver amplifier (preamplifier)

A driver amplifier is usually required to increase the output level of the modulating chain of the transmitter up to the input level of the high power TWT or klystron amplifier. The driver amplifier should be operated under conditions such that the intermodulation occurs mainly in the HPA itself.

So far, the preamplifiers have used a TWT with a saturation output level of a few watts. Solid-state preamplifiers are now available at 6 GHz, 14 GHz and up to 18 GHz and should preferably be used.

The noise figure of the preamplifier must be as low as possible and its intercept point (defined in § 7.4.4) as high as possible.

Some typical values for solid-state preamplifiers are:

- at 6 GHz, a noise figure of 6 dB, and an intercept point at 29 dBm (operating level of 18 dBm);

- at 14 GHz, a noise figure of 8 dB, and an intercept point at 23 dBm (operating level of 12 dBm).

7.4.3.6 Logic unit

This unit, also known as the control, monitoring and protection logic system, provides the interface between the amplifier, its power supplies and protection circuits, and the control, display and alarm circuits.

In addition to providing local control, the logic unit may include facilities for allowing remote control from the centralized monitoring and control systems of the station.

The logic unit performs the following functions:

- placing the amplifier in the required state from the manual local or remote control position;

- display of this state;

- continuous monitoring of amplifier operation and its protection by various measurements.

This logic system is fed with the following inputs:

- the control commands;

- the alarms from the various amplifier circuits.

The logic unit processes these various signals and supplies commands in the form of insulated loops for operating the relays employed in the different controls (blower, filament, high voltage, etc.). When high speed is required, transistors instead of relays are used.

When an alarm appears, it is stored and displayed; the logic system controls the switching off of the various amplifiers' functions.

7.4.4   Solid-state power amplifiers

Solid-state power amplifiers can now be used in small earth stations. In fact, solid-state amplifiers with GaAs FETs (field effect transistors) are capable of producing outputs in excess of 100 W at C band and 20 W at Ku band. Even more power is possible by connecting two or more solid state amplifiers in a combiner unit. IMPATT diode amplifiers with higher power have also been proposed but are scarcely ever used due to their rather poor performance (low efficiency, high noise level, possible unstable operation). GaAs FET power amplifiers feature good efficiency (low power consumption), high reliability and low cost and do not require high voltages needed by the microwave tube amplifiers. They also exhibit good linearity performance, often better than microwave tube amplifiers. Consequently, smaller back-offs are needed in multi-carrier operation, which is the case when several SCPC carriers are transmitted.

The maximum output power of solid-state amplifiers is usually specified at the "1 dB compression point" Pc (i.e. for a gain 1 dB less than the linear gain). Also, the intermodulation performance characteristic is usually specified by the "intercept point" (i.e. the theoretical one carrier output power $\overline{IP}$ at the point where the extrapolated linear one carrier output power is equal to the extrapolated linear two carriers intermodulation power (see Figure 7.38)). The intercept point $\overline{IP}$ of the third order intermodulation product is related to the intermodulation ratio D_3 (see § 7.4.5.2.1) and the single carrier output power P_o by the following formula:

$$(\overline{IP})_{dBw} = \frac{(D_3)_{dB}}{2} + (P_o)_{dBW} \tag{10}$$

from which:

$$(D_3)_{dB} = 2\left[(\overline{IP})_{dBW} - (P_o)_{dBW}\right] \tag{11}$$

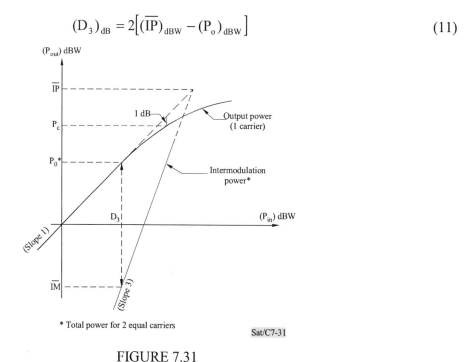

FIGURE 7.31

Amplifier intermodulation characteristics

7.4.5 General performance requirements of HPAs

7.4.5.1 General

The principal performance requirements for the transmitter of an earth station, e.g. in the INTELSAT system, include frequency-amplitude characteristics, intermodulation, delay distortion, AM/PM conversion, out-of-band emission, and residual AM. For these, the characteristics of the HPA play a dominant role, although it is possible to equalize the amplitude and delay distortions in the IF stage.

i) Intermodulation: the amplitude response (output power vs. input power characteristic) of a microwave tube (and of an HPA) is linear for small signals, with a slope equal to the small signal gain, but the response becomes non-linear near the saturation point. This non-linearity causes unwanted carriers (intermodulation products) to appear when multiple carriers are transmitted simultaneously in the same tube with a total power approaching saturation.

The intermodulation performance characteristics of an HPA are usually specified by the intermodulation level of the third-order intermodulation products produced by two equal input carriers. As a typical example, if a TWT HPA with a saturation output power of 1.3 kW is driven by two carriers at frequencies f_1 and f_2, each with an output power of 60 W, two third-order intermodulation products at frequencies $2f_1-f_2$ and $2f_2-f_1$ appear with a level of about –26 dB (relative to the output power of each carrier). Of course, the intermodulation levels depend on the back-off of the tube, i.e. the carriers power level (relative to the saturation). In the example, the back-off (output back-off) is equal to:

$$10 \log \frac{2 \cdot 60}{1\,300} = -10.35 \text{ dB}$$

Intermodulation can be reduced by inserting a linearizer at the amplifier input. A linearizer is a circuit which, due to its own reverse non-linearity, compensates the amplifier non-linearity.

More details on intermodulation calculations and on linearizers will be found in Chapter 2 (§ 2.1.5.2), below in § 7.4.5.2 a) and c) and in Appendix 5.2.

ii) AM/PM conversion: AM/PM conversion occurs when an amplitude variation (AM: amplitude modulation) of the carrier transmitted in the HPA induces a phase variation (PM: phase modulation) in the carrier. AM/PM conversion is also caused by non-linearities in the HPA. In fact, the input-output response of an HPA comprises an input/output amplitude response (AM/AM: see i) above) and an input amplitude versus output phase response (AM/PM). The AM/PM conversion (AM/PM response slope) at saturation is about 4°/dB in klystron HPAs and 7°/dB in TWT HPAs.

Non-linear effects (AM/PM conversion and AM/AM, i.e. amplitude compression near saturation) cause intermodulation (see above) and alter the modulation characteristics (modulation spectrum) of a transmitted carrier. Again, linearizers can reduce these effects. An example of non-linear effects ("spectrum spreading") on a digitally modulated carrier is given in § 7.4.5.2.2.

iii) Gain-frequency response: gain variations with frequency inside the signal bandwidth must be maintained within specified limits.

In fact, when two or more carriers are simultaneously amplified through a circuit which has gain slope and AM/PM conversion, intelligible cross-talk occurs between the carriers. This intelligible cross-talk is proportional to the product of gain slope and AM/PM conversion factor. Gain slope in an HPA is about 0.05 dB/MHz. For earth stations in the INTELSAT system, intelligible cross-talk is specified to be less than –58 dB.

NOTE – Significant gain variations may occur in the operating bandwidth of a TWT HPA (e.g. 5 dB within 500 MHz). Though these gain variations can be compensated, for each individual carrier, at the input level, it is often useful to insert a gain equalizer circuit at the HPA input.

iv) Delay distortions: group delay variations with frequency within the signal bandwidth must also be maintained within specified limits.

In a transmit chain, the delay distortions caused by the HPA are often dominant. Usually, overall delay equalization of each individual transmit chain is performed by adding time delay equalizer (TDE) circuits in the IF stages. However, RF equalizers may also be used for equalizing TWTs throughout their operating band.

v) Residual amplitude modulation: when the transmit signal from the earth station contains AM components, distortion noises are generated in the on-board transponder amplifier due to AM/PM conversion. To prevent this, it is necessary to reduce the residual AM component of the output signal from the earth station. In the INTELSAT system, the allowance for residual AM in the transmit signal at the earth station is specified as follows:

- –20 (1 + log f) dB/4 kHz, for AM components between 4 kHz and 500 kHz (f: centre frequency in kHz of the 4 kHz slot under measurement);

- –74 dB for AM components above 500 kHz.

Since the residual AM is produced mainly by the HPA power supply, ripple of this power supply must be kept within severe limits.

vi) Harmonic generation: microwave tubes generate harmonics, i.e. unwanted signals at radio frequencies which are multiples of the basic wanted RF carrier. The table below shows the typical level of these harmonics. In order to lower this level to about –50 to –60 dB, a harmonic filter is often placed at the HPA output.

Typical harmonic generation in microwave tubes

Tubes	Order of harmonic	Harmonics relative to the fundamental level	
		Saturated	5 dB back-off
Klystron	2nd 3rd	−30 −40	−35 −50
Helix TWT	2nd 3rd	−10 −20	−15 −30
Coupled cavity TWT	2nd 3rd	−20 −30	−25 −40

vii) Unwanted and noise emissions, spurious tones, bands of noise and other undesirable signals, may be transmitted by the earth station throughout the RF operating bandwidth and cause interference. These emissions may be present at the HPA input or may be generated in the HPA itself. In particular, electron fluctuations in a microwave tube generate some RF noise even when no signal is applied at the input. The level of this noise is indicated by the noise figure (F) of the tube.[10] It is to be noted that TWTs generate noise in the whole operating band, while klystrons generate noise only in their actual bandwidth (e.g. 40 to 100 MHz).

It may be useful to divide unwanted and noise emissions into two categories, which correspond to two different performance requirements:

- RF out-of-band and spurious emissions, i.e. the e.i.r.p. outside of the satellite bandwidth unit allocated for each particular carrier: this e.i.r.p. is often specified to be less than 4 dB(W/4 kHz);

- idle state RF emission: in TDMA operation, several earth stations transmit the same RF carrier during different short time periods ("bursts"). The idle state RF emission is the e.i.r.p. which is radiated by a particular earth station outside of its own allocated time periods. This emission causes interference to the other stations and is of particular concern when there are numerous

---

[10] Most HPAs are composed of two amplifying stages: a preamplifier and the power amplifier proper. If the gains and noise figures of these two stages are g_1, g_2 and F_1, F_2, the output noise is given by the classical formula:

$$N = (F_1 \cdot g_1 \cdot g_2 + F_2 \cdot g_2) k \, T_0 b \qquad W \qquad (12)$$

where   k: $1.38 \cdot 10^{-23}$ J/K (Boltzmann's constant)

           T_0:   290 K

           b: bandwidth (Hz)

           (e.g. $k \, T_0 b = -168$ dB(W/4 kHz))

HPA designers must pay attention to the sharing of the overall gain between g1 and g2, taking into consideration both noise and intermodulation products (in case of multiple carriers). TWT noise figures are in the range 25 to 36 dB. In modern HPAs solid-state (FET) amplifiers are to be preferred, as far as possible, due to their better noise figure (e.g. 10 dB).

stations operating in the same TDMA time-frame. The specification in this case may be more severe than the previous one.

7.4.5.2 Non-linearity effects in power amplifiers

Non-linearity in power amplifiers causes such effects as intermodulation, AM-PM conversion, etc. These effects can result in two types of impairment:

i)   Interference: unwanted RF power can be transmitted by the earth station, causing interference. That is why the system's specifications usually put a limit to the unwanted e.i.r.p. (For example, INTELSAT's specifications put a limit of about 20 dB(W/4 kHz) for the out-of-band intermodulation products caused by a particular earth station and reserve an allowance of 500 pW0p in the overall link budget of FDM-FM links for RF out-of-band noise caused by multi-carrier intermodulation from other stations in the system.

ii)   Carrier-to-noise degradation: carrier-to-noise ratio of a given transmitted carrier can be impaired by interfering power arising from other carriers transmitted by the same station.

This section gives some details on non-linear effects in the case of FDMA transmission (intermodulation) and in the case of digital transmission, especially TDMA. The section is also applicable to earth-station HPAs, and to satellite transponder TWT amplifiers. In the latter case carrier-to-intermodulation noise ii) is a major factor in reducing transponder traffic capacity in FDMA.

7.4.5.2.1 Intermodulation

This subsection deals with practical calculation of third-order intermodulation products in a power amplifier, given the intermodulation products in the case of two equal reference carriers (a specification usually given by the tube manufacturer). For generalities on intermodulation effects, the reader is referred to Chapter 2, § 2.1.5.2.

In the case of two carriers (N=2) an approximate formula for the calculation of the power of each third-order intermodulation product $\overline{IM}$ (in watts, first type) is:

$$\overline{IM}_{ij} = \left(\frac{\overline{IM}_o}{P_o} \right) \frac{P_i^2 \, P_j}{P_o^{\,2}} \tag{13}$$

at the two frequencies: $2f_i - f_j$, with i=1, j=2 or i=2, j=1

where:

$\overline{IM}_o$: power of the intermodulation products for two reference carriers (W),

P_o: power of each reference carrier (W),

P_i, P_j: power of each actual carrier (W).

This formula can also be expressed in dB:

$$(\overline{IM}_{ij})_{dBW} = -D_3 - 2\,(P_o)_{dBW} + 2\,(P_i)_{dBW} + (P_j)_{dBW}$$

The term $D_3 = -10 \log \dfrac{\overline{IM}_o}{P_o}$ is the reference third-order intermodulation ratio (see Chapter 2, § 2.1.5.2, Figure 2.8 a)).

This term is obtained from data given by the manufacturer (e.g. $D_3 = 26$ dB in the example given in above in § 7.4.5.1).

In the case of three carriers (N=3), the first type products must be calculated for each carrier pair and an approximate formula for the calculation of the power of each third-order intermodulation product of the second type $\overline{IM}_{1,2,3}$ (in watts) is:

$$\overline{IM}_{1,2,3} = 4 \left(\frac{\overline{IM}_o}{P_o} \right) \frac{P_1 P_2 P_3}{P_o^2}$$

(at the frequencies: $f_1 + f_2 - f_3$, $f_1 - ff_2 + f_3$ and $f_2 + f_3 - ff_1$) or in dB:

$$(\overline{IM}_{1,2,3})_{dBW} = -D_3 - 2(P_o)_{dBW} + 6 + (P_1)_{dBW} + (P_2)_{dBW} + (P_3)_{dBW} \tag{14}$$

These formulae are valid only when there is a sufficient back-off and, more precisely, the power used for each carrier is in the vicinity of the power P_o (which is actually used as a reference for the measurement of D_3).

It should be noted that:

- if each carrier is reduced by 1 dB, the intermodulation products are reduced by 3 dB. In consequence, the carrier power to intermodulation ratio $C/\overline{IM}$ is increased by 2 dB. Near saturation, the intermodulation products are reduced by less than 3 dB, due to amplitude compression near saturation;

- for $N \geq 3$, the amplitude of the second type intermodulation products is dominant, due to the 6 dB factor in ($\overline{IM}_{1,2,3}$).

The above formulas are useful for qualitative or first-order evaluations. Complete and accurate calculations are possible using rather simple computer methods. The most common method is based on an approximation using a Fourier-Bessel development of the actual non-linear characteristic of the microwave tube ($Z(r) = g(r) e^{j\varphi(\gamma)}$) under single carrier operation. Here r is the input signal amplitude, g(r) is the AM/AM characteristic and $\varphi(r)$ is the AM/PM characteristic.

Using the measured characteristic of the tube and the mathematical best fit approximation (with a limited number of Fourier-Bessel terms), it is possible to calculate the intermodulation products and other data under all possible multi-carrier operating conditions. Detailed calculation methods are given in Appendix 5.2. The calculated data prove to be in good agreement with measured intermodulation data. Figure 7.32 gives a typical example of such calculated data.

So far, only unmodulated carriers have been considered. In the actual case of modulated carriers, e.g. FM carriers, the intermodulation products are also modulated and the intermodulation powers (per hertz) are reduced in proportion with the effective energy dispersal of the input carriers. More precisely, in the case of an FM carrier with a large enough modulation index, the spectrum is

Gaussian with a standard deviation σ (σ_i being the r.m.s. frequency deviation of the ith carrier). The intermodulation products keep the same Gaussian spectrum with the following standard deviations:

$$(\sigma_{\overline{IM}_{ij}}) = \sqrt{4\sigma_i{}^2 + \sigma_j{}^2} \qquad \text{for two carriers} \tag{15}$$

$$\text{(3rd order, 1st type)}$$

$$(\sigma_{\overline{IM}_{1,2,3}}) = \sqrt{\sigma_1{}^2 + \sigma_2{}^2 + \sigma_3{}^2} \quad \text{for three carriers} \tag{16}$$

$$\text{(3rd order, 2nd type)}$$

Therefore, the intermodulation power, e.g. that given by the approximate formula for $\overline{IM}_{ij}$ and $\overline{IM}_{1,2,3}$ is reduced by a spreading factor $10 \log [(\sqrt{2\pi/4}) \cdot \sigma_{\overline{IM}}]$ (if the terms are expressed in kHz and $\overline{IM}$ in dB(W/4 kHz)).

In conclusion, due to intermodulation, HPAs must be operated in a sufficiently linear region, well below saturation, whenever they are used for multiple carrier transmission. An output back-off of 7 dB to 10 dB (or even more) is commonly needed when transmitting several FDMA carriers in a common TWT HPA which reduces the overall efficiency of the earth-station transmit subsystem.

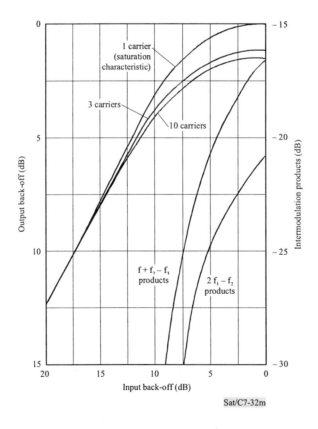

FIGURE 7.32

Typical relative output power and intermodulation products of a TWT

7.4.5.2.2   Non-linearity effects in digital transmission

As already explained in Chapter 2 (§ 2.1.5.3), digitally encoded, phase modulated RF carriers, with a constant amplitude envelope, should theoretically not suffer from AM-AM and AM-PM effects. In the case of high bit rate TDMA, only one RF carrier is usually transmitted through the satellite transponder. A significant example is the INTELSAT 120 Mbit/s 4-PSK modulated TDMA carrier, which occupies the full bandwidth of a 70 to 80 MHz transponder. In this case, it is possible to operate the satellite transponder TWT at saturation, thus getting the full transponder power and traffic capacity. It is also common practice to use a dedicated earth-station HPA for transmitting the single TDMA carrier.

Nevertheless, non-linearity must be taken into account for such a transmission chain (which comprises at least two non-linear circuits: the satellite and the earth-station power amplifiers). This is due to the need for frequency filtering in the satellite transponder and in the earth-station transmit chain. Filtering limits the signal bandwidth (to about 1.2 R for 2-PSK and 0.6 R for 4-PSK, R being the bit rate). As a consequence, the signal no longer has a constant amplitude envelope and is subject to AM-PM conversion (and also to AM-AM) in the non-linear amplifiers. This induces two harmful effects:

i)   a degradation of the overall BER performance (since the digital information is contained in the phase modulation):

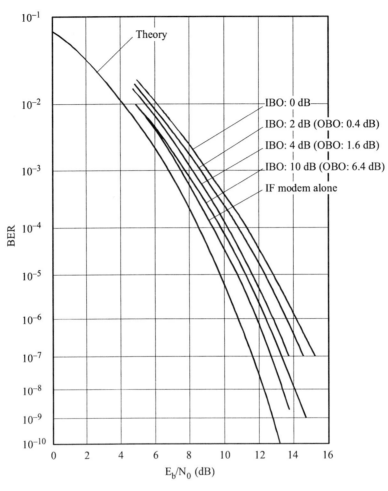

FIGURE 7.33

Typical BER performance of a 120 Mbit/s 4-PSK signal as a function
of E_b/N_0 ratio and earth-station HPA back-off

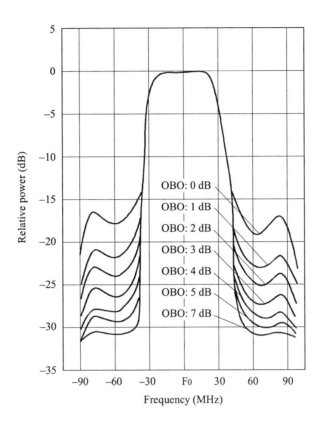

OBO: Output back-off

Sat/C7-34m

FIGURE 7.34

Power spectra of output of HPA (without linearizer)

ii) a modification of the transmitted signal RF spectrum. In fact, when transmitted through a non-linear amplifier (near saturation), the spectrum of a PSK signal (main lobe only after filtering), gives rise to regeneration of its original (sin x/x) spectrum (where $x = \pi(f–f_o)/R$ for 2-PSK and $x = 2\pi(f–f_o)/R$ for 4-PSK, f_o being the carrier frequency). In particular, the first side lobes of the (sin x/x) spectrum reappear ("side lobes regrowth"). Due to this spectrum spreading, some unwanted power is radiated outside of the assigned bandwidth. In order to limit interference into adjacent transponders, the system specifications stipulate the maximum e.i.r.p. level to be radiated by an earth station through its HPA. For example, the INTELSAT TDMA specification states that the e.i.r.p. of the out-of-band emission resulting from spectrum spreading of a PSK carrier (due to earth-station HPA) shall not exceed 23 dB(W/4 kHz) outside a ±44 MHz bandwidth. The EUTELSAT specification is similar but slightly more severe.

Figures 7.33 and 7.34 give typical examples of actual measurement of BER degradation (effect i)) above) and of spectrum spreading (effect ii)) in the case of a 120 Mbit/s 4-PSK signal.

It should be noted that:

- the usual practice with TDMA systems is to operate the satellite TWT amplifier very near saturation (typically 2 dB input back-off, which corresponds to 0.2 or 0.3 dB output back-off);

- INTELSAT operating practice is to limit the earth-station e.i.r.p. to permit more cost-effective HPAs while still obtaining BER performances (under clear-sky conditions) of better than 10^{-7}. The input back-off of the INTELSAT-V satellite is 8.0 dB while that to be used on INTELSAT-VI will be 3.0 dB;

- the earth-station HPA is usually operated with a significant back-off to account for both i) and ii) effects. Typically, output back-offs amounting to 3 dB (INTELSAT specifications) and to 4 to 5 dB (EUTELSAT) are needed. As explained in § 7.4.5.2.3, linearizers can lower these required back-offs;

- amplitude and group delay equalizers must be provided in the transmit and receive chains in order to compensate amplitude and group delay variations (versus frequency) in the complete earth station-satellite-earth-station path. System specifications stipulate the amplitude and group delay "masks" (i.e. tolerances). The measurements shown in Figure 7.34 were made after proper equalization.

7.4.5.2.3   Linearizers

Several methods have been proposed for improving microwave tube linearity near saturation. The most current method is to insert a wideband non-linear predistortion network – called a linearizer – at the input of the amplifier. An example of a linearizer block diagram is shown in Figure 7.35.

This network produces an amplitude expansion and a phase lead when an increasing signal is fed into the input port, thus compensating for the gain reduction and the phase lag (AM-PM) of the HPA when approaching saturation. Of course, this compensation is not perfect but the intermodulation ratio can be improved by about 10 dB (e.g. D_3 increases from 26 dB to 36 dB for 8 dB output back-off) and the phase variation can be minimized (e.g. 20° instead of 60°). Such improvement permits:

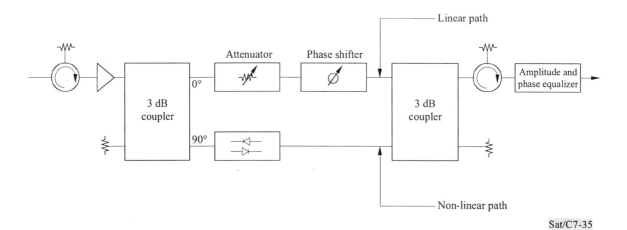

Sat/C7-35

FIGURE 7.35

Example of linearizer block diagram

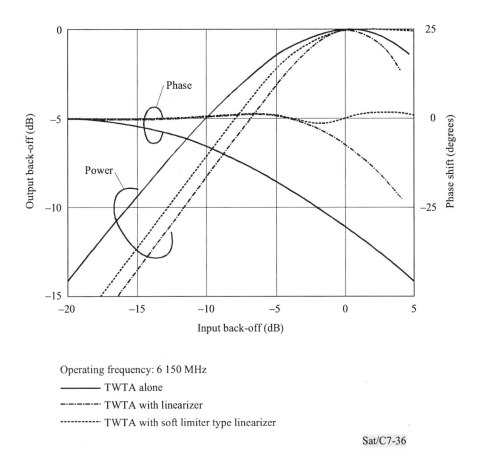

FIGURE 7.36

**Input-output transfer characteristics of TWTA alone,
TWTA with linearizer and TWTA with soft limiter type linearizer**

- an increase of the amplifier traffic capacity in multi-carrier operation by reducing, typically by 3 dB, the output back-off which is required for a given (specified) intermodulation products level;

- a reduction (typically by 1.5 dB) of the required output back-off in digital (PSK) transmission.

The drawback of the simple pre-distortion linearizer is that it exhibits rapid amplitude depression and large phase shift at or beyond saturation, and it is not suitable for use when a TWTA is operated very close to saturation as would be required for the earth-station HPA when used with a single carrier. To overcome this, the use of a soft limiter type of linearizer is promising. A soft limiter type linearizer can be realized using a simple pre-distortion network followed by an amplitude limiter.

Figure 7.36 shows the input-output transfer characteristics of a TWTA alone, a TWTA with a simple pre-distortion linearizer and a TWTA with a soft limiter type linearizer. The amplitude characteristics of a TWTA with a soft limiter type linearizer is linear up to saturation and has a constant envelope beyond saturation. It also exhibits almost zero phase shift, irrespective of the operating point.

Figure 7.37 indicates improvements in carrier-to-intermodulation ratio by the use of TWTA with a soft limiter type linearizer, while Figure 7.38 shows the improvement obtained in the case of 120 Mbit/s 4-PSK signal output spectrum.

Therefore, the installation of a linearizer at the input of an earth-station HPA can either allow an increase in the traffic capacity (for a given HPA) or a reduction in the required HPA size (i.e. the HPA tube saturated power).

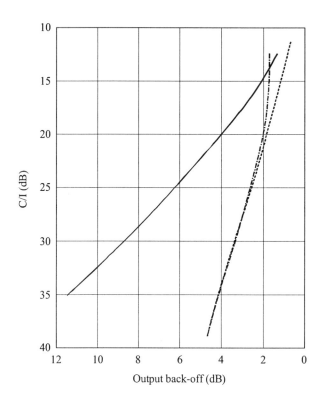

Test carriers: 6 100 and 6 150 MHz (two-carrier intermodulation)

———— TWTA alone

—·—·—·— TWTA with linearizer

············ TWTA with soft limiter type linearizer

Sat/C7-37

FIGURE 7.37

**Carrier-to-intermodulation ratios of TWTA alone,
TWTA with linearizer and TWTA with soft limiter type linearizer**

7.4.6 Output coupling systems and combiners

Three types of coupling systems can be used for connecting the outputs of several transmit chains (i.e. several RF carriers) to the single input port of the antenna (or, possibly to each of the two input ports, in the case of earth stations using dual polarization frequency reuse). They are:

- post-power amplification coupling systems,

- pre-power amplification coupling systems,

- mixed coupling systems.

For clarification, the simple case of single carrier transmission with full redundancy (1+1) is shown in Figure 7.39. RF input and output switches can be operated separately or simultaneously depending on whether or not separate redundancy of IF and RF subsystems is wanted.

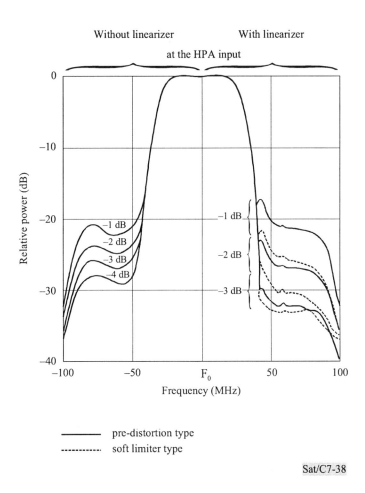

Sat/C7-38

FIGURE 7.38

Typical example of 120 Mbit/s 4-PSK signal output spectrum as a function of earth-station HPA output back-off

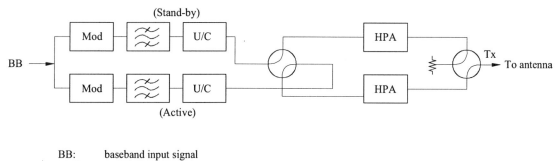

BB: baseband input signal
Mod: modulator
U/C: up converter
HPA: high power amplifier
Tx: transmit RF carrier

Sat/C7-39

FIGURE 7.39

One carrier transmit chain with (1+1) redundancy

7.4.6.1   Post-high-power amplification (post-HPA) coupling systems

In post-HPA systems each RF transmit carrier is amplified in a separate HPA. There are as many HPAs as transmit carriers (n). In addition, m HPAs are provided for redundancy ($1 \leq m \leq n$). Figure 7.40 shows the typical configuration of a post-HPA coupling system with 3 fully redundant transmit chains (3 carriers).

The main post-HPA coupling features are as follows:

- where there is no intermodulation between carriers. Each HPA can be operated at (or near) saturation (of course, in the case of a wideband TDMA carrier, some back-off must be provided, as explained in § 7.4.5.2);

- the overall power efficiency is reduced by the losses in the high power output coupling system (Combiner 1 and Combiner 2 in Figure 7.40);

- the bandwidth of each HPA can be limited to the effective bandwidth of the RF carrier to be transmitted. This is why klystron amplifiers are the usual choice in post-HPA systems, at least in the case of FDMA-FDM carriers.

NOTE – In the case of high bit rate TDMA carriers, klystrons with a sufficient bandwidth (e.g. 80 MHz for 120 Mbit/s TDMA) are not always available. Consequently, a single HPA TWT amplifier, which permits frequency hopping operation (i.e. changing the carrier frequency between the TDMA bursts) is often preferred.

Two types of high power combiners can be used for coupling the HPAs to the antenna:

i)   First type: (hybrid couplers combiner). Combiners of the first type are based on wideband conventional microwave circuits such as the 3 dB coupler, magic-tee, etc. For example, Combiner 2 in Figure 7.40 is a simple 3 dB coupler, featuring equal coupling of the Tx2 and Tx3 carriers with a 3 dB intrinsic loss (which is dissipated in an RF passive load). A more complex combiner, providing adjustable coupling ratio is shown in Figure 7.41. Such a combiner permits optimum selection of the overall system performance and of the power of each individual HPA.

Combiner 1 in Figure 7.40 is an adjustable ratio combiner, which provides a -4.78 dB coupling factor (1/3) from Tx1 to the antenna and a -1.77 dB coupling factor (2/3) from Combiner 2 to the antenna (i.e. 1/3 coupling for each Tx2 and Tx3 carrier). Again, half of the total RF power is dissipated in the fourth port passive load.

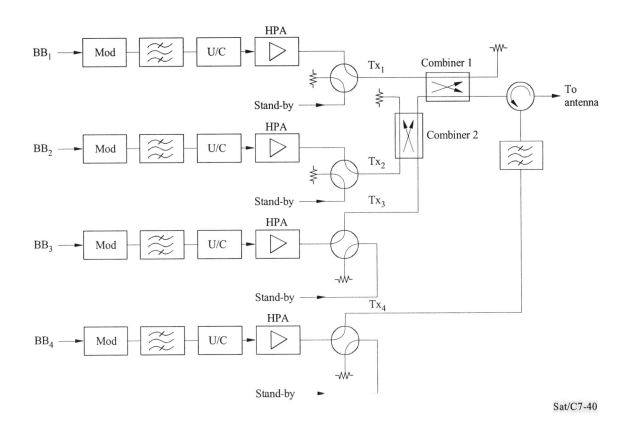

Sat/C7-40

FIGURE 7.40

Post-HPA coupling with first type (hybrid) and second type (filter) combiners

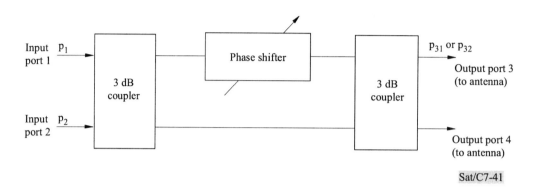

FIGURE 7.41

Adjustable ratio combiner

ii)   Second type: (filter combiner). The first type of combiner relies on a combination of microwave "hybrid" circuits which are frequency independent (in the operating band) but which entail a loss in an "idle" fourth port. The second type of combiner relies on a combination of microwave filters which takes advantage of the difference in frequency of the various transmit channels to connect them with low losses to the common antenna port. The disadvantages using this second type of combiner are:

•   a lack of flexibility in the transmit subassembly;

•   restrictions in the transmit carriers frequency plan (forbidden frequency bands at the filter response cross-over points);

•   thermal dissipation problems in the microwave cavities of the filter(s). Forced air cooling, or even liquid cooling may be needed.

Figure 7.42 shows two typical arrangements for this type of combiner: in both arrangements, one part of the band (e.g. one satellite transponder bandwidth) is transmitted through the filter(s) and the remaining part of the band is reflected by the filter(s). Connection of the filter(s) to the antenna is made by 3 dB couplers (Figure 7.42 a)) or by a ferrite circulator (Figure 7.42 b)).

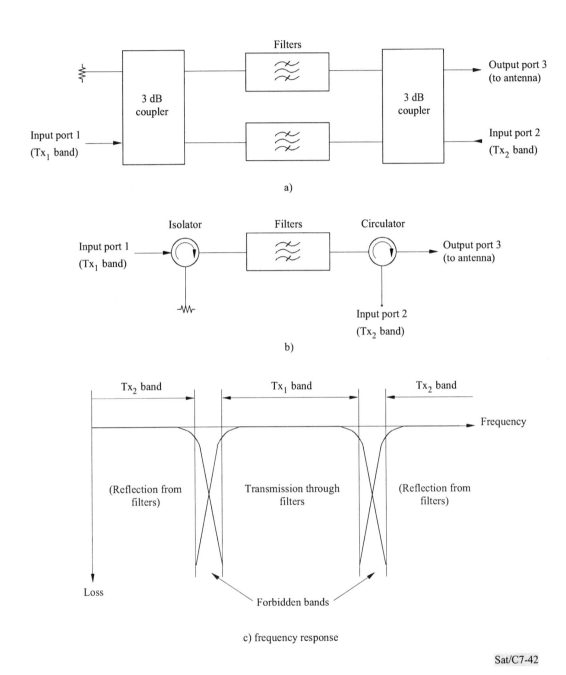

FIGURE 7.42

Filter combiner

It should be noted that, when compared to the first type combiner, there is no intrinsic power loss in the fourth port load, which is only a matching load for absorbing residual reflections.

Combiners of this type are also called diplexers since they couple two RF carriers at different frequencies to a single port. This arrangement can be generalized by cascading multiple diplexers to form an nth order multiplexer with n possible input carriers at n different frequencies (Tx1, Tx2... Txn). Very efficient multiplexers, using coupled cavity filters and featuring low in-band losses and very narrow forbidden bands have been developed. They could be used, for example, to couple 12 klystron HPAs operating in the 12 contiguous 40 MHz satellite transponder bandwidths. However, highly sophisticated and costly filter combiners are needed for such an arrangement. The system shown in Figure 7.43 offers the same capability with less complex filter combiners. In this system, even-order and odd-order transmit channels are first coupled separately through a cascade of filter combiners and then coupled to the antenna through a first type (hybrid) combiner (with a 3 dB loss). In this case, filter combiners are simpler, since contiguous passbands with narrow forbidden bands are avoided. It should be noted, however, that all these multiple channel coupling systems using the second type (filter) combiners lack operational flexibility and may be difficult to implement for the various transponder bandwidths employed in the satellite.

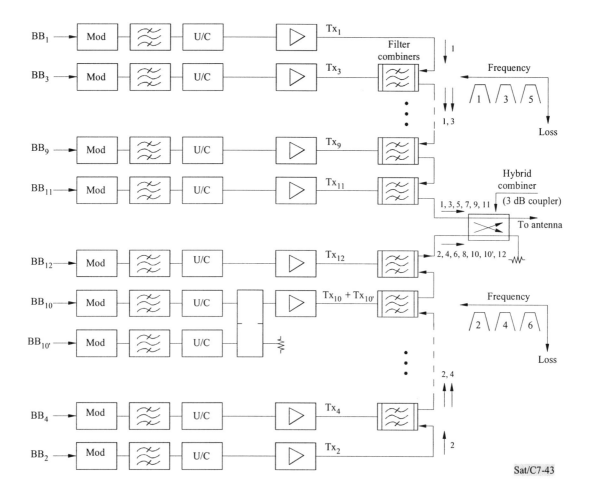

FIGURE 7.43

Post-HPAs coupling system with up to 12 transmit channels using first type (hybrid) and second type (filter) combiner (stand-by chains and HPAs not represented)

Another typical and rather simple system featuring the first and second type combiners is shown in Figure 7.40. Such an arrangement may be used, for example, to couple a TV carrier (Tx4) in addition to the message carriers (Tx1, 2, 3). If TV transmission is only occasional, the Tx4 chain can also be used as a common stand-by chain for the three other chains (by using a proper arrangement of supplementary switches). This provides an efficient and economical redundancy (3+1) scheme.

7.4.6.2   Pre-high-power amplification (pre-HPA) coupling systems

In pre-HPA systems, the various RF transmit carriers are amplified in a common wideband HPA (TWT).

The main pre-HPA coupling features are as follows:

• due to the simultaneous transmission of multiple carriers, the HPA must be "oversized", i.e. it must allow for operation with a significant back-off in order to keep the intermodulation products within their specified limits. This results in a loss of efficiency (by approximately 7 dB to 10 dB) of the HPA system (see § 7.4.5 i)); however, since the transmit carriers are combined before amplification, there are no effective losses in the combiners;

• pre-HPA coupling provides operators with good flexibility for implementing changes. For example, extension inputs to the HPA are available. It should be noted that the usual practice in large earth stations is to locate the transmit chains equipment and the HPAs respectively in the central building and in the antenna building and to connect them by a waveguide inter-facility link (IFL).

7.4.6.3   Mixed-coupling systems

Mixed transmit subassembly configurations, employing both post-HPA and pre-HPA coupling systems are often used. A very simple example is shown in Figure 7.43 which is a post-HPA system, but where two different carriers (Tx10 and Tx10') are to be transmitted in the frequency band of the same satellite transponder. That is why a first type (hybrid) combiner is provided before common amplification in HPA No. 10.

7.4.6.4   Comparison between coupling systems and combiners

Table 7.9 gives a simplified overall comparison of the advantages and disadvantages of the coupling systems and combiners described above. Mixed configurations of systems and combiner types are not included in the table.

TABLE 7.9

Comparison between coupling systems and combiners

Coupling system	Combiners	Advantages		Disadvantages
Post-HPA	Hybrid couplers	- Full power from HPA (no intermodulation products) - Narrow-band channels (lower cost klystron HPAs usable)	Good operational flexibility	High intrinsic losses
	Filter combiners		Low intrinsic losses	- Rather low operational flexibility - Possible forbidden frequency bands - Thermal dissipation problems
Pre-HPA	Low power combiners (e.g. hybrid)	- No combiner losses - Good operational flexibility (extension[1]) - One single HPA		- HPA to be operated with sufficient back-off (intermodulation) - High cost of TWT HPAs
[1] Extensions are limited by the HPA maximum available power.				

7.5 Communications equipment (Generalities and analogue)

7.5.1   General

Communications equipment (also called GCE: ground communications equipment), as shown in the general block diagram of an earth station (see § 7.1.1), comprises the functional units which convert the baseband signal into a modulated RF carrier (at transmit) and conversely at receive. This block diagram applies both to analogue and digital communications. Although this § 7.5.1 is mostly oriented towards analogue equipment, some of the general considerations below can be applied also to digital communications equipment. This is also the case for § 7.5.2, which describes frequency converters. However, specific digital communications equipment is dealt with in § 7.6.

The baseband signal may be:

- a single telephone channel (i.e. one 300 Hz to 3.4 kHz analogue signal) in the case of so-called SCPC (single-channel-per-carrier) systems;

- an FDM (frequency division multiplex) signal (for example: an arrangement of 72 channels – i.e. one group plus one super group extending from 12 to 300 kHz);

- a television signal composed of a video signal (e.g. a 6 MHz wide signal) and of one or several associated audio (sound) channel(s).

The communications equipment (GCE) interfaces with:

- the multiplexing/demultiplexing equipment (MUX) (if any) or the connection point to the terrestrial equipment;

- the power amplifier (HPA) system (for transmit GCE) and the low noise amplifier (LNA) system (for receive GCE).

The major functions of the GCE are:

Transmit side

i)   To provide pre-emphasis and RF energy dispersal for the baseband signal, and to add a 60 kHz pilot to the FDM telephone signals (transmit baseband equipment).

ii)   To convert the baseband signal into an FM-modulated signal at an intermediate frequency (IF), e.g. at 70 MHz (FM modulator).

iii)   To filter the IF signal and to provide group delay equalization (GDE) for the complete transmit path (including earth-station equipment and satellite transponder).

iv)   To convert the IF signal into an RF signal (e.g. at 6 GHz) (up converter: U/C).

v)   To switch and combine the RF signals and to send them to the HPA (HPA input combiner).

Receive side

vi)   To separate and switch RF signals (e.g. at 4 GHz) from LNA (or LNA output divider).

vii)   To convert the RF signals into an IF signal (down converter: D/C).

viii)   To filter the IF signal and to provide group delay equalization (GDE) for the earth-station receive equipment.

ix)   To demodulate the IF signal into the baseband signal (demodulator).

x)   To remove emphasis (de-emphasis) and energy dispersal from the received baseband signal, to detect the 60 kHz pilot in the FDM telephone signal and synchronizing pulses in the TV video signal, and to squelch the GCE by monitoring the out-of-band noise (receive baseband equipment).

Section 7.5.2 describes the frequency conversion equipment (up and down converters) which is almost the same for different modulation methods.

Sections 7.5.3 and 7.5.5 describe frequency modulation (FM) equipment (modulator, demodulator and auxiliary baseband equipment). FM is the most common modulation method currently used in satellite communications systems.

The use of companding techniques (on a channel per channel basis) may be necessary for improving the (subjective) quality (signal-to-noise ratio) of a satellite link. Compandors are commonly used in SSB and SCPC-FM links and may also find applications in FDM-FM links (§ 7.5.3.6).

7.5.2   Frequency converters

7.5.2.1   General description and characteristics

The up converters (U/Cs) translate the IF signals into RF signals, e.g. in the 6 GHz or 14 GHz bands. Conversely, the down converters (D/Cs) translate the RF signals, e.g. in the 4 GHz or 11-12 GHz bands, into IF signals usually at the conventional intermediate frequencies of 70 MHz or 140 MHz. 70 MHz is used for bandwidths up to 36 MHz and 140 MHz for bandwidths up to 72 MHz.

For large RF bandwidths, e.g. 500 MHz, dual conversion converters are used in order to improve the image frequency rejection, i.e. the signal frequency which is symmetrical to the local oscillator frequency.

Up and down converters are usually composed of:

- an RF filter;
- one mixer or two cascaded mixer(s), depending on whether the converter uses single frequency conversion or double frequency conversion (see § 7.5.2.2);
- one or two local oscillator(s) (LOs);
- IF amplifier(s), possibly with automatic gain control;
- IF filters;
- group delay equalizer(s) (GDEs).

The main performance characteristics of the up converters (U/Cs) and down converters (D/Cs) are listed below:

i)   Bandwidth

Three bandwidths have to be considered:

- the RF bandwidth, which defines the capability of the converter to cover the operational RF band, i.e. to transmit (or receive), by adjusting the LO's frequency, the various carrier frequencies which are capable of being operated in the satellite communications system;
- the IF total bandwidth, which defines the capability of the converter to cover all the bandwidths of the various carriers which are capable of being transmitted in the satellite communications system. As a typical example, Table 7.10 gives the various FDM-FM carriers in the INTELSAT-V (and also -VA and -VI) systems;
- the instantaneous IF bandwidth. This bandwidth depends on the number of channels (e.g. as given for FDM-FM by column 2 of Table 7.10. The component which restricts the bandwidth of a given carrier is the IF band-pass filter (IF BPF), the characteristics of which are specified (e.g. by INTELSAT) by the bandwidth unit of each carrier capacity (see § 7.5.3.2).

ii)   Frequency agility

The frequency and channel capacity plan is often altered as the traffic through the satellite is changed and increased. Up and down converters which can be adjusted in frequency over the whole RF bandwidth are specially useful for making these changes. Variable frequency microwave filters and frequency synthesizer local oscillators are used to meet the frequency change requirements. As explained below, frequency agility, i.e. the ability to change the RF carrier frequencies, is improved by the use of double conversion U/Cs and D/Cs.

TABLE 7.10

INTELSAT-V, -VA and -VI FDM-FM carriers

Bandwidth unit (MHz)	Number of channels
1.25	12
2.5	24, 36, 48, 60, 72
5.0	60, 72, 96, 132, 192
7.5	96, 132, 192, 252
10.0	132, 192, 252, 312
15.0	252, 312, 372, 432, 492
17.5	312, 372, 432
20.0	432, 492, 552, 612, 792
25.0	432, 492, 552, 612, 792, 973
36.0	792, 972

iii)   Equalization

The amplitude-frequency response and group delay of the transmit and receive sections of earth stations are equalized in their respective IF sections. The group delay of satellite transponders is usually equalized in the IF section of the U/C (see § 7.5.3.2).

iv)   Linearity

In SCPC systems, a number of carriers are frequency converted by one up or down converter, and intermodulation between carriers can occur. In the transmit section, it is necessary to keep these unwanted intermodulation products negligibly small compared to those in the HPA; therefore, the up converter is required to have good linearity and a sufficiently high intercept point. For a carrier with a large number of channels, good linearity is also necessary to decrease distortion noise caused by the parabolic component of the delay equalizer for the whole system in the IF system and AM-PM conversion occurring in the converter.

NOTE – A definition of the intercept point is given in § 7.4.

v)   Carrier frequency tolerance

The frequency tolerance (i.e. the maximum uncertainty of initial frequency adjustment plus long-term drift) for the transmission of FDM-FM carriers is specified (in the INTELSAT system) from ±40 kHz (1.25 MHz carriers) to ±150 kHz. The tolerance is ±250 kHz for TV carriers. However, the frequency tolerance for the transmission of SCPC carriers is much more stringent, i.e. ±250 Hz. To realize this latter tolerance, the local oscillator has to use a crystal-controlled oscillator with a stability of the order of 10^{-8}.

7.5.2.2   Single and double frequency conversion

Figure 7.44 shows block diagrams of the three most common types of converters:

a) is a single frequency conversion down converter. It is a very simple, rugged and economical equipment, but in order to change the frequency in the operating RF band, it is necessary both to tune the local oscillator RF frequency, and to mechanically adjust the narrow-band microwave band-pass filter. In this up converter, the IF signal is mixed with a local frequency which is 70 MHz lower than the output frequency and converted to the required RF signal. In the RF section, there is a band-pass filter with a 40 MHz bandwidth (centre frequency ±20 MHz) and a rejection filter to suppress any local oscillator frequency leak (below a transmitted e.i.r.p. of +4 dBW). To facilitate frequency change in this type, the microwave filters must have a special mechanical construction.

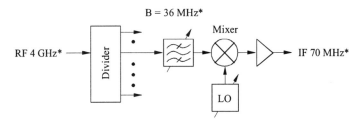

a) Single-frequency conversion

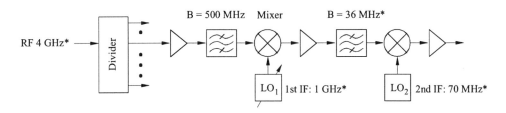

b) Double-frequency conversion (1st type)

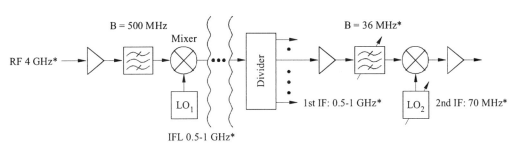

c) Double-frequency conversion (2nd type)

LO: Local oscillator
IFL: Inter-facility link
*Values given as an example

Sat/C7-44m

FIGURE 7.44

Down-converter block diagrams

Up-converter block diagrams would be similar: IF (e.g. 70 MHz) band-pass filter and timer delay equalizer not represented.

b) is a double frequency conversion down converter. This type of converter features high frequency agility since tuning of the first LO (RF oscillator) is sufficient to change the RF frequency in the whole (i.e. 500 MHz) operational RF band. The two LOs can be derived from the same pilot oscillator. This type of converter is the most commonly used in modern earth stations. In this D/C the 4 GHz receive input signal passes through a microwave filter with a 500 MHz bandwidth, enters a mixer, and is mixed with the first local oscillator frequency (variable) and converted into the first intermediate frequency (1st IF). The 1st IF signal passes to a band-pass filter with a 40 MHz bandwidth and is converted into a 70 MHz signal at the 2nd mixer (fixed frequency 2nd LO). In this configuration, making the 1st IF higher than the RF bandwidth (500 MHz) avoids image signals and spurious emission (in the case of up converters). The 1st IF is generally in the range of 800 MHz to 1.7 GHz. Frequency can be changed in the 500 MHz band, by only changing the 1st local frequency, without readjusting any filter. Consequently, combined with a frequency synthesizer, this type of converter is very attractive, and satisfies the requirements for quick frequency change and remote frequency control. It is also effective as a single stand-by unit for multiple converters.

c) is a second type of double frequency conversion down converter. This type does not feature the same frequency agility as the previous type. The frequency of the RF channel is changed in the same way as in the case of a single converter a). However, the adjustable filters and the divider (or combiner for a U/C) are simpler since they operate at a lower frequency (e.g. 1 GHz). This solution can permit a very simple and efficient equipment layout for reception: the RF section (including the microwave wideband filter and the 1st mixer) can be incorporated with the LNA, and directly connected to the antenna receive port. The receive channels can then be connected by a coaxial cable (e.g. at 1 GHz) to the (remote) divider and to the multiple receive chains.

7.5.2.3   Local oscillators

The local oscillators used in converters can be driven either by a pilot crystal oscillator or by a frequency synthesizer. In the first case, changing the frequency requires replacement of the crystal (or switching between multiple crystals). In the second case, changing the frequency can be effected very simply by thumbwheels or even by remote control.

The required frequency stability (long term) may range from some $\pm 10^{-5}$ (FDM-FM and TV) to $\pm 3 \cdot 10^{-8}$ (SCPC).

Local oscillators must feature low phase noise at baseband signal frequencies in order to comply with the general requirements on earth-station equipment noise (see § 7.5.3.1). It should be noted that both low phase noise requirements and frequency stability requirements are specially stringent in the case of SCPC transmission and reception. High performance crystal controlled oscillators or frequency synthesizers must be used in this case.

7.5.3   FDM-FM equipment

7.5.3.1   Introduction

As a typical example, the transmission parameters of some FDM-FM telephony carriers implemented in the INTELSAT system are listed in Table AN2-1 of Annex 2. Such parameters have been selected in order to guarantee the transmission quality which results from the application of Recommendation ITU-R S.353, with the additional condition that the earth-station equipment noise is also included in the total noise of the link. More precisely, the main requirement of Recommendation 353 is that the total system noise must not exceed 10 000 pW0p under certain conditions. This is equivalent to a minimum signal-to-weighted noise ratio of 50 dB. The 10 000 pW0p total noise is shared, as follows, by the INTELSAT specifications:

link noise: 8 000 pW0p
terrestrial interference: 1 000 pW0p
earth station: 1 000 pW0p

This means that:

i) an 8 000 pW0p link noise is allocated for the following sources:

- up- and down-path thermal noise,

- satellite transponder intermodulation noise,

- co-channel interference within the operating satellite,

- interference from adjacent satellite networks,

- unwanted emissions from other earth stations, in particular, multi-carrier intermodulation (for which 500 pW0p is reserved).

Calculation of the first two items is explained in § 7.2.1.2 and 7.4.5.2. Calculation of interference noise from adjacent satellite networks (and also from terrestrial networks) is explained in Chapter 6;

ii) a 1 000 pW0p noise is allocated to the earth-station equipment proper. The INTELSAT specification proposes the following breakdown of this 1 000 pW0p noise as a guide for equipment designers:

- earth station transmitter noise
  (excluding multi-carrier
  intermodulation and group delay noise): 250 pW0p

- noise due to total system group delay
  after equalization by the earth-station
  group delay equalizers (GDE):

  - transmit chain: 200 pW0p
  - satellite intrinsic group delay: 100 pW0p
  - receive chain: 200 pW0p

- other earth-station receive noise: 250 pW0p
  (excluding thermal noise)

  ———

  Total: 1 000 pW0p

Figures 7.45 and 7.46 show block diagrams of a typical earth-station transmit chain (baseband equipment and modulator) and receive chain (demodulator and baseband equipment) for telephony. The main functions and equipment are described in the following sections.

7.5.3.2   IF filtering and equalization

IF filtering and group delay equalization in the earth-station transmit and receive equipment must comply with the communication system specifications. For illustration, Figure 7.47 and the associated tables give the INTELSAT specifications for some typical telephony (and also television) carriers.

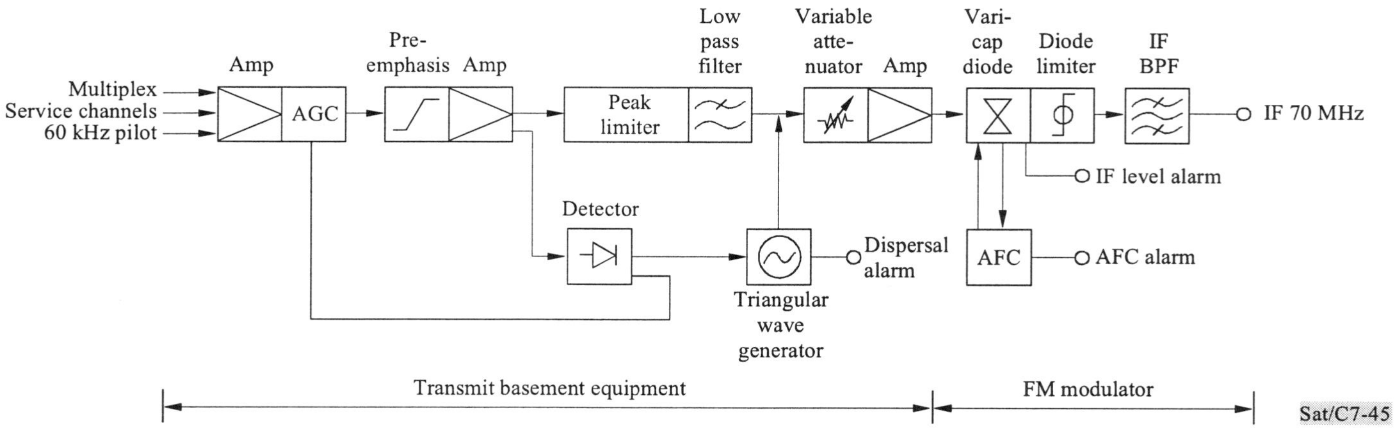

FIGURE 7.45

Transmit baseband equipment and FM modulator

AMP: amplifier

7.5 Communications equipment (Generalities and analogue)

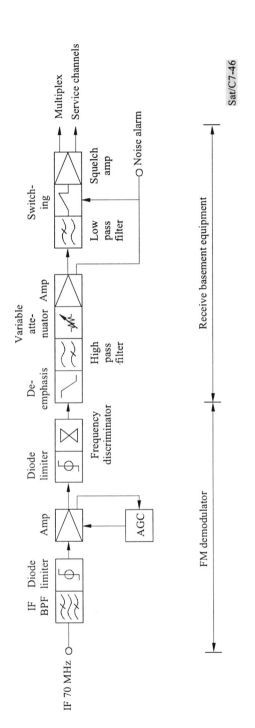

FIGURE 7.46

Receive baseband equipment and FM demodulator

Sat/C7-46

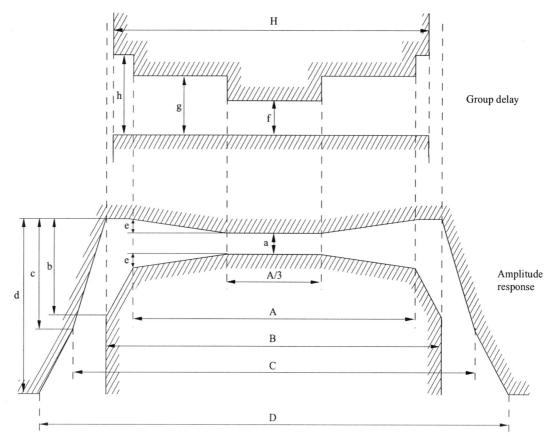

Amplitude and group delay "masks"

Band-width unit (MHz)	A (MHz)	H (MHz)	f (ns)	g (ns)	h (ns)
1.25	0.9	1.13	24	24	30
2.5	1.8	2.1	16	16	20
5.0	3.6	4.1	12	12	20
10.0	7.2	8.3	9	9	18
20.0	14.4	16.6	4	5	15
Video	12.6	14.2	6	6	15
Video	24.0	30.0	5	5	15

a) Transmit and receive equipment
group delay characteristics

Band-width unit (MHz)	A (MHz)	B (MHz)	C (MHz)	D (MHz)	a (dB)	b (dB)	c (dB)	d (dB)	e (dB)
1.25	0.9	1.13	1.50	4.0	0.7	1.5	3.0	25	0.0
2.5	1.8	2.25	2.75	8.0	0.7	1.5	2.5	25	0.0
5.0	3.6	4.50	5.25	13.0	0.5	2.0	3.0	25	0.0
10.0	7.2	9.00	10.25	19.0	0.3	2.5	5.0	25	0.1
20.0	14.4	18.00	20.50	28.0	0.3	2.5	7.5	25	0.1
Video	12.6	15.75	18.00	26.5	0.3	2.5	6.5	25	0.1
Video	24.0	30.00	-	-	2.5	2.5	-	-	0.3

b) Transmit and receive equipment
gain/frequency characteristics

Sat/C7-47

FIGURE 7.47

**Typical examples of INTELSAT specifications
for filtering and group delay equalization**

Typically, a seven-pole Butterworth or Chebyshev filter is convenient for complying with the INTELSAT amplitude/frequency masks.

These filters commonly require low loss inductors and capacitors. This is because, especially for narrow-bandwidths, unloaded Q_g must be high (≈ 400).

In order to meet the requirements of the INTELSAT group delay/frequency mask, two methods can be used:

- a separate group-delay equalizer (GDE);

- a self-equalized filter, with a larger number of L-C components. In this case, the unloaded Q constraint may be relaxed, because the corresponding loss is nearly frequency independent.

Group delay equalizers are used in order to equalize the group delay variation of:

- IF filters (transmit and receive);

- other filters such as RF filters, spurious filtering due to amplifiers, etc.;

- the satellite transponder.

It is generally more convenient to use a separate GDE for each of these three items.

Group delay equalization is obtained by using all-pass filters.

7.5.3.3   Modulation

FM modulation techniques are described in Chapter 4 (§ 4.1.2).

7.5.3.4   Demodulation

FM demodulation techniques are described in Chapter 4 (§ 4.1.3).

7.5.3.5   Other functions

i)   Pre-emphasis/de-emphasis

After demoduation, the parabolic characteristic of the noise spectral density versus baseband frequency would impair the reception quality (S/N) of the higher baseband telephony channels compared to the lower ones. This imbalance is compensated for by using a special baseband transmit filter (pre-emphasis network), the characteristic of which is such that, around a neutral frequency, the relative amplitude (in dB) of the output signal amplitude is positive at high frequencies and negative at low frequencies. At receive, a de-emphasis filter exhibits a symmetrically reversed characteristic. Therefore the relative signal amplitude is recovered while the noise is more attenuated at the higher baseband frequencies. Recommendation ITU-R S.464 specifies the characteristic of the pre-emphasis filter for multiplex telephony. An effective (S/N) improvement of about 4 dB is given by this process.

ii)   RF energy dispersal

When no modulation or a very small modulation signal is applied, the energy concentrates at the carrier frequency and interference can be caused to other terrestrial and satellite systems. Also multiple carrier intermodulation products become excessive in the types of amplifiers that are commonly used.

In order to reduce these effects and to comply with Recommendations ITU-R S.446 and ITU-R S.524, a low-frequency triangular wave is added to the baseband signal prior to the FM modulator.

The level of the triangular wave is set between the following limits:

- lower limit: controls the maximum carrier energy per 4 kHz to a level of 2 dB above (1.58 times) the maximum energy density at full telephone channel load;

- upper limit: determined by distortion noise in the channel and interference to the adjacent channels. It usually makes the carrier energy per 4 kHz equal to the maximum energy density at full telephone channel load.

In practice, the transmit baseband equipment detects the r.m.s. value of the input FDM telephone signal and according to this value, controls the amplitude of the triangular wave so that the energy density meets the above conditions. This triangular wave is then low-pass filtered and added to the baseband signal.

At receive, a circuit (e.g. a low-pass filter) is provided to remove the additional modulation used for energy dispersal.

iii) Carrier transmission monitoring

In order to automatically monitor the operation of the equipment in satellite communications systems, a pilot and noise continuity monitoring are used. A 60 kHz pilot is added to the baseband in the transmit baseband equipment and monitored in the receive baseband equipment. To monitor changes in circuit noise, noise level is measured at a frequency 1.1 times the top baseband frequency. If the noise exceeds the level corresponding to a circuit noise of 50 000 pW0p (S/N = 43 dB), the working system is switched to the stand-by system. If the noise exceeds the level corresponding to 1 000 000 pW0p (S/N = 30 dB), the squelch circuit is operated.

7.5.3.6   FDM-FM with companding

A means for increasing transponder traffic capacity in FDMA-FDM-FM is to insert a compandor at each telephone circuit termination, i.e. a compressor at transmit and an expander at receive. The compandors are usually installed in the telephone exchange premises (and not in the earth station) and companded FDM-FM terrestrial links are established between the telephone exchange and the earth stations.

Companding permits a reduction in the required RF bandwidth for a given carrier size (i.e. number of telephone channels per carrier). This is because the so-called companding gain compensates for the reduction in frequency modulation gain. Due to this companding gain, it may also be possible to reduce the required satellite e.i.r.p. for the carrier, if the (C/N) remains sufficiently above the demodulator threshold.

Some problems in implementing companding on satellite links are listed below:

- the companding gain is determined by subjective assessment of the transmission quality, in comparison with the standard 50 dB S/N (10 000 pW0p) for uncompanded circuits. Figures of 10 to 20 dB are often quoted for the companding gain;

- tandem connections involving multiple compandors result in some quality degradation;

- the impact of signalling transmission (and, more generally, of data) over companded circuits has to be carefully evaluated.

7.5.4 SCPC-FM equipment

For small traffic capacities, telephony communication on a non-multiplexed, channel by channel basis – known as SCPC (single-channel-per-carrier) – is a very popular transmission method because of its simple operation and maintenance, compatibility with multiple access low density ("thin route") traffic and operational flexibility (see Chapter 5, § 5.2.2).

The SCPC equipment of an earth station is composed of:

- the SCPC common equipment (common to all channel units),

- the SCPC channel units (one per telephony channel).

The main features of frequency modulation SCPC systems (SCPC-FM) are described below. However, it should be noticed that these systems tend to become obsolete and to be replaced by digital systems (SCPC-PSK), often based on low bit rate (LRE) encoders (see below, § 7.6.2.1).

i) Channel spacing and modulation parameters

The most commonly used channel spacings are 45 kHz (i.e. 800 possible channels per 36 MHz transponder), 30 kHz (1 200 channels/transponder) and 22.5 kHz (1 600 channel/transponder). Given the channel spacing and allowing for small guardbands between each channel, the available RF bandwidths are approximately 38 kHz, 25 kHz and 19 kHz respectively. The maximum (peak) and r.m.s. frequency deviations are calculated from Carson's rule (see § 3.3.1.2). Whenever sufficient e.i.r.p. is available from the satellite transponder, 22.5 kHz spacing is chosen because of its high traffic capacity (bandwidth limited systems). However, as most systems available at present are power limited, 30 kHz or 45 kHz spacings are more usual.

NOTE – 200 to 400 kHz channel spacings can also be used for the transmission of high fidelity radio programmes (40 Hz - 15 kHz baseband bandwidth).

ii) Voice activation

In FDM-FM systems, the loading factor of the telephone multiplex accounts for the average activity factor of the telephone circuits. To achieve the same improvement in the SCPC transmission mode, a special voice activation circuit is needed in each SCPC channel equipment. This circuit is composed of an on/off switch activated by a speech detector. Therefore, each individual SCPC RF carrier is actually transmitted only when the telephone circuit is active, i.e. when a speech signal is present at the input of the SCPC channel equipment. This means that the average number of SCPC RF carriers which are simultaneously transmitted through the satellite transponder is reduced by the activity factor ($\approx$0.4). For example, among the 800 channels (i.e. "frequency slots") in a 36 MHz transponder with 45 kHz spacing, only 320 are being simultaneously transmitted which reduces the intermodulation products and increases the utilization efficiency of the transponder output power. Similarly, due to the voice activation circuit, an activity factor can be applied to the calculations of the intermodulation products in the earth-station HPA and of the required output power of this HPA. The activity factor to be accounted for in HPA calculations depends on the number of circuits to be transmitted by the station: it varies from 85% for 12 circuits to 40% for more than 60 circuits.

An essential part of the voice activation circuits is the speech detector, of which the voice threshold is critical. A delay line (10 ms) is also included at transmit in order to prevent clipping of the first syllables during carrier "turn-on" and to provide a clean carrier for a 10 ms period for assisting the lock-on of the demodulator at the receive end.

At the receive end, the output of the demodulator is disconnected when there is no carrier at the demodulator input. In the absence of a carrier, the demodulator would operate on input noise only, causing high level output noise which would be very disturbing to the listener.

The delay line at the receive end allows the burst of noise, which would occur when the carrier is turned off at the transmitter and before the signal detector could squelch the demodulator output, to be eliminated.

It should be noted that an echo suppressor can be very easily included in each SCPC channel unit, thus avoiding the need for a separate echo suppressor. This is because most of the echo suppression functions are already included in the voice activation circuits.

iii)   Threshold extension demodulation (TED)

SCPC transmission is usually used in domestic satellite communications systems that have power limited satellite transponders and relatively small earth stations. Under these conditions, the overall carrier-to-noise ratio (C/N) of the received signal can be rather low, especially when using 45 kHz or 30 kHz channel spacing, and a TED must be used.

iv)   Emphasis

Pre-emphasis circuits can be either of the conventional 6 dB/octave type or of the "roll-off" type. The latter is sometimes preferred as it gives better performance for low frequency and impulsive noise.

v)   Companding

Given the C/N and the modulation parameters (and therefore the modulation gain), the transmission quality (i.e. the S/N) obtained from an SCPC link is usually insufficient.

In order to improve the subjective quality, a syllabic compandor with a 1:2 (dB) transfer characteristic in accordance with ITU-T Recommendation G.162 is commonly used.

As explained in § 3.2.1.1, the noise improvement due to the use of companding depends on subjective causes: typical noise improvement may vary from 10 dB to 20 dB (the highest figures being valid for low-level speech). This improvement figure should be included as a supplementary factor in the S/N formulae.

vi)   Pilot generation

SCPC RF frequency pairs (i.e. transmit and receive half-circuits RF frequencies) may be assigned to any frequency "slot" in the satellite transponder. In order to provide convenient frequency agility, the SCPC channel units are equipped with frequency synthesizer local oscillators. In some cases, the transmit and receive frequencies can be paired, allowing the use of a single oscillator for both the transmit and receive sub-unit of each SCPC channel unit. However, in most cases, independent oscillators are preferred. All local oscillator frequencies are usually generated by high stability, very low noise pilots (time and frequency unit: TFU) located in the SCPC common equipment. For the stability and noise performance of local oscillators for SCPC, see § 7.5.2.3.

vii)   Demand assignment

Since SCPC communications are implemented on a channel by channel basis, they are in an ideal medium for operating in the demand assignment mode of multiple access.

By incorporating a demand assigned multiple access (DAMA) system, which is equivalent to an automatic telephone switching exchange, SCPC satellite circuits can be automatically assigned between any two earth stations in the network according to actual traffic demand. DAMA functions and operation is described in detail in Chapter 5 (§ 5.5).

Despite its advantages, it is not the usual practice in domestic satellite communications systems to start operations with DAMA. It is more common to begin using the preassigned mode (while using DAMA compatible SCPC channel units) and to provide DAMA operation when the traffic grows to a point which makes installing DAMA equipment more cost-effective than providing for more space segment capacity (satellite transponder) and for more earth segment equipment (SCPC channel units).

viii)   Terrestrial interface

Whenever small earth stations, equipped with only a few SCPC channels are sited near users, it may be very cost-effective to provide direct interface with the (small) local exchange by incorporating a terrestrial interface module in each SCPC channel unit. This module ensures compatibility with the signalling requirements of the local exchange and of the terrestrial network, in compliance with the signalling system used locally. The terrestrial interface module can also possibly include the DAMA channel control functions.

The complete SCPC equipment of an earth station is composed of:

a)   The SCPC common equipment including:

•   up and down converters,

•   a time and frequency unit,

•   an automatic frequency and gain control (AFC-AGC) unit,

•   an IF unit combiner and divider for connecting the channel units.

The function of the AFC unit is to correct the variations in the received RF carrier frequencies. These variations are mainly due to the satellite transponder instability. The AFC centres the IF received spectrum by receiving a special pilot frequency which is transmitted to all stations of the SCPC network by a dedicated ("master") station.

The AGC function regulates the IF carrier levels.

b)   The SCPC channel units (see Figure 7.48 for a typical block diagram of a channel unit). Note that SCPC channel units can be fabricated in a very compact form (e.g. one or two printed circuit boards). Due to the low baseband frequency, LSI or MSI circuits can be used which allow several functions to be integrated in a small module, e.g. syllabic expander, echo suppressor, squelch.

7.5.5   FM television equipment

7.5.5.1   General

As opposed to the transmission of a telephone circuit, the satellite transmission of a television programme is a one-way point-to-multipoint service. Table 7.11 summarizes some typical characteristics of the various applications of TV satellite transmission. These applications range from i) occasional or regular transmission of TV programmes (e.g. exchange of international programmes between TV studios) by the means of large earth stations, to ii) TV programme transmission towards regional TV broadcast transmitters (e.g. in the framework of a domestic satellite system) and to iii) TV distribution by the means of small TVRO stations to local community re-broadcasting centres or for feeding cable TV networks. Note that two types of service are not included in Table 7.11, i.e. videoconferencing applications (because its parameters can be quite different, e.g. 1.5 Mbit/s or 2 Mbit/s digital TV) and direct broadcast TV (because this application depends on the broadcasting-satellite service, and is outside the scope of the FSS).

The parameters of the baseband signals used in the main current TV systems are defined in Report ITU-R BT.624-4.

The quality objectives for all long-distance TV transmission circuits (terrestrial and/or satellite) are the subject of ITU-T Recommendations J.567 and J.568 and Report 965. Namely the S/N should be equal to or better than 53 dB for 99% of the time and 45 dB for 99.9% of the time. As indicated in Table 7.11 these quality objectives cannot always be met by satellite links (however, it should be noted that reference is only made in ITU-T Recommendation J.567 to transmission between Standard A earth stations).

The signal-to-noise ratio for TV is calculated in § 4.1.1.4. However, it should be noted that the $(S/N)_w$ is not the only factor to be accounted for in assessing the subjective TV quality. In particular, overdeviation noise must be taken into account: overdeviation noise is a special type of impulsive noise which is seen on the TV screen as small horizontal spots occurring at sharp intensity transitions (e.g. white to black transitions). Overdeviation noise occurs whenever a relatively narrow-band IF filter is chosen (in order to increase the (C/N) by limiting the noise bandwidth B_{RF}). In fact, the Carson's bandwidth rule (formula (4), § 4.1.1.1) is difficult to apply to TV transmissions because the frequency deviation $\overline{\Delta F}_{p-p}$ varies with the baseband frequency (pre-emphasis) and also because the high frequency components of the actual video signal can be relatively small. Overdeviation noise can be explained as follows: during a sharp transition, the video signal is differentiated by the pre-emphasis network and the overshoot at the modulator input gives rise to a very high instantaneous frequency deviation. Therefore, the instantaneous IF signal frequency is driven into the attenuated region of the IF filter and the (C/N) is temporarily reduced below threshold, giving rise to impulsive noise peaks.

In conclusion, the main parameters of a satellite TV link (i.e. $\overline{\Delta F}_{p-p}$ at 15 kHz and B_{RF}) should be carefully selected in order to get:

- sufficient (C/N) margin over the demodulator threshold;

- satisfactory $(S/N)_w$, i.e. a sufficient modulation gain, and

- low overdeviation noise.

As shown in Table 7.11, two TV transmission modes are provided in the INTELSAT global system: a so-called full-transponder mode with 30 MHz allocated bandwidth and a half-transponder mode with 20 MHz allocated bandwidth. Table 7.12 gives the main parameters of these two modes. Although the half-transponder mode features a lower quality grade (see Table 7.12) and some overdeviation noise, it is – at present – the most commonly used mode because it allows the simultaneous transmission of two TV programmes in a single transponder.

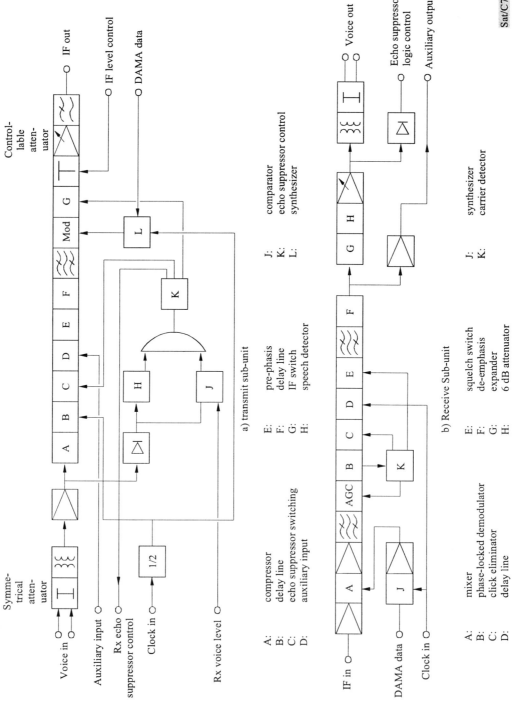

a) transmit sub-unit

A: compressor
B: delay line
C: echo suppressor switching
D: auxiliary input

E: pre-phasis
F: delay line
G: IF switch
H: speech detector

J: comparator
K: echo suppressor control
L: synthesizer

b) Receive Sub-unit

A: mixer
B: phase-locked demodulator
C: click eliminator
D: delay line

E: squelch switch
F: de-emphasis
G: expander
H: 6 dB attenuator

J: synthesizer
K: carrier detector

Sat/C7-48

FIGURE 7.48
SCPC channel unit

TABLE 7.11

Some typical characteristics for TV applications in the FSS

Applications	Frequency band (GHz)	Space segment		Earth segment (Receive earth stations)[5]	Video quality $(S/N)_W$ (dB)
		Allocated bandwidth (MHz)	Available transponder e.i.r.p. (dBW)		
Long-distance TV transmission (e.g. INTELSAT international service or leased transponders for domestic service)	6/4	30	22	Standard A: $G/T = 35$ (dB(K^{-1})) (>15 m antenna) Standard B: $G/T = 31.7$ (dB(K^{-1})) (11 m antenna)	50/53[1] 44/47[1]
		17.5	18	Standard A Standard B (at 55° elevation)	47/48[1] 41.5/43[2]
TV distribution[5]	6/4	30	25	$G/T = 28.7$ (dB(K^{-1})) (7 m antenna)	≈44[3]
			34	TVRO stations: $G/T = 22.7$ (dB(K^{-1})) (4.5 m antenna - FET INA)	≈48[3]
	14/11-12	30	41	TVRO stations: $G/T ≈ 25$ (dB(K^{-1})) (3.5 m antenna - FET INA)	45 to 50[4]
			45	TVRO stations: $G/T ≈ 21$ (dB(K^{-1})) (2 m antenna - FET INA)	45 to 50[4]

[1] Calculated figures for 625/50 and 525/60 TV systems.

[2] It should be noted that the video quality achieved with 17.5 MHz TV reception at Standard B earth stations may be unacceptable, especially when receiving at low elevation angles. This is why earth stations with an improved G/T (using a larger antenna, e.g. 13 m diameter) are often used for such applications.

[3] Typical estimated figures.

[4] Typical estimated figures. Depends on local meteorological conditions.

[5] Transmit earth station (uplink earth station): in order to deliver the required e.i.r.p., the transmit station design is controlled by its antenna diameter (i.e. its antenna gain) and HPA output power. A very important consideration, which often puts a lower limit on the antenna diameter, is the maximum permissible level of off-axis e.i.r.p. (see Recommendation ITU-R S.524).

7.5.5.2 Audio programme transmission

For the satellite transmission of a TV programme, the video signal must be associated with at least one high quality audio (sound) channel and, possibly, with other signals such as a 2nd high quality audio channel (for stereo), commentary channels, order wire, cue, coordination and signalling channels, and data transmission channels for new services (teletext, etc.)

Two methods can be used for the transmission of the associated channel(s), i.e. transmission on one (or a few) separate RF carrier(s), or transmission on a single RF carrier by multiplexing the associated channel(s) with the video. The multiplexing methods fall into the two general classifications of frequency-division (FDM) or time-division (TDM) multiplexing. The FDM techniques employ either an analogue or digital subcarrier to shift the processed programme audio frequency spectrum to a range above that occupied by the TV-video signal. The TDM methods require that the individual programme audio channels be converted to a buffered digital format. The digitized programme channels are time multiplexed into the TV video waveform on a video line-by-line basis.

TABLE 7.12

Main parameters of INTELSAT TV transmissions

	Full-transponder mode		Half-transponder mode	
- TV standard	525/60	625/50	525/60	625/50
- Video bandwidth (MHz)	4.2	6.0	4.2	6.0
- Peak-to-peak frequency deviation (MHz) for 1.0 V peak-to-peak test tone at the cross-over frequency	10.75	9	9.4	9.4
Differential gain	10%		10%	
Differential phase	±3°		±4°	
Reception by a Standard A station (30 m antenna)[1] Receiver bandwidth[2] B_{RF} (MHz)	30		18	
Carrier-to-total noise ratio at 10° elevation) (C/N) (dB)	19		17.3	
Reception by a Standard B station[1] Receiver bandwidth[2] B_{RF} (MHz)	22 to 30		18	
Carrier-to-total noise ratio at 10° or 55° elevation) (C/N) (dB)	12.0 (10°)		11.6 (55°) 9.0 (10°)	

[1] This data is applicable to the INTELSAT-V system.
[2] This is the bandwidth of both IF transmit and receive filters.

Among all the video-associated channel transmission methods, three are currently being used and are discussed below:

i)   Transmission of separate carriers

One (or a few) separate SCPC channel(s) can be used as the transmission medium. Note that the high-quality (i.e. 10 kHz to 15 kHz audio bandwidth) audio channels must be transmitted on relatively wideband SCPC channels (e.g. 150 kHz to 250 kHz bandwidth, see § 7.5.4).

Another method is to use a standard 12- or 24-channel FDM-FM carrier. This was the conventional INTELSAT method until 1978: the audio baseband containing programme, cue and commentary channels was transmitted as a normal 24-channel telephony baseband (with engineering service channels below 12 kHz). The main drawback of these separate carrier methods is that they require additional bandwidth and power from the satellite transponder and that multi-carrier operation causes intermodulation in the transponder.

ii)   "Sound-in-sync" (SIS)

According to ITU-T Recommendation J.66, one high-quality audio signal can be associated with an analogue TV video signal by time division multiplexing (TDM), using the horizontal blanking time slot which is available for synchronization at the beginning of each TV line. This method, called "sound-in-sync" (SIS) is implemented in some terrestrial and satellite TV links and, in particular, in the EBU EUTELSAT system.

The main characteristics of this SIS method are listed below:

*   the audio signal is sampled at twice the TV line frequency, i.e. at about 30 kHz;

*   the audio signal is PCM encoded with 10 bits per sample;

*   therefore, 20 bits (plus 1 cue bit) are inserted (after buffering) in each line synchronization time interval (i.e. about 42 to 5.8 μs);

*   emphasis in compliance with ITU-T Recommendation V.17 and companding are used in order to obtain a signal-to-noise ratio (weighted) of 56 dB. Codecs for SIS encoding-decoding are commercially available. The FM TV modem must be compatible with the transmission of the SIS pulses (see § 7.5.5.3).

iii)   Transmission by an FM subcarrier

The TV-associated audio transmission using an FM subcarrier above the video baseband is, at present, the most commonly implemented method in satellite communication systems. It is now the standard method in the INTELSAT global system and in many domestic systems.

In most cases, only one high quality sound channel is transmitted. However, more complex systems are feasible, e.g. transmission of multiple channels on an FDM subcarrier, transmission of two sound channels on the same TV carrier (but on two separate subcarriers) etc.

The main features of the FM subcarrier technique are listed below:

*   at transmit, the baseband audio and video signals are combined to form a composite signal. This is accomplished by combining an FM subcarrier with the normal video baseband signal. Then, the composite signal is transmitted via the normal earth station video communication equipment. At receive, the composite signal received from the normal earth station video communication equipment is demodulated in order to separate the audio subcarrier from the video baseband signal and to recover both baseband signals. Figure 7.49 shows the block diagram of a typical FM subcarrier TV audio terminal;

*   the IF filters in the transmit and receive equipment must take account of the full bandwidth of the composite signal, i.e. the normal video signal bandwidth plus the subcarrier bandwidth which is centred at the subcarrier frequency F_{sc} (see Figure 7.49). This subcarrier bandwidth B_{sc} can be calculated by Carson's rule (§ 4.1.1.1 formula (4) with $\Delta F = \Delta F_{sc}$ = peak deviation of the

audio subcarrier at the output of the FM modulator of the transmit channel unit and F_A = top baseband frequency of the high quality sound, i.e. 15 kHz);

- the sound transmission quality at receive is obtained by applying twice the general formula (1) of § 4.1.1.1.

First, the carrier-to-noise density ratio $(C/N_0)_{sc}$ of the subcarrier at the output of the video demodulator can be calculated similarly to the signal-to-noise ratio in the highest baseband channel of an FDM signal. Therefore:

$$(C/N_0)_{sc} = (C/N_0) + 10\log \frac{1}{2}(\overline{\Delta F}_{VA}/F_{sc})^2 \quad \text{dB} \tag{17}$$

where:

   C/N_0: video carrier-to-noise density ratio

   $\overline{\Delta F}_{VA}$: peak deviation of the video carrier by the audio subcarrier.

It should be noted that the video pre-emphasis is applied to the composite signal. Therefore, the audio subcarrier, which is in the upper part of the spectrum, is subject to full video pre-emphasis.

Then, the signal-to-noise ratio $(S/N)_A$ of the recovered sound signal is given by:

$$(S/N)_A = (C/N_0)_{sc} + 10\log \frac{3}{2}(\overline{\Delta F}_{sc}^2/F_A^3) + P_A + Q_A \quad \text{dB} \tag{18}$$

P_A and Q_A being the audio emphasis and weighting improvement ratios.

For illustration, Table 7.13 gives the INTELSAT TV-associated sound programme subcarrier parameters.

NOTE – It is common practice to use companding to improve the sound transmission quality, i.e. to use a compressor at the transmit link input and an expander at each receive terminal.

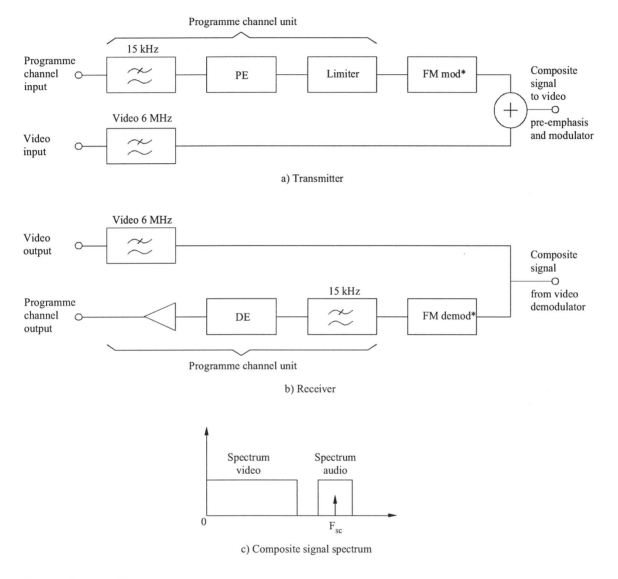

a) Transmitter

b) Receiver

c) Composite signal spectrum

LPF : Low pass filter
PE: Pre-emphasis
DE: De-emphasis

* F_{sc} = 6.60 MHz for video channel 1; F_{sc} = 6.65 MHz for video channel 2

Sat/C7-49m

FIGURE 7.49

Typical FM sound programme terminal block diagram (INTELSAT system)

TABLE 7.13

INTELSAT TV-associated sound-programme parameters
Half-transponder operation

-	Top audio baseband frequency	F_A	15 kHz
-	Occupied bandwidth	B_{sc}	540 kHz
-	Peak deviation for 0 dBm0 test tone at 1.42 kHz	$\overline{\Delta F}_{sc}$	90 kHz
-	Subcarrier frequency	F_{sc}	6.60 or 6.65 MHz
-	Subcarrier peak-to-peak nominal level at the subcarrier modulator input		114 mV (i.e. 1/10th of the video level)
-	Carrier-to-noise density ratio of the subcarrier[1]	$(C/N_0)_{sc}$	74.5 dB(Hz)
-	Audio signal-to-noise ratio[2]	$(S/N)_A$	52 dB
[1]	Reception by a Standard A 30 m antenna.		
[2]	525/60 standard.		

7.5.5.3 Modulation and demodulation of TV carriers

Modulation and demodulation of TV carriers are carried out with the same type of equipment used for FDM-FM telephony (§ 7.5.3.3 and 7.5.3.4). However, some particular features and additional equipment are worth mentioning:

i)   TV threshold extension demodulators (TED): efficient threshold extension demodulation is more difficult to achieve with TV carriers than with low or medium capacity FDM-FM carriers because of the high baseband frequency components in the TV signal; the improvement in thermal noise and impulsive noise can only be appreciated in actual TV pictures. However, TV TEDs are now available, with 2 dB to 3 dB improvement, i.e. they can provide about the same picture quality as a conventional demodulator with a 3 dB lower receive (C/N_0);

ii)   automatic frequency control (AFC) is often provided in the modulator: the AFC circuit is used to stabilize the spectral lines corresponding to the blanking level, irrespective of the video signal levels. This improves the centring of the modulated signal spectrum, thus reducing impulsive noise at sharp image transitions;

iii)   IF filters and group delay equalizers, baseband amplifiers and emphasis networks are different from those used in FDM-FM telephony;

iv)   energy dispersal: energy dispersal circuits are more complex than for FDM-FM (see § 7.5.3.5 ii)):

•   to avoid harmful effects, the energy dispersal signal (usually a sawtooth signal) is synchronized with the field frequency of the video signal;

- in order to comply with Recommendation ITU-R S.524, it may be necessary to increase the energy dispersal level in the absence of video modulation. For example, INTELSAT requires that the peak-to-peak frequency deviation of the energy dispersal sawtooth should be automatically switched from 1 MHz to 2 MHz when there is no video signal. More generally, larger peak-to-peak frequency deviations may be needed (e.g. 4 MHz) when a small antenna is used at the transmit earth station. This is because, for a given e.i.r.p. to be transmitted in the satellite direction, the off-beam e.i.r.p. increases when the antenna gain (i.e. the antenna diameter) decreases;

- at receive, video clamping is the most common technique for removing the energy dispersal waveform and restoring the video d.c. component. Clamping is usually made on the black level of the video signal.

7.6 Communications equipment (digital)

7.6.1 General

Satellite communications are now fully involved in the general trend of the digitalization of telecommunications transmission and switching. In practice, analogue satellite communications are rapidly becoming obsolete and are very generally replaced by the digital counterparts. There are many reasons for this current prevalence of satellite digital communications:

- General factors:
  - increased flexibility in operation and applications;
  - progress in availability, at reduced prices, of digital integrated circuits (up to VLSI) and complete functional units (including microprocessors);
  - operational management, processing and configuration control by software;
  - compatibility with the provision of a broad scope of new services;
  - easy integration in fully digital end-to-end channels and in integrated services digital networks (ISDN);
  - possibility of new services offerings.
- Specific factors:
  - high quality data transmission through powerful error correction schemes;
  - information compression and concentration possibilities in the transmission of data, voice (through signal interpolation and low bit rate encoding – LRE) and of video and television signals (e.g. by MPEG processing);
  - compatibility with multiple access and traffic routing rearrangement (e.g. DAMA operation);
  - and, therefore, increased spectrum efficiency.

At present, digital satellite communications are operationally used for telephony, data transmission and digital TV applications ranging from 64 kbit/s videophone to high-definition television (HDTV) transmission. Table 7.14 illustrates these current applications of digital satellite communications.

Beside the applications shown in the table, services of transmitting high-speed digital traffic such as 139.264 Mbit/s PDH and 155.52 Mbit/s SDH through INTELSAT satellites are available.

Communication equipment design must take into account the transmission quality objectives at the output of the hypothetical reference digital path (see Recommendations ITU-R S.521 and ITU-R S.522 and also § 2.2.3). These objectives which apply to pulse-code modulation (PCM) telephony, are characterized by maximum allowable BER and are as follows:

$\text{BER} \leq 10^{-6}$, 10 min mean value for more than 20% of any month,

$\text{BER} \leq 10^{-4}$, 1 min mean value for more than 0.3% of any month,

$\text{BER} \leq 10^{-3}$, 1 s mean value for more than 0.01% of any year.

For services other than telephony, quality objectives are different.

The design objectives for 64 kbit/s ISDN transmission and digital transmission at or above 1.5 Mbit/s are in Recommendations ITU-R S.614 and ITU-R S.1062 respectively (see § 2.2.3).

The interfaces of the communications equipment (GCE) for digital transmission are the same as for analogue transmission (see § 7.5.1).

Digital communications equipment (GCE) performs part or all of the following functions:

Transmit side

i)   To process the input digital stream in order to convert it into a specific format for transmission. This may include the following functions:

- standard interface data processing;

- telephone channels concentration and interpolation (in the case of digital speech interpolation: DSI);

- compression buffering of the continuous data stream, preamble insertion, frame timing and burst generation (in the case of TDMA transmission);

- error correction encoding (if needed);

- energy dispersal by scrambling.

ii)   To convert the transmit formatted data stream into a modulated signal (e.g. 4-PSK modulated) at an intermediate frequency (IF), e.g. at 70 or 140 MHz (modulator), to filter the IF signal and, possibly, to control the switching operation between multiple transmit carrier frequencies (in the case of "transponder hopping" see § 7.6.3.2 i)).

iii)   To provide group delay equalization and, possibly, path length equalization of the complete transmit path (including earth-station equipment and satellite transponder).

iv)   To convert the IF signal into an RF signal (e.g. 6 GHz) (up converter U/C).

v)   To switch and combine RF signals and to send them to the HPA (HPA input combiner).

TABLE 7.14

Some current applications of satellite digital transmissions

	Multiple access	**Multiplexing**	**Modulation**	**Coding**	**Applications**
One single channel per carrier (SCPC)	FDMA[1]	SCPC	PSK[5]	[6]	Data transmission
	FDMA[1]	SCPC[2]	PSK[5]	PCM[4,6]	Small capacity telephony
Multiplexing	FDMA[1]	TDM	PSK[5]	PCM[3,4,6]	Medium high capacity telephony, data transmission and digital TV transmission
	TDMA, etc.	TDM	PSK[5]	PCM[3,4,6]	VSAT (data, video, and telephony)
	FDMA[7]/ TDMA[1]	TDM	PSK[5]	PCM[3,4,6]	Small capacity telephony and data transmission
	TDMA[1,8]	TDM	PSK[5]	PCM[3,4,6]	Medium to very high capacity telephony and data

a) Techniques for increasing system traffic capacity:

[1] Demand assignment (DAMA)

[2] Voice activation

[3] Digital speech interpolation

[4] Data compression and reduced bit rate telephony (32 kbit/s ADPCM and less, e.g. 3.6 kbit/s LRE)

[5] 2-PSK, 4-PSK to N-PSK.

b) Techniques for improving link budget:

[6] Forward error correction (FEC).

[7] Also called "SCPC-TDMA": a small bandwidth carrier is operated in a time-shared mode between several small earth stations. Occasionally operated in mixed mode: one TDM carrier for transmission from a central (master) station to several small stations with another TDMA carrier for the return channels from the small stations to the central stations.

c) Bandwidth efficiency (4-PSK):

   (64 kbit/s PCM):   20 to 25 voice channels/MHz

[3] (64 kbit/s PCM with DSI)   40 to 50 voice channels/MHz

[3,4] (32 kbit/s ADPCM with DSI)   80 to 100 voice channels/MHz

[8] Note that maximum bandwidth efficiency is obtained when one single TDMA carrier is transmitted through the satellite transponder (since the transponder can, in this case, be operated very near saturation).

Receive side

vi)   To separate and switch RF signals (e.g. 4 GHz) from the LNA (LNA output divider).

vii)   To convert the RF signal into an IF signal (down converter: D/C).

viii)   To provide group delay equalization and possibly path length equalization of the earth-station receive equipment.

ix)   To switch between multiple receive carrier frequencies (in case of "transponder hopping"), to filter the IF signal and to demodulate (demodulator) the IF signal into the receive formatted data stream.

x)   To process the receive formatted data stream in order to convert it into a standard output data stream for terrestrial transmission (or for direct user connection). This may include the following functions:

- removing energy dispersal by descrambling;

- error correction decoding (if encoding provided);

- restoration of a continuous data stream (in the case of TDMA transmission);

- pairing of satellite to terrestrial telephone channels (reverse function of the transmit traffic concentration in the case of DSI operation);

- standard interface data processing.

Functions iii) to viii) are performed using techniques and equipment similar to those used in analogue communications and will not be repeated here. In particular, reference is made to § 7.5.2 for a description of the up and down converters.

Digital communications may be operated either in a continuous or quasi-continuous mode (lines 1 to 4 of Table 7.14) or in time-division mode (lines 5 and 6). The latter mode covers TDMA (including some applications of packet data transmission techniques to satellite communications).

Section 7.6.2 describes communications equipment for the continuous mode and § 7.6.3 describes TDMA communications equipment. Digital communications modulators and demodulators (modems) are described in § 7.6.5.

7.6.2   Communications equipment (SCPC and TDM)

Two types of continuous or quasi-continuous mode digital transmission implementation and equipment will be described in this section:

- FDMA-SCPC-PSK

- FDMA-TDM-PSK

It should be noted that:

- In the case of pure telephony, SCPC is generally operated with voice activation, i.e. the carrier is transmitted only during actual voice (speaking) periods, which means that this is not a fully continuous transmission mode.

- FDMA-TDM-PSK could alternatively be called FDMA-MCPC-PSK (MCPC: Multiple Channels Per Carrier) because, as opposed to SCPC, several channels are multiplexed (in a digital, TDM, frame) on a single carrier.

7.6.2.1 FDMA-SCPC-PSK

It has already been explained that FDMA-SCPC can be operated with analogue (SCPC-FM) modulation. However, in modern equipment, digital SCPC, i.e. digitally encoded and modulated channels, as described in this subsection, is quite generally preferred.

The system general operation and architecture of a digital SCPC network is very similar to those of a SCPC-FM network, as described above in § 7.5.4.

Currently available digital SCPC equipment should enable the transport of voice (telephony) and of data services.

7.6.2.1 a) Telephony

Various types of voice encoding for telephony transmission are described in Chapter 3, § 3.3.1 and 3.3.2). 64 kbit/s Pulse Code Modulation (PCM) voice encoding associated with 4-PSK (QPSK) carrier modulation, as defined by INTELSAT specifications (IESS-303, see below § 7.6.2.1 e) was the most conventional technique used in digital SCPC until the eighties. In particular, it was and remains used in the INTELSAT global system for public telephone services with Standard B earth stations (Standard B with Standard B or with Standard A).

However, due to its rather low efficiency (in terms of bandwidth utilization and transmission power requirement), lower bit rate encoding (LRE) is nowadays generally preferred, especially for domestic and Very Small Aperture (VSAT) systems. LRE algorithms have long been the subject of intensive studies and development in view of combining acceptable voice transmission quality, reduced power requirements and a relatively small additional signal processing delay. The most common techniques currently offered are:

- Adaptive Differential Pulse Code Modulation (ADPCM) at 32 kbit/s, which offers the advantage of being a well-recognized, quality-proven technique standardized by ITU-T (Recommendation G.721).

- Low Delay Code Excited Linear Predictive (LD CELP) vocoder techniques at 16 kbit/s and 8 kbit/s (CELP), which are now also standardized by ITU-T Recommendations (G.728 and G.729).

- Proprietary techniques, usually in the range 9.6 - 4.8 kbit/s. In fact, the utilization of such techniques is fundamentally based on the availability of low-cost, Large-Scale Integration (LSI) chips and of relevant software on Digital Signal Processors (DSP).

These LRE encoding techniques are combined with powerful Forward Error Correction (FEC) schemes, e.g. 1/2 or 3/4, which increase the actual transmitted bit rates in order to improve the link budget (i.e. for lessening the transmitted power requirement for a given transmission quality). Modulation can be either 4-PSK (QPSK) or 2-PSK (BPSK). The latter is less sensible to phase noise and frequency error effects and is generally preferred for low bit rate links.

As mentioned before, voice activation is always provided for in SCPC telephony. This consists of implementing, in each individual voice channel, a voice detector and a carrier control device which activates the carrier only when voice is present.

In consequence, the transmission is no more continuous, but takes the form of bursts. At the beginning of each burst, a preamble must be sent in order to permit carrier and bit timing recovery at reception.

Voice activation brings the same advantages as in the case of analogue SCPC (SCPC-FM, see § 7.5.4).

A comparison of the operational performance of various SCPC voice encoding techniques is given in Table 7.15 below.

TABLE 7.15

Operational performance of various voice encoding and data transmission SCPC techniques

Mode	Bit rate (kbit/s)	FEC ratio	Modula-tion Format	Symbol rate (ksymb/s)	Occupied bandwidth (kHz)	E_b/N_0[1] (dB)	C/N[1] (dB)	Carrier Spacing (kHz)
Voice (companded, FM)[2]			FM		38		8.5	45
Voice (companded, FM)[2]			FM		19		14.2	22.5
Voice[3]	64	No	QPSK	32.0	38.4	11.3	13.5	45
Voice G.721	32	1/2	QPSK	32.0	38.4	6.5	5.7	45
Voice G.721	32	3/4	QPSK	21.3	25.6	8	9	30
Voice G.728	16	1/2	QPSK	16.0	19.2	6.5	5.7	22.5
Voice G.728	16	3/4	QPSK	10.7	12.8	8	9	15
Voice G.729	8	1/2	BPSK	16.0	19.2	6.5	2.7	22.5
Data	64	1/2	QPSK	64.0	76.8	7	6.2	90
Data	64	3/4	QPSK	42.7	51.2	8.5	9.5	60
Data	19.2	1/2	QPSK	19.2	23	7	6.2	27.5
Data	9.6	1/2	BPSK	19.2	23	7	3.2	27.5

[1] E_b/N_0 and C/N are given for a 10^{-4} BER for voice and 10^{-7} for data transmissions.

[2]   FM-SCPC is listed for comparison. The quoted (C/N_0) corresponds to a typical required value without the companding gain (±17 dB).

[3]   INTELSAT SCPC-QPSK (see § 7.6.2.1 e)).

7.6.2.1 b) Data

SCPC-PSK can also be very efficiently implemented to transmit all kinds of low or medium bit rate data, the limitation being usually around 128 kbit/s.

Some typical operational performance of various SCPC data transmission techniques are also given in Table 7.15 above. Note that, although in voice mode a 10^{-4} BER provides a sufficient quality, data transmission generally requires 10^{-7} to 10^{-9} BER. This quality can be obtained by adding extra data block coding.

SCPC data transmissions may be roughly classified in two categories:

- Voiceband data mode (VBD, also called in-band data mode), i.e. transmission of data through the telephony interfaces: These telephony interfaces can, in turn, be either transparent to data transmission or non-transparent.

   The low bit rate voiceband data can be transparently transmitted through the voice codec (up to 4 800 bit/s with the ITU-T Recommendation G.721 codec and up to 2 400 bit/s with ITU-T Recommendation G.728). For higher data rates, it is necessary to demodulate locally the VBD before transmitting the digital information (and to remodulate it at the receive output). With such a technique, it is possible to transmit VBD at a data rate close to the nominal data rate of the codec (i.e. up to 14 400 bit/s for ITU-T Recommendation G.728, the remaining data rate being usable for an additional error correcting coding).

   Moreover for fax transmission, the relevant ITU-T protocols (ITU-T Recommendations T.30 and T.40) must be emulated.

- Digital data mode: In this mode, data are directly transmitted through a digital interface (such as ITU-T Recommendations V.24, V.28, V.11, V.35) to a modem of the earth station terminal.

   In direct data mode, modern digital SCPC equipment should enable the transmission, in a circuit mode, of input data up to 56 kbit/s or even higher.

7.6.2.1 c) Frequency plan

Transmission of SCPC channels is generally supported by a pre-defined RF frequency plan. This means that all the possible carrier frequencies are assigned, in the satellite transponder, to a precise RF frequency.

The frequency spacing between the possible channels depends on the channel bandwidth, which, in turn, is defined by the information rate and the coding and modulation method. For example:

- for 64 kbit/s channels and QPSK modulation, the occupied (noise) bandwidth is BW = 38 kHz and the required Channel Spacing is CS = 45 kHz (BW = 76.8 kHz, CS = 90 kHz with 1/2 FEC);

- for 32 kbit/s (ADPCM telephony) channels with 3/4 FEC encoding and QPSK modulation, the occupied (noise) bandwidth is BW = 25.6 kHz and the required Channel Spacing is CS = 30 kHz.

Note that the number of channels which can be transmitted in a given bandwidth-limited part of a satellite transponder is equal to TB/CS (TB being the total available bandwidth).

Frequency synthesizers are generally provided for generating the assigned frequencies with the required accuracy at transmit and receive (e.g. ±250 Hz at transmit).

In order to compensate the downlink frequency uncertainty (due mainly to the satellite transponder oscillator), automatic frequency control (AFC) is provided at receive. The AFC loop is generally actuated by a pilot signal transmitted by one of the earth stations in the network (e.g. the hub or master station).

In the case of demand assigned operation (DAMA), the transmit and receive frequencies are assigned, during the duration of a given communication, by the DAMA process (see below, § 7.6.2.1 g)).

7.6.2.1 d) Digital SCPC system architecture and earth station terminals operation

SCPC communications are generally used in the case of low-density traffic, but there are two main types of operation (and this is valid for SCPC-FM as well as for digital SCPC):

- Point-to-point links for low-density traffic. A typical case is the INTELSAT SCPC-QPSK system (see below, § 7.6.2.1 e)) which is specifically designed for "thin routes" between the "medium-size" Standard B earth stations and the main Standard A earth stations (or between Standard B stations).

- Dedicated SCPC networks for low-density communications with multiple scattered locations. Examples of such operation are: domestic networks for rural areas, islands, etc., business (corporate) networks. The latter are often categorized as VSAT (very small aperture terminals) networks. As already explained in § 5.6, there are two possible architectures for dedicated SCPC networks:

  - Star networks for communications between a main, central earth station (called the "hub") and small earth stations in remote locations (in this case, communications between two remote locations are possible only via the hub, i.e. through two satellite hops).

  - Meshed networks, in which all remote earth stations are "equal". In meshed networks direct, single-hop, links can be established between any pair of remote earth stations. However, even in meshed networks, a main ("master") earth station is generally implemented for network control and supervision and also for assigning the circuits in the case of DAMA operation, as explained below.

SCPC links may be preassigned (PAMA mode) or demand-assigned (DAMA mode, see § 5.5 and below § 7.6.2.1 g)).

Figure 7.50 represents a typical block diagram of a remote SCPC earth station. It comprises:

- The RF subsystem, including, the antenna, the transmit and receive RF amplifiers (HPAs and LNAs) and the up- and down-converters (U/Cs, D/Cs). In the case of very small earth stations (e.g. VSATs), the RF units, and, in particular the so-called LNC (low-noise converter, i.e. LNA + D/C in a single block), can be included in an outdoor unit directly integrated with the antenna feed and diplexer.

- The SCPC traffic terminal, divided into:

  - the common equipment, which is shared by all channel units and includes, as long as they are needed, an IF distributor, the IF amplifiers, the reference oscillators (TFU, time and frequency unit), the automatic frequency and gain control units (AFC, AGC). In the case of dedicated networks, the common equipment generally includes also a signalling and control unit (SCU);

- the channel units (CUs), each one being assigned, at a given preassigned (PAMA) or demand-assigned (DAMA) carrier frequency, to the transport of one communication channel (voice or data). Each CU includes a terrestrial interface module (TIM) and a digital modulator-demodulator unit (QPSK or BPSK Modem).

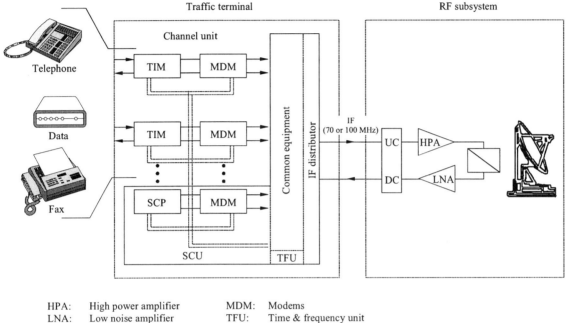

HPA: High power amplifier
LNA: Low noise amplifier
UC: Up convertor
DC: Down convertor
TIM: Terrestrial interface models

MDM: Modems
TFU: Time & frequency unit
SCU: Signalling & control unit
SCP: Signalling & control processor

Sat/C7-50

FIGURE 7.50

General block diagram of an SCPC earth station

NOTE – Possible redundancy configurations in the RF subsystem and in the traffic terminal (including, possibly, supplementary CUs) are not represented in the figure.

The signalling and control unit (SCU) may be composed of:

- a signalling and control processor (SCP) in charge of the general control of the station (configuration, start-up, equipment monitoring and control, etc.) and also of the channel units signalling processing (in the DAMA mode).

- a modem for establishing signalling and data links with the network central station (see § 7.6.2.1 g)).

Each TIM performs functions as follows. Note that this list is only typical and not limitative and may comprise both basic and optional functions:

- telephone interface (usually a standard analogue 2/4-wire + E&M signalling interface). A variety of terrestrial interfaces and signalling systems can generally be accommodated.

For example the possible terrestrial interfaces (2-wire, 2-wire E + M, 4-wire, 4 wire E + M) may include direct connection to standard telephone handsets or connection to public or private exchanges (PABX). Supported signalling systems (for telephony) may include: ITU-T No. 5, R1, R2, subscriber line signalling, PABX signalling, etc.

- digital encoding (usually at 64 kbit/s in accordance with ITU-T Recommendation G.711)

- low bit rate (LRE) encoding (see above § 7.6.2.1 a))

- echo cancelling (ITU-T Recommendation G.165)

- multiplexing and framing for voiceband data mode data (see above § 7.6.2.1 b))

- digital data mode interface (see above § 7.6.2.1 b))

- voice activity detection and carrier control

- control, monitoring and signalling functions (in connection, through the control bus with a SCU, if available).

Each modem unit performs functions as follows (typical list):

- IF interfaces and frequency conversions, including tuning of the traffic frequencies (in case of a wideband, frequency agile, modem)

- scrambling for energy dispersal (see §§ 3.3.4 and 7.5.5)

- forward error correction (FEC, generally using convolutional encoding and soft-decision Viterbi decoding, see Appendix 3.2)

- framing and synchronization of the voice bursts or of the data streams

- modulation and coherent demodulation (QPSK or BPSK), including carrier and clock recovery (see §§ 4.2 and 7.6.4).

7.6.2.1 e) INTELSAT SCPC-QPSK system and earth station terminal

As mentioned above, 64 kbit/s SCPC-QPSK is the conventional SCPC transmission mode in the INTELSAT global system for communications in the 6/4 GHz bands between INTELSAT Standard A and/or Standard B earth stations.

The terminal, the technical parameters of which are specified by INTELSAT document IESS-303, can be operated with four functional configurations of the channel units (as single function or multi-function CUs):

1) Conventional 64 kbit/s voice or voiceband data mode at, or below, 4.8 kbit/s.

2) Same as 1) but, optionally, with voiceband data mode above 4.8 kbit/s using a special FEC encoding (120, 112 modified Hamming code).

3) Digital data mode at 48 or 50 kbit/s using a rate 3/4 convolutional FEC encoding.

4) Digital data mode at 56 kbit/s using a rate 7/8 convolutional FEC encoding.

Table 7.16 below gives the main parameters of the INTELSAT SCPC-QPSK system.

TABLE 7.16

Summary of INTELSAT SCPC-QPSK characteristics
(Excerpt from INTELSAT IESS-303 (Rev.4))

Parameter	Requirement
VOICE	
Audio channel input bandwidth	300-3 400 Hz
Transmission rate	64 kbit/s (including preamble)
Encoding	7 bit PCM, A = 87.6 companding law, 8 kHz sampling rate
Modulation	4-Phase Coherent PSK (QPSK)
Ambiguity resolution	Unique words
Carrier control	Voice activation
Channel spacing	45 kHz
Channel bandwidth	45 kHz
IF noise bandwidth	38 kHz
C/T per channel at nominal operating point	-167.3 dB(W/K)
C/N in IF bandwidth at nominal op. point	15.5 dB
Nominal bit error ratio (BER) at op. point	1×10^{-6}
C/T per channel at threshold	-169.3 dB(W/K)
C/N in IF bandwidth at threshold	13.5 dB
Threshold BER	1×10^{-4}
DATA	
Data rate (3/4 FEC)	48 kbit/s or 50 kbit/s
Data rate (7/8 FEC)	56 kbit/s
Clock recovery	Clock timing must be recovered from the received data stream
Threshold C/N:	
48 kbit/s (Bandwidth, BW = 38 kHz)	13.5 dB
56 kbit/s (BW = 38 kHz)	13.5 dB
50 kbit/s (BW = 38 kHz)	13.6 dB
Threshold BER at operating point without coding or scrambling	1×10^{-6}
Nominal BER at op. point with FEC coding	1×10^{-9} (without scrambling) 3×10^{-9} (with scrambling)

The INTELSAT document specifies all the technical parameters required from the earth stations, in particular: e.i.r.p. level (including reduction by the activity factor due to voice activation) and stability, spurious emissions and intermodulation products (including also reduction by the activity factor), carrier frequency tolerance and phase noise, AFC and pilot functions, telephony codec characteristics, performance of the modulator (including jitter and out-of-band emission), of the demodulator (including filtering, carrier and bit timing recovery, BER) and of the synthesizers, interfaces characteristics (voice and data, including synchronizer, codecs and FEC).

7.6.2.1 f) Digital SCPC earth stations and terminals for domestic communications

The configuration and characteristics (as concerns, e.g. the voice codec and the modem type, usually QPSK or BPSK) of the various digital SCPC earth station terminals currently on the market are dependent on the manufacturer. However, most of them offer different types of earth stations, e.g.:

- small-capacity remote earth stations, usually based on a "VSAT type" configuration with a compact, low-power consumption, generally unattended, terminal (comprising a limited number of channel units, e.g. up to 6);

- medium capacity earth stations, usually for rural applications;

- high-capacity terminals, either for traffic nodes or for main (e.g. hub) earth stations. These terminals may be able to include up to hundreds of channel units and use specific equipment configurations (in general with TWT HPAs).

The earth stations are usually implemented either in the 6/4 GHz bands or in the 14/11-12 GHz bands, although other FSS bands could be envisaged.

An optional configuration with significant advantages in the case of networks with a few medium- or high-traffic links should be mentioned. It consists of combining MCPC operation for these links with the basic SCPC operation. MCPC (multiple channels per carrier) is the mode in which several digital channels are TDM multiplexed on a single carrier (see § 7.6.2.2). In the mixed SCPC-MCPC operation mentioned here, some SCPC channel units of the earth station terminals are replaced by low- or medium-capacity MCPC units (modem unit + TIM). Such a configuration can be an economic means for increasing the traffic capacity of remote earth stations without increasing their complexity. A simple example consists of implementing two 32 kbit/s or four 16 kbit/s telephone channels on a single MCPC carrier.

Note that, in the case of domestic or business applications, vendors may offer complete, turn-key systems, including centralized functions, such as DAMA control and network management (with monitoring and maintenance).

7.6.2.1 g) DAMA implementation

In the demand assignment multiple access mode (DAMA), the satellite channels are allocated and released on a call by call basis (see above § 5.5).

The most significant advantages of DAMA operation are recalled below:

- the various two-way communication channels (trunks) are not fixed assigned to given, preassigned directions, but are pooled for the whole network traffic and are assigned on request to a given direction only during the duration of the concerned communication.

  Due to the concentration effect, the number of channels in the network is determined by the total traffic.

  The calculation (using Erlang tables) of the number of required SCPC channels (and satellite carrier frequencies) shows a significant reduction compared to the preassignment (PAMA) case (examples of DAMA dimensioning calculations are given in § 5.5.7).

- not only is the number of needed satellite channels reduced by DAMA operation, but also the quantity of SCPC channel units in the earth stations;

- the advantage of DAMA is particularly significant in the case of meshed networks where any pair of remote earth stations can communicate via single satellite-hops. A typical case calculation shows that the number of required satellite circuits is only 25% (of the number needed for PAMA) for a meshed network compared to about 70% for a star network with the same traffic capacity. The same calculation shows that the number of SCPC channel units is about 30% for a mesh against 80% for a star network, the latter having furthermore to cope with double-hop delay in communications between remote stations[11, 12].

Mostly for historical reasons, and also for its technical interest, a brief reminder of the INTELSAT SPADE is given here (SPADE stands for: SCPC multiple access demand assignment equipment). Since 1973, INTELSAT, in the framework of its global main earth stations (Standard A) system, operated, under this name, a pool of SCPC-QPSK-PCM channels (IESS 303, see § 7.6.2.1 e)) on a fully demand assigned basis. The purpose of SPADE was to allow Standard A operators to access multiple correspondents with low-density traffic.

Contrarily to most DAMA systems, SPADE was characterized by a distributed assignment control method, in which each earth station, through a demand assignment signalling and switching (DASS) unit, was able to assign its own channels independently. This was accomplished by maintaining, in each DASS, a continuously updated table of active frequencies and by randomly selecting an unused pair of frequencies for establishing a circuit.

A special common signalling channel (CSC) permitted delivery to the other stations in the system of the current status of the concerned station. However, due to the rather restricted utilization and to the terminal cost, INTELSAT decided to close this service.

---

[11] Double-hop delay could be acceptable in some cases, especially when telephony is not the main concern (data links, e.g. for business communications).

[12] However, there are situations where star networks should be preferred to mesh networks. This is the case, in particular, when the main traffic is from the central station (the hub, equipped with a large antenna) towards the remote stations (equipped with small antennas). In any case, given a particular network traffic situation, an analysis of the satellite transponder power utilization must be made (power limitation versus bandwidth limitation). It should also be noted that the advantages of DAMA are really significant in the case of low density traffic links (e.g. 0.1 Erlang).

At present, INTELSAT is beginning to implement a new DAMA system for small earth stations operating in the 6/4 GHz bands. This is a centralized DAMA system which is controlled by (at least) one network management and control centre (NMCC). In this system, the calling station sends a request message (including the called telephone number) to the NMCC on a reserved shared SCPC channel (by a random Aloha access method, see § 5.3.2).

The NMCC broadcasts, on an SCPC channel, messages containing traffic management information (providing, in particular, the transmit and receive frequencies allocated to the calling and called stations for the duration of the requested connection).

Vendors of dedicated (domestic, corporate, etc.) satellite systems often offer their own DAMA system. Since these (proprietary) DAMA systems are generally combined with network management and supervision functions, they may be configured to accommodate the possible particular requirements of the client operator.

Some typical characteristics of currently proposed domestic DAMA systems are listed below:

*   the complete management and supervision functions of the network are centralized in a network control system (NCS), connected to a central computer in the central station of the network;

*   the NCS is in charge of the network management functions:

    *   configuration management

    *   performance management

    *   fault management

    *   security

    *   statistics

    *   billing, etc.

    *   resources management functions in the pre-assigned (PAMA) and demand assigned (DAMA) modes;

    *   the NCS is linked to all traffic stations by a specialized signalling network, often based on TDM/TDMA carriers[13].

NOTE – In the TDM/TDMA network, the central station sends messages to all or part of the remote stations via a (continuous) TDM carrier. The remote stations communicate with the central station via one (or a few, depending on the needed traffic capacity) common TDMA carrier(s).

A TDMA carrier is shared between the remote stations by an Aloha process. Such an operation is very commonly used in VSAT business systems (in these systems, TDM/TDMA links are usually implemented to carry all the traffic between the remote stations and the central – hub – station, Ref: Supplement 3 to the Handbook: VSAT systems and earth stations).

---

[13] In the case of fully preassigned networks, the supervision information can be more simply carried to/from the monitoring and maintenance centre on special overhead bits in the traffic carriers.

7.6.2.2   FDMA-TDM-PSK (MCPC) equipment

7.6.2.2 a) General

FDMA-TDM-PSK carriers are more and more often used on satellite links, instead of the previous FDMA-FDM-FM carriers commonly used during the "analogue transmission" era.

Because in the case of FDMA-TDM-PSK, just as in FDMA-FDM-FM, and contrarily to SCPC, several communication channels can be multiplexed on a single carrier (possibly towards several destinations), this type of transmission is often called MCPC (multiple channels per carrier).

Considering the various possible types of digital transmissions, the implementation of FDMA-TDM-PSK transmissions may be seen as intermediate between, on the one hand, point-to-point SCPC transmissions with small capacity (e.g. 8 kbit/s to 128 kbit/s, see above § 7.6.2.1) and, on the other hand, TDMA transmissions (see below § 7.6.3) with full possibilities for digital multiplexing and multiple accessing, up to very high overall traffic capacities (up to 120 Mbit/s or even more).

In that sense, the preferred range of application of FDMA-TDM-PSK links may typically extend from 64 kbit/s to, e.g. 45 Mbit/s. Often, new FDMA-TDM-PSK communication chains can be easily added in existing earth stations to augment their traffic capacity, using the same (or partially the same) RF equipment. It is even possible, in "old" earth stations, to mix new FDMA-TDM-PSK communication chains with conventional FDMA-FDM-FM chains, thus permitting a progressive implementation of digital satellite communications.

It should be noted that link budget calculations show that the bandwidth and earth station e.i.r.p. required for transmitting FDMA-TDM-QPSK carriers are generally smaller than those needed for transmitting FDMA-FDM-FM carriers of the same communication capacity, especially if digital circuit multiplication (DCME, i.e. low bit rate encoding – LRE – and/or digital speech interpolation – DSI, see § 3.2.2.3) or transcoder techniques[14] are employed.

Most currently available FDMA-TDM-PSK terminals can be used either in open network (according to INTELSAT IDR and IBS standards or EUTELSAT SMS standard) or in closed network (proprietary specifications). FDMA-TDM-PSK carriers should be considered as a means of transmitting any type of information in a digital form, e.g. PCM-encoded telephony (with or without DCME), data, digital video or a multiplexed combination of these (in particular in an ISDN framework).

In order to improve the link budget, and therefore the performance and the power/bandwidth resource utilization, powerful error-correction (FEC) schemes are usually embodied in the transmitted satellite signal. This generally includes convolutional encoding with soft-decision Viterbi decoding plus, possibly, the concatenation of an outer Reed-Solomon code. The utilization of Reed-Solomon codes can allow the quality of digital satellite transmissions to be raised to the level of the best optical cable links.

---

[14] Transcoder techniques are specified in ITU-T Recommendation G.761.

7.6.2.2 b) Terminal description and characteristics

The block diagram of a typical FDMA-TDM-PSK terminal is given in Figure 7.51.

The terminal consists in one transmit channel unit dedicated to the satellite carrier to be transmitted (possibly a multidestination one) and one receive channel unit (or several channel units in the case of a multidestination terminal). It should be noted that, in general, the terminal proper includes only the communication transport means, i.e. the modem unit and that terrestrial equipment, such as multiplexing/demultiplexing (MUX/DEMUX) and terrestrial interface units, DCME, engineering service channels (ESC), frequency reference clock (REF), alarms systems (BWA)[15], etc., as well as up- and down-converters (UC/DC), RF equipment, etc. (in oval boxes on the figure) are to be considered as external units which are parts of the general equipment of the earth station.

On the transmit side, the incoming data from the terrestrial equipment (MUX or transmit part of DCME) enters in the line interface (INT) which performs the data multiplex format conversion and passes to the FIFO (First In, First Out) register to be dejittered (DEJITTER) by an external reference clock (REF)[16] if applicable. This signal is forwarded to the satellite framer (SAT FRAM) to be added with possible housekeeping data (ESC, BWA) according to the transmission system used. For instance, in an open network configuration such as INTELSAT or EUTELSAT, ESC channels (voice for IDR and low rate data for IBS and SMS) and backward alarms generated by the receive side) are included as an overhead in the information data frame. The signal with the overhead information is then scrambled (SCRB), encoded (convolutional encoding – FEC – with or without concatenation of Reed-Solomon outer coding – RS COD) and finally QPSK modulated (QPSK MOD) before transmission as a carrier through satellite via the up- and down-converters (UC/DC) and the RF part of the earth station.

On the receive side, each wanted carrier received from satellite, is QPSK demodulated (QPSK DEM) and regenerated, decoded (sequential or Viterbi decoding – FEC DEC – with or without Reed-Solomon decoding – RS DEC – according to the transmit side) and descrambled (DSCR). This signal is fed to the satellite deframer (SAT DEFRAM) to remove the housekeeping information and passed to a receive buffer (BUFFER) for Doppler effect and plesiochronism compensation by an external reference clock (REF). Finally the line interface (INT) performs data mux format restoration before sending it to the terrestrial equipment (DEMUX or DCME).

---

[15] Backward alarm (BWA) relates to fault conditions from "the remote end" of the circuits. According to ITU-T Recommendations, such alarm indications are often to be supplied to the terrestrial equipment. This is, in particular, a requirement of the INTELSAT specifications (IDR, IESS-308).

[16] The utilization of a high-accuracy external reference clock (REF) (e.g. 10-11) is generally needed to allow synchronization with terrestrial networks. However, an internal back-up clock, with a lesser accuracy, should generally be provided for.

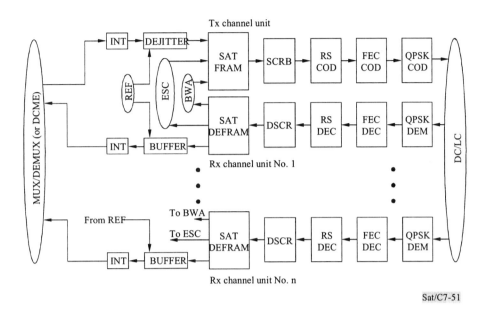

FIGURE 7.51

FDMA-TDM-PSK terminal block diagram

FDMA-TDM-PSK terminals should be designed as modular and flexible units in order to accommodate diversified applications:

i) Multiple bit rates: fully digital, wideband, modems, are now available. They feature compatibility with a wide range of bit rates, e.g. from 64 kbit/s to 8.444 Mbit/s, by changing plug-in filters, or, in modern versions, simply by software control.

ii) Error correction codecs: as mentioned above, FEC codecs using convolutional encoding and soft decision Viterbi decoding are now systematically implemented in most FDMA-TDM-PSK terminals. Various FEC ratios (e.g. 1/2, 3/4, 7/8) should be accommodated. Moreover, additional, outer codecs (usually Reed-Solomon) are often required as options.

iii) Single destination and multi-destination operation: In certain cases, the utilization of a FDMA-TDM-PSK terminal is limited to information exchange between two end users (this is the case, in particular, for the INTELSAT IBS business links). In such cases, a single destination modem and terminal arrangement is sufficient. However, multiple access, a specific characteristic of satellite communication, should find very efficient applications with FDMA-TDM-PSK digital transmissions. In these multi-destination applications, a carrier transmitted by one earth station can be received by more than one earth station (and reciprocally, in a meshed network architecture). Therefore, the implementation of FDMA-TDM-PSK terminals with multidestination capability (such as represented in Figure 7.51) should be recommended. Of course, it remains possible to implement a single – concatenated – multiplexer/demultiplexer equipment: bit streams destined to a particular earth station can be demultiplexed in this station, from higher hierarchy signals transmitted from another station.

iv) Multiplex/demultiplex structures: The FDMA-TDM-PSK terminals must be designed in order to accommodate the great variety of possible multiplexing arrangements of the digital signals to be transparently transmitted through the satellite. These multiplexing arrangements depend on the network architecture (single vs. multidestination), on the traffic requirements and information rates and on the digital hierarchies specified between the corresponding users (often in accordance with the ITU-T Recommendations: G.702 for hierarchies with a primary level of 2.048 Mbit/s or of 1.544 Mbit/s, G.802, for interworking operation between these two hierarchies).

For relatively low traffic between two, or more, destinations, a particular arrangement, known as a "drop and insert mux" can prove very useful: such a device is usable with the first primary order MUX (1 544 or 2 048 kbit/s) when only a part of MUX, i.e. a few 64 kbit/s channels, has to be sent through the satellite. In the transmit side, the drop mux extracts data from time slots in a terrestrial bearer signal and passes the data to the satellite framer.

In the receive side, the insert mux injects time slots from the satellite deframer into the terrestrial bearer signal. Drop and insert muxes can be cascaded if required for multidirection applications.

v) Digital circuit multiplication equipment (DCME) and terrestrial interface equipment: the FDMA-TDM-PSK terminal should be compatible with the utilization of DCME and with various types of terrestrial interface equipment used for connecting the satellite circuits to the terrestrial network. As already explained, the use of DCME (digital speech interpolation – DSI – plus low bit rate encoding – LRE, in general by 32 kbit/s ADPCM) allows the satellite link traffic capacity to be increased (by a factor of 4 to 8) by taking advantage of the normal voice activity factor of each one-way telephone channel coded at 32 kbit/s.

7.6.2.2 c) FDMA-TDM-QPSK transmission parameters and terminal performance

Table 7.17 below gives typical examples of transmission parameters and terminal performance for FDMA-TDM-QPSK carriers. Although this table is mainly derived from INTELSAT specifications for IDR carriers (IESS-308, Rev.8), it can be considered as of general interest and generally applicable for most types of FDMA-TDM-QPSK transmissions.

TABLE 7.17

**Typical examples of transmission parameters and terminal
performance for FDMA-TDM-QPSK carriers**
(INTELSAT IDR carriers, excerpt from INTELSAT specifications -IESS-308, Rev.8)

Information bit rate (bit/s)	Data rate (With overhead) bit rate (bit/s)	FEC ratio	Transmission bit rate (bit/s)	Occupied bandwidth (Hz)	Allocated bandwidth (Hz)	C/N_0 (BER=10^{-10}) (dB-Hz)	C/N (BER=10^{-10}) (dB)
64 k	64 k	3/4	85.33 k	51.2 k	67.5 k	59.1	12
64 k	64 k	1/2	128.0 k	76.8 k	112.5 k	58.1	9.1
384 k	384 k	3/4	512.0 k	307.2 k	382.5 k	66.8	12
1.544 M	1.640 M	3/4	2.187 M	1.31 M	1 552.5 k	73.1	12
2.048 M	2.144 M	3/4	2.859 M	1.72 M	2 002.5 k	74.3	12
2.048 M	2.144 M	1/2	4.288 M	2.57 M	2 992.5 k	73.2	9.1
8.448 M	8.544 M	3/4	11.392 M	6.84 M	7 987.5 k	80.3	12
44.736 M	44.832 M	3/4	59.776 M	35.87 M	41 875.0 k	87.5	12

NOTE 1 – These examples are excerpted from the standard transmission parameters recommended for operation on the INTELSAT VII, VIII and K (from IESS-308, Rev.8, Appendix D, Tables D.3 and D.4).

NOTE 2 – The examples are taken from a list of recommended carrier sizes. These sizes are included in the ITU-T digital hierarchies (Recommendation ITU-T G.802). However, it should be noted that any other information bit rates between 64 kbit/s and 44.736 Mbit/s may be used.

NOTE 3 – Overhead: the standardized overhead size is 96 kbit/s for information bit rates 1.544 Mbit/s.

NOTE 4 – Occupied and allocated bandwidths: The occupied bandwidth (noise bandwidth) and the allocated bandwidth are generally taken as 60% and 70% of the transmission bit rate.

NOTE 5 – (E_b/N_0) performance (Modem and overall channel) (from IESS-308, Rev.8, Appendix D, Tables D.1 and D.2): In order to account for possible performance degradation during a part of the available time, the quoted (C/N_0) and (C/N) have been calculated to provide a clear-sky link BER of better than 10^{-10}. This corresponds to an overall (E_b/N_0) = 11 dB (3/4 FEC) or 9.9 dB (1/2 FEC) (including the satellite channel). The table below gives other INTELSAT values of (E_b/N_0) in order to allow calculations of (C/N_0) and (C/N) for other BER values. The table also includes the INTELSAT modem performance specification (IF back-to-back):

(E_b/N_0) (dB) for BER =	10^{-3}	10^{-6}	10^{-7}	10^{-8}	10^{-10}
Modems back-to-back					
FEC ratio = 3/4	5.3	7.6	8.3	8.8	10.3
FEC ratio = 1/2	4.2	6.1	6.7	7.2	9.0
Through satellite channel					
FEC ratio = 3/4	5.7	8.0	8.7	9.2	11.0
FEC ratio = 1/2	4.6	6.5	7.1	7.6	9.9

TABLE 7.17

(continued)

NOTE 6 – REED-SOLOMON (RS) outer coding (from IESS-308, Rev.8, Appendix H): INTELSAT provides for the optional utilization of a supplementary FEC coding: An "outer" Reed-Solomon (RS) block code (See Chapter 3, § 3.3.5) may be concatenated with the specified FEC (the "inner" code). The purpose of this optional supplementary code is to enhance the clear sky performance and the availability of the links. It should be noted that:

- The occupied bandwidths, when implementing a RS code, are increased by a factor depending on the RS code parameters, practically from 12.5% for small carriers (< 1.544 Mbit/s) to 8.33% for greater size carriers.

- The allocated bandwidths and the required (e.i.r.p.)s remain unchanged.

- The performance table in NOTE 4 above becomes as follows:

(E_b/N_0) (dB) for BER =	10^{-3}	10^{-6}	10^{-7}	10^{-8}	10^{-10}
Modems back-to-back					
FEC ratio = 3/4		5.6	5.8	6.0	6.3
FEC ratio = 1/2		4.1	4.2	4.4	5.0
Through satellite channel					
FEC ratio = 3/4	5.7	6.0	6.2	6.4	7.0
FEC ratio = 1/2	4.6	4.5	4.7	4.9	5.9

7.6.3   Communications equipment (TDMA)

7.6.3.1   General operation of TDMA terminals

The general principles of TDMA operation are given in Chapter 5 (§ 5.3).

A typical TDMA communication system is composed of three types of elements:

- traffic earth stations

  These stations provide the connecting points to the satellite network for the users. They process and buffer (compress) the incoming information signals (PCM telephony, synchronous data, or other) into the required digital format and transmit these signals as data bursts in appropriate time slots inside the TDMA frame. Symmetrically, they receive the data bursts from the corresponding stations, extract from these bursts the appropriate TDM sub-burst, and finally buffer (expand) and restore the outgoing information into conventional format.

- reference earth station

  These stations provide for acquisition, synchronization, monitoring and also general management of the traffic for the whole system. They usually transmit a timing reference signal used by all stations, plus other information needed for the internal operation of the system.

  In some networks, traffic and reference functions are grouped in the same stations, and performed by the same terminals.

- network control centre

  The network control centre provides the operators with a centralized management of network equipment (TDMA terminals and radio), transmission resources and traffic organization.

The main part of the traffic station is the TDMA traffic terminal, which incorporates the functions listed in § 5.3. It comprises:

- an IF subsystem (which includes the modem);

- a signal processing and operation control logic subsystem (common logic equipment);

- a terrestrial interface subsystem.

Two types of TDMA terminals are described: a high bit rate terminal for international traffic (INTELSAT/EUTELSAT system), and a typical example of a medium bit rate terminal implementing FDMA/TDMA access scheme.

7.6.3.2 High bit rate TDMA terminals (INTELSAT/EUTELSAT 120 Mbit/s TDMA-DSI)

i) Main features

The INTELSAT/EUTELSAT 120 Mbit/s TDMA-DSI terminals will permit high capacity multiple access, fully digital telephone (and data) communications to be established via the INTELSAT (and EUTELSAT/ECS) satellites with high efficiency (up to about 3 300 (64 kbit/s) channels per 70 MHz transponder, with DSI concentration).

For data transmission, TDMA terminals without the traffic concentration function (digital non-interpolated: DNI) are used with n x 64 kbit/s interfaces.

Direct digital interface (DDI) which provides connections between terrestrial 2.048 Mbit/s digital bearers through the TDMA system is also available.

The basic transmission parameters of the INTELSAT/EUTELSAT TDMA are given in Table 5.1. Other features which are relevant for terminal design are listed below:

- TDMA frame: the digital signals are arranged in a time-frame of 2 ms duration. This frame contains two types of bursts:

  - the reference bursts (incoming from one primary and one back-up reference station) which provide timing, control and system management information to the earth stations;

  - the traffic data bursts which are allotted to the various stations for handling their traffic. Each traffic burst is divided into one preamble and (up to 8) sub-bursts for accommodating multi-destination (point-to-multipoint) operation. A traffic terminal can transmit up to 32 sub-bursts within up to 16 bursts and receive up to 32 sub-bursts within 32 bursts per frame.

    Figure 7.52 gives details of the frame structure.

- multiple transponders operation: one single terminal can operate through a number of satellite transponders. This remarkable feature is called "transponder hopping".

  The various satellite transponders which can be used for transmission and reception in a single TDMA frame can have different carrier frequencies and/or polarizations. One given burst can be transmitted or received via any of the concerned transponders.

  Transponder hopping is carried out by rapid switching at IF (or RF) as shown in the right-hand part of Figure 7.54 (maximum switching time is 16 symbols, i.e. 26.5 μs). Of course, the electrical length of (i.e. the time delay through) all possible paths from the modulator output to the antenna feed input port (at transmit) and from the antenna feed output to the demodulator input (at receive) must be carefully equalized (time delay differences less than 32 ns). This specification applies to all the various possible IF/RF communication equipments (U/C, D/C etc.) used for transponder hopping at different carrier frequencies (or polarizations) and also to the various IF/RF equipment (U/C, D/C, inter-facility links, HPA, LNA, etc.) which could be affected by any redundancy switching configuration.

- error correction: forward error correction (BCH code of rate 7/8) can be applied whenever necessary for obtaining a satisfactory bit error ratio;

- digital speech interpolation: digital speech interpolation (DSI) is applied to the telephony channels in order to make the most efficient use of the satellite capacity. Telephony input channels are PCM encoded in accordance with ITU-T Recommendation G.711 (A law). Signalling compatibility: ITU-T No. 5 and (EUTELSAT only) R2 systems.

As explained below (see Figure 7.54), each DSI-DNI TDMA interface module (TIM) acts as an interface between the common TDMA terminal equipment (CTTE) and the 2 048 kbit/s digital multiplex equipment (PCM telephony or data) of the terrestrial network (up to ten 2 048 kbit/s PCM lines per module). It also performs either a traffic concentration function (digital speech interpolation: DSI) or only channel multiplexing/demultiplexing (digital non-interpolated: DNI). DNI channels are primarily used for data transmissions.

- redundancy: proper redundancy, i.e. built-in spare functional units, with automatic status monitoring and switching, should be provided in the TDMA terminal (and in the concerned earth-station communication equipment) in order to guarantee the specified availability and to permit maintenance operation;

- peripherals: peripherals (i.e. consoles with visual display units and keyboards) should be provided for:

  - operational control and monitoring of the terminal;

  - burst time plan (BTP) loading;

  - DSI-DNI mapping (i.e. burst and sub-burst pairing with terrestrial circuits).

ii)   TDMA equipment configuration and operation

Figure 7.53 shows the general block diagram of an earth station for operating as a traffic terminal in the INTELSAT or EUTELSAT TDMA-DSI system. The block diagram includes:

- the antenna and RF/IF equipment;

- the TDMA terminal comprising:

  - the common TDMA terminal equipment (including the IF subsystem, the modem and the common logic equipment (CLE));

  - the TDMA interface modules (TIMs) consisting of DSI-DNI units;

- the terrestrial interface equipment.

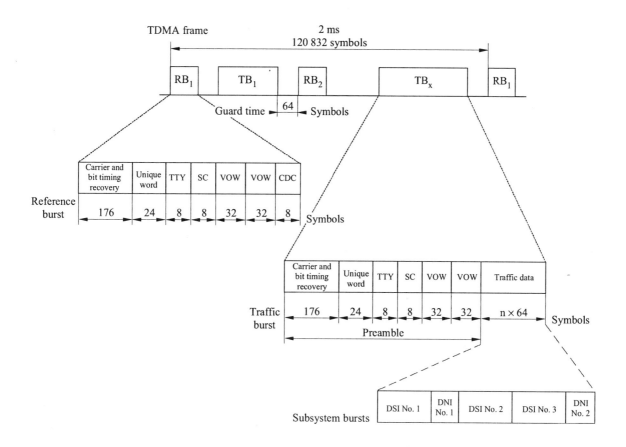

RB$_1$:   Reference burst from reference station 1
TB$_x$:   Traffic burst from station x
Unique word: Special bit pattern in the preamble which permits precise
                     synchronization (start of data) and phase ambiguity
                     resolution (for non-differential decoding) at receive
SC: Service channel (SC) contains alarms and various
                     network management information
CDC: Control and delay channel (CDC) contains the delay
                     information (Dn) for synchronizing the transmit bursts
TTY, VOW: Telegraphy and telephony order wires for inter-station
                     communications

Sat/C7-52m

FIGURE 7.52

INTELSAT/EUTELSAT TDMA frame structure

A typical block diagram of the terminal is shown in Figure 7.54. The digital bit streams generated by the TDMA interface modules are fed to compression buffers where the information bits are accumulated. At the appropriate time, as determined by the transmit synchronizer unit, the preamble generator is activated and the contents of the compression buffers are read out sequentially at the TDMA bit rate.

The high-speed burst signal is then fed to the modulator input for transmission to the satellite. Similarly, on the receive side the preamble decoder and the various expansion buffers will be loaded with the appropriate sections of the received bursts, under the control of the received burst timing unit. The expansion buffers are read out continuously at the terrestrial bit rate.

The received frame timing is obtained by detecting the unique word of the reference burst, once per frame. The timing for burst and sub-burst decoding is derived from the received frame timing and from knowledge of the burst position in the frame and the sub-burst position in the burst. This information is contained in the burst time plan (BTP).

The transmit frame timing is obtained from the received frame timing by adding a delay (Dn) which is periodically adjusted in accordance with the synchronization control information. In the INTELSAT system this information is provided to the TDMA terminals by the reference stations through the reference burst.

The timing for transmission of the terminal's burst and sub-bursts is derived from the transmit frame timing and from the burst time plan.

Each burst carries its own means of carrier and clock synchronization. The burst preamble contains a synchronization pattern for PSK demodulation which permits the receiving modem to recover the carrier and demodulate it correctly for recovering the clock timing, and a unique word to provide reference timing for data reception. The burst may contain a forward error correcting (FEC) code which permits the receiving device to correct transmission bit errors.

In addition, a data scrambling technique which adds an exclusive PN code (pseudo-random noise code) to the data stream is applied to all bursts after the end of unique word, in order to maintain the maximum flux-density at the Earth's surface within the ITU-R limit.

iii)   Terrestrial interlace modules, DSI operation and terrestrial connection

The DNI module accepts PCM voice channels or data channels of different bit rates or any combination of them and arranges the information bits in a predetermined sequence. No traffic concentration is performed by the DNI module.

The DSI module accepts the terrestrial PCM encoded telephone channels and condenses them into a smaller number of channels for transmission over the satellite by interleaving the speech bursts from different terrestrial channels into the same satellite channel (interpolation).

DSI gain is defined as the ratio of the number of incoming terrestrial channels to the number of available normal satellite channels excluding assignment channels and may typically be about 2.0 to 2.2

A schematic representation of digital speech interpolation is shown in Figure 7.55.

The basic configuration of DSI equipment is shown in the left-hand part of Figure 7.54. On the transmit side, a voice detector detects whether or not speech signals are present on each of the N incoming terrestrial channels and then the channels containing active speech signals are connected to M satellite channels. In this case, since on the receive side it is necessary to recognize the exact connection status of terrestrial and satellite channels, an assignment message containing connection information is transmitted through the assignment channel. On the receive side the satellite channels are respectively connected to the corresponding terrestrial channels in accordance with the assignment message sent from a transmit station.

The inputs to the DSI and DNI modules are provided from the terrestrial interface equipment which can be a different type depending on whether the terrestrial link connecting the earth station with the switching centre (CT) is analogue or digital.

If a standard FDM-FM link is employed, conventional demultiplexing equipment can be used to convert from group and supergroup to individual baseband voice channels which are then fed to one or more PCM codecs. The PCM codec is used to convert the analogue voice channel to an 8-bit pulse code modulation (PCM) digital stream. The same equipment (i.e. PCM codecs and FDM multiplexer) is used to perform the reverse function of converting PCM channels to FDM multiplex signals. Alternatively a single piece of equipment called a transmultiplexer can be used to perform the functions of both the FDM multiplexer/demultiplexer and the PCM codecs.

If the link to and from the switching centre carries time division multiplex PCM channels, only TDM demultiplexer equipment is necessary at the earth station to extract the channels to be routed over the satellite link.

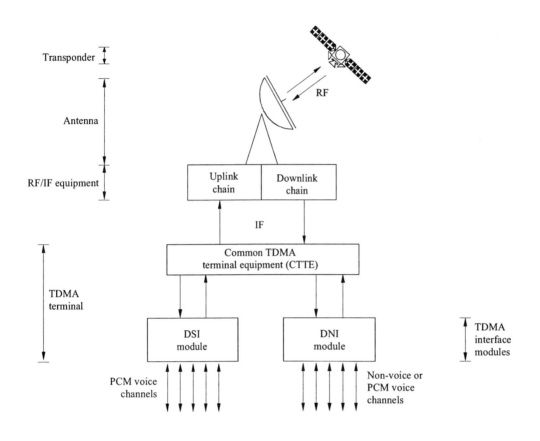

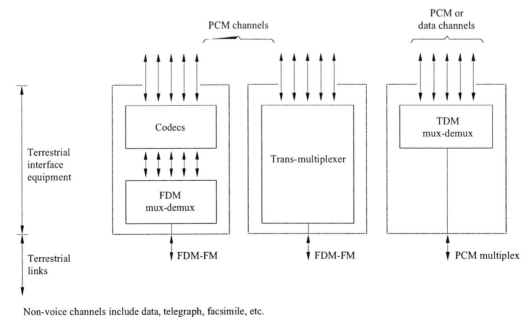

Non-voice channels include data, telegraph, facsimile, etc.
mux-demux: multiplex/demultiplexer

Sat/C7-53m

FIGURE 7.53

Basic TDMA functions

629

7.6 Communications equipment (digital)

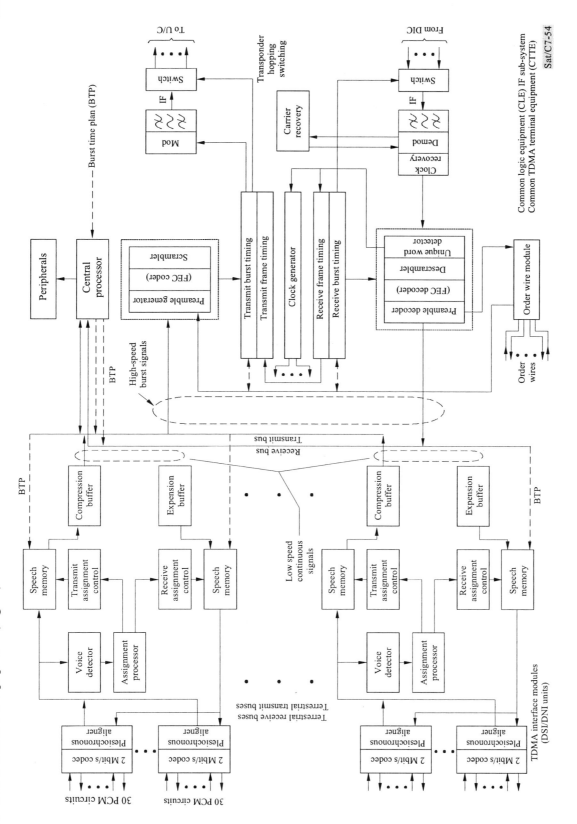

FIGURE 7.54

INTELSAT/EUTELSAT 120 Mbit/s TDMA terminal: typical block diagram

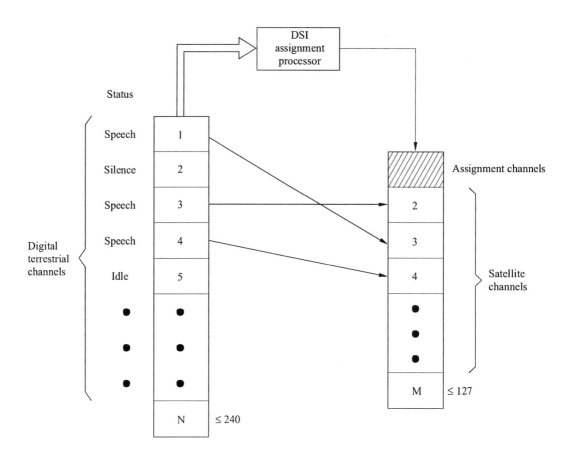

N incoming terrestrial channels are compressed into M outgoing satellite
channels through DSI operation, where the ratio N/M is the DSI gain

Sat/C7-55m

FIGURE 7.55

Digital speech interpolation function

A significant difference still remains between the analogue and digital circuit interfaces which is related to synchronization. Analogue circuits on both the transmit and receive paths which are converted to the digital format, can be operated synchronously with the TDMA terminal transmit and receive sides since the sampling rates of the FDM-TDM converter can be varied to accommodate the rate determined by the TDMA terminal without distorting the sampled analogue signals. However, for digital circuits, the clock rates of the terrestrial digital network may differ from those derived from the TDMA terminal and a digital buffer must be included between the two. The digital interface operates in a plesiosynchronous (nearly synchronous) mode (see ITU-T

Recommendation G.811), and the buffer combines the functions of a plesiochronous aligner and a Doppler buffer.

7.6.3.3   Medium bit rate TDMA terminals

7.6.3.3.1   General

As opposed to 120 Mbits TDMA networks, there is no international open standard available for medium bit rate TDMA systems. As most VSAT networks, these systems rely on "closed" specifications (terminals from different suppliers are not compatible with each other).

Technology evolution has induced a significant reduction in the hardware complexity and cost of medium rate TDMA terminals. Latest products are compact rack-mounting units or autonomous transportable cases, containing typically 3 to 6 boards.

7.6.3.3.2   Access scheme

All medium bit rate TDMA systems use a combination of FDMA and TDMA access schemes. In these systems, the transmission band is divided into carriers (not necessary of identical rates), which are accessed in TDMA mode. The connectivity between all stations is usually guaranteed by a mechanism of frequency hopping (at transmit or at receive side, or both).

In the simplest case, each traffic terminal transmits on a single carrier but can receive traffic coming from any other carrier by modifying its receive frequency in real time (burst per burst). Traffic control software ensures that incoming bursts carrying traffic for the same station in different carriers never overlap in time (see Figure 7.56).

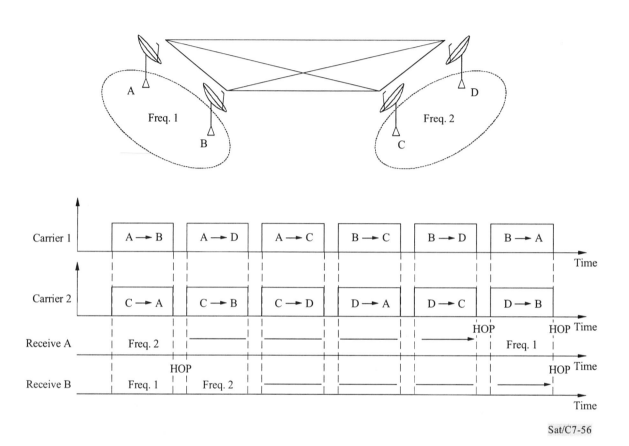

FIGURE 7.56

**FDMA/FDMA access scheme - example
of frequency hopping at receive side**

In more complex systems, each traffic terminal can also transmit on several carriers, modifying its transmit frequency (and even the transmit rate) from any burst to the next burst. This capability is useful when the traffic matrix is very dissymmetrical, or if the capacity allocated per station may vary on a large scale.

Traffic capacity assignment corresponds to the setting of the length and the position in time of each burst transmitted on each carrier. In centralized systems, this process is performed at the network control centre, whilst in decentralized systems it is performed at each traffic terminal. Assignment can be permanent (PAMA mode) or "on demand" (DAMA mode), or both (see § 5.5).

7.6.3.3.3   TDMA traffic terminals architecture

Traffic TDMA terminals are usually organized in 3 main parts (see Figure 7.57):

i)   IF (modem) subsystem

This part includes the modulator and demodulator, IF switches for frequency hopping, and cable connectors (if any). The modem usually works in QPSK mode. Modulation rates per carrier may vary from 256 kbit/s up to 8 Mbit/s. Some systems offer even higher rates (up to 46 Mbit/s) for multi-transponder operation.

The modem subsystem generally comprises one or two boards, making extensive use of digital signal processing technology.

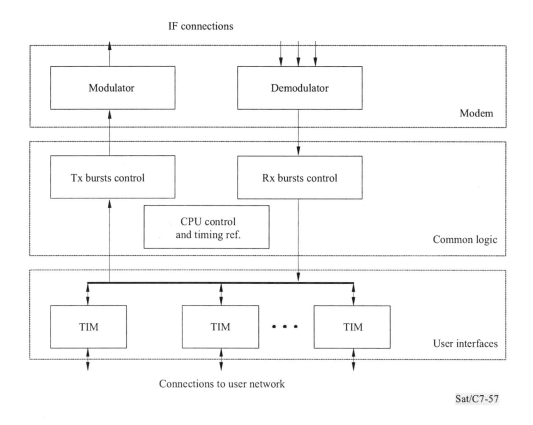

FIGURE 7.57

Typical architecture of a medium rate TDMA traffic terminal

ii)   Common logic subsystem

This part includes bursts transmit and receive control units, a timing reference generator unit and a processor unit in charge of overall terminal management.

The following functions are performed at the transmit side:

- multiplexing of data from various interface units of the terrestrial interface subsystem;

- data scrambling (usually synchronized on TDMA frame or multiframe);

- FEC encoding (usually convolutional FEC);

- addition of preambles and postambles at beginning and end of each burst to transmit;

- bursts transmission to the modulator, at precise time instants.

Similar functions are performed at the receive side:

- unique word detection for each burst;

- preambles and postambles discarding;

- FEC decoding (usually soft decision Viterbi decoding);

- data descrambling;

- data (sub-bursts) switching to appropriate interface units of terrestrial interface subsystem.

The common logic subsystem generally comprises two or three boards.

iii)   Terrestrial interface subsystem

This subsystem comprises one or several terrestrial interface modules (TIM), which connect to each user link and provide the appropriate buffering for data compression at the satellite transmit side (sub-bursts creation) and data expansion at the satellite receive side (sub-bursts extraction).

Several types of TIM may be found depending on the type of terrestrial link to connect:

- 2 Mbit/s or 1.544 Mbit/s PCM links (telephony applications);

- higher rate PCM links (telephony or digital TV applications);

- various types of serial data links (EIA RS-232 or RS-422, ITU-T X.2l, etc.);

- 2- or 4-wire analogue links.

In some cases TIM units include additional processing for on-board voice compression or fax demodulation (telephony applications), or protocol spoofing (data applications). Some experimental systems include also ATM-oriented TIM units.

7.6.4   Modems for digital communications

7.6.4.1   Introduction

Digital modem techniques are described in Chapter 4 of this Handbook. The purpose of this section is to give a better understanding of modern embodiment and technology used in earth station modems. Because of their very wide and common applications, the description applies more particularly, as a typical example, to FDMA-TDM-PSK MCPC multirate modems, as defined by the INTELSAT (IDR and IBS) and EUTELSAT (SMS) specifications (see above § 7.6.2.2, which deals with the system aspects of the modem implementation).

The modem equipment is usually delivered as self-contained units including:

- the modem proper, i.e. the modulator (baseband to IF) and the demodulator (IF to baseband);

- the forward error correction (FEC) encoder/decoder;

- the baseband processing units, including the input/output terrestrial interfaces.

  NOTE – These three units are usually grouped on a single board, often called a transceiver (for transmitter/receiver unit);

- the auxiliary equipment, including supervisory and, control boards and panel.

However, this section is mainly dedicated to the description of the modem proper.

Over the years, this type of equipment has become a mature, highly sophisticated product involving very significant digital signal processing operations. In this section, the various technical solutions will be discussed and typical equipment will be described. Some views about future developments and technological evolution will be expressed at the end of the section.

7.6.4.2 Main features of a state of the art modem

Beside the clear requirements imposed by the INTELSAT (IESS-308) specifications, some technical features that should be provided in modern equipment, either in the basic version or including optional features, are listed hereafter:

- an operational range of any data rate from 64 kbit/s up to 8 Mbit/s;

- both 70 and 140 MHz IF interfaces within the same units, selectable by a software command;

- both BPSK, QPSK and possibly 8-PSK modulation formats;

- FEC encoding by convolutional codes k=7, at rate 1/2, 3/4, 7/8 with Viterbi decoding, or sequential decoding;

- concatenation of a Reed-Solomon code with the convolutional one;

- possible implementation of trellis coded modulation (TCM) for 8-PSK modulation.

In addition, the following practical aspects are to be considered:

- a high degree of integration, e.g. less than two mechanical units (1 unit = 2 inches) for one complete transceiver;

- overall compactness, i.e. several transceivers (e.g. up to 10) can be placed in a single cabinet, which should include also a few boards for the auxiliary equipment and control panel(s).

7.6.4.3 Design options

The purpose of this section is to explain the essential technological problems found in the design of up-to-date modems and to select up-to-date solutions.

7.6.4.3 a) Baseband data processing and terrestrial interfaces

Baseband data processing for converting the transmission formats of the terrestrial network into the satellite link formats (and reciprocally at receive) should be provided by the modem equipment. The following terrestrial interface functionalities may be included in a framing unit:

- physical interface for the incoming/outcoming data streams;

- engineering service channel (ESC) equipment facilities;

- redundancy switching for secured links;

- data framing and synchronization;

- energy dispersion by scrambling.

The physical interfaces must comply with common standards (V.11, V.24, X.21 and HDB3, ...) depending on the data rate.

The ESC functions provide operation such as insertion/extraction of the dedicated channel into/from the normal traffic channel. Synchronous multiplexing of overload bits for the provision of ESCs and alarms is also accomplished in the framing unit.

The clock of the data transmission along the satellite link may be provided by a network clock, a local earth station clock, or from the incoming signal. A plesiochronous buffer copes with asynchronism, jitter and doppler effects.

The last function of the framing unit is to give some minimal stochastic characteristics to the data stream sent to the link. The two main purposes are scrambling for energy dispersion of the transmitted spectrum and of the receiver synchronization, and interleaving that permits spreading out bursts of errors generated by loss of synchronization in decoders. This latter technical approach is also used in the implementation of Reed-Solomon codes.

7.6.4.3 b) Modulation

i)   Baseband signal generation

The input data provided by the framing unit must be translated into analogue vectors of modulation to carry the information on the carrier. Coding is easy to achieve, because the convolutive code is simple and the alphabet of symbols is rather limited (2 for BPSK up to 8-PSK). The symbol rate may be deduced from the incoming rate of framed data, incremented by the error correcting code redundancy, and divided by the spectral efficiency of the chosen modulation. Each symbol is determined by its I and Q coordinates in the modulation plan[17].

An interpolating digital filter permits the generation of samples at a high frequency, that is relative to the maximum symbol rate in order to keep the same analogue smoothing filters, whatever be the modulation format. In order to fit the expected spectrum shape at the transmitter, a length of less than 8 symbols is sufficient to translate the pulse response of the filter. In addition, the selection of

---

[17] In this section, P and Q are used to designate the digital information paths, while I and Q are used for the modulation vectors.

the binary states for each I or Q path can be performed very simply, especially for QPSK, using a look-up table where all the possible sequences of samples are registered.

Digital-to-analogue converters with 8 bits of resolution are sufficient to achieve the definition of the filtered I and Q waveforms.

ii)   Carrier synthesis

The INTELSAT specifications on spurious emissions and phase noise are rather difficult to implement over all the IF bandwidth, i.e. 52-88 MHz or 104-176 MHz.

Moreover, a frequency resolution of a few Hertz is required. The specifications can be only achieved by the technique of direct digital synthesis (DDS). Such a process method combines a digitally controlled oscillator, which feeds a digital-to-analogue converter (D/C) to generate a reference sine wave, with a phase locked loop (PLL), which permits delivery of an harmonic of the reference. The major difficulty comes from the fact that this frequency multiplication process (by a factor n) also multiplies the spurious level by the same factor. In fact, any DDS system generates some spurious frequencies in a wide band due to the phase sampling and to the impairments of digital to analogue converters.

DDS could be enhanced by a direct digital modulation technique which achieves the generation of the modulated carrier directly at the intermediate frequency (IF), without generating two analogue baseband I and Q signals. This copes with quadrature or carrier rejection problems. Anyway, there is, for the moment, a technological limit in digital to analogue conversion that does not allow generating a modulated signal with the required characteristics at a frequency higher than 20 MHz.

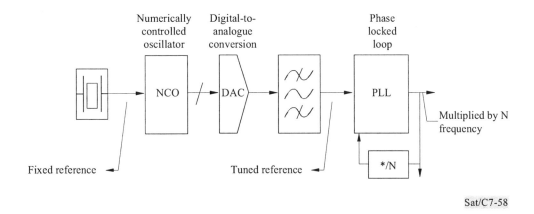

Sat/C7-58

FIGURE 7.58

Block diagram of a synthesizer using the DDS method

This fact imposes some complexity on frequency conversions and filtering. Figure 7.58 shows a conventional block diagram of a synthesizer using the DDS method.

iii)   I, Q modulation

As mentioned above, the direct generation of the modulated spectrum is currently possible at low frequency only, although this should probably be used for future products.

In consequence, a more conventional method with a quadrature modulator is generally implemented in currently available equipment: it uses either an active component or an association of mixers and coupler. The mixer process in the modulator can directly transfer the signal to the carrier frequency. The major advantage of this technical approach is to avoid image spectrum problems but, on the other hand, precise control of the phase over the whole bandwidth is needed. Another technical approach consists in modulating at a first, almost fixed, IF frequency in order to ensure a well-tuned quadrature, and then to shift the modulated signal to the normal IF frequency by a second mixing.

7.6.4.3 c) Demodulation

i)   Amplification, gain control and filtering

Due to the multiple access technique (FDMA), the RF signal can be received in a wideband (usually 36 MHz for a 70 MHz IF, 72 MHz for a 140 MHz IF). The relevant carrier occupies a reduced part of this bandwidth.

The automatic gain control (AGC) must compensate the differences both of the spectral density (coming from the uplink) and of the global level (due to the downlink). The variable rate of carried data involves a variation of spectrum bandwidth which increases the complexity of the gain control somewhat. An AGC is based on a bandpass filter, a variable gain amplifier, a detector and a feedback loop.

In modern receivers, variation of the bandwidth of the filters should be implemented by digital filtering, taking into account the constraints on linearity, the noise factor of the complete receiver and the limitation in the resolution of A/D converters.

Typically, the control of gain is achieved in two steps. The first is done in an analogue way and controls the power inside a bandwidth determined by the tightest filter before A/D conversion. The second operates after the reduction of band by the digital filter.

ii)   I, Q demodulation

On the receive side, the process can be considered the same as the I, Q modulation described above. That means that either the complex IF signal or two baseband I and Q signals must be sampled before processing. Just as in the case for the transmitter, this is merely a question of filters and of availability of fast A/D converters. The current trend is to use direct IF sampling but this requires good analogue rejection filters and more digital processing, which has so far been relatively cumbersome and expensive.

iii)   Signal regeneration

This part includes the Nyquist filtering, the carrier and clock recovery, and the decoding.

The latter function is achieved by a Viterbi decoder, which provides high performance for convolutive codes. As explained in Appendix 3.2, it computes metrics, i.e. distances from the received sequence to some ideal coded sequences, in order to determine which one offers the maximum likelihood. The convolutive code may be concatenated with a powerful Reed-Solomon code that requires specific synchronization. Another method for decoding is the sequential suboptimal technique that permits the use of a higher length of code constraint (e.g. $K = 36$), which, in the case of low data rate links, permits improvement of the transmission quality by a simple process.

The regeneration itself consists of interpreting the received analogue signal at sampled instants and deciding what data word each symbol represents. Ensuring the best likelihood of a right decision consists of reducing the uncertainties and distortions around the received signal. Among them, the sampling instant is optimized by the clock recovery and the phase of the received complex signal is coherently demodulated by the carrier recovery. But some more loops like AGC, DC offset compensation, quadrature correction and possibly adaptive filtering should be added to minimize the distortion.

Contrarily to burst modems for TDMA or packet transmissions, FDMA/TDM (e.g. IDR) modems have to cope only with a continuous flow of data. This makes it possible to make extensive use of feedback loops for carrier, clock, gain recoveries, etc., which permits very fine adaptive tuning of receiver parameters and results in a BER curve very near to the theoretical one at any rate, any carrier frequency, or any environment condition.

7.6.4.4 Equipment description

7.6.4.4 a) Equipment technology and layout overview

An IDR modem comprises analogue and digital signal processing units. Clearly, the analogue part of the hardware may be mainly identified with the intermediate frequency section, modulation and demodulation, frequencies synthesis functions. These parts require special cares in the layout and some shielding in order to ensure conformity with the severe specifications on spurious emissions. In the embodiment described, the analogue part is implemented on a specific board to isolate it from the digital part. This latter part gathers the data processing, including multiplexing and framing, and the digital signal processing, generation and regeneration.

Physically, the digital electronics occupy up to 70% of the space in the modem while the analogue electronics occupy 30%. This figure should be compared with an equal share of the technologies in earlier generations. Moreover, the digital part may be split one third for data processing, one third for data signal processing, and one third for the interface and supervisory functions.

Data processing is based on the use of field programmable gate arrays (FPGAs), which incorporate an equivalent of more than ten thousand logical gates and present the significant advantage of being entirely configurable by software. This technology is well-adapted to the multiple data rates and formats in IDR and other similar (e.g. IBS and SMS) applications.

Digital signal processing is generally implemented with FPGAs or with ASICs (application specific integrated circuits) for the digital filtering. DSPs (digital signal processors) are, in principle, good candidates for these functions but they fail to meet current performance requirements for more than 2 Mbit/s data rate.

7.6.4.4 b) Framing unit

As explained above, the baseband data processing and terrestrial interfaces functions can be implemented almost entirely with digital hardware. The currently available FPGA technology permits a dramatic reduction of the amount of components while complying with the constraints of the different standards. The supervisory system of the equipment loads from memory the specific combination of gates and gate array cells suited for each application.

7.6.4.4 c) The modulator

In the particular layout described below, direct modulation is used, involving the generation of two quadrature baseband analogue signals. This solution spares costly filters that are required by the rejection of image spectra in other architectures.

i)   Baseband signal generation

On the transmit side, the baseband signal generation unit receives the framed data signal and converts it into two baseband analogue signals. It includes therefore the functions listed as follows:

• convolutional encoding;

• symbols mapping (especially for 8 PSK mode);

• variable oversampling digital filter with its associated sampling clock generation;

• digital to analogue conversion of samples;

• analogue smoothing filters.

These functions are organized as described on Figure 7.59 below.

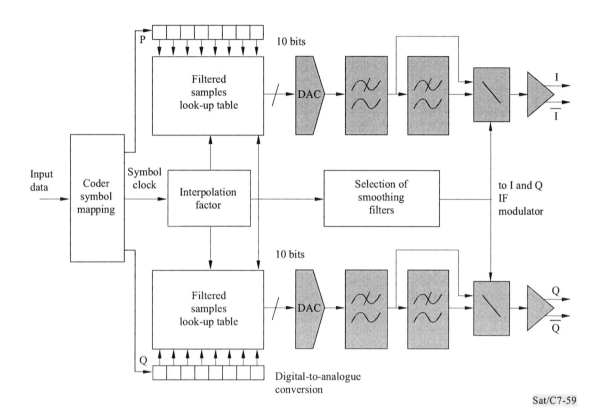

FIGURE 7.59

Baseband signal generation

ii)   IF modulation

After the generation of the I and Q signals, these baseband signals are transferred to the carrier by direct modulation. After this stage, a bandpass filter eliminates the harmonics and other products of the mixing. The level of transmitted power can be finely tuned depending on the spectrum bandwidth and the transmit power budget. These functions are organized as described in Figure 7.60 below.

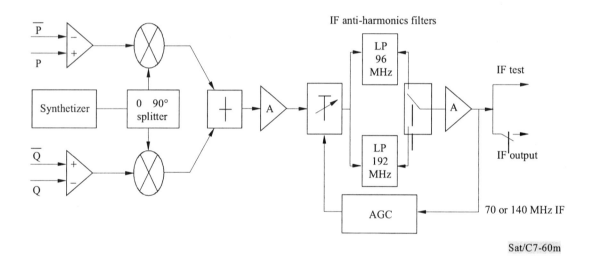

FIGURE 7.60

IF modulation

7.6.4.4 d) The demodulator

i)   IF Demodulation

Direct demodulation is achieved in the example given in Figure 7.61 below.

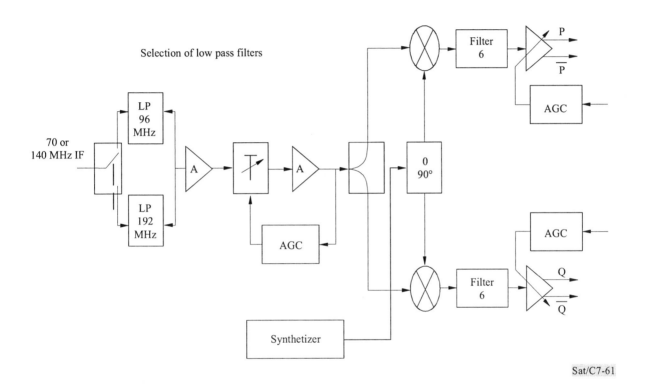

FIGURE 7.61

IF demodulation

ii)   Signal regeneration

The regeneration unit includes the digital filtering, the carrier and clock synchronizations, amplitude and quadrature loops, and the Viterbi decoding. This is represented in Figure 7.62 below:

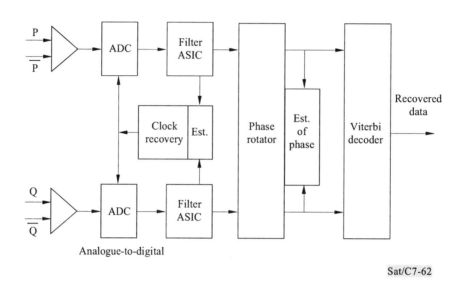

FIGURE 7.62

Signal regeneration embodiment

7.6.4.5   Future developments

Trends in modulation techniques

Combining coding and modulation like in trellis coded modulation already offers new attractive solutions for radio links. The 8-PSK (with 2/3 error correction) TCM is already optionally included in the INTELSAT IDR Standard and other coding schemes like 8-PSK, 5/6, 8/9 may be used, based on 16-PSK or QAM.

Moreover, the versatility of codes and modulations available in a flexible modem with the same hardware opens the way to adaptive coded modulations. This technique consists in ensuring an always error-free transmission by modulating the data rate versus the coded rate at a fixed spectrum occupancy. It is an attractive response to the rain attenuation providing variation of the link budget margin. But this technique is more suited for packet transmission or for TV speech broadcasting which allows a control of the message source. Anyway, it should be underlined that today modems may be designed to provide switching, during operation, between various modulation and/or coding processes.

Trends in modem technology

It has been shown that today digital modems are mainly based on specific or programmable gate arrays, data and signal processors. The technology of digital integrated circuits shows continuous progress. This should soon permit the concentration of the digital part, as a first step, in a group of gate arrays and processors and eventually into only two or three devices. Beside this progress in digital technology which permits the concentration or addition of more digital processing, the analogue parts may benefit from the new integrated circuits designed for terrestrial radiocommunications and mobile phone technology.

Significant results of the technical progress should also provide:

- more compact equipment at reduced price and ensured reliability;

- improved customer facilities, ease of use, higher synchronization speed and also compatibility with powerful network management means. However, it should be noted that progress in modems is not only related to the product itself but also to the definition of the whole communication system.

7.6.5 Digital television equipment

7.6.5.1 General

Analogue satellite TV transmission has existed since the use of satellites began.

However due to the advancement of digital signal compression techniques digital TV transmission has been rapidly developed recently.

The analogue links use a complete transponder (20-36 MHz) and need a high value for the carrier-to-noise (C/N) ratio (10-14 dB). Digital TV transmissions with high compression technologies require only 10 to 20% of the bandwidth needed for analogue which makes possible the transmission of 6-10 video signals into the same transponder. A smaller required C/N ratio (4-8 dB) and robust error code correction improve the receiving conditions.

Digital TV transmission applications range from long-distance TV transmission (e.g. INTELSAT services) to satellite news gathering (SNG; see § 7.9) to TV distribution (see § 7.10) and to low bit rate videophone.

Digital TV image signal compression requires rapid processing of vast amounts of data and has been made feasible only during the last few years due to the availability of fast and powerful signal processing computers. Much of the initial technical and standardization work in TV image signal processing has been done by the moving picture expert group (MPEG) association, and the digital video broadcasting (DVB) project office (under the European Broadcasting Union (EBU)).

7.6.5.2 Digital TV transmission

At present, digital TV applications range from low bit rate videophone to high-definition television (HDTV) transmission. According to these various applications many TV transmissions with various compression levels and bit rates are available.

The international MPEG-2 standard is flexible and may be implemented using various degrees of image and sound qualities and numerous features. The MPEG-2 standard is a family of systems, each having an arranged degree of commonality and compatibility. It allows four source formats, or levels (low level, main level, high-1440 level, and high level), to be coded, ranging from limited definition (about today's video recording quality), to full high definition TV (HDTV) – each with a range of bit rates.

In addition to this flexibility in source formats, MPEG-2 allows currently 5 different profiles (simple profile, main profile, SNR scalable profile, spatially scalable profile, and high profile).

Each profile offers a collection of compression tools that together make up the coding system. A different profile means that a different set of compression tools is available. More than 20 combinations of levels and profiles have been approved. The main profile and main level (MP@ML) is the most used for satellite transmission.

In general digital TV satellite transmission employs some form of forward-error-correction coding (FEC), for example, Reed-Solomon (RS) block coding to minimize the occurrences of random bit errors at the input to the Viterbi decoder which can result in picture degradation.

7.7 Monitoring, alarm and control

7.7.1 General

The monitoring, alarm and control (MAC) facilities are of prime importance to the proper operation and management of an earth station. This subsystem can simplify the operation of a station by providing easy identification of problems. MAC functions are an essential part of satellite earth stations whether they be small (1 m) VSAT, TVRO (television receive only) earth stations or an INTELSAT Standard A type.

In order to choose a MAC system which is suited for a particular earth station application, careful consideration should be given to the following:

- complexity of the MAC function;
- expected future growth of the earth stations;
- cost;
- size/type of earth station;
- size of network;
- user-friendliness;
- use of standard interfaces/protocols;
- operator/MAC interface;
- possibility of interfacing with the existing MAC.

Three different types of MAC systems can be considered each with varying degrees of complexity, e.g. analogue, computerized, or unattended MAC system. These types of systems will be discussed in the following sections.

The size of a MAC system depends on the complexity of the monitoring and control function and on the level of detail that the earth station manager wishes to exercise. Obviously the greater the degree of control and monitoring that one wishes to exercise the more complex the MAC becomes.

The expected future growth of the earth station plays a significant role. A MAC system which can be easily expanded at a nominal cost should be considered if considerable expansion is expected.

The cost is an important parameter. In choosing a MAC system which is cost-effective, one should realize that a good system can reduce significantly the personnel requirements which translate into annual operational cost savings and greater utilization of skilled manpower resources. Thus, a good system will pay for itself in a short time if the cost/performance trade-offs are carefully weighed against each other.

The size and type of earth station plays a significant role in choosing the proper MAC system to use. Large earth stations require MAC systems with greater handling capacity than small unattended earth stations.

Many organizations have the intention of eventually centralizing and extending all their stations to a network control centre (NCC). The role of an NCC is to monitor, measure and, if necessary, control the flow of traffic and thus ensure maximum use at all times of the equipment and facilities in order to complete as many paid calls as possible and prevent congestion. The above point should be considered when choosing a MAC system if this is the eventual direction which the organization intends to follow.

The operator/MAC interface is another criterion which is often neglected in the design objectives.

The operator is often the most critical control component. His efficiency and performance can be optimized through the use of "human engineering" (the study of man's interrelationship with his working environment).

Efficiency in the MAC operation can and should be extended to the operator functions just as this is applied to any other portion of the system.

7.7.2 Major objectives of a MAC

The station monitoring, alarm and control subsystem performs the following three major functions:

i) provides the station staff with a centralized monitoring, alarm and control system. The distribution monitoring points and the dispersed equipment alarms could be centralized at the operator's desk. Aids such as CRT console, audio and visual alarm signals, memory storage and hard-copy printer are some of the usual peripherals. Control commands should also allow him to turn on/off equipment, switch over transmission paths and other such functions;

ii) provides the network control centre with real-time monitoring of transmission and traffic performance of individual stations. Information from several stations could be combined to predict immediate or short-term network degradation, so that preventive and corrective network management action could be applied accordingly. Station equipment alarms, pilot and noise levels as well as alarms from switching equipment and traffic data should be used for this purpose;

iii) provides the integrated test centres with remotely controlled testing capabilities. These centres could access the automated testing equipments in the station for fault location, transmission measurements or any other routine testing.

7.7.3   Functional arrangement of a MAC

The control and monitoring facilities can be divided into three functions, namely monitoring, alarm, and control. Monitoring is the obtaining of information concerning the condition or status of equipment. Alarm provides either aural or visual information (e.g. a buzzer, change of colour or a flashing light) for alerting personnel to abnormal operation of earth-station equipment, at the same time indicating which section of the equipment is at fault.

Control is principally the capability for (automatic or manual) transfer of signals from a path through their normal on-line equipment to a path through appropriate stand-by equipment. Control also includes the ability to set variable parameters (priority) and operating modes (e.g. automatic, manual, supervisory).

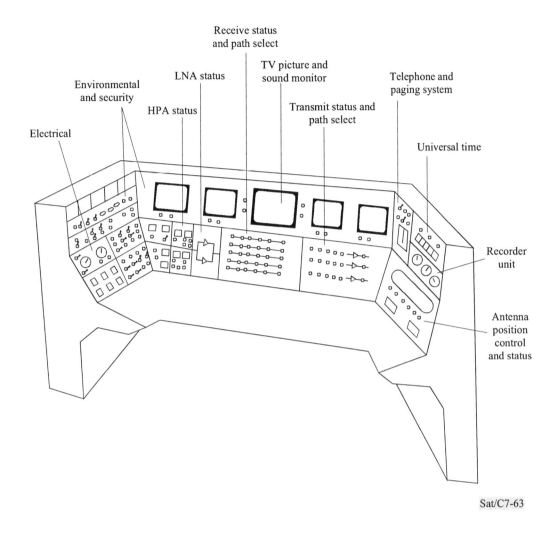

Receive status
and path select

TV picture and
sound monitor

Telephone and
paging system

LNA status

Environmental
and security

HPA status

Transmit status and
path select

Universal time

Electrical

Recorder
unit

Antenna
position
control
and status

Sat/C7-63

FIGURE 7.63

Conventional MAC system control centre

The MAC facilities are located in the station's control centre. The relevant parameters of all the different subsystems are brought to the control centre so as to centralize the MAC facilities at one location for ease of operation.

MAC facilities can be classified into three categories: local, remote, supervisory.

Local MAC – Local MAC facilities are those which are located close to the equipment that is being monitored or controlled.

Remote MAC – Remote MAC facilities are located at some distance from the equipment being monitored or controlled. Remote facilities usually provide duplications of at least some of the information provided by the local MAC facilities.

Most of the remote facilities of each major subsystem of an earth station are centralized in one or two respective control units located in the control centre. From the control centre the operator can monitor and control the operation of the station (see Figure 7.63).

Supervisory MAC – The MAC parameters are made available to computer type supervisory equipment. Monitoring of the parameters would be automatic and pre-programmed corrective action(s) to specific failures can be included in the software, subject to alteration by the station staff if required (see Figure 7.64 for the input/output relationship of the station monitoring system).

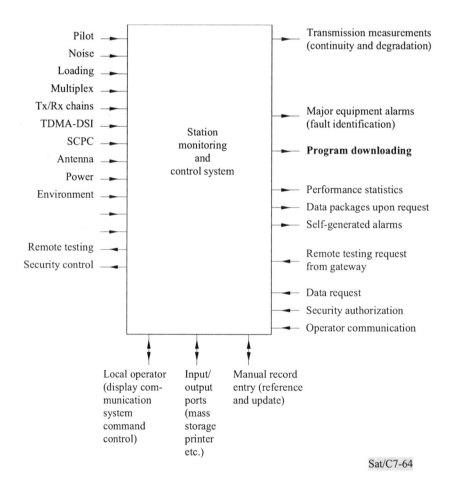

FIGURE 7.64

Input/output functions of station monitoring and control system

7.7.4 MAC systems

MAC systems can be classified into three general categories, namely analogue, computerized and unattended MAC systems.

Analogue MAC systems

In the past, analogue MAC systems have been characterized by analogue indicators, flashing light indicators, or audible alarms. Many of the parameters are usually located locally on the equipment itself with duplication of at least some of the parameters at a central control centre (CCC) where an operator acknowledges all alarms and takes appropriate action to remedy the problems. This usually includes switching stand-by equipment on-line (many MAC systems have automatic switch-over) and alerting maintenance personnel of the equipment failures.

Computerized MAC systems

Recent advancements in computers, microprocessors, and integrated circuits have led to MAC systems with a large capability. These computerized MAC systems have new features which have significantly changed the whole aspect of monitoring and control.

With the ever-increasing complexity of earth stations, it is essential that a computerized monitoring system be implemented if efficient utilization and logging of the monitored parameters is to be effected.

A computerized monitoring system, as compared to an analogue monitoring system, has the following advantages:

- monitoring of the parameters would be automatic and pre-programmed corrective action(s) to specific failures would be included in the software, subject to alteration by the station staff as required;

- minimizes the routine work of periodic meter readings;

- a video screen is generally used to display the status of the operational carriers or channels and of the equipment (at different levels of detail and including on-line and off-line stand-by circuits) are usually displayed under the form of diagrams with programmed menus in order to give clear, real-time, indications of all earth station parameters under surveillance;

- performs alarm filtering and diagnostic evaluation of the source of failure;

- in most cases, software changes rather than hardware changes are required to alter any parameter or to upgrade the capabilities;

- can be used to print out hard copies of parameter values for further processing, e.g. failure reports, availability reports, outage, etc.;

- it manages and updates the database of the earth station (equipment composition, traffic, events, etc.);

- can be expanded as the system grows and extensions to a network control centre can be easily accomplished (see Figure 7.65).

Unattended MAC system

Changing system constraints and developments in microwave technology have led to new satellite communication earth terminals designed for unattended operation. With the number of earth terminals increasing, greater operational economy can be achieved with terminals left unattended for long periods of time, and which require only periodic maintenance by visiting technical personnel.

A control and monitoring system is essential for the proper operation of these unattended earth terminals (UET). The essential parameters which are monitored and the operations which can be commanded are shown in Figure 7.66.

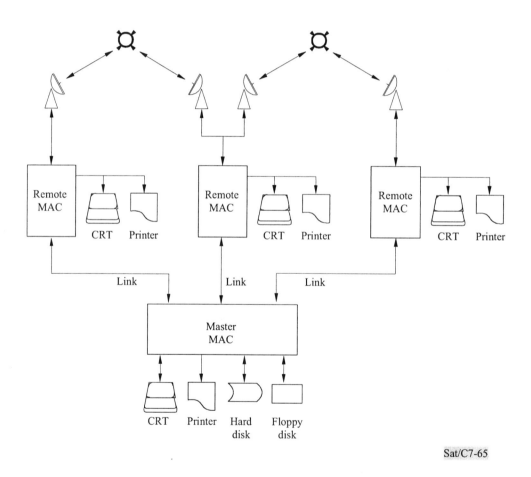

Sat/C7-65

FIGURE 7.65

Remote MAC system for satellite earth-station network

7.7.5   Parameters monitored

The main parameters that should be monitored (and possibly controlled) at the earth station can be classified as follows:

i)   Communication carrier parameters which can be monitored at different levels (RF, IF and BB) to measure the quality and the availability of the channel. The following parameters are included.

  •   Analogue transmission: monitoring of multiplex pilot level, pilot frequency[18] and channel noise[19].

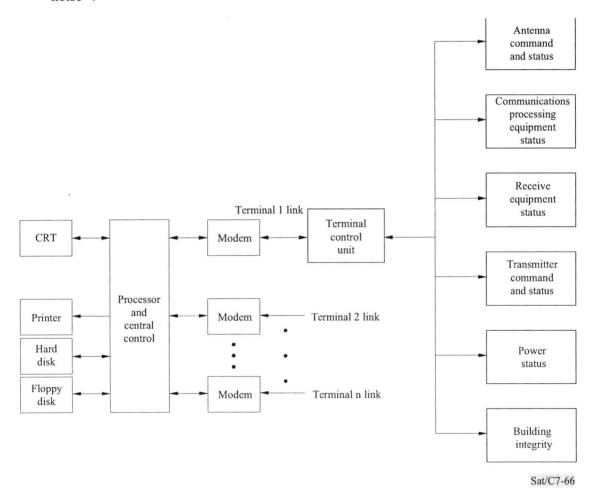

Sat/C7-66

FIGURE 7.66

---

[18] The pilot level and frequency indicates the trend and the degree of degradation.

[19] In analogue, the noise power is measured and converted to its equivalent in a psophometrically weighted 3.1 kHz channel. An alarm is set off if the mean noise power exceeds a pre-assigned threshold level for a pre-determined amount time (quality/availability).

Unattended earth terminal system diagram

- Digital transmission: monitoring of BER, E_b/N_0[20] or errored blocks in modern systems (B-ISDN, ATM, etc.).

ii) Communication equipment status monitoring: this includes the indication of individual equipment status (on-line, stand-by, maintenance, fault, etc.) of both transmit and receive chains and also performs automatic switching operation in case if any of the on-line equipment fails[21].

iii) Tracking subsystem: monitoring of tracking parameters such as beacon level, angle indications, etc. Control of antenna motors through the servo system and selection of an available back-up mode if the main mode of tracking fails.

iv) Support subsystem: this includes all other monitoring and controls which are not directly involved with communication and can be categorized as follows:

- power generation;

  this includes an uninterrupted power system (UPS), power generators, battery banks and associated equipment.

- environmental and security;

  this includes outdoor/indoor facilities and general equipment.

- miscellaneous;

  various other equipment and facilities need monitoring and control (terrestrial links, TV transmission facilities, internal telephone system, etc.).

7.7.6   Summary

A monitoring, alarm and control (MAC) system offers several very distinct operational advantages that justify the initial cost. Present and future station owners should give serious consideration to a MAC system for the following reasons:

- it reduces significantly the quantitative and qualitative personnel requirements which translates into annual operational cost savings and greater utilization of skilled manpower resources;

- it reduces significantly those earth-station traffic interruptions which are related to human error;

- its provides detailed traffic statistics and should therefore permit enhancement of the system performance and the quality of service;

- detailed breakdown of equipment failure status allows the operator to minimize the downtime;

---

[20] The BER measurement provides the station operator with an indication of the quality performance of the received carrier. An alarm is set off if the BER exceeds a pre-assigned threshold level for a pre-determined amount time (quality/availability).

[21] Receive and transmit chains are normally provided with off-line, hot stand-by, equipment. The redundant equipment has all the necessary functions, such as automatic adjustment of frequency, level and parameters needed to replace any on-line chain failure.

- it may incorporate a control system (including billing function) for the centralized management of local networks (e.g. domestic or business networks).

7.8 General construction of main earth stations

Introduction

Satellite earth stations are similar to terrestrial microwave stations to the extent that they consist of two main elements: antennas and equipment buildings. They have, however, the following differences.

Terrestrial microwave stations are usually located at high elevations and consist of a tall tower on which antennas, usually not more than 4 m in diameter, are mounted, and a small unmanned electronic equipment building near the base of the tower.

Satellite earth stations are usually located in low-lying areas that are shielded from terrestrial RF interference by low hills surrounding the area. Note that this section deals only with earth stations operating with significant traffic capacities. These are equipped with medium or large antennas (typically 10 m to 18 m, but possibly even more[22] and are implemented with relatively large, manned, buildings. In fact, the case for small stations used in domestic or business operation (e.g. VSATs) is quite different and is not included in this section.

They often include within the station proper, a terrestrial microwave link connecting the station to the central area it is serving.

The earth station should be designed to provide shelter and a suitable internal environment for the telecommunications equipment, control and monitoring equipment, the station support equipment and the personnel operating the station. The basic components are:

- the civil works necessary to provide shelter and a working environment;
- the power supply providing the energy to the electronic equipment and the building services;
- the antenna system (antenna civil works).

General remark

All descriptions given in this section are to be considered only as representative of current trends in technology. In fact, project engineers and designers should, in each particular case, consider fully the local conditions and comply with the regulations, standards and requirements which are locally in force as concerns environment, construction, security, human engineering, etc.

---

[22] Up to the eighties, the INTELSAT specifications for Standard A earth stations (G/T $\geq$ 40.7 dBi at 4 GHz) asked the use of very large antennas (up to 33 m diameter). Since then specifications have been relaxed (G/T $\geq$ 35.0 dBi at 4 GHz) to permit the utilization of smaller, less expensive Standard A antennas. However, older stations remain operating with those previous very large antennas.

7.8.1   Civil works

The design of the station has to cater for the following main features:

*   antenna;
*   telecommunication equipment;
*   power supply equipment;
*   mechanical equipment (heating, ventilating and air-conditioning);
*   administration;
*   station support services.

Additional supplementary features that should be considered are:

*   access roads and parking;
*   security;
*   water supply and treatment;
*   fire protection system;
*   power entrance;
*   terrestrial link telecommunication facilities.

The characteristics of the areas designated for the above features will depend upon the type of station being considered and are given below.

7.8.1.1   Large multi-antenna earth stations

The main features mentioned previously, apart from the antennas, should be housed in discrete areas (individual areas isolated from their neighbours by fire-resistant walls, door openings, etc.) with a central building so located as to provide the shortest possible inter-facility links between the telecommunications equipment room and the antennas. Figure 7.67 shows the layout of a typical multi-antenna station.

The location, height and orientation of the various buildings, structures and antennas should be such that no obstacles obstruct the radiation of the antennas wherever they are pointed during operations.

The main building should be designed to withstand the severest seismic shocks anticipated in the locality. After such a shock the building should have suffered only superficial damage and its support and communications systems should continue to maintain service.

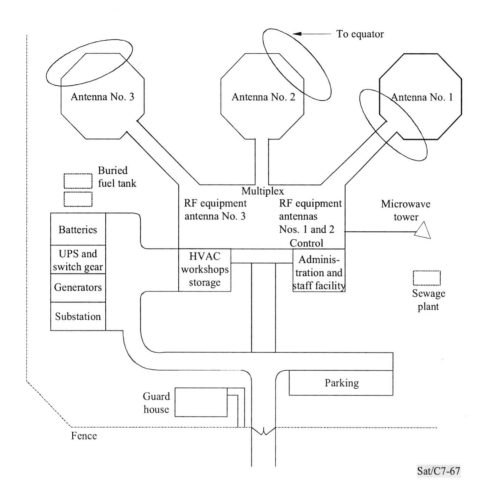

FIGURE 7.67

Layout of typical multi-antenna station

7.8.1.1.1 Telecommunication equipment area

The telecommunications equipment should be housed in a discrete area having adequate floor-to-ceiling height to permit the use of overhead inter-rack cabling. Elevated air-conditioning ducts are usually installed in the space provided between the concrete ceiling and a false ceiling. Alternatively, the area should have a technical elevated floor (computer type) with sufficient underfloor depth (0.6 m) to provide space for air-conditioning ducts and for inter-rack wiring, including power distribution cabling.

The environment of the area should be controlled to meet the requirements specified by the equipment suppliers. The air-conditioning equipment should be of sufficient capacity to adequately meet all heating/cooling loads and have ample redundancy to allow for individual unit failure.

7.8.1.1.2   Power supply equipment area

The power supply equipment (i.e. transformers, switchgear, uninterruptible power supply (UPS), stand-by generators, d.c. power supplies, etc.) should be housed in a discrete area somewhat adjacent to the telecommunications equipment room. This area should have amply resistant fire walls to prevent the spread of combustion to and from the other areas. The stand-by generators, batteries, switchgear and UPS should be housed in separate interconnected rooms within the area.

Self-closing, anti-panic, fire-resistant doors should be used throughout and all cable routes through walls or underfloor should be sealed with fire-resistant materials to prevent the spread of combustion. Sound-absorbing materials should be used in the construction of the generator room, particularly if this is the only source of continuous electrical power. If possible this room should be cooled by outside ambient air using natural convection or fans. The generator control equipment should, if possible, be housed in an adjacent room with a sound-proof window between rooms.

The battery room should have adequate ventilation to disperse hydrogen fumes. Adequate curbs, sloping floors and a cesspool for removing and retrieving accidental spillage should be provided. The floor and lower part of the walls should be protected with anti-acid paint. For security reasons, the battery room should also be equipped with a shower/eye douche.

The d.c. power and uninterruptible power supply (UPS) rooms should be air-conditioned. The use of heat pumps in the UPS room is recommended if other station support areas required heating in the winter months as the excess heat from this equipment can thus be utilized.

The power supply area should be laid out so that the d.c. power and UPS rooms are as close as possible to the electronic equipment room to avoid long power cable runs.

An adequate underground reservoir for fuel oil of sufficient size to meet the needs of the station's power generating equipment, should be provided. The reservoir should consist of two or more interconnected tanks linked by dual fuel lines, valves, etc. to dual daytanks located in the generator room. All fuel lines should be equipped with fusible link stop valves at their entry to the generator room.

7.8.1.1.3   Mechanical equipment (heating, ventilating and air-conditioning: HVAC)

The mechanical equipment (heating, ventilating, air-conditioning) should have a separate area adjacent to the power equipment area with fire-resistant walls, doors, etc. The type of HVAC system will depend on the location of the station. It is usual for large stations to have heat exchangers (cooling towers) with a central circulating plant using either chilled water or water at a controlled temperature to individual air-conditioning units or heat pumps distributed throughout the station.

If heating is required in the winter season, the use of heat pumps between the UPS room, the electronic equipment room and the station support areas should be considered. The excess heat from UPS and equipment can thus be utilized.

The design of the building envelope (walls, roof, windows, etc.) should include a study of the local climatic conditions and environmental requirements of the station to determine the optimum design of both the envelope and mechanical system. Natural (free) cooling should be used whenever possible and consideration should be given to energy-reducing features to improve the building energy efficiency.

The cooling of the microwave HPAs should be carried out by means of either an air-cooled system or a water-cooled system depending on the size of the microwave tube that is used, as described in § 7.4.

In the case of an air-cooled system, cooling of the HPA should be by means of outside air connected directly to the intakes and exhausts of the individual units by insulated ducts. (A plenum can be used for adjacent multiple HPAs.) Supplementary fans should be considered to assist the flow of cooling air and to overcome duct friction. In climates with large variations of seasonal temperatures the intake and exhaust ducts should be interconnected and the hot exhaust air mixed with the cold outside air by means of modulating duct dampers to give the required intake temperatures at the HPA inlet.

In the case of a water-cooled system, de-ionized pure water should be used for cooling the HPA in order to prevent blockages of the water paths in the collector and to provide low water conductivity. It is also necessary to keep the amount of diluted oxygen absorbed in the water flow as small as possible to avoid oxidization. Materials such as copper, brass and stainless steel are commonly used for water path. However, recently, water-cooled systems are seldom used.

7.8.1.1.4   Administration and station support area

The administration and station support areas can be combined in one discrete area or be in two separate discrete areas. The area should contain the offices (manager, supervisors, clerical staff), the staff personnel facilities (lunchroom, lockers, toilets, classroom/conference room, etc.), receiving, storage/warehouse area and mechanical and electronic workshops.

7.8.1.1.5   Supplementary features

Apart from the preceding main features the following supplementary features should also be considered:

- access roads and parking: access to the station, particularly during the construction stage when large heavy structural components of the antenna have to be transported, must be reviewed. Adequacy of the road bed, bridges, tunnels, etc. must be determined and corrections made where necessary. Turning areas for delivery vehicles and space for parking should be allowed.

- water supply and treatment: investigations should be undertaken before the site has been selected to ascertain the availability of potable water, or water requiring the minimum amount of treatment, adjacent to the site. The availability of this water will have a direct bearing on the design of the HVAC system, the stand-by generator cooling as well as the personal needs of the station staff. If wells have to be drilled, their capacity should be determined for all seasons of the year and a redundant system should be provided to cater for failure or maintenance of pumps and equipment.

An adequate sewage system should be provided. This system should be connected to a main line system or consist of a septic tank with drainage field.

- fire protection system: the buildings should be protected by a zone-designated fire protection alarm system including local fire detectors. In equipment areas the protection system could consist of an automatic release of CO_2 gas. A portable system, associated with proper alarm announcement, could be sufficient in attended areas.

Where the stations are located in heavily-wooded areas, a clear space to act as a fire break should be maintained between the buildings and foliage.

A water hydrant system connected to a reservoir of sufficient capacity and with pumping facilities to maintain an adequate flow of fire-fighting water should be considered.

- power entrance: if a high voltage or commercial power source is available it should be fed via dual cables (underground and/or overhead) to a high voltage substation which can be located within the main building or exterior to it. The exterior substation should be protected by a 2.5 m chain link or equivalent type fence.

- terrestrial link telecommunication facilities: provision should be made for providing terrestrial telecommunication facilities from the station. Protected space should be provided for the electronic equipment and, if required, the location of a terrestrial microwave transmission tower should be determined so as not to interfere with the operation of the main antennas. If communication is by cable then an adequate duct or burial system should be provided.

7.8.1.2   Large single-antenna station

This type of station is assumed to have a maximum of one antenna only, the antenna pedestal acting as the central building (see Figure 7.68 for a typical layout). The main features (antenna, telecommunication equipment power supply, administration and support services) as given for a multi-antenna station apply equally for this type of station. The characteristics and locations of these features should be followed, except for the following:

- mechanical equipment (HVAC): for a station of this size a separate HVAC room may not be necessary. A split unit (internal evaporator, external condenser) system should be considered for the conditioned areas with individual fan coil (evaporator) air handling units plus back-up units in each area. The condenser units should be outside and adjacent to the pedestal in an area with an adequate wind flow. The air handling units can be provided with a heating coil if seasonal changes necessitate the use of heating in the winter months.

- power generating equipment: the stand-by generators and their control equipment should be housed in a separate building some distance from the antenna. This building should be cooled by natural and/or mechanical (fans) air flow using ambient outside air. Ventilation openings should be provided with automatic louvres that will close in the event of a fire, the generators and fans should shut off automatically in this event and fire extinguishing devices should be provided. The requirement for an underground fuel reservoir remains as before.

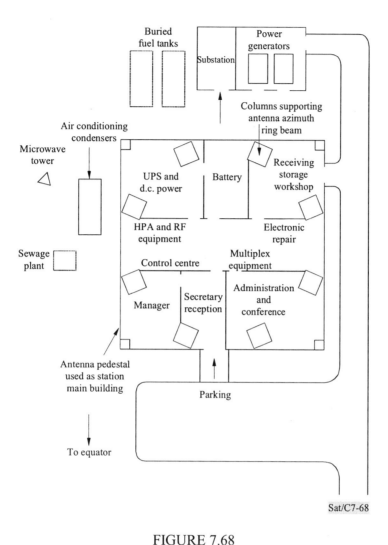

Buried
fuel tanks

Power
generators

Substation

Columns supporting
antenna azimuth
ring beam

Air conditioning
condensers

Microwave
tower

UPS and
d.c. power

Battery

Receiving
storage
workshop

HPA and RF
equipment

Electronic
repair

Sewage
plant

Control centre

Multiplex
equipment

Manager

Secretary
reception

Administration
and
conference

Antenna pedestal
used as station
main building

Parking

To equator

Sat/C7-68

FIGURE 7.68

Layout of typical single-antenna station

7.8.1.3   Medium-sized stations

The stations considered in this subsection are domestic network type stations with a single medium-sized antenna (10-15 m). The main and supplementary features detailed previously apply in general but modifications and omissions can be tolerated depending on the reliability required for the station (Figure 7.69 shows a typical layout).

The construction of this station should consist basically of the antenna, a radio equipment shelter and a main building. The 10-15 m antennas are usually of the limited motion variety consisting of the reflecting surface mounted on an Az-El mechanism fabricated from medium steel sections. The radio equipment shelter should be an abutment from the main building located between the support members of the antenna pedestal thus bringing the RF equipment and LNAs as close to the antenna feed as possible. The RF shelter should be well-constructed and transportable, having a removable chassis, thus permitting the RF equipment to be assembled and tested at the manufacturer's premises and transported to site ready for operation with a minimum of field labour.

The main building can be constructed of local material on a concrete slab or be of the steel frame type with a prefabricated hung panel which can be readily transported and assembled on site.

The supplementary features (access roads, parking, fire protection, etc.) as given for a multi-antenna station apply equally for this type of station.

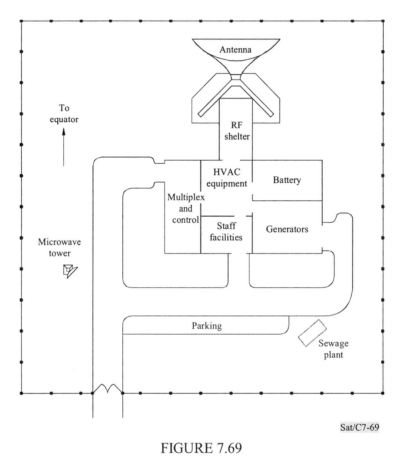

Sat/C7-69

FIGURE 7.69

Layout of typical medium size antenna station

7.8.2   Power system

The power system should be designed to deliver the electrical power which is required for the communications equipment, the buildings services and auxiliary facilities.

The requirements are as follows:

- critical loads: high power amplifiers, telecommunications equipment, multiplex equipment, etc. Critical loads are normally supplied through the battery bank of the uninterruptible power supply system (UPS);

- essential loads: antenna servo system, antenna anti-icing, air-conditioning, ventilation, lighting, etc. Essential loads are normally supplied by a commercial power line and, in case of its failure, by a stand-by motor/generator at the station;

- non-essential loads: external lighting. Non-essential loads are normally supplied by a commercial power line only.

The main power subsystems are as follows:

- the power entrance facility; including the high voltage feed and transformer substation with high voltage protection;

- the a.c. power distribution; including switchgear, metering, low voltage distribution panel boards with integral low voltage protection;

- the emergency stand-by generator(s); including related auxiliary equipment and control facilities;

- the uninterruptible power supply (UPS); including rectifier, inverter and battery bank;

- the d.c. power supplies; including rectifiers, battery banks and associated distribution;

- the station grounding facilities.

The characteristics of the subsystem indicated above will depend upon the type of station being considered as given below.

A block diagram of a typical power system is shown in Figure 7.70.

7.8.2.1 Large earth stations

The main design criteria for large multi-antenna earth stations and large single-antenna stations are listed below:

i)   The initial point of entry of the power supply system should be connected to a dual HV feeder each having an independent fused or circuit breaker protection fed from a common utility or commercial power source. The incoming feed to each fuse or circuit breaker should be equipped with lightning arrestors. Two main power transformers should be used to step down the incoming voltage. The transformers should have star (Y) secondary windings with the neutral grounded at the transformer and connected to the station ground. An isolated neutral can also be used (depending on local regulations). The transformers should be rated so that either transformer is capable of feeding the station load. It is preferable to utilize both transformers to feed the station via separate input circuit breakers for additional redundancy.

ii) The main switchgear should incorporate at least two main bus circuits to permit the essential loads to be separated from the non-essential loads. The bus configuration should also provide diversity of operation in that the routing of power may be derived from either or both transformers without paralleling the transformer outputs. Metering and protection facilities should be provided. Coordinated selective trip facilities are of prime importance. Coordinated ground fault protection is recommended on all three-phase circuit breakers. An isolated generator bus should be provided to permit testing of the stand-by generators. Use of a tie breaker and a dummy load feeder breaker facilitate generator operation. The rating of the main bus circuits should allow for future growth. Switchgear control power should be fed from an independent d.c. power supply backed up by batteries.

The requirement for voltage regulation is primarily dependent on the quality of the commercial power source. The majority of the station equipment will operate on the nominal voltage within $\pm 10\%$. If the required voltage tolerance cannot be met by the commercial power source, voltage regulators should be provided. Power factor correction should be incorporated in the distribution facilities; however the amount of correction required should depend upon the cost saving in operation versus the installation cost.

7.8 General construction of main earth stations

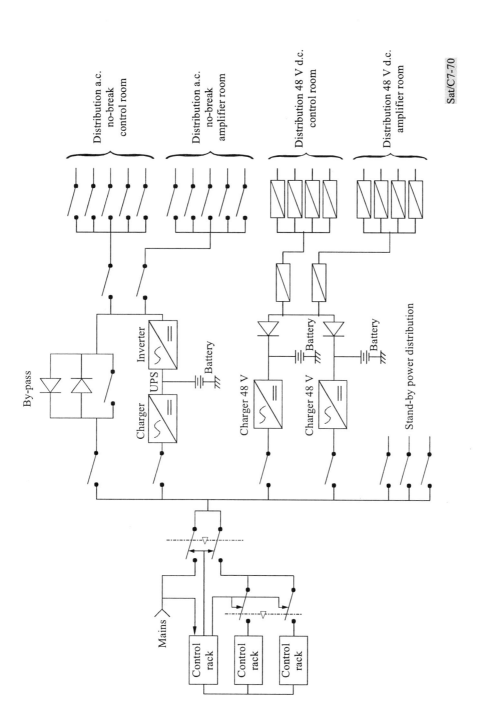

Sat/C7-70

FIGURE 7.70

Typical block diagram of a power system

iii) Emergency stand-by generators are required to back up at least the essential loads. The rating of the diesel generators can be based on stand-by use but the sets must be capable of continuous use at the stand-by rating. The location of the site above sea level and the ambient temperature coupled with the choice of radiator are of prime importance in determining the continuous output capability. The requirement for multiple diesel generators is dependent on the availability and reliability of the commercial power source. The advantages of utilizing more than one stand-by generator set are numerous and if multiple sets are used, synchronizing facilities should be provided to parallel the sets. The fuel system should include dual electric transfer pumps to keep the day tanks full. The generator output voltage regulation should be $\pm 2\%$ for steady-state conditions for no load to full load and the frequency regulation should be $\pm 1\%$ for the same steady-state conditions.

The use of an electronic voltage regulator and an electronic governor is recommended. The generator and/or voltage regulator should incorporate an exciter current boost circuit to prevent exciter field collapse during short circuit conditions, to ensure circuit breaker fault clearing.

iv) Uninterruptible power supply systems (UPS) may employ rotary or static generation. The preferred scheme utilizes a static inverter which under normal conditions feeds the critical a.c. loads of the station and provides total isolation from commercial power interruptions and line disturbances including voltage or frequency excursions. The system employs a rectifier charger to supply current to a d.c. bus which feeds a battery and an inverter. An automatic transfer switch is used to transfer the critical load from the UPS to the UPS bus in the case of an inverter failure. The use of a static transfer switch for this function is preferable over an electromechanical type switch in cases where computer equipment is being fed. The rectifier should incorporate a delayed start and a current control facility to coordinate and limit power inrush to the working bus upon restoration of a.c. input power. The system should also incorporate adjustable output current limit facilities, the capability to parallel with additional future UPS modules and a d.c. logic power supply backed up by a separate battery. The UPS battery should be of long lifetime type (lead antimony, lead calcium, nickel cadmium).

v) d.c. power supplies consisting of rectifier chargers feeding a battery bank and its associated distribution via circuit breakers or fuses are required to feed communications equipment. The use of multiple conservatively-rated rectifier chargers suitable for load sharing is recommended. The battery bank should have local disconnect facilities with cable protection. The bank should consist of two sets of cells in parallel to permit isolation of one set for maintenance.

vi) Station grounding facilities should be provided as follows:

An important requirement of station grounding is to maintain an equal potential across all portions of the system and the equipment to which it is connected, particularly during electrical power faults and strokes of lightning. In order to distribute the ground throughout the station, a perimeter ground system is used. This consists of a large diameter conductor (No. 4/0 AWG stranded copper) buried adjacent to the walls of the building with conductors connected at intervals to the perimeter conductor and carried to the interior where they are connected to an interior copper bus bar. The bus extends internally around the building walls of rooms where equipment requires grounding. The extension of the perimeter ground to the antenna and the transformer substation is normally done using two buried ground conductors. All ground connections should utilize a cad-welded process. Ground rods are normally driven in adjacent to the perimeter conductor and connected to it. The number of ground rods required will depend upon soil conductivity.

The antenna grounding is of special concern with regard to lightning protection. Antenna structure grounding down leads should be welded to the antenna perimeter ground with a ground rod at the same point. If the overall ground resistance is poor, consideration should be given to utilizing deep earth electrode grounds around the antenna with connections to the perimeter ground. The nature of the soil conditions determines the methods which must be used to achieve a low-resistance ground. A design goal of 5 Ω should be used.

In an area which is predominantly rocky and soil conductivity is low, it may be necessary to utilize deep earth electrodes to meet the design goal of 5 Ω.

7.8.2.2   Medium-sized stations

Power subsystem characteristics for domestic network type stations are listed below:

i)   the initial point of entry of the power supply system should use a single high voltage fused disconnect switch feeding a step-down transformer. The transformer secondary star (Y) point should be connected to the station ground;

ii)   the switchgear could consist of a CDP type panel board with dual bus circuits to permit essential loads to be separated from non-essential loads. The bus configuration should also permit feeding a dummy load from the stand-by generator while the working bus is fed from the commercial power source;

iii) the characteristics of the uninterruptible power supply system are essentially the same as those used on large earth-station complexes;

iv)   the characteristics of the emergency stand-by generator system are essentially the same as those used on large earth-station complexes. However, less emphasis should be given to the use of redundant or multiple diesel generator sets;

v)   the characteristics of the d.c. power supply system are the same as those used on large earth-station complexes;

vi)   the characteristics of the station ground system are essentially the same as those used on large earth-station complexes.

7.8.3   Antenna system (antenna civil works)

A general description of antenna systems is included in § 7.2.3. The civil works associated with the antenna are described below.

Two types of antennas are considered in this section: large antennas (>15 m) used in high traffic capacity earth stations and medium-sized antennas (15 m to 10 m) used in domestic network stations.

The pointing accuracies of the antennas require that the antenna should be built on ground with stable soil conditions. The soil should be analysed in depth to determine the actual condition on which the foundations are to be laid. These analyses are usually conducted by the antenna supplier/erector and the type of foundation is determined after a study of the soil report.

The foundation must be designed to meet the operational and survival loads imposed by the antenna, the support structure and the pedestal. These include wind loads, gravitational loads and seismic loads.

7.8.3.1   Large antennas

Most large antennas are mounted on an azimuth-over-elevation (Az-El) mechanical assembly of the wheel and track type. In this design, a large diameter circular rail supports the mechanical assembly and allows its rotation in azimuth, while the horizontal stresses are transmitted to a central bearing. The foundation pedestal can consist of eight columns of reinforced concrete supporting the azimuth rail tied to four columns supporting the central bearing (pintle), connected by radial beams and a top slab. The pedestal can also consist of a continuous polygonal concrete shell with a self-bearing dome-shaped roofing embedded in the circular beam, the pintle being included in this roofing. In this latter case the foundation is an integral part of the antenna building.

As the diameter of the azimuth rail ring is about 15 m to 18 m for large (30 m) antennas, the area under the top slab or the roofing is an ideal place to locate RF equipment. The internal construction and finish of this area should follow the general guidelines previously given for the main building.

The antenna foundation and pedestal should have a good drainage system which is capable of gathering and dispersing stormwater captured by the antenna and support structure.

7.8.3.2   Medium-sized antennas

Medium-sized antennas (10-15 m) are generally mounted on limited motion assemblies (Az-El or X-Y type) attached to a fixed support structure secured directly to the foundation by means of embedded anchor bolts.

The foundation should be designed after completion of a soil analysis. In general, it consists of footings or of a raft with piles tied in the upper part by groundsels, the whole being constructed in reinforced concrete. The foundation should have a sufficient mass, conveniently distributed to compensate for the winds, gravitational and seismic loads imposed by the antenna. A less stringent soil condition can thus be tolerated.

Small footings should also be provided for supporting an electronic equipment shelter situated, if possible, between the rear legs of the antenna support structure.

7.9   Satellite News Gathering (SNG) and Outside Broadcasting (OB) via satellite

7.9.1   General

7.9.1.1   Definition and main functions

SNG refers to temporary and occasional transmission at short notice of television or sound for broadcasting purposes, using highly portable or transportable uplink earth stations operating in the framework of the fixed-satellite service.

The definition of the equipment is that it should be capable of uplinking the video programme with its associated sound or sound programme signals, and capable of providing two-way coordination and communication circuits. The equipment may provide for data transmission and should be capable of being set up and operated by a crew of not more than two (2) people within a reasonably short time (such as within one hour).

Transportable earth stations may also fall under the SNG requirements when logistics dictate the use of such systems and the systems meet the basic functional characteristics of the SNG systems. SNG sound may also be used in the mobile-satellite service.

The main functions of the SNG system are:

- to transmit with a minimum of impairments, a visual and associated sound or sound programme signal;

- to provide limited receiving capability to assist in aiming the antenna and to monitor the transmitted signals, where possible;

- to provide two-way communication channels.

SNG link budget calculation examples will be found in Annex 2 to the Handbook.

7.9.1.2   Configuration of the SNG system

SNG transmission equipment consists of a vehicle mounted or highly portable uplink earth station, space segment (e.g. communication satellite) and main earth station equipped with a large receiving antenna.

A general outline of the SNG transmission system is given in Figure 7.71.

The space segment and main earth station are generally the same systems as these used in the fixed-satellite service.

The general operational diagram of an SNG uplink earth station is shown in Figure 7.72.

SNG terminals using a flat (usually phased array antenna) or parabolic antenna consist of the following main units:

- antenna and feed system with polarization adjustment;

- antenna mount with azimuth/elevation adjustment;

- high-powered amplifier/solid state power amplifier (HPA/SSPA) for visual/sound and auxiliary (voice/data) communication channels in the case of analogue transmission or for multiplexed visual/sound/data channels in the case of digital transmission;

- receiver unit to assist antenna pointing;

- baseband/modulation equipment and IF to RF upconverter;

- two-way voice/data communication equipment;

- local/remote control panel;

- optional power generator.

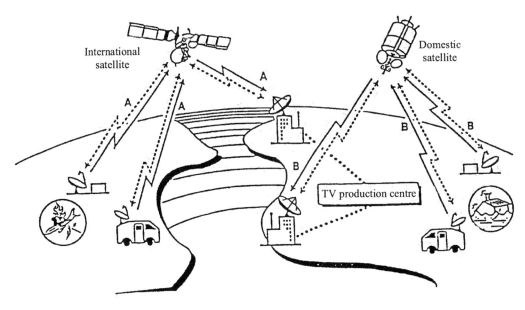

A: international SNG transmission (TV or sound)
B: domestic SNG transmission (TV or sound)
◄······► two-way communication channels (voice and data)

Sat/C7-71scm

FIGURE 7.71

A general outline of the SNG transmission

7.9 Satellite News Gathering (SNG) and Outside Broadcasting (OB) via satellite

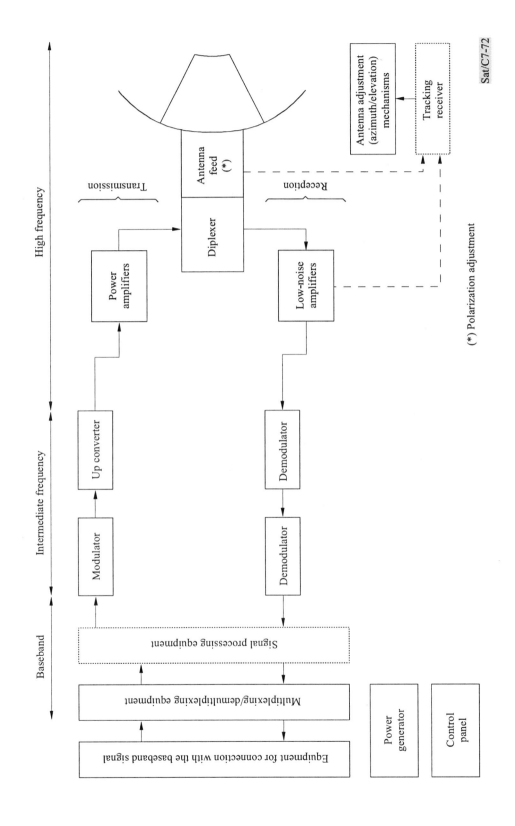

FIGURE 7.72

General operational diagram of an SNG uplink earth station

7.9.2   Antenna system

The antenna system for an SNG earth station should be small, lightweight and easy to assemble. Regarding e.i.r.p., however, there is a trade-off between antenna gain and RF output power. Size, weight and power consumption have to be well-balanced for mobile applications. Depending on the modulation scheme (analogue or digital) and frequency band (6/4 or 14/11 GHz band), the diameter of a parabolic antenna may vary between 3 m (6/4 band) and 2 m (14/11 GHz band) in an analogue system and between 1.2 m and 0.75 m in a 14/11 GHz band digital system. In a digital SNG terminal with a flat antenna, the antenna is square and measures 60 cm by 60 cm.

The antennas of SNG terminals are used for both transmission and reception. According to Recommendations ITU-R SNG.722-1 (for analogue) and SNG.1007 (for digital), the RF transmission performance requirements are specified as follows:

- The off-axis e.i.r.p. density complied with Recommendation ITU-R S.524 or the satellite operator's requirements, whichever is more stringent.

- The cross-polarization design for linearly polarized antennas should be better than 30 dB within the −1 dB points of the main beam axis and 25 dB elsewhere.

From the mechanical point of view, an SNG terminal must be capable of rapid deployment to transmit video programs and the associated sound with two-way communication circuits. Repeated assembly and disassembly of the antenna should not affect the radiation and cross-polarization discrimination performance. Extreme structural accuracy is required for suitable RF performance: the surface of the main reflector of the parabolic antenna must be accurately shaped, and the surface of the flat antenna (planar array antenna) composed of microstrip patch antenna elements, must be similarly accurate because errors in the physical location of array elements cause phase errors. Moreover, the antenna beam must remain pointed towards the satellite under all environmental operating conditions. The antenna mount with azimuth/elevation adjustment and polarization adjustment should be designed to deploy the antenna accurately and to keep it pointed at the satellite even in high winds.

The antenna system for a parabolic antenna consists of the following components:
- the reflector;
- the illuminating horn with polarization adjustment;
- the feed system, comprising an OMT to separate the transmitted and received radio waves;
- the antenna mount, to adjust the azimuth and elevation.

Example of parabolic and Gregory antennas for SNG are shown in Figures 7.73 and 7.74.

In the newly-developed transportable digital SNG terminal for 14/11 GHz band applications, the reflector is an offset parabolic antenna of 1.2 ~ 0.75 m diameter.

The antenna system using a flat antenna consists of the following:
- the planar array antenna and feed system with polarization adjustment;
- the antenna mount with azimuth/elevation adjustment;
- the antenna tracking system.

An example flat antenna is shown in Figure 7.75.

The small digital SNG terminal with flat antenna for 14/11 GHz band applications is designed for easy portability, rapid deployment and prompt activation. The antenna is a square microstrip array comprising 1 024 elements (32 x 32 elements). To attain low side-lobe characteristics for the geostationary orbit, the flat antenna is installed so that its diagonal plane is oriented in the plane of the orbital arc as seen from the earth station. This is because the square array antenna with uniform aperture excitation has tapered excitation in the diagonal plane. The flat antenna uses electronic tracking techniques, receiving a beacon signal transmitted, via satellite, by a "hub" station. It is only necessary to orient the flat antenna to within a few degrees (3 or less) of the satellite's orbital position; the system then adjusts for optimum orientation automatically.

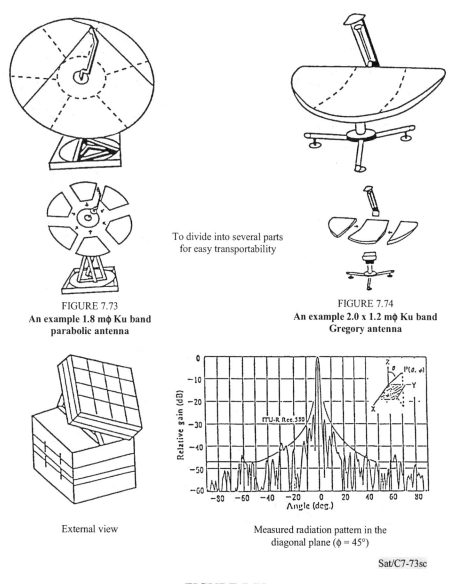

FIGURE 7.73
**An example 1.8 mφ Ku band
parabolic antenna**

To divide into several parts
for easy transportability

FIGURE 7.74
**An example 2.0 x 1.2 mφ Ku band
Gregory antenna**

External view

Measured radiation pattern in the
diagonal plane ($\phi = 45°$)

Sat/C7-73sc

FIGURE 7.75

An example flat antenna of 60 cm x 60 cm for Ku band digital SNG

7.9.3   Baseband signal processing

Recently, the changeover is being made from the analogue to the digital baseband signal for SNG transmission, due to the improvement of digital video compression techniques which makes the transmission system smaller and lighter. Digital SNG also has the following characteristics:

1) better portability of transmission equipment;

2) effective use of the transponder;

3) lower cost;

4) lower transmission e.i.r.p.;

5) lower power consumption;

6) less interference between channels;

7) scrambling of the transmission signal.

Digital SNG services with these characteristics will wholly replace the analogue ones. The video coding method is already recommended in ITU-T Recommendation J.81 (34 ~ 45 Mbps) for long-distance circuits. The most recent video coding method, based on ITU-T Recommendation H.262 (MPEG-2), although not yet recommended, will probably be recommended for digital SNG services in the future.

Analogue and digital baseband signals are indicated as follows.

7.9.3.1   Analogue baseband signal

The analogue baseband signal is based on Recommendation ITU-R SNG.722-1. For analogue television broadcasting, the video signal is based on ITU-T Recommendation J.61 and the audio signal is based on ITU-T Recommendation J.21. For analogue sound broadcasting, the baseband signal is based on ex-CCIR Recommendation 504-2 (CMTT).

7.9.3.2   Digital baseband signal

The digital compressed baseband signal is based on Recommendation ITU-R SNG.1007. 40 Mbps (at least 34 Mbps) is needed for contribution quality and 10 Mbps for lower quality. However, lower bit rate ranges are admitted under bad conditions. Compatibility between the satellite link and terrestrial links also need to be considered.

1)   Video coding method

There are several coding methods. One combines motion compensated adaptive prediction between frames or in the field with DCT (discrete cosine transformation). The other uses the WHT (Walsh Hadamard transform) to code and compress composite signals directly without separating the luminance signal from the chrominance signal. The delay arising from the time disposition between frames or fields in the encoder and the decoder is then the most important problem. Though the quality of video and the delay are related by trade-off, the delay is more important than the quality of video for SNG services.

2) Audio coding method

The method based on Layer II, which is one of MPEG standard layers, and DOLBY AC2, which is used in the United States, are adopted.

3) Error correction method

Dual error corrections with Reed-Solomon code and convolution code are adopted in order to receive signals of sufficient quality by lower C/N than with analogue.

4) Scramble

A quasi-random signal is inserted in the television signal for energy dispersal in communication between specified partners.

5) Digital SNG system

First the video signal, the audio signal and the data signal are coded in each coder. The fixed output bit streams from the coders are multiplexed in the multiplexer. Then the baseband signal is constructed after energy dispersal and dual error corrections.

The digital SNG system configuration is indicated in Figure 7.76.

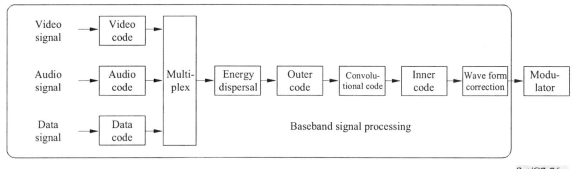

Sat/C7-76m

FIGURE 7.76

7.9.4 Carrier modulation for SNG

For satellite news gathering (SNG), one can choose from the same set of modulation techniques as for any other type of television or data transmission. Modulation methods can be divided into analogue and digital. Analogue modulation techniques are discussed in considerable detail in Chapter 4 (§ 4.1) of this handbook. Digital modulation techniques are explained in Chapter 4 (§ 4.2) of this handbook. The following paragraphs summarize some of the characteristics of the most popular modulation techniques for SNG, and also give some examples of how the SNG signals, along with the coordination channels, can be transmitted over a satellite transponder.

Most – if not all – new SNG network startups throughout the world have opted to go digital with ease. Since few of these ventures have existing investments in equipment and systems to deal with, many justify their investments in digital compression on the first year's savings on the international space segment alone. In contrast, the widespread use of digital SNG in the United States appears to be a number of years down the road because of the amount of capital invested in existing SNG trucks, fixed low-power uplinks, communications hubs and scheduling systems. Therefore, the most likely scenarios for the SNG market in the United States is the continued use of analogue technology, particularly frequency modulation (FM). Both analogue and digital modulation will be addressed in the next sections.

7.9.4.1   Analogue modulation

Frequency modulation (FM) is the predominant analogue transmission method used for satellites because it has the advantage of being quite robust (not as susceptible to noise and interference), especially compared to amplitude modulation or AM. In FM, the modulating signal varies the frequency of the carrier on a continuous basis. Bandwidth can be traded to reduce satellite transponder power requirements, if necessary.

In addition to the video carrier, the transponder must accommodate various other carriers for auxiliary communications. Figure 7.78 shows an arrangement where a 36 MHz transponder is loaded with one TV/FM carrier, and 10 SCPC/FM carriers. In this case, the 10 SCPC/FM carriers provide five full duplex circuits; one between the SNG terminal and the satellite operating entity, and four between the SNG terminal and the broadcaster's facility.

7.9.4.2   Digital modulation

In selecting a digital modulation system, the two primary factors for comparison are the transmitted power and bandwidth that are required to achieve a specific bit error ratio (BER) performance. To measure power efficiency, the parameter Eb/No is used. For spectrum efficiency, the ratio of transmission rate to transmission bandwidth is used. Other factors that may influence the choice of modulation technique include effects of fading, interference, channel or equipment nonlinearities, and implementation complexity and cost.

The most popular digital modulation technique for TV transmission over satellite is phase modulation, more commonly called phase shift keying or PSK, where the phase of the carrier is varied according to the source signal. Other techniques are amplitude shift keying (ASK), where the amplitude of the signal is varied, or frequency shift keying (FSK), where the frequency of the signal is varied. ASK is not a popular choice for SNG applications because of its relatively poor error performance and because of its susceptibility to fading and non-linearities. FSK is more commonly used for low data rate applications.

QPSK (or 4-PSK) is the modulation scheme most popular for SNG, because it offers the best compromise between the bandwidth available on satellite transponders, and the required power to offer a particular signal quality.

As with analogue SNG, there is the problem of how to accommodate the coordination channels between the SNG terminal and the hub. Traditional solutions entailed the purchase of separate equipment for the transmission of these signals. Figure 7.77 shows a possible alternative. The SNG setup is such that the coordination channel can operate with the video channel or alone. The channel is either multiplexed with the video data, when there is video, or transmitted independently on an SCPC channel when there is no video signal. One 64 kbit/s channel will provide six voice lines, one 9.6 kbit/s data line, and one line for fax transmission.

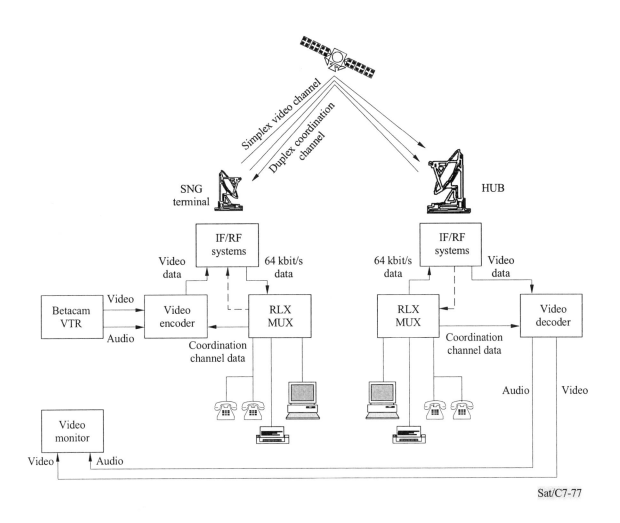

Sat/C7-77

FIGURE 7.77

Packaged digital C-band SNG system configuration

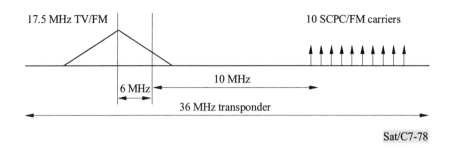

FIGURE 7.78

Possible analogue SNG configuration in a 36 MHz transponder

7.9.5   Power amplifiers

The typical block diagram of an earth station (Figure 7.79), shows various input chains which are connected to up converters, the outputs of which pass through an input which feeds the final high power amplifiers. The output power (Pt) of a high power amplifier must be sufficient to deliver the equivalent isotropically radiated power (e.i.r.p.) which is required for each transmit carrier, taking into account the antenna gain (Gt). The e.i.r.p. = Pt · Gt.

The necessary e.i.r.p. of the SNG terminal is dependent on the required uplink carrier-to-noise ratio C/N and the satellite G/T. The e.i.r.p. can however be limited by the off-axis e.i.r.p. density limits as set in Recommendation ITU-R S.524 or the satellite operator's requirements. The SNG terminal may consist of a single active high power amplifier, which is usually associated with a standby high power amplifier, through an input and an output switch (1 + 1 redundancy) arrangement as shown in Figure 7.79.

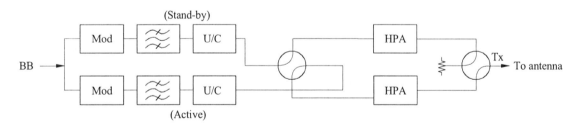

BB: Baseband input signal
Mod: Modulator
U/C: Up converter
HPA: High power amplifier
Tx: Transmit RF carrier

Sat/C7-79m

FIGURE 7.79

One carrier transmit chain with (1+1) redundancy

The types of high power amplifiers used in the SNG terminals is dependent on the mode of the transmitted signal. SNG terminals used for analogue sound and vision transmissions use microwave tube amplifiers due to power and bandwidth requirements whereas SNG terminals used for digital transmission tend to use either low power microwave tube or solid state amplifiers.

Microwave tube amplifiers: In an analogue SNG system a 400 watt travelling wave tube in a single amplifier configuration is capable of delivering +71 dBW e.i.r.p. when coupled with a 1.5 metre antenna. The maximum output power is achieved by 1 + 1 active standby by switching or by phase combining two high power amplifiers in a power combiner system. A combined power level of 670 watts can be delivered to the antenna using two 400 watt travelling wave tubes connected via a combiner unit. Section 7.4 of Chapter 7 gives details on microwave tube high power amplifiers and the output combiner technologies and the problems associated with using these devices.

Solid state amplifiers: Solid state power amplifiers are increasingly used in digital and, where power requirements permit, for analogue SNG terminals for transmission of signals via high gain satellite transponders. The Gallium Arsenide Field Effect Transistors (FET) are capable of producing outputs in excess of 100 watts at 6/4 GHz band and 20 watts at Ku band. The output powers in excess of those specified above can be produced by connecting two or more solid state amplifiers in a combiner unit. The solid state high power amplifiers have an added advantage that they do not require high voltages needed by the microwave tube amplifiers and tend to be more reliable in use.

7.9.6 Equipment for auxiliary communications

SNG uplink often originates from remote areas. In such cases, communication using the public switched telephone network (PSTN) is difficult or impossible. The SNG terminal should therefore be equipped to provide all of its own communications through the satellite to both the satellite operator's communication control centre and the broadcaster's premises, with the equipment detailed below.

Some methods for implementing these communication channels have been described above in § 7.9.4.1 and § 7.9.4.2.

7.9.6.1 Communication channels for supervision and coordination

SNG terminals require two-way communication channels, in addition to the pictures and associated sound, to provide communications between the satellite operator's communications control centre and the broadcaster's facilities.

It should be noted that several domestic systems are presently in operation using various communication techniques. However Recommendation ITU-R SNG.771-1 recommends as follows:

- two or more duplex circuits should be provided, whenever possible within the same transponder as the program pictures and associated sound or sound program signal;

- these communication circuits should be in compliance with ITU-T Recommendation G.703, i.e. 64 kbit/s;

- the modulation of the communication channel carriers shall be fully compatible with the current INTELSAT Standard IESS-308.

There will be occasions when return communications to the SNG terminal cannot be provided in the same transponder as the video. In these circumstances it may be possible to provide the return communications in another transponder in the same frequency band on the same satellite. In this case, the return communications should be provided with the appropriate polarization to avoid the need for a dual polarization feed to the SNG terminals. If it is not possible to provide communications in the same frequency band as the video signal, a number of other solutions are proposed.

a)   6/4 GHz band

In one such solution, the communication channel could be provided on a 6/4 GHz band transponder with global coverage. One scheme currently under investigation involves the use of small (circa 0.8 m) micro terminals employing spread-spectrum modulation. This would support a single 64 kbit/s duplex circuit, which may itself include multiple low bit-rate voice duplex circuits.

A second solution to this problem, which has been proposed, is the implementation of "partial" dual band feed (i.e. 14/12-11 GHz band Tx/Rx, 6/4 GHz band Rx) on the SNG terminal. This solution would augment the normal 14/12-11 GHz band transmit/receive capability of the SNG terminal with a 6/4 GHz band receive-only capability, thereby enabling the reception of 6/4 GHz band transmissions. Initial studies in this area indicate that multiple 64 kbit/s carriers could be supported by this approach using existing commercially-available digital modem technology.

b)   Land mobile earth station (LMES)

In some situations, a solution could be provided by using a land mobile earth station operating at 64 kbit/s or an LMES which can support voice-grade communication capability.

7.9.6.2   Circuits between the SNG terminal and satellite operator

Liaison with the satellite operator's communications control centre should be available at all times and should not be restricted to the duration of the transponder booking. For this purpose it is desirable that a minimum of one, two-way narrow-band voice/data coordination circuit be provided in each direction, preferably in the same transponder as the program vision and sound.

If it is necessary to provide these carriers elsewhere on the satellite, and linear polarization is being used, they should be provided with the same polarization in order to avoid the need for a dual polarization feed on SNG terminals.

7.9.6.3   Channels between the SNG terminal and the broadcaster's premises

To communicate with the broadcaster's premises, typically up to four (4) two-way (duplex) voice/data circuits are required per broadcaster. These circuits generally operate for brief periods before and after transponder bookings and during actual program transmissions.

These "two-way" circuits between the SNG terminal and the broadcaster's premises can be used for:

- production coordination;

- engineering coordination;

- program-related data transmission;

- more than one broadcaster;

- more than one language.

7.9.7 Operating procedures required for temporary authorization for SNG

Uniform operating procedures for SNG are indicated in Recommendation ITU-R SNG.770-1. The following is a summary.

SNG differs from most other forms of satellite transmission in a number of ways. For example, the requirement for SNG is typically identified only a few days, or even hours, before transmission. It typically lasts for no more than several days, or at the most, a few weeks. Nevertheless, the SNG operator has to comply with the regulations of the host country and with a number of procedures which are designed to ensure the proper management and protection of the space segment and frequency spectrum.

The regulatory framework in which an SNG operation takes place has a dual effect on its operational effectiveness. In order to carry out its intended function, the SNG operator must have access to temporary agreements and/or authorizations in a timely and cost-effective manner. The operator's needs range from frequency authorization and coordination with the space segment entity to tariffs and administrative costs, and the necessary supporting lines of communication.

Given that SNG requirements are occasional and/or temporary and that coverage of unplanned fast-breaking news is a valuable worldwide service, expeditious approval for the activation of portable earth stations is essential.

7.9.7.1 Earth station approval

Earth station approval is necessary to allow the body responsible to ensure the compatibility of the SNG terminal with the space segment. To meet this requirement, administrations are required to consider procedures to permit the SNG terminal to be brought into service as quickly as possible. Administrations are urged to investigate whether an SNG terminal whose technical performance is approved by the space segment provider can be accepted on a uniform basis. They are encouraged to complete administrative procedures in close cooperation with SNG operators as expeditiously as possible. A technical report demonstrating the measured performance characteristics should be prepared and made available to the administration. The following technical characteristics should be documented as the minimum:

- transmit gain as a function of frequency;

- transmit off-axis gain;

- transmit main beam e.i.r.p.;

- transmit beam width and polarization;

- transmit main beam spectral density for the worst 4 kHz;

- transmit off beam spectral density for the worst 4 kHz;

- maximum energy dispersal (where required);

- reception of G/T as a function of frequency;

- cross-polarization isolation;

- pointing accuracy performance;

- frequency agility for reception and transmission within the operating bands;

- spurious emissions (in-band and out-of-band);

- manufacturers' model numbers, modulation characteristics and frequency stability;

- other technical characteristics which are part of the SNG standard used in the country concerned.

7.9.7.2   Frequency assignment and coordination

Frequency coordination procedures are derived from international and national regulations. In order to assess the acceptability of an SNG terminal in this respect, the body responsible may require the information detailed in § 7.9.7.1, together with details of the geographical location of the SNG terminal and the anticipated transmission times.

7.9.7.3   Space segment booking

The SNG operator needs to have a quick and clear understanding of what space segment will be available in a timely manner (e.g. within less than 24 hours) for this purpose. This information needs to include:

- transponder characteristics (satellite identifier);

- amount of bandwidth and power;

- earliest available time of access.

The SNG operator may require direct contact with the space segment provider on a continuous basis.

7.9.7.4   Auxiliary coordination circuits

Auxiliary coordination circuits are required between both the satellite operator's communication control centre and the broadcaster's premises.

7.9.7.5   Additional support communications/transmission facilities

Support communication facilities may be required to facilitate the effective operation of the SNG terminal. These facilities may include point-to-point microwave, telephone communication systems, two-way simplex/duplex radio, wireless microphones and mobile satellite terminals for voice and data.

7.9.7.6 Radiation hazards

It is essential to protect the public and personnel from hazardous radiation. Many administrations have established standards for safe exposure to radio (non-ionizing) radiation, which are a function of the frequency, power and duration of exposure.

SNG operators should comply with permitted radiation (health and safety) standards established by the host country. Where the host has not established its own standards, the World Health Organization (WHO) standards should be used. (The WHO develops health criteria in conjunction with the International Non-Ionizing Radiation Committee of the International Radiation Protection Association.)

7.9.7.7 Importation and customs

The SNG operator should have a sufficient understanding of the importation and customs system of the host country. This is particularly important when there is frequent news gathering and where facilities in that country cannot be used.

7.9.7.8 Contact point for information, guidance and approval

Each administration or relevant organization should, if possible, establish a designated point of contact (DPC) for SNG, which should be available 24 hours per day, seven days per week.

This contact point should be available for assistance to facilitate the temporary authorization of SNG earth stations owned by foreign operators by mediating the exchange of information necessary for authorization procedures and frequency coordination, and providing guidance on the administrative procedures of the host country.

The "SNG User's Guide" (ITU BR 1996) provides information for foreign broadcasters on how to bring the terminal to a given country (or area) and obtain the temporary licence to operate it.

7.10 Earth stations for direct reception of TV and audio programmes

7.10.1 Introduction

TV transmissions in analogue have been used from the beginning of satellite system for contribution and broadcast applications. These links use a complete transponder (20-36 MHz) and need a high value for the carrier-to-noise (C/N) ratio (10-14 dB). Digital TV transmissions with high compression technologies require only 10 to 20 % of the bandwidth needed for analogue which makes possible the transmission of 6-10 video programs (called a "bouquet") into the same transponder. A smaller required C/N ratio (4-8 dB) and robust error code correction improve the receiving conditions and hence, improve the satellite coverage.

Due to the technically advanced features, digital TV satellite transmissions should progressively make analogue transmissions obsolete. This is the reason why this section is entirely devoted to digital TV transmission.

Until late 1990, digital television broadcasting to the home was thought to be impractical due to the technologies used for coding. By 1993 an expert European group launched the DVB (digital video broadcasting) team. By 1997 the development of the DVB project had successfully followed the initial plans, and extended its open standards globally, making digital television a reality.

The digital satellite system which can be used in frequency bands up to 11/12 GHz and which is configurable to suit a wide range of transponder bandwidths and powers is called DVB-S. It was the first standard available. The DVB-C is a digital cable delivery system, compatible with DVB-S. A lot of standards are under study to define the complete broadcasting chain.

DVB standards are open and interoperable, the specifications were offered for standardization to the relevant standards body (ETSI, Cenelec, ITU-R, ITU-T and DAVIC forum, Digital Audio-Visual Council).

7.10.1.1   Definitions

The digital vision and sound-coding systems adopted use sophisticated compression techniques. The MPEG standard (see § 3.3.4) specifies a data stream syntax. For audio, MPEG Layer II (Musicam, masking pattern adapted universal subband integrated coding and multiplexing) is a digital compression system which achieves 6:1 bit-rate reduction in the digital audio data stream, using adaptive PCM in 24 subbands. It takes into account the psychoacoustic principle of spectral masking which quantifies human hearing's inability to perceive some frequencies of sound when they are in close spectral proximity to other louder sounds. This means that the frequencies produced by one sound may render other sounds inaudible. This system can achieve sound quality which is very close to compact disc.

The international MPEG-2 video standard is a family of systems each having an arranged degree of commonality and compatibility which allows four source formats (levels) to be coded ranging from limited definition (VCR, videocassette recorder) to full HDTV.

There are currently five profiles, each one is progressively more sophisticated. The main profile and main level (MP@ML) is the most used for satellite transmission. A current "NTSC/PAL/SECAM" quality requires a system which can operate with a bitstream between 2.5 and 6 Mbit/s.

The DVB-S system is a single carrier system carrying either one or several programs in a multiplexed data stream (TS or MPEG-2 transport stream).

Note that digital signals other than TV programmes can be included in the data stream (typically from 350 kbit/s to 6 Mbit/s). In this way, various digital transmissions, such as multimedia services, Internet etc. can be offered to the users.

In simple systems, a fixed data bit rate is allotted to each program. In more advanced systems, statistical multiplexing can be applied.

In statistical multiplexing, the data bit rate of a given program is instantaneously modified and optimized to its current content. For example, rapidly moving subjects (e.g. in sport events) can be provisionally allotted with a higher bit rate, the total bit rate of the multiplex being kept constant.

On-board multiplexing is another advanced system which offers the ability to implement independent direct uplinks from different locations with on-board assembly of these various uplink signals to form a single downlink multiplex in the DBV format. For example, in the EUTELSAT "Skyplex^{TM}" system, the uplinks can be established by means of small earth stations and can be carried out either in SCPC (e.g. a single station accessing a 6 Mbit/s channel) or in a shared mode. In the shared mode, the stations can access simultaneously an uplink channel in TDMA (e.g. six stations, each one transmitting at 350 kbit/s and accessing a 2 Mbit/s channel).

The modulator uses a QPSK modulation with two error correction systems (outer coding and inner coding, as mentioned below).

7.10.1.2 System configuration

Figure 7.80 below illustrates the current capabilities of TV transmission satellite systems spanning from contribution links, i.e. distribution to local broadcast transmitters, up to direct broadcast to home TV sets (DTH).

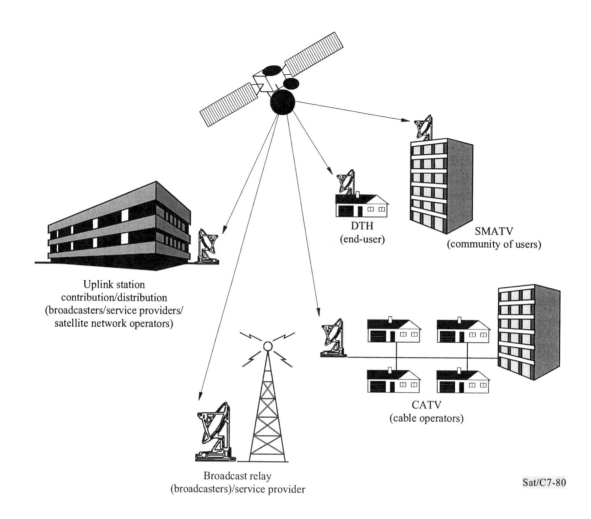

FIGURE 7.80

DVB networks

7.10.2 Transmission chain

A typical transmission chain, which performs the adaptation of the baseband TV signals from the output of the MPEG-2 multiplexer, to the satellite channel characteristics is shown by Figure 7.81.

The following processes are applied to the data stream:

*   transport multiplex adaptation and randomization for energy dispersal;

*   outer coding (i.e. Reed-Solomon);

*   convolutional interleaving;

- inner coding (i.e. punctured convolutional code);
- baseband shaping for modulation;
- QPSK modulation;
- upconverters and amplifiers (same as for analogue transmission).

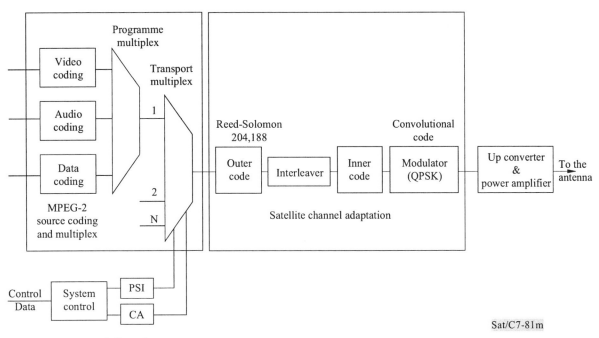

PSI: Program system information
CA: Control access

FIGURE 7.81

DVB transmission chain

7.10.3 Receive station

The typical receive earth station (TVRO) shown by Figure 7.82 uses a very low-cost architecture. It is composed of:

- a small antenna (dish from 40 to 90 cm diameter);
- a low noise block converter(s) (LNB) with a very good phase noise;
- an interconnection cable (75 ohm) with very low losses;
- an integrated receiver decoder (IRD).

NOTES

• The antenna should generally be designed to be able to receive in either one of the two orthogonal polarizations (e.g. vertical and horizontal) on which the various programs can be transmitted. This is generally operated, under control of the IRD, through either a switch or a polarization rotation device located in the antenna primary horn.

• It is a rather usual practice to equip the antenna with two (or even more) receiving systems (primary horn + LNB) properly located around the reflector focus. This permits to receive the emissions from two (or more) satellites located at sufficiently close positions. The angular separation between the primary horns (as seen from the reflector summit) is approximately equal to the angular separation of the satellites on the GSO.

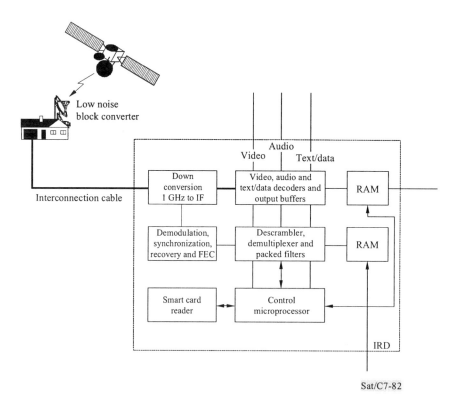

FIGURE 7.82

Receive station

7.10.3.1 Outdoor unit (ODU)

The outdoor unit includes the antenna and the low noise block (LNB) converter. It is linked to the indoor unit by a cable operating usually at a first IF frequency around 1 GHz. On the market there is a lot of choice concerning antenna types and sizes.

The most common are offset antenna but some manufacturers can provide Gregory, Cassegrain, pure parabolic dish and flat antenna (phased array). They can be fixed or turntable mounted on an azimuthal support and driven by an electrical motor.

The sizes range from 40 cm to 90 cm for home installation in 14/11 GHz band or from 1 to 2.5 m for SMATV or CATV. The gain for a 0.60 m offset antenna is about 36 dB.

Horizontal, vertical or circular polarization are used and the more common sets use double polarization. In the United States and European countries, the 11 GHz band is usual. Due to the high rain attenuation in bands above 10 GHz, the 4 GHz band is currently used in equatorial countries.

The noise figure of the LNB is about 0.7 dB in 11 GHz band. The bandwidth is 10.7 to 12.75 GHz with the possibility to switch the local oscillators by a 22 kHz signal and the polarization from a d.c. voltage.

7.10.3.2 Digital receiver

This integrated receiver decoder (IRD) of the indoor unit, also called a set top box can include the following functions:

- down conversion from 1 GHz band to IF (e.g. 450 MHz);

- demodulation;

- error correction;

- energy dispersion removal;

- de-encryption;

- baseband processing to transform MPEG program in analogue video (R, G, B and PAL/SECAM systems) and sound;

- software functions giving the possibility to download new applications such as an electronic program guide and menus that allow the user to make an informed choice about which program to watch;

- phone modem to purchase pay per view programs or interactive games, to order from home shopping etc.;

- data output for downloading software to a local PC.

Figure 7.83 shows an advanced version of possible IRD applications. In this version, the local personal computer (PC) performs the following functions:

- receives satellites data stream;

- reassembles original files and stores them on hard disk or plays data in real time;

- retrieves and executes files via PC (+ video/sound card);

- creates diagnostics and statistics log files;

- regularly transmits log files to processing centre.

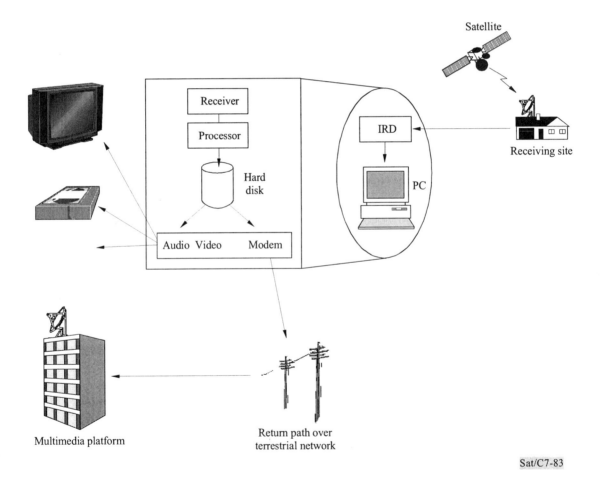

Sat/C7-83

FIGURE 7.83

Advanced IRD applications

These set top boxes will become more and more intelligent to provide new services including video on demand, videoconferencing, home shopping, virtual travel agencies and remote banking services. The multimedia capabilities of a set top box will introduce a new generation of value added applications.

7.10.3.4   Conditional access

There are advantages to a pay TV system. For the broadcaster this simplifies the copyright negotiations because the number of consumers watching his program is known and he has to pay less than for a very large audience. The direct effect for the consumer is that the quality of programs will improve.

Almost all new broadcasters propose only bouquets with conditional access.

Before the transmission the clear signal is scrambled given that normal viewing has become impossible. The reverse operation, descrambling is only possible with a secret key. Systems such as Eurocrypt are able to update the authorizations by sending securely encrypted authorization messages (EMM, entitlement management messages) which are transmitted together with the scrambled television signal and after reception and de-encryption the authorization is stored inside the chip of a smart card. For reasons of security the key changes every 5 or 10 seconds. Another message is transmitted which contains a description in parameter form of the program. After having received an ECM (entitlement control messages), the smart card checks if it is authorized to descramble the program. If this is the case, then the smart card activates the descrambler by issuing the key.

These systems are very sophisticated, they make use of various addressing schemes and allow different forms of authorization such as normal subscription and pay per view for near video on demand (the same movie is broadcasted in a repetitive scheme, each series is staggered by, for instance, 20 minutes).

In the case where the consumer wishes to receive several bouquets from several operators he needs to have an access for each conditional access (for example SYSTER or VIA ACCESS systems). With the aim to solve this problem, the DVB group propose two possibilities which are termed Simultcrypt and Multicrypt.

Simultcrypt is based on a mutual agreement between various program service providers. They authorize a common utilization of their own control access (CA) and the providers broadcast at the same time on each bouquet all EMMs and ECMs.

Multicrypt or common interface: this interface is a PCMCIA card currently in use for computer applications. It may contain more than the descrambler chip and the control access system, for instance software for an electronic program guide (EPG).

7.10.4   Digital audio broadcast

DAB is the future technology for radio, providing multiplexes or packages of about six stereo channels or twelve mono or a combination of the two where currently there would be a single analogue broadcast. This technology was developed by European broadcasters team "Eureka 147".

The first sets are expected to be in car models, where listening in AM or FM has been something of a problem. In cities with high rise buildings AM transmission can fade to near zero and FM transmissions are also susceptible to multipath distortion. Another advantage is to use a single HF frequency to cover a complete country.

Today radio channels are transmitted by satellite as a part of DVB-S stream, with an MPEG-1 Layer II coding (Musicam) with a bit stream from 128 kbit/s to 384 kbit/s for a stereo channel with a near CD quality sound.

The Eureka 147 DAB system is based on a frequency multiplex of QPSK carriers employing error correction coding. One important feature is the presence of a well defined "guard interval" between each individual digital pulse of the transmission signal. (COFDM system: orthogonal frequency division multiplex used in conjunction with channel coding.)

For the future specific satellite would be launched either in geostationary (GEO) or in multi-regional highly-inclined elliptical orbit (M-HEO). WARC-92 allocated 40 MHz of spectrum in the range 1 452 to 1 492 MHz for the Broadcasting-Satellite Service (BSS) for transmission of digital sound programs on a worldwide basis (except the United States). The introduction on the market will start from the end of 1998.

7.10.5 Future developments (see Figure 7.84)

Digital satellites have the potential to download megabits of data to a PC instantly. The main advantage is speed. A 10 Mbytes file, which would possibly hold five minutes of video, would take nearly 100 minutes to download over a 14.4 kbit/s modem. On the ISDN digital phone lines, it would take a lengthy 22 minutes. Over satellite link at 38 Mbit/s this is just 2.2 seconds. Unfortunately, most desktop PCs need time to catch up and have to reduce the transmission rate to 6 Mbit/s, but that still only takes 14 seconds to download. Today adding a modem (in PC side) link to the satellite uplink, enables users to control what was transmitted from their own PC. This would work with the Internet system, where most of the information is downloaded to the PC, the only information going back to the uplink would be the location of the web page to download.

The next phase will be to use a return channel (uplink) through the satellite link. This will require a small transmitter attached to a larger dish to transmit back to the satellite. This technology should offer a bit rate from 20 to 40 Mbit/s for the downlink and up to 2 Mbit/s for the uplink.

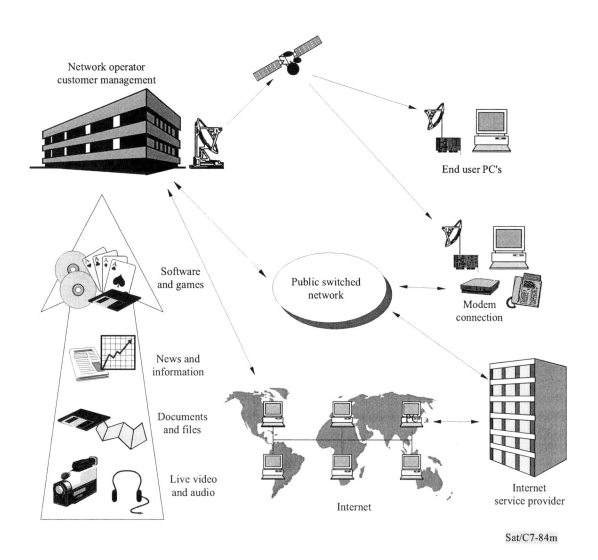

FIGURE 7.84

Worldwide data network

APPENDIX 7.1

This Appendix contains four tables as follows.

TABLE AP7.1-1

INTELSAT Standard A, B and D earth stations

No.	Parameter	Performance characteristics		
		A	**B**	**D-1 and D-2**
1	RF bandwidth	Transmit (Tx) 5 925 to 6 425 MHz Receive (Rx) 3 700 to 4 200 MHz For Standard A, B and D-2 new earth stations INTELSAT VI and VIII satellites Transmit 5 850 to 6 425 MHz Receive (Rx) 3 625 to 4 200 MHz INTELSAT VIII A satellites Transmit 5 850 to 6 650 MHz Receive (Rx) 3 400 to 4 200 MHz		
2	$G/T(dB(K^{-1}))$	$G/T \geq 35.0 + 20\log f/4$ f = frequency in GHz	$G/T \geq 31.7 + 20\log f/4$	D-1: $G/T \geq 22.7 + 20\log f/4$ D-2: $G/T \geq 31.7 + 20\log f/4$
3	Antenna diameter (m)	15 – 18	11 - 13	D-1: 4.5 - 6, D-2: 11
4	Antenna side-lobe pattern (see Recommendation ITU-R S.465)	Standard A: after 1986, Standard B and D: after 1995 $\leq 29 - 25\log_{10} \phi$ dBi, $1° \leq \phi \leq 20°$ ≤ -3.5 dBi, $20° < \phi \leq 26.3°$ $\leq 32 - 25\log_{10} \phi$ dBi, $26.3 < \phi < 48°$ ≤ -10 dBi, $48° \leq \phi$		
5	Polarization	• Dual circular polarization (Tx polarization orthogonal to Rx polarization) INTELSAT VIII A satellites also use dual-linear polarization • Tx and Rx axial ratio ≤ 1.06		D-1: axial ratio ≤ 1.3 D-2: axial ratio ≤ 1.06
6	Antenna pointing and tracking	• Pointing capability: in any direction of geostationary orbit (above 5° elevation angle) • Manual and autotrack	– Manual and autotrack	D-1: Fixed antenna D-2: Manual and autotrack
7	Modulation and access characteristics	FDM/FM, CFDM/FM, SCPC/QPSK, TV/FM, TDMA, QPSK/IDR, IBS, TCM/IDR and DAMA	CFDM/FM, SCPC/QPSK, TV/FM, QPSK/IDR, IBS, TCM/IDR and DAMA	SCPC/CFM

694

TABLE AP7.1-2

INTELSAT Standard F earth stations

No.	Parameter	Performance characteristics	
		.	F-1, F-2 and F-3
1	RF bandwidth	Transmit (Tx) 5 925 to 6 425 MHz Receive (Rx) 3 700 to 4 200 MHz	
2	G/T(dB(K^{-1}))	F-1: $\geq 22.7 + 20\log f/4$ F-2: $\geq 27.0 + 20\log f/4$ F-3: $\geq 29.0 + 20\log f/4$ f = frequency in GHz	
3	Antenna diameter (m)	F-1: 4.5 to 7, F-2: 7 to 8, F-3: 9 to 10	
4	Antenna side-lobe pattern	Standard F (after 1995) $\leq 29 - 25\log_{10} \phi$ dBi, $1° \leq \phi \leq 20°$ ≤ -3.5 dBi, $20° < \phi \leq 26.3°$ $\leq 32 - 25\log_{10}\phi$ dBi, $26.3° < \phi < 48°$ ≤ -10 dBi, $48° \leq \phi$	
5	Polarization	• Dual circular polarization (Tx polarization orthogonal to Rx polarization) INTELSAT VIII A satellites also use dual-linear polarization • Tx and Rx axial ratio ≤ 1.09	
6	Antenna pointing and tracking	• Pointing capability: in any direction of geostationary orbit (above 5° elevation angle) • Fixed antenna	
7	Modulation and access characteristics	QPSK/IDR, IBS, TCM/IDR and DAMA	

TABLE AP7.1-3

INTELSAT Standard C and E earth stations

No.	Parameter	Performance characteristics	
		C	**E-1, E-2 and E-3**
1	RF bandwidth	Transmit (Tx) 14 to 14.5 (GHz) Receiver (Rx) 10.95 to 11.2 (GHz) and 11.45 to 11.70 (GHz) Transmit (Tx) 14 to 14.25 (GHz) or 14.25 to 14.50 (GHz) Receiver (Rx) 11.70 to 11.95 (GHz) or 12.50 to 12.75 (GHz) or 11.45 to 11.70 (GHz)	
2	$G/T(dB(K^{-1}))$	$\geq 37.0 + 20\log f/11.2$ f = frequency in GHz	E-1: ≥ 25.0 (but < 29.0) $+ 20\log f/11$ E-2: ≥ 29.0 (but < 34.0) $+ 20\log f/11$ E-3: $\geq 34.0 + 20\log f/11$
3	Antenna diameter (m)	11 - 14	E-1: 3.5 - 4.5, E-2: 5.5 - 7, E-3: 8 - 10
4	Antenna side-lobe pattern	Standard C (after 1988) and Standard E (after 1995) $\leq 29 - 25\log \phi$ dBi, $1° \leq \phi \leq 20°$ ≤ -3.5 dBi, $20° < \phi \leq 26.3°$ $\leq 32 - 25\log \phi$ dBi, $26.3° < \phi < 48°$ ≤ -10 dBi, $48° \leq \phi$	
5	Polarization	• Linear • Tx and Rx axial ratio > 31.6	
6	Antenna pointing and tracking	• Pointing capability: in any direction of geostationary orbit (above 10° elevation angle) • Autotrack	E-1 and E-2: Fixed antenna E-3: Autotrack
7	Modulation and access characteristics	FDM/FM, CFDM/FM, QPSK/IDR, IBS and TCM/IDR	QPSK/IDR, IBS and TCM/IDR

TABLE AP7.1-4

EUTELSAT SMS earth stations

No.	Parameter	Performance characteristics		
		SCPC/SMS[1]		
1	RF bandwidth	Transmit(Tx) 14.0 to 14.5 GHz Receive (Rx) 12.5 to 12.75 GHz		
2	G/T(dB(K^{-1}))	Standard 1	Standard 2	Standard 3
		G/T > 30 +20log (F/12.5)	G/T > 27 + 20log (F/12.5)	G/T > 23 + 20log (F/12.5)
3	Antenna diameter (m)	5.0 - 5.4	3.7	2.4
4	Antenna side-lobe pattern	Recommendation ITU-R S.465 + objective: G > 29 – 25log ϕ $25° > \phi > 7°$		
5	Polarization	• Linear polarization (Tx polarization orthogonal to Rx polarization) • Many stations use dual-polarized operation • Polarization isolation of antenna system > 35 dB (in –1 dB contour), mandatory for Tx recommended for Rx		
6	Antenna pointing and tracking	• Full steering capability desired		
		• Tracking so as to meet the ±0.5 dB e.i.r.p. stability performance		• No tracking
7	Receive noise temperature	160 K	160 K	140 K
8	e.i.r.p.	e.i.r.p. > P – Δ + 20log (f/14) • P = 53 + 10log(n) where n x 64 kbit/s is the customer bit rate n = 1, 2, 3, 4, 6, 12, 18, 24 and 30 • f: carrier frequency (GHz) • Δ: adjustment factor for geographical advantage • Higher quality grade channels can be obtained by transmitting with higher e.i.r.p.		
9	e.i.r.p. stability	±0.5 dB	±0.5 dB	±0.5 dB
10	Modulation multiplexing and multiple access	• Modulation 4 - PSK differential encoding differential decoding • Rate: 64 kbit/s SCPC and multiples of n with n = 1, 2, 4, 6, 12, 18, 24, 30 • 1/2 rate FEC or 3/4 rate FEC • Encryption (optional) • Access FDMA		
11	Spurious[2] emission (excluding multicarrier intermodulation products	e.i.r.p. < 4 dB(W/4 kHz) (outside the nominated carrier bandwidth) e.i.r.p. 55 dB below the total e.i.r.p. of the transmitted carrier (within the transponder which is accessed)		
12	Intermodulation products	e.i.r.p. < 12 dB (W/4 kHz)		
[1]	This service is provided using EUTELSAT II satellites and will later be extended to W-satellite and SEASAT.			
[2]	In the 14.0-14.5 GHz band.			

CHAPTER 8

Interconnection of satellite networks with terrestrial networks and user terminals

8.1 Interconnection of telephony networks

8.1.1 General interfacing aspects

International telephony traffic still represents one of the most important applications of satellite networks. Apart from conventional voice services, satellite networks can also convey through telephony channels, telegraphy/telex/data services (AVD: alternative voice/data services). Whenever digital terrestrial links are available, high bit rate data (e.g. 64 kbit/s or more) can also be carried. A few channels are also used to convey leased circuits for private telephony and data (low bit rate or, if possible, high bit rate). In the present fully end user-to-end user digital (i.e. ISDN) environment, all channels may be customized and may carry voice and/or data.

Figure 8.1 shows a typical arrangement for establishing an international FSS link for providing high traffic telephony connections between country A and country B. In this example, countries A and B are divided in telephony regions (two regions are shown). Each region is equipped with its own automatic international switching centre (ISC) which is connected to the international earth station by a terrestrial link carrying the international channels specific to the region. Often, one of these regional ISCs is dedicated as the nation's main ISC and is used for connecting the incoming channels which have not been dedicated by the sending ISC to a particular regional ISC of the receiving country. The figure also shows other types of international links, using microwave relays, cables (coaxial or optical fibres, terrestrial or submarine) or even using an alternative satellite link. In the example, the traffic between countries A and B is shared between the satellite link and a submarine cable link, in proportions which result from a bilateral agreement and which depend on economic considerations, availability and general management policy. In the routing plan, provision is taken to avoid, as far as possible, double satellite hops. Each means (satellite and cable) is also considered as a restoration facility in case of failure of the other.

As shown in Figure 8.1, connections between an ISC and the various international transmission means are made through a facility called the ITMC (international transmission maintenance centre). The ITMC forms the interface between the terrestrial links carrying the international channels and the ISC. The main equipment and functions usually provided in an ITMC are:

- line terminating equipment (LTE) for connecting terrestrial cables or transmitter/receiver for connecting radio relay links;
- multiplexing/demultiplexing;

- possibly, circuit concentration on the transmission links, e.g. by providing 64 kbit/s PCM to 32 kbit/s ADPCM conversion, digital circuit multiplication;

- echo suppression or cancellation;

- possibly, signalling (e.g. ITU-T No. 5). However, it should be noted that conversion between international signalling and national signalling is a major function which is incorporated in the ISCs.

It should be emphasized that the configuration shown in Figure 8.1 is only given as an example and that the switching and routing arrangements are particular to each country and are, in each case, the subject of bilateral agreements. In particular, for smaller traffic connections, only one ISC will be provided on one or both sides of the link.

Separate facilities have to be provided for inserting/extracting the leased circuits in/from the transmitted/received multiplexed channels and for routing them to their subscribers, possibly via specialized switches and networks.

For national (domestic) telephony traffic, the various possibilities due to using the FSS have already been summarized in § 1.4.5.1. Basically, there are two main categories of satellite national telephony links:

8.1 Interconnection of telephony networks

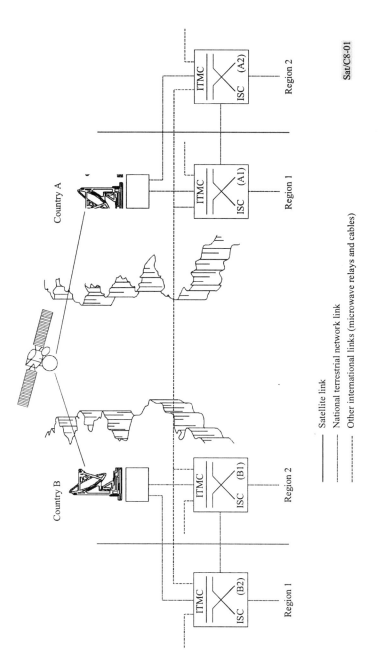

Country B

Country A

Region 2

Region 1

ITMC

ISC (B1)

ITMC

ISC (B2)

ITMC

ISC (A1)

ITMC

ISC (A2)

Region 1

Region 2

Sat/C8-01

——————— Satellite link

- - - - - - - National terrestrial network link

– – – – – – Other international links (microwave relays and cables)

FIGURE 8.1

A typical international FSS telephony link

- the first category is very similar to an international link; it consists of interconnecting the local terrestrial networks of the regions of a country which are separated by very significant geographical obstacles and/or distances. In the case of certain overseas territories, the link, if not available on a dedicated domestic satellite, can be established via an international satellite (INTELSAT explicitly provides for these cases). The only difference from the international connections previously described concerns switching centres and numbering operations which normally remain in a national framework;

- the second category covers the cases where a satellite constitutes a major means for establishing a telephony network in a country. As has already been explained in this handbook, these types of domestic satellite networks are currently implemented in many countries and especially in developing countries. They use either dedicated satellites or transponders leased or bought on available satellites (e.g. INTELSAT satellites). They can provide high or medium traffic arteries between main centres and/or thin route networks (often called rural telephony) for remote areas. Details concerning the design of these networks are treated in § 5.6.

Delay problems

Using modern analogue and more especially digital transmission techniques, most of the salient characteristics of connections established via satellite can be designed to be indistinguishable from connections established via any other media. The main difference between satellite connections and those of any other media, is the propagation delay associated with the long distances (36 000 km) to the satellite and back. Typically, the mean one-way propagation delay solely due to the satellite path can be expected to be 235 ms. In practice, this delay will be larger at earth stations located at higher latitudes (perhaps as much as an additional 20 ms delay). Of course, these considerations apply only to GSO satellites. The delay is much reduced in the case of low-orbit satellites.

Many experiments have been performed to determine the customer response to the amount of delay experienced in a telephony link. The general consensus holds that delays considerably in excess of 300 ms are well tolerated in the absence of echo. However, echo return losses of better than 36 dB are required for this condition to be met and the establishment and maintenance of such low levels of return loss is technically difficult. However, modern echo cancellers (ITU-T Recommendation G.165) are capable of achieving the necessary return losses under most conditions. It may be concluded therefore, that with the careful selection (and subsequent maintenance) of good echo cancellers, connections established via satellite can be designed to achieve the same or similar levels of customer acceptance as any other long-haul media.

The trend towards digital signal processing brings about significant additional path delays. For example, 16 kbit/s ADPCM-based DCMEs, cordless handsets and mobile radio designs can add in excess of 100 ms for each included device. On a single end-to-end link, two or even three of these devices could be encountered. The inclusion of some of these elements will be known *a priori* (for example, the existence of a DCME device will be known to the circuit planners). However, other delay contributing devices such as cordless handsets can not be predicted since their existence is a variable resulting from many different customers accessing the public switched networks.

Given these problems, the establishment of a telephony service via more than one geosynchronous satellite may have difficulties. It should be noted that, although the typical one-way propagation delay for two simple satellite connections in tandem is about 470 ms, ITU-T Recommendation G.114 specifically states that the use of transmission paths having a one-way propagation delay in excess of 400 ms is not recommended.

A possible exception would be for private networks which would not be as concerned with the 400 ms mean one-way propagation delay recommendation of ITU-T.

In the case of data transmissions, given that the data communication protocols (more especially the block numbering range) are capable of accommodating the 600 ms or more of delay, efficient data exchanges can be conducted via two satellite networks providing that the bit error performance is acceptable.

8.1.2 Digital networks interfacing aspects

8.1.2.1 Interfacing problems between terrestrial networks and digital satellite systems

The equipment used for interfacing a satellite network with a terrestrial network provides a set of functions which makes the conversions between the transmission formats required in the two networks to enable the connection of terrestrial channels to satellite channels to be made.

Different problems have to be solved if the two networks are both analogue or both digital or if the terrestrial network is analogue and the satellite network is digital. Some of these problems and in particular those relevant to analogue transmission are discussed in § 8.4, 4.1 and 7.5.

Additional information on interfacing with digital satellite networks is given below with reference to the current specifications of INTELSAT and EUTELSAT systems using TDMA and FDMA. The synchronization aspects are addressed in particular.

Further information can also be found in Chapter 3 ("Evaluation of digital satellite system") of ITU-T GAS 3 Handbook ("Methods for evaluating new digital inter-exchange transmission systems as a guide to national network planning").

8.1.2.1.1 Terrestrial interfacing with TDMA systems

The interface between the TDMA system and the terrestrial network must perform the following functions:

- encode and decode analogue circuits, if they exist;

- synchronize the digital signals as necessary;

- carry the signalling and alarm signals associated with individual terrestrial circuits, or groups of circuits.

The INTELSAT and EUTELSAT 120 Mbit/s TDMA/DSI systems require a 2.048 Mbit/s terrestrial interface. This interface corresponds to a module of the traffic terminal called TIM (terrestrial interface module) which is a DSI/DNI unit.

When an existing analogue terrestrial link is interconnected with a TDMA/DSI satellite system, analogue-to-digital conversion is required, using primary order multiplexers or transmultiplexers, in order to provide 2.048 Mbit/s streams at the interface. Also, if the terrestrial link is digital the incoming stream at the interface has to be configured in a primary order PCM multiplex format.

The main problem in the interfacing is the synchronization of the digital stream on the terrestrial side with the clock timing of the satellite network.

In fact the TDMA/DSI system is an independent synchronous network driven by a high stability clock with an accuracy of the order of 10^{-11}, generated by the TDMA reference station and distributed to all the TDMA transponders for network synchronization.

If the terrestrial link is analogue, the synchronization is obtained by slaving the A/D converter to the TDMA clock. A simple method called loop synchronization (see Figure 8.2) consists in extracting the clock from the incoming signal at the satellite receive side of the interface and using it to synchronize the clock of the transmit signal. In this case the interface should include a buffer for compensation of the Doppler effect on the transmit side.

If the terrestrial link is digital, synchronization requires the lining-up of the clocks in the terrestrial and in the satellite networks.

When the digital PCM link on the terrestrial side is asynchronous, a slaved synchronization method, similar to the loop synchronization of A/D converters shown in Figure 8.2 can be adopted. In this way the PCM digital multiplexer on the satellite receive side is slaved to the clock of the incoming TDMA signal and this clock is transferred to the terrestrial link.

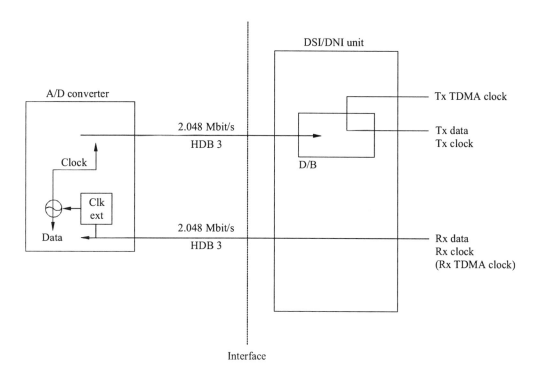

Clk ext: clock extractor
D/B: doppler buffer
HDB: high density bi-polar code

Sat/C8-02

FIGURE 8.2

Loop synchronization method between an analogue terrestrial link
and a TDMA digital satellite network

When the digital terrestrial link uses a synchronous digital PCM link which is synchronized with a national clock source or equivalent, plesiochronous alignment is necessary. The national clock source of the terrestrial link and the master clock of the TDMA system both have a highly stable master clock source of the order of 10^{-11}. When these stable and different links are directly connected to each other, data slip will occur at approximately every 72 days. For this slip, plesiochronous alignment is necessary at some interface; in practice a plesiochronous buffer is provided in the DSI/DNI unit for this purpose.

Since in this case the Doppler effect has also to be removed in both transmit and receive links, the actual buffer circuit in the TIM includes plesiochronous buffer and Doppler buffer functions.

8.1.2.1.2   Terrestrial interfacing with FDMA digital systems

In systems using FDMA for the transmission of continuous digital carriers, some special considerations are required since multiple destinations and different bit rates can be included in the time division multiplexed (TDM) arrangement of each carrier.

The fact that a given carrier can transmit data towards multiple destinations usually means that the earth station terminal should be able to receive the carriers incoming from these destinations. Therefore, as is commonplace in FDMA, symmetrical modems cannot be used and the earth station terminal should include separate transmit units (baseband equipment and modulator) and receive units (demodulator and baseband equipment). Each transmit unit is usually associated with a few receive units.

The fact that carriers at different bit rates can be accommodated, means that flexible modulators and demodulators with variable bit rate capability should preferably be used. For illustrative purposes, the examples given below refer to the INTELSAT IBS, EUTELSAT SMS and the INTELSAT IDR, which are well specified systems.

a)   INTELSAT IBS and EUTELSAT SMS carriers

At the customer interface (often called interface A), customer bit rates either lower than 64 kbit/s (2.4, 4.8, 9.6 kbit/s) or multiples of it (n x 64 kbit/s, n = 1 to 32) can be provided.

Since the minimum bit rate accepted for an IBS or SMS carrier is 64 kbit/s, user bit rates lower than 64 kbit/s (2.4 to 9.6 kbit/s) have to be multiplexed into 64 kbit/s aggregated signals by means of lower level order data multiplexers.

If the earth station is located at the end user's premises each n x 64 kbit/s data channel accesses the FDMA terminal directly. If the earth station and the user's premises are at different places the data have to be transmitted from the customer interface to the terminal interface (also called interface H) using a terrestrial link. Since the n x 64 kbit/s data rates are not standardized for terrestrial transmission for any value of n, it becomes necessary to combine the data channels into a 2.048 Mbit/s stream, which is suitable for terrestrial means. This requires the use of flexible data multiplexers which accept input data channels at different bit rates, as shown in Figure 8.3.

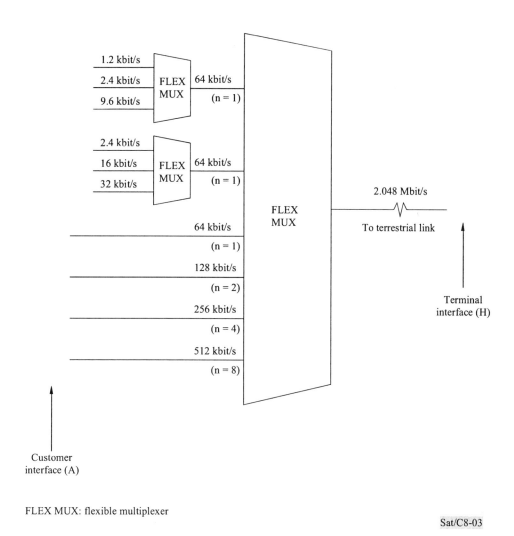

Sat/C8-03

FIGURE 8.3

Lining up the terrestrial side data

The opposite arrangement is however not needed at the earth station to restore the independent data channel as the terminal itself includes an opposite demultiplex scheme, as shown in Figure 8.4, to obtain independent n x 64 kbit/s data channels. Such a scheme is compatible with multidestination networks.

The timing of the digital signals at interface H in both directions of transmission can be derived in one of three ways:

i)   from a national clock with an accuracy of one part in 10^{11} as recommended in ITU-T Recommendation G.811;

ii) from a local earth station clock with an accuracy of at least one in 10^9 over any 40 day period;

iii) from an incoming clock received from a remote earth station by satellite.

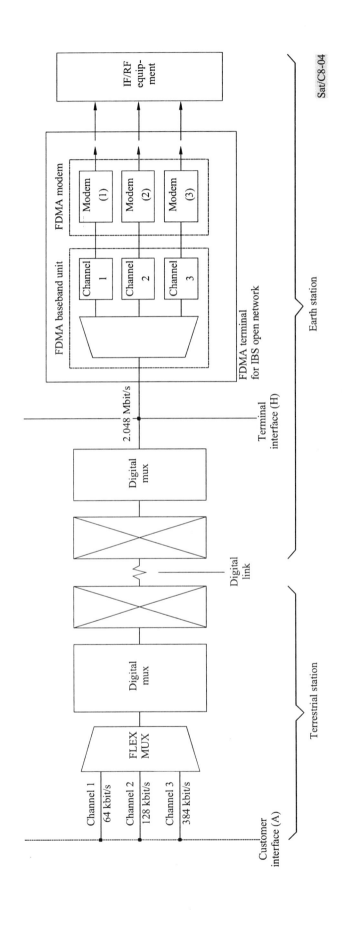

FIGURE 8.4

Overall line up of IBS E/S with terrestrial link

Depending on which of the three synchronization methods is adopted at each end of the satellite link, different alternative synchronization arrangements can be established.

Figure 8.5 shows possible synchronization schemes. Buffering at the receive earth station is required to compensate for the effects of satellite movements (Doppler compensation) and for disparity between clocks at originating and receiving ends of a data channel. The location of the buffer depends upon the configuration of each channel and upon the point where transitions from one clock to another occur. The required buffer capacity depends on the sources of timing, the satellite delay variations, the interval between frame slips and the configuration of each particular channel.

b) The INTELSAT IDR system

Intermediate data rate carriers in the INTELSAT system utilize coherent QPSK modulation operating at information rates ranging from 64 kbit/s to 44.736 Mbit/s and offer single destination and multidestination services. Unlike IBS and SMS, both voice and data can be carried on an IDR carrier. This introduces some slight differences in the synchronization schemes. In this case the clock at interface H is derived either from a plesiochronous clock with an accuracy of one part in 10^{11} or from the incoming clock received from a remote earth station. However, in cases where there is not a synchronous digital network at either end, but the channels are converted to analogue voice circuits, the internal clock of the PCM multiplex equipment is considered to be of sufficient accuracy (about 50 parts in 10^6).

There is an increasing trend at present to adopt digital circuit multiplication techniques in transmission links using IDR. This involves the introduction in the terrestrial interfaces of digital circuit multiplication equipment (see § 8.1.4).

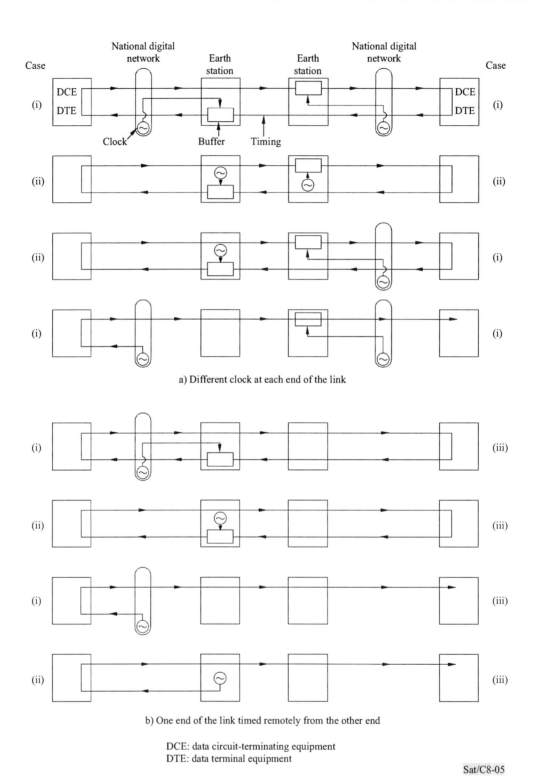

a) Different clock at each end of the link

b) One end of the link timed remotely from the other end

DCE: data circuit-terminating equipment
DTE: data terminal equipment

Sat/C8-05

FIGURE 8.5

Possible schemes for clocking arrangements between two earth stations

8.1.3   Earth station multiplex equipment

8.1.3.1   Analogue satellite communications (FDMA-FDM)

As already explained (see § 7.1.1.5), multiplexed telephony (FDM) baseband assemblies arriving at the earth station from the terrestrial link must be rearranged prior to satellite transmission. This is because:

i)   it may be necessary to distribute the telephony channels between several satellite RF carriers and also, inside a given carrier, to distribute the telephony channels according to their respective destinations;

ii) it is the usual practice (e.g. in the INTELSAT satellite system) that the baseband channel arrangements used for satellite transmission, though normally having a 4 kHz spacing and complying with ITU-T Recommendations, show some differences with the conventional arrangements used for terrestrial transmission. In particular, the first group (group A) containing 12 channels is systematically included in the baseband between 12 kHz and 60 kHz. This arrangement has been adopted to increase the channel capacity and to raise the effective modulation index of the frequency modulation.

Table 8.1 gives the baseband compositions used in the INTELSAT system.

At receive, a third factor arises:

iii) whereas one single FM-FDM carrier can be used to transmit a number of telephone channels from the earth station towards several destinations (i.e. several corresponding earth stations), the same earth station must be able to receive each individual incoming carrier from each of these destinations, in order to operate in normal duplex telephony. These incoming carriers are generally multidestination carriers and the receiving earth station has to extract from these carriers only those supergroups, groups and/or telephone channels which correspond explicitly to its own transmit signals. The terrestrial multiplex arrangement is then reassembled from these incoming supergroups, groups and/or telephone channels. This asymmetric structure of the earth station multiplex equipment is specific to multiple access in satellite communications.

The complete set of the earth station multiplex equipment is generally divided into two sub-assemblies:

•   the so-called "terrestrial multiplex equipment" which demultiplexes (at transmit) the terrestrial FDM baseband assembly into the minimum number of supergroups, groups and/or telephone channels and which multiplexes (at receive) the (incoming) supergroups, groups and/or telephone channels in order to restore the (symmetrical) terrestrial FDM baseband assembly;

•   the so-called "satellite multiplex equipment" which multiplexes (at transmit) the supergroups, groups and/or telephone channels in order to form the FDM signals (for modulating the RF carriers) and which demultiplexes the received demodulated carriers into the minimum number of supergroups, groups and/or telephone channels.

It must be emphasized that, in general, it is not necessary to demodulate completely the transmit (terrestrial) baseband and the receive (satellite) baseband down to the telephone channel level, but only to select the multiplex equipment "building blocks" in order to interface the two sub-assemblies ("terrestrial multiplex" and "satellite multiplex") by the minimum number of supergroups, groups and/or telephone channels, as explained above. However, such an optimum design should remain flexible and should not preclude future extensions of the station traffic capacity.

TABLE 8.1

INTELSAT carrier baseband composition

No. of channels	Baseband composition	Frequency band (kHz)
12	Group A (e)[1]	12-60
24	Group A (e)[1] plus Group 5 of SG 1 (i)	12-108
36	Group A (e)[1] plus Groups 5 and 4 of SG 1 (i)	12-156
48	Group A (e)[1] plus Groups 5-3 of SG 1 (i)	12-204
60	Group A (e)[1] plus Groups 5-2 of SG 1 (i)	12-252
72	Group A (e)[1] plus SG 1 (i)	12-300
96	Group A (e)[1] plus SG 1 (i) plus Groups 1 and 2 of SG 2	12-408
132	Group A (e)[1] plus SG 1 (i) plus SG 2 (e)[2]	12-552
192	Group A (e)[1] plus SG 1 (i), SG 2 (e) and SG 3 (i)[2]	12-804
252	Group A (e)[1] plus SG 1 (i), SG 2 (e), SG 3 and SG 4 (i)[2]	12-1 052
312	Group A (e)[1] plus SG 1 (i), SG 2 (e) and SG 3-5 (i)[2]	12-1 300
372	Group A (e)[1] plus SG 1 (i), SG 2 (e) and SG 3-6 (i)[2]	12-1 548
432	Group A (e)[1] plus SG 1 (i), SG 2 (e) and SG 3-7 (i)[2]	12-1 796
492	Group A (e)[1] plus SG 1 (i), SG 2 (e) and SG 3-8 (i)[2]	12-2 044
552	Group A (e)[1] plus SG 1 (i), SG 2 (e) and SG 3-9 (i)[2]	12-2 292
612	Group A (e)[1] plus SG 1 (i), SG 2 (e) and SG 3-10 (i)[2]	12-2 540
792	Group A (e)[1] plus SG 1 (i), SG 2 (e) and SG 3-13 (i)[2]	12-3 284
972	Group A (e)[1] plus SG 1 (i), SG 2 (e) and SG 3-16 (i)[2]	12-4 028
1 092	Group A (e)[1] plus SG 1 (i), SG 2 (e) and SG 3-16 (i)[2] plus SG 16 and 15 (e) (modulated)[2]	12-4 892
1 332	Group A (e)[1] plus SG 1 (i), SG 2 (e) and SG 3-16 (i)[2] plus SGs 16-11 (e) (modulated)[2]	12-5 884

NOTE 1 – Group: an assembly of 12 telephone channels derived from the basic group by frequency translation (the basic group is obtained by frequency conversion of 12 telephone channels in the frequency range from 60 to 108 kHz).

NOTE 2 – Supergroup (SG): an assembly of 60 telephone channels derived from the basic supergroup by frequency translation (the basic supergroup is obtained by frequency conversion of five groups in the frequency range from 312 to 522 kHz).

(e): erect

(i): inverted

[1]   Refer to ITU-T Recommendation G.322.

[2]   Refer to ITU-T Recommendation G.423.

The main "building blocks" of the multiplex equipment are:

• channel translating units, each one converting 12 telephone channels into a group (channel modems and filters);

• group translating units, each one converting five groups into a supergroup (group modems);

• supergroup translating units for locating the supergroups into the baseband;

• through group and through supergroup filters;

• carrier generating units for providing (from a main pilot) the carrier frequencies for the various translating units;

• distribution frames for providing proper connections between the various units.

The translating units should include conventional pilots, automatic level regulation and alarm circuits.

Furthermore, it is usually required that a 60 kHz continuity pilot signal be inserted in the baseband of each transmit carrier for monitoring the satellite link.

8.1.3.2 Digital satellite communications

At present, there is no standard practice for the traffic capacities of digital satellite links nor for their terrestrial interfaces. However, some of the main trends are as follows:

i) Low bit rate digital communications (e.g. 32 kbit/s to 1.5 or 2 Mbit/s) are mainly used for point-to-point, user-to-user data transmissions (digital SCPC links). Their interfaces are standardized by ITU-T Recommendations V.35, V.32, X.24 and X.21. Very low bit rate communications (e.g. 2.4, 4.8, 9.6 kbit/s, V.24/X.24 ITU-T interface) should be multiplexed prior to their transmission on satellite links.

ii) If multiple access is to be provided for these types of communications, several digital SCPC links can be combined through a digital multiplex equipment, using, for example, 2 Mbit/s (or 1.5 Mbit/s) standard interfaces.

iii) Higher bit rate digital communications (e.g. 2 Mbit/s, 8 Mbit/s or 34 Mbit/s in the European hierarchy) are used in the FDMA-TDM-PSK satellite transmission mode. The most effective way of connecting these satellite communications to the terrestrial network is by direct coupling through digital terrestrial links (cable or microwave links). It should be noted that the satellite transmission efficiency can be enhanced (i.e. at least doubled) by traffic concentration through digital speech interpolation (DSI) and even more by the use of digital circuit multiplication equipment (DCME).

iv) In the TDMA mode of satellite communications, very high bit rates and traffic capacities can be provided, with full multiple access capability. In most cases, the normal interface with the terrestrial network is at 2 Mbit/s (e.g. in the INTELSAT and EUTELSAT TDMA systems) or at 1.5 Mbit/s (e.g. in the United States' domestic TDMA systems). This interface is usually provided through DSI units for traffic concentration: for example, in the INTELSAT and EUTELSAT TDMA systems, the standard telephony DSI interface provides connection to ten PCM (2 Mbit/s) lines (300 telephone channels), but owing to the DSI process, only 120 (approximately) active channels are instantaneously transmitted through the satellite link. DCME can also be used for increasing the traffic capacity of TDMA systems.

v) The connection between the direct users and the digital satellite link may also be made on a demand assignment basis (either delayed demand assignment – i.e. in reservation mode – or instantaneous demand assignment). This can be easily implemented by using a digital switching equipment: such a system is used, for example, in the French TELECOM 1 system.

vi) In the common case where the terrestrial standard interface is at 2 Mbit/s (HDB 3 interface) – or at 1.5 Mbit/s – several types of terrestrial connections may be envisaged depending on whether the terrestrial link is analogue or digital and also on whether the switching equipment is an analogue switch or a digital time division switch. Some typical cases are shown in Figure 8.6.

vii) When interfacing between an analogue multiplex assembly (e.g. a 60 channels FDM supergroup) and a digital multiplex assembly (e.g. two 30 channel PCM groups) – or vice versa – conversion equipment must be used. Two types of process can be used for such a conversion:

- conversion through multiplexing/demultiplexing at the telephone channel level. This process uses conventional analogue and digital equipment;

- direct conversion: equipment for direct conversion, called transmultiplexers, is now available. The most common technique for creating a transmultiplexer is based on frequency domain to time domain conversion by using the fast Fourier transform algorithm.

8.1 Interconnection of telephony networks

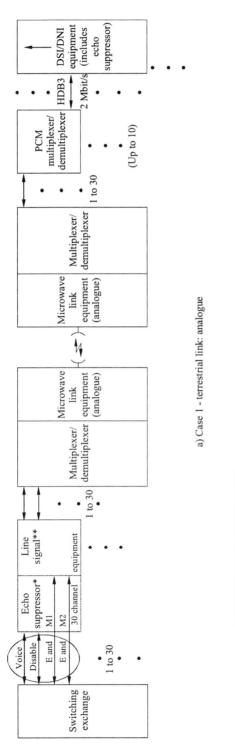

a) Case 1 - terrestrial link: analogue

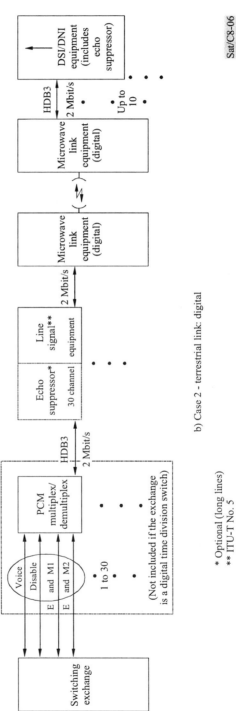

b) Case 2 - terrestrial link: digital

* Optional (long lines)
** ITU-T No. 5

Sat/C8-06

FIGURE 8.6

Typical cases for connection of digital satellite links to the terrestrial network

8.1.3.3   SCPC satellite communications (telephony)

SCPC (i.e. non-multiplexed) links, either with digital (PCM/PSK) or analogue (companded FM) transmission, are commonly used for low-density, scattered traffic satellite communications with small stations. Multiplex equipment for SCPC is much simpler than that for multiplexed (FDM or TDM) communications. This is due not only to the small capacity, but also to the fact that only the "terrestrial multiplex equipment" is needed in this case.

8.1.4   Digital circuit multiplication

8.1.4.1   General

Digital circuit multiplication equipment (DCME) had been used for some time in some domestic satellite systems and in submarine cable systems. However, the main impetus to its development was due to its standardization by INTELSAT and by its future standardization by EUTELSAT and by ITU-T. All the specifications either in existence or under preparation are very similar and are based on the principles listed in §§ 8.1.4.2 to 8.1.4.4 below. It should be noted that operation with non-standard DCMEs (sometimes called DCMs) is possible in closed networks, i.e. when the correspondents agree on specific characteristics and/or use identical equipment (from the same manufacturer) at both ends of the system.

8.1.4.2   Circuit multiplication operation

Circuit multiplication is performed by combining digital speech interpolation (DSI) and low bit rate encoding (LRE) normally at 32 kbit/s. In consequence, as a result of the combination of the factor 2 due to LRE (telephony at 32 kbit/s in lieu of 64 kbit/s) and of a factor of about 2.5 due to the telephone conversation activity factor used in DSI operation, the total circuit multiplication gain provided by DCME utilization will be about 5 or even more, depending on actual operational conditions (and, in particular on the percentage of data transmissions using the network). Figure 8.7 gives typical DCME multiplication gains.

8.1.4.3   Low bit rate encoding

Various algorithms have been proposed for encoding voice at 32 kbit/s. The selection of a common i.e. standard algorithm is essential for interworking in a satellite network with equipment provided by different manufacturers. The system already adopted by INTELSAT and EUTELSAT uses the 5/4/3-bit ADPCM encoding chosen by ITU-T on the basis of Recommendation G.721 and also ITU-T Recommendation G.723. With this technique, voice channels are normally encoded as 4-bit samples, but during periods of overloading, the fourth bit is stolen from enough 4-bit samples to produce additional 3- or 4-bits samples. All bit stealing operations are randomly chosen in order to distribute any circuit degradation across all pooled channels.

It is possible to use the bit stealing technique for both voice and data traffic within each pool of channels (the term "pool" is defined in § 8.1.4.5 below). Voiceband data traffic will be provided with high-speed 40 kbit/s ADPCM channels, by assigning a 5-bit sample obtained by adding a supplementary bit to a normal 4-bit sample. An available 4-bit channel is temporarily used as a bit pool which is able to provide this fifth bit to as many as four data channels. 64 kbit/s PCM preassigned channels and 64 kbit/s PCM demand assigned channels are used to accommodate ISDN requirements and these channels will not be subject to DSI and LRE operation.

An important feature is that a built-in dynamic load control (DLC) is provided to prevent excessive degradation of the voice quality due to an exceptionally high traffic load. Whenever a predetermined encoding rate limit is reached (typically 3.6 bits/sample), the DCME will send a message to the telephone exchange (generally the ISC) to prevent additional calls being placed on the trunks served by the DCME until a normal load is re-established.

8.1.4.4 DCME interfaces

On its terrestrial (trunk) side, the DCME interfaces with primary multiplex signals either of the 24-channel T1 type at 1.544 Mbit/s, or of the 30-channel CEPT carrier at 2.048 Mbit/s. The nominal capacity of the INTELSAT DCME is 150 channels. Therefore, an interface capable of accommodating at least seven T1 or five CEPT carriers is required. In fact, more input carriers can be connected to the available DCMEs (up to eight 30-channel carriers) to permit advantage to be taken of actual traffic configurations.

On its satellite transmission side, the interface consists of either a 2.048 Mbit/s or a 1.544 Mbit/s bearer.

The DCME is provided for operation either with FDMA-TDM-PSK carriers, e.g. the so-called intermediate data rate (IDR) carriers defined by INTELSAT, or with TDMA-TDM-PSK carriers, e.g. the 120 Mbit/s TDMA carriers defined by INTELSAT and EUTELSAT.

8.1.4.5 Multiple destination operation of DCME

The DCME specification provides considerable operational flexibility as multiple destination operation is provided for by inserting several possible destinations in the DCME bearer frame structure. Two modes of operation with multiple destinations are possible, called multiclique and true multidestination:

- in the multiclique mode, the frame is divided into a small number of separate interpolation (DSI) pools, each pool being called a "clique". Each clique contains its own assignment channel for interpolating conversations (speech) within the pool and all channels within a clique are associated with one separate destination. The INTELSAT specification provides for up to two cliques, but other systems may provide for more. The multiplication gain provided by DSI is lower when the number of channels within the pool is reduced;

- in the multidestination mode, a single pool is used to communicate with the multiple destinations (also equipped with multidestination DCMEs). In the INTELSAT specifications one additional pool can be placed in the frame to communicate with one multiclique DCME, but the total number of destinations served in the frame (by the two pools) must not exceed four. The single pool, multidestination mode provides higher multiplication gain than the multiclique mode.

In the case of FDMA-TDM-PSK carrier multiclique operation, the received pools must be processed by a special PCM time slot interchange function (cross-connect or drop-and-insert circuit) called by INTELSAT a clique sorting facility (CSF). This CSF is not normally included in the DCME and must be installed at the earth station. Its function is to extract each pool from its respective bearer and to combine them into a single bearer to be processed by the DCME (see Figure 8.8).

In the case of TDMA, the DCME processed input/output digital signal must not be subject to the DSI operation which is often associated with TDMA. In consequence, their interface with the common TDMA terminal equipment (CTTE) can be either through a special transparent direct digital interface (DDI) module, or through a DNI (digital non-interpolated) module, or through a DNI port of a DSI module (see Figure 8.10).

8.1.4.6 The effects of the installation of DCME on earth station equipment

The DCME is a baseband equipment which processes the trunks switched by the telephone exchange (e.g. the international switching centre (ISC)). This is why it should very often be installed in the telephone exchange premises, with the echo suppressors/cancellers and the various signalling facilities. Moreover, locating the DCME at the telephone exchange and a CSF in the earth station allows the benefit of circuit multiplication to be also obtained on the exchange-to-earth station microwave link. This is one advantage of selecting the multiclique mode for multiple destinations operation of the DCME. In consequence, users with one or two destinations per DCME could select the multiclique mode, and place the DCMEs at the exchange. Users with a smaller amount of traffic per destination (and with more than two destinations) could select the multidestination option.

Three typical examples of DCME and earth station equipment configurations (which can be either new and complete earth stations, or retrofitted units used for equipping existing stations with digital/DCME communications) are given in the following figures:

Figure 8.8 shows a simple configuration for implementing a 2 Mbit/s FDMA-TDM-PSK/DCME link with two destinations in the multiclique mode. The equipment consists of three subsystems:

- the RF and converters subsystem (antenna, LNA, HPA, U/C and D/C). In the usual case where the two received carriers are transmitted through the same satellite transponder, a common down converter can be used;

- the "signal processing terminal" which comprises the modulator and two demodulators (with, if required, forward error correction encoders/decoders) and the transmit (Tx) and receive (Rx) baseband modules. In these modules are included various baseband digital functions, such as buffering, framing, scrambling/descrambling, engineering service channels addition/extraction, etc. The receive "master" module also includes the CSF which extracts from the 2 Mbit/s bearers received from the corresponding stations (B and C) the cliques destined to the concerned station (A) and recombines them in a 2 Mbit/s bearer with the same format as the transmit bearer;

- the DCME, which can be located at the earth station or, more frequently, at the telephone exchange.

It must be emphasized that the configuration shown can, of course, be extended, e.g. with higher bit rate carriers. However, this simple configuration demonstrates that, when operating with circuit multiplication, even small stations can carry a significant traffic capacity.

Figure 8.9 shows a similar configuration with about the same potential traffic capacity as above, but with four corresponding stations (B, C, D, E) in the multidestination mode. In the example shown the "signal processing terminal" must be equipped with four demodulators and four baseband modules (not including, as in the previous case, supplementary units if redundancy is required). In this example, no CSF is needed.

Figure 8.10 shows an example of an earth station working in TDMA (INTELSAT/EUTELSAT 120 Mbit/s TDMA) and retrofitted with DCMEs. It demonstrates that TDMA appears to be a very good solution in the case of high traffic capacity with numerous destinations. This is because the time division operation permits a single common TDMA terminal equipment (CTTE), to process the multiple destinations bearers in lieu of all the units involved, especially on the receive side, in FDMA operation. The figure shows three examples of DCME utilization in association with a TDMA equipment conforming to the INTELSAT or EUTELSAT specifications:

•   the first (upper) DCME is assumed to carry high capacity traffic to/from a single destination (station B). Here a single DNI port for DSI/DNI terrestrial interface module (TIM) is used and up to seven ports remain available for DSI only, DNI only, or other DCME operation;

•   the second DCME is assumed to be corresponding with two stations (C and D) in the multiclique mode. Again, thanks to the multiple destination operation of the TDMA equipment, a single DNI port is needed in the TIM;

•   the third DCME is assumed to link four corresponding stations (E, F, G, H) in the multidestination mode. In this case, the multidestination function is carried out by the DCME. In consequence, the TIM must be operated in full DNI mode. Then, in accordance with the requirements of the INTELSAT and EUTELSAT specification, the four received 2 Mbit/s bearers must be connected to the four available DNI ports and consequently no spare port is available on this TIM. In fact, a simplified TIM, the so-called DDI could be used in this example.

8.1.4.7   Conclusion

It is evident that the implementation of DCME in satellite communications will greatly contribute to a more efficient utilization of the satellite transponders and, more generally, of the orbit/spectrum resource. This is why DCMEs should find, in the coming years, more and more applications, not only in international satellite communications (e.g. INTELSAT or EUTELSAT), but also in domestic satellite communications systems. In fact, taking the example of a satellite transponder with a nominal traffic capacity of 1 200 channels (at 64 kbit/s), this same transponder could carry 6 000 telephone channels (half-circuits) by implementing DCMEs. This increase in transponder efficiency should be reflected in a decrease in space segment costs.

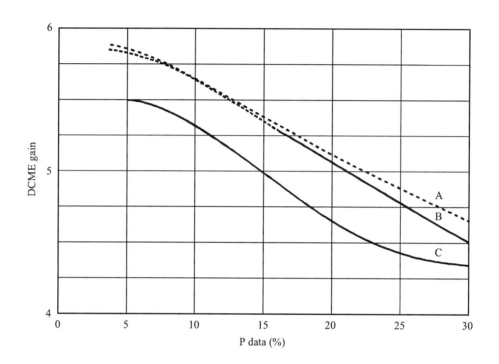

A, B, C:	30, 24, 15 transmission channels (8-bit channels)
———	gain limited by clipping (2% probability)
- - - - -	gain limited by encoding rate (3.6 bits/sample)
Data links:	FAX (50% Tx and 50% Rx)
Preassigned channels:	none
Unrestricted 64 kbit/s channels:	none

Sat/C8-07

FIGURE 8.7

DCME gain curves

8.1 Interconnection of telephony networks

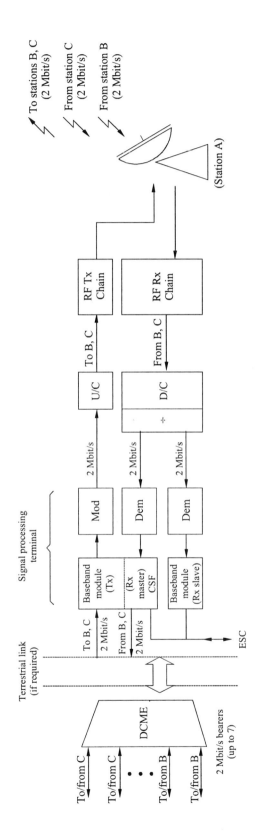

FIGURE 8.8

An example of simple earth station with FDMA-TDM-PSK and DCME: equipment configuration (multiclique operation)

CHAPTER 8   Interconnection of satellite networks with terrestrial networks

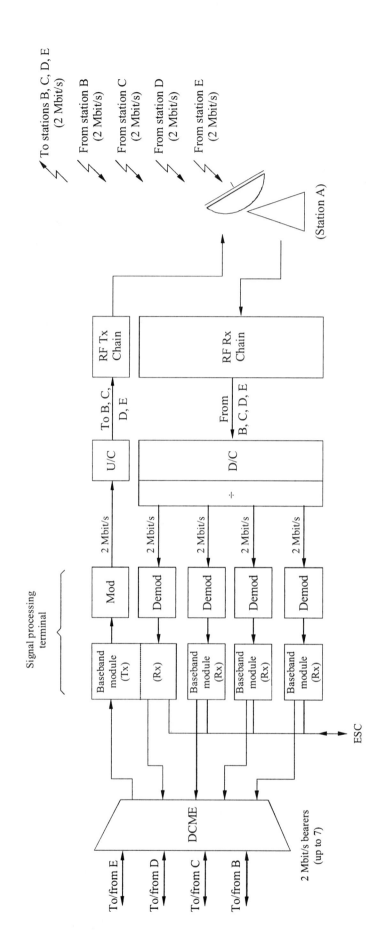

FIGURE 8.9

An example of simple earth station with FDMA-TDM-PSK and DCME: equipment configuration (multidestination operation)

Traffic capacity: 150 to 216 telephone circuits
ESC: Engineering service centre

Sat/C8-09

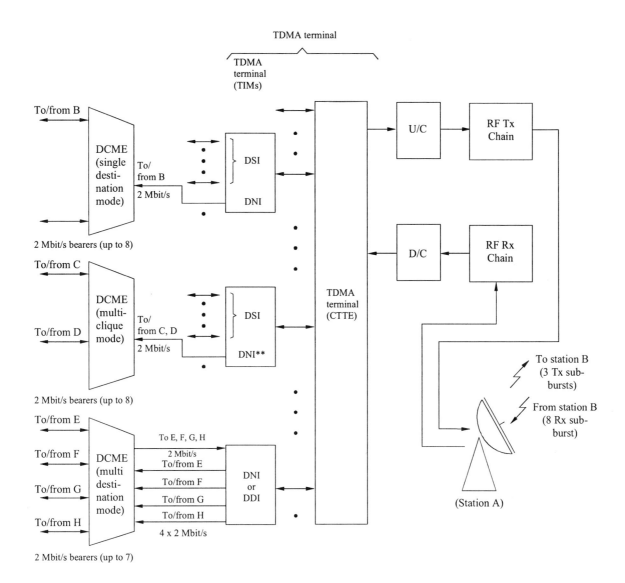

FIGURE 8.10

An example of high capacity earth station with TDMA-TDM-PSK
and DCME: equipment configuration

8.1.5   Engineering service circuits

In most cases, an engineering service circuit (ESC) network should be provided as part of a satellite communications system in order to allow communications, for service purposes, between the various traffic earth stations inside the system and also between any traffic earth station and the technical and/or operations centre(s) which are in charge of managing and monitoring the system.

In general, a small number of specialized telephone and data channels are specifically used for this ESC function.

Consequently, special ESC equipment should be provided in the earth station. This equipment often comprises two different parts:

i)   ESC multiplex equipment which inserts (at transmit) and extracts (at receive) the specialized telephone and data channels into/from the normal traffic signals;

ii)   ESC telephony and data switching equipment which permits automatic access to the satellite system ESC network.

Following the INTELSAT specifications, the common practice for providing ESC channels is as follows:

a)   In the case of FDMA-FDM-FM operation, two specialized 4 kHz telephone channels can be included in the so-called sub-baseband of each transmitted carrier. This sub-baseband (4 to 12 kHz) comprises two inverted sidebands of virtual carrier frequencies of 8 kHz and 12 kHz. Each of the specialized telephone channels is shared between one voice channel and one to five data channels (at 2.7, 2.82, 2.94, 3.06 and 3.18 kHz) on a speech plus duplex voice frequency telegraphy basis.

Consequently, the ESC multiplex equipment (in i) above) usually provides for two functions, i.e.:

• inserting (and extracting) the 4 kHz channels into (from) the multiplexed baseband;

• inserting (and extracting) the data channels into (from) each of the 4 kHz channels.

b)   In the case of SCPC operation, one or two SCPC channel units (SCPC modems) are usually provided in the earth station for ESC in addition to the normal traffic channel units. These specialized channel units are used in the same telephony/data mode as in the previous case. Of course, the ESC multiplex equipment is much simpler and has only to provide for the second function outlined above.

c)   In the case of FDMA-TDM-PSK operation, also called intermediate data rate (IDR) carriers, synchronous multiplexing of overhead bits for the provision of ESC and alarms is accomplished. Figure 8.11 shows a system currently in use in which 96 kbit/s of overhead are added to a 2 048 kbit/s information bit stream.

d)   In the case of TDMA operation, two voice channels and eight data channels are digitally encoded and are inserted in the preamble of each transmitted traffic burst of the TDMA frame.

In each earth station TDMA traffic terminal, a special terrestrial interface module (TIM), the so-called order wire module (OWM), is used and constitutes the ESC multiplex equipment (see i) above).

With regard to ESC telephony and data switching (see ii) above) modern equipment is currently available. This equipment is usually offered with all the additional services of a modern switching exchange and with supplementary circuits, e.g. for local communications and for communications with other centres (telephone traffic exchange, TV facilities, etc.) via terrestrial links.

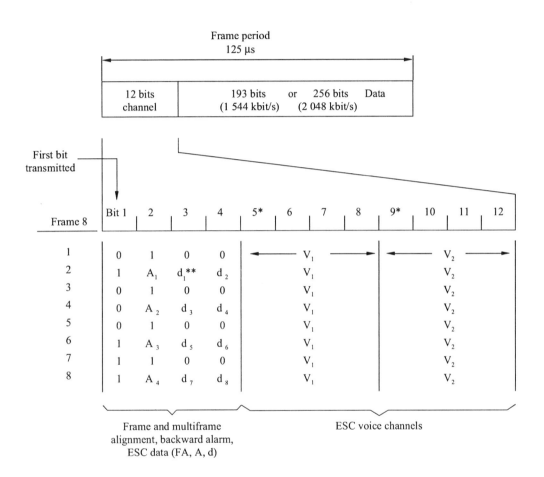

V_i: ESC voice channel 1 bits (i = 1, 2); (set to 1 if not used)
A_i: Backward alarm to destination i (i = 1, 2, 3, 4); no alarm = 0, alarm = 1
d_i: ESC digital data (i = 1 to 8); (set to 1 if not used)
8 frames = 1 multiframe period (period = 1 ms)
Channel rate = 12 bits/125 µs = 96 kbit/s
*   Bits 6 and 9 in the overhead frame correspond to the first bits transmitted
         in the ESC voice channels
** d_i corresponds to the first bit transmitted in the ESC data channel

Sat/C8-11m

FIGURE 8.11

Overhead structure of 1 544 and 2 048 kbit/s carriers

8.1.6   Echo problems (echo controllers)

8.1.6.1   Echo problems

As previously stated, geostationary satellites are situated at an altitude of 36 000 km. Thus, when a speaker has finished his conversation, he has to wait a minimum of 600 ms for an answer. In practice, it is not the propagation delay itself which poses a problem for a telephone conversation, as the speaker becomes used to the delay, but rather its combination with the echo phenomenon. Of course, echo problems are lessened in the case of lower orbit satellites.

Echo has always existed in telephone circuits. Taking the simplified case of the link in Figure 8.12 between the client and the exchange, only two wires are generally used for the two transmission directions in order to reduce the cost of the link. However, it is usually necessary to separate the two directions of transmission between exchanges separated by large distances.

In order to carry out this operation, a differential (or hybrid) coupler is used i.e. a network with four ports a, b, c and d (see Figure 8.12) so that all signals coming from port a reappear entirely at port c and all signals coming from port d reappear at port a. This is obtained when the impedances at ports a and b are balanced. If this condition is not achieved, a part of the signal coming from port d is found at port c and is re-emitted to its origin where it appears as an echo.

In practice, the impedance observed in port a varies according to the 2-wire circuit that connects the exchange to the link. Since it is not possible to design a balancing network valid for all connections, balancing is provided for an average link. Thus, an echo will always exist for a long-distance link, the volume of which depends on the degree of imbalance of the differential coupler. When the length of the link is less than 2 500 km, the propagation delay is sufficiently short so that the person speaking does not realize that an echo exists. However, when an echo exists with a large delay, as in the case of a satellite link, it disturbs the speaker so much that conversation becomes practically impossible.

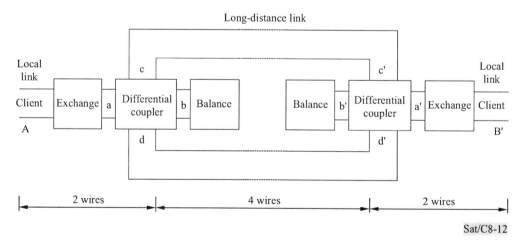

FIGURE 8.12

Simplified diagram of a long-distance link

This defect is eliminated in satellite links by using special devices. Two types of devices are available: the so-called echo suppressors, which have been used for many years on most satellite links, and the more recently developed echo cancellers.

Echo suppressors and echo cancellers are connected to the telephone interface in a circuit-per-circuit basis. They are often located, not at the earth station itself, but in the framework of the switching exchange facilities. This is because it may be necessary to disable the echo suppression (or echo cancellation) function during telephone signalling phases. This disabling operation may be facilitated if the echo suppression (cancellation), the signalling and the switching equipment are co-located. A more detailed description of echo suppressors and echo cancellers is given below.

a)   Echo suppressor

The principle of echo suppressor operation is shown in Figure 8.13. When B speaks, the echo suppressor nearest to A detects the speech on the receiving channel and blocks transmission on the emission channel to prevent the return of the echo. However, it often happens that, during a conversation, one speaker interrupts another, and for a short instant, double talk may be heard. To prevent both transmissions from being simultaneously blocked in these cases, a decision circuit detects this situation and, instead of cutting emission, inserts an attenuation line of 6 dB on reception. Therefore, the direct speech signal is attenuated by 6 dB but the echo passing through two lines is attenuated by 12 dB.

In most cases, echo suppressors based on this principle are satisfactory. However in certain extreme cases, conversation may be troublesome. In particular, it is possible that the sound level of one of the speakers is such that his echo returns at a higher volume than the speech of his correspondent. In these cases it is difficult to distinguish single talk from double talk. Untimely cuts in speech from one of the speakers may result from this effect and momentarily make conversation impossible. Other situations may also degrade the quality of echo suppression. These difficulties are aggravated by the length of propagation delay which is why links using more than one satellite are avoided as far as possible. Moreover, two pairs of echo suppressors in cascade may cause unacceptable clipping of speech signals (this would be the case of an undersea cable link prolonged by a satellite link), and in this case the intermediate pair of echo suppressors must then be inhibited.

However, much progress has recently been made in the technique of echo suppressing, thereby overcoming most of the problems listed above, and high-quality equipment is now available. Modern echo suppressors must comply with ITU-T Recommendation G.164. This recommendation improves the operating conditions significantly when double talk occurs i.e. in the case of the "break-in" of the talker on the near-end side (A), while the far-end side (B) is normally talking. Two conditions are now defined: partial break-in and full break-in. The blockage in the emitter path is first removed and the "on" position is maintained during a short duration (short hangover time < 26 ms): this is the partial break-in condition. If the double talk condition is confirmed (full break-in) the "on" position is maintained with the normal hangover time (48 to 66 ms) and the 6 dB loss is introduced. In this way, the clipping of the first syllables in case of break-in is avoided.

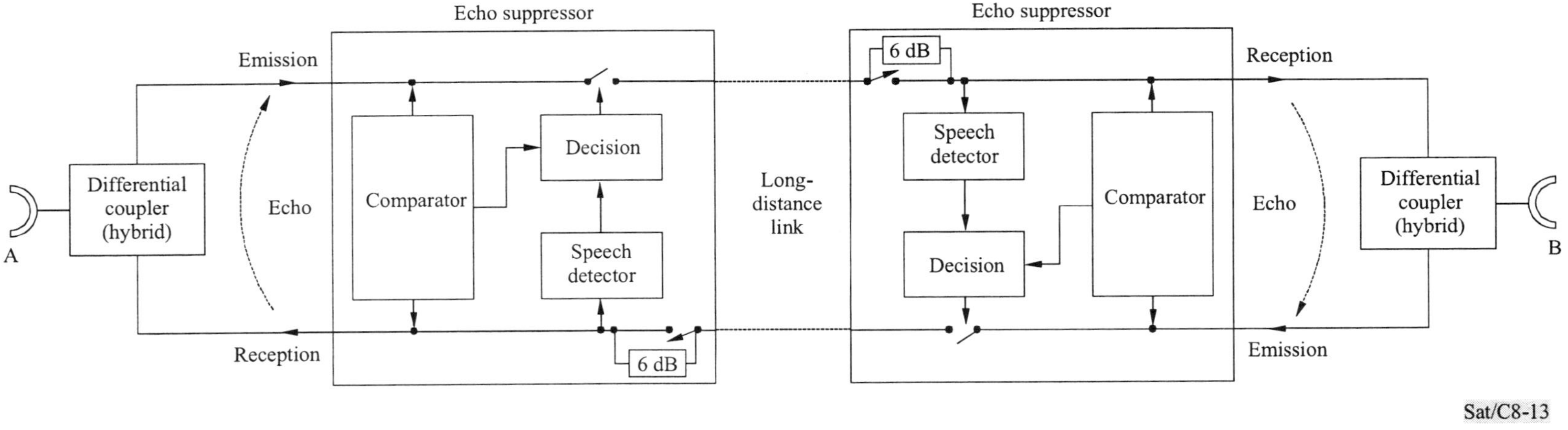

Sat/C8-13

FIGURE 8.13

Principle of a link with echo suppressor

Furthermore, ITU-T Recommendation G.164 proposes that the decision for removing the blockage condition in the emitter path and for introducing a 6 dB loss should be controlled by an adaptive process. This process is based on the comparison between the emission and reception level taking into account the actual echo transfer loss.

It must be emphasized that the actual performance of an echo suppressor can only be measured subjectively by statistical tests between users under variable operational conditions, the user being asked to evaluate the grade of quality of the telephone conversation.

ITU-T Recommendation G.164 also classifies echo suppressors into four types, depending on whether or not analogue or digital links and analogue or digital echo suppression circuits are used:

- type A: echo suppressors of this type interface with analogue telephone channels and are fully realized in analogue techniques. In general, they only comply with the previous ITU-T Recommendation G.161. They should be considered as obsolete for use in new earth stations;

- type B echo suppressors also interface with analogue telephone channels. Logic operations are digitally processed after signal sampling. Blocking and attenuation circuits are analogue. This type of echo suppresser can be made very compact by using multiplexed digital logic circuits. For example, 30 echo suppressors can be contained in a standard 19 inch shelf;

- type C echo suppressors interface with digital telephone channels and are based entirely on digital circuits. They are usually used in multiplexed units interfacing with conventional digital multiplex groups (30 or 24 circuits);

- type D is similar to type C, but codecs are incorporated in the equipment in order to allow interfacing with analogue telephone channels.

It should also be noted that echo suppression can be obtained from functions which are normally included in other types of terrestrial interface equipment. This may be the case, in particular, for speech interpolation equipment (e.g. DSI units). This may also be the case for SCPC telephony modems, which normally include a voice activation device. In these cases, the use of separate echo suppressors in the switching exchange can be avoided if some supplementary conditions are met: i) the "short echo path", i.e. the path between the echo suppressor and the final user, has a delay of less than 24 µs; ii) the availability of the telephone circuits should be maintained in case of failure of the equipment which includes the echo suppression function, and iii) the echo suppression disabling connection (for signalling and/or data transmission purposes) should continue to be able to be made. In cases where these conditions cannot be met, separate echo suppressors should be provided at the switching exchange and the echo suppression function should be disabled in the interface equipment.

b) The echo canceller

The disadvantages discussed above have given rise to much research on a device theoretically capable of functioning correctly under all conditions, including double talk (the echo canceller). Positioned in the same place as an echo suppressor, it synthesizes a replica of the echo and subtracts it from the returned signal. Several algorithms with a number of refinements have been studied for the cancelling process, but the basic principle in every case is the following:

If the transmission channel over which the echo travels is linear and invariable in time, where $x(t)$ represents the signal received and $y(t)$ the resulting echo, the following relation is obtained:

$$y(t) = \sum_0^\infty h_j \; x(t - \tau_j)$$

where the coefficients h_j are the samples taken at time intervals T of the impulse response of the four terminal networks constituting the echo path.

In practice, the impulse response has finite duration and it is thus sufficient to consider a finite number N of samples large enough to cover the whole duration of the response.

The echo canceller synthesizes a replica $y(t)$ of $y(t)$ represented by the expression:

$$\hat{y}(t) = \sum_0^\infty c_j \; x(t - \tau_j)$$

The calculation of the c_j coefficients is carried out continuously according to a reiterative procedure so that the absolute value of $e(t) = y(t) - \hat{y}(t)$ is minimal. At each period of sampling k, the c_j coefficients are corrected by the addition of the value:

$$\Delta c_j - f(e_k) \cdot g(x_k^{\,j})$$

where e_k is the kth sample of $e(t)$ and $x_k^{\,j}$ is the jth sample of $x(t)$ considered during the kth sampling period. The functions f and g are non-decreasing odd functions. Their choice results from a compromise between the need to reduce the convergence time of the process and the necessity of not overcomplicating the equipment.

When there is double talk, the correction of the c_j coefficients is stopped and they keep constant values equal to those they had just before the period of double talk.

The number of samples N depends on the duration of the impulse response of the echo path. An example of such a response is shown in Figure 8.14. It is characterized by an initial delay T_o (e.g. 15 ms), and a period of activity T_A which can last up to 8 ms. Consequently, to obtain a good approximation of this response, it suffices that the number of samples covers at least 23 ms. In practice, 30 ms are taken.

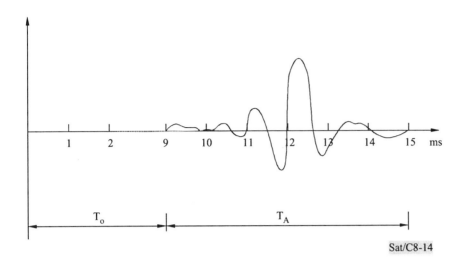

Sat/C8-14

FIGURE 8.14

Example of the impulse response of an echo path

Figure 8.15 shows an example of an echo canceller in which the algorithm is given in digital form. The signals x(t) and y(t) are sampled in PCM at 8 kHz. A duration of 30 ms will then need 240 c_j coefficients.

- 240 successive samples are stocked in a closed-circuit shift register where they circulate at the rate of one complete cycle per sampling period of 125 μs. At the end of a period, a new sample is introduced, replacing the previous one which will then disappear;

- the 240 c_j coefficients circulate in another closed-circuit shift register with an adder incorporated in the loop;

- the adder's other input receives the quantity $\Delta c_j = f(e_k) \cdot g(x_k^{\,j})$ with which the c_j coefficients are corrected;

- at each step j of the registers, a multiplier calculates the product $c_j \cdot x_k^{\,j}$ and the result is summed up in an accumulator over the 240 values obtained during the kth sampling period, which gives:

- a subtracter carries out the operation $\hat{y}_k - y_k = e_k$.

Echo cancellers are the subject of ITU-T Recommendation G.164. In practice, echo cancellers have now replaced echo suppressors in modern equipment.

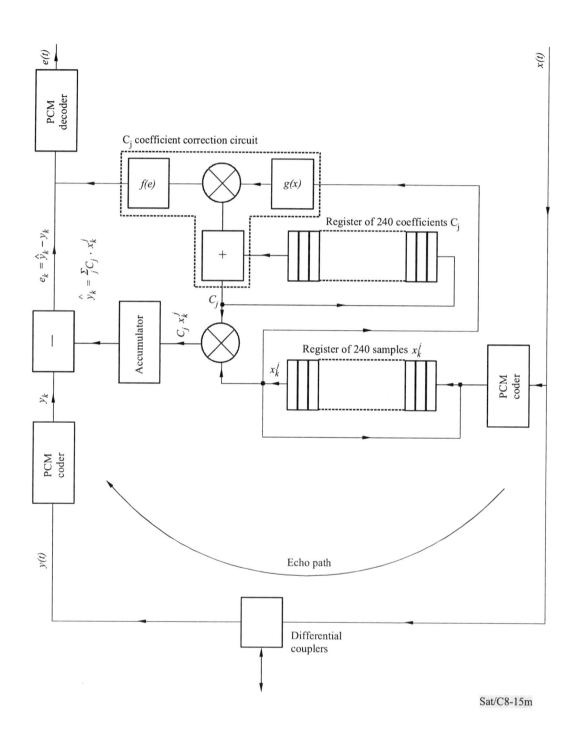

FIGURE 8.15

Diagram of a digital echo canceller

8.1.7 Signalling transmission aspects

8.1.7.1 General

Signalling is the exchange of information (other than by speech) specifically concerned with the establishment, release and other control of calls, and network management, in automatic telecommunications operation. Some basic information of signalling systems are given below in Appendix 8.1.

Signalling, which is necessary to establish communication, is also affected by propagation delay. In particular, this is the case for compelled signalling systems. In these systems, the emitter sends a continuous signal until it receives an acknowledgement signal from the far receiver indicating to it to stop emitting. Therefore, the duration of the signal corresponds to the total return path length delay of the satellite link, approximately 600 ms. To reduce the waiting time after dialling and the occupation time of circuits, it is important that signalling systems used in satellite links make the least possible use of this type of signalling.

For systems which use voice channels for the transmission of signalling, care must be taken that the echo suppressors do not disturb signalling. It is possible that during the signalling phases, signals could be emitted simultaneously from both ends. This is equivalent to double talk and a 6 dB loss is applied to the emission by the echo suppressor. In certain cases the signal could then be under the signal detection threshold on reception, which would entail the loss of the corresponding signal and therefore the impossibility of establishing the communication.

The echo suppressor is generally placed between the switching exchange and the signalling emitter-receiver, so that it has no action on signalling signals. When this arrangement cannot be made, the echo suppresser must be neutralized by command from the switching exchange during the signalling phase.

When speech interpolation devices are used, signalling systems using speech channels must not be affected by the clipping of signals caused by these devices. Indeed, it is of no advantage to use a traffic concentration method which gives good quality of speech if the resulting clipping speed entails a rate of loss of signalling that is too high.

8.1.7.2 Signalling, analogue

Two main analogue signalling systems have been standardized and are applicable within an FSS satellite network:

- Signalling System No. 5, standardized by ITU-T in 1964;

- Signalling System R2, standardized by ITU-T in 1968.

ITU-T No. 5 code is the signalling code most widely used in international satellite links. This is a system using voice channels which was specially designed to operate with satellite links and with speech interpolation devices. However, the corresponding signalling equipment is relatively expensive and the time to establish communications is fairly long.

The ITU-T R2 code (already used extensively in terrestrial links) is being used in the EUTELSAT system. Although this code was not initially designed to be compatible with satellite links or with speech interpolation, modifications have been made to adapt it to the new transmission conditions.

8.1.7.3   Signalling, digital

Two main digital signalling systems have been standardized and are applicable within an FSS satellite network:

- Signalling System No. 6, standardized by ITU-T in 1968;

- Signalling System No. 7, standardized by ITU-T in 1980.

Signalling Systems Nos. 6 and 7 are common channel signalling systems. This signalling method uses separate channel(s) to transmit signalling information related to circuits.

Common channel signalling is a signalling method in which a separate channel conveys, by means of labelled messages, signalling information relating to a multiplicity of circuits, or other information such as that used for network management. Common channel signalling can be regarded as a form of data communication that is specialized for various types of signalling and information transfer between processors in telecommunication networks.

The signalling system uses signalling links for transfer of signalling messages between exchanges or other nodes in the telecommunications network served by the system. Arrangements are provided to ensure reliable transfer of signalling information in the presence of transmission disturbance or network failures. These include error detection and correction on each signalling link. The system is normally applied with redundancy of signalling links and it includes functions for automatic diversion of signalling traffic to an alternative path in case of link failures.

The main advantages of the common channel signalling method are:

- signalling is completely separate from speech. Therefore, the likelihood of misoperation due to imitation by speech signals is greatly reduced;

- the exchange of information between nodes is not subject to physical circuit seizure;

- signalling is significantly faster, therefore speech circuit utilization is maximized;

- there is potential for a very large number of signals;

- it is possible to handle management and administration signals during speech;

- there is flexibility to add or change signals (without requiring changes to hardware);

- it is possible to carry user-to-user or access-to-access information in case of ISDN service.

8.1.8   Problems of data transmission

In a data transmission link using a normal telephone channel, both transmission directions are often used simultaneously. Disturbances to transmission caused by untimely cuts or insertions of losses due to echo suppressors must therefore be avoided. In order to prevent this, a device called a tone disabler is added which inhibits the function of echo suppression when data transmission takes place. This is done by detecting the emission of a 2 100 Hz tone by the data transmission terminal before the beginning of transmission.

In other respects, propagation delay is also a cause of operational difficulties: in principle, all messages must arrive at the receiving end with no error. When errors are caused by the link they must be detected by the receiver and the emitter must re-transmit the erroneous part of the message. It is clear that the necessity of repeating these erroneous parts entails a loss of transmission efficiency which increases in proportion with the bit error ratio.

Numerous operational procedures have been developed for terrestrial links to take this factor into account, but they are not all suitable for satellite transmission. The most widely used, at present, consists in the transmission of data in blocks. After sending each block, the emitter waits for an acknowledgement of receipt. If this is positive, the next block is sent, if negative, the block is repeated. It can be seen that this technique can be very inefficient if propagation delay is long compared to the duration of the block. On the other hand, the emission of very long blocks would require very large memory capacities on transmission and would entail very high repetition times since a single error means the repetition of the whole block.

This difficulty is reduced by lowering the bit error ratio as far as possible by the use of error correction codes (Appendix 3.2). Moreover, operational procedures are being studied to permit the best possible transmission efficiency to be obtained for a given bit error ratio. These generally consist in the continuous transmission of blocks, i.e. without waiting for an acknowledgement of reception and repeating erroneous blocks on request.

8.1.9   Facsimile transmission

8.1.9.1   General

Facsimile (also simply called "fax") is the largest non-voice service operating on the world's public switched telephone network PSTN and this is particularly significant to satellite communications. This gives it the maximum geographic reach with the lowest call charges. There are a large number of suppliers producing terminal equipment so it is difficult to be precise about the current total number of dedicated facsimile terminals but it is believed to be over 50 million. In addition there are a large number of personal computers which can now emulate facsimile terminals.

Facsimile is a high resolution raster scanning image transfer system that uses PSTN modem technology for its communications. It was patented in 1843 by a Mr Alexander Bain as a development of the master-slave-clock time distribution system. Thus it predates the telephone by several years.

In 1850 the cylinder and screw scanning mechanism improved the synchronization significantly and in 1902 the photo-electric sensor was incorporated which allowed plain paper originals to be used instead of metallic type. By 1910 there was an international photo-telegraphy service between London, Paris and Berlin.

The size of the service stayed small until international standards were established via ITU-T for the terminal equipment in the late 1960s which opened up the possibility of a large volume market. The first ITU-T Recommendation for facsimile terminal equipment was agreed in 1968. This was called group 1 facsimile as new developments were already visible (see Table 8.2).

Group 1 facsimile was an analogue system that employed frequency modulation for its communications and could transfer a grey-scale image on one page at a resolution of 96 lines per inch in 6 minutes.

Group 2 facsimile was also an analogue system but it employed an amplitude modulated vestigial sideband communication system which improved the transmission time to 3 minutes a page using the same resolution. The group 2 recommendation was agreed in 1976.

Both of these analogue systems suffered from echoes in the PSTN which produced visible defects on the received image.

8.1.9.2   Group 3 and group 4 facsimile

Group 3 facsimile was the first all digital terminal equipment system and it employed a data modem to communicate the bits across the telephone network. This avoided the problem of echoes appearing on the received image and it allowed image compression techniques to be applied to the quantized black and white image at the transmitter. This combination of techniques reduced the page transmission time to 30 seconds per page when it was operated over a connection that could support a modem speed of 9 600 bits per second. A double scanning resolution option was also introduced to improve the quality of the image.

In 1984 a new group of facsimile equipment was standardized for the Integrated Service Digital Network ISDN. It was called group 4 and was intended for 64 kbps operation, the ISDN basic rate, although the speed was not actually limited to this figure as the technology could have been used over 12-channel-group-band modems or high speed private data circuits.

Group 4 facsimile equipment had more and higher resolutions than group 3 for improved image quality, 200, 240, 300 and 400 lines per inch in both dimensions and it had a better, page based instead of scan line based, compression algorithm plus a seven layer error correcting protocol. Most group 4 terminals employed plain paper laser printers to ensure the highest quality of received image.

The page transmission time was a nominal 5 seconds depending upon the amount of information needed to represent a page after a page had been through the compression process.

TABLE 8.2

Facsimile characteristics

Facsimile group	Transmission time per page	Communication system	Date of standardization
1	6 minutes	FM	1968
2	3 minutes	VSB-AM	1976
3	30 seconds	V.29 + V.27*ter* modem technology	1980
4	4 seconds	baseband digital 64 kbps	1984

The geographical penetration of the ISDN was much less than that of the PSTN so the major market place remained with the group 3 equipment which thanks to the digital nature of its operation had been rapidly converted into solid state and thereby achieved a very low cost of production. This continued interest in PSTN facsimile has lead to the group 3 standard being enhanced with most of the group 4 features, i.e. higher resolutions, better compression algorithm and plain paper printers but not the heavyweight protocol.

8.1.9.3   Group 3 protocol

The group 3 protocol has to deal with the analogue characteristics of the worldwide PSTN which is still an analogue network end to end even if certain parts are digitized and it therefore includes echoes.

The signal processing parts of the modem function deals with echoes of short duration, of the order of a few tens of milliseconds but the protocol has to operate correctly in the face of echoes with duration up to 1.5 seconds and possibly larger amplitudes than the wanted signal.

The basic operation of the protocol is defined in the T-series of ITU recommendations and is illustrated in Figure 8.16 which shows the uncorrupted transmission of one page.

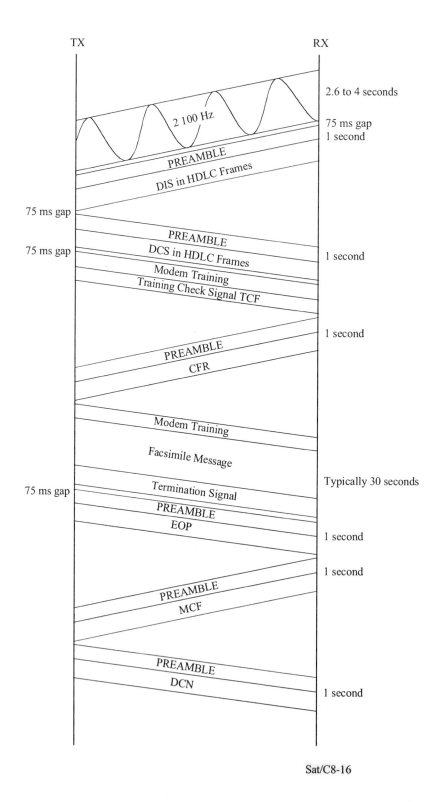

Sat/C8-16

FIGURE 8.16

Group 3 facsimile protocol operation

for a single transmitted page

a) **Link establishment**

Silence is the first signal on the line after the connection has been set up. This lasts for 1.8 to 2.5 seconds and then a 2 100 Hz answering tone is transmitted from the answering terminal. This is known as the called station identification signal CED and is mandatory in most countries to indicate to human callers that they have made contact with a non-voice terminal. It also disables any echo suppressors or echo cancellers in the circuit and may enable data-bypass or facsimile demod-remod operation in any DCME that might have been included in the connection.

The answering terminal also sends the next signal to line. This is the preamble signal which is a 1 second stream of HDLC flags at 300 bits per second modulated on to the upper band of a V.21 modem signal. This provides octet synchronization and prepares the calling station to receive the next signal which is the digital identification signal (DIS) again transmitted from the answering terminal. Thus all the initial signals are from the answering station which saves time as the 2 100 Hz signal has to be sent from the answering station first.

The digital identification signal (DIS) is also a 300 bps upper V.21 channel signal and it contains a menu of the capabilities of the answering terminal. This menu is coded in a very compact format with each bit representing the availability or otherwise of an item on a standard list of features and functions.

The preamble and DIS are repeated at 3 second intervals until a signal is received from the calling station. The line signal operates in a half duplex mode using only the upper channel of a V.21 300 bps modem signal.

V.21 is normally a 300 bps FM duplex modem with frequency separation of the transmit and received signals.

The first signal from the calling station is a preamble followed by a digital command signal (DCS) which is also a 300 bps signal. This signal contains a selection from the menu offered in the DIS signal using the same compact bit coded selection format as the DIS.

Thus at the end of this exchange of short 300 bps signals the two terminals know what parameters, and their values, will be used for the rest of the communication. The only outstanding issue at this point is the selection of the high speed modem operating speed and this depends upon the quality of the channel.

b) **High speed image transmission modem selection**

The highest speed modem available at both ends has been agreed in the initial handshake and the transparency of the channel is established by the simple process of trying each available modem speed in descending order until a satisfactory error rate is discovered. This process is illustrated in Figure 8.16 with the signal exchanges labelled "modem training", the high speed modem training pattern, followed by a training check format (TCF) which is a known pattern with a duration of 1.5 seconds. The positive response is a confirmation to receive signal (CFR), a negative response would be failure to train (FTT). These are 300 bps signals as they are very short and high speed operation is only required in the opposite direction for half duplex facsimile traffic. A CFR indicates that only a low number of errors had been detected in the TCF pattern. The exact number of errors that are acceptable is a variable that is selectable at the receiver.

c)   Page transmission

Facsimile is a page-based system. The page sizes include the Japanese, European and American standard paper dimensions.

Once CFR has been received at the calling station the transmission of the page image may commence. A modem training signal precedes the page image traffic to make sure that the modem is still adjusted to its optimum setting.

In the original version of group 3 facsimile, the page transmission system did not include any error correction and the run-length coding algorithm was terminated at the end of each scan line. Later versions included selective repeat error correction and compression algorithms which encompassed the whole page.

After the page image has been sent there is a page closing signal to inform the receiver that the next signal from the transmitter will be a 300 bps V.21 signal. The 300 bps signal is an end of procedure (EOP) signal which indicates that there are no more pages awaiting transmission.

The receiving station confirms reception with a 300 bps message confirmation (MCF) signal and the transmitting station closes the protocol with a disconnection DCN signal.

This is normally followed by a disconnection of the connection.

If a multiple page transmission is required then there is a different command sent from the transmitter after each page at 300 bps, the multipage signal MPS. The same response signal MCF is sent from the receiver and the closing sequence is the same as that shown in Figure 8.16.

d)   Unique employment of modem technology

From the previous description of the operation of the facsimile protocol it should be noted that this service employs modem technology in a rather unique manner, i.e. it uses more than one modem signal type during a single connection. This has led to problems with the design of low rate encoding algorithms in the past and it seems to be a little known feature of facsimile traffic that might be a cause of future problems in the development of new systems if it is not more widely appreciated.

e)   Terminal identification

Terminal identities may be exchanged at the start of the protocol by sending a called subscriber identification (CSI) frame, from the answering station, immediately before the DIS and a calling subscriber identification (CIG) frame, from the calling station, immediately before the DCS. Note that the bit sequences in the octets representing the addresses are in the reverse order in these identification frames.

8.1.9.4   Non-standard capabilities

Provision was made in the protocol to permit proprietary features to be negotiated between terminals produced by the same manufacturer. This negotiation is realized by the exchange of two 300 bps frames, the non-standard facilities (NSF) frame and the non-standard setup (NSS) frame which are positioned immediately in front of the identification frames.

The ability to add some proprietary features to individual manufacturers products and consequently to be able to compete on features other than raw cost has been a great incentive for manufacturers to enter the facsimile terminal market.

Optional standard facilities

Some of the optional facilities of facsimile terminals are not well known and are not implemented on all terminals, for example, polling. This is a mode of operation where a document may be left in the input hopper of a facsimile machine and be sent to a caller only when an incoming connection is established by that caller. Other facsimile messages may be left as normal. This feature is useful for the unattended distribution of the same document to many callers some of whom may be mobile and not have their own facsimile machine.

A recent extension of the polling feature has been the introduction of "selective polling" where a 20 digit number can be added to the polling handshake to identify individual documents from a large library/database. Another mode of operation that exploits the polling function has been developed in the last few years with the incorporation of DTMF signals to select documents from a database after an initial poll has recovered a catalogue of the documents held in the database. This has proved to be commercially popular as a means of distributing sales literature with minimal personnel support, e.g. for out-of-hours access.

Other new features are:

- sub-addressing capability: to identify individual terminals on a LAN, for example;

- password exchange: for extra security, usually coupled with some inhibition on the printing of a received message until the correct password has been entered at the receiver;

- encryption: of the line signal and/or to verify the identity of the author of a document. This feature is still under study.

8.1.9.5   Incorporation of faster modem technology

As modem technology has developed and the quality of the PSTN has improved as a transmission medium for non-voice signals, the speed achievable by modem technology has stepped up from 9.6 to 14.4 kbit/s then 19.2 kbit/s and now up to 28.8 kbit/s and it is still increasing with V.34 promising speeds above 30 kbps. This new technology has been incorporated into facsimile terminals as soon as the chips have become available and ITU-T Recommendations have just been enhanced to support these higher speed options.

V.34 modems also incorporate a useful terminal selection feature, defined more fully in V.8, which supports the separate identification of modems from facsimile machines. This is a more sophisticated approach to the general problem of terminal selection.

8.1.9.6   Terminal selection and standardization aspects

A facsimile machine is only one of a multitude of items which may be connected to a single PSTN line and there are various schemes for selecting/identifying the desired terminal when an incoming call is received. For example, the calling tone of facsimile machines is different to that of modems so the simplest scheme is to place the answering machine on-line first to give its announcement then if a calling tone burst is received during the announcement the correct non-voice terminal can immediately be connected and the announcement suppressed. This works best if all the terminal functions are in one box or at least under the control of one intelligence.

There is a similar terminal selection problem in the ISDN but there are more network functions available to "solve" the problem in this environment, e.g. sub-addressing, MSN, DDI, bearer capability indication, higher layer compatibility indicator and, not least, the interworking indicator. All these terminal selection features allow a call to be rejected before it is answered and thus saves the caller the expense of a call charge. However, it is not really an ideal solution because it prevents the caller, who wishes to communicate something, from making any adjustments to his requirements to match the facilities at the receiving terminal/s, e.g. a fall-back to facsimile transmission will solve most document transmission requirements.

Many facsimile machines can be programmed to send documents at some given time of day, say, when the call tariffs are decreased after office hours. This raises the possibility of calls being made by mistake to normal telephones and there is usually a connection approval requirement which insists that calls are abandoned after 60 seconds if they have not been answered by another facsimile machine.

Unfortunately 60 seconds is not long enough to reach all the facsimile machines in the world, especially if they are fitted behind terminal selection boxes which insist on sending out their full answering machine announcement before looking for calling tone. The worst-case post-dialling delay through the international telephone network is not known although the issue is under study. A figure of 90 seconds is the best guess at the present time.

Satellite transmission usually receives most of the blame for such delays.

From the preceding discussion it is apparent that an international standard cannot dictate the length of a timer for post-dialling delay but implementors should be aware that the first timer mentioned in the recommendation, T1 = 35 seconds, is not a post-dialling delay timer!

Another problem which is difficult to cover in a standard is a method for dealing with echoes. Satellite transmission systems also receive most of the blame for long duration echoes. They are particularly awkward to handle when they occur at a change of modem speed and it is an implementation matter as to how much information is available from the signal processing chips at such a boundary.

Therefore the whole matter is left for the implementor and it is not mentioned in the text of the standard. This lack of text has misled to some implementors into not realizing that there was a problem that needed a solution.

Terminal testing

A set of tests for group 3 facsimile terminal equipment may be found in the European standards ETS on group 3 facsimile.

Voice plus fax

Studies are in progress on various schemes for the simultaneous sharing of one PSTN connection between a voice signal and a modem or facsimile signal.

8.2 General considerations on protocols and terrestrial interfaces

Section 8.1 above dealt with the interconnection of satellite links with conventional telephony networks (PSTNs). However, many new applications of satellite communications, are implemented in the framework of autonomous satellite systems and imply the connection of satellite links or networks either directly with end users or with private communication networks (or, more generally with so-called "closed user groups (CUGs)". Such applications include, *inter alia*: private data networks (PvtDNs), often used for dedicated business services, rural communications networks, etc. These applications generally make used of small, low-cost earth stations, often called "VSATs" (an acronym standing for: very small aperture terminal, i.e. an earth station with a small antenna). A very important feature of such systems is that the earth station design comprises, not only the earth station hardware, but also the complete software for allowing end-to-end (user-to-user) operation, including protocols and interfacing functions.

In consequence, § 8.2 and the following § 8.3 and 8.4 are devoted to the general principles which are the basis of this software design. This is also the reason why the terms "VSAT" and "VSAT network" are very commonly used in these sections although the principles are more generally applicable in any satellite systems with similar interfaces[1].

8.2.1   VSAT network protocol model

Packet-oriented communications are generally used within VSAT networks. In packet-data communications, information is transmitted by grouping the data into packets. However, the fact that VSAT networks operate internally in the packet mode does not limit users to packet mode communications because packetization functions can be performed in the user interface units of the VSAT terminals.

In data communications, open systems communicate with each other with communication functions which are separated into layers. The International Organization of Standardization (ISO), in cooperation with the ITU Telecommunication Standardization Sector (ITU-T), developed the Open System Interconnection (OSI) protocol reference model, which consists of seven layers [Ref. 1]. The upper four layers contain end-to-end communication protocols between the communicating systems. The three lowest layers contain network, and network interface, protocols for virtually error-free transmission of users' data packets across networks. Packet switched data networks use communication protocols in these three layers for transferring the user data across the network and providing service to the four upper layers which contain end-to-end protocols [Ref. 9].

- The physical layer (layer 1) is the lowest level in the OSI model, which contains physical characteristics and specifications of connections for bit level transmission through the network and across network interfaces.

- The data link layer (layer 2) comprises communication procedures and protocols between end user terminals and the network, or within networks between two nodes. These protocols

---

[1] In fact, because of their general significance, § 8.2, 8.3 and 8.4 are reproduced from "VSAT systems and earth stations" (ITU-R, Geneva, 1995, Supplement 3 to Edition 2 of the Handbook on Satellite Communications").

typically perform error detection and correction of framed data packets. If errors cannot be corrected, an error message is sent to layer 3. These protocols may also have capabilities for addressing and flow control of the data. Layer 2 also provides the synchronization between the user terminal and the network.

- The network layer (layer 3) establishes, maintains and terminates data connections through the network. In layer 3 the data packets are provided addressing information for routing through the network, errors are corrected and the flow of the packets is controlled. Long data packets can be split up and reassembled.

VSAT network packet mode communications involves functions and protocols which generally belong to these three lower OSI layers, used both internally in the network, as well as across its external network interfaces.

VSAT networks are used mainly as independent private data networks, interconnecting a number of user data terminals (or terminal groups) which interface with remote VSATs, and host computers which interface with the VSAT network hub. More recently, VSAT networks are also being used to connect remote VSAT users to terrestrial-based (private and public) data networks and possibly to ISDN in the future. Interconnection is either through the hub, or another VSAT.

8.2.2 Internal VSAT network architecture and protocol implementation

8.2.2.1 General

In terms of communication protocols and procedures, a VSAT network can be divided into a network kernel and network interfaces as shown in Figure 8.17.

Network interfaces are situated at the extremities of the network through which VSAT network users are connected with the VSAT network. A network interface is also provided at the network hub, connected to a host computer or alternatively to other terrestrial-based networks. Each VSAT network interface can be configured to support one of several different user interface types independently of other VSAT network interfaces. The network interfaces rely on the network kernel to provide a certain level of service.

8.2.2.2 Network kernel

The VSAT network kernel has its own communication structures and protocols to transfer data across the satellite transmission medium in the most efficient way. The network kernel provides either reliable delivery of data, or an indication that data has been lost either due to multiple errors or equipment failure. The network kernel includes a set of functions:

- satellite access protocol(s);
- packet addressing mechanism;
- satellite channels congestion control procedures;
- packet routing and switching;
- network management.

Network management functions are used for network configuration and operation, e.g. to alert a network operator to exceptions in the user interface such as unexpected link disconnection or excessive retransmissions.

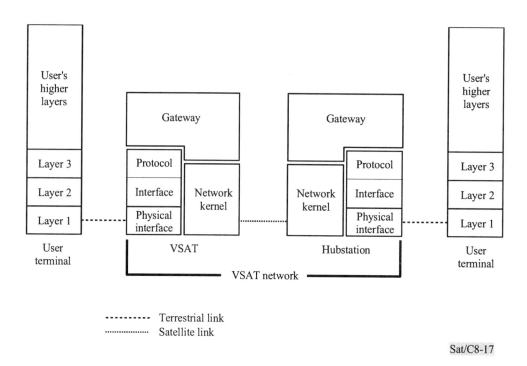

FIGURE 8.17

Protocol architecture of a VSAT network

8.2.2.3　System architecture and satellite access protocols

Most business application oriented VSAT networks use a star architecture (see Chapter 5, § 5.6.2 (iii)) implementing a central earth station (the "hub") and small remote user earth stations (the VSATs proper). In such networks, the satellite access protocol is generally asymmetric. There are a few common forms of VSAT-to-hub access in use, i.e. Slotted Aloha (see § 5.3.6), TDMA or reservation TDMA. In the hub-to-VSAT direction the mode of access is generally TDM.

8.2.2.4　Intranetwork data communication protocols

Above the access protocol, point-to-point, or point-to-multipoint protocols may be used for reliable communications through the network, which include error recovery and data flow control. These are intranetwork communication protocols specifically designed for satellite transmissions in the VSAT network [Ref. 2]. Factors that need to be taken into account in the design of internal VSAT network protocols include the special characteristics of the network, such as the star topology of VSAT networks as well as the multiple access method, which have an impact on data throughput and call setup time performance of the VSAT network.

The packetized information is structured in formats containing error detection codes used to either acknowledge correct reception, or to reject erroneously received packets and to request retransmission. In VSAT networks using TDMA/RA for VSATs-to-hub transmission of data packets, the acknowledgement process and packet retransmissions are under the control of the VSAT network management software.

In general, the satellite link BER should be low enough to avoid too frequent repetitions of messages. For example, a 10^{-7} BER corresponding to a packet error rate of about 10^{-4} (for a packet size of 128 bytes) can be considered low enough to avoid network congestion due to retransmissions. Above this layer, the messages are practically error-free. If a VSAT network does not implement layer 2 gateway-to-gateway protocol, and only provides interfaces between the internal satellite access procedures and standard user protocols, the user's communications at OSI layer 3 are supported transparently through the VSAT network. Users' protocols would have to be adapted to the particular features of satellite transmissions.

If end-to-end error recovery mechanisms in the higher layers are used, e.g. application layer protocols between the host and terminals, a very low information throughput could result since erroneous data would be repeated only after a very long delay. Without error correction measures at the lower layers, the satellite link BER would have to be much lower, e.g. 10^{-10} to ensure acceptable throughput performance.

8.2.2.5 Packet switching function

VSAT networks with a star configuration are essentially packet switching networks with a central packet switch performing routing and relay functions. Packet switching functions are carried out by the baseband processing and control equipment in the VSAT and hub earth stations.

There are two fundamental packet switching mechanisms: datagram and virtual circuit. With datagrams, the packets are delivered with certain limited reliability (with low but finite probability of out-of-sequence or duplicate packet delivery). Virtual circuits guarantee sequential delivery of packets without duplication. In a VSAT network, each mechanism offers advantages and disadvantages. Virtual circuit switches require less overhead on a per packet basis, but the need to maintain state information on each connection in a network may be severe in a large VSAT network with potentially a large number of connections to be supported. At the cost of a higher overhead per packet, a datagram-based switch can provide higher throughput and offer a significant additional advantage: the ability to be restarted without requiring re-establishment of network connections [Ref. 3, 4].

The internal design of a packet switching network can be considered as a problem of internetworking between switching systems and processing elements. Thus OSI network layer protocols can be used as an architectural basis for the internal structure of a VSAT packet switching network.

The packet switch function forms the heart of the network kernel. Since many VSAT network functions are similar to those found in terrestrial packet switching networks, some networks use a modified packet switch to provide X.25 PSDN functions as a VSAT packet switch. Other VSAT networks may use proprietary protocols to perform the same function. The principle functions required in a VSAT packet switch are:

- satellite multiple access control;

- reliable link level transfer to/from VSATs (also a local host if there is no protocol gateway at the hub);

- routing of data between VSATs and the host;

- connection to the network management system;

- connection to other networks.

VSAT packet switches have some unique requirements not found in traditional X.25 packet switches:

i)   The star topology of most VSAT networks requires that all traffic pass through a single switch, whereas in typical terrestrial PSDNs a number of packet switches form a complete meshed network with each switch carrying only a portion of the total traffic. With only a single switch in the network, a single VSAT switch may need to provide in excess of 1 000 packets/s in networks with hundreds of VSATs carrying typical interactive traffic.

ii) High data rates are required in most VSAT networks which typically operate with satellite channel data rates of greater than 56 kbit/s. Outroute (hub-to-VSAT) speeds are typically 128, 256, 512 kbit/s or higher, which are not usually encountered in terrestrial packet switch designs.

iii) Most PSDN designs provide for alternate routes between switches so that the failure of a single switch will not interrupt a significant portion of network traffic. VSAT networks typically depend on the operation of a single switch to carry traffic between hundreds or thousands of nodes. Redundancy considerations, therefore can be crucial to the design of a VSAT switch [Ref. 4].

There are many possible hardware architectures suitable for constructing VSAT packet switching functions. Most modern packet switches use a multiprocessor implementation with a central control or executive processor and multiple input/output (I/O) processors performing protocol conversion and routing functions in parallel. Most commercial packet switches use proprietary internal architectures or some variation on the X.25 access protocol. As the original protocol lacks some critical functions for VSAT operation, additional protocols are added to X.25 to build a workable internal switch protocol. The internal switch architecture implementation is vendor specific. Some switch architectures have most of the network management hardware and software integrated in the packet switch, other manufacturers have a separate network management computer which is connected to the packet switch.

8.2.2.6   VSAT network gateways

Each network interface type includes a gateway function which performs protocol conversion and, if necessary, protocol emulation.

Protocol conversion

When interconnecting data communications networks, a gateway generally performs conversion between dissimilar network communications protocols at higher OSI layers (e.g. email networks and message gateways).

The VSAT network gateway function essentially performs conversion at the lower layers between the network user communication protocols and the internal VSAT network protocols. The VSAT network gateway gives access to the network kernel, performs data packetization (if required), and addresses translation. In all cases, each type of user protocol interface has its own associated gateway functions. Both the hub and VSATs may provide network interfaces with this structure, which can be configured to support any one of several types of user protocols. This solution has the advantage of having an intranetwork protocol (in the kernel) independent from the user protocols, which makes the VSAT network easily adaptable to support different types of common as well as proprietary user interfaces.

Protocol emulation

Most data communication protocols, such as those based on the high-level data link control (HDLC) protocol defined by ISO, use some form of acknowledgement scheme to ensure correct transmission of data. A receiving station must send an acknowledgement packet after having received a certain predetermined number of packets. The sending station may not proceed to transmit the new set of data packets before it has received the acknowledgement. A parameter called window size defines the maximum number of unacknowledged packets.

The value of the window size depends on the protocol operation. In terrestrial networks this value is often small, for example 7 (modulo 8 operation). Because of the long round-trip delay of the satellite link the protocol may become ineffective (depending on the packet size and bit rate) as there could be a long waiting period for acknowledgement after each set of packets (i.e. one message unit) has been transmitted. This leads to sub-optimum throughput at the pertinent protocol layer. Larger window sizes are needed in the case of satellite links to ensure effective data transmission. Other protocol features such as retransmission timers, which specify the maximum time to wait for an acknowledgement before retransmitting the frame, should have timeout values which are long enough to accommodate a satellite round-trip delay.

The default parameter values of packet network protocols (timers, window sizes, etc.) are usually chosen for terrestrial (shorter delay) network operation. Selecting other parameter values for optimum protocol performance for satellite operation may not always be possible, for example due to built-in inflexibility of data transmission equipment.

The concept of protocol emulation is used to overcome the protocol performance degrading effects due to the satellite delay. The principle is shown in Figure 8.18 for a layer 2 protocol. The emulation function in the VSAT terminal at B' emulates the host computer protocol functions of B by sending locally the acknowledgement frames to the sending terminal A, e.g. after each seven data packets have been received. Thus transmission by the sending terminal A does not have to be interrupted due to the emulated acknowledgements. Error-free transmission of the packets over the satellite to the opposite end (hub) is then the task of the VSAT network, using intranetwork satellite efficient protocols (e.g. higher modulo operation). At the other end terminal A' (e.g. the hub) locally emulates the protocol functions of terminal A towards the host computer B.

Protocol emulation is also required for protocols which are based on polling (SDLC, BiSync), where a host computer polls (invitation to send) remote terminals in turn to send their data packets. This will be discussed later with the description of the particular protocol.

In VSAT systems, the protocol emulation process is performed cooperatively by the user protocol interface and the protocol gateway. The latter has it as an inherent part of its protocol conversion function.

8.2.2.7   ITU recommendation on VSAT network interconnection

ITU-R prepared a special recommendation entitled: "Connection of VSAT systems with Packet Switched Public Data Networks (PSPDNs) based on ITU-T Recommendation X.25" in 1995. For details on this draft recommendation, the reader is referred to Supplement 3 ("VSAT Systems and Earth Stations", ITU, Geneva, 1995), Appendix 5.3-2.

This draft recommendation was transferred to ITU-T, and was finally approved as ITU-T Recommendation X.361.

8.3 Interconnection with user data terminating equipments

Typically VSAT networks have several network interface types. Each type consists of the physical layer interface and user protocol interface for proper interfacing with the local user data terminating equipment (DTE). Similar provision may also be present between the VSAT hub equipment and the host computer.

8.3.1   Physical layer interface

The physical interface performs the physical connection of the user DTE to the VSAT network interface (for example V.24). Each VSAT system usually has several independent and reconfigurable physical interfaces, supporting synchronous and asynchronous physical standards at various data rates. The user interfaces typically support user data rates up to 64 kbit/s [Ref. 3, 4].

8.3.2   User protocol interface

The user protocol interface is associated with each physical interface, in order to properly terminate the user access to the VSAT network with a complete data circuit-terminating equipment (DCE) functionality at layers 2 and 3. Customized user protocol interfaces allows user equipment to connect to the network with their specific protocol.

Network configuration information, which includes user protocol interface information, is maintained and updated in the network configuration databases at the hub. Each interface in the network can be configured remotely from the hub, where the parameters of the interfaces are stored in the database. This versatility allows the network to adapt to the user interface and greatly simplifies the task of users when replacing existing data networks by VSAT networks.

Most VSAT systems support at least BiSync, SDLC and X.25 user protocols. Aside from supporting these most frequently used user interfaces, a VSAT network can easily be adapted to support proprietary interfaces as modifications are restricted to the user side of the network interfaces and not the entire network. Other types of supported user protocols are Async-X.25 (PAD X.25), Burroughs Pool/Select, DDCMP, Ethernet, HDLC frame pass through (transparent), etc.

a)   BiSync protocol

BiSync (or BSC) is one of the first communications protocols to connect remote terminal equipment to host computers. It is based on polling, selection and data transfer operation between a BiSync master station (the host or front end processor) and the remote terminals (tributaries). The master station continuously solicits data from the remotes by sending the polling sequence which contains the unique address assigned to a particular remote. When the remote has data to send to the host (i.e. a user hits the enter key), data is sent in response to the next poll. Whenever a master has data to send to a remote, a select sequence is sent followed by the data string. A block check is appended to the end of each message for error detection. If an error is detected, the receiving station responds with a NAK (negative acknowledgement) or does not respond at all, and the selection process is repeated.

BiSync does not allow multiple unacknowledged frames and enforces a stop and wait line discipline for each individual frame. In order to allow BiSync data to be sent efficiently through a VSAT network (or any other long delay network) the stop and wait nature must be removed from the long delay part of the connection by a protocol emulation process, and replaced by a satellite efficient VSAT network protocol [Ref. 4].

b)   SDLC protocol

SDLC is a bit-oriented link level protocol, which is a subclass of the high-level data link control (HDLC) protocol standardized by ITU-T and ISO. In normal response mode (NRM) of SDLC, all data transfers are initiated by the host (primary station) and a polling process is used to solicit data from the terminal (secondary station).

SDLC allows multiple unacknowledged frames (usually seven, but in some cases 127), which lead to better throughput over satellite connections than BiSync. Nevertheless, the polling mechanism still needs to be emulated for efficient operation in a VSAT network.

The interface between an SDLC primary and secondary station connected across a VSAT network involves synchronization of the opposite links and sending only information fields and important link events through the network [Ref. 4].

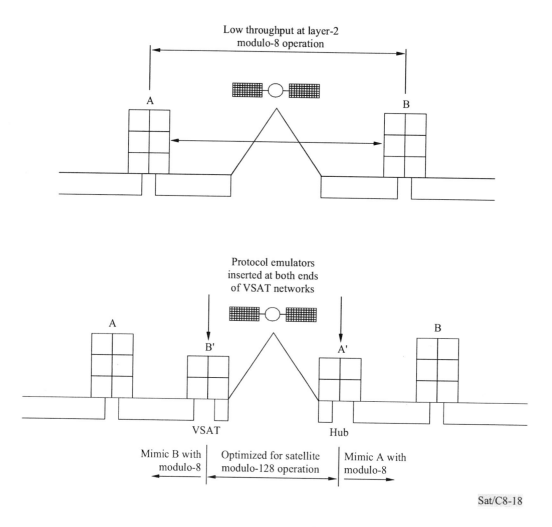

FIGURE 8.18

Principle of protocol emulation [Ref. 9]

An example of the SDLC protocol support in the VSAT network is in Figure 8.19. At the VSAT user side, an SDLC secondary device is assumed. The SDLC function within the VSAT terminal provides an SDLC primary station function, and polls the user's secondary device. Packets received from the secondary device are routed to the satellite network protocol function (gateway), which uses the link level service provided between the VSAT and hub terminals. Similarly, the packets received at the hub are routed to the appropriate SDLC secondary emulator at the hub, and then to the user's SDLC primary equipment attached to the hub.

c)   X.25 protocol

In terrestrial X.25 networks, each node of the network performs local acknowledgement of received data packets. Therefore X.25 user protocol interfacing to the VSAT network is more straightforward than SDLC or BiSync. The physical interface between the user X.25 DTE and the network is based on the ITU-T Recommendation V.24 standard. The user protocol interface of the VSAT network interface is fully compliant with the ITU-T Recommendation X.25 levels 2 and 3. The gateway performs protocol conversion between the X.25 access protocol and the internal VSAT network protocol, and controls the end-to-end virtual circuit [Ref. 2].

The VSAT user protocol interface performs local levels 2 and 3 acknowledgements to the user equipment, as would the nodes and DCEs in terrestrial X.25 networks. Given the absence of satellite delay across this local X.25 interface, the timer values and window sizes of the X.25 levels 2 and 3 protocols in the user equipment do not have to be modified when used with VSAT networks. The timer values of those level 3 procedures requiring end-to-end changes (e.g. call connect, call clear, etc.) are significantly greater than the round-trip satellite delay. Therefore they also remain unchanged in the user equipment [Ref. 4].

8.4 Interconnection with data networks

8.4.1 Interconnection with packet switched terrestrial data networks

Again, this section applies primarily to VSAT networks.

In PSPDNs, user DTEs (data terminating equipment) are interconnected to PSPDN DCEs (data circuit-terminating equipment) using X.25 interfaces. Asynchronous DTEs can be connected to the network using X.28 interface and a PAD (packet assembly/disassembly) function defined in ITU-T Recommendation X.3. PSPDNs (A and B) are interconnected with each other through network gateways with an X.75 interface between them). Data is transmitted through the network in packets between the packet switches (nodes) responsible for routing the packets. Each packet carries a header which contains the address information. Unlike circuit switched networks, no permanent physical connection is maintained through the PSPDNs between the communicating DTEs.

A number of interconnection scenarios between VSAT networks and PSPDNs are conceivable:

i)   VSAT network as a replacement of part of the terrestrial PSPDN;

ii) VSAT network as a transit subnetwork between PSPDNs;

iii) a VSAT network accessing PSPDN through a network interface at the hub or at one of the VSATs.

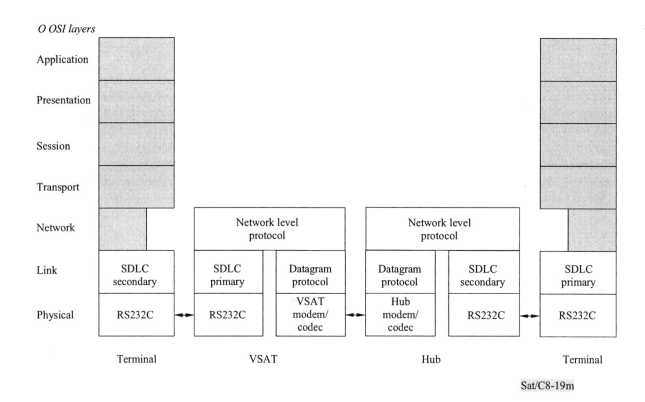

FIGURE 8.19

Example of SDLC protocol support [Ref. 3]

The third scenario is the most common amongst the three, and is formalized through the development of international standards. In this scenario the VSAT network is considered as a particular implementation of a private data network (PvtDN) connected to PSPDN through the standard user-network interface of the PSPDN. The VSAT network is perceived by the PSPDN as a normal X.25 DTE. For the VSAT network, the connection point to the PSPDN is typically at the hub [Ref. 5]. Alternatively, a VSAT terminal could be the interconnection point to PSPDN.

The general structure of PSPDN networks with a VSAT network, interconnection is shown in Figure 8.20.

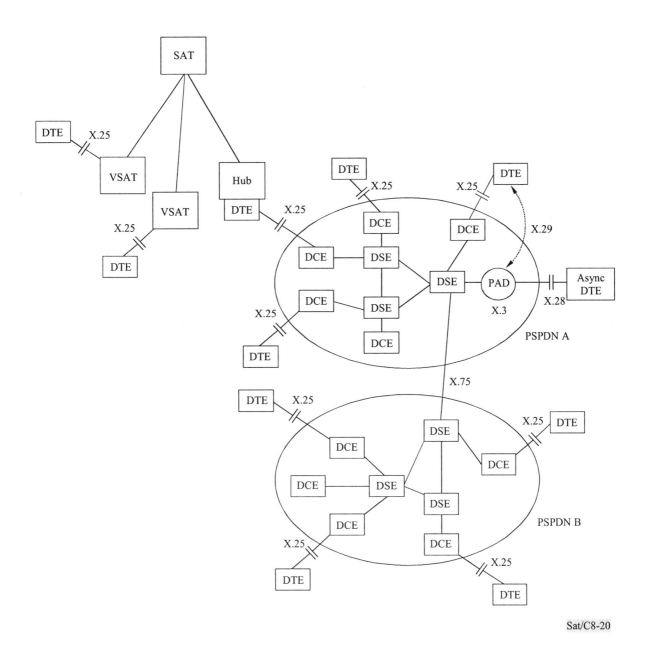

Sat/C8-20

FIGURE 8.20

General structure of PSPDNs and a possible interconnection with a VSAT network

The X.25/VSAT gateway performs functions such as address translation, packet routing, virtual circuit management, data exchange, and flow control between its local level 3 and the remote gateways. When a call connect level 3 packet is received at a gateway from its network interface, the gateway performs its virtual circuit control and management functions, transforms the packet to the internal network format, and appends the corresponding network address before passing it to the network kernel for transmission to the remote gateway. The remote gateway, on receipt of this packet, strips the information appended by the sending gateway, transforms the packet and passes it to its local network interface. The packet is then sent to the user terminal via the lower layers. If the remote user terminal accepts the call it responds with a call accept, which is sent back to the originator, and both gateways allocate system resources for the virtual circuit.

Once a virtual circuit has been established, information frames can be freely exchanged between user equipment subject to flow control restrictions. Level 3 protocols locally acknowledge information frames across the network interfaces so that the satellite transmission delay is not apparent to the user equipment. When a virtual circuit is cleared either as a result of a call clear packet or an incident in the network, all resources allocated to the virtual circuit are released [Ref. 4].

Performance aspects

VSAT/PSPDN interconnection requirements depend on the private VSAT network's access to the PSPDN network. In addition to the mandatory interfacing requirements, a certain level of network quality objectives must be met. Quality requirements include, for example, throughput performance and delay performance (call setup delay, data packet transfer delay and call clear delay). The private VSAT networks introduce additional performance degradation of the overall connection. The private VSAT network appears at the PSPDN interface as a DTE, and no quality allocations have been made in ITU-T Recommendation X.134 for the performance of DTEs beyond the public data network hypothetical reference connection.

Views on how this issue can be addressed are contained in the annexes to the draft new ITU-R recommendation on "Connection of VSAT systems with Packet Switched Public Data Networks (PSPDNs) based on ITU-T Recommendation X.25" and also in the European Standard section on the subject (ETS 300 194). VSAT systems which are not restricted in their connectivity may have to meet, in addition to the basic interfacing requirements, connection quality requirements also [Ref. 5]. Resolution of this matter is still pending. For call setup delay, an estimate of the additional transmission delay over a VSAT network experienced by a call connect packet is 0.8 seconds, assuming one satellite hop (i.e. the connection to PSPDN is at the hub) and a Slotted Aloha satellite access under a normal system load (10-15%). The additional transmission delay for the call accept packet in the VSAT part of the connection is approximately 0.3 seconds (TDM). Therefore the total call setup duration will be extended by an average of 1 second due to the VSAT system. If the interconnection scenario involves two hops, i.e. the connection point to the PSPDN is at another VSAT remote terminal, the call setup extra delay in the VSAT network will be of the order of 2 seconds.

For comparison, in terrestrial PSPDN networks under normal load conditions, the network transit time for both call connect and call accept packets is approximately 0.6 seconds. Although the node-to-node transmission delay in terrestrial networks is significantly less, the higher number of packet nodes encountered results in values of connection setup times comparable to VSAT networks.

During the data transfer phase of a call, both types of networks perform local acknowledgements and the data packet transfer delay can be considered one-way. In the VSAT network the transit delay is about 0.5 seconds and 1 second for the one-hop and two-hop configurations respectively. In terrestrial data networks the transit delay is approximately 0.4 seconds [Ref. 4, 5].

8.4.2 Interconnection with asynchronous and synchronous clear channels

Clear channel operation refers to a connection scenario where the user protocol is connected transparently through the VSAT network, i.e. no protocol conversion or emulation is performed in the VSAT network interface.

A common application of asynchronous clear links for VSATs is in replacing traditional dial-up data networks, where occasional data transfers are performed using the switched telephone network and asynchronous modems. Examples of such applications are reservation networks, or personal computer networks where data transfer is needed infrequently. Data is received on an RS-232 interface, and packetized in the gateway upon the occurrence of conditions such as expiration of a time interval between receiving characters, receipt of a terminating character, or the accumulation of a preset number of characters.

In the case of synchronous clear channels, permanent circuits are maintained between VSATs and the hub, without any protocol support in the user interfaces. Examples of such applications are voice transmission or distribution and real-time measurement collection.

Unlike asynchronous applications, synchronous transmission requires a continuous minimum capacity on satellite frames for the duration of the call (random access or dynamic demand assignment protocols are generally not suitable for such communications, since a constant network transit time is required, therefore extensive buffering and flow control must be used in the absence of dedicated satellite channels). A centralized reservation programme, part of the network management system, must be used to open and close the clear links.

Typically, synchronous channels are established between V.24, physical interfaces. Data is packetized in the corresponding gateway, and packets are transmitted via the kernel, using addresses given by the reservation system.

The role of the receiving gateway is to regenerate the synchronous data flow and forward it to the physical interface. It is important to have packets as short as possible in order to limit the buffer length in the gateway (and therefore to limit the time spent waiting in the packet disassembling queues). Since fixed-satellite channel reservations are used, the transmission delay is constant, and data buffering in the gateway can correct slight variations that may occur due to clock drift [Ref. 2].

8.4.3 Interconnection with synchronous multipoint protocol interfaces

Many potential applications for VSATs are as an alternative to existing communication networks and involve the use of synchronous point-to-point and multipoint protocols such as BiSync or SDLC, which are link level protocols designed to allow multiple communicating devices to share a common communications facility, typically a multipoint digital or analogue circuit.

In long delay networks, such as VSAT networks, these protocols cannot be efficiently used with clear channels, since protocols use a polling process to share a communication line. A protocol emulation function must be implemented in the network access interfaces, one at the host end and the other at the terminal end of the network, in order to carry the information across a satellite network efficiently [Ref. 2].

8.5 Interconnection with ISDN

8.5.1   Satellite transmission in the ISDN

An integrated services digital network (ISDN) is a general purpose digital network capable of supporting or integrating a wide range of services (voice and non-voice) on a switched basis and using a set of standard multipurpose user network interfaces.

The great promise of ISDN is the cost savings (equipment and more efficient use of the resources) and the universality and flexibility of service that it offers.

The increasing use of ISDN terminal equipment and the availability of satellite networks make their interconnection an interesting proposition. Particularly for those areas where terrestrial ISDN connections cannot be provided. VSAT networks can be used for connections between ISDN user terminals and ISDN local switching centres. The feasibility of such network interconnection may critically depend on the functioning of the ISDN protocols over the satellite link [Ref. 7].

There is nothing in the ISDN definition which prevents the use of satellites as part of it. In fact, since the transmission system is only one portion of the ISDN, the question that must be answered for the satellite radio engineer is: how will the transmission system required for the ISDN differ from the traditional system used for analogue traffic? Radio transmission systems are affected by thermal noise, interference and the propagation medium. The question thus reduces to: how do the above factors affect the performance of a transmission system carrying digital messages? Thermal noise, interference and the propagation medium together affect the ratio of the received carrier to the total noise at the receiver input. The propagation medium will make this ratio a time varying quantity. The distance covered by the radio signal will determine the delay incurred by the signal in passing from the transmitting terminal to the receiving terminal and vice versa.

The carrier-to-noise ratio can be controlled to any degree required by increasing power or through the use of error correction coding and/or the use of diversity. Each of these techniques requires paying a penalty of increased transmission system cost. Propagation delay can be managed by using properly implemented data transfer protocols.

8.5.2   ISDN studies in ITU

8.5.2.1   General

The concept of the ISDN has been the subject of study by ITU for over 20 years. More specifically, the CCITT, then ITU-T has been very active in preparing a new series of recommendations known as the I-series, to cover ISDN-related topics, such as:

- the ISDN concept and associated principles;
- service capabilities;
- overall network aspects and functions;
- user-network interfaces;
- internetwork interfaces.

Following the ITU-T effort, ITU-R, through Study Group 4, has been active in the definition of relevant requirements on conditions and performance for satellite links carrying ISDN channels and has issued recommendations translating the ITU-T Recommendations in terms which are significant for the satellite portion of the overall ISDN connections.

8.5.2.2 ITU-T definitions

The initial definitions about transmission in the ISDN came from ITU-T Recommendation G.821 which defines an "ISDN hypothetical reference connection" or HRX (ISDN) as schematically represented in Figure 8.21.

Use is made of the HRX to develop the performance requirements of each of the major transmission segments of the overall connection. As indicated by Figure 8.21, an ISDN service can include a connection of 27 500 km, the longest possible connection between subscribers (the distances being defined along the Earth's surface).

Three basic segments are identified with distances that are expected to be typically covered by each portion of the connection. These are: the local grade segment which includes the first switch, the medium grade segment, similar to inter-office service, and the high grade segment. In ITU-T Recommendation G.821, these three grades are allocated 30%, 30% and 40% of the performance degradation respectively. Figure 8.21 illustrates the longest HRX as given in ITU-T Recommendation G.821 and includes a single satellite link. In this HRX the local and medium grade portion apportionments are divided among the two ends of the connection (1 250 km), while the high grade portion is taken to cover the remainder of the connection or 25 000 km.

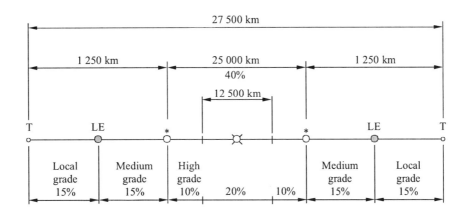

T: reference point
LE: local exchange

Sat/C8-21

FIGURE 8.21

ISDN hypothetical reference connection based upon
ITU-T Recommendation G.821

NOTE 1 – It is not possible to provide a definition of the location of the boundary between the medium and high grade portions of the HRX (see ITU-T Recommendation G.821 which provides further clarification of this point).

In apportioning the performance degradation to a fixed-satellite service hypothetical reference digital path (FSS HRDP), ITU-T considered the allocation that would have to be given to the local network, terrestrial radio-relay, cable and satellite systems, etc., each of which forms a part of the HRX and has certain performance constraints. As a result of this study ITU-T determined that the allocation for an FSS HRDP should be 20%.

Note, however, that this 20% allocation may be different from other ITU-T Recommendations.

8.5.2.3   ITU-R definition of the hypothetical reference digital path (HRDP)

Recommendation ITU-R S.521 clearly addresses the possibility that a fixed-satellite service HRDP may form part of an ISDN HRX as defined by ITU-T.

The HRDP, which should consist of one Earth-space-Earth link with the possible inclusion of one or more inter-satellite links (ISL) in the space portion, should include the equipment indicated in Figure 8.22 a) and should interface with the terrestrial network at a suitable digital distribution frame (DDF) at the lowest bit rate appropriate to the HRDP. However, the bit rate at the terrestrial network interface may be any value, depending upon the application.

The HRDP should accommodate different types of access as single channel or TDMA and allow for the use of techniques such as digital speech interpolation (DSI) or low rate encoding (LRE).

Additionally, the earth stations should include facilities to compensate for the effects of satellite link transmission time variation introduced by satellite movements which are of particular significance in a plesiochronous network.

The ISDN HRX for a 64 kbit/s circuit switched connection, which is defined in ITU-T Recommendation G.821, encompasses a total length of 27 500 km and is composed of three circuit grades (local, medium and high) which have different performance requirements. The model adopted by ITU-R as a reference for the development of performance and availability objectives of the satellite portion of the HRX, assumes that one FSS HRDP may form part of the high grade portion and may replace an equivalent terrestrial connection covering a distance of 12 500 km. This reference model is illustrated in Figure 8.29 b).

It should be noted that the 12 500 km distance used in ITU-T Recommendation G.821 is close to the figures resulting from ITU-R studies for various network configurations. More specifically, in these studies it was noted that satellite performance is generally distance independent, with a maximum single hop covering an equivalent terrestrial distance of approximately 16 000 km. Consequently, in the vast majority of cases where a satellite is in the international portion of the connection, the satellite hop forms the entire international portion with the two ends being located within the countries being connected, and usually less than 1 000 km from the end users. When satellites are used in the national portion of a connection, it would appear to be reasonable to expect that terrestrial extensions from the earth stations to the end user will be even shorter. In several existing or planned satellite systems, particularly with business services, the earth terminals are in many cases located at the end users' premises, virtually eliminating the terrestrial extensions.

In the various HRX configurations considered by ITU-R, although a distance related approach did not appear to be applicable, it was found that a distance could be postulated for the satellite portion resulting in the development of the concept of a "satellite equivalent distance" (SED) that could be assigned to the satellite HRDP. As seen before, a value of 12 500 km has been assigned to the SED.

Models involving two satellite hops and inter-satellite links were also developed but it was recognized that connections of this type would be used only in exceptional cases. The use of dual satellite hops may lead to problems in telephony transmission due to propagation delay and the synchronization of digital equipment due to variation in delay.

8.5.2.4 ITU-R performance requirements

As explained above, following the issuing of ITU-T Recommendations, ITU-R, through Study Group 4, has prepared recommendations translating these ITU-T Recommendations in terms which are significant for the satellite portion of the overall ISDN connections. Although these works are still continuing, the following ITU-R recommendations have been issued in the field of ISDN:

- Recommendation ITU-R S.614 on quality objectives for a 64 kbit/s ISDN circuit (translating ITU-T Recommendation G.821);

- Recommendation ITU-R S.1062 on "Error performance for a HDRP operating at or above the primary rate" (i.e. 1.5 Mbit/s) (translating the new ITU-T Recommendation G.826).

Details on these recommendations are given above in the section on Quality and Availability in Chapter 2 (§ 2.2.3.3).

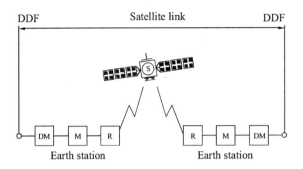

a) Hypothetical reference digital path

S:　　　space station in the fixed-satellite service
DM:　　digital multiplex equipment, including TDMA,
　　　　DSI and LRE equipment if used
M:　　　modem equipment
R:　　　IF/RF equipment
DDF:　　digital distribution frame

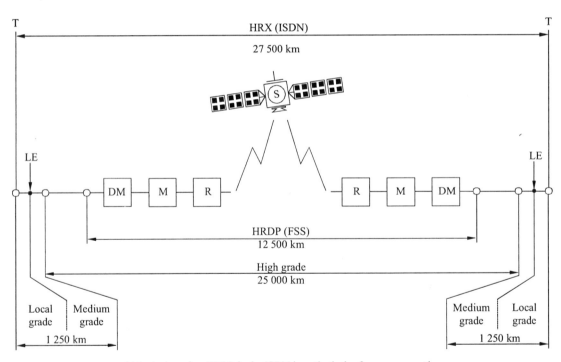

b) Inclusion of an HRDP in the ISDN hypothetical reference connection
based upon ITU-T Recommendation G.821

LE:　　Local exchange
T:　　T reference point

Sat/C8-22

FIGURE 8.22

HRDP and its inclusion in the ISDN HRX

8.5.3 ISDN structure and services

As already explained, the ISDN network architecture is divided into three major segments which are recalled in Figure 8.23 as user network, local network and transit network.

The user network may consist of a variety of different customer premises equipment such as data terminals, host computers, facsimile, telephones, and video equipment. These are connected to the local network through network termination equipment (NT) at the subscriber S and T reference points. Reference point S is for connection of terminal equipment to a user premises PABX, LAN or some type of time division multiplexer. The functional group is referred to as network termination NT2. Reference point T provides essentially a layer 1 termination (NT1) of the ISDN local loop at the user premises. A user ISDN terminal can be connected directly at reference point T for direct access to the local network.

Two ISDN user-network access interfaces are defined. The basic access has an interface structure consisting of two B-channels and one D-channel (2B+D). The bit rate of the B-channel is 64 kbit/s and of the D-channel is 16 kbit/s. The primary rate access interface structure is 23B+D (1 544 kbit/s) or 30B+D (2 048 kbit/s), where the bit rate of the D-channel is 64 kbit/s. The D-channel, which is primarily intended to carry signalling information, may also be used to carry packet switched data [Ref. 8].

A satellite network interconnected to the ISDN should be capable of supporting one or more of the bearer services, teleservices and supplementary services defined for ISDN in the relevant ITU-R recommendations. The VSAT internal network design is determined by the intended ISDN services to be supported.

As a minimum the satellite network needs to support ISDN circuit mode bearer services, which requires adequate capacity for channels ranging from 64 kbit/s to 1 920 kbit/s plus either 16 or 64 kbit/s D-channels. In addition, since satellite networks are primarily intended for data communications, support of ISDN packet mode bearer services appears logical. To ensure efficient use of the circuit supporting the packet bearer service, appropriate extended frame and packet windows should be provided.

Some of the ISDN supplementary services such as sub-addressing, direct dialling, multiple subscriber number and closed user groups can be used to satellite networks' and ISDN users' advantage [Ref. 7].

8.5.4 Satellite network to ISDN interconnection scenarios

A common view is of the satellite network as a part of the user network connection to ISDN through a NT2 termination. Several interconnection scenarios are conceivable, as well as several ways in which the satellite network may be involved in the communication. The following scenarios have been identified so far, with particular application to VSAT networks (although more general application could be envisaged).

a) Single node distributed customer network

A single node distributed ISDN customer network is shown in Figure 8.24. The ISDN may offer at the T reference point either a basic or a primary rate interface structure by means of a network terminal NT1. The NT2 forms part of the customer network and is therefore part of the VSAT system. The relevant ITU-T specifications of the T interface at the physical, data link and layer 3 level need to be satisfied. It would be cumbersome for VSAT systems to assume the functions of an integrated NT2-NT1, as it would require special interfaces to the public network at reference point U which is not standardized by the ITU Telecommunication Standardization Sector.

The NT2 can be envisaged as the node of a distributed PABX, while the S interface represents the standard for the interface between terminal equipment (TE) and the PABX node. If a VSAT system is used to connect remote terminals (TEs) to the distributed PABX, the remote TEs are connected to the VSATs and the NT2 is located at the hub (or one of the VSATs). Therefore the VSAT system conceptually translates the S interface from the NT2 site to the TE site.

8.5 Interconnection with ISDN

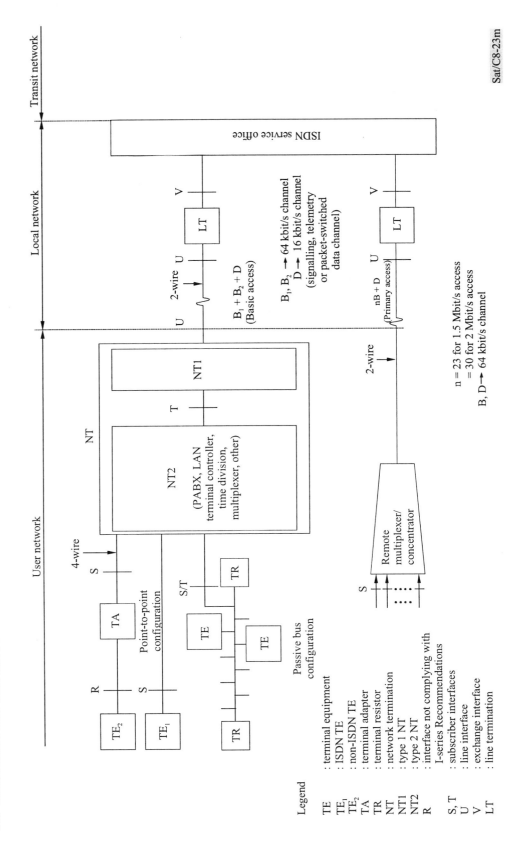

Legend

TE	: terminal equipment
TE₁	: ISDN TE
TE₂	: non-ISDN TE
TA	: terminal adapter
TR	: terminal resistor
NT	: network termination
NT1	: type 1 NT
NT2	: type 2 NT
R	: interface not complying with I-series Recommendations
S, T	: subscriber interfaces
U	: line interface
V	: exchange interface
LT	: line termination

FIGURE 8.23

ISDN definitions [Ref. 8]

Sat/C8-23m

b)   Multiple node distributed customer network

In this architecture of multiple node distributed ISDN customer networks, shown in Figure 8.25, several private (ISDN) networks (nodes) can be interconnected by satellite but can also be interconnected with the public ISDN via a gateway (interworking function (IWF)). It is assumed that these private networks are operated according to an agreed protocol.

The interconnection of private networks with the public ISDN network may take place over standard subscriber interfaces or may otherwise require some IWF to adapt the procedures for defining the connection requirements within the ISDN to those applicable to the private network. Currently such an interworking function between the public ISDN and private networks is not specified, but ITU-T Recommendation I.500 identifies the need for such a function. Some form of SS No. 7 signalling system may also be required. The DSS 1 (ISDN local access signalling) is not applicable in this case and therefore it must be reviewed against the private network signalling [Ref. 7].

c)   Interface through message switch

One design approach to interconnect dissimilar networks is to implement a gateway function between them using a store and forward switch. Such a system would be limited to ISDN teleservices that do not require real-time interaction. With this approach, the operation on the VSAT system side is decoupled from the ISDN, and compatibility with the ISDN interface requirements should be sufficient for interoperation of the systems. The terminal equipment need not be ISDN-compatible if the gateway structures the information for both high and low layers in accordance with ISDN teleservices and bearer services [Ref. 7].

8.5.5   VSAT network ISDN interface

The terrestrial interface at the VSAT terminal can be the basic rate interface (2B+D), and the terrestrial interface at the hub can be both the basic rate and the primary rate (23B+D) or (30B+D). The link layer (layer 2) signalling protocol for ISDN access is LAPD. The D-channel rate is 16 kbit/s for basic rate access and 64 kbit/s for primary rate access.

ITU-T Recommendation I.430 and I.431 define the physical characteristics of the S interface. The physical separation between the terminal equipment (TE) and NT2 is less than 1 000 metres, and the corresponding round-trip delay is less than 43 microseconds (point-to-point wiring configuration in ITU-T Recommendation I.430). The round-trip delay specification is necessary for timing consideration of the D-channel echo bits. Timer values in the I.430 and I.431 protocols may not be suitable for a satellite link, and revisions to the recommendations would be necessary. Alternatively, local emulation at both ends of several S interface functions can be considered to ensure compatibility with standard terminal equipment. The VSAT system must make internal provision for the required channels and the correct operation of the physical layer including clock, byte and frame alignment, activation/deactivation of the TEs, D-channel bit contention and echo schemes, maintenance and status indications, etc.

LAPD is the protocol used in the D-channel, which normally operates between ISDN terminals and the local exchange. It is important to ensure that this protocol works properly over a satellite link (VSAT network) which is part of the connection between the terminal and the local exchange (e.g. in reference configuration 1 described above). At layer 2, parameters which are critical to satellite operation are retransmission timer T200 and k, the maximum number of outstanding frames in the LAPD (link layer) protocol. It would be desirable to have the value of T200 equal to 2.5 seconds on the interface entities involved (TE/NT2). Also the value of k should be chosen higher than its default value to improve performance over the D-channel, i.e. a value at least 5 and 19 for 16 kbit/s and 64 kbit/s D-channels respectively [Ref. 7].

For ISDN packet mode bearer service support, the gateway at the hub sets up an appropriate virtual connection with the corresponding gateway at the destination VSAT terminal. An appropriate ISDN protocol is reintroduced at the VSAT terminal to communicate over the terrestrial ISDN interface to the end user. As a result, an end-to-end ISDN packet communication is conducted over the VSAT network in a cost-effective manner. In this scenario, the gateway performs a number of important functions to provide an efficient packet communication service, namely, the interface to the VSAT multiple access protocol and error/flow control, the virtual connection setup and release, and removal of interframe idle pattern data in LAPD protocol.

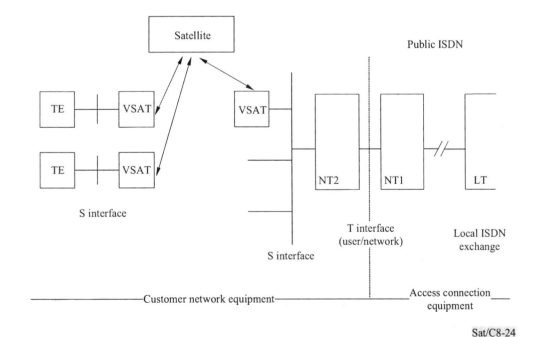

Sat/C8-24

FIGURE 8.24

Single node distributed ISDN customer network [Ref. 7]

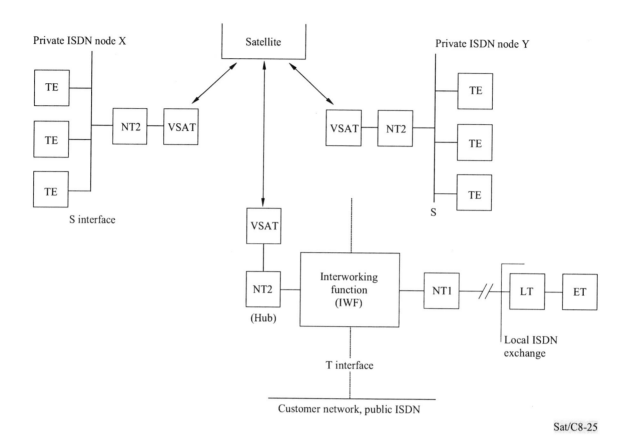

FIGURE 8.25

Multiple node distributed ISDN customer network [Ref. 7]

By isolating the call control and LAPD protocol procedures at the terrestrial interface one is free to choose a suitable multiple access and gateway-to-gateway protocol. One choice could be a selective acknowledgement protocol that buffers the out-of-sequence packets. Typically, a VSAT network uses a random access technique or one of its variants as a multiple access technique scheme resulting in a finite probability of packet collision. As the LAPD protocol will not accept out-of-sequence packets, any collision will cascade into the loss of many uncollided packets [Ref. 6].

8.6 Interconnection with ATM networks

8.6.1   Introduction

As explained in Chapter 3 (§ 3.5.4) and in Appendix 3.4, to which the reader is referred for a summary of the basic principles, the asynchronous transfer mode (ATM) is a specific packet oriented mode which uses asynchronous time division transmission and switching of fixed size blocks (53 bytes long) called cells.

This section deals with the specific problems met when satellite links are implemented in ATM networks.

In the conclusion, reference will be made to recent works on future satellite architectures for advanced ATM applications.

8.6.2 Transport of ATM multiplex

Transport of ATM cells was initially defined at the synchronous digital hierarchy (SDH) rates, i.e. at 155 Mbit/s and above (see ITU-T Recommendations G.707 and I.432-series). SDH is still considered as the basis for the transport of ATM cells. However, at least for a transition period, transport of ATM cells using the plesiochronous digital hierarchy (PDH) has been defined. The ITU-T Recommendation G.804 defines the ATM cell mapping in the PDH at rates 1.544, 6.312, 44.736 and 97.728 Mbit/s (1.544 Mbit/s hierarchy) and at rates 2.048, 34.368 and 139.264 Mbit/s (2.048 Mbit/s hierarchy). Cell based systems (see again ITU-T Recommendations G.707 and I.432-series) and MPEG-2 based systems can also be supported by ATM (for MPEG-2, see Chapter 3, Appendix 3.3, § AP3.3-2.3 and also Chapter 7, § 7.10.1.1).

8.6.3 Effect of satellite transmission on ATM

As already explained in Appendix 3.4 (§ AP3.4-2), the satellite link introduces a delay in the transmission between the transmit and the receive earth station. This delay is generally between 240 ms and 280 ms for a given link in the case of a geostationary (GSO) satellite. Almost all services can tolerate such a delay, although it can significantly increase, in particular, the latency of feedback mechanisms which are essential for congestion control.

Other delays exist, like the forward error correction (FEC) encoding and decoding delay, the interleaving and de-interleaving delay (see below), the buffering delay for access (including queuing, switching and/or routing delays). All these delays are constant, except the last one, and so they have no influence on the cell delay variation parameter.

Of course, the delay is much shorter, but includes a variable component, in the case of a low-Earth orbit (LEO) satellite constellation. There are various methods for compensating, at least partially this variable component (e.g. "dating" the cell in order to restore it correctly at receive).

In order to reduce the earth station transmitted power and/or antenna size, satellite transmission is generally associated with FEC, the most frequently used code being the convolutional encoding (rate ½ or ¾) associated with Viterbi decoding. Residual errors after such decoding are not independent but appear in the form of bursts (even more with rate ¾ than with rate ½). This presence of error in bursts impacts ATM transmission, because ATM cells header formats and the ATM adaptation layer (AAL) have been designed assuming a low bit error rate and isolated errors.

In fact, the header error correction (HEC) of ATM cells can correct single errors and detect almost all multiple errors in the 5 bytes that contain the header of an ATM cell. When the HEC detects errors that it cannot correct, the whole cell is discarded and its payload is lost for an end-to-end connection[2].

The two main processes currently used for dealing with the error bursts, are:

Interleaving: An interleaver is used which rearranges the encoded bits over a span of several block lengths. The amount of error protection, based on the length of burst errors encountered on the channel, determines the span length of the interleaver. The overall effect of interleaving is to spread out the effects of long burst errors so that they appear to the decoder as independent random bit errors. Of course, a de-interleaver re-establishes the normal order of the bits at its output. The interleaving method can be implemented only in the case of relatively long signals involving a sufficient number of blocks.

Concatenation of an outer code: The implementation of an outer code in addition to an inner FEC coding can considerably reduce the bit error ratio. The Reed-Solomon (RS) block code is among the most efficient codes which can be implemented using state-of-the-art hardware and software technology (see Chapter 3, §3.3.5 and Appendix 3-2). In block codes, the input signal being partitioned into blocks and each block is processed as a single unit by both the encoder and the decoder. Concatenated RS coding schemes are used with convolutional inner codes to achieve high quality, cost effective ATM satellite transmission links with "fibre-like" quality. An example of the performance of the utilization of a concatenated RS scheme is shown by Figure 8.26. In this example of simulation, the BER and the cell error ratio (CER, see below) are shown as a function of E_b/N_0 for the particular case of an RS (126, 106) outer code extending over two cells (without interleaving): for $E_b/N_0 = 6$ dB, the BER is nearly 10^{-9} and the CER is about 3.10^{-8}.

If possible, these two processes are combined to optimize the link quality. Note again that the use of an RS code and interleaving increases the delay.

---

[2] Moreover, some patterns of more than two errors in the header are recognized by the HEC as a single error and therefore an improper correction may take place over a cell that should be discarded. In this case, the ATM is either eliminated or transmitted to a wrong direction by the ATM node (misrouted, see the "cell misinsertion rate – CMR-" below in § 8.6.5).

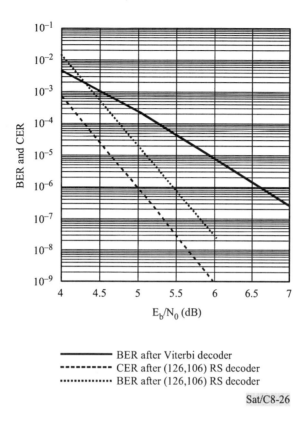

BER and CER

E_b/N_0 (dB)

——————— BER after Viterbi decoder
----------- CER after (126,106) RS decoder
•••••••••••• BER after (126,106) RS decoder

Sat/C8-26

FIGURE 8.26

BER and CER performance for a concatenated code RS (126,106)

8.6.4   Role of satellite transmissions in ATM networks

8.6.4.1   Satellite as a transmission medium

Satellites need to be used as a transmission medium for ATM data, just like any other classical mediums. The transport of ATM must be taken into account on satellite links in order to provide a transmission quality comparable to that obtained with fibre optic cables.

8.6.4.2   Satellite as transmission and switching element

The satellite can also play a role in ATM switching. Several studies have shown the feasibility of satellite on-board processing (OBP) including a complete ATM virtual path (VP)/virtual circuit (VC) switch, or, at least a VP (cross-connect) switch.

A transparent satellite coupled with TDMA can also play an active role in ATM switching. This has been demonstrated, in particular, by the European RACE CATALYST project which is described below (§ 8.6.7.2).

8.6.5   ITU-R works on ATM satellite transmissions

8.6.5.1   General

Following ITU-T recommendations on ATM, and, in particular, Recommendation I.356, (on B-ISDN ATM layer cell transfer performance) and Recommendation I.357 (on B-ISDN availability performance for semi-permanent connections), ITU-R is currently issuing recommendations which translate the performance objectives in terms applicable to connection portions including satellite links. Such ITU-R recommendations should ensure compliance of the satellite links with overall end-to-end performance objectives and, in turn, of end-user quality objectives. It should be noted however that:

- ATM-based satellite transmission systems may exist which are outside the scope of I.356 and I.357 and which, in particular, do not operate in the framework of international links in B-ISDN networks;

- the ITU-R recommendations currently in preparation will apply to GSO and, possibly, to NGSO satellites. This means that the satellite path may comprise the earth stations and a single transparent ("bent-pipe") satellite or a series of satellites. Some satellite systems may include on-board processing (OBP) transponders, ATM switching, and inter-satellite links (ISL);

- the demarcation point between the domestic ATM network and the international ATM network is known as the measurement point international (MPI). The portion between the two MPIs is known as the international inter-operator portion (IIP).

8.6.5.2   ITU-R recommendation on satellite B-ISDN ATM performance

The first ITU-R recommendation to be issued on the subject will deal with "Performance for B-ISDN asynchronous transfer mode (ATM) via satellite".

The parameters accounted for in this recommendation are based on those which are accounted for by ITU-T in Recommendation I.356 and which are listed below (see also Appendix 3.4, § AP3.4-2):

- the cell loss ratio (CLR);

- the cell error ratio (CER) which is defined as the ratio of errored cells to the total number of successfully transferred errored and non-errored cells;

- the severely errored cell block ratio (SECBR) which is the ratio of total severely errored cell blocks[3] in a population of interest;

- the cell misinsertion rate (CMR, see footnote[3]);

---

[3] A severely errored cell block outcome occurs when more than M errored cells, lost cells or misinserted cells are observed in a received cell block. A cell block is a sequence of N cells transmitted consecutively on a given connection.

- the cell transfer delay (CTD);
- the cell delay variation (CDV)[4].

The principles for translating parameters CLR, CER and SECBR in terms of bit error ratio (BER) in the satellite channel and then translating the ITU-T recommendation in terms of satellite performance objectives are somewhat similar to those used for translating ITU-T Recommendation G.826 ("Error performance parameters and objectives for international constant bit rate digital paths at or above the primary rate") into Recommendation ITU-R S.1062 ("Allowable error performance for a HRDP operating at or above the primary rate") (see Chapter 2, § 2.2.3.3 ii)). The starting point is the allocation of the ATM performance objectives for the international portion of ATM satellite connections forming part of a hypothetical reference connection (HRX) for B-ISDN ATM systems.

Table 8.3 below shows the performance objectives which are currently envisaged for services of Class-1 (the most stringent class, which is a delay-sensitive class and it is intended to support constant bit rate and real-time variable bit rate services such as telephony and videoconference).

TABLE 8.3

ATM performance objectives for satellites (Class-1 services)

Performance parameters	ITU objective end-to-end	ITU objective satellite
CLR	3×10^{-7} (NOTE 1)	7.5×10^{-8}
CER	4×10^{-6}	1.4×10^{-6}
SECBR	1×10^{-4}	3.5×10^{-5}
CTD	400 ms	320 ms (max)
CDV	3 ms	Negligible
CMR	1/day	1 per 72 hours (NOTE 2)

NOTE 1 – It is possible that in the future, networks will be able to commit to a CLR = $1 \cdot 10^{-8}$ for Class-1. This is for further study.

NOTE 2 – The allocation for on-board ATM processing equipment is for further study.

It should be noted that, while the ATM layer performance parameters and objectives are specified by ITU-T Recommendation I.356, the physical layer performance parameters and objectives for connections that will carry ATM traffic are given in the ITU-T Recommendation G.826 which, as mentioned above was translated in Recommendation ITU-R S.1062.

---

[4] The cell delay variation (CDV) or jitter that may arise in a satellite link depends on several aspects. First, CDV depends on the traffic load structure. CDV also depends on the switch buffering capacity and mechanism. As concerns systems based on LEO satellite constellations, the CDV will depend on the handover. It will increase with the number of ATM nodes within a connection (this may be a critical component of satellites that use OBP and ISL). Finally, CDV will depend on operations resulting from satellite specific ATM equipment. The use of multiple access schemes may have an impact on CDV.

The parameters dealt with in G.826, i.e. errored second ratio (ESR), severely errored second ratio (SESR) and background block error ratio (BBER), are different to those of I.356 and studies and measurements have shown that a facility designed to just meet G.826 may not meet the I.356 objectives for ATM Class-1 services. Thus, it is suggested that satellite links that will carry ATM traffic should be designed to meet I.356 requirements with enough margin to compensate for operational impairments.

8.6.5.3   ITU-R recommendation on satellite B-ISDN ATM availability

Another ITU-R recommendation will deal with "Availability objectives for a hypothetical reference digital path (HRDP) when used for transmission of B-ISDN asynchronous transfer mode (ATM) in the FSS by geostationary orbit satellite systems using frequencies below 15 GHz[5].

This recommendation will correspond to ITU-T Recommendation I.357 (B-ISDN ATM semi-permanent connection availability). The availability of B-ISDN ATM over satellite is determined by the aggregate effects of network congestion, transmission errors, equipment failures and propagation characteristics. Based on previous studies for 6/4 GHz and 14/10-11 GHz bands transmission paths and for spacecraft and earth station availability, a value of 99.86% is proposed for the yearly availability of a satellite HRDP, i.e. the percentage of time during which ITU-T Recommendation I.356 is met.

8.6.6   Some applications of ATM satellite transmissions

Of course, the main application of ATM satellite transmissions are and will be in the transport of voice and, more generally of integrated voice/data, ISDN type, streams. However, two more specific applications, i.e. Internet and television/video compressed programmes are described below.

8.6.6.1   Internet traffic over satellite ATM

The transport of Internet traffic over ATM networks is an important satellite application area. ATM-based satellite systems can be used to provide high-speed backbone transport as well as access connectivity in many regions of the world, including access by LEO satellite systems.

Although most Internet applications run successfully over satellite ATM, there is a concern that high-speed applications may not work efficiently over satellite links due to the latency (delay) effects on current versions of the transmission control protocol of the Internet (TCP/IP) used by the Internet, specially in the case of high-speed data streams over long-delay links. This problem is not unique to satellites and is also of concern in some data services transmitted at gigabit rates over fibre networks.

There are two broad categories of problems limiting enhancement of the current TCP efficiency over satellite: the available bandwidth and the manner in which packets are acknowledged and re-

---

[5]   Note that the availability objectives for B-ISDN ATM switched connections is for further study. However, until the specific availability requirements of these type of connections are adopted by ITU-T, the objectives given in this recommendation may be used.

transmitted. The throughput of a single application using a TCP connection may be limited by the "window size" of TCP. One way of increasing the throughput is to increase the window size, since more data will be sent before awaiting for an acknowledgment message. However, other aspects need to be considered as well. These include the retransmission algorithms and the congestion control mechanism of TCP. Some potential solutions have been proposed and simulation studies have been conducted, especially to determine the buffering requirements.

The following cases were considered in the quoted studies:

- LEO systems (e.g. 700 km altitude) with a single link (earth station-to-earth station latency ≈ 5 ms) and with multiple LEO hops (latencies up to 50 ms);

- GEO systems (typical earth station-to-earth station latency ≈ 275 ms);

- configurations with 5, 15, 50 sources (i.e. TCP connections) on a single bottleneck link (and 100 sources for a single LEO hop);

- system bit rate and configuration: The entire satellite network is assumed to be a 155 Mbit/s ATM link without any on-board processing or queuing (all processing or queuing are performed at the earth stations);

- buffer size multiples or sub-multiples of the round trip time (RTT).

The simulation results (throughput efficiency Vs buffer size) show a high asymptotic value of the efficiency for a sufficient buffer size. In all cases of latency and even for a large number of sources, the TCP throughput is over 98 % for buffer sizes of 0.5 to 1 RTT.

8.6.6.2   Video and multimedia applications

These applications are generally based on MPEG-2 video compression protocol. The MPEG-2 transport stream is a complicated time multiplexing protocol allowing multiple programmes of video, audio and user specific data to be transmitted in a single data stream. Because of the complexity of the MPEG-2 video/audio encoding and transport multiplexing, it is extremely difficult to determine the video quality resulting from random errors inserted in the transport stream., i.e. to determine how the programme content degrades or how the decoder looses synchronization when the BER increases on the satellite link.

The MPEG-2 transport stream (composed of 188 byte packets) can be segmented and placed into ATM cells using ATM application layer AAL-1 or AAL-5.

Another subject of interest is the joint transport of interactive services based on ATM and of broadcasting services on satellite systems, based on MPEG-2 (see Chapter 3, Appendix 3.3, § AP3.3-2.3 and also Chapter 7, § 7.10.1.1). In order to avoid the use of separate carriers for these classes of services in the forward (broadcasting) direction, ATM cells can be encapsulated into well-identified streams of MPEG-2 packets. In this case, MPEG-2 acts as the physical layer for the transport of ATM cells.

8.6.7 ATM satellite transmission tests

8.6.7.1 Measurement of performance at physical layer and ATM layer

Several test were conducted to evaluate the physical layer and ATM performance of satellite links. Two examples of these tests are given below. Other tests were performed by INTELSAT and NASA (United States). The measurement results of these tests are used by ITU-R in the preparation of the recommendations on ATM.

Tests between AT&T (United States) and KDD (Japan)

These tests are a part of a complete test campaign conducted in 1995 between the Salt Creek (California), the Ibaraki (Japan) and the Sidney (Telstra, Australia) earth stations with the assistance of INTELSAT. A 45 Mbit/s IDR link was operated. The trial involved a mix of satellite and terrestrial fibre links (DS-3). The results presented pertain to the AT&T-KDD link only since this link was tested for a longer time and it exhibited greater rain events than the AT&T-Telstra link. Table 8.4 shows the main measured performance parameters, compared with the ITU objectives (those have been adjusted according to the allocation given to the GSO satellite portion in an international end-to-end connection). The results show that the performance in both directions met the ITU-T Recommendation G.826 objectives (and also the Recommendation ITU-R S.1062 "masks") by a margin of 1-to-2 orders of magnitude. However, the results also show that this level of performance may not be sufficient for ATM since the upper bound objectives of Class-1 services (ITU-T Recommendation I.356) will require BER thresholds near 10^{-9}.

Consequently, the AT&T-KDD results may have barely met the ITU-T Recommendation I.356 objectives. Link enhancement techniques, such as RS outer coding and/or interleaving can provide improved performance which may be more consistent with ATM requirements.

TABLE 8.4

**Physical and ATM layer performance of 45 Mbit/s links
between AT&T and KDD**

Parameters	Physical layer April-June 1995					ATM layer August-December 1995	
	% ES (G.826)	% SES (G.826)	BBER (G.826)	Avg. BER	Threshold BER (0.2% of Time) (S.1062)	CLR I.356 (Class-1)	CER I.356 (Class-1)
ITU objectives	2.62	0.07	$7 \cdot 10^{-5}$	-	$4.0 \cdot 10^{-6}$	$7.5 \cdot 10^{-8}$	$1.4 \cdot 10^{-6}$
KDD to AT&T	0.014	0.008	$1.5 \cdot 10^{-8}$	$4.8 \cdot 10^{-10}$	$4.5 \cdot 10^{-9}$	$1.9 \cdot 10^{-10}$*	$5.4 \cdot 10^{-10}$*
AT&T to KDD	0.0056	0.0027	$7.6 \cdot 10^{-9}$	$2.6 \cdot 10^{-10}$	$3.0 \cdot 10^{-9}$	$3.9 \cdot 10^{-10}$*	$8.8 \cdot 10^{-10}$*

* These are the average values and not the upper bounds specified by ITU-T Recommendation I.356.

ATM tests by EUTELSAT

EUTELSAT conducted ATM and physical layer measurements on TDMA (2.048 Mbit/s input to 120 Mbit/s TDMA) and IDR (34.468 Mbit/s) in order to characterize the relationship between ITU-T Recommendations G.826 and I.356 parameters as a function of the link performance. Figures 8.27 a) and b) below show typical performance and objective comparison.

8.6.7.2   Satellite ATM switching system demonstration

As explained above (§ 8.6.4.2), a transparent satellite coupled with TDMA can also play an active role in ATM switching, as demonstrated by the European CATALYST project.

The CATALYST system implements a full mesh network between earth stations acting as traffic nodes. It behaves as a distributed ATM switch: input cell streams are routed from inputs to outputs (i.e. between the ATM ports at the earth stations) according to their virtual path identifiers (VPIs).

The CATALYST system thus allows to interconnect geographically scattered broadband networks (ATM networks or local area networks – LANs – via an interworking unit).

As shown in Figure 8.28, a traffic terminal maps the valid ATM cells through a small ATM VP switch (the ATM adaptation module – ATMAM). The TDMA burst time plan is managed by the NCC, which allocates to each traffic terminal, via immediate or short term reservation, a capacity function of the estimated peak traffic. As concerns the ATM interfaces, the system performs functions of the ATM layer and of the physical layer. As concerns the LAN interfaces, the system performs also AAL functions, through the interworking unit, to adapt the LAN traffic to ATM.

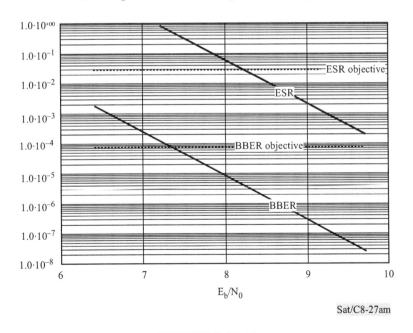

Sat/C8-27am

FIGURE 8.27 A)

EUTELSAT tests: G.826 (physical layer) performance vs. E_b/N_0 at the demodulator input for an IDR modem (with 3/4 FEC) at 34.368 Mbit/s

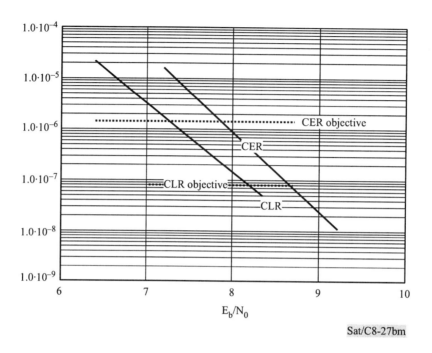

FIGURE 8.27 B)

EUTELSAT tests: I.356 (ATM layer) performance vs. E_b/N_0 at the demodulator input for an IDR modem (with 3/4 FEC) at 34.368 Mbit/s

Note that TDMA allows easy reconfiguration of the capacity. However, a drawback is to introduce a cell delay variation (CDV), which is due to time compression and expansion and which is thus equal to the TDMA frame period. It is possible to greatly reduce this CDV by dating the cells at the input, by transmitting this time information and reconstructing the cell flow at the output with the best time approximation. Of course, this increases the complexity and reduces the useful satellite throughput.

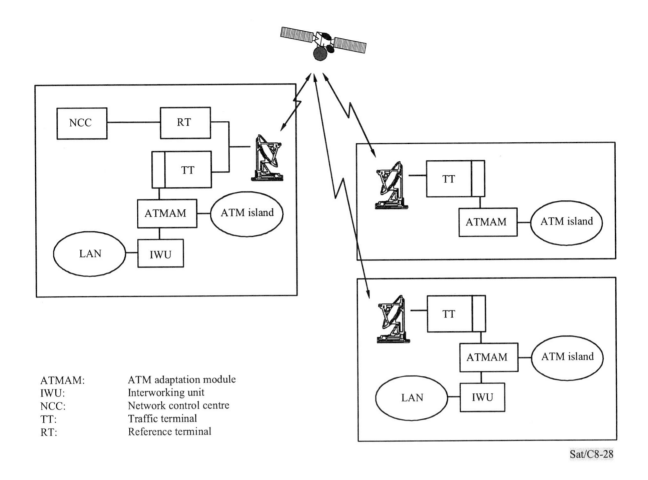

ATMAM: ATM adaptation module
IWU: Interworking unit
NCC: Network control centre
TT: Traffic terminal
RT: Reference terminal

Sat/C8-28

FIGURE 8.28

The CATALYST network

8.6.8 Conclusion: Future architectures for advanced ATM satellite networks

In order to allow satellites to play a major role in the evolution of digital telecommunications at the national level (national information infrastructure – NII) and at the worldwide level (global information infrastructure – GII), much effort is being spent in various organizations to define and standardize the architectures and implementation processes of satellite networks using ATM, either at the user access or as an interconnection medium, or both[6].

In the case of transparent satellites and of fixed ATM networks, two scenarios for high speed data transmissions can be envisaged:

---

[6] This conclusion is based on the following references: IEEE Communications Magazine, March 1999, Vol. 37, No. 3, which includes five papers on satellite ATM network architectures (pages 28 to 71), in particular: "Next-generation satellite networks: Architectures and implementation", by P. Chitre and F. Yegenoglu (COMSAT Laboratories).

- access by user terminals: The VSAT type user terminals (UTs) are connected through a standard ATM user network interface (UNI) to a gateway earth station (GES). This GES can be directly connected to a terrestrial broadband network through a UNI or it could perform switching or concentration, being therefore connected through a network-network interface (NNI);

- interconnection of remote ATM networks, i.e. high-speed interconnection (e.g. up to 1.2 Mbit/s) by GESs among ATM islands: In this scenario, the GESs do not perform switching or concentration which are effected in the switches of the local network(s). Of course, the traffic is less bursty than in an access from a UT. The interface is normally through an NNI.

It is usually assumed that resource management functions are implemented to allow a dynamic assignment of the bandwidth at the UTs and at the GESs, this function being centralized in a network control centre (NCC).

Implementation procedures are also developed to support terminal mobility functions in these satellite ATM networks. Mobility functions can be needed due to two possible circumstances: mobility of the UT (portable terminal) and mobility due to the satellite motion in the case of non-geostationary (NGSO) satellite systems (e.g. LEO constellation). The former case requires only UT location management procedures, while the latter case requires handover support. Handover can be needed at the UT or at the GES from one satellite to another or from one satellite beam to another. In these cases, the handover does not require rerouting at the ATM layer and can be performed at the satellite physical layer. However, handover can also be needed from one GES to another when the satellite leaves the line of sight of one GES and enters another. In this case, rerouting of the ATM connection is needed since the two GESs are connected to the terrestrial ATM network via two separate switches.

Finally, mobility can also be envisaged at the satellite network level for providing interconnectivity between a mobile network (e.g. an aeronautical ATM transmission network) and a fixed network or between two mobile networks.

In the case of on-board processing satellites, the satellite performs ATM switching functions, the control function being possibly distributed between the on-board ATM switch and the NCC on the ground. Various types of connectivity should be envisaged, depending on the implementation of the network and of the satellite system operation (in particular when the satellite system forms a mesh ATM network in the sky via intersatellite links (ISLs)).

8.7 Interconnection of television networks, television distribution

8.7.1 Historical overview

Historically, the introduction of satellites for transmitting and distributing television programmes can be seen as the means by which the television service has conquered its natural medium. Before the advent of communication satellites, television was distributed from studios to the regional or local television transmitters by microwave link networks which were at that time the only available long distance transmission medium with a sufficient bandwidth. International transmission of TV programmes by radio means (and in consequence, live long distance transmission) did not exist. In

1964, the first generation of communication satellites made it possible to transmit TV over intercontinental distance, but they needed rather expensive earth stations to enable them to be used to replace regional or local microwave networks. Since the seventies, domestic satellites with higher e.i.r.p. and even INTELSAT leased transponders have allowed, not only the transmission of TV programmes to isolated TV stations, but a generalized replacement, on a cost/effectiveness basis, of terrestrial microwave links by satellite links for distributing TV programmes to local TV stations, broadcast transmitters, cable distribution (CATV) head-ends, etc. Moreover, the availability of powerful satellites (at 6/4 GHz and, more promisingly, at 14/11-12 GHz), and also of mass produced, low cost, small receiving earth stations (TVROs), have made it possible to feed directly by satellite community TV distribution systems such as CATVs, SMATVs, i.e. satellite master TV antennas, hotels, etc. and even home receivers (DTH, i.e. direct-to-home satellite terminals). And now, a new era is opening, with the advent of digital TV which permits the broadcasting of multiplexed arrangements of high quality programmes (often called "bouquets"). All these achievements have been accomplished in the framework of the FSS, without resorting to the broadcasting satellite service (BSS). Note, however, that direct broadcast satellites (DBS), operating in the BSS, are also employed in some countries for direct-to-home service with very small TVRO antennas.

8.7.2   Interconnection of TV networks and distribution via satellite

Today's regional and international satellites are owned and operated by large corporations or consortiums. Transponder time is then sublet to users who provide programming. For example, in the United States, satellite television transponders are used by cable programmers to relay programmes to their cable TV affiliates across the country. Some satellite-delivered TV programmes are created for home dish owners.

a)   TV networks use satellite transponders to provide programming to their affiliate stations across the country. Several hundred local TV broadcasting stations have their own dishes (as shown in Figure 8.29) and they can choose many TV programmes on satellite transponders to supplement or replace some of the regular network programmes.

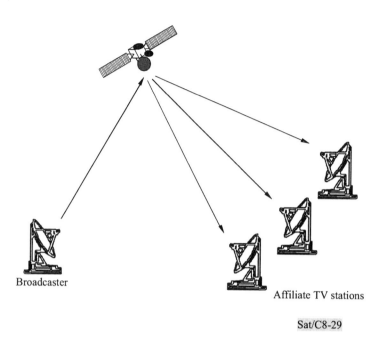

Broadcaster

Affiliate TV stations

Sat/C8-29

FIGURE 8.29

Television network distribution

b)   Cable television systems (CATV) headend distribution: satellite earth receiving stations or televisions which receive only (TVRO) stations are most commonly used by CATV systems. Television picture and sound programme material may originate from local or distant broadcasting stations relayed by satellite to cable headends as shown in Figure 8.30. Signals received will then be processed at the cable headends and redistributed to their subscribers. Some pay television services and a number of advertiser-supported networks serving CATV are distributed by satellite. More and more cable networks use satellite to expand their international markets. Some of their programmes are modified to match local interests. To reach subscribers globally a "double hop" communication as shown in Figure 8.31 may be required.

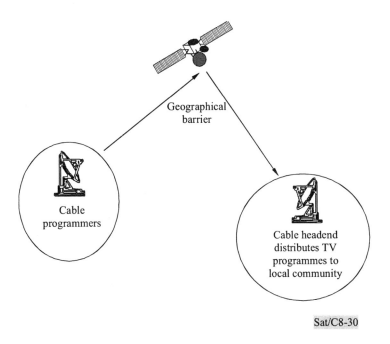

Sat/C8-30

FIGURE 8.30

Cable headend distribution

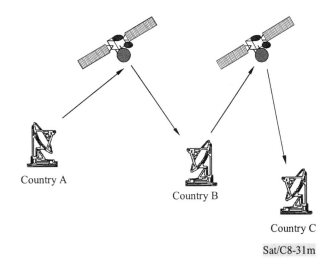

Sat/C8-31m

FIGURE 8.31

International television distribution - double hopped by satellites

The broadcast cable networks use satellite transponders for the relay of news programmes as shown in Figure 8.32. For instance, several transponders may be used simultaneously, with one used as the main feed to other affiliates and others as feeds from different parts of the countries or overseas.

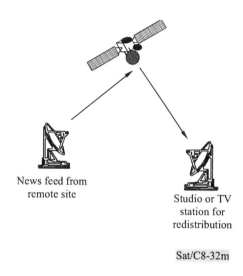

News feed from
remote site

Studio or TV
station for
redistribution

Sat/C8-32m

FIGURE 8.32

Satellite news gathering

c)   Direct broadcast satellite (DBS) and fixed-satellite system (FSS)

The term DBS referred exclusively to high-powered Ku-band satellites operating within the 12.2 to 12.7 GHz broadcasting-satellite service (BSS) band – a range of frequencies assigned by the World Administrative Radio Conference (WARC) for direct satellite broadcasting within the western hemisphere. However, FCC has permitted non-broadcasting frequencies 11.7 to 12.2 GHz to be used for interim DBS ventures. At present, a number of companies have made preliminary moves to establish themselves as DBS services in the United States. A true DBS system offers 9 degree spacing.

In Europe, Australia and Japan however, there are several Ku-band frequency ranges in use: 10.95-11.70, 11.70-12.50 and 12.50-12.75 (Europe), 11.90-12.10 GHz (Japan) and 12.25-12.75 GHz (Australia and Japan). Circular polarization is used by the European and Japanese high-powered DBS satellites, while linear-polarization is used by all European, Australian and Japanese medium-powered Ku-band satellites. It should be noted that circular polarization makes receiving equipment much easier to align and install. Also, true DBS systems typically use smaller dishes than FSS systems.

Various satellite TV programming from these broadcasting satellites is available, such as sports, movies, pay-per-view, home shopping, and news channels, etc. Some of these programmes are advertiser supported while the use of a descrambler and the payment of a monthly subscription fee are required.

Present developments of DBS and actions by regulatory bodies worldwide promise more channels by a number of broadcasting and communication organizations in the Ku band. The services will include signals on both conventional 525- and 625-line video standards and HDTV standards using a substantially greater number of lines and wider video baseband.

8.7.3 Satellite transmission and broadcasting of digital television

The drive towards digital broadcasting has effectively come from both the broadcaster and the consumer. Broadcasters have been seeking easier editing and the capacity to introduce pay-per-view, video-on-demand and online games, while consumers have been striving for wider programme choice and improved picture quality.

8.7.3.1 Driving opportunities for digitization

a) Wider programme choice

Prior to the introduction of cable and direct satellite broadcasting there was a demand for multiple TV programmes and for new television delivered services. This increased demand was reinforced by a request for more specialized programmes. It appeared that consumers were no longer satisfied with channels that show a wide selection of programmes but prefer channels that accurately respond to their specific requirements, e.g. channels devoted to films, to sports, to music, to telemarketing, to education, etc. or even to very specialized hobbies. Digital editing being simpler than previously current processes, costs are reduced and specialist channels can become commercially viable as required returns are proportionally lower.

b) Improved picture quality

In addition to a wider choice of programming, digital television gives improved picture quality. Television reception is constant, regardless of physical barriers. Improved picture quality facilitates the introduction of new applications such as online games. Similarly, consumers will be more willing to pay for video-on-demand if picture quality is assured. The prospect of films in wide-screen format and with cinema-quality digital sound is likely to encourage some subscribers to invest in digital receiving equipment.

c) New applications and services

Digital broadcasting opens up possibilities for new applications and services such as pay-per-view, video-on-demand (VOD), interactive games and displays (similar to video disks), educational programme downloading, etc.

The development of new applications has a number of effects. Pay-per-view and VOD are significant in that they offer new revenue streams. Consumers are paying for the programming they watch as opposed to a flat rate regardless of viewing. VOD will be a direct competitor to video rental and so is of added importance. The use of television for the downloading of games changes the role of the television from being a receiving device to a focus for home-based leisure activities. This is reinforced by the use of television as an aid to training and education. Each of these new applications has a direct effect on the intricacies of market development.

d)   Digital programme production

With the advent of digital broadcasting and the proliferation of cable, satellite and terrestrial television channels, the quest for content will become ever more fierce. Here digitization offers a further benefit, the development of digital production.

Digital cameras, recently developed, store picture and sound in digital form on a computer disk as opposed to on film. This aids the process of editing. Historically, the process of editing film footage required highly complex and specialized machinery to copy and splice video pictures into a finished programme. With material stored digitally, it is possible to move this process away from highly specialized equipment and onto the desktop. The advantages of the newer systems will not be primarily hardware and software costs, as the technical requirements of desktop editing systems are still highly complex, but rather the ability to quickly and easily create broadcast quality programming.

8.7.3.2   Effects of digitization

a)   Multiplication of channels

The principal effect of digitization is to allow the multiplication of channels. For example, eight to ten channels can be broadcast in the same bandwidth as an analogue broadcast, i.e. in a conventional (say 40 MHz) satellite transponder. This is permitted by a bit-rate compression, which is carried out according to MPEG standards. Currently the transmitted bit rate generally implemented is 4 Mbit/s (compressed, e.g. from a 270 Mbit/s stream from a digital camera).

b)   Variable bit rate and dynamic bandwidth allocation

In picture processing and encoding, the use of "discrete cosine transform (DCT)", by which only changes in a picture are transmitted, drastically reduces the amount of data, i.e. the effective bit rate, that needs to be coded. DCT is particularly relevant in live events broadcasting because the amount of on-screen motion affects the rate at which data can be transmitted. For example, live broadcast from a newsroom requires little change in picture (the background is usually fixed), whereas a live sporting event is highly dynamic and requires more information to be transmitted. In consequence, whilst some programmes require greater bandwidth than others, it is possible to compensate for this effect. A process called statistical multiplexing allows bandwidth to be dynamically assigned to video channels, for example, a sport programme would have more bandwidth than a news broadcast within the same multiplex.

c)   Picture improvement and transmission quality

As explained above, digital television gives an improved picture quality, although this quality depends, of course, on the compression factor. The picture improvement in quality and clarity results from the fact that compression technologies remove the distortions which cause blurring or ghosting in conventional analogue transmission.

As concerns the quality of the satellite link proper, provision is generally made in the satellite link budget to guarantee a very good transmission quality, even in poor meteorological conditions and with small TVRO antenna dishes. This is because, due to the implementation of powerful error correcting codes (e.g. Reed-Solomon codes), the system performance objective is "quasi error free" operation, corresponding to less than one uncorrected error event per transmission hour, i.e. BER $\leq 10^{-10}$ or 10^{-11}, depending on user bit rate (see Chapter 2, § 2.2.3.3).

8.7.3.3   Connection of user terminals

a)   Source information and coding

Digital cable, satellite and terrestrial broadcasting have a number of common elements, particularly the processes of encoding, multiplexing and modulating digital signals. In all cases, source information consisting of video, audio, text and data is digitized and coded using MPEG standards as shown in Figure 8.33.

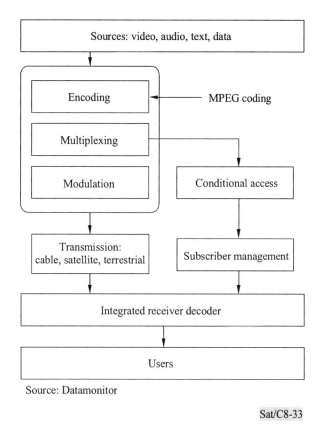

Source: Datamonitor

Sat/C8-33

FIGURE 8.33

Digital broadcasting structure

Once programming information has been coded, multiplexing and modulation allows channels to be overlaid and data prepared for transmission. Although this process differs slightly between cable, satellite and terrestrial, the result in all cases is a transmission bit stream.

b)   Multiplexing and conditional access

At the multiplexing stage, access to services can be restricted and the possibility of conditional access emerges. Conditional access refers to the process whereby signals are encrypted, allowing pay-per-view or subscription broadcasting with decoding, generally through a decoder card (a form of smart card) restricted to those homes which pay their charge. Conditional access services require subscriber management and control because they need to account for management operations, such as billing, by the operator.

c)   User equipment

The user equipment basically comprises an antenna with its low noise block (LNB) receiver, the distribution set (which can be a simple connection cable, but which can also include a distribution frame in the case of a community of subscribers) and an integrated receiver decoder terminal which converts the received bit flow into signals usable by the television set. This terminal is generally integrated in a set-top box. In the future, fully integrated digital television sets will become commercially available. More information on the user equipment will be found in Chapter 7 (§ 7.10).

d)   Hybrid analogue and digital services

Until digital services become commonplace, the old analogue services will be run concurrently with the new digital services. Public-sector broadcasters (PSBs) cannot "switch off" or discontinue access to the current analogue programmes. Hence, full conversion to digital broadcasting will certainly take years. PSBs may need to transmit services in digital format, using currently spare frequencies or satellite capacity, and simultaneously broadcast analogue services, a process known as simulcasting.

e)   The importance of standards and regulations

With cable, satellite and terrestrial broadcast services all moving toward digital transmission, the need for universal standards increases. Standards like those provided by the digital video broadcasting (DVB) project provide consistency and compatibility of products and services.

Consistency and compatibility are important both to the consumer entering the digital marketplace and to the developers wishing to supply digital services to the market. Standards also create an umbrella for component production, which allows all companies involved to exploit economies of scale, thereby reducing the purchase price of new digital technology.

The delivery mechanism that channel packages use is likely to alter significantly. The requirement for a decoder for each medium and the perceived unwillingness of consumers to have a "stack" of set-top boxes means that the first digital set-top box will be the only digital receiving equipment installed.

The role of national or regional regulatory agencies is very important. In Europe, for example, the European Commission (EC) has already stipulated that access should be "fair, open and non-discriminatory", this may mean that a single decoder must be able to receive all digital broadcasts; terrestrial, satellite and cable. National governments may add more stringent legislation to ensure that PSBs do not lose out to commercial broadcasters.

APPENDIX 8.1

Introduction to signalling systems

AP8.1-1 Signalling systems to be used within satellite networks

AP8.1-1.1 General

Signalling is the exchange of information (other than by speech) specifically concerned with the establishment, release and other control of calls, and network management, in automatic telecommunication operations.

Four main signalling systems have been standardized and are applicable within an FSS satellite network:

- Signalling System No. 5, standardized by ITU-T in 1964;

- Signalling System R2, standardized by ITU-T in 1968;

- Signalling System No. 6, standardized by ITU-T in 1968;

- Signalling System No. 7, standardized by ITU-T in 1980.

Signalling Systems No. 5 and R2 are analogue systems. In these systems, signalling is transmitted through the same transmission channel as the user's communications, by means of pulses or voice frequencies.

Signalling Systems Nos. 6 and 7 are common channel signalling systems. This signalling method uses separate channel(s) to transmit signalling information related to circuits.

A summary of the signalling methods is given in Table AP8.1-1.

AP8.1-1.2 Analogue signalling systems

Two types of signalling are considered:

- line signalling, which deals with the circuit handling or clearing, and the supervision of the overall transmissions;

- interregister signalling, which is exchanged during the call establishment (e.g. numbering signal).

TABLE AP8.1-1

Suitability of signalling systems for specific criteria

	Used on satellite links	**Used with speech interpolation systems**	**Used with echo suppressor systems**
System No. 5	Signals are not provided for indicating whether the connection already includes a satellite link	Suitable	Signals are not provided for echo suppressor control
System R2	Contains signals indicating whether the connection already includes a satellite link A non-compelled version can be used in order to speed up the exchange of signals via satellite	Suitable if the DSI system is transparent to pulsed interregister signals	Interregister signals for echo suppressor control are included
System No. 6	Contains signals indicating whether the connection already includes a satellite link	Suitable	Contains signals for echo suppressor control
System No. 7	Contains signals indicating whether the connection already includes a satellite link	Suitable	Contains signals for echo suppressor control

i)   Line signalling

Analogue line signalling systems can be divided into in-band (in the speech frequency band) and out-band systems. In addition, two signalling techniques may be employed: pulsed signalling or continuous signalling.

In-band systems require the use of a guarding method to prevent false operation of the signalling equipment by speech interference. Even so, occasional receiver misoperation by speech can occur, and thus in the speech phase a suitable minimum signal recognition time should be chosen.

Out-band signalling requires the use of a transparent 4 kHz bandwidth between the two signalling equipments. Part of the signalling equipment is usually provided within the transmission equipment.

Continuous signalling is arranged to operate with "tone-on-idle" signal. It consists in sending a tone whenever the line is free. Such systems have the inherent advantage of allowing immediate identification of circuit availability. Only two signal states are available in each direction, thus recognition arrangements are simple.

Pulsed signalling allows a greater signal variety ("the repertoire") than continuous signalling, but requires more complex signal recognition arrangements. In general, the signalling tone is recognized by the signal receiver but requires persistent checking and correlation with the circuit state before the signal is validated.

ii)   Interregister signalling

Two types of interregister signalling may be suitable: the decadic signalling and the multifrequency signalling.

Decadic signalling uses the same frequency and sender/receiver equipment as the line signalling. Signals are composed of a sequence of tone pulses analogous to subscriber line signalling employing rotary dials.

Multifrequency signalling has the advantage of greater speed and signal repertoire than decadic systems. To provide both an adequate repertoire and signalling reliability, signals are composed of two frequencies from a set of 4, 5, 6 or 8 frequencies.

AP8.1-1.3 Common channel signalling systems

Common channel signalling is a signalling method in which a separate channel conveys, by means of labelled messages, signalling information relating to a multiplicity of circuits, or other information such as that used for network management. Common channel signalling can be regarded as a form of data communication that is specialized for various types of signalling and information transfer between processors in telecommunication networks.

The signalling system uses signalling links for transfer of signalling messages between exchanges or other nodes in the telecommunications network served by the system. Arrangements are provided to ensure reliable transfer of signalling information in the presence of transmission disturbance or network failures. These include error detection and correction on each signalling link. The system is normally applied with redundancy of signalling links and it includes function for automatic diversion of signalling traffic to an alternative path in case of link failures.

The main advantages of the common channel signalling method are:

*   signalling is completely separate from speech. Therefore the likelihood of misoperation due to imitation by speech signals is greatly reduced;

*   the exchange of information between nodes is not subject to physical circuit seizure;

*   signalling is significantly faster, therefore speech circuit utilization is maximized;

*   there is potential for a very large number of signals;

*   it is possible to handle management and administration signals during speech;

*   there is flexibility to add or change signals (without requiring changes to hardware);

*   it is possible to carry user-to-user, or access-to-access information in case of ISDN service.

AP8.1-1.4 Specific criteria for satellite networks

The factors that influence the selection of a given signalling system for a satellite application (see Table AP8.1-1), are the following:

i)   Propagation delay

Some signalling systems will not work correctly over satellite links since the long propagation delay (240 to 280 ms one way) exceeds that assumed by the line signalling specifications. This is the case for compelled signalling systems in which each forward signal is acknowledged by a backward signal. As the propagation time has to be included twice in one signalling cycle, the exchange of signals is rather slow if the propagation delay is long.

In addition, the inclusion of two satellite links in a speech connection can be avoided by means of signalling that informs the transit centre that such a satellite link is already included in the connection and that a terrestrial link should be selected during the following routing process.

ii)   Echo control

Satellite links require the insertion of echo suppressor (or echo canceller) devices. Arrangements should be incorporated in the switching equipment to prevent echo suppressor action from disturbing forward and backward signalling. Typical arrangements are:

*   locating the echo suppressors on the switching side of the signalling equipment;

*   inhibiting the action of echo suppressors located on the line side of the signalling equipment by means of an appropriate control signal sent from the signalling equipment to the echo suppressor while signalling is in progress.

iii)   Speech interpolation

Some signalling systems may not be compatible with speech interpolation systems, e.g. the speech interpolation equipment may cause excessive clipping of pulse signals resulting in their non-recognition by the distant signalling equipment.

AP8.1-2 Signalling System No. 5

AP8.1-2.1 General

Signalling System No. 5 is an analogue system and all the signalling is transmitted though the user's communication path.

It is suitable for both-way operation and for terminal and transit working; in the latter case two or three circuits equipped with Signalling System No. 5 may be switched in tandem.

The signalling equipment is in two parts: the line signalling for the supervisory signals and the register signalling for the numerical signals (call establishment).

AP8.1-2.2 Line signalling

The line signal coding arrangement is based on the use of two in-band frequencies (2 400 Hz and 2 600 Hz) transmitted individually or in combination as shown in Table AP8.1-2.

The signalling equipment must operate in a sequential manner retaining memory of the preceding signalling state and the direction of signalling in order to differentiate between signals of the same frequency content. All signals except the forward-transfer signal are acknowledged in the compelled-type manner as indicated in Table AP8.1-2.

Recognition time is defined as the minimum duration which a direct-current signal, at the output of the signal receiver, must have in order to be recognized by the switching equipment. All recognition times are the same (125 ms) except for the seizing and proceed-to-send signals (40 ms) as these two signals are not subject to signal imitation by speech and fast signalling is desired in particular to minimize double seizings.

AP8.1-2.3 Register signalling

This is a link-by-link two out of six multifrequency in-band en block pulse signalling system, with forward signalling only. En bloc register signalling is the transmission by a register, of all the call information as a whole in a regular timed reference of signals. The signal consists of a combination of two out of six frequencies (700, 900, 1 100, 1 300, 1 500 and 1 700 Hz). The address information necessary to route the call to its destination is assembled into one block which is transmitted as a rapid sequence of multifrequency pulse code signals.

Automatic access to the international circuits must be used for outgoing traffic and the numerical signals from the operator or subscriber are stored in an outgoing international register before an international circuit is seized. The register signals are presented in Table AP8.1-3. As soon as the "end-of-pulsing" condition is available to the outgoing register, a free international circuit is selected and a seizing line signal transmitted. On receipt of a proceed-to-send line signal the seizing signal is terminated and a "start-of-pulsing" pulse, followed by the numerical signals, is transmitted by the register. The final register signal transmitted is an "end-of-pulsing" pulse.

TABLE AP8.1-2

Line signal code

Signal	Direction[1]	Frequency[2]	Sending duration	Recognition time
Seizing	→	f1	continuous	40 ± 10 ms
Proceed-to-send	←	f2	continuous	40 ± 10 ms
Busy-flash	←	f2	continuous	125 ± 25 ms
Acknowledgement	→	f1	continuous	125 ± 25 ms
Answer	←	f1	continuous	125 ± 25 ms
Acknowledgement	→	f1	continuous	125 ± 25 ms
Clear-back	←	f2	continuous	125 ± 25 ms
Acknowledgement	→	f1	continuous	125 ± 25 ms
Forward-transfer	→	f2	850 ± 200 ms	125 ± 25 ms
Clear-forward	→	f1 + f2 (compound)	continuous	125 ± 25 ms
Release-guard	←	f1 + f2 (compound)	continuous	125 ± 25 ms

(1) ——————→ forward signals   ←—————— backward signals

(2) f1 = 2 400 Hz f2 = 2 600 Hz

Seizing: transmitted at the beginning of a call to initiate circuit operation and to seize equipment for switching.

Proceed-to-send: to indicate that the equipment is ready to receive the numerical signals (see register signalling).

Busy-flash: to show that either the route, or the called subscriber, is busy.

Answer: to show that the called party has answered the call.

Clear-back: to indicate that the called party has cleared.

Forward-transfer: send when the outgoing exchange operator wants the help of an operator at the incoming exchange.

Clear-forward: sent when the calling subscriber clears.

Release-guard: sent in response to the clear-forward signal. It serves to protect circuits against subsequent seizure as long as the disconnection operations have not been completed.

TABLE AP8.1-3

Register signals

Signal	Function
Start-of-pulsing	Sent on receipt of a proceed-to-send signal and used to prepare the incoming register for the receipt of the numerical signal
Numerical signal	Provides information necessary to affect the switching of the call in the desired direction
End-of-pulsing	Sent to show that there are no more numerical signals to follow

AP8.1-3 Signalling System R2

AP8.1-3.1 General

Signalling System R2 is an analogue system used as an international signalling system within international regions (world numbering zones). Moreover, System R2 can be used for integrated international/national signalling if it is employed as a signalling system in the national networks of the region concerned.

System R2 is suitable for terminal and transit working. It is specified for one-way operation on analogue and digital transmission systems and for both-way operation on digital systems.

Distinction is made between line signalling (supervisory signals) and interregister signalling (call set-up control signals). Two versions of line signalling are specified for use on carrier systems (analogue) and PCM systems (digital). The interregister signalling is a compelled multifrequency code system.

System R2 offers a high reliability for the transmission of information necessary for setting up a call. It provides sufficient signals in both directions to permit the transmission of numerical and other information relating to the called and calling subscriber lines and to increase routing facilities.

AP8.1-3.2 Line signalling

The analogue version is a link-by-link using the out-band, low level continuous tone-on-idle signalling method. When the circuit is in the idle state, a low-level signalling tone is sent continuously in both directions over the signalling channels. The tone is removed in the forward direction at the moment of seizure and in the backward direction when the called subscriber answers. The connection is released when the signalling tone is restored. The six characteristic states of the circuit are shown in Table AP8.1-4.

This signalling method, requiring only simple equipment, provides rapid signal recognition and transmission. The signal transfer speed provided by continuous type signalling compensates for the need for signal repetition inherent in link-by-link transmission.

The digital version of System R2 line signalling is a link-by-link using two signalling channels in each direction of transmission of the speech circuits. Primary PCM multiplexes economically provide such channels (for 2 048 kbit/s PCM systems, the signalling information of the 30 speech channels is transmitted in time slot No. 16. By making use of the increased signalling capacity, simplification of the outgoing and incoming switching equipment can be achieved. Table AP8.1-5 shows the signalling code on the PCM line under normal conditions.

The analogue version and the digital version of the line signalling can be converted to each other by a transmultiplexer. Such an equipment forms a conversion point between the analogue (FDM) transmission and digital (PCM) transmission systems.

TABLE AP8.1-4

Line signalling code (analogue version)

State of the circuit	Line signalling condition	
	Forward	Backward
1) Idle	Tone-on	Tone-on
2) Seized	Tone-off	Tone-on
3) Answered	Tone-off	Tone-off
4) Clear-back	Tone-off	Tone-on
5) Release	Tone-on	Tone-on or off
6) Blocked	Tone-on	Tone-off

TABLE AP8.1-5

Line signalling code (digital version)

State of the circuit	Line signalling condition			
	Forward		Backward	
	a_f	b_f	a_b	b_b
Idle/Released	1	0	1	0
Seized	0	0	1	0
Seizure acknowledged	0	0	1	1
Answered	0	0	0	1
Clear-back	0	0	1	1
Clear-forward	1	0	0	1
			or	
			1	1
Blocked	1	0	1	1

"a_f" channel identifies the operating condition of the outgoing switching equipment and reflects the condition of the calling subscriber's line.

"b_f" channel provides a means for indicating a failure in the forward direction to the incoming switching equipment.

"a_b" channel reflects the condition of the called subscriber's line. "b_b" channel indicates the idle or seizing state of the incoming switching equipment.

AP8.1-3.3 Register signalling

The interregister signalling is performed end-to-end using a two out of six in-band multifrequency code with forward and backward compelled signalling. It uses six signalling frequencies (1 380, 1 500, 1 620, 1 740, 1 860 and 1 980 Hz) in the forward direction and six signalling frequencies (1 140, 1 020, 900, 780, 660 and 540 Hz) for the backward direction.

The outgoing R2 register starts call set-up as soon as it has received the minimum requisite information. Therefore signal transfer starts before the complete address information is received (i.e. before the caller finishes dialling). This overlapping interregister signalling is in contrast to the en bloc register signalling (see System No. 5) in which the transmission of all the address information is made as a whole in one sequence starting only after it has been completely received.

The System R2 interregister signalling is a compelled signalling system (the basic compelled cycle) is shown in Figure AP8.1-1. The acknowledging backward signals, besides being a functional part of the compelled procedure, serve to convey special information concerning the required forward signals – to indicate certain conditions encountered during call set-up or concerning the state of the called subscriber's line to be transferred.

In addition to the capability of transferring address information, the use of the 15 forward and 15 backward combinations of the multifrequency code provides several operational features such as the control of echo suppressors, information concerning the nature and the origin of the call (national or international, from an operator or from a subscriber, from data-transmission, maintenance or other equipment, etc.) information on the nature of the circuit (satellite link), information on congestion, unallocated number, etc.

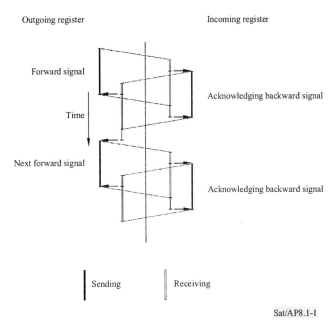

Outgoing register   Incoming register

Forward signal

Acknowledging backward signal

Time

Next forward signal

Acknowledging backward signal

Sending   Receiving

Sat/AP8.1-1

FIGURE AP8.1-1

Compelled signalling cycle

AP8.1-3.4 Compatibility of System R2 with satellite links

The R2 signalling system, which was not initially designed for satellite links, is being used by EUTELSAT with some adaptations, as explained below:

Signalling System R2 is an analogue system in which signals are transmitted through the user's communication transmission channel. Therefore, all the signalling goes through the terrestrial switched network and is handled at each point of this network (local, transit and international switching centres, see Figure AP8.1-2).

System R2 has been designed to allow end-to-end interregister signalling over several links in tandem without signal regeneration in the intermediate exchanges. However, in circumstances where the transmission conditions do not comply with the requirement specified for System R2, or in case of using this system via a satellite link, the overall multilink connection is divided into sections, each with its own interregister signalling (signals being then relayed and regenerated by a register at the point where the division is made).

System R2 is a fully-compelled signalling system. That means that each forward signal is acknowledged by a backward signal. Operation of such a compelled signalling system may be affected by the satellite propagation delay which increases for instance the holding times and the post-dialling delay. In order to avoid a loss in service quality due to this adverse effect, a non-compelled signalling system can be used on the satellite link (see Figure AP8.1-3). Such a system is designed to speed up the process of interchange of signals via the satellite. The basic differences with the specifications of Signalling System R2 are the way of sending signals in both directions (signals are sent in the form of pulses) and the non-existence of acknowledgements.

The typical end-to-end call set-up procedure is shown in Figure AP8.1-4.

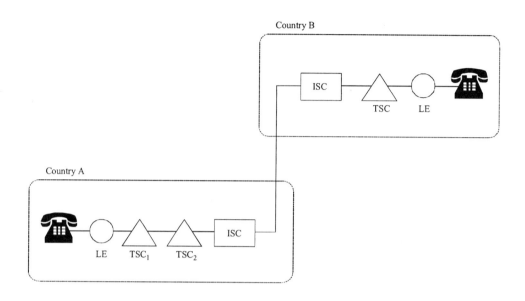

ISC: international switching centre
TSC: transit switching centre
LE: local exchange

Sat/AP8.1-2

FIGURE AP8.1-2

International signalling network architecture

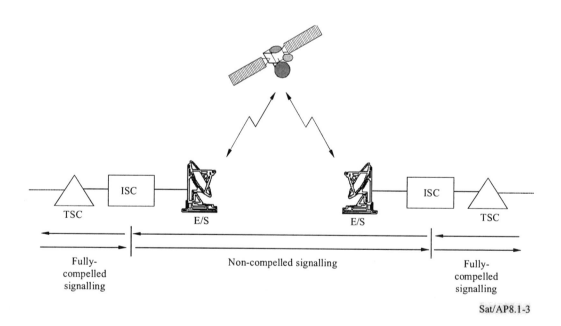

Sat/AP8.1-3

FIGURE AP8.1-3

Signalling network with a satellite link

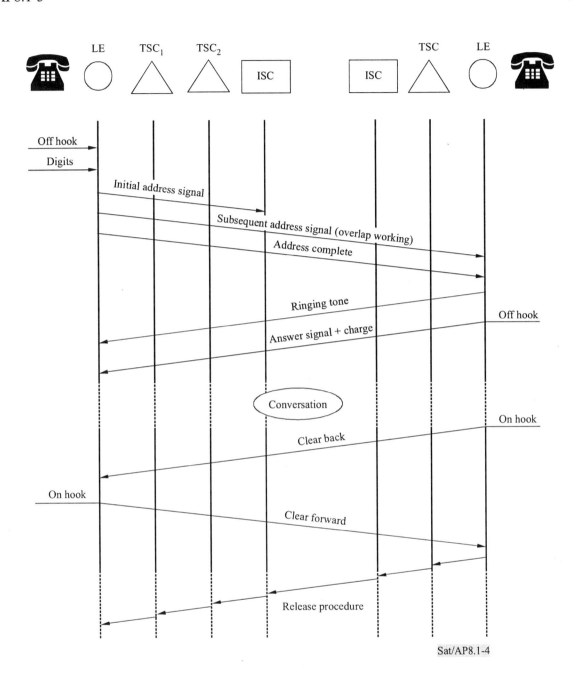

FIGURE AP8.1-4

Typical end-to-end call set-up

AP8.1-4 Signalling System No. 6

AP8.1-4.1 General

Signalling System No. 6 can be used to control the switching of all types of international circuits and be used in a world-wide connection, including speech interpolation circuits and satellite circuits.

It is designed for both-way operation and is suitable for terminal or transit working.

The system can also be used for regional and national applications, and a large part of the signal code capacity is reserved for this purpose.

Moreover, a large unused signal code capacity will allow the addition of new signals to cater for unknown future requirements. This spare capacity may be used for increasing the number of telephone signals as well as for introducing other signals, e.g. network-management or network-maintenance signals.

The system features are obtained by entirely removing the signalling from the speech paths and introducing the concept of a separate common signalling link over which all signals for a number of speech circuits are transferred. A number of these common signalling links interconnected by a number of signal transfer points will form a coherent signalling network which can transfer all signals for all speech circuit groups within that network area.

The signalling system may be operated both in an associated mode and in a non-associated mode. In the associated mode of operation, the signals are transferred directly between the two exchanges which are the end points of a group of speech circuits. In the non-associated mode of operation, the signals are transferred via two or more common signalling links in tandem associated with other groups of circuits, intermediate nodes acting only as signal transfer points. Non-associated mode of operation makes the signalling channel economically suitable for use with small circuit groups by sharing the capacity of the signalling link among several groups.

This signalling system is not widely implemented. It is still used in the North American common channel inter-office signalling (CCIS) but it will eventually be replaced by Signalling System No. 7.

AP8.1-4.2 Description of the common channel signalling system

The common signalling link is capable of operation over analogue circuits (its use over digital circuits is also possible).

Each signalling channel of the system (shown in Figure AP8.1-5) is operated synchronously: a continuous stream of data flows in both directions. The data stream is divided into signal units of 28 bits each, of which the last eight are check bits. These signal units are grouped into blocks of 12 signal units. The first 11 signal units carry either telephone, management or synchronization signals. The twelfth and last signal unit of each block is an acknowledgement signal unit coded to indicate the number of the block being transmitted, the number of the block being acknowledged and whether or not each of the 11 signal units of the block was received without error.

The transmission of a signal in System No. 6 starts in the processor as shown in Figure AP8.1-6. Signals corresponding to the information to be transmitted, which may be one-unit messages or multi-unit messages, are stored in the output buffer according to their priority level. In the coder, each signal unit is encoded by the addition of check bits in accordance with the check bit polynomial.

In the analogue version of the signalling system, the signal is then modulated and sent to the outgoing voice frequency channel for transmission to the distant receiving terminal. The stream of pulses is normally transmitted at a rate of 2 400 bit/s using the four-phase modulation method.

In the digital version, the signal is passed through the interface adaptor before entering the outgoing digital channel. In the case of 2 048 kbit/s PCM primary multiplex, the stream of pulses is transmitted at 64 kbit/s.

The receiving function starts with the acceptance of the serial data from the transmission path. The output of the demodulator or the interface adaptor is delivered to the decoder where each signal unit is checked for errors on the basis of the associated check bits. Signal units received with detected errors are discarded. Signal units which are error-free are transferred to the input buffer after deletion of the check bits. The input buffer delivers the signal to the processor which analyses the signals and takes appropriate action.

The error control necessary for a common signalling link is based on error detection by redundant coding and on error correction by retransmission of those signal messages found to be in error.

In addition, provision is made for automatic transfer to an alternative link in the event of failure caused by breakdown or excessive bit-error ratio.

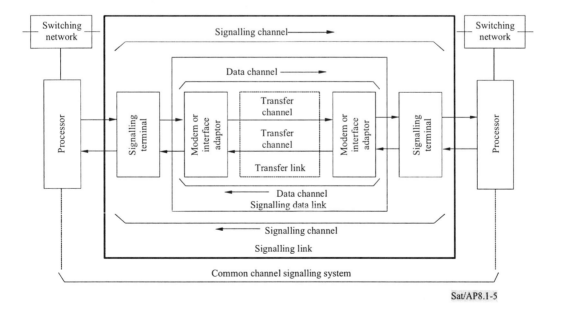

Sat/AP8.1-5

FIGURE AP8.1-5

Basic diagram of the common channel signalling system

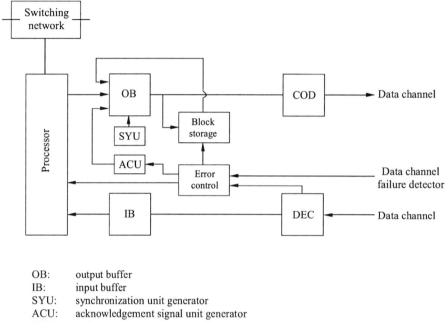

OB: output buffer
IB: input buffer
SYU: synchronization unit generator
ACU: acknowledgement signal unit generator
COD: coder
DEC: decoder

Sat/AP8.1-6

FIGURE AP8.1-6

Functional block diagram of a System No. 6 terminal

AP8.1-5 Signalling System No. 7

AP8.1-5.1 General

The overall objective of Signalling System No. 7 is to provide an internationally standardized general purpose common channel signalling (CCS) system:

*   optimized for operation in digital telecommunication networks in conjunction with stored programme controlled exchanges;

*   that can meet present and future requirements of information transfer for inter-processor transactions within telecommunication networks for call control, remote control, management and maintenance signalling.

Signalling System No. 7 is a modern system based on the concept of levels, or layers (see Appendix 8.2), which meets requirements of all control signalling for telecommunication services such as the telephone and circuit switched data transmission services. It can also be used as a reliable transport system for other types of information transfer between exchanges and specialized centres in telecommunication networks (e.g. for management and maintenance purposes). The system is thus applicable for multipurpose uses in networks that are dedicated for particular services and in multiservices networks. The signalling system is intended to be applicable in international and national networks.

AP8.1-5.2 Signalling system structure

The fundamental principle of the signalling system structure is the division of functions into a common message transfer part (MTP) and a separate user part for different users. This is illustrated in Figure AP8.1-7.

The overall function of the message transfer part is to serve as a transport system providing reliable transfer of signalling messages between the locations of communicating user functions.

The term user in this context refers to any functional entity that utilizes the common transport capability provided by the message transfer part. The user part comprises those functions of a particular type of user that are part of the common channel signalling system. A degree of commonality exists between certain user parts, e.g. the telephone user part (TUP) and the data user part (DUP).

The elements of the signalling system are specified in accordance with a level concept – illustrated in Figure AP8.1-8 – in which:

• the functions of the message transfer part are separated into three functional levels; and

• the user parts constitute parallel elements at the fourth functional level.

i) Signalling data link functions (level 1) define the physical, electrical and functional characteristics of a signalling data link and the means to access it. Level 1 provides a bearer for a signalling link. In a digital environment, 64 kbit/s digital paths (a time slot of PCM stream) are normally used, but other types of data links, such as analogue links with modem, can also be used.

ii) Signalling link functions (level 2) define the functions and procedures for the transfer of signalling messages between two points. The signalling message is transferred over the signalling link in variable length signal units which also include transfer control information for proper operation of the link. The major signalling link functions are:

   • delimitation of signal unit by means of flags;

   • error detection by means of check bits included in each signal unit;

   • error correction by retransmission and signal unit sequence control by means of explicit sequence numbers and explicit acknowledgements;

   • signalling link failure detection by means of signal unit error ratio and signalling link recovery by means of special procedures.

iii) Signalling network functions (level 3) define the transport functions and procedures that are common to the operation of individual signalling links. As illustrated in Figure AP8.1-9 these functions fall into two major categories:

a) signalling message handling functions which direct the message to the proper signalling link or user part;

b) signalling network management functions which control the current message routing and configuration of signalling network facilities.

iv) User part functions (level 4) consist of the different functions and procedures of the signalling system that are particular to a certain type of user. The extent of the user part functions may differ significantly between different categories of users, such as:

- users for which most user communication functions are defined within the signalling system, e.g. telephone and data call control functions;

- users for which most user communication functions are defined outside the signalling system. An example is the use of the signalling system for transfer of information for some management or maintenance purpose with an "external user".

AP8.1-5.3 Signalling message

Signalling messages are an assembly of information, defined at level 3 or 4, pertaining to a call, management transaction, etc., that are transferred by the message transfer function. Each signalling message is packed into a message signalling unit (MSU) which contains additional information from level 2 (see Figure AP8.1-9)

The beginning and end of a message signalling unit are indicated by a unique 8-bit pattern, call flag (F). Measures are taken to ensure that the pattern can not be imitated elsewhere in the unit.

A length indicator (LI) is used to indicate the integer number of octets of the signal unit.

Each message contains a service indicator octet (SIO) identifying the source user part and giving additional information such as an indication of whether the message relates to national or international application of the user part.

The signalling information of the message is given by the signalling information field (SIF) which includes the actual user information, such as telephone or data code control signals, management and maintenance information, identification of the type and format of the message. It also includes a label that provides information enabling the message to be routed to its destination.

For the purpose of acknowledgement and signal unit sequence control, each signal unit carries two sequence numbers. The signal unit sequence control is performed by means of the forward sequence number (FSN). The acknowledgement function is performed by means of the backward sequence number (BSN).

The error detection function is performed by means of 16 check bits (CK) provided at the end of each message signalling unit. The check bits are generated by the transmitting signalling link terminal by operating on the preceding bits of the message signalling unit following a specified algorithm. If consistency is not found between the received check bits and preceding bits according to this algorithm, then the message signalling unit is disregarded.

Two forms of error correction methods are provided: the basic method and the preventive cyclic retransmission one. The second method is the only one which is suitable for long transmission delay links (e.g. for satellite links). It is a positive acknowledgement, cyclic retransmission, forward error correction system. A signal unit which has been transmitted is retained at the transmitting signalling link terminal until a positive acknowledgement is received. During the period where there are no new signal units to be transmitted, all the signal units which have not yet been positively acknowledged are transmitted cyclically.

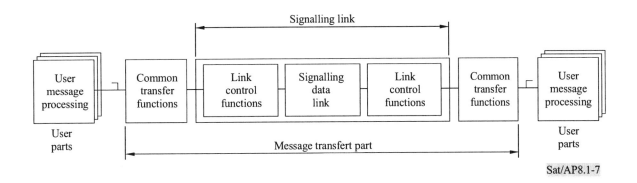

FIGURE AP8-1.7

Functional diagram for the common channel signalling system

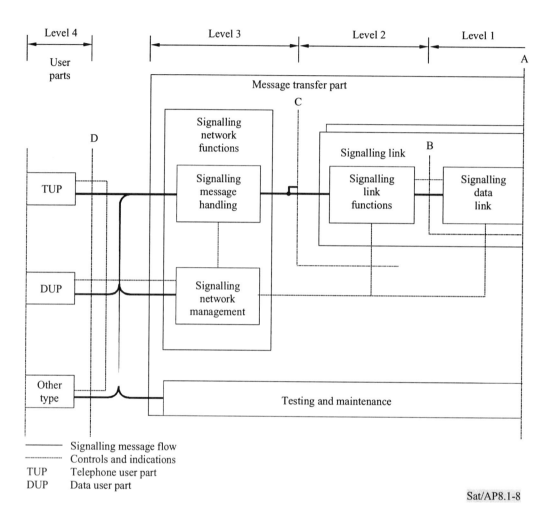

Sat/AP8.1-8

FIGURE AP8.1-8

General structure of signalling system functions

F	BSN	B I B	FSN	F I B	LI		SIO	SIF	CK	F
8	7	1	7	1	6	2	8	8n, n>2	16	8

PAN

F: flag   LI: length indicator
BSN: backward sequence number   SIO: service information octet
BIB: backward indicator bit   SIF: signalling information field
FSN: forward sequence number   CK: check bits
FIB: forward indicator bit

Sat/AP8.1-9

FIGURE AP8.1-9

Basic format of a message signalling unit (MSU)

APPENDIX 8.2

Introduction to the open systems interconnection (OSI) model

AP8.2-1 Introduction

In order to implement a universal compatibility between computers, computerized equipments, and communication networks, the International Organization for Standardization (ISO) in Geneva, has developed the open systems interconnection (OSI) model, which is a framework for defining the communication process between systems. This reference model defines the functions involved in communicating and the services and the protocols required to perform these functions.

Many of the protocols associated with the OSI model have been defined to incorporate existing standards (i.e. the digital ITU-T X.21, the ITU-T X.25, and the IEEE 802 for local area network (LAN), etc.) which are already widely implemented. Others are aimed at meeting the likely future requirements of international telecommunications. In particular, ongoing work on the integrated services digital network (ISDN) is closely related to work now taking place on OSI.

AP8.2-2 Overview of the OSI reference model

AP8.2-2.1 A layer structure

The OSI reference model, the structure of which is presented in Figure AP8.2-1, breaks the process of communicating into an orderly sequence of seven layers. It was adopted by the ISO in 1984.

Viewed as a system, the layers of the reference model can be split into two groups. The bottom three layers – physical, data link, and network – cover the components of the network used to transmit the message – satellite, X.25 network, or local area network (LAN). The top three layers – application, presentation, and session – reflect the characteristics of the communicating end systems, regardless of the physical means actually used. the transport layer acts as a liaison between the end systems and the network

AP8.2-2.2 Service concept

At each layer of the reference model, there are services to carry out the functions. For instance, a service such as requesting initialization is needed to initiate the control function for a conversation between two end systems.

When an application process initiates a communication, it passes its message down through each layer. The functions of each layer add value by providing services which enable the communication to be completed.

The main feature of OSI is that these service definitions are independent. In other words, every service can be implemented regardless of the method used to implement services in the layers above and below it. Error checking, for instance, may be provided in different systems by dissimilar devices or software. As long as the devices or software use the same approved protocol, it will perform the same function for the end user.

The ISO specifies protocols for each service definition within the layers of the model. These are a description of the bit coding formats in which specific information is passed between processes, as well as the procedure to interpret it. Protocols operate between the same entities (the parts of a system providing services for a given layer) of the end systems. For example, information about network layer protocol in a message sent by a system are used only by the network layer in the receiving system.

AP8.2-2.3 Data processing

The message from the application process, plus information added by each layer below, form the frame that is sent out over the network (see Figure AP8.2-2). At each layer, header control information is appended to the data unit received from the layer above. This information identifies the protocol options used and gives other data about the message and its routing.

At the receiving end, the header information is removed and processed by each layer. The remaining data unit is then passed up to the next layer, where a similar operation takes place.

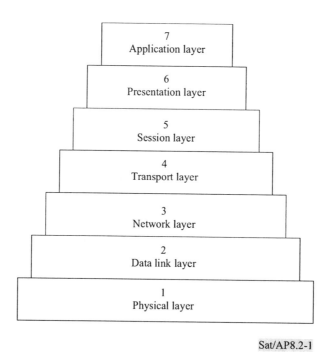

Sat/AP8.2-1

FIGURE AP8.2-1

Structure of the OSI reference model

AP8.2-2.4 Message routing

Data routed between networks, or from node to node within a network, requires only the functions of the three lower layers – network, data link, and physical (see Figure AP8.2-3). The network node is called a relay system.

A message may pass through many relay systems on its way between the application processes. In an OSI application, the path taken is invisible to the end-users. Only relay systems know what route a message is using.

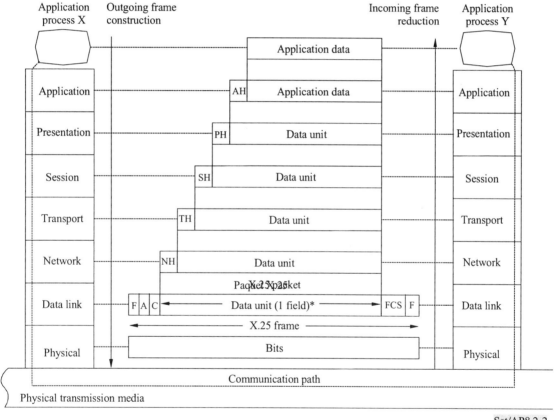

Sat/AP8.2-2

FIGURE AP8.2-2

Data processing

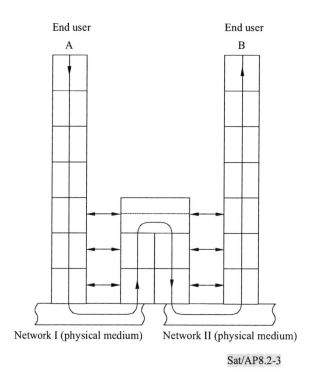

FIGURE AP8.2-3

Message routing through the OSI layers

AP8.2-3 Description of the seven layers

AP8.2-3.1 Physical layer

The physical layer, layer 1, covers the physical interface between devices and the rules by which bits are passed from one to another. The physical layer has four important characteristics:

- mechanical
- electrical
- functional
- procedural

Examples of standards at this layer are RS-232-C. RS-449/422/423 and portions of X.21.

AP8.2-3.2 Data link layer

Although the physical layer provides only a raw bit stream service, the data link layer, layer 2, should ensure the reliability of the physical link from node-to-node within the network or from the end user to the network and provide the means to activate, maintain, and deactivate the link. The principal service provided by the data link layer to the higher layers is that of error detection and control. Thus, with a fully functional data link layer protocol, the next higher layer may assume virtually error-free transmission over the link. However, if communication is between two systems that are not directly connected, the connection will comprise a number of data links in tandem, each functioning independently. Thus the higher layers are not relieved of an error control responsibility.

Examples of standards at this layer are HDL, LAP-B, LAP-D and LLC.

AP8.2-3.3 Network layer

The basic service of the network layer, layer 3, is to provide for the transparent transfer of data between transport entities. It relieves the transport layer of the need to know anything about the underlying data transmission and switching technologies used to connect systems. The network service is responsible for establishing, maintaining, and terminating connections across the intervening communications facility.

It is at this layer that the concept of a protocol becomes a little fuzzy. The layer 1 and 2 protocols are station/node protocols (local). Layers 4 through 7 are clearly protocols between (N) entities in the two stations. Layer 3 is a little bit of both.

The principal dialogue is between the station and its node; the station sends addressed packets to the node for delivery across the network. It requests a virtual circuit connection, uses the connection to transmit data, and terminates the connection. All of this is done by means of a station-node protocol. However, because packets are exchanged and virtual circuits are set up between two stations, there are aspects of a station-station protocol as well.

There is a spectrum of possibilities for intervening communications facilities to be managed by the network layer. At one extreme, the simplest, there is a direct link between stations. In this case, there may be little or no need for a network layer, as the data link layer can perform the necessary functions of managing the link. Between extremes, the most common use of layer 3 is to handle the details of using a communication network. In this case, the network entity in the station must provide the network with sufficient information to switch and route data to another station. At the other extreme, two stations might wish to communicate but are not even connected to the same network. Rather, they are connected to networks that, directly or indirectly, are connected to each other. One approach to providing for data transfer in such a case is to use an Internet protocol (IP) that sits on top of a network protocol and is used by a transport protocol. IP is responsible for internetwork routing and delivery, and relies on a layer 3 at each network for intranetwork services. IP is sometimes referred to as layer 3.5.

The best known example of layer 3 is the X.25 layer 3 standard. The X.25 standard refers to itself as an interface between a station and a node (using our terminology). In the context of the OSI model, it is actually a station-node protocol.

AP8.2-3.4 Transport layer

The purpose of layer 4 is to provide a reliable mechanism for the exchange of data between processes in different systems. The transport layer ensures that data units are delivered error-free, in sequence, with no losses or duplications. The transport layer may also be concerned with optimizing the use of network services and providing a requested quality of service to session entities. For example, the session entity might specify acceptable error rates, maximum delay, priority, and security. In effect, the transport layer serves as the user's liaison with the communications facility.

The size and the complexity of a transport protocol depends on the type of service it can get from layer 3. For a reliable layer 3 with a virtual circuit capability, a minimal layer 4 is required. If layer 3 is unreliable, the layer 4 protocol should include extensive error detection and recovery. Accordingly, ISO has defined five classes of transport protocol, each oriented toward a different underlying service.

AP8.2-3.5 Session layer

The session layer, layer 5, provides the mechanism for controlling the dialogue between presentation entities. At a minimum, the session layer provides a means for two presentation entities to establish and use a connection, called a session. In addition it may provide some of the following services:

- dialogue type: This can be two-way simultaneous, two-way alternate, or one-way.

- recovery: The session layer can provide a checkpointing mechanism, so that if a failure of some sort occurs between checkpoints, the session entity can retransmit all data since the last checkpoint.

AP8.2-3.6 Presentation layer

The presentation layer, layer 6, offers application programmes and terminal handler programmes a set of data transformation services. Services that this layer would typically provide include:

- data translation: code and character set translation.

- formatting: modification of data layout.

- syntax selection: initial selection and subsequent modification of the transformations used.

Examples of presentation protocols are text compression, encryption, and virtual terminal protocol. A virtual terminal protocol converts between specific terminal characteristics and a generic or virtual model used by application programmes.

AP8.2-3.7 Application layer

The application layer, layer 7, provides a means for application processes to access the OSI environment. This layer contains management functions and generally useful mechanisms to support distributed applications. Examples of protocols at this level are virtual file protocol and job transfer and manipulation protocol.

AP8.2-4 Perspectives on the OSI model

Figure AP8.2-4 provides two useful perspectives on the OSI architecture. The annotation along the right side suggests viewing the seven layers in three parts. The lower three layers contain the logic for a host to interact with a network. The host physically is attached to the network, uses a data link protocol to reliably communicate with the network, and uses a network protocol to request data exchange with another device on the network and to request network services (e.g., priority). The X.25 standard for packet-switched networks actually encompasses all three layers. Continuing from this perspective, the transport layer provides a reliable end-to-end connection regardless of the intervening network facility. Finally, the upper three layers, taken together, are involved in the exchange of data between end users and making use of a transport connection for reliable data transfer.

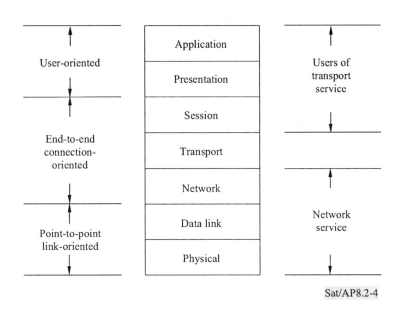

Sat/AP8.2-4

FIGURE AP8.2-4

Perspectives on the OSI architecture

Another perspective is suggested by the annotation on the left. Again, consider host systems attached to a common network. The lower two layers deal with the link between the host system and the network. The next three layers are all involved in transferring data from one host to another. The network layer makes use of the communication network facilities to transfer data from one host to another; the transport layer assures that the transfer is reliable; and the session layer manages the flow of data over the logical connection. Finally, the upper two layers are oriented to the user's concerns, including considerations of the application to be performed and any formatting issues.

REFERENCES

[1]   Handbook on Satellite Communications (FSS) (ITU, Geneva, 1988), Appendix 8-II.

[2]   User protocols for multi-service VSAT networks (E. Denoyer, J. Bousquet, ICDSC 8, April 1989).

[3]   VSAT networks Architectures, Protocols and Management (D.M. Chutre, J.S. McCoskey, IEEE Comm. magazine, July 1988, Vol. 26 No. 7).

[4]   Packet switch architectures and user protocol interfaces for VSAT networks (J. Stratigos, R. Mahindru, IEEE Comm. magazine, July 1988, Vol. 26, No. 7).

[5]   European standard for the interconnection of VSAT systems to PSPDNs (ETSI, draft Standard ETS 4.1, February 1991).

[6]   Network architectures for satellite ISDN (D Chitre, ICSDSC 8, April 1989).

[7]   Connection of very small earth station networks (VSATs) with the ISDN (ad hoc ITU-R/ITU-T – ISDN/Sat. meeting contribution of ESA, October 1990).

[8]   Small aperture earth station networks and their relationship to ISDN (L. Golding, Globecom Conf., November 1987).

[9]   Technical overview of ISDN and its network impact on satellites (W.S. Oei, INTELSAT seminar, December 1988).

[10]   Voelcker, J. – March 1986 – Helping computers communicate. *IEEE Spectrum.* Vol. 23, 3, 61-70.

[11]   Stallings, W. – 1988 – Data transmission systems – Standards, Vol. 1 OSI Standards.

CHAPTER 9

Frequency sharing, interference and coordination

This chapter provides a summary – for information only – of regulations and procedures in the field of frequency sharing and interference. It is not intended to replace the Radio Regulations (RR), the pertinent ITU-R recommendations and the applicable provisions of international agreements. When planning or implementing new stations in the FSS, the reader should refer to the appropriate portions of ITU and other documentation issued by the relevant international organizations.

The 1992 Additional Plenipotentiary Conference restructured ITU into three Sectors to improve coordination and the interface with users and to provide for a continuous review of strategy and planning. The Radiocommunication Sector (ITU-R Sector) was created on 1 March 1993, from the previously existing CCIR and IFRB. The two other Sectors are the Telecommunication Standardization (ITU-T) and Telecommunication Development (ITU-D) Sectors. The aim of the Radiocommunication Sector is to ensure rational, equitable, efficient and economical use of the radio-frequency spectrum and satellite orbits.

The Radio Regulations (RR), which complement the provisions of the International Telecommunication Convention constitute an international treaty on radiocommunications. World Radiocommunication Conferences, WRC (previously called World Administrative Radio Conferences, WARC), normally convened every two years, may revise the RR and any associated Frequency Assignment and Allotment Plans. The last WRC was held in 1997 and due account of the works of this conference is taken in this chapter. The Radiocommunication Bureau (BR) organizes and coordinates the work of the Radiocommunication Sector and also records and registers frequency assignments and orbital characteristics of space services and maintains the Master International Frequency Register (MIFR).

The reader should be advised that many RR texts analysed in this chapter could be revised by the next, incoming WRC, which will take place in the year 2000 (WRC-2000).

9.1 Radio Regulations provisions and ITU-R procedures

9.1.1 Introduction

The present Radio Regulations (RR, Geneva, 1990 revised in 1994 and 1996, and by the Final Acts of WRC-97) are intended to establish procedures and limits to prevent harmful interference from affecting the proper operation of services sharing the same frequency bands or networks of a certain service operating in the same frequency bands.

The "Simplified Radio Regulations", the principles of which were adopted by WRC-95, have been reviewed by WRC-97 to ensure consistency between all of the provisions of Articles S4, S7, S8, S9,

S11, S13 and S14 (S for simplified provisions) and new Appendices S4 and S5 concerning advance publication, coordination, notification and registration procedures of a satellite network.

It is important to note that, at the present day, the new RR, applying these new principles, could be not fully available and that, in consequence, mention of the previous RR references could remain in some documents[1].

WRC-97 adopted a revision of the RR and appendices thereto, including the review and revision of existing resolutions and recommendations and the adoption of various new resolutions and recommendations as contained in the Final Acts[2, 3].

The revised RR includes four volumes: Volume 1 relating to Articles, Volume 2 relating to Appendices, Volume 3 relating to Resolutions and Recommendations, Volume 4 relating to Recommendations ITU-R incorporated by reference plus a booklet relating to Maps to be used in relation to Appendix S7.

It is very important to note that the sections relating to non-geostationary orbit (non-GSO) networks are provisional, the final decisions will be taken by WRC-2000 (or the next one). In consequence, the texts on this subject in this chapter should be considered only as indicative.

9.1.2 Frequency allocations

Many frequency bands are allocated to more than one service in the ITU-R Table of Frequency Allocations and its footnotes (RR, Article S5); such bands are said to be shared. Allocations contained in the table relate to the whole of one or more of the three ITU Regions of the world[4] if not qualified in some respect by footnotes. Footnotes may prescribe limitations on the use to be made of the frequency band. More usually, footnotes add services to, withdraw services from or change the category of the allocations in the table for application in specified countries.

WRCs take decisions on the allocation of the various frequency bands to the many radiocommunication services. Telecommunication satellite networks are implemented in the framework of the fixed-satellite service (FSS), the broadcasting-satellite service (BSS) and the mobile-satellite service (MSS). The MSS may involve land, maritime and aeronautical

---

[1] *In certain cases, double references are used in this chapter* to provide correspondence between the former numbers of the provisions and the S-numbers assigned to the simplified provisions. Certain references to S-numbers do not have a corresponding reference or are either new or a combination of a number of former provisions.

[2] It is to be noted that the newly published version of the RR is an updated document incorporating the four volumes (V1-2-3, Edition 94 and V4, Edition 96) and the Final Acts of WRC-95 and WRC-97. The majority of the provisions revised or adopted by WRC-97 is applied as from 1 January 1999. The other provisions shall apply as from the special dates of application indicated in the revised or adopted texts.

[3] Meaning of the following symbols: MOD97: change to the substance of the text by WRC-97; SUP97: deletion of a provision; ADD97: addition of a new provision; DNR: draft new recommendations adopted by the Radiocommunication Assembly (RA) 1997.

[4] Region 1 (R1): Africa, Arab nations, Europe, Russian Federation (including the part in Asia); Region 2 (R2): The Americas; Region 3 (R3): Asia and Oceania.

mobile-satellite services. The FSS is defined as involving links between points on the Earth which are fixed when transmitting or receiving signals, as opposed to MSS, which involves links which may be in motion when transmitting or receiving. BSS involves radio and TV broadcast transmissions intended to be received by the general public. For both the MSS and the BSS, part of the link may involve a feeder link transmission from/to a fixed point of the Earth and therefore be operated within the FSS. Any frequency allocation to the FSS may be used for feeder links. However, there is an increasing trend to designate specific frequency bands for this purpose, as for the feeder links to non-geostationary satellites in the MSS.

As explained above, in order to promote the efficient use of the spectrum, most of these frequency allocations are shared by several services. Since the geostationary orbit arc is generally well above the local horizon, sharing with the fixed service (FS, i.e. radio-relay links) is readily implementable and is therefore generally present. There is only one significant exception worldwide, at 29.5-30 GHz/19.7-20.2 GHz where the allocation to the FSS (and/or MSS) is not shared with the FS. Moreover, in most countries in Region 1, the band 12.5-12.75 GHz is also allocated to the FSS on a primary exclusive basis.

There are three categories of allocations, primary, permitted, and secondary as defined in Section II of Article S5 of the Radio Regulations. Primary and permitted services have equal rights, except that, in the preparation of frequency plans, the primary service, as compared with the permitted service, shall have prior choice of frequencies.

Secondary services have no rights against primary or permitted services concerning the potential for harmful interference transmitted or received; they can only claim protection from other secondary services to which frequencies may be assigned at a later date.

Table AP9.1-1 of Appendix 9.1 to this chapter contains a list of frequency bands allocated to the FSS, BSS, MSS and ISS and indicates the other services which share these bands on an equal primary basis (services listed in lower case characters are allocated on a secondary basis).

RR Resolution 2 presents the principles relating to "equitable use, by all countries, with equal rights, of the GSO and of frequency bands for space radiocommunication services"[5].

All countries have equal rights in the use of both the radio frequencies allocated to various space radiocommunication services and the GSO orbit for these services. The radio-frequency spectrum and the GSO orbit are limited natural resources and should be most effectively and economically used. The registration of frequency assignments for space radiocommunication services and their use should not provide any permanent priority for any individual country or groups of countries and should not create an obstacle to the establishment of space systems by other countries.

Where a frequency band is allocated to a single service, it is necessary to ensure that interference between different networks of that service operating in that band does not exceed acceptable limits. Where a band is shared by two or more services, similar methods are used to ensure that stations of secondary services do not interfere with stations of primary services, and that interference between

---

[5] According to RR S4.8, equality of rights to operate when a frequency band is allocated in different regions to different services must be observed, i.e. limits concerning interregional interference are to be observed.

stations of services with allocations of equal status does not exceed acceptable limits. Typically these methods involve the following:

a)   Minimum acceptable performance standards are defined for links of each service; typically this process defines a maximum noise level measured under specified conditions or a maximum bit error ratio at the receiving end of a circuit or radio link. For example, for broadcasting services these conditions may apply at the edge of the service area and for international telephone circuits they apply to a defined hypothetical reference circuit (HRC) or a hypothetical reference digital path (HRDP). Specific fractions of this maximum level of degradation are allotted to signal degradations occurring within the system, to interference from other networks of the same service and to interference from networks of other services. These last two components are called "permissible interference".

b)   Where it can be foreseen that most of the interference will come from a limited number of identifiable interfering stations, frequency coordination is carried out to determine the level of interference from each source, and changes in the characteristics of the two networks are negotiated, if necessary, to reduce interference to an acceptable level.

c)   Where the number of interfering stations is potentially large or their location is indeterminate, it becomes necessary to apply constraints to the characteristics of all such stations using the frequency band in question in order that the aggregate level of interference will be acceptably small, regardless of the number of interfering stations or their location.

9.1.3   Possible interference modes

It is useful to consider the various modes of interference between stations in the space and terrestrial services. These are shown in Figure 9.1.

9.1.3.1   Modes of interference between space and terrestrial services

A1   Terrestrial-station transmissions possibly causing interference to reception by an earth station.

A2   Earth-station transmissions possibly causing interference to reception by a terrestrial station.

C1   Space-station transmissions possibly causing interference to reception by a terrestrial station.

C2   Terrestrial-station transmission possibly causing interference to reception by a space station.

9.1.3.2   Modes of interference between stations of different space systems in frequency bands with separated Earth-to-space and space-to-Earth allocations

B1   Space-station transmission of one space system possibly causing interference to reception by an earth station of another space system.

B2   Earth-station transmissions of one space system possibly causing interference to reception by a space station of another space system.

9.1.3.3   Modes of interference between stations of different space systems in frequency bands for bidirectional use

The modes above should be extended as follow:

E   Space-station transmission of one space system possibly causing interference to reception by a space station of another space system.

F   Earth-station transmission of one space system possibly causing interference to reception by an earth station of another space system.

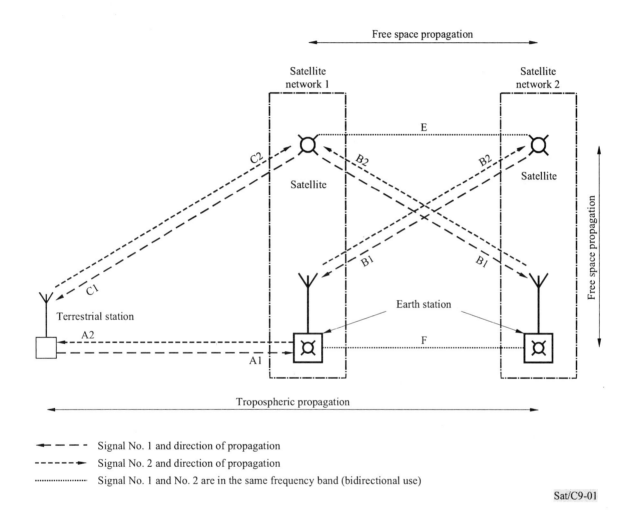

FIGURE 9.1

**Modes of interference concerning the FSS in the frequency bandsallocated
with equal rights for terrestrial radiocommunications**

9.1.3.4   Propagation conditions over the interference paths

In studying the effects of the above-mentioned modes of interference it is necessary to consider the following worst-case propagation conditions over the interference paths:

- free-space propagation for paths between Earth and space, for modes B1, B2, C1, C2 and E;

- tropospheric propagation for paths which effectively follow the surface of the Earth, for modes A1, A2 and F.

9.1.3.5   Applications of the Radio Regulations

The Radio Regulations (RR) were positively designed to deal with each of these situations to allow the development of all services without the risk of harmful interference in either sense.

Modes A2, C1 and C2 have been regulated by imposing suitable limitations on the radiated energy, especially in a critical direction.

Modes A1 and A2 related to interference paths between earth stations and terrestrial stations are dealt with by a concept of coordination, which has to be effected between administrations, in cases when the "coordination area" around an earth station (either transmitting or receiving) includes the territory of any other country.

Modes B1 and B2 are usually combined to evaluate the apparent increase in equivalent satellite link noise temperature caused by another satellite link. When this increase exceeds the threshold value (RR Appendix S8) the two administrations responsible for operation of the two space systems shall effect their coordination. A similar approach applies to mode E.

Mode F is not subject to any provisions of the Radio Regulations.

9.1.4   Radiation limitations

9.1.4.1   Stations of the FSS

(See RR Articles S21 and S22.)

9.1.4.1.1   Transmitting earth station

The limits imposed on radiation by transmitting earth stations (mode A2) are defined by the Radio Regulations in Article S21, Sections III and IV. Specifically RR S21.8-S21.13, concern the maximum permissible e.i.r.p. in any direction towards the horizon when the elevation of the horizon (δ) viewed from the centre of the earth-station antenna is less than or equal to 5° (Table 9.1A below).

In addition RR S21.14 provides that the earth-station antennas shall not be employed for transmission at elevation angles of the main beam axis of less than 3° above the horizontal plane. The purpose of these limitations is to prevent the production of such high values of interference as would unreasonably restrict the ability to choose frequencies and sites for terrestrial stations in neighbouring countries.

TABLE 9.1A

Maximum permissible e.i.r.p. (δ) produced by a transmitting earth station (e/s)
(no restriction for $\delta > 5°$)

Frequency bands (F, GHz)	e/s e.i.r.p. (dBW)	Bandwidth (kHz)	$0° < \delta \le 5°$ e.i.r.p. (δ) (dBW)
$1 < F < 15$	40	4	$40 + 3. \delta$
$F > 15$	64	1 000	$64 + 3. \delta$

Through interference mode B2, earth-station transmissions may cause interference to other geostationary-satellite networks. No. S22.26 of Article S22 of the RR requires that the level of e.i.r.p. emitted by an earth station at angles in the direction of the GSO off the main-beam axis should be minimized and that administrations are encouraged to achieve the lowest values practicable for off-axis emissions bearing in mind the latest ITU-R recommendations. Recommendation ITU-R S.524 prescribes the recommended level of off-axis radiations from earth stations (see Table 9.1B below). More detailed considerations related to off-axis e.i.r.p. density limits are given in Annex 1 to this recommendation.

The special case of very small aperture terminals (VSATs) is the subject of Recommendation ITU-R S.728.

TABLE 9.1B

Off-axis e.i.r.p. limits (φ) emitted by an earth station in the FSS in any direction
within 3° of the GSO (see RR Article S22 Section VI)

Frequency bands (F, GHz)	e/s off-axis e.i.r.p. (dBW) e.i.r.p. (φ)	Bandwidth (kHz)	$\varphi°$
6	$32 - 25\log(\varphi)^{(*)}$	4	$2.5° < \delta \le 7°$
14	$39 - 25\log(\varphi)^{(**)}$	40	$2.5° < \delta \le 7°$
	$32 - 25\log(\varphi)^{(***)}$		$2.0° < \delta \le 7°$

NOTES:

$^{(*)}$ For earth stations using emissions other than SCPC (see ITU-R S.524).

$^{(**)}$ The off-axis e.i.r.p. for FM-TV emissions at 14 GHz with energy dispersal (or properly modulated) should not exceed the following value: $53 - 25\log(\varphi)$ dBW (see ITU-R S.524).

$^{(***)}$ For VSAT earth stations operating with GSO FSS satellites at 14 GHz (see ITU-R S.728).

However, the provisions of Section VI of RR Article S22 concerning earth station off-axis power limitations in the FSS are suspended pending the review of the values in Notes S22.26, S22.27 and S22.28 by WRC-2000 (see No. S22.29: e.i.r.p. limits applicable to the following frequency bands assigned to FSS Earth-to-space: 12.75-13.25, 13.75-14, 14-14.5 GHz).

9.1.4.1.2   Transmitting space station

The interference potential created by a transmitting space station can affect receiving terrestrial stations (mode C1), receiving earth stations of another satellite network (mode B1) and, in addition, receiving space stations when operating in a frequency band allocated for bidirectional use (mode E).

The interference potential to terrestrial stations (mode C1) is restricted by limitation of the maximum power flux-density produced by a space station on the Earth's surface.

An important aspect of these limits is that they vary as a function of the angle of wave front arrival on the Earth in such a way as to relax the limit with increasing angle. Although it could be thought at first glance that, for higher angle of wave front arrival the slant range is shorter and therefore the power flux-density is higher, the decrease in the spreading loss is negligible in comparison with the predominant antenna discrimination of the terrestrial stations for increasing elevation angle, and thus the relaxation is justified.

The power flux-density, pfd, limits are given in Table S21.4 of Section V of Article S21 of the RR and are summarized in Table 9.2. The limits relate to the pfd which would be obtained under free-space propagation conditions and apply to emissions in the space radiocommunication services where the frequency bands are shared with equal rights with the fixed or mobile service (FS or MS).

For the identification of the pertinent downlink frequency bands allocated to the fixed-satellite services reference should be made to Table AP9.1-1 of this handbook.

The limits of power flux-density established for protection of the line-of-sight radio-relay links would not be sufficient in case of sharing the frequency band 2 500-2 690 MHz with tropospheric scatter systems with more sensitive receivers. For this particular situation see RR S21.16.3.

9.1.4.2   Terrestrial stations

The interference potential created by a transmitting terrestrial station can affect receiving space stations (mode C2) sharing the same frequency bands and is restricted by limitation of the maximum equivalent isotropically radiated power (e.i.r.p.) and by limitations of the direction of maximum radiation towards the geostationary-satellite orbit as indicated in Sections I and II of Article S21 of the RR (see Tables S21.1 and S21.2).

The maximum e.i.r.p. of a station in the fixed or mobile service shall not exceed in any case the limit of 55 dBW while more stringent limitations are applicable if the direction of maximum radiation is confined towards the geostationary orbit.

The e.i.r.p. limits, as function of pointing angle towards the geostationary orbit, and the limits of power delivered to the antenna of a station in the fixed or mobile services are summarized in Table 9.3 below.

TABLE 9.2

**Maximum allowable power flux-density produced by a space station
on the Earth's surface
(see RR Article S21)**

Frequency range (GHz)	Maximum power flux-density as function of arrival angles δ (dB(W/m^2))				Service
	$0° < \delta \le 5°$	$5° < \delta \le 25°$	$25° < \delta \le 90°$	Reference bandwidth	
1.525-2.290	−154	$-154 + 0.5\,(\delta - 5)$	−144	In any 4 kHz band	TT and C space operation
2.500-2.690	−152	$-152 + 0.75\,(\delta - 5)$	−137	In any 4 kHz band	GSO FSS (and MSS)
2.520-2.670					GSO BSS
3.400-4.200 4.500-4.800 7.250-7.750	−152	$-152 + 0.5\,(\delta - 5)$	−142	In any 4 kHz band	GSO FSS
5.150-5.216	−164	−164	−164	In any 4 kHz band	
6.700-6.825	−137	$-137 + 0.5\,(\delta - 5)$	−127	In any 1 MHz band	
6.825-7.075	−154	$-154 + 0.5\,(\delta - 5)$	−144	In any 4 kHz band	FL non-GSO MSS (FSS)
	−134	$-134 + 0.5\,(\delta - 5)$	−124	In any 1 MHz band	(S9.11A)
10.7-11.7 (NOTE 1)	−150	$-150 + 0.5\,(\delta - 5)$	−140	In any 4 kHz band	GSO and non-GSO FSS (WRC-97 Res. 130 and 131)
11.7-12.5 (R2) 12.2-12.7 (R2) 11.7-12.2 (R2 and 3)	−148 (NOTE 1)	$-148 + 0.5\,(\delta - 5)$ (NOTE 1)	−138 (NOTE 1)	In any 4 kHz band	BSS Plan or GSO and non-GSO FSS (WRC-97 Res. 131, 138 and 538)
12.2-12.5 (R3) 12.50-12.75 (some R1 countries)	−148 (NOTE 1)	$-148 + 0.5\,(\delta - 5)$ (NOTE 1)	−138 (NOTE 1)	In any 4 kHz band	GSO and non-GSO FSS (WRC-97 Res. 130 and 131)
15.430-15.630	−127	5°-20°: −127 20°-25°: $-127 + 0.56$ $(\delta - 20)$	25°-29°: −113 29°-31°: $-136.9 + 25\log\,(\delta - 20)$ >31°: −111	In any 1 MHz band	FL non-GSO MSS (FSS) (S9.11A) (WRC-97 Res. 123)

17.7-19.3 (NOTES 2, 3)	−115	−115 + 0.5 (δ − 5)	−105	In any 1 MHz band	GSO FSS (Res. 130 and 538, S9.11A)
	−125	−125 + (δ − 5)	−105		Non-GSO FSS (provisionally for >100 satellites) (WRC-97 Res. 131)
19.3-19.7	−115	−115 + 0.5 (δ − 5)	−105	In any 1 MHz band	GSO FSS FL non-GSO MSS (S9.11A) non-GSO FSS (S9, S11)
37.5-40.5	−115	−115 + 0.5 (δ − 5)	−105	In any 1 MHz band	GSO FSS and MSS (provisional pfd limits)

NOTES – The limits in this table relate to the pfd which would be obtained under assumed free-space propagation conditions.

NOTE 1 – These values require further study (see WRC-97, Res. 131).

NOTE 2 – Equality of rights to operate when a frequency band is allocated in different regions to different services of the same category (see No. S4.8). Limits concerning interregional interference be observed.

NOTE 3 – The risks of interference caused by non-GSO and GSO satellites systems to the EESS and space research (passive sensors) must be reduced to a to minimum (see Recommendation ITU-R SA.1029).

TABLE 9.3

Maximum transmissible equivalent isotropically radiated power (e.i.r.p.) delivered to the antenna of a station in the fixed service (FS) or mobile service (MS) towards the GSO (see RR Article S21, Sections I and II)

Frequency range (GHz)	Maximum transmissible e.i.r.p.[*] as a function of avoiding angle δ from the GSO (dBW)		Maximum power delivered to the antenna (dBW)
1-10	+35	δ > 2°	+13
	+47 + 8(δ − 0.5)[**]	0.5° < δ < 1.5°	
10-15	+45	>1.5°	+10
25.250-27.500	+24	>1.5°	+10
Other bands above 15 GHz	+55	No limit	+10

[*] The maximum e.i.r.p. of a station in the FS or MS shall not exceed +55 dBW.

[**] These values apply if compliance with the other is impracticable.

For the identification of the pertinent uplink frequency bands allocated to the fixed-satellite service reference should be made to Table AP9.1-1 of this handbook.

The interference potential created by a transmitting terrestrial station can affect receiving earth stations (mode A1) but in this case sites and frequencies for terrestrial stations shall be selected having regard to the relevant ITU-R recommendations with respect to geographical separation from earth stations.

9.1.5   Procedure for effecting coordination

9.1.5.1   General

The provisions of RR Articles S9 and S11 apply to space radiocommunication services for the coordination and notification and registration of frequency assignments (see also RR Resolution 51 for provisional application of certain provisions of the RR).

Following WRCs, two types of frequency bands have been identified: planned ones, i.e. formally assigned to (a) given service(s) in (a) given region(s) and unplanned ones. The procedure for effecting coordination in these two cases is summarized below:

For satellite networks or satellite systems using frequency bands covered the FSS Plan, see also RR Appendix S30B, for those using frequency bands covered the BSS and BSS FL Plans, see also Article 4 and Annexes 1 and 2 to RR Appendices S30 and S30A, and for frequency bands used by the non-GSO FSS systems, see also RR Resolutions 130 and 538.

i)   In the unplanned frequency bands

In the unplanned frequency bands, satellite networks, except in the BSS, are generally required to meet the provisions of Articles S9 and S11 of the RR.

- Spectrum for GSO networks is provided by application of procedures of Articles S9 and S11 (RR1042 and RR1060).

- Spectrum for GSO BSS networks is provided (Special Sections RES33/A and RES33/C) by application of RR Resolution 33 (§§ 3.1 and 3.2.1) and RR Articles S9.7, S9.11 and S11.

- Spectrum for non-GSO networks in certain bands is provided (see also RR Resolution 130):

  - by application of the provisions of Resolution 46 (WRC-95/97, RR S9.11A),

  - and by application of procedures of Article S11.

ii)   In the planned frequency bands

- In the FSS Plans (6/4 GHz and 13/10-11 GHz), spectrum for GSO networks is provided by application of RR Appendix S30B (see Annex 2 to this appendix), and also of Articles 6, 8, 10 and 11 of this appendix. See also: SUP97 RR Resolution 107, RR Article S9.1 and Article S11.1.

- In the BSS plans (and associated feeder links (FL) Plans), spectrum for GSO networks is provided by application of RR Appendices S30 and S30A (see Annex 2 of these appendices (RR S9.8 and S9.9)). See also in Appendix S30: Articles 4, 5, 7, 10, 11 and in Appendix S30A: Articles 4, 5, 7, 9, 9A, and 11.

- Spectrum for non-GSO networks in the planned frequency bands is provided by application of procedures of Article S11 of the RR. RR Resolution 538 relates to the use of the frequency bands covered by the BSS and BSS FL Plans by non-GSO systems in the FSS and RR Resolution 130 relates to the use of the frequency bands covered by the FSS Plan (RR App. S30B) by non-GSO systems in the FSS.

Various other RR resolutions and recommendations which concern the procedures for allocating frequencies in the planned frequency bands, especially as concerns the BSS, are simply listed below:

- RR Resolution 46, Interim procedures for the coordination of non-GSO systems in certain bands (see RR App. S5 Table S5-1A relating to the applicability of RR No. S9.11A to space services);

- RR Resolutions 33 and 507 (BSS and BSS FL Plans);

- RR Recommendation 35 (FSS Plans and BSS and BSS FL Plans) and RR Resolution 42 (Interim systems in R2 in the bands of the BSS Plans).

9.1.5.2   Table of Frequency Allocations and particular provisions (RR S9.21)

i)   No. S9.21 of RR Article S9:

Frequency allocations are contained in a table given by RR Article S5. This table contains a number of footnotes which provide for the allocation of a given frequency band to a given service or application, subject to the application of RR S9.21 (ex-Article 14) provisions.

RR S9.21 (ex-Article 14) provisions are, in contrast with the other procedures described in this section, not a coordination procedure in the sense that its successful application to a given frequency assignment only means that this frequency assignment is in conformity with RR Article S5. In addition to this conformity, the applicable coordination procedure (such as RR Article S9 or Resolution 46, which can be initiated before or at the same time as RR S9.21), needs to be carried out successfully in order to obtain protection for this assignment.

Under these provisions, the requesting administration sends the RR Appendix S4 information to the Bureau. This information is published by the Bureau in a special section (AR14/C) of its weekly circular, with a list identifying the administrations which might be affected, based on the relevant sharing criteria.

Any administration believing that the planned assignment may affect its services operated in accordance with the Table of Frequency Allocations or planned to be so operated, shall, within four months of the date of the weekly circular, so inform the administration requesting agreement and the Bureau. Any administration not having commented within this period shall be regarded as unaffected by the planned assignment.

When a disagreement from an administration is based on a planned assignment to a space radiocommunication station not yet in operation, only assignments for which RR Appendix S4 has been received, or for which RR S9.21 procedure has been initiated, and which are planned to be operated within the next three years (Rule of Procedure December 1994) can be considered. In the case of a disagreement based on a planned assignment to a terrestrial radiocommunication station, a three year period is also applicable (Rule of Procedure December 1994).

ii)   No. S4.4 of RR Article S4:

If agreement cannot be obtained from some administrations, the assignment might be recorded in the Master International Frequency Register (MIFR) with the mention that it is operated under No. S4.4 of RR Article S4 (ex-R.342) with respect to the corresponding administrations. Under this regulation the administration responsible for the assignment is under obligation that it shall not cause interference to, and shall not claim protection from, harmful interference caused by other assignments in accordance with RR.

9.1.5.3   Advance publication of information (API)

An administration that intends to establish a satellite network, under Articles S9 and S11 of the RR (general case: FSS networks in the unplanned bands), has to send to the Radiocommunication Bureau (BR), not earlier than five years and preferably not later than two years before the date of bringing the network into service, the information listed in RR Appendix S4 (Annexes 2A and 2B).

The date of bringing into use will be extended at the request of the notifying administration by not more than two years only under the conditions specified in addenda to Article S11.

The provisions of RR Article S9 (S9.1 or S9.2) apply to space radiocommunication services for the advance publication of information (API) on satellite networks or satellite systems.

The recommended format for submitting this information is given in forms of new notice APS4/V or VI.

The major items of information relating to GSO or non-GSO satellite system subject to coordination under RR Article S9 Section II to be provided with the form of notice APS4/VI (see RR Article S9.3) are:

*   for GSO satellites only: orbital location of space station;

*   for GSO satellites only: inclination angle, period, apogee, perigee, number of satellites;

*   frequency bands;

*   service areas on the Earth;

*   class of station and nature of service;

*   date of bringing into use and period of validity.

For a non-GSO satellite system not subject to coordination under RR Article S9 Section II, additional characteristics for reception at the space station and for transmission from the space station (i.e. information to be provided for the receiving antenna beam, and information related to the space station emissions and associated receiving stations and related to the associated transmitting stations) must be provided with the form of notice APS4/V (see RR Article S9.1).

The BR publishes this information within three months in special sections of its weekly circular (as of 22 November 1997, this publication became fortnightly) and if any administration is of the opinion that interference that may be unacceptable could be caused to its existing or planned space radiocommunication services, it shall, within four months after the date of the weekly circular, send its comments to the administration concerned, and attempts should then be made to resolve these potential interference problems.

9.1.5.4   Coordination procedure (see Section II of RR Article S9, and Appendices S8 and S5)

Some time after the publication of the advance information, usually six months, coordination of the individual satellite network frequency assignments, under RR Article S9 (S9.1/1058E) (general case: FSS networks in the unplanned bands), should be initiated with any other administration whose frequency assignments, for a space station on a geostationary satellite or for an earth station that communicates with a space station on a geostationary satellite, are already in the coordination or notification process and might be affected.

The provisions of RR Article S9 apply to space radiocommunication services for the requirement and request for coordination with other administrations identified under No. S9.27 (see Table S5-1 to RR Appendix S5 relating to technical conditions for coordination).

If the information (coordination request) under S9.30 (RR Appendix S4) has not been received by the RB within a period of two years after the date of receipt by the RB of the relevant information (API), the advance publication shall be cancelled.

For effecting coordination under RR Article S9 and for identifying the administrations with which coordination is to be affected, RR Appendix S5 gives the different cases where coordination is required.

The problems that may be resolved at this stage are of a general nature, mainly involving calculations of increase in equivalent satellite link noise temperature $\Delta T/T$ (see § 9.3.1.2).

The criterion for deciding whether coordination is required between assignments in different networks is given in Appendix S8 to the Radio Regulations, which is based upon a threshold criterion of 6% increase in the equivalent satellite link noise temperature. Coordination of frequency assignments in the FSS is governed by provisions in Section II of Article S9 of the Radio Regulations.

To effect coordination the information listed in Annex 2A of Appendix S4 of the RR should be sent to the other administrations concerned, with a copy to BR.

The recommended format for submitting this information is given in RB forms of new notice APS4/II.

The information to be provided for coordination is more detailed than that provided at the advance information stage. The additional main information includes (see Forms of Notice APS4/II):

- assigned frequencies (usually taken to be the transponder centre frequencies and bandwidths);
- classes of emission (Appendix S1 of the RR);
- maximum power density and total peak envelope power for each class of emission and for each assigned frequency;
- earth station and space station antenna radiation patterns;
- for GSO satellites only: longitudinal tolerance, visible arc and service arc of the GSO;

- for GSO space stations using simple frequency-changing transponders and operating with earth stations, values of equivalent satellite link noise temperatures and transmission gains for each link and its associated receiving earth stations.

The BR publishes this information (coordination request) within three months in a special section of the weekly circular (now fortnightly) and the administrations concerned are required to notify the requesting administration, and BR, of their agreement, or disagreement within four months. If, at that time, agreement has not been reached, then detailed coordination by correspondence or by meetings continues. Sections 9.3.2, 9.3.3 and 9.3.4 provide information on the calculations and methods that may be used to facilitate coordination.

In the case of BSS and BSS FL, the procedures to be followed when an administration intends to make a modification to one of the Regional Plans are indicated in RR Appendices S30 and 30A. For the application of the provisions of RR Article S9 with respect to stations in a space radiocommunication service using frequency bands covered by the BSS and BSS FL Plans, see also RR Appendices S30 and S30A (Article 4).

9.1.6 Notification and registration of frequency assignments (RR S11.2)

For a satellite network under Article S11 of the RR, (general case: FSS networks in the unplanned bands), this Article stipulates that, when a frequency assignment is to be brought into use, notices relating to this assignment shall reach BR not earlier than three years before the assignments are brought into use (see RR Article S11.25).

The provisions of RR Article S11 apply to space radiocommunication services for the notification and registration of frequency assignments in the Master International Frequency Register (MIFR).

This notification should be drawn up as prescribed in RR Appendix S4. If the RB finds that the notice conforms with the Table of Frequency Allocations and with the provisions concerning coordination, then the frequency assignment is recorded in the Master Register.

The notified date of bringing into use of any assignment to a space station of a satellite network shall be no later than five years following the date of receipt by the BR of the relevant information. It may be extended by not more than two years only under the conditions specified in RR S11.44 and S11.44B to I, as follows:

- if due diligence information is provided for the satellite network;

- if the procedure for effecting coordination in accordance with Section II of RR Article S9 as applicable has commenced; and

- if the notifying administration certifies that the reason for the extension is one or more of the following specific circumstances: launch failure, launch delays outside the control of the administration or the operator, delays caused by modifications of satellite design necessary to reach coordination agreements, problems caused by the satellite design specifications, delays in effecting coordination after the assistance of BR, financial circumstances outside the control of the administration or the operator and force majeure.

NOTE – Due diligence information procedure is detailed in WRC-97 Resolution 49. If the due diligence information is not received before the expiry date, the requests for coordination or for a modification to the Plans shall be cancelled and any recording of the frequency assignments in the MIFR shall be deleted by the Bureau.

Moreover:

- RR Resolution 4 indicates that if a notifying administration wishes to extend the period of operation originally shown on the assignment notice of a frequency assignment of an existing space station, it informs the Radiocommunication Bureau (BR) accordingly more than three years before the expiry of the period in question. The BR shall amend as requested the period of operation originally recorded in the Master Register;

- RR Article S11.49 provides for the case without any complaint of harmful interference (for at least four months).

- Article RR S11.49 provides for the case of suspension of the regular use (for a period not exceeding 18 months).

9.1.7   Evolution of the procedures

The evolution of the RR is taken into account in this Edition 3, as concerns, in particular WRC-95 and WRC-97 issues relating to the frequency sharing between the MSS, BSS, non-GSO FSS and non-GSO MSS feeder links.

As indicated in Table AP9.1-1 of Appendix 9.1 to this chapter, WARC Orb-85 and 88 had divided the frequency bands allocated to the fixed-satellite service into three categories (see Supplement 1 to Edition 2 of the Handbook: "Effects of WARC Orb-88 decisions"):

i)   allotment plan (FSS planned bands: see RR Appendix 30B/S30B);

ii)   improved procedures (MPM: multilateral planning meeting: see RR Article S9 and Resolution 110);

iii) simplified procedures (procedures concerning the global satellite network in the standard bands: see RR Articles S9 and S11).

The main subjects studied by WRC-97 were the following:

- review of the Articles of and Appendices to the Simplified Radio Regulations adopted by WRC-95;

- review of BSS Plans in Regions 1 and 2. The new Plans are given in the new Article 11 of Appendix S30 and in the new Article 9A of Appendix S30A. However, the incorporation of the outcome of these reviews of these Appendices into the Articles of RR has not yet been agreed. The adoption of RR Article S10/T10 relating to possible procedure for modification of a frequency allotment or assignment has been postponed. An Inter-conference Representative Group (IRG) is established to study the feasibility of increasing the minimum capacity for countries in Regions 1 and 3 (see WRC-97 Resolution 532);

- possible use of the GSO FSS and BSS planned bands by introducing non-GSO satellite systems: regulatory matters and compatibility analysis between GSO and non-GSO satellite systems operating in these planned bands (see WRC-97 Resolution 130 and Rev.WRC-97 Resolution 506);

- possible use of certain GSO FSS bands (between 10-30 GHz) by introducing non-GSO satellite systems (see WRC-97 Resolution 130);

- effects of frequency allocations and regulatory aspects for the non-GSO FSS and non-GSO MSS FL.

The coordination, modification, notification and registration procedures to be applied to those new cases have either been determined at WRC-97, or are expected to be determined at the incoming WRC-2000.

Note that, according to WRC-97 Resolution 72, the success of a conference depends on a greater efficiency of regional coordination and interaction at interregional level prior to future conferences and that there is a need for an overall coordination of interregional consultations and in order to help the regional telecommunication organizations in their preparations for future WRCs.

9.2 Other coordination procedures

Note that the procedures described below may change in the future, due to the possible evolution of the status of the organizations.

9.2.1   Coordination with INTELSAT based upon Article XIV of its Agreement

9.2.1.1   General

INTELSAT, the International Telecommunications Satellite Organization, owns and operates the global communications satellite system used by most of the countries around the world (see Annex 3 to this handbook).

The States, Parties to the INTELSAT Agreement, desiring to continue to develop this telecommunications satellite system with the aim of achieving a single global commercial system, decided that in order to foster and protect the INTELSAT system, they will assume certain obligations regarding the establishment of satellite systems separate from the INTELSAT system.

Article XIV of the INTELSAT Agreement defines the rights and obligations of INTELSAT members regarding this subject.

9.2.1.2   Article XIV intersystem coordination

This Article provides that certain requirements should be met in case an INTELSAT Party or Signatory or person within the jurisdiction of an INTELSAT Party intends to establish, acquire, or utilize space segment facilities separate from the INTELSAT space segment facilities to meet its telecommunications service requirements. These provisions can be summarized as follows:

i)   Domestic public telecommunication services

(telephony, telegraphy, telex, radio and television distribution, etc.) (Article XIVc))

To the extent that any Party or Signatory or person within the jurisdiction of a Party intends to establish, acquire or utilize space segment facilities separate from the INTELSAT space segment facilities to meet its domestic public telecommunications services requirements, such Party or Signatory, prior to the establishment, acquisition or utilization of such facilities, shall consult the Board of Governors, which shall express, in the form of recommendations, its findings regarding the technical compatibility of such facilities and their operation with the use of the radio-frequency spectrum and orbital space by the existing or planned INTELSAT space segment.

ii)   International public telecommunication services

(telephony, telegraphy, telex, radio and television distribution, etc.) (Article XIVd))

To the extent that any Party or Signatory or person within the jurisdiction of a Party intends individually or jointly to establish, acquire or utilize space segment facilities separate from the INTELSAT space segment facilities to meet its international public telecommunications services requirements, such Party or Signatory, prior to the establishment, acquisition or utilization of such facilities, shall furnish all relevant information to, and shall consult with, the Assembly of Parties, through the Board of Governors (and in this case its Technical Committee), to ensure technical compatibility of such facilities and their operation with the use of the radio-frequency spectrum and orbital space by the existing or planned INTELSAT space segment and to avoid significant economic harm to the global system of INTELSAT.

Upon such consultation, the Assembly of Parties, taking into account the advice of the Board of Governors, shall express, in the form of recommendations, its findings regarding the considerations set out in this paragraph, and further regarding the assurance that the provision or utilization of such facilities shall not prejudice the establishment of direct telecommunication links through the INTELSAT space segment among all the participants.

iii)   Specialized services

(space research, meteorological, etc.) (Article XIVe))

To the extent that any Party or Signatory or person within the jurisdiction of a Party intends to establish, acquire or utilize space segment facilities separate from the INTELSAT space segment facilities to meet its specialized telecommunications services requirements, domestic or international, such Party or Signatory, prior to the establishment, acquisition or utilization of such facilities, shall furnish all relevant information to the Assembly of Parties, through the Board of Governors. The Assembly of Parties, taking into account the advice of the Board of Governors, shall express, in the form of recommendations, its findings regarding the technical compatibility of such facilities and their operation with the use of the radio-frequency spectrum and orbital space by the existing or planned INTELSAT space segment.

9.2.1.3   Technical compatibility

With regard to consultations on the technical compatibility of the separate facilities, technical information, including the system transmission parameters, is submitted by the Party or Signatory for consideration in two steps:

a) Informal consultations which begin as early as possible and preferably before the advance publication of information by BR. This first step serves to identify possible problems relating to the eventual establishment of the technical compatibility between the INTELSAT system and the planned separate system, and would provide the opportunity for consideration and discussion of alternatives for resolving any problems that may arise.

During the informal consultation, the increase in the equivalent link noise temperatures of the INTELSAT system due to the separate system will be calculated first. If such an increase exceeds the relevant criterion, a detailed interference evaluation will be performed.

b) Formal consultations which commence subsequent to the informal consultations, but well in advance of the proposed establishment, acquisition, or utilization of the facilities of the separate system. The formal consultation process, under Article XIV, includes an examination of the potential for interference from the planned system into the existing or planned INTELSAT system. Factors to be examined are technical compatibility of such facilities and their operation with the use of the radio-frequency spectrum and orbital space by the existing or planned INTELSAT system.

Discussions are held with the Party or Signatory requesting the consultations with a view to achieving the necessary clarifications and/or resolutions of any potential problems concerning technical compatibility. After commencement of the formal consultations, INTELSAT has six months to provide its findings about the technical compatibility of the network with the INTELSAT system.

There has been a progressive relaxation in INTELSAT interference criteria over the past years. For example, the INTELSAT criteria have been aligned to ITU-R criteria where these are available. This has made it progressively easier to coordinate new systems and to obtain the specific orbital position desired, despite the marked increase in orbit occupancy. The parameters of the INTELSAT system as well as the applicable interference criteria, guidelines and procedures are published in the INTELSAT intersystem coordination manual (IICM).

The application of these procedures has proved to be beneficial to INTELSAT as an organization as well as to its members who wish to establish separate satellite systems. It ensures technical compatibility between the systems which utilize the geostationary-satellite orbit.

9.2.2 Coordination with Inmarsat

Inmarsat, the International Maritime Satellite Organization, operates the global maritime communications satellite system used by most seafaring nations (see Annex 3 to this handbook).

Any Party to the Inmarsat Convention must undergo coordination with Inmarsat if it intends to use the non-Inmarsat space segment to provide maritime satellite services. Coordination covers both technical and economic aspects.

9.2.3 Coordination with other organizations

In addition to INTELSAT and EUTELSAT, other multi-user satellite communications organizations might establish similar rules to ensure that separate systems established by their members would have to be coordinated with the multi-user system prior to their implementation.

Therefore, if a country belongs to such an organization, it should check the provisions of the organization's charter before undertaking any activities relating to the establishment of a separate satellite system.

9.3 Frequency sharing between GSO FSS networks

9.3.1 Frequency sharing between GSO FSS networks in unplanned bands

(See Recommendation ITU-R S.738.)

9.3.1.1 General: Procedure for effecting coordination with or obtaining agreement of other administrations

Before initiating any action under RR Article S9 or S11 in respect of frequency assignment for a satellite network or a satellite system, an administration shall provide, prior to the coordination procedure (Article S9, Section II), to BR a general description of the network or system for advance publication (API) as already explained above in § 9.1.5.3.

The coordination information may also be communicated to BR at the same time as the API.

a) In a first case API, according to RR Articles Nos. S9.3 to S9.5A, applies to satellite networks or satellite systems that are not subject to coordination procedure under Section II. In this case, when the data on the new network is published by BR in its weekly circular (API Form of notice APS4/V, for non-GSO satellites), if any administration believes that unacceptable interference may be caused to its existing or planned satellite systems or networks, it may send to the publishing administration within four months its comments on the particulars of the anticipated interference to its existing or planned system.

Both administrations shall endeavour to cooperate in joint efforts to resolve any difficulties and shall exchange any additional relevant information that may be available. In the case of difficulties, the administration responsible shall explore all possible means to resolve the difficulties by means of mutually acceptable adjustments to their networks

b) In another case API, according to RR Articles Nos. S9.5B to S9.5D, applies to satellite networks or satellite systems that are subject to coordination procedure under Section II. In this case, when the data on the new network is published by BR in its weekly circular (API form of notice APS4/VI, for non-GSO and GSO satellites), if any administration considers its existing or planned satellite systems or networks to be affected, it may send its comments to the publishing administration, so that the latter may take those comments into consideration when initiating the coordination procedure. Both administrations shall endeavour to cooperate in joint efforts to resolve any difficulties and shall exchange any additional relevant information that may be available.

Before an administration notifies BR (see RR Article S11) or brings into use a frequency assignment in any cases listed under RR S9.7 to S9.21, it shall effect coordination, as required, with other administrations identified under RR S9.27 (frequency assignments to be taken into account in effecting coordination are identified using RR Appendix S5).

As concerns notification and recording of frequency assignments in the Master Register, any frequency assignment to a transmitting station and to its associated receiving stations shall generally be notified to BR.

When notifying a frequency assignment, the administration shall provide the relevant characteristics listed in RR Appendix S4. When the examination with respect to its conformity with the different mandatory points leads to a favorable finding, the assignment shall be recorded in the Master Register.

No. S11.17 indicates that frequency assignments relating to a number of stations or earth stations may be notified in the form of the characteristics of a typical station or a typical earth station and the intended geographical area of operation.

9.3.1.2   Calculations to determine if coordination is required

Recommendation ITU-R S.737 gives a diagram indicating the relationship of technical coordination methods within the FSS (see these methods given in Recommendations ITU-R S.738/739/740/741).

In compliance with the above described procedures, when the data on the new network is published by BR in its weekly circular under "Coordination Request", the other concerned administrations will perform certain calculations to determine whether the new network will cause interference or receive interference that is higher than a certain "threshold level", into or from their own existing or planned networks.

If this threshold level is exceeded, coordination would be necessary between the administration planning to establish the new network and each of the other administrations that are subject to be affected by this new network.

Since the number of parameters characterizing a system is so large, it is useful to devise a simple method to determine whether there is any risk of interference between two given satellite networks. ITU has prepared such a method as part of the Radio Regulations which is described in RR Appendix S8. Furthermore, ITU-R presents similar information in Recommendation ITU-R S.738.

This method is applicable whenever two networks share a common portion of the assigned frequency band on at least one of their paths. Figures 9.2 and 9.3 show the geometry of interference between two satellite networks. The first is applicable for the case when the uplinks and downlinks of both links share the frequency bands in the same transmission direction, while the second is applicable for frequency bands sharing in opposite transmission directions.

The method is based on the concept that the noise temperature of the system, subject to interference, undergoes an apparent increase, due to the effect of the interference.

The interfering signals are treated as thermal noise, whose spectral power density would be equal to the maximum spectral power density of the signals. Although this would result, in most cases, in a pessimistic result, the method is simple, and can be used irrespective of the modulation characteristics of both satellite networks (interfering and interfered-with) and the precise carrier frequencies employed.

The "apparent increase in the equivalent satellite link noise temperature", resulting from the interfering emission, caused by a given system is calculated by the method of RR Appendix S8 or of Recommendation ITU-R S.738. This is called ΔT. The "equivalent satellite link noise temperature" of the interfered-with network is also calculated using the same Appendix S8 or Recommendation ITU-R S.738. This is called T. The ratio of ΔT to T (called $\Delta T/T$, expressed in percentage) is compared to a threshold value of 6%. If the $\Delta T/T$ is less than, or equal to 6%[6], coordination between the two networks would not be necessary. Otherwise coordination would be required.

The calculation of $\Delta T/T$ requires the predetermination of the "transmission gain" (called γ, in RR Appendix S8 and in Recommendation ITU-R S.738), as well as the increases in the noise temperature of the uplinks and/or the downlinks of the interfered-with system. These in turn depend upon the spectral power densities in each of the uplinks and downlinks of the system, the gains of the earth station and space-station antennas of each network in the direction of the stations of the other network, and the path losses between the interfering and interfered-with stations. Furthermore, if administrations agree to take account of any polarization isolation between the two networks, this additional factor is included in the calculations.

Other methods could be considered to determine whether coordination is necessary.

The normalized $\Delta T/T$ method outlined in Annex 1 to Recommendation ITU-R S.739 could represent an improvement over the previous method because it would reduce the amount of required coordination. It is based on the method outlined in Appendix S8 to the Radio Regulations which has been modified in order to obtain more accurate results.

Therefore, the 6% threshold value has been replaced by a series of values which depend on the types of wanted and interfering carriers and which conform to ITU-R criteria. The $\Delta T/T$ threshold values are shown in Table 3 of Annex 1 to Recommendation ITU-R S.739.

9.3.1.3   Detailed coordination calculations

If coordination is necessary as a result of the previous phase, the administrations concerned take account of more precise data concerning their respective networks so as to be able to make detailed calculations.

These data should be those stipulated in RR Appendix S4. On the basis of this information, the administrations determine the interference which will be caused on each network using the calculation methods described in Recommendations ITU-R S.740, S.741, SF.766 and SF.675. The calculated interference levels are then compared with the permissible levels (given in the case of FSS in Recommendations ITU-R S.466, S.483 and S.523). If the calculated levels are greater than the permissible values, a study must be made to decide how this situation can be remedied.

When the administrations have agreed on the measures to be taken, coordination is completed and the transmission frequencies of the planned network can be notified and registered.

---

[6] In the particular case of narrow-band SCPC carriers being interfered with by slow-swept television carriers, the 6% criterion may not afford enough protection to the interfered network. This should be taken into account by administrations when using this method (see Table 3 of Annex 1 to Recommendation ITU-R S.739).

The data for the API and those for the Coordination Request are related to the network as a whole, and are not given for each station (earth or space).

In the case of a transmitting earth station, the most significant characteristics are: location of the station, antenna radiation pattern, polarization, transmitting power of each carrier and corresponding maximum power density. In addition, for each carrier, more detailed characteristics are given regarding the type of modulation used (FDM-FM, PCM-PSK, FM-TV, etc.). In the case of a receiving earth station, the most significant characteristics are the location of the station, the antenna radiation pattern and polarization.

In the case of a transmitting space station, the most significant characteristics are: information relating to the orbit, such as antenna radiation pattern, polarization, transmitting power of each carrier and corresponding power flux-density. In the case of a receiving space station, the most significant characteristics are the information relating to the orbit, the antenna radiation pattern, polarization and noise temperature of the entire receiving system.

To calculate the ratio of carrier power to interference power (C/I), Recommendation ITU-R S.740 explains the procedures for such calculations and should be referred to. For telephony-type interfered-with carriers, the C/I ratios are converted into baseband noise power utilizing the procedures of Recommendation ITU-R SF.766 (Annex 1). For television-type interfered-with carriers, reference should be made to Recommendation ITU-R SF.766 (Annex 2).

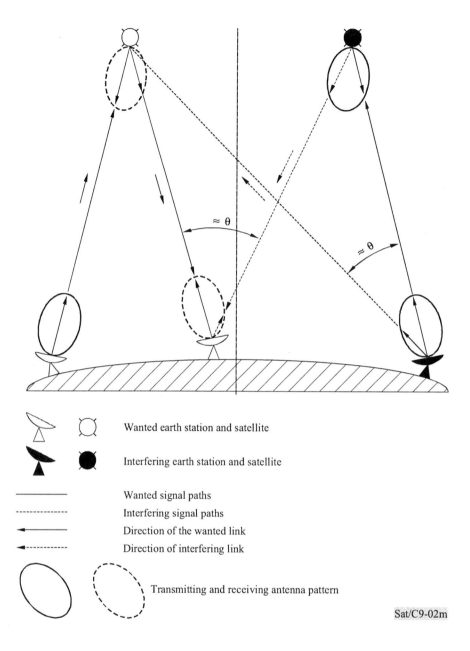

Wanted earth station and satellite

Interfering earth station and satellite

Wanted signal paths
Interfering signal paths
Direction of the wanted link
Direction of interfering link

Transmitting and receiving antenna pattern

Sat/C9-02m

FIGURE 9.2

Interference geometry between two satellite networks
(uplink of wanted network sharing frequencies with uplink
of interfering network (and similarly in the downlink)

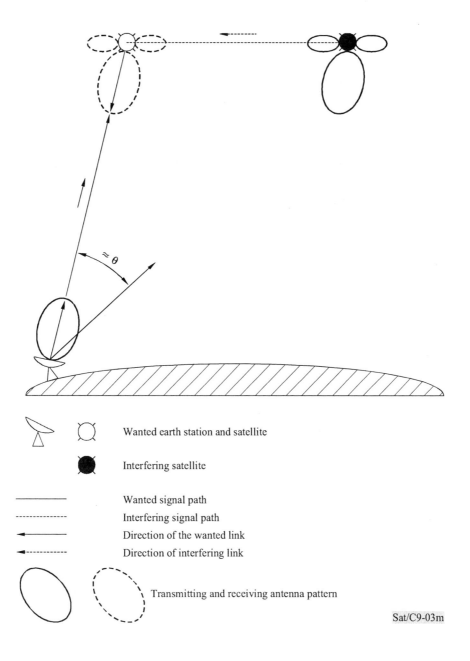

Wanted earth station and satellite

Interfering satellite

Wanted signal path

Interfering signal path

Direction of the wanted link

Direction of interfering link

Transmitting and receiving antenna pattern

Sat/C9-03m

FIGURE 9.3

Interference geometry between two satellite networks
(uplink of wanted network sharing frequencies
with downlink of interfering network)

9.3.1.4   Applicable C/I criteria: single entry interference criterion

(See Table 2 of Recommendation ITU-R S.741 relating to C/I calculations between FSS networks. The case for aggregate interference is dealt with below in § 9.3.1.6.7.)

The permissible levels of interference are given, in the case of FSS, in Recommendations ITU-R S.466 (for FM telephony), S.483 (for FM-TV), S.523 (for PCM telephony), S.735 (for ISDN) and S.671 (for the protection of narrow-band SCPC transmission interfered by FM/TV).

These can be summarized as in Table 9.4 below.

TABLE 9.4

Applicable C/I criteria

(δ: ratio of the wanted signal useful bandwidth to the peak-to-peak energy dispersal
of the interfering signal produced by a transmitting earth station)

Type of wanted signal	Type of interfering signal	Permissible level (dB)	Recommendation ITU-R
FM-TV	Noise	C/N + 14	S.483
Digital	Noise	C/N + 12.2	S.735, S.523
SCPC-FM	Noise	C/N + 14	
SCPC-FM	TV-FM	$12.6 + 2 \log(\delta)$	S.671
SCPC-Digital	TV-FM	$C/N + 7.6 + 3.5 \log(\delta)$	S.671

9.3.1.5   GSO FSS networks and feeder links of GSO MSS networks

ITU-R reports give some provisional permissible interference criteria for maritime MSS circuits. These cover analogue telephony circuits. The types of emission used for the MSS, that is, narrow-band analogue and digital emissions, are likely to require careful protection from certain types of emission commonly used in the FSS, and in particular FM-video emissions.

9.3.1.6   Methods that may be used to facilitate coordination

9.3.1.6.1   General

Several methods may be used to facilitate the coordination of satellite networks, i.e. increasing the angular separation between satellites; reducing or eliminating the overlapping between the most critical carriers; improved earth station or satellite antenna patterns; acceptance of interference levels higher than those recommended by ITU-R, etc. The possibilities of applying these methods and their particular implications are set out below. However, the methods are not arranged in terms of any specific order of priority for the purpose of their application. Additional methods are mentioned in Recommendations ITU-R S.739 and S.740.

9.3.1.6.2 Increase in angular separation

The effect of this increase is to reduce the level of interference in a ratio equal to that of the gain of the earth-station antenna in the direction of the satellite of the other network. If θ_1 and θ_2 are the initial and final values of the angular separation, and I_1 and I_2 are the corresponding interference levels, in dBW, we can write:

$$I_2 - I_1 = 25 \log (\theta_1/\theta_2) \tag{1}$$

assuming that the radiation pattern of the earth station is in accordance with Recommendation ITU-R S.465.

This increase in angular separation may be obtained, first of all, by reducing the tolerance around the nominal longitude: for example, if two satellites are 2° apart, a reduction of the tolerance from ±0.5° to ±0.1° will increase the minimum angular separation from 1° to 1.8°, which will produce an interference level 6.4 dB lower. In certain cases, this effect can be used to greater advantage by shifting one (or both) of the satellites within its service arc.

The limits within which this method may be used depend on the congestion of the orbit as a whole. If the area in which the satellite network is located is congested, the shifting of a satellite to move it further away from another satellite will increase the problems of interference with other networks; the available margin in this respect is often very small.

9.3.1.6.3 Adjustment of network parameters

When two satellite networks are coordinated from the standpoint of mutual interference, a number of the parameters in one or both of the networks may be adjusted to obtain mutually acceptable interference levels.

These adjustments may consist in modifying the link parameters, leading to variations in the power density and the sensitivity. ITU-R recommendations draw attention to four basic factors which limit the power density:

- the up-path power density (P_u);
- the interfering power density against which a satellite receiver is protected (I_u);
- the power density on the downlink (P_d);
- the interfering power against which an earth station receiver is protected (I_d).

If it is possible to reduce the values of the ratios P_u/I_u and P_d/I_d, the interference level between the two systems will be reduced. It is perhaps possible to effect coordination on this basis, i.e. on the basis of the limit values accepted for these four parameters.

It is also possible to effect coordination on the basis of the four parameters in question by readjusting access to transponders to reduce to a minimum the effects due to differences between the antenna gains of earth stations or between the G/T ratios.

If during coordination between two systems, it is found that the interference criteria cannot be met in the whole band of frequencies, it might be desirable to split up the band into two or more sub-bands, for the various types of carriers.

9.3.1.6.4   Reduction or elimination of overlapping between carriers

The method most commonly used – because it is simple, effective and selective – is that which consists in separating the carriers involved so as to reduce or eliminate mutual interference. As a general rule, it is found, in fact, that only certain particular pairs of carrier types cause interference. The best known and the most frequent of these pairs is the case of an FM-TV carrier causing interference to a very narrow-band carrier. RR Appendix S8 refers to this case: Section 4 of the appendix indicates that administrations should exchange information on their use of these carriers in order to facilitate coordination between the satellite systems and to reduce the number of administrations involved in this procedure.

For particularly critical carrier pairs (such as those referred to above), the method used for solving the problem consists in avoiding overlapping between these carriers. This method should be used when the interference level which would arise if the carriers were of the same frequency is markedly in excess of the permissible level (typically, 10 dB or more).

When, for a certain pair of carrier types, it is found that interference exceeds the permissible level by a relatively small amount (typically, from 1 to 5 dB), it is not usually necessary to eliminate completely the overlapping between carriers and it may be sufficient to arrange for a quite small frequency separation between the carriers, which still overlap, but with an increased interference reduction factor compared to the case of "co-frequencies".

The limits within which this method can be applied are due to the operational constraints of administrations in terms of frequency plans. However, where the whole of the spectrum sector affected by the most difficult cases of interference is not very large, it is usually possible for one of the networks to avoid using certain types of carrier in certain parts of the spectrum.

9.3.1.6.5   Improvement of earth-station antenna patterns

The characteristics of earth-station antenna radiation patterns affect interference levels directly. Any improvement in the pattern will therefore be fully reflected in the interference level and such improvement constitutes a very effective means of solving interference problems.

To improve the pattern, one can either increase the antenna diameter or, with a constant diameter, use a specific technique for reducing the side lobes.

This method is applicable primarily when the satellite network is in the initial stages of development.

9.3.1.6.6   Improvement of satellite antenna patterns

Unlike the earth-station antenna pattern, the satellite antenna pattern does not linearly affect the level of interference received. Thus while this method can be employed in certain very special cases, it is not in general use. Moreover, its application also presupposes that the network is in a very early stage of development.

9.3.1.6.7   Acceptance of interference levels higher than those recommended by ITU-R

In the case of analogue and digital telephony in the fixed-satellite service, ITU-R has fixed levels of maximum permissible interference from one network in the FSS (single entry) and from all the other networks in the FSS (aggregate interference).

According to Recommendation ITU-R S.466, the maximum permissible levels in the case of FDM-FM telephony expressed as interference noise power in a telephone channel is 800 pW0p (single entry) and 2 000 to 2 500 pW0p (aggregate with or without frequency reuse).

According to Recommendation ITU-R S.523, the maximum permissible levels in the case of digital 8-bit PCM encoded telephony expressed as a percentage of the noise power level at the demodulator input which would give rise to a bit error ratio of 10^{-6} is 6% (single entry) and 20 to 25% (aggregate). Note that Recommendation ITU-R S.523 applies to networks not forming part of the ISDN.

The ISDN interference limits are given (at the demodulator input under clear-sky conditions on the interference paths in Recommendation ITU-R S.735), these limits are the same as the limits given in Recommendation ITU-R S.523. Concerning the apportionment of this single entry criterion of 6% into the aggregate interference criterion recommended by the ITU-R for FSS systems, it can be noted that the present situation for FSS frequency reuse systems allows 6% noise increase for single entry and 20% for aggregate in the case of digital transmissions and 4% noise increase for single entry and 10% for aggregate in the case of FM/TV transmissions.

In certain cases these values can be exceeded by mutual agreement between the administrations concerned. For instance an administration could tolerate an increase in the aggregate interference level if this is compatible with the noise budget of its network or, in congested arcs of the geostationary-satellite orbit, could tolerate an increase in the level of one or more single entries if the aggregate interference level is not exceeded.

However, in the latter case establishing single entry levels higher than the permissible levels reduces the margin left for other entries.

While such acceptance may be compatible with a given satellite configuration, there is a risk that networks planned in the future may be penalized by the relatively small allocations.

To avoid this drawback Note 3 of Recommendation ITU-R S.466 and Note 4 of Recommendation ITU-R S.523 state that any single entry interference in excess of the recommended value should be disregarded in calculating whether the recommended aggregate interference level is exceeded.

9.3.1.6.8   Other methods

Other methods can be envisaged, but their feasibility is closely dependent on the stage of development of the network: changing the pointing angle of the satellite antenna beam, greater pointing accuracy, changing the polarization used, etc. (see Recommendation ITU-R S.740 for more information on these methods).

9.3.1.7 Conclusion

Coordination between space networks is only necessary if the preliminary procedure which follows the advance publication of the information has revealed an increase of noise temperature greater than 6%.

In this case, if more detailed calculations indicate interference levels higher than the permissible values, an attempt must be made to reduce these levels by various means, such as the coordination of frequencies, which consists in separating the carriers so as to reduce or eliminate overlapping between frequencies, or such as increasing the angular separation between satellites or by other means discussed above. If this is not sufficient, one method that may be considered is the acceptance of interference levels higher than the permissible values, but in that case care must be exercised as stated above.

9.3.2 Frequency sharing between GSO FSS networks in planned bands

9.3.2.1 WARC Orb-88 FSS Plan[7]

The FSS frequency Plan resulting from the WARC Orb-88 decisions is defined in Appendix 30B of the Radio Regulations. It is relevant to the following frequency bands:

- in the 6/4 GHz bands: 6 725-7 025 MHz (Earth-to-space)/4 500-4 800 MHz (space-to-Earth);

- in the 12-13/10-11 GHz bands: 12.75-13.25 GHz (Earth-to-space) 10.7-10.95 GHz, 11.2-11.45 GHz (space-to-Earth), resulting in a total bandwidth of 800 MHz in each direction.

This Plan consists of two parts: (see Article 10 of RR Appendix S30B)

- Part A: the national allotments, such that each administration has at least one allotment (800 MHz, having access to an orbital position within a predetermined orbital arc (PDA))[8]; and

- Part B: the networks using the planned bands already communicated to ITU-R BR before the date of plan development ("existing systems").

Table 9.5 below gives the main technical characteristics of Part A of the FSS Plan.

The Plan was developed, with the assistance of the BR "ORBIT II" computer program, on the basis of a set of "standardized parameters" which approximated the technical characteristics of a typical FSS network. However flexibility in the design of specific satellite networks is kept. A set of generalized parameters (A, B, C and D) have been defined in view of determining if the assignments

---

[7] For more details on the subject, refer to Supplement 1 to Edition 2 of the "Handbook on Satellite Communications (FSS)" (Geneva 1991) entitled: "Effect of WARC Orb-88 decisions".

[8] The predetermined orbital arc (PDA) is the position interval on the GSO within which the nominal position can be moved while still providing coverage of a service area. This is limited to ±10° during the pre-design phase and to ±5° during the design phase.

of a satellite network are in conformity with the Plan[9]. In order to conform to the Plan, an allotment should not exceed the values of its relevant A, B, C and D parameters. Some care is required in designing in satellite networks to avoid excessive receiver sensitivities. To help in applying this principle, a "macrosegmentation" concept was included in the Plan by stipulating that the upper 60% of each allotment band are to be used for high-density carriers[10] and the lower 40% for low-density carriers. This approach would greatly ease the problem of separating frequencies for analogue TV and narrow-band digital carriers.

---

[9] The A and C parameters represent the interference producing capability of the wanted carriers in the uplink and downlink directions respectively, whereas the B and D parameters represent the interference sensitivity of these carriers in the uplink and downlink directions respectively.

[10] High-density carriers are defined as those carriers whose ratio of peak power spectral density (psd) to average psd is greater than 5 dB.

TABLE 9.5

Main technical characteristics of Part A of the FSS Plan for Regions 1, 2, 3
(see Annex 1 to Appendix S30B)

Main parameters	
Period of validity of the Plan	In force until a revision by a competent conference WRC (at least 20 years from Orb-88)
Frequency bands (Uplink/Downlink):	
• **6/4 bands: 6/4 GHz (300/300 MHz)**	6 725-7 025/4 500-4 800 MHz
• **14/11 bands: 12-13/10-11 GHz (500/250+250 MHz)**	12.75-13.25/10.70-10.95 and 11.20-11.45 GHz
Orbital spacing	Non-uniform
Orbit position	See concept and application of predetermined orbital arc (PDA) in Article 5 and Annex 5 of RR App. S30B
	See service arc and PDA in Article 10 of RR App. S30B
Service area	National coverage (possibility of subregional systems: see Article 6 of RR App. S30B)
Reference bandwidth	1 MHz (the Plan is based on power density by dBW/MHz)
C/I Overall protection ratios:	
Single entry	30 dB
Cumulative	26 dB
C/N with rain attenuation	
(C/N↑, C/N↓, C/N overall)	(23 dB↑, 17 dB ↓, 16 dB)
(at 6/4 GHz 99.95% of the year, C/N is exceeded)	
(at 13/10-11 GHz 99.90% of the year, C/N is exceeded)	
Propagation loss	
• **rain attenuation model**	Model in ITU-R Rec.
• **max. rain attenuation margin (6/4 and 13/10-11 GHz)**	8 dB
Transmitting satellite	
Type of beam	Cross-section: circular or elliptical (with orientation angle)
Pointing accuracy	0.1° (in any direction)
Angular rotation of elliptic beams	±1°
Minimum half-power beamwidth:	
• at 6/4 GHz	1.6°
• at 13/10-11 GHz	0.8°
Efficiency	55%
Antenna gain (with a and b: beamwidth of elliptical cross section at half-power)	
	$27\,843/(a \cdot b)$ or (in dBi): $44.45 - 10.\log(a \cdot b)$

TABLE 9.5 (continued)

e.i.r.p. density (mini, maxi) **(see RR Appendix S30B Article 10)** **Edge of coverage (EOC) reference pattern of satellite antenna**	• At 4 GHz: (–42.2 dBW/Hz, –35.6 dBW/Hz) • At 10-11 GHz: (–32.5 dBW/Hz, –21.0 dBW/Hz) See curves A and B of Figures 1 and 2 in Annex 1 of RR Appendix S30B
Receiving earth station	
Antenna diameter ∅	7 m at 6/4 GHz 3 m at 13/10-11 GHz
On-axis antenna gain	$10.\log(\eta(\pi \cdot \varnothing/\lambda)^2)$ in dBi
Reference pattern of earth station antenna:	See Table 1 of RR Appendix S30B
System noise temperature T:	
• **at 4 GHz**	140 K
• **at 10-11 GHz**	200 K
Minimal elevation angles (generally)	10°, 20°, 30°, 40° respectively for climatic areas (A to G), (H to L), (M and N), and P
Transmitting earth station	
e.i.r.p. density (mini, maxi) **(see RR Appendix S30B Article 10)**	• At 6 GHz: (–7.5 dBW/MHz, 2 dBW/MHz) • At 12-13 GHz: (–9.3 dBW/MHz, 14.6 dBW/MHz)
Antenna diameter ∅	Same as for receiving earth station
On-axis antenna gain	Same as for receiving earth station (with η = 70%)
Reference patterns	Same as for receiving earth station
Minimal elevation angle	
Minimal power density	–60 dBW/Hz if rain attenuation
Receiving satellite antenna	
Type of beam	Same as for transmitting satellite
Pointing accuracy, angular rotation of elliptic beams	
Minimum half-power beamwidth	
Antenna gain	Same as for transmitting satellite (with η = 55%)
Noise temperature T:	
• **at 6 GHz**	1 000 K
• **at 13 GHz**	1 500 K
Reference pattern of satellite antenna	Same as for transmitting satellite

Concerning the compatibility analysis, the C/I calculations are based on power densities averaged over the carrier bandwidth and the results are compared to single-entry and aggregate criteria. The applicable single-entry and aggregate criteria are 30 dB and 26 dB respectively, or the values resulting from the Plan whichever is the lower.

However, if the macrosegmentation concept is not observed by a system that has high density carriers of peak factor k>5 dB over the band reserved for low density carriers, the single-entry criterion is changed to 25 + k dB.

A wanted allotment of an administration shall be considered as being affected by another administration if, at its nominal orbital position within the PDA, the calculated single-entry and aggregate C/I are less than the above mentioned criteria or the values given in the Plan for this wanted allotment.

Concerning the sharing with the other services in the FSS planned bands, it has to be treated with the same criteria as for the non-planned FSS bands.

A specific procedure is applied for the introduction of subregional systems in the Plan. The definition of a subregional system (§ 2.5 of Article 2 of Appendix S30B) limits the group of administrations in it to neighbouring countries only. These administrations need to suspend their allotments for the duration of the subregional system unless these allotments can be used simultaneously with the subregional system without affecting the Plan. A separate assessment of the specifics of this procedure is provided in Section II of Article 6 of Appendix S30B.

9.3.2.2   Use of the FSS planned frequency bands by non-GSO satellite systems

(See § 9.5.5 below.)

Several studies show the possible use of GSO FSS planned bands for non-GSO satellite systems.

WRC-97 Resolution 130 of the RR asks ITU-R to study frequency sharing mechanisms among GSO, non-GSO and terrestrial systems.

RR Article S5.441 provides that the use of the bands 10.7-10.95 GHz (space-to-Earth), 11.2-11.45 GHz (space-to-Earth), 12.75-13.25 GHz (Earth-to-space) by non-GSO satellite systems in the FSS shall be in accordance with the provisions of WRC-97 Resolution 130.

RR Article S5.484A provides that the use of the bands 10.95-11.20 GHz (space-to-Earth), 11.45-11.7 GHz (space-to-Earth), 11.7-12.20 GHz (space-to-Earth) in Region 2, 12.2-12.75 GHz (space-to-Earth) in Region 3, 12.5-12.75 GHz (space-to-Earth) in Region 1, 13.75-14.5 GHz (Earth-to-space), 17.8-18.6 GHz (space-to-Earth), 19.7-20.2 GHz (space-to-Earth), 27.5-28.6 GHz (Earth-to-space), 29.5-30 GHz (Earth-to-space) by non-GSO and GSO satellite systems in the FSS is subject to the provisions of WRC-97 Resolution 130. The use of the band 17.8-18.10 GHz (space-to-Earth) by non-GSO FSS systems is also subject to the provisions of WRC-97 Resolution 538.

9.3.3 Frequency sharing between GSO FSS networks and feeder links of BSS in planned and unplanned in planned bands

(See RR Appendices S30A BSS Plan, S30A BSS FL Plan and S30B FSS Plan.)

It should be remembered that the BSS feeder links belong to the FSS.

In sharing between FSS networks and BSS feeder links (BSS FL), interference can occur as follows:

i) For bands allocated bidirectionally to the FSS in which the Earth-to-space allocation is exclusively for use by feeder links to the BSS (e.g. 17.7-18.1 GHz), 10.7-11.7 in Region 1 and 18.1-18.4 GHz:

   • feeder link earth station interference into FSS receiving earth station;

   • FSS space station interference into BSS receiving space station, both nearby and near-antipodal.

ii) For bands allocated to the FSS Earth-to-space usable for feeder links to the BSS (14-14.5 GHz and 24.75-25.25 GHz in Regions 2 and 3):

   • Feeder link earth station into FSS receiving space station;

   • FSS uplinks into BSS feeder link receiver.

(Case 1) BSS feeder link earth station into FSS receiving earth stations

(FSS, space-to-Earth and BSS FL, Earth-to-space)

Procedures and criteria are required for sharing between BSS feeder links and FSS receiving earth stations. Two kinds of interference interactions need to be accounted for:

Mode a)   coupling along a great circle tropospheric interference horizon path;

Mode b)   coupling through scattering from hydrometeors.

Recommendation ITU-R SF.1005 provides a method for determination of the bidirectional coordination area (see also RR Appendix S7 and Recommendations ITU-R IS.848 and IS.850).

(Case 2) FSS space station into BSS feeder links receiving space station

(FSS, space-to-Earth and BSS FL, Earth-to-space)

Procedures and criteria are required for nearby and near-antipodal FSS and BSS space stations.

For instance, in the 17.7-18.1 GHz for instance, procedures and criteria are required for nearby and near-antipodal FSS and BSS space stations. WARC Orb-88 adopted, for the 17.7-17.8 GHz band, an approach requiring coordination when the space stations are separated by less than 3° or greater than 150° and when a $\Delta T/T$ increase (case II of the method given in RR Appendix S8) of 4% might occur. However, in the case where the separation exceeds 150°, coordination is required only if the pfd towards the equatorial limb of the Earth exceeds -137 dB(W/m^2/24 MHz) on the Earth's surface (see Annex 4 to RR Appendix S30A of the Radio Regulations).

(Case 3) BSS feeder links into FSS space stations

(FSS, Earth-to-space and BSS FL, Earth-to-space)

RR Article S5.506 provides that the 14.0-14.5 GHz band may be used, within the FSS (Earth-to-space) for BSS feeder links, subject to coordination with networks in the FSS. Such use of feeder links is reserved for countries outside Europe. Such use of the band could lead to a potential interference between the two services, taking account of the fact that the BSS orbital assignments are fixed in accordance with the Region 1, 2 and 3 BSS Plans, and the fact that use of 14.0-14.5 GHz for the FSS is growing rapidly. Large angular separations are generally required to protect both services.

Recommendation ITU-R S.1063 provides criteria between BSS feeder links and other Earth-to-space or space-to-Earth links of the FSS.

9.4 Frequency sharing between the GSO FSS networks and other space services

The subsections which follow discuss the important situation of sharing between FSS networks and various other space services, e.g.:

- FSS and the broadcasting-satellite service (BSS downlink Plan);

- FSS and the earth exploration-satellite service (EESS) (passive and active);

- FSS and the mobile-satellite service (MSS).

9.4.1   Sharing between GSO FSS and the GSO BSS Plan

9.4.1.1   GSO BSS Plans and GSO BSS FL Plans

(RR Appendix S30, BSS downlink Plan and Appendix S30A, BSS feeder-link Plan)

9.4.1.1.1   General: Frequency sharing between BSS networks in planned bands

It should be remembered that the BSS frequency Plans for all three regions are defined in term of assignments for each administration respectively for the downlinks (11.7-12.5 GHz for Region 1, 12.2-12.7 GHz for Region 2 and 11.7-12.2 GHz for Region 3) and for the feeder-links (14.5-14.8 GHz for Regions 1 and 3 outside Europe, 17.3-18.1 GHz for Regions 1 and 3 and 17.3-17.8 GHz for Region 2) emissions and that an assignment of a notifying administration corresponds mainly to:

- an orbital position;

- a channel number (on five channels average by country for Region 1) or a centre frequency;

- a channel bandwidth (27 MHz in Regions 1 and 3 and 24 MHz in Region 2);

- a polarization (circular polarization is generally used);

- one or several beams;

- a transmitting power;

- transmitting and receiving antenna parameters;

- a set of test-points;

- and fundamental criteria called the reference situation which can be a reference equivalent protection margin EPM or a reference overall equivalent protection margin (OEPM).

The parameters adopted for planning purposes during the relevant conferences are defined in Annex 5 to RR Appendix S30 and in Annex 3 to RR Appendix S30A, they correspond to the standard parameters.

Articles 10 and 11 of Appendix S30 contain respectively the Region 2 BSS Plan and Regions 1 and 3 BSS Plans. Articles 9 and 9A of Appendix S30A contain respectively the Region 2 BSS FL Plan and Regions 1 and 3 BSS FL Plans.

The procedures for modifications to the Plan (see Article 4 of Appendices S30 and S30A) have proven to prescribe constraints difficult to respect while the requests were very numerous before WRC-97. With a view to improving the efficiency and flexibility of the Plans for R1 and R3 contained in RR Appendices S30 and S30A, WRC-97 proceeded with the revision of Appendices S30 and S30A. But certain countries requested that a re-planning be undertaken in order to increase the Plan capacity by a future WRC.

9.4.1.1.2 Proposed revisions and adoption of new BSS and BSS FL Plans by WRC-97 (for Regions 1 and 3)

WARC-92 and WRC-95 recommended various modifications to the procedures and to the Plans in view of improving their efficiency. These modifications were worked on by WRC-97, involving: a general reduction of 5 dB from the levels of e.i.r.p. planning values, the use of an improved receive earth station reference antenna pattern with a figure of merit G/T equal to 11 dB/K, the simultaneous planning of feeder link (FL) and downlink with calculation of OEPM, an aggregate overall C/I ratio of co-channel interference of 23 dB with no single-entry co-channel lower than 28 dB.

WRC-97 revisions involved modifications or additions in various RR articles relating to the coordination, notification and registration procedures and the revision of Appendices 30 and 30A of the RR for Regions 1 and 3 (while the integrity of the Region 2 Plan was maintained).

ITU-R studies in this field were carried out for the revision of the Appendix S30 and S30A Plans with the aim of avoiding their obsolescence by using a combination of updated technical parameters and revised procedures in view of ensuring the long-term flexibility of the Plans.

The revision of Appendices S30 and S30A by WRC-97 is based on the following principles:

- use the revised planning parameters adopted in WRC-95 RR Recommendation 521;

- provide for new countries, and those countries having less than the minimum number of channels assigned by the BSS WARC-77 (in Region 1, this was five channels);

- be based on national coverage;

- take account of the increased requirements of subregional systems in order to facilitate the development of multi-administration and subregional systems, and take account of systems communicated to BR under Articles 4 of RR Appendices S30 and S30A and protect the assignments which are in conformity and have been notified;

- provide the simultaneous operation of analogue and digital systems.

In conclusion, WRC-97 has adopted a revision of the BSS Plans for Regions 1 and 3 providing capacity for all new countries in accordance with WARC-92 Resolution 524 and WRC-95 Resolution 531. But certain countries requested that a replanning be undertaken in order to increase the Plan capacity and to provide a channel capacity large enough (ten channels) to permit the economical development of a BSS system, and the increasing number of applications for modifications under Article 4 of RR Appendices S30 and S30A involves too many additions to the Plans.

In consequence, WRC-97 resolved that an Inter-conference Representative Group (IRG) shall be established. The results shall be presented to WRC-2000 regarding the minimum assigned capacity for countries in Regions 1 and 2 to around ten (10) analogue equivalent channels.

9.4.1.1.3 Possible introduction of non-GSO satellite systems: use of the BSS planned frequency bands by non-GSO systems (also see § 9.5.5 below)

Resolution 506 (Orb-88) of the RR was finally revised by WRC-97 in order to introduce the possibility, under hard limits, of introducing non-GSO FSS systems in the frequency band of the BSS Plan, while not permitting the introduction of non-GSO BSS systems.

The regulatory matters as well as the compatibility analysis between GSO and non-GSO satellite systems operating in these planned bands of the RR shall be subject to further studies within ITU-R study groups. WRC-97 Resolution 538 of the RR asks ITU-R to study the frequency sharing between GSO and non-GSO systems and between non-GSO systems.

RR Article S5.487A provides that the use of the 11.7-12.5 GHz band in Region 1, the use of the 12.2-12.7 GHz band in Region 2 and the use of the 11.7-12.2 GHz band in Region 3 are also allocated to the FSS (space-to-Earth) on a primary basis, limited to non-GSO systems and subject to the provisions of WRC-97 Resolution 538 of the RR.

Also, RR Article S5.516 provides that the use of the 17.3-18.1 GHz band in Regions 1 and 3 and the use of the 17.8-18.1 GHz band in Region 2 by non-GSO satellite systems in the FSS (Earth-to-space) is subject to the provisions of WRC-97 Resolution 538 of the RR.

RR Article S5.484A provides that the use of the 17.8-18.6 GHz band by non-GSO and GSO satellite systems in the FSS (space-to-Earth) is subject to the provisions of WRC-97 Resolution 130 and the use of the 17.8-18.1 GHz band by non-GSO satellite systems in the FSS (space-to-Earth) is also subject to the provisions of WRC-97 Resolution 538.

9.4.1.2 Frequency sharing between the FSS and the BSS FL Plans

The difference in the frequency allocations in the frequency bands around 12 GHz in the three Regions gives rise to several areas where sharing occurs between the BSS and the FSS. These areas are identified in Tables 9.6A (BSS Plans) and 9.6B (BSS FL Plans) and it should be noted that these tables have been compiled taking into account the revisions to RR Article S5 adopted at WARC Orb-85/88, WARC-92 and WRC-95/97.

i) Intraregional sharing in the BSS Plans

(See Recommendations ITU-R S.1065 and S.1067.)

In Region 1 there are no shared allocations between the BSS and the FSS at 12 GHz.

In Region 2, the 11.7-12.2 GHz band is allocated to the FSS, however, transponders on such satellites may, by virtue of RR Article S5.485, be used for transmissions on the BSS provided that they do not have a maximum e.i.r.p. greater than 53 dBW per television channel and do not cause greater interference, or require more protection than coordinated FSS assignments. Both the FSS and BSS are limited to national and subregional systems in these bands.

Also in Region 2, the BSS assignments in the 12.2-12.7 GHz band may, under RR Article S5.492, be used for transmissions in the FSS (space-to-Earth), provided that such transmissions do not cause more interference or require more protection than BSS transmissions operating in accordance with the Plan for Region 2 in RR Appendix 30.

In Region 3, the 12.5-12.75 GHz band is shared between the FSS (space-to-Earth) and the BSS. The BSS is limited by RR Article S5.493, to a maximum power flux-density of -111 dB(W/m^2). Under Resolution 34, coordination between the two services is required on the basis of RR Article S9 and RR Resolution 33, until such time as a Region 3 BSS Plan is developed for this band.

ii)   Interregional sharing

(See Recommendation ITU-R S.1066 and also RR Articles 5, 6 and Annex 4 of RR Appendix S30.)

Interregional sharing between the Region 3 BSS (community) and the Region 1 FSS (space-to-Earth) links, occurs in the 12.5-12.75 GHz band. At present there is no BSS plan for this band, and the BSS is permitted to operate under RR Article S5.493, subject to a power flux-density limit of -111 dB(W/m^2).

In the bands between 12.5 and 12.75 GHz, sharing occurs also between Region 3 BSS (community) and FSS (Earth-to-space) in Regions 1 and 2. Interference is possible both between closely spaced and antipodal satellites.

The interregional sharing problems between the BSS and the FSS (space-to-Earth) are briefly discussed in Recommendation ITU-R S.1066. Potential interference from networks established in accordance with the Regions 1 and 2 BSS Plans may count as difficulties for the location and operation of FSS networks in the GSO unless interference mitigation techniques (i.e. advanced methods for reducing interference) are employed for present sensitive FSS transmitters.

Note that the band 12.2-12.5 GHz is allocated on a primary basis to the BSS in Region 1 and FSS in Region 3 and both services should have equitable access to the orbit and spectrum. At present, the procedures of RR Appendix S30 applicable to the Region 3 FSS in respect of the Region 1 BSS Plan are such that only the BSS plan is protected.

WRC-97, by Resolution 73, asked administrations to make all possible mutual efforts to solve the interference problems.

TABLE 9.6A

Interference situation of the FSS and the BSS in the 11-12 GHz band (NOTE 1)

Interference situation	Where interference occurs (GHz)			Available criteria
	Region 1	**Region 2**	**Region 3**	
FSS ↓ into BSS Plan	11.7-12.2 from R2 12.2-12.7 from R3	12.5-12.7 from R1 12.2-12.7 from R3	11.7-12.2 from R2	RR Appendix S30, Annex 4 (NOTE 2)
FSS ↑ into BSS			12.5-12.75 from R1 12.7-12.75 from R2	RR Resolution 33
FSS ↓ into BSS			12.5-12.75 from R1 and intraregional	RR Resolution 33
GSO and non-GSO FSS ↓ into BSS			12.5-12.75 from GSO and non-GSO FSS	RR Resolution 130
non-GSO FSS ↓ into BSS Plan	11.7-12.5 from non-GSO FSS	12.2-12.7 from non-GSO FSS	11.7-12.2 from non-GSO FSS	RR Resolution 538
BSS Plan into FSS ↓	12.5-12.7 from R2	11.7-12.2 from R1 and R3	12.2-12.5 from R1 12.2-12.7 from R2	RR Appendix S30, Annex 1, § 6 (NOTE 3)
BSS into FSS ↓	12.5-12.75 from R3		Intraregional	RR Resolution 33
BSS Plan into FSS ↑	12.5-12.7 from R2			RR Appendix S30 Annex 1, § 7 ΔT/T threshold 4% (App. S7, case II)
BSS into FSS ↑	12.5-12.75 from R3	12.7-12.75 from R3		RR Resolution 33
BSS Plan	11.7-12.5	12.2-12.7	11.7-12.2	RR Appendix S30

NOTES – ↑: Uplink, ↓: Downlink

NOTE 1 – See also (table in App. S30 Annex 6) protection requirements (C/I or N) for sharing between services in the 12 GHz band.

NOTE 2 – (pfd limit for coordination of a FSS space station with respect to the BSS Plans, Interregional interference case, θ = orbital separation):

-147 dBW/m²/27 MHz if $\theta < 0.44°$;

$-138 + 25.\log\theta$ dBW/m²/27 MHz for $0.44° \leq \theta < 19.1°$; and

-106 dBW/m²/27 MHz for $\theta \geq 19.1°$.

NOTE 3 – (limits to change in the pfd of assignments in the BSS Plan to protect the FSS):

FSS in R2 to be protected from R1 and R3 BSS Plan: pfd increase ≤ 0.25 dB or pfd ≤ -138 dB/m²/27 MHz;

FSS in R1 and R3 to be protected from R2 BSS Plan: pfd increase ≤ 0.25 dB or pfd ≤ -160 dB/m²/27 MHz.

Relevant RR footnotes:

S5.487A: The band 11.7-12.5 GHz in Region 1, the band 12.2-12.7 (R2) and the band 11.7-12.2 GHz (R3) of the BSS Plans are also allocated to the FSS (space-to-Earth) on a primary basis, limited to non-GSO systems and subject to the provisions of RR Resolution 538 (see Section II of RR Article S22 for interference to GSO systems).

S5.484A: The use of the band 12.5-12.7 GHz (R1), the band 11.7-12.2 (R2) and the band 12.2-12.75 GHz (R3) by GSO and non-GSO systems in the FSS (space-to-earth) is subject to the provisions of RR Resolution 130 (see also Section II of RR Article S22).

S5.491: In R3, the band 12.2-12.5 GHz is also allocated to the FSS (space-to-earth) on a primary basis, limited to national and subregional systems.

TABLE 9.6B

Interference situation of the FSS and the FFS uplink (BSS feeder links) in the 17 and 18 GHz band

Interference Situation	Where interference occurs (GHz)			Available criteria
	Region 1	Region 2	Region 3	
non-GSO FSS ↑ into FSS ↑(BSS FL or BSS FL Plan)	17.3-18.1	17.8-18.1	17.3-18.1	• RR Art. S22.5D • S5.516: RR Res. 538
GSO FSS ↑ into FSS ↑(BSS FL Plan)	17.7-17.8 (spanning)			RR Appendix S30A, Annex 4, § 1
	17.8-18.1 from GSO or non-GSO FSS		17.8-18.1 from GSO or non-GSO FSS	RR S.484A: Res. 130 and 538
FSS ↓ into BSS ↓		17.3-17.8 (no interference from FSS after 2007)		S5.517 (17.7-17.8: BSS after 2007)
non-GSO FSS ↑ into GSO BSS ↓		17.3-17.8 (for further study)		S22.5C
GSO (BSS or FSS) ↓ (transmit space station) into FSS ↑(BSS FL Plan)	17.3-18.1 (FSS)	17.3-17.8 (BSS) 17.7-17.8 (FSS)	17.3-18.1 (FSS)	• Appendix S30A, Annex 4, § 1 for receiving space stations • ΔT/T threshold 4% (see RR App. S8 and Art. S5.515: 17.3-17.8, BSS)
non-GSO FSS ↑ into FSS ↑ (BSS FL)		17.8-18.1		RR S5.516: Res. 538
GSO and non-GSO FSS ↓ into FSS ↑ (BSS FL)	18.1-18.4	17.8-18.1 18.1-18.4	18.1-18.4	• Res. 130 (17.8-18.6) • Res. 538 (17.8-18.1) • S5.484A
FSS ↑ (BSS FL Plan) into GSO FSS ↓	17.7-18.1 (spanning)			App. S30A, Annex 1, § 1 and Annex 4, § 3 for receiving earth stations (see App. S7)
FSS ↑ (BSS FL) into FSS ↑ (BSS FL Plan)	17.3-18.1 (from BSS FL of R2)	17.3-17.8 (from BSS FL of R1 and R3)	17.3-18.1 (from BSS FL of R2)	• App. S30A, Annex 1, § 5 for receiving space stations. • ΔT/T threshold 3% (see RR App. S8)
(BSS FL Plan)	17.3-18.1 (spanning)			App. S30A

Relevant footnotes:

S5.516: The use of the band 17.3-18.1 GHz in Regions 1 and 3 (Earth-to-space) and the band 17.8-18.1 in Region 2 by GSO and non-GSO systems in the FSS is subject to the provisions of RR Resolution 538 (see Section II of RR Article S22 for interference to GSO systems).

The use of the band 17.3-18.1 GHz by GSO systems in the FSS (Earth-to-space) is limited to the BSS FL.

S5.484A: The use of the band 17.8-18.6 GHz (space-to-Earth) by GSO and non-GSO systems in the FSS is subject to the provisions of RR Resolution 130 (see § II of RR Article S22 for interference to GSO systems).

The use of the band 17.8-18.1 GHz (space-to-Earth) by non-GSO systems is also subject to the provisions of RR Resolution 538.

9.4.1.3   Frequency sharing between the BSS and BSS FL Plans and the other services

The feeder-link and downlink frequency bands allocated to the BSS are also shared with terrestrial services (fixed, broadcasting and mobile, except aeronautical mobile in the BSS downlink bands).

The sharing criteria applicable in these situations are based on pfd limits or coordination areas and are defined in RR Appendices S30 and S30A.

In dealing with interference from a BSS space station into terrestrial services, particular attention should also be paid to provision RR Article S23.13.

Criteria defined for the FSS are used to check the compatibility of a modification or an addition to a Plan with the other assignments in the same regional BSS Plan.

In addition, sharing criteria are required for both feeder links and downlinks in the allocated frequency bands shared with other services.

i)   BSS compatibility within the same region

In both Regions 1 and 3 feeder-link and downlink Plans, a modification or an addition to a Plan is compatible with the other assignments of this Plan if it does not degrade below –0.25 dB the reference equivalent protection margin (EPM) of the other assignments which are positive and if it does not degrade by more than 0.25 dB the reference overall equivalent protection margin (OEPM) of the other assignments which are negative.

In the Region 2 feeder-link and downlink Plans, the same degradation limit is applied with the OEPM instead of the EPM.

ii)   Sharing with other services

The frequency bands allocated to the feeder links associated with the BSS require criteria for sharing with the following services, as shown in Table 9.7 below.

TABLE 9.7

Criteria for sharing between FSS ↑(BSS feeder links) and other services

Region	Frequency band (GHz)	Service to be protected	Applicable criteria
R1 and R3 outside Europe	14.5-14.8	Fixed service (FS)	RR Appendix S8 (also RR App. S30A, Article 6 and Annex 1, § 2)
All regions	17.7-17.8	FS	
		FSS ↓ (receiving earth stations)	RR Appendix S7 type coordination procedure (also RR App. S30A, Annex 1, § 1 and Annex 4, § 3)
R1 and R3	17.8-18.1	FS	RR Appendix S8 (also RR App. S30A, Article 6 and Annex 1, § 2)
		FSS ↓ (receiving earth stations)	RR Appendix S7 type coordination procedure (also RR App. S30A, Annex 1, § 1 and Annex 4, § 3)
R2	17.3-17.8	In R1 and R3: FSS ↑ (BSS FL Plan) (receiving space stations)	• RR Appendix S30A, Annex 1, § 5 • ΔT/T threshold 3% (see RR App. S8)
R1 and R3	17.3-17.8	In R2: FSS ↑ (BSS FL Plan) (receiving space stations)	• Appendix S30A (Annex 1, § 5) • Orb-88, Resolution 42

Downlink frequencies allocated to the BSS may also require sharing with other services (FS, BSS of another region, FSS) as shown in Table 9.8 below.

TABLE 9.8

Criteria for sharing between BSS downlinks and other services

Region and BSS	Frequency band (GHz)	Case No.	Service to be protected in region	Applicable criteria (Annex 1 to RR App. S30): pfd limits[(*)]
R1 and R3 BSS Plans (App. S30)	11.7-12.2	1s	FSS↓ in R2	§ 6: pfd increase ≤ 0.25 dB or pfd ≤ −138 dB/m²/27 MHz
		1fs	FS in R2 (except for 12.1-12.2 GHz)	§ 4: −125 dBW/m²/4 kHz for circular polarization −128 dBW/m²/4 kHz for linear polarization
		1t	Terrestrial services in R1 and R3	§ 8a: pfd increase ≤ 0.25 dB or § 5a and b
R1 BSS Plans (App. S30)	12.2-12.5	2s	• BSS Plan in R2	§ 3: −147 dBW/m²/27 MHz if θ<0.44° (θ = orbital separation) −138+ 25.logθ dBW/m²/27 MHz for 0.44°≤θ< 19.1° −106 dBW/m²/27 MHz for θ≥19.1°
			• FSS↓ (national and subregional) in R3[(**)]	§ 6: pfd increase ≤ 0.25 dB or pfd ≤ −160 dBW/m²/4 kHz
		2t	• Terrestrial services in R1 and R3	§ 8a: See Case No. 1t
			• Terrestrial services in R2	§ 4: See Case No. 1fs
R2 BSS Plans (App. S30)	12.5-12.7	3s	• BSS Plan in R1	§ 3: See Case No. 2s
			• FSS↓ (national and Subregional) R3[(**)]	§ 6: See Case No. 2s
		3t	• Terrestrial services in R2	§ 4 or § 5: See Case No. 1t
			• Terrestrial services in R3	§ 8a or § 8b: See Case No. 2t
R2 BSS Plans (App. S30)	12.7-12.75	4s	• BSS in R3	§ 3: See Case No. 2s
			• FSS↓ in R1 and R3	§ 6: See Case No. 2s
			• FSS↑ in R1 and R2	§ 7: ΔT/T threshold: 4% (RR App. S8, case II)
		4t	• Terrestrial services in R2	§ 8b: See Case No. 3t
			• Terrestrial services in R3	§ 5: See Case No. 1fs
R3 BSS[(***)] **(See RR Res. 34 and 33)**	12.5-12.7	5s	• BSS Plan in R2	§ 3: See Case No. 2s
			• FSS↓ in R1 and R3	§ 6: See Case No. 1s
			• FSS↑ in R1 and R2	§ 7: See Case No. 4s
		5t	• Terrestrial services in R2	§ 4: See Case No. 1fs
			• Terrestrial services in R3	§ 8a: See Case No. 1t

TABLE 9.8 (continued)

R3 BSS[(***)] (See RR Res. 34 and 33)	12.7-12.75	6s	• FSS↓ in R1 and R3	§ 6: See Case No. 1s
			• FSS↑ in R1 and R2	§ 7: See Case No. 4s
		6t	• Terrestrial services in R2	§ 4: See Case No. 1fs
			• Terrestrial services in R3	§ 8a: See Case No. 1t

[(*)]   FSS: see § 6 and BSS: see § 3. BSS in R1 and R3 into terrestrial services in R2: see § 4. BSS in R2 into terrestrial services in R1 and 3: see § (5a/b/c/d). BSS into terrestrial services in R1 and R3: see § 8a. BSS into terrestrial services in R2: see § 8b. NOTE – Terrestrial services: FS, MS, BS …

[(**)]   S5.491: The band 12.2-12.5 GHz in R3 is also allocated to the FSS (space-to-Earth) on a primary basis, limited to national and subregional systems.

[(***)]   S5.493: The BSS in the band 12.5-12.75 GHz in R3 is limited to a pfd not exceeding –111 dBW.

9.4.2   Frequency sharing between the FSS and the earth exploration-satellite service (EESS)

i)   Passive EESS

The 18.6-18.8 GHz band is allocated to the earth exploration-satellite service EESS (passive) and Space Research SR (passive) services on a primary basis in Region 2 and on secondary basis in Regions 1 and 3.

In Region 2 the EESS (passive) and the fixed-satellite service FSS (space-to-Earth) share the 18.6-18.8 GHz band on an equal primary basis. The fixed service FS, the mobile service MS and the FSS (space-to-Earth) are allocated worldwide on a primary basis. RR Article S5 (footnotes S5.522 and S5.523) request administrations to limit the power of FS transmitters and the pfd produced by FSS space stations as far as possible in order to reduce the risk of interference to passive sensors.

Measurements by spaceborne passive sensors in the 18.6-18.8 GHz band are needed because the band has unique characteristics that are important to obtaining information regarding environmental conditions on the Earth's land and ocean surfaces.

Studies by the EESS users have shown that loss of data could occur when the passive sensor satellite passes through the fixed satellite FSS main beam or from FS signals reflected from the Earth's surface. A reduction of 22 dB in maximum pfd from that permitted in the Radio Regulations has been proposed as a suitable level to allow sharing of this band, but not accepted by users of the FSS.

In the FSS, systems are currently developed in this band and suitable standards which would assure that this band can be successfully utilized with such a 22 dB reduction in this maximum pfd level has not yet been established.

In Regions 1 and 3 the passive satellite sensor services have a secondary allocation at 18.6-18.8 GHz, whereas the FSS allocation is primary. It is foreseen that the pfd power flux-density set up for the Earth's surface in Regions 1 and 3 by some FSS space stations will exceed the limits for a satisfactory sharing with passive satellite sensors.

In conclusion, urgent studies are required on a suitable pfd limit.

ii)   Active EESS

The FSS (Earth-to-space) and the EESS (space-to-Earth) services share the 8 025-8 400 MHz band in Region 2 on an equal-primary basis and RR Article S5.464 provides that the band may be used by the EESS service on a primary status in Regions 1 and 3 countries subject to agreement obtained under Article S9.21 (Article S9/14 procedures). Elsewhere the EESS is secondary and the FSS is primary.

Protection criteria for EESS links are contained in Recommendation ITU-R SA.514.

9.4.3   Frequency sharing between the GSO FSS and the GSO mobile-satellite service (GSO MSS)

The GSO FSS is shared on a primary basis with the GSO MSS in the 7.250-7.375 GHz (downlink), 7.900-8.025 GHz (uplink), (19.7-20.1 GHz downlink in Region 2), 20.1-21.2 GHz (downlink), 29.5- 29.9 GHz (R2) and 29.9-31 GHz (uplink), 39.5-40.5 GHz (downlink) bands and other higher frequency bands (71-74 GHz uplink and 81-84 GHz downlink). A number of systems are in existence or planned to operate in the 8/7 GHz bands. Many of these systems operate in both the FSS and MSS so that intranetwork sharing occurs. It could be expected that this combined service use will also occur at the higher frequency bands as systems are developed. From information published to date in BR circulars, some basic factors which are common to a number of systems using the 8/7 GHz bands are:

- large service areas, i.e. approaching the visible areas;
- Earth coverage, hemispheric coverage and redirectable narrow beam satellite antennas;
- capabilities for changing satellite antenna/transponder configurations;
- circular polarizations; no frequency reuse within a network;
- large differences in earth-station antenna sizes, the smallest are in the order of 1 to 3 m;
- relatively high maximum transmission gains (see RR Appendix S27), which coupled with high up-path satellite antenna gains, result in relatively high up-path sensitivity.

These factors are consistent with satellite networks which could operate in either or both of the fixed-satellite and mobile-satellite services.

Conversely, there is no uniformity in transponder arrangements, frequency translations, satellite antenna configurations, or types of modulation, carriers, and satellite accessing.

The factors which affect the use of the geostationary satellite orbit are those related to frequency reuse within and among satellite networks operating in the FSS and MSS. Some of the most significant factors are as follows:

- satellite antenna discrimination;
- earth-station antenna discrimination;
- orthogonal polarization;
- modulation.

Because of these factors, the minimum satellite spacing tends to be limited by up-path interference under co-coverage conditions. Generally, the characteristics of the MSS will determine the minimum spacing from satellites in other services.

Recommendation ITU-R S.132 specifies frequency sharing between systems of the bands 19.7-20.2 GHz between systems in the MSS and systems in the FSS.

9.5 Frequency sharing between the GSO and non-GSO FSS and other GSO and non-GSO space services (except the GSO/GSO case)

(See CPM-97 Report and WRC-97 Final Acts.)

9.5.1   General

Non-geostationary satellite (non-GSO) systems, in particular using low orbiting satellites (LEOs) are becoming a well recognized medium for establishing both mobile and fixed communication networks. A few systems are already in construction and should enter operation and many others are projected (see Chapter 6, § 6.5 and Appendix 6.1 of the handbook). These systems are a new challenge for ITU, in terms of potential interference and frequency sharing.

The subsections which follow discuss the important situation of frequency sharing between GSO and non-GSO FSS networks and other GSO and non-GSO space networks (except the GSO/GSO case) e.g.:

*   sharing between GSO and non-GSO FSS;

*   sharing between non-GSO FSS;

*   sharing between non-GSO FSS and non-GSO MSS FL;

*   sharing between non-GSO MSS FL (FBW and RBW);

*   sharing between non-GSO FSS and GSO (FSS and BSS) Plans.

RR Article S22.2 summarizes as follows the general regulation context concerning the interference situation from the non-geostationary (non-GSO) space services into GSO geostationary satellite systems in the FSS:

*   Before WRC-97, it was said that "non-GSO space stations shall cease or reduce to a negligible level their emissions, and their associated earth stations shall not transmit to them, whenever there is unacceptable interference to geostationary satellite space systems in the FSS operating in accordance with these Regulations".

*   The WRC-97 modified this wording as follows: "non-GSO satellite systems shall not cause unacceptable interference to GSO satellite systems in the FSS and the BSS operating in accordance with these Regulations".

Revised WRC-95 Resolution 46 (Article S9.12) presents "Interim procedures for the coordination and notification of frequency assignments of satellite networks in certain space services and the other services to which certain bands are allocated" (see Annex 2 to Resolution 46 (Rev.WRC-97), also see § 9.8 below concerning sharing between non-GSO space stations and FS stations).

WRC-95 made certain frequencies available for the purpose of providing feeder links to space stations in the non-GSO satellite networks of the mobile-satellite service (FSS (non-GSO-MSS (FL))).

Pending a permanent procedure, the mutual use of certain bands by non-GSO satellite networks in relation to other GSO satellites networks or to non-GSO satellite networks or terrestrial stations (including the case of terrestrial stations in relation to the earth stations of non-GSO satellites networks) shall be regulated in accordance with the interim procedures and the associated provisions and criteria annexed to Resolution 46.

These interim procedures apply in addition to those of Articles S9 and S11 of the RR for non-GSO satellite networks in the bands identified by footnote to the Table of Frequency Allocations in Article S5 of the RR and must be applied since 17 November 1995.

RR Resolution 46 invites ITU-R to study and develop recommendations on the coordination methods, the necessary orbital data relating to non-GSO satellite systems and the sharing criteria.

WRC-95 Recommendation 104 of the RR asks ITU-R to study the possibility of developing e.i.r.p. and pfd limits for non-GSO MSS FL to protect GSO FSS networks in accordance with RR S22.2 (ex-RR 2613) in bands where RR Resolution 46 (WRC-95) does not apply.

WRC-97 Resolution 130 advises ITU-R that technical, operational and regulatory studies concerning the use of non-GSO systems in the FSS in certain frequency bands are required, in time for consideration by WRC-2000, in order to review the conditions under which sharing of the frequency bands 10-30 GHz which are allocated to the planned (RR APS30B) or unplanned FSS is feasible between GSO and non-GSO systems, between non-GSO systems and between non-GSO and terrestrial systems.

Similarly WRC-97 Resolution 538 applies to the request for an appropriate protection of the BSS Plans and their future modifications and do not place unreasonable constraints on the development of non-GSO systems in the bands concerned.

Tables S22-1 to S22-4 of RR Article S22 contain provisional limits of the interference caused by one non-GSO FSS system in the frequency bands considered in the above resolutions.

9.5.1.1 Background

Systems using non-GSO are proposed in the FSS and the frequency bands 18.9-19.3 and 28.7-29.1 GHz were allocated to the non-GSO FSS on a primary basis at WRC-95.

WRC-95 identified bands near 5, 7, 15, 20 and 30 GHz in which feeder links (FL) for constellations of non-geostationary spacecraft in the mobile-satellite service non-GSO MSS (FL) may be implemented.

Studies have been made in order to facilitate frequency sharing with the other services to which those bands are allocated and also to develop recommendations for estimating interference between non-GSO MSS (FL), between non-GSO MSS (FL) and GSO FSS, and between GSO and non-GSO FSS.

The frequency sharing is generally co-directional but is also bidirectional in certain bands.

Under the present Radio Regulations, a non-GSO system has to follow the following procedure:

- Section I of RR Article S9: Advance publication procedures

- Sections I and II of RR Article S11: Notification and registration procedures.

Unlike GSO systems, coordination is not required, except for the non-GSO satellite systems in the bands subject to coordination under Rev.WRC-97 Resolution 46 procedures. However, consultations between administrations are usually carried out in order to resolve potential difficulties, either between a non-GSO and a GSO or between non-GSO systems.

A non-GSO system has to comply with RR S22.2 provisions. In FSS bands which are not subject to Resolution 46, there is a need for a procedure to provide for protection of non-GSO systems from previous and future GSO systems.

9.5.1.2   Interference statistics between GSO and non-GSO networks

9.5.1.2.1   Technical characteristics of non-GSO/MSS feeder links, GSO/FSS and non-GSO/FSS satellite systems in the 30/20 GHz bands

The technical characteristics of non-GSO/MSS feeder links, GSO/FSS and non-GSO/FSS satellite systems in the 30/20 GHz bands under consideration in the current ITU-R sharing studies are given in the new Recommendation ITU-R S.1328.

For these studies, the assumptions are as follows:

- non-GSO systems are considered with different inclinations, altitudes, numbers of satellites, numbers of orbital planes and frequency bands and use three different types of access/modulation: CDMA/FDMA, TDMA/QPSK and FDMA/QPSK;

- GSO systems use spot beams and some global beams and three different types of carriers (FM-TV, IDR, SMS);

- the non-GSO heights (near 800 km, 1 400 and 10 000 km) and the main pairs of GSO FSS frequency bands (4/6, 11/14 and 20/30 GHz) are covered by the non-GSO network carriers chosen.

Sharing between non-GSO and GSO networks and sharing between multiple non-GSO networks are analysed in terms of interference and performance criteria for forward band working (FBW), and reverse band working (RBW) in which the uplink frequency and downlink frequency of one system with respect to another system are reversed.

9.5.1.2.2   In-line interference and worst-case interference between GSO and non-GSO networks

i)   **In-line interference between GSO and non-GSO networks for co-directional co-frequency sharing (in forward band working (FBW))**

The situation is illustrated by Table 9.9 and Figure 9.4.

When one of the networks employs non-GSO orbits operating in the same frequency band, whilst the angular separation between wanted and interfering satellites will be significant for the great majority of the time, the separation will be very small for short periods during which there will be little or no earth station antenna pattern discrimination.

These in-line instances occur when a satellite of one network is momentarily in line with a satellite of another network and an earth station of either network.

In-line interferences occur on the up-path and on the down-path and from each network to the other and they occur from time to time for every point on the Earth's surface visible to the higher of the two satellites. Each point on the Earth's surface which is in-line with any pair of satellites will move as the satellites progress around their orbits.

GSO earth stations suffer severe excess interference from non-GSO satellite downlinks when an in-line situation occurs (non-GSO satellite within the main beam of the GSO earth station). In order to have valid statistical results, non-Earth coherent orbits are used for simulations. Every earth station within a certain range of latitude receives hits at one time or another.

The idea of using coherent orbits may be a possible way to reduce the number and duration of in-line situations, such orbits are synchronized with the Earth rotation in order to provide a very accurate ground track which will repeat itself after a certain amount of time.

However, the large number of spacecraft orbiting on several determined orbital planes makes the use of a single ground track almost impossible, because with a short cycle, the total number of revolutions for all ground tracks may be relatively large due to the presence of several ground tracks.

In general, it is concluded that coherence or near-coherence of the non-GSO orbit is undesirable from the point of view of resulting interference statistics into GSO networks, because some locations suffer increased interference in order to reduce the interference at other locations.

TABLE 9.9

Worst interference cases for FBW between GSO and non-GSO networks

	GSO satellite (1)	GSO earth station (1)
Non-GSO satellite (2)	None	uplink (1) → uplink (2) (Case c) downlink (2) → downlink (1) (Case b)
Non-GSO earth station (2)	uplink (2) → uplink (1) (Case a) downlink (1) → downlink (2) (Case d)	None
NOTE – The GSO and non-GSO satellite systems are designated respectively by (1) and (2).		

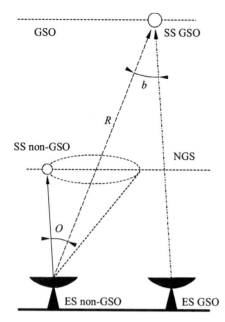

Case a: Emission from non-GSO earth station
affecting GSO receiving space station

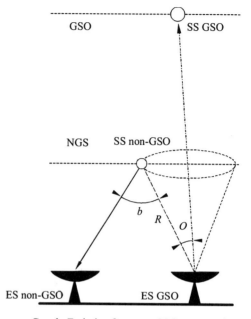

Case b: Emission from non-GSO space station
affecting GSO receiving earth station

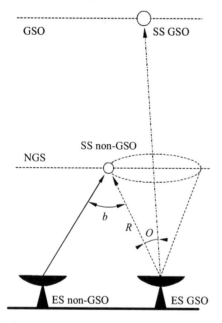

Case c: Emission from GSO earth station
affecting non-GSO receiving space station

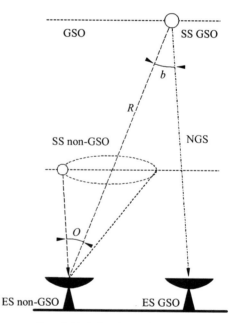

Case d: Emission from GSO space station
affecting non-GSO receiving earth station

⎯⎯⎯→ Wanted signal   - - - - -→ Interference   Sat/C9-04

FIGURE 9.4

Interference geometry between GSO and non-GSO satellite networks (FBW case)

ii) **In-line interference between GSO and non-GSO networks for co-frequency sharing in reverse band working (RBW)**

The situation is illustrated by Table 9.10 and Figure 9.5.

TABLE 9.10

Worst interference cases for RBW between GSO and non-GSO networks

	GSO satellite (1)	**GSO earth station (1)**
Non-GSO satellite (2)	downlink (1) → Satellite (2) downlink (2) → Satellite (1)	None
Non-GSO earth station (2)	None	uplink (2) → earth station (1) uplink (1) → earth station (2)
NOTE – The GSO and non-GSO satellite systems are designated respectively by (1) and (2).		

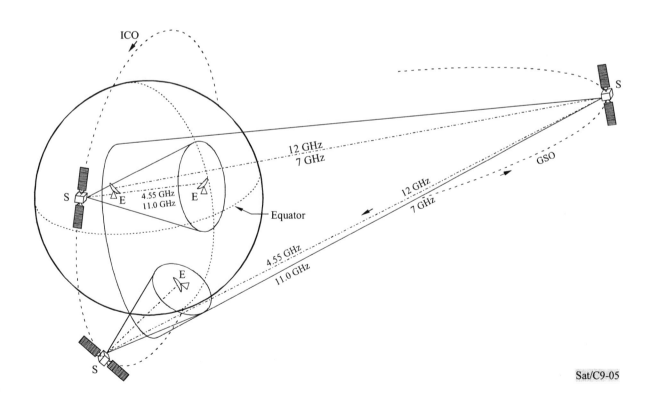

Sat/C9-05

FIGURE 9.5

Interference geometry between GSO and non-GSO satellite networks (RBW case)

- ### Earth station-to-earth station interference

The worst case will occur when the transmitting and receiving earth stations are pointing at each other in the azimuth direction. For every possible azimuth of a non-GSO earth station, the worst-case interference will occur in two specific azimuths.

The percentage of time for a receiving earth station during which interference from one source may exceed the threshold value (see Recommendation ITU-R IS.847) is provisionally assumed to be the same in the reverse band working as in the forward band working.

Recommendation ITU-R IS.849 is used to calculate the cumulative distribution of earth station antenna gain in a specified direction with the basic transmission loss that converges on a coordination distance for the worst percentage of the time. The worst-case geometrical alignment is that which results in the largest coordination distance.

- ### Satellite-to-satellite interference

The worst-case interference between GSO and non-GSO satellites will occur along the antipodal path between two such satellites or if there is no antenna discrimination from either satellite. The worst-case alignment however is infrequent and of relatively short-term duration. But these antennas are usually designed to cover elevation angles greater than a minimum positive value, Therefore, main beam-to-main beam satellite coupling should never appear. The worst-case interference will be encountered when the aggregate interfering power of the whole non-GSO constellation into a single GSO satellite is maximized.

iii)   Mutual interference between non-GSO networks

The situation is illustrated by Table 9.11 and Figure 9.6.

The worst-cases interference between non-GSO networks are similar to the above-mentioned cases for FBW and RBW.

TABLE 9.11

Worst mutual interference cases for FBW between non-GSO networks
(FSS, 30/20 GHz bands)

	Space station FSS, non-GSO satellites constellation (1)	Earth station FSS, non-GSO satellites constellation (1)
Space station FSS, non-GSO satellite, constellation (2)	None	uplink (1) → uplink (2) downlink (2) → downlink (1)
Earth station FSS non-GSO satellite, constellation (2)	uplink (2) → uplink (1) downlink (1) → downlink (2)	None
NOTE – The GSO and non-GSO satellite systems are designated respectively by (1) and (2).		

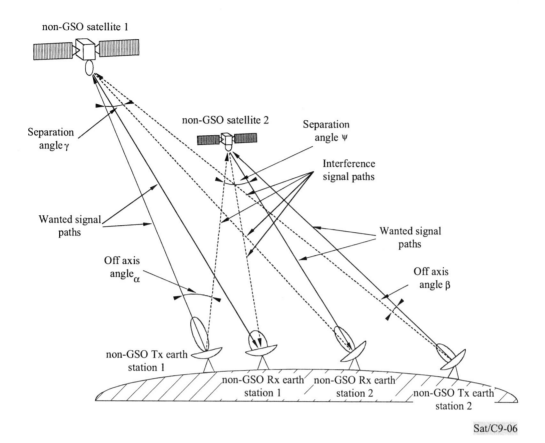

α:	Angular separation between the non-GSO satellite 1 and the non-GSO satellite 2, as viewed by the non-GSO earth station 1;
β:	Angular separation between the non-GSO satellite 1 and the non-GSO satellite 2, as viewed by the non-GSO earth station 2;
ψ:	Angular separation between the non-GSO earth station 1 and the non-GSO earth station 2, as viewed by the non-GSO satellite 1;
γ:	Angular separation between the non-GSO earth station 1 and the non-GSO earth station 2, as viewed by the non-GSO satellite 2.

FIGURE 9.6

Interference geometry between non-GSO satellite networks (FBW case)

9.5.1.2.3   Criteria for acceptable interference

Recommendation ITU-R S.1257 describes "Analytic method to calculate visibility statistics for non-GSO satellites as seen from a point on the Earth's surface".

Downlink and uplink simulations were made over a very large period of time in order to have statistically valid results. Some simulations ran a complete constellation while others were made for one non-GSO satellite and a multiplying factor was used to take into account the number of non-GSO satellites in the whole constellation.

Excess interference values, duration of interference (max and mean), mean time between interference events, and aggregate interference durations were the main data gathered from the various simulations.

There are simulations involving C/I computations and simulations involving visibility and geometric interference statistics. Results were expressed in terms of C/I values not exceeded for a given percentage of total simulation time or in terms of interference statistics.

i)   Criteria for acceptable interference for co-directional sharing (forward band working (FBW) – see Recommendations ITU-R S.1323 and S.1325)

These interferences are of intermittent nature.

Criteria for acceptable interference are expressed in terms of short-term interference allowances.

Long-term interference between non-GSO and GSO will be inherently of a time varying nature just as the short-term interference, for a GSO-FSS will provisionally be limited to 6% of the total link noise.

A methodology for calculating mutual long- and short-term interference limits between non-GSO FSS networks and between GSO FSS and non-GSO FSS is given in Recommendation ITU-R S.1325. Short-term performance objectives refer to those BER or C/N associated with 1% of the time or less.

Recommendation ITU-R S.1323 gives the maximum permissible of interference in a satellite network (GSO FSS; non-GSO FSS; non-GSO MSS FL) for a HRDP in the FSS caused by other co-directional networks below 30 GHz.

Annex 1 to Recommendation ITU-R S.1323 includes three methodologies A/B/C for deriving interference allowances (simultaneous effects due to fading and interference/interference effects considered separately from fading/aggregate interference).

•   Interference from non-GSO networks into GSO networks

For high quality digital transmissions, the BER requirement is 10^{-9} for 0.04% of the year.

Short-term interference criteria can be in general related to short-term performances and it is convenient to specify short-term interference criteria in terms of allowances of excess noise during allowable percentages of time.

Ten per cent of the allowance percentage of time, corresponding to degraded performance of that wanted, may be allocated to the interference from non-GSO links.

Using Methodology A of Recommendation ITU-R S.1323, the following single entry interference allowances were derived (GSO FSS carriers) on the assumption that two non-GSO networks interfere with the GSO network as follows (Table 9.12). These criteria could be used provisionally.

TABLE 9.12

**Derivation of single entry interference allowance
(from Recommendation ITU-R S.1323)**

Frequency	Single entry interference level (% of total link noise N_T)	% of time of the year
6/4 GHz band (fade depth 3 dB)	I non-negligible	<2.154%
	I > 0.26N_T	<0.153%
	I > 0.58N_T	<0.014%
	I > N_T	<0.0008%
14/10-12 GHz band (fade depth 3-10 dB or more)	I non-negligible (0.06 N_T)	<0.87%
	I > 0.26N_T	<0.119%
	I > N_T	<0.029%
	I > 2.16N_T	<0.0004%
30/20 GHz band (clear sky) (fade depth 12 dB)	I non-negligible	<0.87%
	I > 0.78N_T	<0.119%
	I > 2.98N_T	<0.0294%
	I > 14.8N_T	<0.0004%
NOTE – The interference allowances are estimated by essentially regarding non-GSO networks interference events as equivalent to fade events.		

But interference from non-GSO networks is characterized by individual short and strong events which are likely to occur much more frequently than similar individual events resulting from fading. Therefore, the frequency of occurrence of non-GSO networks interference events may present a significant limitation to the service quality of GSO networks, it is appropriate to limit the occurrence of severe interference events to one in 14 days per non-GSO network (as indicated in grey colour in the table) and to define such events as those exceeding the shortest-term single entry protection criteria with this limitation.

- **Interference to non-GSO networks**

There is not yet any ITU-R recommendation on the permissible short-term interference objectives for interference to non-GSO networks.

- **Possible revision of the RR Appendix S8**

By the application of the $\Delta T/T$ method in RR Appendix S8 to the worst geometrical alignment (shortest interference path length) when a non-GSO satellite and a GSO satellite are instantaneously in line with a receiving earth station of the existing network or a transmitting earth station of the incoming network, it is possible to determine the need for coordination in the case of a non-GSO network with an existing GSO network, or of a GSO network with an existing non-GSO network. There are four cases in which the need for coordination may have to be established:

- incoming GSO system → existing GSO system;

- incoming non-GSO system → existing GSO system;

- incoming GSO system → existing non-GSO system;

- incoming non-GSO system → existing non-GSO system.

However, such an application is likely to lead to very conservative results.

A modification to Recommendation ITU-R S.738 should also take account of non-GSO systems.

ii)   Criteria for acceptable interference for bidirectional sharing in reverse band working (RBW)

- **Earth station-to-earth station interference: coordination area**

Recommendations ITU-R IS.847, IS.848 and IS.849 provide the basis of existing coordination methods for GSO and non-GSO earth stations. Mode 2 (hydrometeor scatter) can be neglected, mode 1 (great-circle propagation) is only important in calculating the coordination distances for a non-GSO earth station.

These coordination distances result either in the calculated interference level being equal to the interference threshold (general approach) or in the calculated C/I being equal to the protection ratio (specific approach). For example: several studies use interference criteria of 6% or C/I of 30 dB, and a percentage % of time of 0.1% for the overland path loss Lb(p).

- **Satellite-to-satellite interference**

In the satellite-to-satellite mode, interference is acceptable provided that the calculated C/I into each satellite exceeds the required protection ratio. A long-term interference of 6% of the total noise (as specified by Appendix S8) into the GSO satellite is used.

The very nature of operation of a non-GSO satellite would suggest that only a short-term objective is needed for interference into non-GSO satellites. Presently, there is no long- or short-term objective for interference into non-GSO satellites. The percentage of time for which the acceptable level of long-term interference into a non-GSO satellite is exceeded also has not been established.

In the 6 700-7 075 MHz frequency band, RR Article S22.5A specifies that the pfd produced at the GSO by a non-GSO in the FSS shall not exceed -168 dBW/m^2 in any 4 kHz band.

Simulation and analysis results for both $\Delta T/T$ and C/I can be used and, in most cases, the satellite-to-satellite interference in RBW mode is acceptable. The earth station-to-earth station interference can also be acceptable with some appropriate constraints.

9.5.1.2.4   Coordination methods and acceptance of interference

With the exception of Resolution 46 of the RR in certain bands, there is no regulatory provision for coordination involving non-GSO and GSO networks. Two main features will be needed, i.e.:

- a method for determining whether coordination with existing GSO and non-GSO networks needs to be considered;

- a procedure to facilitate coordination when necessary and the establishment of techniques to assist coordinating parties to reach agreement.

The ΔT/T method described in RR Appendix S8 can be extended for non-GSO networks, given certain additional information to that presently required by RR Appendix S4.

The following techniques might be used to reduce frequency sharing difficulties and a further study is needed to determine their respective effectiveness: earth station beam switching, satellite beam switching, use of satellite spot beam tracking (for tracking non-GSO earth stations), implementation of fade-compensation mechanisms, optimization of non-GSO satellite handover strategy for minimum in-line interference, etc. The techniques employed in coordination between GSO networks could also be envisaged, e.g. transponder frequency interleaving, macro- and micro-segmentation, carrier level adjustments, etc.

However, given the large number of coordinations potentially involved and the need for the non-GSO networks to incorporate interference mitigation facilities into their initial design, many people consider that some form of pfd limitation is a more satisfactory means of frequency sharing than coordination when non-GSO systems share with GSO systems.

In fact, non-geostationary satellites require a new approach for spectrum sharing and new techniques must be used to allow spectrum sharing between non-GSO satellite service networks. Several sharing methodologies have been presented but further studies are required to determine the feasibility, practicability and economic impact of some of these techniques.

9.5.2   Interference between non-GSO-FSS and GSO-FSS networks and between non-GSO--FSS networks operating co-frequency in the same direction (30/20 GHz band)

WRC-95 allocated for non-GSO satellites in the fixed-satellite service FSS under the provisions of RR Resolution 46 (RR Article S9.11A), the following frequency bands: uplink: 28.7-29.1 GHz/ downlink: 18.9-19.3 GHz, these bands being usable by both GSO FSS networks and non-GSO FSS networks on a equal basis (RR Article S22.2).

Studies were performed to determine if the sharing of the frequency bands 20/30 GHz is feasible between GSO FSS and non-GSO FSS systems, between non-GSO FSS systems and between non-GSO FSS systems and terrestrial systems (to protect existing terrestrial services) (see WRC-95 Resolution 118).

By RR S5.523A, WRC-97 provided that the use of the bands 18.8-19.3 GHz (space-to-Earth) and 28.6-29.1 GHz (Earth-to-space) by GSO and non-GSO FSS is subject to the application of the provisions of Article S9.11A/Resolution 46 (Rev.WRC-97).

9.5.2.1   Interference between GSO-FSS and non-GSO-FSS networks

Refer again to Table 9.9 and Figure 9.4.

i)   Analytical methods: worst-case approach (interference calculated for in-line events without the application of mitigation techniques)

The results show unacceptable mutual interference: calculation of I_0/N_0 and C/I margins for the worst case (i.e. when earth stations are colocated and the satellites corresponding are in-line from these co-located earth stations), calculation of C/I statistics of excess in-line interference events for earth stations near the Equator and near longitude of GSO satellite, calculation of C/I and of I/N_T (as a function of interference duration and for earth stations at 0° and 60° latitudes), calculation of C/I for earth stations at 0° latitude (highest GSO earth station = 90°) and at 50° latitude (highest GSO earth station = 32.69°).

In the case of LEO constellations, at latitudes below approximately 50°, there will be severe interference from GSO FSS earth stations into non-GSO FSS satellites, but interference from non-GSO FSS satellites into GSO FSS earth stations is also quite severe.

The mutual interference is dependent on the latitude of the earth station and in some cases at higher latitudes can become negligible. At low latitudes, the worst-cases events are associated with in-line events of a duration in the order of a minute, while at latitudes 50° the worst-cases events are shorter and less severe.

ii)   Computer simulation: realistic and time-varying models of the interference

The results of the simulations are provided both in terms of I_0/N_0 and C/I: curves and tables of the percent time interference exceeding each of the noise level thresholds as a function of earth stations latitude (grid centre latitude or cell), tables of interference statistics using a C/I threshold to calculate interference duration and interval statistics as a function of site latitude and curves of C/I cumulative probability distribution functions (CDF) as a function of earth stations latitude. The interference levels change as a function of earth station latitude.

Co-directional frequency sharing is not feasible between such systems unless mitigation techniques can be applied.

iii)   Reduction of interference

As already explained, different techniques may reduce the interference between GSO and non-GSO FSS networks operating co-directionally e.g.: adaptive power control, use of high gain antennas, geographic isolation of earth stations, restricted operational elevation angle, satellite diversity, site diversity, signal design and network traffic management, satellite footprint shift, GSO arc avoidance, residual interference limitation, optimization of the constellation, hybrid systems, exclusion zones.

Studies are undertaken on interference mitigation techniques for non-GSO and GSO FSS techniques operating in the 18.800-19.300 GHz and 28.600-29.100 GHz and other bands.

In conclusion, several studies have demonstrated that co-directional co-frequency sharing between the GSO FSS and non-GSO FSS may be feasible if appropriate interference mitigation techniques are employed.

9.5.2.2   Mutual interference between non-GSO-FSS networks

Refer again to Table 9.11 and Figure 9.6.

The feasibility of frequency sharing between non-GSO-FSS in this band may be possible with some constraints.

i)   Feasibility of frequency sharing between homogeneous non-GSO-FSS systems

Assuming identical parameters for non-GSO-FSS networks based on the first network submitted to ITU-R (LEO-SAT 1), some studies have been performed for orbit constellations separated spatially by either interleaving orbital planes or interleaving satellites within the same planes.

In a first example, simulation results relating to interference between identical non-GSO FSS operating at latitudes of 45° and 60° (interference is not modelled at lower latitudes since the interference statistics are worse for higher latitudes), except that the orbit planes of the two systems are interleaved. The C/I values on the up and downlinks of an earth station were computed at one second intervals over a simulated period of several days. Results show that C/I values substantially below the mean were occurring infrequently but at fairly regular intervals when the geocentric spacing between adjacent constellation 1 and constellation 2 becomes smaller.

In another example, simulation results relating to interference between identical non-GSO FSS, except that the orbital planes of a system 2 are interleaved halfway between the orbital planes of the system 1.

Since the interference is symmetric, the following cases are considered:

* network 1 satellites interfering into network 2 earth stations;
* network 1 earth stations interfering into network 2 satellites.

The interference levels change as a function of earth station latitude and are driven by the minimum topocentric separation of any two satellites from the two systems which could service the same location on the Earth.

The results of these studies indicates that co-directional co-frequency sharing between several such homogeneous non-GSO-FSS systems appears to be feasible with the application of mitigation techniques at high latitudes.

ii)   Feasibility of frequency sharing between two or more non-homogeneous non-GSO-FSS systems

These studies are in progress. Early indications are that sharing will depend on the ability of the networks to employ satellite diversity to avoid main beam-to-main beam coupling of interference.

iii)   Use of mitigation techniques

There are a number of mitigation techniques which may help to reduce the interference between non-GSO FSS networks operating in the 20/30 GHz bands and in particular the following sharing techniques may allow for two or more non-homogeneous systems to share the same frequency co-directionally: satellite diversity, space division (alternating orbital planes or alternating satellites within a common orbital plane), polarization division, frequency division, time division, code division.

If these techniques can be implemented, then two or more constellations may be able to operate simultaneously in the same frequency band with an acceptable level of interference to each other.

9.5.2.3   Interference between the non-GSO-FSS and the GSO FSS Plan (RR Appendix S30B)

Some administrations are investigating further possible use of the GSO FSS planned bands by introducing non-GSO satellite systems bands.

Studies are being performed on a "Methodology for deriving criteria to protect GSO networks operating in RR Appendix S30B planned bands from interference of non-GSO networks" with long-term and short-term interference criteria.

9.5.3   Interference between GSO-FSS networks and non-GSO-MSS feeder links

WRC-95 made provisions for the non-GSO-MSS/feeder links (FL) and for the GSO-FSS to operate on a shared primary basis in portions of the 20/30 GHz FSS bands and allocated the frequency bands 19.300-19.600 GHz and 29.100-29.400 GHz for the feeder links of non-GSO MSS in co-direction with the GSO FSS (RR Resolutions 120 and 121).

Note that WRC-95 also allocated the band 6.700-7.075 GHz downlink for the non-GSO MSS FL while band 6.725-7.075 GHz uplink was already allocated to the GSO FSS Plan (Allotment Plan of RR Appendix S30B). Recommendation ITU-R S.1256 describes the methodologies for determining the maximum aggregate pfd at the GSO orbit in the downlink 6.700-7.075 GHz from non-GSO MSS/FL.

Studies were performed for the use of bands 19.300-19.700 GHz and 29.100-29.500 GHz by the feeder links for non-geostationary mobile-satellite service networks (non-GSO-MSS (FL)).

Due to the non-application of RR S22.2 in the bands 19.6-19.7 and 29.4-29.5 GHz for coordination between GSO FSS networks and non-GSO-MSS (FL) networks, Rev.WRC-97 (RR Resolution 121) invited to undertake further studies on sharing between non-GSO-MSS (FL) networks and GSO FSS networks in the full bands 19.3-19.7 GHz (space-to-Earth) and 29.1-29.5 GHz (Earth-to-space) on a equal basis.

MOD97 of Article S5.541A shows that the non-GSO MSS FL systems and GSO FSS systems operating in the band 29.1-29.5 GHz (Earth-to-space) shall employ uplink adaptative control or others methods of fade compensation, such that the earth station transmissions shall be at a power level required to meet the link performance while reducing the level of mutual interference between both networks.

i)   Co-directional FBW sharing between GSO-FSS and non-GSO-MSS FL

Reference is made again to Table 9.9 and Figure 9.4.

Recommendation ITU-R S.1324 gives a methodology for estimating interference between non-GSO-MSS (FL) networks and GSO FSS networks operating co-frequency and co-directionally and Recommendation ITU-R S.1255 describes the use of adaptative uplink power control to mitigate interference between GSO FSS networks and non-GSO MSS FL networks.

Some of these mitigation techniques which may reduce the interference between GSO FSS networks and non-GSO MSS (FL) networks operating in the 20/30 GHz are again: adaptive power control, use of high gain antennas, geographic isolation of earth stations, restricted operational elevation angle, satellite diversity, site diversity, signal design and network traffic management, satellite footprint shift, GSO arc avoidance, residual interference limitation, optimization of the constellation, hybrid systems, exclusion zones.

In conclusion, several studies have demonstrated that co-directional co-frequency sharing GSO FSS networks and non-GSO MSS (FL) networks may be feasible if appropriate interference mitigation techniques are employed.

ii)   Reverse band (RBW) frequency sharing between non-GSO-MSS FL and GSO-FSS

Reference is made again to Table 9.10 and Figure 9.5.

Interference is negligible between non-GSO and GSO satellites. For coordination between GSO and non-GSO earth stations the method described in Annex 1 of Recommendation ITU-R IS.847 can be used. The worst cases of the Earth-to-Earth interference will occur when the transmitting and receiving earth stations are pointing at each other in the azimuth direction and when both are operating at their lowest elevation angles.

9.5.4   Mutual interference between non-GSO/non-GSO or non-GSO/GSO MSS (FL)

Rev.WRC-97 Resolution 215 of the RR identified the need to develop criteria for determining necessity to coordinate and calculation methods for determining the interference levels and the required protection ratios between networks in the MSS.

9.5.4.1   Mutual interference between co-directional non-GSO MSS (FL)

WRC-95 identified bands near 5, 7, 15, 20 and 30 GHz for which feeder links can be used for constellations of non-geostationary in the mobile-satellite service non-GSO MSS (FL).

WRC-95 and WRC-97 allocated the following frequency bands for the feeder links of non-geostationary satellites in the mobile-satellite service non-GSO MSS (FL) under the provisions of RR Resolution 46:

- uplink: 5.091-5.150 GHz   (RR Resolution 114)   and   downlink: 6.700-7.075 GHz* (RR Resolution 115);

- uplink 15.430-15.630 GHz (RR Resolution 117) (15/17 GHz);

- downlink: 6.700-7.075 GHz* (RR Resolution 115);

- uplink 5.150-5.250 GHz and downlink 5.150-5.216 GHz;

- uplink   19.310-19.700 GHz   (RR Resolution 119)   and   downlink   15.430-15.630 GHz* (RR Resolution 116);

- uplink 29.100-29.500 GHz* (RR Resolution 120) and downlink 19.310-19.700 GHz* (RR Resolution 120) (but only 29.100-29.400 GHz/19.300-19.600 GHz according to RR Resolution 46, see § 9.5.3 above).

Note that the above-mentioned pairing is not compulsory. With the exception of the bands marked by an asterisk (which are also available for general FSS use), these allocations (on a primary basis) are limited to FL of non-GSO systems in the MSS and are subject to coordination under Rev.WRC-97 Resolution 46/No. S9.11A.

Recommendation ITU-R S.1325 describes a common methodology for simulating short-term interference between co-frequency, co-directional non-GSO/non-GSO or non-GSO/GSO FL networks. The methodology uses a geometrical analysis and includes simplifying assumptions for systems employing automatic power control and issues a set of common output statistics for I_0/N_0, C/I, C/(N+I).

The situation is again illustrated by Table 9.11 (where "FSS" is to be replaced by "MSS") and Figure 9.6.

9.5.4.2 Mutual interference between non-GSO-MSS (FL) in reverse band working (RBW)

WRC-95 or WRC-97 allocated certain frequency bands (5.150-5.216 GHz, 15.430-15.630 GHz, 19.300-19.700 GHz) for simultaneous operation of non-GSO MSS (FL) in reverse band working (RBW).

A new ITU-R recommendation (under approval) describes "Methodologies for estimating mutual interference between FL of non-GSO satellite networks co-frequency operating in RBW".

These methods may be used to determine the need for further coordination between non-GSO MSS (FL) co-frequency operating in RBW: mutual interference in space-to-space direction and Earth-to-Earth direction (these methods may be used for constellations having different orbital altitudes).

Reference is made again to Table 9.11 (where "FSS" is replaced by "MSS") and Figure 9.6.

9.5.5 Control of interference from non-GSO satellite systems into GSO satellite systems

(See RR Article S22 and WRC-97 Resolutions 130 and 538.)

9.5.5.1 General: Possible use of the GSO bands (planned or unplanned, BSS or FSS) by non-GSO satellite systems

As it was already pointed out, different studies show the possible use of the GSO BSS or GSO FSS bands by non-GSO satellite systems, if the power levels radiated by the non-GSO satellites, as defined below, are assigned limits and corresponding proportions of time for which those limits must be met.

- The equivalent pfd (epfd), at any point on the Earth's surface visible from the GSO, produced by emissions from all the space stations of a non-GSO system operating in the FSS frequency bands, is defined as the sum of the pfd produced at a point of the Earth's surface by all space stations within this non-GSO satellite system, taking into account the off-axis discrimination of a reference receiving antenna assumed to be pointing towards a GSO network (see S22.5C1).

- The aggregate pfd (apfd or $epfd_{up}$) produced at any point in the GSO by the emissions from all the earth stations in a non-GSO system operating in the FSS frequency bands, is defined as the sum of the pfd produced at a point in the GSO by all earth stations of this non-GSO satellite system (see S22.5D).

Provisional epfd and apfd limits corresponding to interference caused by one non-GSO FSS system are given in RR Article S22. Tables S22-1 to S22-4 (see below Tables 9.13 and 9.14) and Nos. S22.26 to S22.29 of RR Article S22 contain **provisional** limits corresponding to an interference level caused by one non-GSO FSS system in the frequency bands to be applied in accordance with WRC-97 Resolutions 130 and 538. These provisional limits are subject to review by ITU-R and are subject to confirmation by WRC-2000.

Moreover, in the bands 6 700-7 075 MHz, RR Article S22.5A provides that the maximum apfd produced by non-GSO at the GSO orbit and within ±5° of inclination around the GSO by a non-GSO system in FSS shall not exceed -168 dBW/m² in any 4 kHz band (see Recommendation ITU-R S.1256).

9.5.5.2   Use of the unplanned and planned GSO FSS bands for non-GSO FSS satellite systems

(See RR AP S30B: Planned GSO FSS bands.)

RR WRC-97 Resolution 130 presents the possible use of non-GSO in the FSS in certain bands, i.e.:

- in the planned bands (RR S5.441): 10.7-10.95 GHz (space-to-Earth), 11.2-11.45 GHz (space-to-Earth), 12.75-13.25 GHz (Earth-to-space);

- in the unplanned bands: 10.95-11.20 GHz (space-to-Earth), 11.45-11.7 GHz (space-to-Earth), 11.7-12.20 GHz (space-to-Earth) in Region 2, 12.2-12.75 GHz (space-to-Earth) in Region 3, 12.5-12.75 GHz (space-to-Earth) in Region 1, 13.75-14.5 GHz (Earth-to-space), 17.8-18.6 GHz (space-to-Earth), 19.7-20.2 GHz (space-to-Earth), 27.5-28.6 GHz (Earth-to-space), 29.5-30 GHz (Earth-to-space).

In these bands, there is a need for the provision of services on a competitive basis between GSO FSS and non-GSO FSS as well as between non-GSO FSS and non-GSO FSS.

Further studies are required on sharing conditions. WRC-97 Resolution 130 requests ITU-R to conduct them in time for consideration by WRC-2000.

TABLE 9.13

Provisional epfd limits
(see also RR Tables S22.1 and S22.3)

Frequency band GHz	Equivalent pfd (epfd)	Time during which epfd level may not be exceeded (%)	Earth station antenna diameter (m), (Reference radiation pattern)
BSS bands (WRC-97 Res. 538 (S5.487A. BSS Plan))			
	dBW/m²/4 kHz		
11.7-12.5 (Region 1), 11.7-12.2 (R3), 12.5-12.75 (R3)	−172.3/−169.3 −183.3/−170.3 −186.8/−170.3	99.7	0.30 0.60 0.90
12.2-12.7 (R2)	−174.3/−165.3 −186.3/−170.3 −187.9/−170.3 −191.4/−170.3	99.7	0.45 1.00 1.20 1.80
17.3-17.8 (R2)	For further study		
FSS bands (WRC-97 Res. 130 and S5.484A: unplanned FSS, S5.441: FSS Plan)			
	dBW/m²/4 kHz		
10.7-11.7 (R2) 10.7-10.950 (FSS Plan) 11.2-11.450 (FSS Plan) 10.950-11.2 11.450-11.7 11.7-12.2 (R2) 12.2-12.5 (R3) 12.5-12.75 (R1)	−179 −192 −186 −195 −170 −173 −178 −170	99.7 99.9 99.97 99.97 99.999 99.999 99.999 100	(Rec. ITU-R S.465) 0.60 3 10 0.60 3 10 ≥0.60
	dBW/m²/40 or /1 000 kHz		
17.8-18.6 17.8-18.1 (Res. 538)	−165/−151 −165/−151 −167/−153 −180/−166 −184/−170 −180/−174 −165/−151	99 99.5 99.8 99.9 99.9 99.9 100	0.30 and 0.70 (Rec. ITU-R S.465) 0.90 (Rec. ITU-R S.465) 1.5 5 7.5 12 0.3 to 12
19.7-20.2	−154/−140 −164/−150 −167/−153 −174/−160 −154/−140	99 99.9 99.8 99.9 100	0.30 (Rec. ITU-R S.465) 0.90 (Rec. ITU-R S.465) 2 5 0.30 to 12

TABLE 9.14

Provisional apfd (or epfd$_{up}$) limits
(see also RR Tables S22.2 and S22.4)

Frequency band GHz	Aggregate pfd (apfd)	Time during which apfd level may not be exceeded (%)
BSS bands (uplink) (WRC-97 Res. 538) (see S5.516: BSS LC Plan)		
	dBW/m²/4 kHz	
17.3-18.1 (R1 and 3: FL Plan) 17.8-18.1 (R2: FL Plan)	−163 «	100 «
FSS bands (uplink) (WRC-97 Res. 130) (see S5.441 and 484A)		
12.5-12.75	−170	100
12.75-13.25 (FSS Plan)	−186	«
13.75-14.5	−170	«
27.5-28.6	−159	100
	DBW/m²/1 000 kHz	
29.5-30 (484A)	−145	«

9.5.5.3   Use of the planned GSO BSS bands by introducing non-GSO FSS satellite systems

(See RR AP S30 and S30A planned GSO BSS bands.)

RR WRC-97 Resolution 538 deals with the utilization of the frequency bands covered by RR Appendices S30 and S30A (Planned bands):

- 17.3-18.1 GHz in Regions 1 and 3 and 17.8-18.1 GHz band in Region 2 (Earth-to-space, RR S5.516);

- 11.7-12.5 GHz in Region 1, 12.2-12.7 GHz in Region 2 and 11.7-12.2 GHz band in Region 3, also allocated to the FSS on a primary basis, limited to non-GSO (space-to-Earth, RR S5.487A).

WRC-97 also decided to introduce in RR Article S5 a new allocation to the FSS in the frequency bands 11.7-12.5 GHz in Region 1, 12.2-12.7 GHz in Region 2 and 11.7-12.2 GHz in Region 3, limited to non-GSO FSS downlinks. The integrity of the RR AP S30 and S30A and their future modifications is to be ensured.

These non-GSO FSS systems operating in the frequency bands covered by RR Appendices S30 and S30A shall comply with the provisional limits specified in RR Article S22 and in the annex to Resolution 538 (WRC-97). These limits shall be revised, if appropriate, by WRC-2000.

These non-GSO FSS systems shall also apply the procedures of RR Articles S9 and S11 and shall be subject for the coordination with other non-GSO FSS systems, to the application of Resolution 46/S9.12 provisionally.

In Regions 1 and 3, such non-GSO FSS systems shall conform to a epfd threshold of -185.3 dBW/m²/4 kHz for 99.7% of the time, calculated with a reference 90 cm diameter antenna pattern of the earth station.

9.5.6   Conclusion

In view of WRC-2000, further studies concerning non-GSO and GSO satellite systems are required with a view to promoting efficient use of spectrum/orbit resources and equitable access to these resources by all countries. The following important points need particular attention:

- short duration interference peaks;

- short-term interference;

- power flux-density (pfd, and epfd);

- frequency sharing, and, in particular:

  - frequency sharing between non-GSO FSS networks using circular orbits and networks using slightly-inclined GSO orbit, and also between non-GSO FSS networks and networks using Quasi-GSO orbit;

  - techniques for the mitigation of interference between GSO and non-GSO networks, and between non-GSO networks;

  - criteria for sharing between GSO and non-GSO FSS networks and systems in the fixed, radiolocation and space science services which use the same frequency bands.

The complete situation of frequency allocations to the non-GSO systems and networks is summarized in the Tables 9.15 and 9.16 below. In these tables, the cells in the first lines illustrate the general situation (FSS, BSS, MSS, terrestrial) and the cells in the following lines refer to the particular situation in the various ITU regions for the satellite services only.

TABLES 9.15A/1 AND A/2

Bands allocated to the non-GSO FSS for the downlink (space-to-Earth)

Downlink (10-12 GHz band)	10.7-10.95	10.95-11.2	11.2-11.45	11.45-11.7	11.7-12.2	12.2-12.5	12.5-12.7	12.7-12.75
FSS, BSS or MSS	FSS Plan (App. S30B)	FSS	FSS Plan (App. S30B)	FSS	-BSS Plan (R1, R3)(App. S30B) -FSS (R2),	-BSS Plan (R1, R2) (App. S30B) -FSS National/ Subregional (R3) (Art. S5.492)	-BSS Plan (R2)(App. S30) -BSS (R3), -FSS (R1, R3) (Art. S5.492)	-FSS (R1), -BSS (R3)
Terrestrial serv. RR97 (Res. 131)	Fixed (F) and Mobile (M)				-F (NOTE 2) -M(R1 & R3) -B(R1 & R3)	F, M and B	-F (R2, R3) -M(R2, R3) -B(R2 at 12.5-12.7)	
Satellite, Region 1 RR97 Res. RR Art. S5	FSS Plan + non-GSO-FSS Res. 130 S5.44 (FSS Plan) or S5.484A (FSS)				BSS Plan (NOTE 1) + non-GSO-FSS + (NOTE 4) Res. 538 S5.487A		GSO FSS + non-GSO-FSS Res. 130 S5.484A	GSO FSS + non-GSO-FSS Res. 130 S5.484A
Satellite, Region 2 RR97 Res. RR Art. S5	FSS Plan or FSS + non-GSO-FSS Res. 130 S5.441 (FSS Plan) or S5.484A (FSS)				GSO-FSS Nat/Subreg. (S5.448) + non-GSO-FSS Res. 130 S5.484A	-BSS Plan (NOTE 1) +GSO FSS Nat/Subreg. (S5.448) + non-GSO-FSS (NOTE 4) Res. 538 S5.487A		
Satellite Region 3 RR97 Res. RR Art. S5	FSS Plan + non-GSO-FSS Res. 130 S5.441 (FSS Plan) or S5.484A (FSS)				BSS Plan (NOTE 1) + non-GSO FSS (NOTE 4) Res. 538 S5.487A	-GSO-FSS Nat/Subreg. (S5.491) +non-GSO-FSS Res. 130 S5.484A	-BSS + non-GSO-FSS Res. 130 S5.484A	GSO-FSS +non-GSO-FSS Res. 130 S5.484A

Downlink (20 GHz band)	17.8-18.6 (NOTE 3)	18.8-18.9-19.3	19.7-20.2
FSS, BSS or MSS	FSS		FSS, MSS, mss (R1 & R3 at 19.7-20.1)
Terrestrial services **RR97 Res.**	F and M Res. 131		F
Satellite **R1, R2, R3** **RR97 Res.** **RR Art. S5**	GSO FSS + non-GSO FSS Res. 130 S5.484A	GSO FSS + non-GSO FSS Res. 46/S9.11A SUP97 Res. 118 Res. 132 S5.523A	GSO FSS + non-GSO FSS Res. 130 S5.484A

NOTES – Lower case characters indicate allocation on a secondary basis.

NOTE 1 – Additional Allocation: the use of the BSS PLANS is limited to the non-GSO FSS and shall not be operated before the end of WRC-2000.

NOTE 2 – Except in Region 2 between 12.1 and 12.2 GHz.

NOTE 3 – The use of the band 17.8-18.1 GHz (space-to-Earth) by the non-GSO FSS is also subject to the provisions of WRC-97 Res. 538.

NOTE 4 – GSO-FSS (space-to-Earth) in these bands may also be used in conformity with BSS Plans.

• RR Resolution 131 gives the pfd limits applicable to non-GSO FSS systems in the bands 10.7-12.75 and 17.7-19.3 GHz for the protection of terrestrial services.

TABLE 9.15B

Bands allocated to the non-GSO MSS feeder links (FL) in the FSS for the downlink (space-to-Earth)

Downlink (4, 11, 20 GHz)	5.150-5.216	6.700-7.075	15.43-15.63	19.3-19.6-19.7
FSS, BSS or MSS	FSS	FSS	FSS	FSS
Terrestrial services	Radionavigation, Aeronautical (RN, AN)	F and M	RN, AN	F and M
Satellite R1, R2, R3	Non-GSO MSS FL S5.447B	Non-GSO MSS FL	Non-GSO MSS FL	GSO FSS + non-GSO MSS FL
RR Res. **RR Art. S5**	Res. 46/S9.11A S5.447B	Res. 46/S9.11A S5.458B	Res. 46/S9.11A Res. 116, Res. 123 S5.511A	Res. 46/S9.11A Res. 121 S5.523C/D/E

TABLE 9.16A

Bands allocated to the non-GSO FSS for the uplink (Earth-to-space)

Uplink (14 and 30 GHz bands)	12.75-13.25	13.75-14.5	17.3-17.8	17.8-18.1	27.5-28.6	28.6-28.7-29.1	29.5-30.0
FSS, BSS or MSS	FSS Plan (App. S30B)	-FSS (S5.506) -bss fl at 14-14.5 outside Europe	-FSS, -BSS FL Plan (App. S30A)	-FSS, -BSS FL -BSS FL Plan (App. S30A) (R1 & R3)	-FSS, -bss fl (S5.539)	-FSS bss fl (S5.539)	-FSS, bss fl -MSS -mss (R1 & R3 at 25.5-29.9)
Terrestrial services	F and M	-F and M (14.3-14.5) -RL (13.75-14) -RN (14-14.3)	-F and M (17.7-18.1) -m (R2)	F and M			
Satellite, R1	GSO FSS +non-GSO FSS		BSS FL Plan +non-GSO FSS		GSO FSS + non-GSO FSS		
Satellite, R2			BSS FL Plan	BSS FL +non-GSO FSS			
Satellite, R3			BSS FL Plan +non-GSO FSS				
RR Res.	RR Res. 130	RR Res. 130	RR Res. 130 and 538		RR Res. 130	Res. 46/ S9.1RS Res. 132	RR Res. 130
RR Art. S5	S5.441	S5.484A	S5.516		S5.484A	S5.523A	S5.484A

TABLE 9.16B

Bands allocated to the non-GSO MSS feeder links (FL) in the FSS for the uplink (Earth-to-space)

Uplink (6, 14, 30 GHz)	5.091-5.150	5.150-5.250	15.430-15.630	19.3-19.6	29.1-29.4-29.5
FSS, BSS or MSS	FSS		FSS	FSS	FSS
Terrestrial services	RN AN		RN AN	F and M	F and M
Satellite, R1, R2, R3	Non-GSO MSS FL	Non-GSO MSS FL	Non-GSO MSS FL	Non-GSO MSS FL	GSO FSS + non-GSO MSS FL
RR Res.	Res. 46/S9.11A, Res. 114	Res. 46/S9.11A	Res. 46/S9.11A Res. 117	Res. 46/S9.11A	Res. 46/S9.11A Res. 121
RR Art. S5	S5.444A	S5.447A	S5.511A	S5.523B	S5.535A

9.6 Frequency sharing between the FSS networks and other radio services in bands also allocated to the FSS

9.6.1   Sharing with ARNS and RDSS in the band 5 000-5 250 MHz (Earth-to-space for FSS)

The situation in this band is currently as follows: the full band is allocated to the aeronautical radionavigation service (ARNS) on a primary basis, but:

- the band 5 000-5 250 MHz is also allocated, on a co-primary basis, to the FSS (Earth-to-space, i.e. uplink) for non-GSO MSS feeder links (FL). Moreover, there is also an allocation on a primary basis to the FSS (space-to-Earth, i.e. downlink), in the portion 5 150-5 216 MHz, again for the non-GSO MSS FL and this same portion is also allocated to the radio determination satellite service (RDSS, downlink), on a primary basis in Region 2 and on a secondary basis in R1 and R3 (except some allocations on a primary basis);

- the band 5 000-5 150 MHz is to be used for operation of the International Standard Microwave Landing System (MLS), the requirements for this system taking precedence over uses of this band. Notwithstanding these requirements the portion 5 091-5 150 MHz is allocated, on primary basis (up to January 2010) to the FSS (uplink) for non-GSO MSS FL.

Most of these allocation are subject to coordination under various articles of the RR.

As a result, RR WRC-95 Resolution 114 invited ITU-R to undertake studies on the use of the band 5 091-5 150 by the FSS (uplink, to be limited to feeder links for non-GSO MSS) and, following a question concerning the "Sharing between the non-GSO MSS FL in the band 5 091-5 250 MHz and the ARNS in the band 5 000-5 250 MHz", Recommendation ITU-R S.1342 on "A method for calculating coordination distances in the 5 GHz band, between the MLS stations operating in the ARNS and feeder uplinks of the non-GSO MSS FL stations (Earth-to-space)" was issued.

9.6.2   Sharing with ARNS in the band 15.430-15.630 GHz (Earth-to-space and space-to-Earth for FSS)

RR S5.511A provides that the use of the band 15.43-15.63 GHz by the FSS (space-to-Earth) and (Earth-to-space) is limited to non-GSO MSS FL subject to coordination under Rev.WRC-97 Resolution 46/RR S9.11A. The earth stations (elevation angle, gain and coordination distances) shall be in the space-to-Earth direction in accordance with Recommendation ITU-R S.1341.

In the space-to-Earth direction, harmful interference shall not be caused to stations of the radio astronomy service. The threshold levels of interference and associated pfd limits are given in Recommendation ITU-R RA.769.

According to WRC-97 Resolution 123, the band 15.43-15.63 GHz (space-to-Earth) is allocated to the FSS on a primary basis for use by non-GSO MSS FL while taking into account the protection of the radio astronomy, the earth exploration-satellite service (EESS) and the space research (passive) which are using the nearly adjacent band 15.35-15.40 GHz. The band 15.43-15.63 GHz is shared with ARNS on a primary basis in agreement with RR Article S4.10.

The levels of interference detrimental to the radio astronomy are given in Recommendation ITU-R RA.769 and may not be easily met by non-GSO MSS FL operating in the space-to-Earth. The ITU-R was invited to study, as a matter of urgency, for WRC-2000 the interference potential of non-GSO MSS FL to the radio astronomy service (RAstS) in the 15 GHz and the feasibility of implementing non-GSO MSS FL in the band 15.43-15.63 GHz.

Two ITU-R recommendations give information for operation of the FSS:

- Recommendation ITU-R S.1340: Sharing between non-GSO MSS FL and ARNS systems in the Earth-to-space direction in the band 15.4-15.7 GHz;

- Recommendation ITU-R S.1341: Sharing between non-GSO MSS FL and ARNS systems in the space-to-Earth direction in the band 15.4-15.7 GHz and the protection of the radio astronomy service in the band 15.35-15.40 GHz.

9.6.3 Sharing with radionavigation (RN) and radiolocation (RL) services in the band 13.75-14 GHz (Earth-to-space for FSS)

This band is allocated to the FSS and also to RN and RL services, both on a primary basis.

Footnote S5.502 of the RR gives the constraints between the FSS (space-to-Earth) and the radiolocation (RL) and radionavigation (RN) services (see Recommendation ITU-R S.1068).

Footnotes S5.503 and 503A of the RR give the constraints between the FSS (space-to-Earth) and the space research (SR) or earth exploration-satellite services (EESS) (see Recommendation ITU-R S.1069).

9.6.4 Sharing in the band 40.5-42.5 GHz (space-to-Earth for FSS)

WRC-97 has added a primary allocation in the band 40.5-42.5 GHz to the FSS in Regions 2 and 3 and in certain countries in Region 1 and to the fixed service (FS). The date of the provisional application of the allocation to the FSS in Regions 1 and 3 band 40.5-42.5 GHz is 1 January 2001. WRC-97 resolved that WRC-2000 should review this allocation under WRC-97 Resolution 129 taking full account of the requirements of the other services.

Space service networks (FSS and BSS) will share the band 40.5-42.5 GHz on a primary basis with the FS and BS and are in the band adjacent to the band 42.5-43.5 GHz which is the subject of a study under WRC-97 Resolution 128. There is a need to undertake and to establish, as matter of urgency, studies of sharing criteria and methodologies, including pfd limits, to facilitate the coexistence of the space and terrestrial services with allocations in this band.

This band is adjacent to the band 42.5-43.5 GHz which is allocated to the radio astronomy service. Unwanted out-of-band emissions from space stations of the FSS (space-to-Earth) may result in harmful interference to the radio astronomy service. WRC-97 resolves that administrations shall not implement FSS systems in the band 41.5-42.5 GHz until technical and operational measures have been identified and agreed to protect the radio astronomy service from harmful interference.

9.7 Inter-satellite service (ISS)

9.7.1 General

The inter-satellite links (ISLs) are employed to provide connections between earth stations in the service area of one satellite-to-earth station in the service area of another satellite, when neither of the satellites covers both sets of earth stations. If there is an overlap between the two services, an alternative to employing an ISL is to link an earth station in the non-overlapping part of one area to an earth station in the non-overlapping part of the other area by a double satellite hop via an intermediate earth station in the overlap region. If the satellites are separated by angles large enough to make parts of the Earth's surface visible to only one or other of the two, then an ISL may be used to extend the service area available to users of the network. Constellations of satellites in low-Earth orbits sometimes incorporate a pattern of ISLs to enable long overland routes to be covered without recourse to multiple satellite hops via intermediate earth stations.

9.7.2 In the band 32-33 GHz

Since WARC-79, the band 32-33 GHz was allocated to the ISS, shared with the radionavigation (RN) service. Recommendation ITU-R S.1151 deals with the sharing between the ISS involving GSO satellites in the FSS and the RN service at 33 GHz. An ITU-R recommendation (in preparation) presents "A method for calculating single entry carrier-to-interference for GSO ISS systems".

9.7.3 In the band 25.5-27.5 GHz

At WARC-92, the frequency band 25.25-27.5 GHz was allocated on a primary basis to the ISS inter-satellite service, and the band 25.5-27 GHz on a secondary basis to the EESS (passive).

Recommendation ITU-R SA.1278 concludes that sharing between transmitting EESS and receiving data relay satellites (DRS) in the ISS near 26 GHz is feasible provided that the EESS does not produce a pfd greater than -155 dBW/m^2 in a 1 MHz bandwidth at any location in the GSO for more than 1% of the time and that sharing between transmitting satellites operating in the ISS and receiving earth stations in the EESS is also feasible provided that the probability of receiving brief periods of interference from data relay satellites DRS operating in the ISS is less than 0.1%.

9.7.4 In the band 50.2-71 GHz

Following WRC-95 which invited ITU-R to undertake studies to identify the bands most suitable for the ISS between 50 and 70 GHz in order to enable WRC-97 to make appropriate allocations to that service, several new recommendations were prepared:

- Recommendation ITU-R S.1327 which recommends the following bands as the most suitable for the ISS in the range 50.2-71 GHz: 54.25-58.2 GHz limited to GSO satellites, 59-64 GHz and 65-71 GHz and which summarizes the sharing potential of the ISS with other services in the frequency range 50.2-71 GHz;

- Recommendation ITU-R S.1339: "Feasibility of sharing between spaceborne passive sensors of the EESS and ISLs of geostationary satellite networks in the range 50 to 65 GHz";

- Recommendation ITU-R SA.1279: "Spectrum sharing between spaceborne passive sensors and inter-satellite links in the range 50.2-59.3 GHz";

- Recommendation ITU-R S.1326: "Feasibility of sharing between the ISS and the FSS in the frequency band 50.4-51.4 GHz".

Taking into account the need for protecting the EESS operating in the frequency range from about 50 GHz to about 70 GHz (in the bands 50.2-50.4 passive EESS, 51.4-54.25 passive, 54.25-58.2 passive, 58.2-59 passive, 64-65 passive, 65-66 GHz), ITU-R studies have concluded that the sharing is feasible under some constraints on the pfd in the range 50 to 65 GHz between the EESS and the ISLs of the GSO satellite service. On the other hand, sharing is not feasible in the same range 50-65 GHz between the EESS and the ISLs of non-GSO satellite systems.

In the band 50.4-51.4 GHz, sharing between ISS and GSO FSS may be feasible with certain constraints, but sharing between ISS and non-GSO FSS is likely to be difficult.

Sharing studies among GSO ISS networks led to the conclusion that sharing is feasible, given sufficient orbital separation, and sharing studies between non-GSO and GSO ISS systems demonstrated that such sharing is also feasible. But co-frequency sharing between non-GSO ISS networks is anticipated.

9.8 Frequency sharing between systems of the GSO and non-GSO FSS and the terrestrial services

9.8.1 Frequency sharing between systems of the GSO FSS and the terrestrial services

9.8.1.1 Introduction

(See Articles S9 and S11 of the RR for coordination, notification and registration procedures.)

Frequency sharing between systems of the FSS and terrestrial services involves four possible cases of interference. These are between:

- transmitting space station and receiving terrestrial stations;
- transmitting terrestrial stations and receiving space stations;
- transmitting terrestrial stations and receiving earth stations;
- transmitting earth stations and receiving terrestrial stations.

Protection from interference of the first type is provided by the adoption of maximum permissible power flux-densities at the Earth's surface due to a space station in the fixed-satellite service as provided in RR Article S21 as discussed previously (see § 9.1.4.1.2). Protection from interference of the second type is provided by certain power (e.i.r.p.) and pointing angle restrictions on radio-relay systems as provided in RR Article S21, also previously discussed (see § 9.1.4.2).

The latter two types are resolved by detailed coordination on a case-by-case basis. To identify specific cases requiring detailed coordination. RR Appendix S7 provides a procedure for constructing a coordination contour around an earth station. Terrestrial stations located within this contour are then subject to detailed coordination.

Recommendation ITU-R SF.1008 indicates a possible use by space stations in the FSS of orbits slightly inclined with respect to the GSO orbit in bands shared with the FS.

Details on procedures for earth stations are given in a Special Section of the Weekly Circular (now fortnightly), the recommended format for submitting this information is given in form of notice APS4/III.

The following sections provide a brief summary of the various steps involved.

9.8.1.2   Calculation of contours around an earth station based on RR Appendix S7

(See Recommendation ITU-R IS.847.)

The purpose of the procedures provided in RR Appendix S7 is the determination of the coordination area around an earth station within which a terrestrial station might cause, or be subject to, excessive interference. Its purpose is to identify cases where detailed coordination is needed. Therefore, the calculations are based on the most unfavorable assumptions as regards interference.

In general, for an earth station having both transmitting and receiving capability, two separate coordination contours will be necessary, one for the earth station transmitter and the other for the earth station receiver.

The basis of the method of Appendix S7 to determine the coordination contour is to determine for each azimuth direction the coordination distance (RR Article S1 No. 173). Determination of the coordination area is based on the concept of the permissible interference power at the antenna terminals of a receiving terrestrial or earth station. The attenuation required to limit the level of interference between a transmitting terrestrial or earth station and a receiving terrestrial or earth station to the permissible interference power for p% of the time is given by the minimum required loss.

Recommendation ITU-R IS.847 describes the determination of the coordination area of an earth station operating with a GSO space station and using the same frequency band as a system in a terrestrial service and Recommendation ITU-R IS.848 describes the determination of the coordination area of a transmitting earth station using the same frequency band as receiving earth stations in bidirectionally allocated frequency bands.

A new recommendation, currently under preparation, will address the determination of the coordination area around a transmitting or receiving earth station, sharing spectrum in frequency bands between 0.1 and 105 GHz with terrestrial radiocommunication services or with earth stations operating in the opposite direction of transmission. This recommendation should be updated based on changes in the RR which will result from decisions of WRC-2000 and the following WRCs. It will cover a very wide number of cases, including GSO, non-GSO systems, mobile earth stations, etc. (see § 9.8.2.4 below).

9.8.1.3   Permissible level of the interfering emission

The permissible level of the interfering emission in the reference bandwidth to be exceeded no more than p% of the time at the output of the receiving antenna of a station is calculated for both the earth station and the appropriate terrestrial station.

The basic equation from which to obtain the permissible level of the interfering emission is given in equation (3) of § 2.3.1 and Tables I and II of Appendix S7 to the Radio Regulations. There are detailed notes to explain and quantify some of the terms in the equation and tables. Some guidance is also given in § 2.3.2 of the same appendix on coordination parameters for very narrow-band transmissions.

The specific value of permissible level of the interfering emission varies depending on a number of variables. One specific example is given in Appendix 9.2 to this chapter.

9.8.1.4   Minimum permissible transmission loss

The amount of attenuation required between an interfering transmitter and an interfered-with receiver is the minimum permissible transmission loss.

In practice, the variation in transmission loss due to propagation effects is recognized by specifying a minimum transmission loss over a given percentage of time which ranges from 99.95% to 99.999%.

In order that generalized tropospheric propagation curves may be used, it is convenient to work in terms of a "minimum permissible basic transmission loss" which is the attenuation between isotropic antennas.

In general, two specific modes of radio propagation should be considered to determine which results in the largest coordination distance: tropospheric or rain scatter modes.

i)   Tropospheric propagation calculations (mode (1))

Propagation mode (1) is concerned with the attenuation of signals subject to tropospheric propagation via near-great-circle paths. Determination of coordination distance for this case is covered in § 3 of RR Appendix S7.

There is a choice between two methods. The numerical method is intended principally for use with the aid of a computer. It is covered in detail in § 3.2.2 of RR Appendix S7. The alternative graphical method given in § 3.2.3 of this appendix is equally valid and all necessary graphs are given in the appendix.

By either method, the starting point is the value of minimum permissible basic transmission loss for a given percentage of the time and the coordination distance is derived taking into account the radio-climatic zone or zones traversed, the frequency, percentage of time and horizon angle from the earth station.

ii)   Rain scatter propagation calculations (mode (2))

Propagation mode (2) is concerned with the attenuation of signals subject to scatter due to hydrometeors (rain). Determination of coordination distance for this case is covered in § 4 of RR Appendix S7. Again there is the choice between a numerical method (§ 4.3.1) and a graphical method (§ 4.3.2).

In this case it is necessary to calculate a "normalized transmission loss" as given in equation (20) of Appendix S7 and to take into account the rain-climatic zones shown in Figure 8 and Table IV of Appendix S7. The minimum required loss should be exceeded by the predicted path loss for all but p% of the time.

9.8.1.5   Determination of coordination contours

(See also Recommendations ITU-R IS.847 and IS.848.)

For each azimuth direction the coordination distance is the greater of the two calculated for mode (1) and mode (2). If the greater of these is less than 100 km then the coordination distance is to be taken as equal to 100 km. These distances are then plotted from the earth station on a map to an appropriate scale and the contour drawn. Auxiliary contours, if desired, are also plotted on the map using the same method and the additional factors S (equation (32) and Table I of RR Appendix S7) and E (equation (33) and Table II). The auxiliary contours may be of use in elimination of certain existing or planned terrestrial stations falling within the coordination contour thus avoiding more precise and tedious detailed calculations.

Appendix 9.2 of this chapter presents an example of the calculation results for the coordination contours.

Recommendation ITU-R IS.850 states coordination areas using predetermined coordination distances.

9.8.1.6   Performance and interference criteria

For those terrestrial stations located within the coordination contour, more detailed calculations are necessary to establish compatibility. The first step is to establish performance and interference criteria.

Both fixed-satellite systems and terrestrial-service systems are designed to satisfy certain overall service requirements in conformance with particular performance objectives (e.g. hypothetical reference circuits (HRC) and hypothetical reference digital path (HRDP)). These criteria are related to the type of traffic (e.g. multichannel telephony, television, etc.) and the type of modulation (e.g. analogue, digital) in use within the system.

Several ITU-R recommendations have been developed.

a)   On performance and availability

For the fixed-satellite service (FSS) (see also, in this handbook, Chapter 2, § 2.2 "Quality and availability"):

- Recommendations ITU-R S.352 and ITU-R S.353 (FDM telephony),

- Recommendations ITU-R S.522 and ITU-R S.579 (8-bit PCM),

- Recommendations ITU-R S.521 and ITU-R S.1062 (Digital),

- Recommendations ITU-R S.614 and ITU-R S.579 (ISDN),

- Recommendation ITU-R S.354 (analogue video),

and for the fixed service FS using radio-relay systems:

- Recommendation ITU-R F/Part 1. 393 (analogue),

- Recommendation ITU-R F/Part 1. 594 (digital).

b)   On criteria of maximum permissible interference

For the fixed-satellite service FSS:

- analogue telephony: Recommendation ITU-R SF.356 (FM),

- 8-bit PCM telephony: Recommendation ITU-R SF.558.

and for the fixed service FS using radio-relay systems:

- analogue telephony: Recommendation ITU-R SF.357 (AM),

- digital telephony: Recommendation ITU-R SF.615 (ISDN).

c)   On constraints of sharing conditions for each service

For the fixed-satellite service FSS: Recommendations ITU-R SF.358, SF.674 and SF.1004, and for the fixed service FS using radio-relay systems: Recommendation ITU-R SF.406.

9.8.1.7   Sharing criteria: coordination and interference calculations

Documentation relevant to detailed assessment of interference between earth stations and terrestrial stations is contained in Recommendations ITU-R IS.847, SF.766, SF.1006 and SF.1193. These ITU-R recommendations describe procedures for assessing whether interference can be expected to exceed a predetermined permissible level. They are, however, intended only as a guide for administrations since the method for determining the possibilities of interference is subject to agreement between the administrations concerned.

The antenna patterns used are important in determining both the coordination area and the assessment of interference. ITU-R has adopted a reference pattern for the side lobes and these are given in Recommendations ITU-R S.465 and S.580.

As concerns more particularly the band 37-40 GHz, WRC-97 (Resolution 133) allocated this band, on a primary basis to the FSS and to the FS. An increasing number of FSS systems and of stations in the FS are deployed or being planned for use. The deployment of high-density systems in either the FS or the FSS may result in interference to the FSS from stations in the FS. Sharing could be facilitated by the adoption of appropriate frequency sub-bands. WRC-97 requests to study for WRC-2000 the pfd limits included in Article S21 which adequately protect the FS stations from FSS networks and requests WRC-2000 to consider the identification of spectrum in the band 37-40 GHz for high-density applications in the FS.

9.8.2   Frequency sharing between systems of the non-GSO FSS and the terrestrial service systems

9.8.2.1   General

Review of Appendix S7 of the RR concerning the coordination area for the non-GSO earth stations is postponed until WRC-2000. Recommendation ITU-R IS.849 presents a determination of the coordination area of an earth station operating with a non-GSO spacecraft station in bands with terrestrial services.

9.8.2.1.1   Power flux-density limits for protection of terrestrial services

WRC-97 Resolution 131 of the RR resolved that emissions from a space station in non-GSO FSS networks in the bands 10.7-12.75 GHz and 17.7-19.3 GHz shall comply with the pfd limits contained in Table 21-4 of RR Article S21 (see Table 9.2 in Section 9.1.4) for the protection of terrestrial services and that further studies are required on the appropriate pfd values to be applied to non-GSO networks for the review the provisional limits by WRC-2000.

So as to facilitate sharing between non-GSO FSS networks and systems in the FS, reductions in the pfd or reductions in the number of satellites in non-GSO FSS shall be considered within the spirit of No. S9.58.

9.8.2.1.2   Non-GSO satellite systems sharing with FS stations

(See also Recommendations ITU-R IS.848 and IS.849.)

Studies were performed on sharing between the fixed-satellite service (FSS) and the fixed service (FS) in the 19.3-19.6 GHz band when used by the FSS to non-GSO-MSS FL (see WRC-95 Resolution 119).

The 19.3-19.6 GHz band is allocated in the direction space-to-Earth to the GSO FSS on a primary basis and WRC-95 also allocated this band to provide for non-GSO MSS FL. WRC-95 also allocated the 19.3-19.6 GHz band in the Earth-to-space direction to the non-GSO MSS FL on a primary basis. The 19.3-19.6 GHz band is also allocated to the FS on a primary basis. The coordination and notification procedures set forth in (Rev.WRC-97) RR Resolution 46 apply to services with equal rights in the 19.3-19.6 GHz band. Pfd limits in this 19.3-19.6 GHz band (space-to-Earth) for non-GSO systems have been adopted by WRC-97.

Recommendation ITU-R SF.1005 is relating to sharing between the FS and the FSS with bidirectional usage in bands 10 GHz currently unidirectionally allocated and states that the criteria of maximum permissible interference from earth stations operating bidirectionnally in the 19.3-19.6 GHz band to stations in the FS are preliminary and require further study.

Studies are being performed concerning the determination of the coordination area around earth stations operating with GSO FSS and earth stations operating with non-GSO MSS FL in opposite directions of transmission (see WRC-95 Recommendation 105).

WRC-95 has identified certain frequency allocations to the FSS for the non-GSO MSS FL which are also used by stations in the FSS operating with GSO satellites and which are also used in the opposite direction of transmission from non-GSO MSS FL. In order to avoid mutual interference between GSO and non-GSO earth stations operating in opposite directions of transmission, there is a need to determine the coordination area of such stations (see Recommendations ITU-R IS.849 and IS.847). WRC-2000 will review the procedures set forth in RR Appendix S7.

9.8.2.1.3   Frequency sharing and coordination for non-GSO earth stations

(Annex 1 to RR App. S5 and Recommendation ITU-R IS.849. Note that certain FSS bands can be used by non-GSO MSS FL systems.)

i)   Frequency sharing between non-GSO and GSO MSS space stations and terrestrial services

a)   Coordination thresholds for determining the need for coordination between MSS (space-to-Earth) and terrestrial services in the same frequency bands and between non-GSO MSS FL MSS (space-to-Earth) and terrestrial services in the same frequency bands:

- below 1 GHz: in the bands 137-138 MHz and 400.120-401 MHz, the coordination of a space station of the MSS is required if the pfd produced by emissions from a space station exceeds (–125 dBW/m²/4 kHz);

- between 1 and 3 GHz: the coordination of GSO and non-GSO space stations of the MSS is required if the pfd produced by emissions from a space station or the FDP of a station in the FS exceed the coordination thresholds value shown in the table of Annex 2 to Resolution 46;

- above 3 GHz: in the band 15.45-15.65 GHz, the coordination of a non-GSO space station MSS is required if the pfd exceeds (–146 dBW/m²/MHz).

b)   Limits for sharing between non-GSO MSS FL (space-to-Earth) and terrestrial services in the same frequency bands:

- in the band 5 150-5 216 MHz, the pfd produced by emissions from a space station shall in no case exceed –164 dBW/m² in any 4 kHz band for all angles of arrival;

- in the bands 6 700-6 825 MHz, 6 825-7 075 MHz, 15.43-15.63 GHz, emissions from a non-GSO space station shall not exceed the pfd limits shown in the table of Annex 2 to Resolution 46;

- in the band 6 700-7 075 MHz, the maximum apfd produced at the GSO orbit by a non-GSO FSS system shall not exceed –168 dBW/m² in any 4 kHz band;

- in the band between 17.7 GHz and 27.5 GHz, the pfd produced by emissions from a space station shall not exceed in any 1 MHz the following values:

    - –115 dBW/m² for arrival angles δ: between 0° and 5° above the horizontal plane,

    - –115+0.5(δ-5) dBW/m² for δ between 5° and 25°,

- -105 dBW/m² for δ between 25° and 90°.

c)   pfd limits produced by non-GSO FSS in the 20-30 GHz

The pfd produced by emissions from a space station shall not exceed the same limits as given above for the band 17.7-27.5 GHz.

However, the following pfd limits in 1 MHz shall apply provisionally to emissions of space stations on non-GSO satellites in networks operating with a large number of satellites, that is systems with more than 100 satellites (see WRC-97 Resolution 131):

- -125 dBW/m² for arrival angles δ between 0° and 5° above the horizontal plane;

- $-125+0.5(δ-5)$ dBW/m² for δ between 5° and 25°;

- -125 dBW/m² for δ between 25° and 90°.

ii)   Coordination areas for non-GSO earth stations

The coordination contour associated with the coordination area is drawn to scale on an appropriate map in order to depict the coordination area and the extent to which it overlaps the territory of administrations that may be affected.

Two types of coordination distances are specified (Tables 1 to 4 of Rev.WRC-97 Annex 2 to Resolution 46): 1) predetermined distances and 2) distances calculated on a case-by-case basis, taking into account specific parameters of the earth station for which the coordination area is being determined. Neither of these distances indicate required separation distances.

The coordination area of a mobile earth station is determined as the service area in which it is intended to operate typical earth stations, extended in all directions by the coordination distance (see Tables 1 and 2 respectively below 1 GHz and in the 1-3 GHz band).

In the case of non-GSO MSS FL or non-GSO FSS earth stations, the coordination contour is determined as the end points of coordination distances measured from the earth station location. Coordination distances for FL earth stations operating above 5 GHz are specified in Table 3 with respect to stations in terrestrial services and, where applicable, earth stations, of other satellite networks operating in the opposite direction of transmission. Coordinations distances for non-GSO FSS earth stations are specified in Table 4.

9.8.2.2   Utilization of the space-to-Earth direction by the non-GSO MSS (FL) on a shared basis with the FS

WRC-95 allocated certain frequency bands (6.700-7.075 GHz and 19.3-19.6 GHz) to the use in the space-to-Earth direction by feeder links for non-geostationary satellite systems in the mobile-satellite service non-GSO MSS (FL) on a shared basis with the fixed service FS.

The band 6.700-7.075 GHz (space-to-Earth) is subject to coordination under RR Resolution 46/RR S9.11A. Recommendation ITU-R S.1257 recommends in the range 6.7-19.6 GHz shared between FL of non-GSO MSS and FS a maximum pfd produced at the surface of the Earth by a satellite non-GSO MSS (FL) and proposes in Annex 1 guidance concerning the interference from non-GSO MSS FL satellites to stations in the FS.

i)   Highest level of interference and statistics of worst-case interference

The highest level of interference occurs when a non-GSO MSS satellite is within the main beam of a terrestrial system antenna. For a given non-GSO constellation, percentage of interference, duration of interference and mean time between interference events will be very dependant on FS latitude and FS link azimuth.

The worst-case interference is from azimuth directions where the probability of exceeding a certain interference level is at its maximum. Depending on orbital parameters (altitude, inclination) and FS station parameters (latitude, elevation) there are 1 to 4 worst-case azimuths: the smaller the geocentric angle between the FS station and the satellite, the greater the off-axis angle and hence the better the discrimination against interference. Most of those terminals will not point to the worst-case azimuths and in some cases there are azimuth directions where the satellite would not appear within the main beam of an FS station. The worst-case interference also depends on geometrical considerations.

ii)   Methodology to determine the fractional degradation of performance (FDP)

Generally, pfd thresholds are used to determine the need for coordination between space stations of the MSS (space-to-Earth) and terrestrial services. However, to facilitate sharing between digital FS stations and non-GSO MSS space stations, the concept of FDP has been adopted. This concept involves methods described in Annex 2 to Resolution 46 (Rev.WRC-95/97).

The FDP is calculated by positioning the FS station at a certain latitude (at an elevation angle of $0°$) and by calculating for each pointing azimuth (az) varying between $0°$ and $360°$ at each instant in time of the simulation, the aggregate interference from all visible space stations received at the FS station and the FDPaz for the azimuth (az) and by the following formula:

$$FDP = max(FDPaz).$$

Recommendation ITU-R F.1108 provides a method to calculate the FDP:

For non-diversity digital FS systems, the FDP, for an azimuth considered, is given as:

$$FDP = \sum f_i (I_i/N_T)$$

the summation being taken over the entire simulation time and where:

- N_T $(= kTB)$ is the FS thermal noise,

- T is the FS station receiving system effective noise temperature,

- B is the reference bandwidth (1 MHz),

- f_i is the fraction of time that a digital FS system experiences an interference level I_i.

iii)   FS interference criteria

Recommendation ITU-R F.1094 recommends a performance degradation criterion of 10% for services shared on a co-primary basis, and Recommendation ITU-R SF.357 specifies interference criteria into terrestrial radio-relay links.

Recommendation ITU-R SF.1005 recommends that for shared bands allocated in both directions to the GSO FSS, the pertinent interference criterion (long term and short term) from earth stations operating in RBW to stations in the FS should be lower by the following values than those calculated by the methods described in Recommendations ITU-R IS.847 and SF.1006: 7 dB in the band 10-15.4 GHz, 5 dB in the band 15.4-20 GHz and 3 dB above 20 GHz.

Recommendation ITU-R SF.1005 pertains to frequency bands above 10 GHz due to the fact that most bands below 10 GHz are heavily used by the FS and then RBW mode is not feasible in these bands.

It is recognized that similar concepts would be applicable to non-GSO FSS and that some tightening of the FDP criterion of 10% would be required when considering interference from RBW non-GSO MSS FL.

iv)   Feasibility of sharing: pfd limits of Recommendation ITU-R SF.358

Simulations were performed using the characteristics of various possible satellite constellations in order to determine the feasibility of co-frequency sharing with FS systems.

It was concluded that the satellite constellations can share with most FS systems because satellite constellations produced pfd levels at the Earth's surface which are much lower than the limits specified in Recommendation ITU-R SF.358 for non-GSO FSS satellites.

In the case of non-GSO MSS, the sharing with the FS will take place successfully with the application of pfd limits, which in fact generate interference that far exceed the thermal noise for short periods of time, because the affected receiver will not usually experience fading at the same time that it is receiving this high level of interference. As a result, even though this boresight interference event cannot be avoided, the FDP criterion of 10% or a smaller value can be satisfied.

The interference power from each satellite is calculated as follows:

$$It = pfd + 10\log(\lambda^2/4\pi) + G_{FS} + \text{Feeder Loss}$$

Various factors (FS latitude, FS elevation, arrival angle, …) have different effects on the sensitivity to the pfd limits derived from an FDP calculation: generally, the higher the latitude, the more severe is the interference.

The FDP is a function of the amount of time the satellite is within the beam of the FS antenna, the discrimination of the FS antenna towards the satellite as a function of time and the satellite interference power. To a large extent, the FDP is dominated by these short-term high-interference events. The pfd limits are proposed to be applicable to each non-GSO MSS satellite operating in a RBW mode (with the following assumptions: peak FDP criterion of 4% and 4 dB relaxation).

A balance which provides adequate protection to the majority of FS systems plus flexibility to future non-GSO MSS FL networks can be reached.

9.8.2.3 Frequency sharing between satellites of non-GSO FSS networks and FS stations in the 18.8-19.3 GHz (space-to-Earth) and 28.6-29.1 GHz (Earth-to-space) bands

In response to WRC-95 Resolution 118 of the RR, several studies related to the feasibility of frequency sharing between a non-GSO FSS network and FS stations operating in the 18.8-19.3 and 28.6-29.1 GHz bands have been performed.

Interference in both directions between a non-GSO FSS satellite and an FS station can occur in one of three modes: side lobe into main beam, main beam into side lobe and side lobe into side lobe. Main beam-to-main beam coupling does not occur because the modelled non-GSO FSS satellites are planned to operate substantially above the horizon, whereas the FS station antennas are generally directed horizontally or at small elevation angles.

In the studies, the majority of FS station parameters are taken from Recommendation ITU-R SF.758 and the LEOSAT-1 network (the first non-GSO FSS network) was used in all of the studies as the example of a non-GSO FSS network.

i) Interference from FS transmitters into non-GSO FSS satellite receivers

Non-GSO satellite receivers operating in these bands will experience interference levels well below the interference power spectral density threshold level of −212.5 dBW/Hz. The conclusion of these studies is that a reasonably large density of FS stations would not cause unacceptable interference to the non-GSO FSS satellites in the case of the LEOSAT-1 system.

E.i.r.p. levels from FS stations as high as 55 dBW are permitted by the RR (see Article S21 and Recommendation ITU-R SF.406). There is also a need to review the e.i.r.p. limit, considering bandwidth and elevation angle, for FS transmitters operating in the band 28.6-29.1 GHz.

ii) Interference from non-GSO FSS satellite transmitters into FS receivers

Pfd levels of −115/−105 dBW/m²/MHz are the limits permitted in the RR (Article S21.16) for a single non-GSO FSS satellite beam, depending on the elevation angle of the interference. Calculations have been made in the case of a single non-GSO FSS satellite beam (calculation of I_0/N_0 in the three modes) and in the case of the entire non-GSO FSS satellite constellation (interference simulation) into a FS station at an elevation angle located at a certain latitude, calculation of the FDP, FML (fade margin loss), long-term interference with maximum and minimum values for I_0/N_0 during the entire simulation, and cumulative distributions of I_0/N_0 for different values of FS receiver gain.

It can be concluded that in the case of LEOSAT-1 system, significant interference will not be caused to typical terrestrial FS stations from a network of non-GSO FSS satellites and that sharing appears to be feasible. The minimum operational elevation angle restriction of LEOSAT-1 is instrumental in promoting a favorable sharing environment.

In the 18.8-19.3 GHz band, it has been shown that a non-GSO satellite transmitting to elevations below 20° could cause unacceptable interference into FS receivers.

Therefore, there may be a need to review the pfd limits for non-GSO satellites operating at elevations below 20°. Recommendation ITU-R SF.1320 gives the "Maximum allowable values of pfd at the surface of the Earth produced by non-GSO in the FSS used in FL for the MSS and sharing the same frequency bands with radio-relay systems (relating to 19 GHz)".

9.8.2.4 Determination of coordination area for earth stations of the non-GSO satellites in the FSS or in the feeder links of the MSS (FL) and the terrestrial stations in the FS; Review of RR Appendix 28/S7 by WRC-2000 (see also Recommendations ITU-R IS.847/848/849)

The review of RR Appendix S7 has been postponed from WRC-97 until WRC-2000.

WRC-97 Resolution 721 presents the agenda for the WRC-2000 and requests in topic 1.3 to consider the results of ITU-R studies in respect of Appendix S7/28 on the method for the determination of the coordination area around an earth station in frequency bands shared among space services and terrestrial radiocommunications services and to take the appropriate decisions to revise this appendix.

A complete new version will be proposed. This new version will be mainly based on Recommendations ITU-R IS.847, IS.848, IS.849 which were written with a view to the modernization of the appendix.

WRC-95 had allocated frequency bands for use by MSS (FL) and by non-GSO FSS systems in bands shared on a primary basis with the FS, and also some of these bands are shared with the GSO FSS in the same direction or in reverse direction. The bands 18.8-19.3 GHz and 28.6-29.1 GHz were identified for use by non-GSO FSS systems.

The new procedure must be appropriate for the determination of the coordination area as well in frequency bands in which the space service has a unidirectional (Earth-to-space or space-to-Earth) allocation as in frequency bands which are bidirectionally (i.e. Earth-to-space or space-to-Earth) allocated to space services. This new procedure must also be applicable as well to the case of earth stations operating with GSO space stations in the reverse direction as to the case of earth stations operating with non-GSO space stations with the exception of non-GSO scientific systems where special provisions apply.

The new procedure must then be directly applicable to the following cases both for earth station operating with GSO space station and for earth station operating with a constellation of non-GSO space stations.

The worst-case interference between a non-GSO earth station and an FS station will occur when the transmitting and receiving stations are pointed towards each other along reciprocal azimuth directions. The two main parameters concerning the level of interference between the two stations are the geometric alignment and the cumulative duration of the non-GSO earth station antenna pointing towards the FS antenna.

ITU-R is in the process of developing a future recommendation concerning "The determination of the coordination area of FSS earth stations operating with non-GSO satellites (whether or not they are for use in MSS (FL)) and stations in the FS". It is expected that co-frequency sharing between non-GSO FSS earth station terminals and FS stations will be feasible.

As already explained above (§ 9.8.1.2), a new recommendation under preparation presents a determination of the coordination area around a transmitting or receiving earth station in frequency bands between 0.1 and 105 GHz. For the case of earth stations operating to non-GSO space stations that use a directional antenna which tracks the space station, the antenna gain in the direction of the horizon on any azimuth varies with time. Two methods are available to take this effect into account:

- the time-invariant gain (TIG) method;

- the statistical method.

The TIG method provides ease of implementation but may overestimate the coordination. In order to reduce the coordination burden and on the basis of bilateral and multilateral agreements, administrations can use the statistical method to obtain less conservative results. The TIG method uses fixed values of antenna gain based on the maximum assumed variation in horizon antenna gain on each azimuth under consideration.

The statistical method requires a knowledge of the statistics of the time-varying horizon antenna gain of an earth station operating with a non-GSO space station. Finally, the TIG method, which is independent of antenna horizon gain statistics, will be recommended because of its ease of implementation but the statistical method, which gives shorter realistic distances, could also be used between administrations on a bilateral or multilateral basis.

APPENDIX 9.1

TABLE AP9.1-1

Allocation of frequencies to the FSS, BSS, MSS and ISS services

FSS/BSS/MSS/ISS Allocations (121 MHz -275 GHz) (ARTICLE S5 of RR)

Frequency	decisions: WARC 85&88	decisions: WARC 92 &WRC 95&97	Regions R1	R2	R3	services to which the band is allocated on a primary basis (services written in lower case are allocated on a secondary basis) R1	R2	R3	footnotes : previous numbering of Final Acts of the WRC95 R2	R3
30.005 - 30.010 MHz										
117.975 - 136 MHz	Art14/S9.21					**Space Operation (satellite identification)** amss (R) AM (R)				
117.975 - 136 MHz						AM (R),			S5.198	
121.500 &123.100 MHz	Mob83: Aeronautical Emergency frequency (main and auxiliary) Distress and safety frequency for MSS (AppS31&S13/ArtN38&38)					mss			S5.200	
121.450 - 121.550 MHz	RadioBeacons (AppS13/Nos3259&3267)		u	u	u	mss			S5.199	
136.000 - 137.000 MHz	Mob87: Res408								S5.111/198/200/201	
137.000 - 138.000 MHz		Add92&Mod95: NGSO MSS &Art S9.11A(Res46)	d	d	d	AM (R), f, mxa sop, meteosat, sres			S5.203	
			d	d	d	NGSO MSS & ngso mss, MeteoSat, SOp, SRes			S5.208A/208A/209	
		NGSO MSS & Res46 &Res 714	d	d	d	m(xa)			S5.204/205/206/207	
137.000 - 137.025		Mod95: ArtS9.11A(Res46)	d	d	d	NGSO MSS, MeteoSat, SOp, SRes			S5.208A/209	
137.050 - 137.175		NGSO MSS &Res46 &Res 714 Mod95: ArtS9.11A(Res46)	d	d	d	ngso mss MeteoSat, SOp, SRes			& S5.204/205/206/207/208 S5.208A/209	
137.175 - 137.825		ngso mss &Res46 &Res 714 Mod95: ArtS9.11A(Res46)	d	d	d	NGSO MSS, MeteoSat, SOp, SRes			& S5.204/205/206/207/208 S5.208A/209	
137.825 - 138.000		NGSO MSS &Res46 &Res 714 Mod95: ArtS9.11A(Res46)	d	d	d	ngso mss MeteoSat, SOp, SRes			& S5.204/205/206/207/208 S5.209	
148.000 - 149.900 MHz		ngso mss &Res46 &Res 714 Mod95: ArtS9.11A(Res46)	u	u	u	NGSO MSS, M, F	MxA, F		S5.218/219/221 S5.209	
		Mod95: ArtS9.21(Art14) NGSO MSS - Res46	u	u	u	SOp			S5.218 S5.218/219/221	
149.900 - 150.050 MHz		Mod95: ArtS9.11A(Res46) &Res715	u		u	NGSO LMSS (until 2015) NGSO MSS (after 2015) RNSS (until 2015)			S5.209/224A S5.209 S5.224B S5.220/222/223	
		NGSO LMSS (until 2015) : WRC97 Res46 NGSO MSS (after 2015) : WRC97 Res46							Add97: S5.224A&B and Mod 97: S5.209	
235.000 - 322.000 MHz		Mod95: ArtS9.21/Art14	d	d	d	mss F, M			S5.254	
		Mod95: ArtS9.21/Art14	d	d	d	sop (267-272 MHz) & SOp (272-273 MHz)			S5.257 (TM on primary basis in the country)	
242.950 - 243.050		RadioBeacons (AppS13/Nos3259&3267) Mob83 : Emergency position-indicating radiobeacons	u	u	u	mss, F, M			S5.199 S5.111/248/254/256	
312.000 - 315.000		Mod95: ArtS9.11A(Res46)	u	u	u	gso & ngso mss (see 387 - 390 MHz) F, M			S5.255	

FSS/BSS/MSS/ISS Allocations (121 MHz -275 GHz) (ARTICLE S5 of RR)

Frequency	decisions: WARC 85&88	decisions: WARC 92 &WRC 95&97	Regions R1	R2	R3	services R1	services R2	services R3	footnotes R2	R3
335.400 - 399.900 MHz	NGSO &GSO MSS - Res46								S5.255	
387.000 - 390.000		Mod95: Art59.21(Art14)	d	d	d		mss / F, M		S5.254	
	Mod95: Art59.11A(Res46)						gso & ngso mss (see 312 - 315 MHz)		S5.255	
	NGSO &GSO MSS - Res46									
399.900 - 400.050 MHz	Mod95: Art59.11A(Res46)&Res715(Com5/9)		u	u	u		NGSO LMSS (until 2015) / NGSO MSS (after 2015) / RNSS (until 2015)		S5.209/224A / S5.209 / S5.224B / S5.220/222/223 / Add97: S5.224A&B and Mod 97: S5.209	
	NGSO LMSS (until 2015) : WRC97 Res46 / NGSO MSS (after 2015) : WRC97 Res46									
400.150 - 401.000 MHz	Mod95: Art59.11A(Res46)		d	d	d		NGSO MSS, MeteoSat, SRes Met sop		S5.208A/209/264 / S5.263	
	NGSO MSS : Res46		d	d	d				S5.262	
406.000 - 406.100 MHz	RadioBeacons: ApS13(exArt38) & ArtS31(exArtN38) (emergency position-indicating radiobeacons)		ux	ux	ux		MSS		S5.266, 267	
454.000 - 455.000 MHz	Add95: Art59.11A(Res46)		u	u	u	(NGSO MSS) : (Additional allocation for some countries) F, M			Add97 S5.286E / Add97 S5.286D / Add97 S5.286E	
									S5.209/271 / Mod97 S5.286B/286C	
455.000 - 456.000 MHz	Add95: Art59.11A(Res46)		u	u	u	(NGSO MSS)	NGSO MSS F, M	(NGSO MSS)	Add97 S5.286E / Mod97 S5.286A/B/C / Add97 S5.286E	
	NGSO LMSS (until 2015) : WRC97 Res46 / NGSO MSS (after 2015) : WRC97 Res46								S5.209/271 / Mod97 S5.286A/286B/286C	
459.000 - 460.000 MHz	Add95: Art59.11A(Res46)		u	u	u	(NGSO MSS)	NGSO MSS F, M	(NGSO MSS)	Add97 S5.286E / Mod97 S5.286A/B/C / Add97 S5.286E	
	NGSO LMSS (until 2015) : WRC97 Res46 / NGSO MSS (after 2015) : WRC97 Res46								S5.209/271 / Mod97 S5.286A/286B/286C	
608.000 - 614.000 MHz			d	u	d		mss (xA) bss (fm)			
620.000 - 790.000 MHz			d	b	d	B (470-790 MHz)	B (614-806 MHz)	B (610-890 MHz), F, M	S5.311(Res 33 et 507)	
806.000 - 890.000 MHz	Mod95: Art59.21(Art14)		u	b	b	F, B, MxA (862-890)	national MSS F, B, M	national MSS (xA&R) F, B, M	S5.317 (national) ... S5.320 (national)	
806.000 - 840.000			u			national mss (xA&R)			S5.319	
849.000 - 851.000				d			AM			
856.000 - 890.000			d			national mss (xA&R)			S5.319	
894.000 - 896.000	Art14			u			AM		S5.318	
942.000 - 960.000 MHz	Mod95: Art59.21(Art14)				b			MSS (xA&R)	S5.320 (national)	
	Art14									
1.390 - 1.400 GHz	WRC97 Res 127: studies for NMSS FL (up) with services links operating below 1GHz									
1.427 - 1.429 GHz			u	u	u		SOp F, MxA		S5.341	
1.427 - 1.432 GHz	WRC97 Res 127: studies for NMSS FL (down) with services links operating below 1GHz									
1.452 - 1.492 GHz	WARC92 Res528 (BSS(sound))		d	d	d	F, MxA,	BSS(DAB) F, M, B(DAB)		S5.345/347	
1.492 - 1.525 GHz	Mod&Add95: S9.11A(Res46)			d		MxA	MSS, F, M		S5.343 / S5.345/347 / S5.341/344 / S5.348/348A / S5.343	

FSS/BSS/MSS/ISS Allocations (121 MHz -275 GHz) (ARTICLE S5 of RR)

Frequency	decisions: WARC 85&88	decisions: WARC 92 &WRC 95&97	Regions R1	R2	R3	services (primary basis) R1	R2	R3	footnotes R1	R2	R3
1.525 - 1.559 GHz	Res46		d	d	d	MSS (see 1.6265 - 1.6605 GHz; Inmarsat Bands)			S5.341/342	S5.341/344/348	S5.341/348A
1.525 - 1.530		Add92 &Mod95: S9.11A(Res46) & S9.13	d	d	d	MSS	MSS		S5.354 & S5.352A / Add97 S5.352A(after 1/4/98)	S5.354 / S5.341	S5.354
		Add95: ArtS9.11A(Res46)	d	d	d		SOp / F	F	S5.347/350	Add97 S5.352A(after 1/4/98)	
1.530 - 1.559		Add95: ArtS9.11A(Res46)	d	d	d	MSS				S5.341/351 / S5.354	
1.530 - 1.544		Add97: priority for GMDSS (see WRC97 218)	d	d	d	MSS & MSS (GMDSS)				Add97 S5.353A	
1.530 - 1.533		Add97: priority for GMDSS (see WRC97 218)	d / d	d / d	d / d	MSS & MSS (GMDSS) SOp				S5.354 & Add97 S5.353A	
1.533 - 1.535		Add97: priority for GMDSS (see WRC97 218)	d / d	d / d	d / d	MSS & MSS (GMDSS) SOp			S5.341/347/351	S5.341/351 / S5.354 & Add97 S5.353A	
1.535 - 1.544		Add97: priority for GMDSS (see WRC97 218)	d	d	d	MSS & MSS (GMDSS) SOp			S5.341/347/351	S5.341/351 / S5.354 & Add97 S5.353A / S5.341/351/355	
1.544 - 1.545		Mod95: ArtN38/S31: Distress &Safety Com	d	d	d	MSS (only GMDSS)				S5.354 / S5.341/355/356	
1.545 - 1.555		Add97: priority 1-6 in ArtS44 for AMS(R)S (see WRC97 218)	d	d	d	MMS & AMMS (R)				S5.354 & Add97 S5.362A / S5.341/351/355/357/358/359	
1.550 - 1.555			d	d	d	(F) (Additional allocation for some countries)				S5.359	
1.555 - 1.559			d	d	d	MMS & AMM(R)S-USA				S5.354 & Add97 S5.362B / S5.341/351/355/359	
1.6100 - 1.6455 GHz			u	u	u	F (Additional allocation for some countries)				S5.359	
1.6100 - 1.6265 GHz		Mod95: S9.11A(Res46) / Mod95: S9.21(Art14)	u	u	u	MSS AMMS(R) ANav				S5.364(up) / S5.367 / S5.366	
		Mod92: Res46	u	u	u (rdss)	RDSS(see 5.150-5.216 GHz)			S5.364/368/371	S5.364/368 (see S5.446: 5.150 -5.216 GHz) / S5.341/364/366/367/368/372	
1.6106 - 1.6106			u	u	u	MSS					
			u	u	u (rdss)	ARN / RDSS					
1.6106 - 1.6138			u	u	u (rdss)	(Radiodetermination bands: see 2483.5 - 2500 MHz)			S5.355/359/363/369 / S5.355/359/363/369	S5.341/364/366/367/368/372 / S5.370	S5.355/359/369
1.6138 - 1.6265		WRC97 Res 125: MSS/SRAst	u	u	u (rdss)	MSS RASt, ARN RDSS			S5.355/359/363/369	S5.149/341/364/366/367/368/372 / S5.370	S5.355/359/369
1.6265		Mod95: ArtS9.11A(Res46)	u / d	u / d	u	MSS(up) mss(d) ARN RDSS			S5.355/359/363/369 / S5.371	S5.364(up) / S5.365(d) / S5.341/364/366/367/368/372 / S5.370	S5.355/359/369
1.6265 - 1.6605 GHz		Mod&Add95: Art S9.13 / Art S9.11A(Res46)	u	u	u	MSS (see 1.525 - 1.559 GHz)			S5.355/359/363/369	S5.354	S5.355/359/369
1.6265 - 1.6315		Add97: priority for GMDSS (see WRC97 218)	u	u	u	MSS & MSS (GMDSS)				S5.354 & Add97 S5.353A / S5.341/351/355/359	
1.6315 - 1.6345			u	u	u	MSS & MSS (GMDSS)				S5.354 & Add97 S5.353A	
1.6345 - 1.6455			u	u	u	MSS & MSS (GMDSS)				S5.341/351/355/359 &Mod97 S5.374 / S5.354 & Add97 S5.353A	

FSS/BSS/MSS/ISS Allocations (121 MHz-275 GHz) (ARTICLE S5 of RR)

Frequency	decisions: WARC 85&88	decisions: WARC 92 &WRC 95&97	R1	R2	R3	Services R1	Services R2	Services R3	footnotes : previous numbering of Final Acts of the WRC95 R2	R3
									S5.341/351/355/359	
1.6455 - 1.6465			u	u			MSS		S5.341/354/375	
1.6465 - 1.6600			u	u			F (Additional allocation for some countries)		S5.359	
1.6465 - 1.6565			u	u			MMS & AMMS (R)		S5.354 & Add97 S5.362A; S5.341/351/355/359/376	
1.6565 - 1.6600			u	u			MMS & AMM(R)S-USA		S5.354 & Add97 S5.362B; S5.341/351/355/359 & Mod S5.374	
1.6600 - 1.6605		WRC97 Res 125: MSS/SRAst	u	u			MMS & AMM(R)S-USA RAst		S5.354 & Add97 S5.362B & 376A	
1.675 - 1.710 GHz		Mod95: Art S9 11A(Res46) Mod 95 Res 213		u			MSS		S5.149/341/351	
1.675 - 1.690		Mod 95 Res 213	d	u	d		MSS, MeteoSat, F, MtA, Met		S5.377	
1.690 - 1.700		Mod 95 Res 213 : MSS in 1675-1710 MHz	d	u	d		MSS, MeteoSat, Met		S5.341 | S5.341/377	S5.341
1.700 - 1.710		Mod 95 Res 213	d	u	d		MSS, MeteoSat, F, MtA		S5.289/341/382 | S5.289/341/377/381	S5.289/341/377/381
1.930 - 1.970 GHz		Rev 97Res212(IMT2000:1.885-2.025 GHz)		u			mss		S5.289/341 | S5.289/341/377	S5.289/341/384
1.980 - 2.010 GHz		Add95: Art S9 11A(Res46)&Res716	u	u			MSS (after 2000)(see 2.170 - 2.200 GHz),		S5.389A (after 2000) (see 2.170 - 2.200 GHz), S5.389A (after 2005) & S5.389B	
1.980 - 1.990		WRC95 Res716	u	u			MSS (after 2005), F, MtA		S5.388/389A/B/F	
1.980 - 2.010 (sat component)		Rev 97Res212(IMT2000:1.885-2.025 GHz)					FPLMTS-IMT-2000		S5.388 (satellite component)	
2.010 - 2.025 GHz		Add95: Art S9 11A(Res46)&Res716; Rev 97Res212(IMT2000:1.885-2.025 GHz); WRC95 Res716		u			MSS (after 2005) (see 2.160-2.170 GHz) F, M		Mod97 S5.389C (after 2002); S5.388 (satellite component); Add97 S5.390(2005) & S5.389D(after 2000) & S5.389E	
2.025 - 2.110 GHz			u u s s	u u s s			SOp, EESS, SRes, SOp, EESS, SRes, F, M		S5.392	
2.120 - 2.160 GHz		Rev97Res212(IMT2000:2.110-2.200 GHz)		d			mss F, M FPLMTS IMT2000		S5.388	
2.160 - 2.170 GHz		Add95: Art S9 11A(Res46)&Res716; WRC95 Res716 & Rev. WRC97 Res 46		d			MSS (after 2005)(see 2.010-2.025 GHz), F, M		S5.388 (satellite component); Mod97 S5.389C (after 2002); S5.388 (satellite component); Add97 S5.390(2005) & S5.389D/E	
2.170 - 2.200 GHz		Add95: Art S9 11A(Res46)&Res716	d	d	d		MSS (after 2000)(see 1.980 - 2.010 GHz), F, M		S5.392A | S5.389A (after 2000)	S5.389A (after 2000)
2.170 - 2.200 (sat component)		Rev97Res212(IMT2000:2.110-2.200 GHz) Res716	d	d	d				S5.388/392A; S5.389A/F	
2.200 - 2.290 GHz			d d s s	d d s s			SOp, EESS, SRes, SOp, EESS, SRes,		S5.392	

FSS/BSS/MSS/ISS Allocations (121 MHz -275 GHz) (ARTICLE S5 of RR)

Frequency	decisions: WARC 85&88	decisions: WARC 92 & WARC 95&97	Regions R1	R2	R3	services to which the band is allocated on a primary basis (services written in lower case are allocated on a secondary basis) R1	R2	R3	footnotes : previous numbering of Final Acts of the WRC95 R2	R3
2.300 - 2.450 GHz							F, M			
2.4835 - 2.500 GHz	WARC 92 Res528 (BSS(sound))						F, M, RL, amat BSS(sound) (2.310 - 2.360 GHz)		S5.393/396	
2.500 - 2.500 GHz	Mod95: S9.11A(Res46)		d	d	d		MSS, F, M, RL,		S5.398 (see S5.446: 5150 -5216 MHz)	
			d	d	d rdss	rdss	RDSS(see 5.150-5.216 GHz) (Radiodetermination bands: see 1610 - 1626,5 MHz)	&S5.371	S5.150/402	
2.500 - 2.690 GHz			b				FSS		S5.415	
2.500 - 2.535 GHz				u	d		FSS		S5.415	
2.500 - 2.535 GHz				u	d		FSS		S5.415	
2.520 - 2.670 GHz	Mod95: S9.21(Art14) &ArtS21		d	d	d		BSS		S5.416/413 (Nat and Reg systems: community reception)	
2.500 - 2.520 GHz	Mod95: S9.21(Art14) &ArtS21 / Mod95: Art S9.11A(Res46) / Mod95:S9.21(Art14)&S9.11A(Res46)		d	d	d		FSS, MSS(after 2005) mxass (until 2005), F, MxA		S5.415 (national and regional systems) S5.414 (after 2005) & S5.403 (national) (S9.21)	
	Mod92: Res46, Art14							S5.409/410/411	S5.409/411	
2515 - 2535					d	amss (J from 2000)			S5.403/407/414 S5.404 (national & S9.21) Add97 S5.403A (S9.21)	
2.520 - 2.535 GHz	Mod95: S9.21(Art14) &ArtS21 / Mod95: S9.21(Art14) &ArtS21		d	d	d		FFS, BSS, F,		S5.415 (national and regional systems) S5.416/413	
	Mod95:S9.21(Art14)&S9.11A(Res46) / Mod92: Res46, Art14		d	d	d		MxA mxass		S5.409/410/411	S5.409/411 Add97 S5.403A (S9.21)
2.535 - 2.655 GHz	Mod95: S9.21(Art14) &ArtS21 / Mod95: S9.21(Art14) &ArtS21		d	d	d		FFS, BSS, F,		S5.405/408/418/417/412 S5.415 (national and regional systems) S5.416/413	
								S5.409/410/411	S5.405/408/418/417/412 S5.409/411	
2.655 - 2.670 GHz	Mod95: S9.21(Art14) &ArtS21 / Mod95: S9.21(Art14) &ArtS21		d	b	u		FFS, BSS(d), MxA, F		S5.415 (national and regional systems) S5.416/413	
	Mod95:S9.21(Art14) &S9.11A(Res46) / Mod92: Res46, Art14		u	d u	u		mxass eess (passive), rast,		S5.409/410/411 S5.420 (national)	
2.670 - 2.690 GHz	Mod95: S9.21(Art14) &ArtS21 / Mod95: S9.11A(Res46)		b	u	u		FSS, MSS(after 2005),		S5.149/417/412/420 S5.415 (national and regional systems) S5.419 (after 2005)	
	Mod95:S9.21(Art14) &S9.11A(Res46)		u	u	u		mxass(until 2005), MxA, F		S5.420 (national) (until 2005)	
	Mod92: Res46, Art14				u	amss (J from 2000)	eess & sres (passive), rast,		S5.409/410/411	S5.149/419/420 Add97 S5.420A (S9.21)
3.400 - 3.700 GHz			d	d			FSS			
3.400 - 3.600			d		FSS, F				S5.431/434	
3.400 - 3.500				d	d		FSS, F		S5.282/432	
3.500 - 3.700				d	d		FSS, F, MxA		S5.435	
3.600 - 4.200			d		FSS, F					

FSS/BSS/MSS/ISS Allocations (121 MHz - 275 GHz) (ARTICLE S5 of RR)

Frequency	decisions: WARC 85&88	decisions: WARC 92 &WRC 95&97	R1	R2	R3	Services R1	Services R2	Services R3	footnotes : previous numbering of Final Acts of the WRC95 R2	R3
3.700 - 4.200 GHz			d	d	d		FSS, F, MxA	F		
4.500 - 4.800 GHz	APS30B (d) (A): FSS Plan		d	d	d		FSS Plan (see 6.725 - 7.025 GHz), F, M		S5.441	
5.000 - 5.150 GHz		Mod95: Art S9.21(Art14)					ARN (MLS) AMSS(R)		S5.444/444A / S5.367	
5.091 - 5.150 GHz		Add95: NGSO-MSS(FL) &Art S9.11A(Res46) &Res114	ux	ux	ux		FSS (NGSO-MSS(FL))(primary until 2010) (see 6.700 - 7.075 GHz)		S5.444/444A (until 2010)	
5.150 - 5.250 GHz		Add95: NGSO-MSS(FL) &Art S9.11A(Res46)	ux	ux	ux		ARN FSS (NGSO-MSS(FL))		S5.447A / S5.447/447A/447C (nov 95)	
5.150 - 5.216 GHz		Add95: NGSO-MSS(FL) &Art S9.11A(Res46)	dx	dx	dx	rdss(fl)	FSS (NGSO-MSS(FL))	rdss(fl)	S5.447B / S5.446 (see S5.446: 1610.0 - 1626.5 or 2483.5 - 2500 MHz)	rdss(fl) S5.446
5.725 - 5.850 GHz			u	d	d	FSS	RL	FSS	S5.150/451/453/455	S5.150/453/455
5.830-5.850			d	d	d	amatss	amatss	amatss		
5.850 - 5.925 GHz			u	u	u		FSS, F, M		S5.150	
5.925 - 6.700 GHz			u	u	u		FSS, F, M		S5.440/458/149 / S5.440/458/149	
6.700 - 7.075 GHz			b	b	b		FSS, F, M		S5.441	
		Add95: Art S9.11A (Res46) & Art S22.5A (Res115A) (not subject to S22.2)	dx	dx	dx		FSS (NGSO MSS(FL)) (see 5.091 - 5.250 & 15.450 - 15.650 GHz)		S5.458/458A/458B/458C / S5.458B	
6.725 - 7.025	APS30B(u) (A): FSS Plan		u	u	u		FSS Plan (see 4.500 - 4.800 GHz),		S5.441	
7.025 - 7.075		Add95	u	u	u		GSO FSS & other NGSO Systems		S5.458C (18 nov 95)	
7.250 - 7.750 GHz			d	d	d		FSS(d) (see 7.900 - 8.400 GHz)		S5.441	
7.250 - 7.300 GHz			d	d	d		FSS, M			
7.250 - 7.375 GHz		Mod95: Art S9.21(Art14)	d	d	d		MSS(d) (see 7.900 - 8.025 GHz)		S5.461	
7.300 - 7.450 GHz			d	d	d		FSS, MxA			
7.450 - 7.550 GHz		Add 97: NGSO MetSat (S5.461A &462A / between 7450-7550 MHz & 7750-7850 MHz)	d	d	d		FSS, MetSat GEO, F, MxA		S5.461A (after 30 nov 97)	
7.550 - 7.750 GHz			d	d	d		FSS, F, MxA			
7.750 - 7.850 GHz			d	d	d		MetSat NGEO			
7.900 - 8.400 GHz			u	u	u		FSS(up) (see 7.250 -7.750 GHz)			
7.900 - 8.025 GHz	Art14/MSS		u	u	u		MSS(up) (see 7.250 -7.375 GHz)		S5.461	
7.900 - 8.025 GHz	Art14/MSS	Mod95: Art S9.21(Art14)	u	u	u		FSS(u), F, M		S5.461	

FSS/BSS/MSS/ISS Allocations (121 MHz -275 GHz) (ARTICLE S5 of RR)

Frequency	decisions: WARC 85&88	decisions: WARC 92 &WRC 95&97	Regions R1	R2	R3	services to which the band is allocated on a primary basis (services written in lower case are allocated on a secondary basis) R1	R2	R3	footnotes : previous numbering of Final Acts of the WRC95 R2	R3
8.025 - 8.175 GHz			u	u	u		FSS(u), F, M EESS(d)		Add97 S5.462A	Add97 S5.462A
8.175 - 8.215 GHz		Add97 EESS limit= S5.462A&Res124	u	u	u		FSS(u), F, M MetSat(u), EESS(d)		Add97 S5.462A	Mod S5.463
8.215 - 8.400 GHz			u	u	u		FSS(u), F, M EESS(d)		Add97 S5.462A	Add97 S5.462A Mod S5.463
10.700 - 10.950 GHz	AP30B(d) (A): FSS Plan	NGSO FSS: WRC97 Res 130	b	d	u		FSS Plan (see 12.75 - 13.25 GHz), NGSO FSS(d) MxA, F,		Mod S5.463 S5.441 Mod 97 S5.441	Add97 S5.462A
10.950 - 11.200 GHz			ux			FSS (BSS(FL))	FSS, NGSO FSS(d) MxA, F		S5.484	Mod 97 S5.484A
11.200 - 11.450 GHz	AP30B(d) (A): FSS Plan	NGSO FSS: WRC97 Res 130	b	d	u		FSS Plan (see 12.75 - 13.25 GHz), NGSO FSS(d) MxA, F		S5.484 S5.441	Mod 97 S5.441
11.450 - 11.700 GHz		NGSO FSS: WRC97 Res 130	ux	b	d	FSS (BSS(FL))	FSS, NGSO FSS(d) MxA, F		S5.484	Mod 97 S5.484A
11.700 - 12.500 GHz	APS30/R1: BSS Plan Space Op: Guardbands at the lower &upper bands (14MHz)	Add97-NGSO FSS:S5.487A&Res538	ux d d	d		FSS (BSS(FL))	BSS Plan (see 14.5 - 14.8 outside Europe & 17.3 - 18.1 GHz) NGSO FSS(d) F, B fss(d)		S5.484 Add97 S5.487A S5.487 Mod97 S5.492	
11.700 - 12.200	APS30/R3: BSS Plan Space Op: Guardbands at the lower &upper bands (14MHz)	NGSO FSS(d): WRC97 Res130 NGSO FSS(d): WRC97 Res538	d	d	d d		FSS(d) NGSO FSS(d) NGSO FSS(d) fss(d) F,MxA, B	BSS Plan (see 14.5-14.8 outside Europe & 17.3-18.1GHz)	S5.485/488 (Nat &Reg FSS & BSS systems) Add97 S5.484A	Add97 S5.487A Mod97 S5.492 S5.487
12.200 - 12.500				d			FSS(d) (National & sub-Reg) F, MxA, B		S5.487	
11.700 - 12.100					d		FSS(d), F		S5.486 S5.485/488	
12.100 - 12.200				d			FSS(d)		S5.485/488/489	
12.200 - 12.700 GHz	APS30/R2: BSS Plan Space Op: Guardbands at the lower &upper bands (12MHz)	Add97-NGSO FSS:S5.487A&Res538	d d d				BSS Plan (see 17.3 - 17.8 GHz) NGSO FSS(d) fss(d) F, MxA, B		S5.488/490/492 (Nat and Reg FSS &BSS systems) Add97 S5.487A Mod97 S5.492	
12.200 - 12.750 GHz		NGSO FSS: WRC97 Res 130	d	d		FSS NGSO FSS (d)		NGSO FSS(d)		Add97 S5.484A
12.500 - 12.750 GHz	WARC79 Res34	NGSO FSS: WRC97 Res 130	b d					FSS(d), F, MxA	S5.494/495/496 Add97 S5.484A	
12.500 - 12.700			d					BSS F, MxA	S5.493	

FSS/BSS/MSS/ISS Allocations (121 MHz - 275 GHz) (ARTICLE S5 of RR)

Frequency	decisions: WARC 85&88	decisions: WARC 92 & WRC 95&97	Regions R1	R2	R3	services — R1	services — R2	services — R3	footnotes R2	R3
12.700 - 12.750				u			FSS(u), F,MxA			
12.750 - 13.250 GHz	AP30B(u) (A): FSS Plan & Res 108	NGSO FSS : WRC97 Res 130	u	u	u		FSS Plan (see 10.7 - 10.950 & 11.2 - 11.450 GHz), NGSO FSS(up), F,M		S5.441, Mod 97 S5.441	
13.750 - 14.500 GHz		NGSO FSS : WRC97 Res 130								
13.750 - 14.000 GHz		Mod95: RL&RN&Sres	u	u	u		NGSO FSS(up), FSS, RL, sres		S5.33?/499/500/501/502/503/503A	
14.000 - 14.250 GHz			u	u	u		FSS & mxas &fss(bss(fl)) (outside Europe), RN		S5.506 (outside Europe), S5.504, S5.505	
14.250 - 14.500 GHz			u	u	u		FSS & mxas &fss(bss(fl)) (outside Europe)		S5.506 (outside Europe), S5.504/S5.505/508/509	
14.250 - 14.300							RN		S5.504	
14.300 - 14.400							F,MxA	F,MxA	S5.506	
14.400 - 14.470							F,MxA		S5.149/506	
14.470 - 14.500						F,MxA	F,MxA			
14.500 - 14.800 GHz	AP30A/R1&3: BSS(FL) Plan / Space Op: Guardbands at the lower &upper bands (11.8/11.80MHz)		u / ux	u	u / ux	Plan (BSS(FL)) (see 11.7 - 12.5 GHz)	FSS(BSS(FL)) (outside Europe), Plan (BSS(FL)) (see 11.7 - 12.2 GHz)		S5.510, S5.510 (outside Europe)	S5.510
15.400 - 15.430 GHz			d	d	d		F,M, ARN		Add97 S5.511D / FSS (NGSO MSS(FL)) published before 21 Nov 97	
15.430 - 15.630 GHz		Add95: NGSO-MSS(FL) &Art S9.11A(Res46) &Res116, Mod97 S5.511A&Res123(down/R.Axii), NGSO-MSS(FL): WRC97 Res46/S9.11A	b	b	b		FSS (NGSO MSS(FL)) (see 19.3 - 19.6 GHz), ARN		Mod97 S5.511A	
~~15.450 - 15.650~~		~~NGSO-MSS(FL): WRC95 Res46/S9.11A Art69.11A/Res46 & S.4.10 & Res117~~	tt	tt	tt		~~FSS (NGSO-MSS(FL)) (see 6.700 - 7.075 GHz), ARN~~		~~5HG~~	
15.630 - 15.700 GHz / 15.630-15.650 / 15.650-15.700			u	u	u		ARN		Add S5.511D / FSS (NGSO MSS(FL)) published before 21 Nov 97	
17.300 - 18.100 GHz	GSO FSS BSS(FL) / BSS(FL) Plan: see APS30A (17.8-18.1GHz: unplanned in R2) / Space Op: Guardbands at the lower &upper bands (14/11MHz)	Mod97 S5.516 & Res 538 : NGSO FSS (u)	ux	ux	ux	GSO BSS(FL) Plan, NGO FSS(u)	GSO FSS(BSS(FL)), BSS(FL) Plan (only 17.3-17.8 GHz), NGSO FSS(u) (only 17.8-18.1 GHz)	GSO BSS(FL) Plan, NGSO FSS(u)	Mod97 S5.516	
17.300 - 17.800 GHz	APS30A: BSS(FL) Plan & Art S11 / APS30A Annex4		ux	ux / d	ux / d	NGSO FSS	GSO FSS(BSS(FL)) Plans, BSS, NGSO FSS		Mod97 S5.516, Mod97 S5.515/517 & Mod97 S5.514	
17.300 - 18.100 (R1&3)	APS30A: BSS(FL) Plan/R1 &3 &Art S11(ex.Art15.A)	WRC97 Res 538 : NGSO FSS (u), Mod 97: S5.516& Res 538 (u)	ux	u	u	FSS (BSS(FL)) Plan (see 11.7 - 12.5 GHz), NGSO FSS(up)	FSS (BSS(FL)) Plan (see 11.7 - 12.2 GHz), NGSO FSS(up)		S5.516 (see 11.7 - 12.5 GHz), Mod97 S5.516, S5.516 (see 11.7 - 12.2 GHz), Mod97 S5.516	
17.300 - 17.800 (R2)	APS30A: BSS(FL) Plan/R2 &Art S11(ex.Art15.A) / Space Op: Guardbands at the lower &upper bands (12/12MHz) / APS30A Annex4	Mod92: Res 525 & 526 (HDTV)		ux	d	GSO FSS (BSS(FL)) Plan (see 12.2 - 12.7 GHz), BSS (after April 2007)			S5.515/516 (see 12.200 - 12.700), S5.517 (after April 2007)	
17.700 - 18.400 GHz			b	b	b		FSS(d) & FSS(BSS(FL)(u)),			
17.800 - 18.600 GHz		NGSO FSS : WRC97 Res 130	d	d	d		NGSO FSS (d),		Add97 S5.484A	
17.700 - 18.100 GHz	FSS(BSS(FL) Plan :APS30A		d	ux	ux		FSS(d), GSO FSS (BSS(FL)) Plan		S5.516	

FSS/BSS/MSS/ISS Allocations   (121 MHz -275 GHz) (ARTICLE S5 of RR)

Frequency	decisions: WARC 85&88	decisions: WARC 92 &WRC 95&97	Regions R1	R2	R3	services to which the band is allocated on a primary basis (services written in lower case are allocated on a secondary basis) R1	R2	R3	footnotes : previous numbering of Final Acts of the WRC95 R2	R3
								F, M	S5.518: M until April 2007 for 17.7-17.8 GHz	
17.700 - 17.800	APS30A: BSS(FL) Plan/R1&2&3		d	d	d		FSS(d),	F, M	S5.516	
			ux	ux	ux		GSO FSS (BSS(FL)) Plan		S5.517	
					d		BSS(after April 2007 & priority on FSS(d)) M(until 31/03/2007)		S5.517(after April 2007) S5.518 (until 31/03/2007)	
17.800 - 18.100	APS30A: BSS(FL) Plan/R1&3 and unplanned in R2		b	b	b	BSS(FL) Plan	FSS(d) & GSO FSS(BSS(FL)(u)),	F, M	S5.515/518	
		WRC97 Res 538 : NGSO FSS (u)	u	u	u		BSS(FL), NGSO FSS (u) F, M	BSS(FL) Plan	Mod97 S5.516	
18.100 - 18.400 GHz		Mod92: FSS(BSS(FL))	d	d	d		FSS(d) FSS(BSS(FL)) M		S5.520	
18.100 - 18.300			ux	ux	ux				S5.519/521	
		Art S21							S5.519	
18.400 - 18.800 GHz			d	d	d		GSO MeteoSat			
18.400 - 18.600 GHz			d	d	d		FSS(d)			
18.600 - 18.800 GHz			d	d	d		FSS, F, M			
							FSS, F, MxA, EESS & SRes (passive)		S5.523	
18.800 - 19.300 GHz		Add95: ex-Res 118 (ex : Teledesic) Mod97: WRC97 Res 46/S9.11A Add97: WRC97 Res 132	d	d	d		GSO & NGSO FSS (see 28.6 - 29.1 GHz),		S5.522	
18.800 - 18.900		GSO &NGSO FSS: WRC97 Res 46 Mod97: ArtS22.2(ex-2613) does not apply	d	d	d		NGSO FSS (see 28.6 - 28.7 GHz)		S5.523A (NGSO FSS) (see 28.600 - 29.100) (18 Nov 97)	
18.900 - 19.300		GSO &NGSO FSS: WRC95 Res 46	d	d	d		NGSO FSS (see 28.7 - 29.1 GHz) F, M		frozen from 17/02/96 until the end of WRC 97	
19.300 - 19.700 GHz		Add95: ex-Res120 (ex : Iridium &Odyssey)	b	b	b		FSS (see 29.1 - 29.5 GHz), GSO FSS & FSS(NGSO MSS(FL))			
19.300 - 19.600		Add95: ArtS9.11A(Res46) (18 nov 95) Art S9&S11&S22.2(exRR2613)	d	d	d		GSO FSS & FSS(NGSO MSS(FL)) (see 29.1 - 29.4 GHz) other NGSO FSS		S5.523C (NGSO MSS(FL))	
		Mod97: S22.2 applies for networks before 18 Nov 95							S5.523D (GSO-FSS & NGSO MSS(FL)(after 18 Nov 97) S5.523D	
19.600 - 19.700		S22.2(exRR2613)	d	d	d		FSS (NGSO MSS(FL)) (see 29.4 - 29.5 GHz),			
19.300 - 19.600		Add97:S22.2 applies for networks before 21 Nov 97	ux	ux	ux		FSS (NGSO MSS(FL))			
19.300 - 19.700			ux	ux	ux		FSS (NGSO MSS(FL)) F, M		S5.523B (NGSO MSS(FL))	
19.7 - 20.2 - 21.2 GHz			d	d	d		FSS(d) (see 29.5 - 30.0 - 31.0 GHz)			
19.7 - 20.2 GHz			d	d	d		FSS NGSO (d)			
19.700 - 20.100 GHz		NGSO FSS : WRC97 Res 130	d	d	d	mss	FSS MSS		Add97 S5.484A Mod97 S5.524	
20.100 - 20.200 GHz			d	d	d		FSS, MSS		S5.525/526/527/528/529	
20.200 - 21.200 GHz			d	d	d		FSS, MSS		S5.524/525/526/527/528	
21.400 - 22.000 GHz		Mod92 : Res 525 & 526 (HDTV)	d		d		F, M	BSS(after April 2007),	S5.524	
								BSS(after April 2007)	S5.530/531	S5.530 (after April 2007)

FSS/BSS/MSS/ISS Allocations (121 MHz.-275 GHz) (ARTICLE S5 of RR)

Frequency	decisions: WARC 85&88	decisions: WARC 92 &WRC 95&97	Regions R1	R2	R3	services (R1)	R2	R3	footnotes : previous numbering of Final Acts of the WRC95 R2	R3
22.550 - 23.550 GHz							ISS, F, M		S5.149	
24.450 - 24.750 GHz							ISS,			
24.450 - 24.650							ISS, RN	F, M, RN	S5.533	
24.650 - 24.750				u		F	ISS, RLSS	F, M	S5.533 & 534	
24.750 - 25.250 GHz	Mod92: BSS(FL)			u	u	F	FSS(BSS(FL)(priority) & fss(u)	F, M	S5.535 (ex-882G)	S5.534
25.250 - 27.500 GHz							ISS		S5.536	
25.250 - 25.500							ISS, F, M,		S5.536	
25.500 - 27.000							ISS, F, M, EESS		S5.536	
27.000 - 27.500 GHz			d	d	d				Add S5.536A & 536B	
27.000 - 27.500 GHz				u	u	ISS, F, M	FSS, ISS, F, M		S5.536	S5.536/537 (for non-geo: exempt from S22.2)
27.500 - 30.000 GHz		Add92 : bss(f)	u	u	u		FSS & fss(bss(f)),		S5.539	
27.501 - 29.999		Add92 : Radiobeacons	d	d	d		fss(d)		S5.540 (ex-882B)	
27.500 - 28.600 GHz		NGSO FSS : WRC97 Res 130	u	u	u		NGSO FSS (u)		S5.484A	
27.500 - 28.500 GHz		Add92 : bss(f)	u	u	u		FSS & fss(bss(f)), F, M		S5.539	
27.500 -27.501		Add92 : Radiobeacons	d	d	d		FSS(d)		S5.538/540	
28.500 - 29.100 GHz			u	u	u		FSS & fss(bss(f)), F, M		S5.539/523A	
				u	u		eess		S5.541 S5.540/541	
28.600 - 29.100	Add91-Res 118 Mod97: WRC97 Res 46/S9.11A GSO &NGSO FSS: WRC97 Res 46 S22.2(ccR R2613) until WRC97 GSO &NGSO FSS: WRC95 Res 46		u	u	u		GSO & NGSO FSS (see 18.8 - 19.3 GHz)		S5.523A	
28.600 - 28.700							NGSO FSS (see 18.8 - 18.9 GHz)		frozen from 17/02/96 until the end of WRC97	
28.700 - 29.100							NGSO FSS (see 18.9 - 19.3 GHz)			
29.100 - 29.500 GHz	Add91-Res 120 Add97: S9.11A(Res46) & Res 121 GSO &NGSO FSS: WRC97 Res 46		u	u	u		GSO FSS & FSS (NGSO MSS(FL) (see 19.3 - 19.7 GHz) FSS (NGSO MSS(FL)) (see 19.3 - 19.7 GHz)		S5.539/535A/541A/523C&E Mod 97 S5.535A & Mod 97 S5.541A (after 18 Nov 95)	
29.100 - 29.400	Mod97:S22.2 applies for networks before 18 Nov 95 S22.2(ccR R2613)						FSS (NGSO MSS(FL)) (see 19.3 - 19.6 GHz)		Mod 97 S5.523C	
29.400 - 29.500	Add97 S22.2 applies for networks before 21 Nov 97						FSS (NGSO MSS(FL)) (see 19.6 - 19.7 GHz)		Mod 97 S5.523E	
29.100 - 29.500		Add92 : bss(f)	u	u	u		FSS(bss(f)), F, M		S5.539/535A	
			u	u	u		eess		S5.541 S5.540/541	
29.5 - 30.0 - 31.0 GHz			u	u	u		FSS(u) (see 19.7 - 20.2 - 21.2 GHz)		S5.484A	
29.500 - 30.000 GHz		NGSO FSS : WRC97 Res 130	u	u	u		FSS NGSO (up)		S5.539	
29.500 - 29.900 GHz		Add92 : bss(f)	u	u	u	mss	FSS &FSS(bss(f)), MSS		S5.525/526/527/529	
							cess		S5.541	

FSS/BSS/MSS/ISS Allocations   (121 MHz -275 GHz) (ARTICLE S5 of RR)

Frequency	decisions: WARC 85&88	decisions: WARC 92 &WRC 95&97	Regions			services to which the band is allocated on a primary basis (services written in lower case are allocated on a secondary basis)			footnotes : previous numbering of Final Acts of the WRC95	
			R1	R2	R3	R1	R2	R3	R2	R3
29.900 - 30.000 GHz		Add92: bss(fl)	d	d	d		FSS &FSS(bss(fl), MSS, eess		S5.540/542 S5.539	
29.999 - 30.000			d	d	d				S5.541 S5.525/526/527/543/538/540/542	
30.000 - 31.000 GHz		Add92 Radiobeacons	d	d	d		FSS(d) FSS, MSS		S5.538 S5.542	
32.000 - 33.000 GHz							ISS			
31.800 - 33.400							F (HD applications)			
32.000 - 32.300 GHz		Add97: Res 126 & 726					ISS, F, RN,		Add97 S5.547 &547A	
32.300 - 33.000 GHz		Add97: Res 126 & 726	d	d	d		SRes (deep space) ISS, F, RN		S5.548 & Add97 S5.547C Add97 S5.547 &547A	
37.500 - 40.500 GHz		Add97: Res 126 & 726	d	d	d		FSS, eess		S5.548 &Add97 S5.547D	
37.500 - 38.000			d	d	d		FSS, eess, SRes (deep space), F, M			
38.000 - 39.500			d	d			FSS, eess, F, M			
39.500 - 40.000			d	d	d		FSS, eess, MSS, F, M			
40.000 - 40.500			d	d	d		FSS, eess MSS, F, M, EESS, SRes			
40.500 - 42.500 GHz		Add97: Res 128 & 134	d	d	d		BSS, B, F, FSS, m	BSS, B, F, m	Add S5.551D Add S5.551D	Add S5.551B/E & C/F Add S5.551C&F
42.500 - 43.500 GHz			d				FSS, F, RAst, MtA		Add S5.551C S5.552	
43.500 - 47.000 GHz							MSS, RN, RNSS, M		S5.149	
47.200 - 50.200 GHz			d				FSS, F,M, Rast		S5.553 S5.554 S5.552	
48.940 - 49.040							to reserve for bss fl (see BSS: 47.2-49.2GHz)		S5.555	
47.200 - 49.200									S5.149/340/555	

FSS/BSS/MSS/ISS Allocations (121 MHz -275 GHz) (ARTICLE S5 of RR)

Frequency	decisions: WARC 85&88	decisions: WARC 92 &WRC 95&97	Regions			services to which the band is allocated on a primary basis (services written in lower case are allocated on a secondary basis)			footnotes : previous numbering of Final Acts of the WRC95	
			R1	R2	R3	R1	R2	R3	R2	R3
47.200 - 47.500		Add97: Res 122					F (high altitude platform)		Add97 S5.552A	
47.900 - 48.200		"					"		"	
50.400 - 51.400 GHz				u	u		FSS, mss F,M			
54.250 - 58.200 GHz		Mod95: Res 643 (50 - 70 GHz)					ISS, EESS & SRes (passive),			
54.250 - 55.780		Add97: ISS : limited to GSO satellites					ISS		Add97 S5.556A Add97 S5.557A	
55.780 - 56.900		Add97: ISS : limited to GSO satellites					ISS, M, F		Add97 S5.556A Mod97 S5.558 Add97 S5.547 Mod97 S5.557	
56.900 - 57.000		Add97: ISS: limited to links between GSO and to transmissions from NGSO HEO to LEO					ISS M, F		Add97 S5.556B Mod97 S5.558 Add97 S5.547 Mod97 S5.557	
57.000 - 58.200		Add97: ISS : limited to GSO satellites					ISS M, F		Add97 S5.556A Mod97 S5.558 Mod97 S5.557 Mod97 S5.557	
55.780 - 59.000		Add97: Res 126 & 726					F (HD applications)		Mod97 S5.557	
59.000 - 71.000 GHz							ISS			
59.000 - 64.000 GHz		Mod95: Res 643					ISS, M, F RL		Mod97 S5.558 S5.559	
59.000 -59.300		Add97: ISS : limited to GSO satellites					ISS, EESS & SRe (passive)		Add97 S5.556A	
59.300 - 64.000							ISS		S5.138	
64.000 - 65.000							ISS, F (HD applications), MxA,		Add97 S5.547	
65.000 - 66.000							ISS, F (HD applications), MxA, EESS & SRe		Add97 S5.547	
66.000 - 71.000							ISS, MSS, M RNSS, RNSS,		S5.138	
71.000 - 74.000 GHz			u u u	u u u	u u u		FSS, MSS, F, M		S5.553 & Mod97 S5.558 S5.554	
74.000 - 75.500 GHz			u	u	u		FSS, F, M SRes		S5.149/556	
81.000 - 84.000 GHz			p p	p p	p		FSS, MSS, F, M SRes			

FSS/BSS/MSS/ISS Allocations (121 MHz -275 GHz) (ARTICLE S5 of RR)

Frequency	decisions: WARC 85&88	decisions: WARC 92 &WRC 95&97	Regions			services to which the band is allocated on a primary basis (services written in lower case are allocated on a secondary basis)			footnotes : previous numbering of Final Acts of the WRC95	
			R1	R2	R3	R1	R2	R3	R2	R3
84.000 - 86.000 GHz							BSS B, F, M			
92.000 - 94.000 GHz & 94.100 - 95.000 GHz				u	u		FSS F, M, RL		S5.561	
95.000 - 100.000 GHz							MSS, RNSS, RN, M		S5.149/556 S5.553	
102.000 - 105.000 GHz			p	p	p		FSS, F, M		S5.149/554/555	
116.000 - 134.000 GHz		*Mod95*					*ISS,*		S5.341	
116.000 - 119.980 GHz		Mod95					ISS,			
119.980 - 120.020 GHz		Mod95					ISS, EESS & SRes (passive), F, M		S5.558 S5.138 & 341	
120.020 - 126.000 GHz		Mod95					ISS, EESS & SRes (passive), F, amat, M		S5.558 S5.138 & 341	
126.000 - 134.000 GHz		Mod95					ISS, EESS & SRes (passive), F, M		S5.558 S5.138 & 341	
134.000 - 142.000 GHz			p	p	p		ISS, F, M, RL		S5.558 S5.559	
149.000 - 164.000 GHz			p	p	p		FSS, F, M		S5.340	
149.000 - 150.000 GHz			p	p	p		*FSS*			
150.000 - 151.000 GHz			p	p	p		FSS, F, M			
151.000 - 156.000 GHz			p	p	p		FSS, EESS & SRes (passive), F, M			
156.000 - 158.000 GHz			p	p	p		FSS, F, M		S5.149, 385	
158.000 - 164.000 GHz			p	p	p		FSS, EESS & SRes (passive), F, M			

FSS/BSS/MSS/ISS Allocations (121 MHz -275 GHz) (ARTICLE S5 of RR)

Frequency	decisions: WARC 85&88	decisions: WARC 92 &WRC 95&97	Regions			services to which the band is allocated on a primary basis (services written in lower case are allocated on a secondary basis)			footnotes : previous numbering of Final Acts of the WRC95	
			R1	R2	R3	R1	R2	R3	R2	R3
170.000 - 182.000 GHz							*ISS*			
170.000 - 174.500 GHz							ISS, F, M		S5.558 S5.149 & 385	
174.500 - 176.500 GHz							ISS, EESS & SRes (passive), F, M		S5.558 S5.149 & 385	
176.500 - 182.000 GHz							ISS, F, M		S5.558 S5.149 & 385	
185.000 - 190.000 GHz							ISS, F, M		S5.558 S5.149 & 385	
190.000 - 200.000 GHz							MSS, RNSS, RN, M		S5.553 S5.341 & 554	
202.000 - 217.000 GHz			u	u			FSS, F, M		S5.341	
231.000 - 241.000 GHz			*p*	*p*			*FSS(d)*			
231.000 - 235.000 GHz			p	p			FSS, F, M, rl			
235.000 - 238.000 GHz			p	p			FSS, EESS & SRes (passive), F, M			
238.000 - 241.000 GHz			p	p			FSS, F, M, rl			
252.000 - 265.000 GHz							MSS, RNSS, RN, M			
265.000 - 275.000 GHz			u	u			FSS, F, M, RAst		S5.149/385/554/555 & Mod97 S5.564	
275.000 - 400.000 GHz							not allocated		S5.149	
									S5.565	

NOTE - Read this table line per line; for a same colour (grey or white) the services allocated at each region and the footnotes corresponding are given.

Glossary

footnotes: summary (relating to the FSS or BSS services)

NOTE - S-numbers are used for the simplified provisions of the RR

S5.43	Where it is indicated in RR that a service may operate in a specific frequency band subject to not causing harmful interference, this means that this service cannot claim protection from interference caused by other services to which the band is allocated
S5.138	These (Industrial, Scientific and Medical) ISM applications shall be subject to special authorization by the administration concerned.
S5.149 (Mod95&97)	Take all practicable steps to protect the RAstS from harmful interference of others services to which the bands are allocated (see the list of these bands in RR)
S5.150	ISM equipment operating in these bands is subject to the provisions of n° S15.13. (see the list of these bands in RR)
S5.199	The bands 121.450-121.550 MHz and 242.950-243.050 MHz are also allocated to the MSS but are limited to emergency position-indicating radiobeacons. (see ApS13)
S5.208 (Add92+Mod95&97)	The use of the band 137-138 MHz by the MSS is subject to coordination under n° S9.11A and to Rev WRC95 Res 46 . The provisions of WRC95 Res 714/COM5 apply for the pfd limitations.
S5.208A (Add95)	Take all practicable steps to protect the RAstS from unwanted emissions from harmful interference of the MSS (see the list of these bands in RR)
S5.209 (Add92+Mod95&97)	The use of the bands 137-138 MHz, 148-149.900 MHz, 400.150-401 MHz, 455-456 MHz and 459-460 MHz by the MSS and the bands 149.900-150.050 MHz and 399.900-400.050 MHz by the LMSS is limited to non-GEO satellites.
S5.218	The band 148-149.9 MHz is also allocated to the Space Operation Service (Earth-to-space)
S5.219 (Add92+Mod95&97)	The use of the band 148-149.900 MHz by the MSS is subject to coordination under n° S9.11A/Res46
S5.220 (Add92+Mod95&97)	The use of the bands 149.900-150.050 MHz and 399.900-400.050 MHz by the LMSS is subject to coordination under n° S9.11A and shall not constrain the use of the F, M, and Space Services in the band 148-149.900 MHz
S5.221 (Add92+Mod95)	The use of the bands 148-149.900 MHz by the MSS shall not cause harmful interference to the F and M Services in the band 148-149.900 MHz
S5.224 (Add92+Mod95+Sup97)	The allocation of the bands 149.900-150.050 MHz and 399.900-400.050 MHz for the LMSS shall be on a secondary basis until 1 January 1997.
S5.224A (Add97)	The use of the bands 149.90-150.05 MHz and 399.90-400.05 MHz by the MSS (Earth-to-space) is limited to the LMSS (Earth-to-space) until 1 January 2015.
S5.224B (Add97)	The allocation of the bands 149.90-150.05 MHz and 399.90-400.05 MHz to the RNSS shall be effective until 1 January 2015.
	The bands 235-322 MHz and 335.400-399.900 MHz may be used by the MSS, subject to agreement obtained under n°S9.21 without harmful interference to others services.
S5.254 (Mod95)	The band 312-315 MHz (Earth-to-space) and 387-390 MHz (space-to-Earth) may also be used by the non-GEO MSS, subject to coordination under n°S9.11A.
S5.255 (Add92+Mod95)	The band 267-272 MHz may be used by administrations for space telemetry in their countries on a primary basis subject to agreement obtained under Article 14/S9.21
S5.257	The use of the band 400.150-401 MHz by the MSS is subject to coordination under n° S9.11A/Res46.
S5.264 (Add92+Mod95)	The band 406-406.100 MHz is also allocated to the MSS but is limited to emergency position-indicating radiobeacons. (see also Art S31 and ApS13)
S5.266	Any harmful interference in the band 406-406.100 MHz is prohibited
S5.267	The use of the bands 455-456 MHz and 459-460 MHz by the MSS is subject to coordination under n° S9.11A/Res46
S5.286A (Add95+Mod97)	The use of the bands 455-456 MHz and 459-460 MHz by the MSS shall not cause harmful interference to, or claims protection from the F and M services.
S5.286B (Add95+Mod97)	The use of the bands 455-456 MHz and 459-460 MHz by the MSS shall not constrain the use of the F and M services.
S5.286C (Add95+Mod97)	Assignments may be made to TV-FM in the BSS within the frequency band 620-790 MHz, subject to agreement with the administrations concerned. (see Res 33 & 507, and Rec 705).
S5.311	In Region 2, (except B and USA), the band 806-890 MHz is also allocated to the MSS within national boundaries,on a primary basis and is subject to agreement obtained under n°S9.21 /Art14
S5.317 (Mod95)	In Region 1 (BLR, RUS, UKR), the bands 806-890 MHz(Earth-to-space) and 942-960 MHz (space-earth) are also allocated to the MSS (except AMSS(R))
S5.319 (Mod95)	and the use of these bands shall not cause harmful interference to, or claim protection from other services.

S5.320 (Mod95) (Art14)	The bands 806-890 MHz and 942-960 MHz are also allocated to the MSS (except AMSS(R)) in Region 3, on a primary basis and for national applications and are subject to agreement obtained under n°S9.21
S5.340 (Mod95&97)	All emissons are prohibited in these bands (see the list of these bands in RR)
S5.345	Allocated to the BSS at the WARC-92
S5.347 (Mod95)	The use of the band 1452-1492 MHz by the BSS is limited to (Digital Audio Broadcasting) DAB and subject to Res 528 (WARC92)
	Allocated to the BSS at the WARC-92
S5.348 (Mod95)	The use of the band 1452-1492 MHz by the BSS is limited on a secondary basis until 1 April 2007 in some countries (mainly in Regions 1 and 3) (WARC92)
S5.348A (Add95)	The use of the band 1492-1525 MHz by the MSS is subject to coordination under n° S9.11A
S5.351	In the band 1492-1525 MHz, the pfd , in application of Rev WRC95 Res 46 shall be -150 dBW/m2 in any 4 kHz band for angles of arrival.
Sup97 S5.352	The bands 1525-1544 MHz, 1545-1559 MHz, 1626.500-1645.500 MHz and 1646.500-1660.500 MHz shall not be used for FL of any service.
S5.352A (Add97)	The use of the bands 1525-1544 MHz, 1545-1559 MHz, 1626.500-1645.500 MHz, 1626.500-1645.500 MHz by the LMSS is limited to data transmissions
	In the band 1525-1530 MHz, stations in the MSS (except in the MMSS) shall not cause harmful interference to, or claim protection from, stations of the Fixed Service in certain countries in R1&3.
Sup97 S5.353 (Mod95)	In several countries (mainly in Region 2) , the band 1530-1544 MHz is also allocated to the MSS (space-to-Earth) and the band 1631.5-1645.5 MHz is also allocated to the MSS (Earth-to-space) on a primary basis ,
S5.353A (Add97)	priority of GMDSS distress and safety communications.
	In applying the procedures of No S9.11 A to the MSS in the bands 1530-1544 MHz and 1626.5-1645.5 MHz, priority shall be given to the GM DSS (priority 1 to 6 in RR Article S44) over all other MSS communications.
	which shall not cause unacceptable to, or claim protection from, AMS(R)S communications.(see WRC97 Resolution 218)
S5.354 (Mod95)	The use of the bands 1525-1559 MHz and 1626.500-1660.500 MHz by the MSS is subject to coordination under n° S9.11A /Res46 except S9.13
S5.356	The use of the band 1544-1545 MHz by the MSS (space-to-earth) is limited to distress and safety communications (see Article N38/S31)
S5.357	In the band 1545-1555 MHz, transmissions are also used to extend or Supplement the satellite-to-aircraft.
S5.360	In the bands 1555-1559 MHz and 1656.5-1660.5 MHz aircraft and ship earth stations may also communicate with space stations in the LMSS (Mob87 Res 208)
S5.361	Alternative allocation: In Australia, Canada and Mexico, the band 1555-1559 MHz, is allocated to the MSS (space-to-Earth) and the band 1656.5-1660.5 MHz is allocated to the MSS (Earth-to-space)
	and the band 1660-1660.5 MHz is allocated to the MSS (Earth-to-space) and the RAstS, on a primary basis.
	Alternative allocation: In Argentina and the USA, the band 1555-1559 MHz, is allocated to the MSS (space-to-Earth) and the band 1656.5-1660.5 MHz is allocated to the MSS (Earth-to-space)
Sup97 S5.362	and the band 1660-1660.5 MHz is allocated to the MSS (Earth-to-space) and the RAstS, on a primary basis.
	The AMSS(R) shall have priority access over all other MSS communications. In account of the priority of MMSS distress and safety communications.
S5.362A (Add97)	In applying the procedures of No S9.11 A to the MSS in the bands 1545-1555 MHz and 1646.5-1656.5 MHz, priority shall be given to the AMS(R)S (priority 1 to 6 in RR Article S44) over all other MSS communications.
	which shall not cause unacceptable to, or claim protection from, AMS(R)S communications.(see WRC97 Resolution 218)
S5.362B (Add97)	In the USA, in the bands 1555-1559 MHz and 1656.5-1660.5 MHz, priority shall be given to the AMS(R)S (priority 1 to 6 in RR Article S44) over all other MSS communications,
	which shall not cause unacceptable to, or claim protection from, AMS(R)S communications.
S5.364 (Mod95)	The use of the band 1610-1626.500 MHz by the MSS (Earth-to-space)and by the RDSS (Earth-to-space) is subject to coordination under n° S9.11A/Res46 and shall make all practicable efforts to ensure protection of stations operating in accordance with the provisions of n° S5.366
	Stations of the MSS shall not claim protection from stations in the ARNS. (see S5.356)
S5.365 (Mod95)	The use of the band 1613.800-1626.500 MHz by the MSS (space-to-Earth) is subject to coordination under n° S9.11A/Res46
S5.372	In the band 1610.600-1613.800 MHz, earth stations of the RDetS and MSS, shall not cause harmful interference to the stations in the RAst service.
Sup97 S5.373A (Add95)	In Argentina and the USA, the use of the band 1626.500-1631.500 MHz by the MSS (space-to-Earth) is subject to the condition of Sup97 n° S5.353.

S5.374	In the bands 1631.500-1634.500 MHz and 1656.500-1660 MHz ,land and ship earth stations in the MSS, shall not cause harmful interference to the stations in the FS.
S5.375	The use of the band 1645.5-1646.5 MHz by the MSS (Earth-to-space) and the ISL is limited to distress and safety communications.
S5.377 (Mod95)	In the band 1675-1710 MHz, stations in the MSS shall not cause harmful interference to the MetSS and shall be subject to coordination under n°S9.11A.
S5.388 (Mod95&97)	The bands 1885-2025 and 2110-2200 MHz are intended for use, on a worldwide basis, for IMT2000 (ex-FPLMTS) in accordance with Revised WRC95&97 Resolution 212
S5.389A (Add95)	The use of the bands 1980-2010 MHz and 2170-2200 MHz by the MSS is subject to coordination under Rev WRC97 Res 46/n° S9.11A and to the provisions of WRC95 Res 716
	The use of these bands shall not commence before 1 January 2000 and for the band 1980-1990 MHz in Region 2 shall not commence before 1 January 2005
S5.389B (Add95)	The use of the band 1980-1990 MHz by the MSS shall not cause harmful interference to the FS and MS in many countries in Region 2 .
S5.389C (Add95+Mod97)	The use of the bands 2010-2025 MHz and 2160-2170 MHz by the MSS shall not commence before 1 January 2002 and is subject to coordination under Rev WRC97 Res 46/n° S9.11A and to the provisions of WRC95 Res 716
S5.389D (Add95)	In Canada and the USA, the use of the bands 2010-2025 MHz and 2160-2170 MHz by the MSSshall not commence before 1 January 2000
S5.389E (Add95)	The use of the bands 2010-2025 MHz and 2160-2170 MHz by the MSS in Region 2 shall not cause harmful interference to the FS and MS in Regions 1and 3.
S5.389F (Add95)	In several countries of Southern Region 1 , the use of the bands 1980-2010 MHz and 2170-2200 MHz by the MSS shall cause harmful interference to the FS and MS, prior to 1 January 2005.
S5.392	The use of the bands 2025-2110 MHz and 2200-2290 MHz by the space-to-space transmissions between 2 or more NGSO satellites shall not impose any constraints on transmissions of SR, SOp,and EESS and between GEO and NON-GEO satellites.
S5.393 (WARC92+Mod97)	Additional Allocation: In India and Mexico, and in the USA, the use of the band 2310-2360 MHz is also allocated to the BSS(sound) and terrestrial BS(sound) on a primary basis and is limited to DAB (Digital Audio Broadcasting) and is subject to the provisions of WARC92 Res528.
S5.396 (WARC92)	Space stations of the BSS(sound) in the band 2310-2360 MHz and the RDetSS is subject to the condition of WARC92 Res 528
S5.402 (WARC92+Mod95)	The use of the band 2483.5-2500 MHz by the MSS and the RDetSS is subject to the coordination under n°S9.11A .
	Take all practicable steps to protect the RAst service from unwanted emissions from harmful interference of the MSS and especially into the 4990-5000 band by second-harmonic radiation.
S5.403 (Mod95) (Art14&RevWRC95Res46)	The band 2520-2535 MHz (and until 1 January 2005 the band 2500-2535 MHz) may also be used for the MSS (space-to-Earth), except AMSS within national boundaries, and is subject to the provisions of n° S9.11A and Res 46.
S5.403A (Add97)	Additional Allocation:In Japon, subject to agreement obtained under No S9.21, the band 2515-2535 MHz may also be used for the AMSS (space-to-Earth) for operation limited to within its national boundary from 1 January 2000.
S5.407 (WARC92)	In the band 2500-2520 MHz, the pfd for the MSS (space-to-Earth) shall not exceed -152dBW/m2/4kHz in Argentina, unless otherwise ageed by the administrations concerned.
S5.414 (Mod95) (Rev.WRC95 Res46)	The allocation of the band 2500-2520 MHz to the MSS (space-to-Earth) shall be effective on 1 January 2005 and is subject to coordination under n° S9.11A
S5.416 (WARC92+Mod95)	The use of the band 2520-2670 MHz by the BSS is limited to national and regional systems for community reception, and is subject to agreement obtained under n°S9.21(see Art S21 for the values of pfd)
S5.417 (Mod92+Mod95)	In Germany and Greece, the band 2520-2670 MHz is allocated to the FS on a primary basis
S5.419 (WARC92+Mod95) (RevWRC95Res46)	The allocation of the band 2670-2690 MHz to the MSS shall be effective on 1 January 2005 and is subject to coordination under n° S9.11A
S5.420 (Mod92+95) (Art14&RevWRC95Res46)	The band 2655-2670 MHz (and until 1 January 2005 the band 2655-2690 MHz) may also be used for the MSS (Earth-to-space), except AMSS within national boundaries, and is subject to agreement obtained under n° S9.21 and to coordination n° S9.11A.
S5.420A (Add97)	Additional Allocation:In Japon, subject to agreement obtained under No S9.21, the band 2670-2690 MHz may also be used for the AMSS (space-to-Earth) for operation limited to within its national boundary from 1 January 2000.
S5.441 (Mod95&97)	The use of the bands of FSS Plans 4500-4800 MHz(space-to-Earth) , 6725-7025 MHz (Earth-to-space), 10.7-10.95 GHz (space-to-Earth) and 11.2-11.45 GHz (space-to-Earth) and 12.75-13.25 GHz (Earth-to-space) shall be in accordance with the provisions of APS30B.

S5.441A (Add97)	The use of these bands of FSS Plans at 10-11-12-13 GHz by GSO FSS systems shall be in accordance with the provisions of RR Appendix S30B and by NGSO FSS systems shall be in accordance with the provisions of WRC97 Resolution 130.
	The use of the bands 10.95-11.20 and 11.45-11.70 GHz (space-to-Earth), 11.7-12.2 GHz (space-to-Earth) in Region 2, 12.2-12.75 GHz in Region 3, 12.5-12.75 GHz in Region 1,
	13.75-14.5 GHz (Earth-to-space), 17.8-18.6 GHz and 19.7-20.2 GHz(space-to-Earth), 27.5-28.6 GHz and 29.5-30 GHz (Earth-to-space) by NGSO and GSO FSS systems is subject to the provisions of WRC97 Resolution 130.
S5.444A (Add95)	The use of the band 17.8-18.1 GHz (space-to-Earth) by NGSO FSS systems is also subject to the provisions of WRC97 Resolution 538
S5.446	The band 5091-5150 MHz is also allocated to the FSS (Earth-to-space) on a primary basis; This additional allocation is limited to FL of non-GEO MSS and is subject to coordination under WRC95 Res46/S9.11A
	The band 5150-5216 MHz is also allocated to the RDSS (space-to-Earth) on a primary basis and in some countries of R1&3 on a secondary basis.
	This additional allocation is limited to FL of RDSS operating in the bands1610-1626.5 and 2483.5-2500 MHz
S5.447A (Add95)	The band 5150-5250 MHz is allocated to the FSS (Earth-to-space) on a primary basis but is limited to FL of non-GEO MSS and is subject to coordination under WRC95 Res 46/S9.11A.
S5.447B (Add95)	The band 5150-5216 MHz is also allocated to the FSS (space-to-Earth) on a primary basis; This additional allocation is limited to FL of NON-GEO MSS and is subject to provisions of WRC95 Res46/S9.11A
S5.447C (Add95)	In the band 5150-5250 MHz the FSS networks under S5.447A&B shall coordinate on an equal basis with NON-GSO RDSS prior to 17 Nov 95 in accordance with WRC95 Res46/S9.11A
	and must not be interfered by this service after 17 Nov 95.
S5.458A (Add95)	The use of the band 6700-7075 MHz by the FSS involve to take all practicable steps to protect the RAstS in the band 6650-6675.2 MHz, from unwanted emissions from harmful interference of the FSS
S5.458B (Add95)	The use of the band 6700-7075 MHz is limited to FL for the non-GEO MSS and is subject to coordination under S9.11A.
S5.458C (Add95)	In the band 7025-7075 MHz (Earth-to-space), a consultation shall be necessary to facilitate shared operation of both GEO in the FSS and non-GEO in this band.
S5.461 (Mod95)	The bands 7250-7375 MHz (space-to-Earth) and 7900-8025 MHz (Earth-to-space) are also allocated to the MSS on a primary basis subject to agreement obtained under n°S9.21.
S5.461A (Add97)	The use of the band 7450-7750 MHz by the Meteosat (space-to-Earth) is limited to GSO systems. NGSO Meteosat notified before WRC97 in this band may continue to operate on primary basis until the end of their lifetime.
S5.461B (Add97)	The use of the band 7750-7850 MHz by the Meteosat (space-to-Earth) is limited to non-GSO systems.
S5.462 (Add97)	In the band, the pfd limits specified in RR Article S21 Table S21-4 shall apply in Regions 1 and 3 to the EESS.
S5.462A (Add97)	In Regions 1 & 3, in the band 8025-8400 MHz, the EESS using GSO systems shall not produce a pfd in excess of the following provisional limits for angles θ of arrival: -174 dBW/m² +0,5(θ-5) in a 4KHz/m²°
	Aircraft stations are not permitted to transmi 8025- 6 8400 MHz in the band
S5.484	In Region 1, the use of the band 10.7-11.7 GHz (Earth to space) by the FSS up-links is limited to the BSS feeder links
S5.485	In Region 2, the use of the BSS band 11.7-12.2 GHz (space-to-Earth) by the FSS may be used additionally for transmissions in the BSS provided that such transmissions do not cause more interference than the coordinated FSS transmissions.
S5.487	In Regions 1 and 3, the use of the band 11.7-12.5 GHz (space to Earth) by the fss shall be in accordance with the provisions of RR APPENDIX S30. without interference into BSS stations operating in accordance with the provisions of RR APPENDIX S30.
S5.487A (Add97)	Additional Allocation: In each Region, the respective bands (space-to-Earth) of the BSS Plans are also allocated to the FSS (space-to-Earth) on a primary basis, limited to the NGSO FSS and subject to the provisions of WRC97 Resolution COM5-19.
	(the band 11.7-12.5 GHz in Region 1, the band 12.2-12.7 GHz in Region 2 and the band 11.7-12.2 GHz in Region 3)
S5.488 (Orb85)	The use of the bands 11.7-12.2 GHz (space to Earth) in Region 2 by the FSS and 12.2-12.7 GHz by the BSS in Region 2 is limited to national and subregional systems.
S5.491 (Mod95)	In Region 3, the use of the band 12.2-12.5 GHz (space to Earth) is also allocated to the FSS on a primary basis, limited to National and sub-Regional systems.
S5.492 (Orb85+Mod97)	In Region 2, The use of the bands 12.2-12.7 GHz of BSS Plans (space to Earth) by the FSS may also be used for transmissions in the fss. provided that such transmissions do not cause more interference than the planned BSS transmissions.
S5.493 (Orb85+Mod97)	In Region 3, the use of the band 12.5-12.75 GHz (space to Earth) by the BSS is limited to a PFD that shall not exceed the value of -111 dBW/m²/27MHz .

S5.502 (Mod92+95)(ex855A)	In the band 13.75-14 GHz: protection of RL and RN Services with the FSS ((Earth to space): 68<eirp e/s<85 dBW and diameter ≥4.5m
S5.503 (Mod92+95)(ex855B)	In the band 13.75-14 GHz, the geo space stations of the SR for which Advance Publication has been received by the Bureau prior to 31 January 1992 shall operate on an equal basis with stations of the FSS.
S5.503A (Add95)	Until 1January 2000, the FSS shall not cause interference to no-geo space stations in the SR and EESS.
S5.506	This band may be used for feeder links of the BSS for countries outside Europe (Orb88)
S5.510	The use of this band is limited to feeder links of the BSS for countries outside Europe (Orb88)
S5.511A (Add95+Mod97)	The use of the band 15.43-15.63 GHz by the FSS (space-to-Earth: see WRC97 Res COM5-8 and WRC95 Res 116) and (Earth-to-space: see WRC95 Res 117) is limited to the NGSO SMS feeder links and and is subject to coordination under n°S9.11A / Rev WRC97 Res 46.
S5.511C (Add95+Mod97)	In the space-to-Earth direction, the minimum earth station elevation angle above and gain towards the local horizontal plane and the minimum coordination distances to protect an earth station from harmful interference shall be in accordance with ITU-R Rec S.1341: Harmful interference shall not be caused to stations of the RA using the band 15.35-15.40GHz, the PFD limits are given in ITU-R Rec RA.769.
S5.511D (Add95+Mod97)	Stations operating in the ARNS shall limit the effective eirp in accordance with ITU-R Rec S.1340 . In the Earth-to-space direction, the maximum eirp transmitting by a FL earth station towards the local horizontal plane and the minimum coordination distances to protect the ARNS (No S4.10 applies) from harmful interference from the FL earth stations shall be in accordance with ITU-R Rec S.1340,
S5.515 (Add92)	Before WRC97, the band 15.4-15.7 GHz (space-to-Earth) and the band 15.45-15.65 GHz (Earth-to-space) were allocated to the FSS on a primary basis, . and this use is limited to feeders links of NGSO SMS with certain limits and protection (see WRC95 Resolution 116 & 117) In Region 2, the sharing in the band 17.3-17.8 GHz between the FSS (Earth-to-space) and the BSS (space-to-Earth) shall also be in accordance with the provisions of ApS30A
S5.516 (Add92+Mod97)	The use of the band 17.3-18.1 GHz (Earth to space) is limited to the BSS feeder links (WARC92). For the use of the band 17.3-17.8 GHz in Region 2 by the BSS FL with the band 12.2-12.7 GHz for the BSS, see RR Article S11.
S5.517 (Add92)	The use of the bands of GSO BSS FL Plans 17.3-18.1GHz (Earth-to-space) in Regions 1&3 and the use of the GSO BSS FL 17.8-18.1 GHz (Earth-to-space) in Region 2 by NGSO FSS systems is subject to the provisions of WRC97 Resolution COM5-19.
S5.520 (Add92)	The band 17.7-17.8 GHz (Earth-to-space) is allocated to the MSS on a primary basis until 31 March 2007.
S5.520 (Add92)	Additional Allocation: The use of the band 18.1-18.3 GHz (space-to-Earth) is also allocated to the GSO Meteosat and shall be in accordance with the provisions of RR Article S21 (Table S21..4)
S5.523	The use of the band 18.1-18.4 GHz (Earth to space) is limited to the BSS feeder links (WARC92)
S5.523A (Add95+Mod97)	The pfd at the at the Earth's surface in the band 18.6-18.8 GHz (space-to- Earth) is limited as far as practicable in order to reduce the risk of interference in the EESS and Space Research. The use of the bands 18.8-19.3 and 28.6-29.1 GHz by the GSO 1 NGSO FSS is subject to the application of the provisions of No S9.11A/ Rev WRC97 Res 46 and No S22.2/No 2613 (see also WRC97 Resolutions 131 and 132 and Sup97 WRC95 Res 118).
S5.523B (Add95)	GSO systems under coordination prior to WRC95 shall cooperate to coordinate pursuant to No S9.11A / Res 46 with NGSO systems for which notification information has been send prior to WRC95. NGSO systems shall not cause unacceptable interference to GSO FSS systems.under coordination prior to WRC95.
S5.523C (Add95+Mod97)	The use of the band 19.3-19.6 GHz (Earth-to-space) is limited to the non-GSO-SMS feeder links and is subject to the application of the provisions of NGSO-MSS systems S9.11A. The use of the band 19.3-19.6 GHz (space-to- Earth) and 29.1-29.5 GHz (Earth-to-space) by GSO FSS systems and by the feeder links of NGSO-MSS systems for which coordination information has been send prior to WRC95, is subject to the application of the provisions of RR No S22.2. (see Sup97 WRC95 Res 120).
S5.523D (Add95+Mod97)	The use of the band 19.3-19.7 GHz (space-to- Earth) by GSO-FSS and by the feeder links of the non-GSO-MSS is subject to the application of the provisions of WRC95 Res46 / n°.S9.11A but not subject to the provisions of No S22.2.
S5.523E (Add97)	The use of this band for other non-GSO-FSS shall continue to be subject to Art S9 (except S9.11A) and S11 procedures and to the provisions of No.22.2. The use of the band 19.6-19.7 GHz (space-to- Earth) and 29.4-29.5 GHz (Earth-to-space) by GSO FSS systems and by the feeder links of NGSO-MSS systems for which coordination information has been send prior to WRC95, is subject to the application of the provisions of RR No S22.2. (see Sup97 WRC95 Res 120).

S5.524 (Add97)	Additional Allocation: The use of the band 19.7-21.2 GHz in certain countries is also allocated to the F &M services on a primary basis. This additional use shall not impose any limitation on the pfd of space stations in the FSS and MSS.
S5.525 (Add92)	In order to facilitate inter-regional coordination between networks in the MSS and FSS, interfering carriers in the MSS shall be located in the higher of the bands 19.7-20.2 GHz and 29.5-30 GHz.
S5.526 (Add92)	In the bands 19.7-20.2 GHz and 29.5-30 GHz in R2, and in the bands 20.1-20.2 GHz and 29.9-30 GHz in R1 & 3, networks which are both in the FSS and in the MSS may include links between earth stations through one or more satellites for p/p and p/mp com.
S5.527 (Add92)	In the bands 19.7-20.2 GHz and 29.5-30 GHz, the provisions of No.S4.10 do not apply with respect to the MSS
S5.528 (Add92)	The use of the band 17.7-20.1 GHz in R2 and of the band 20.1-20.2 GHz by the MSS is intended for use by networks with narrow spot-beam antennas and other advanced technology at the space stations.
S5.529 (Add92)	In Region 2, the use of the bands 19.7-20.1 GHz and 29.5-29.9 GHz by the MSS is limited to satellite networks which are both in the FSS and in the MSS
S5.530 (Add92)	This allocation to the BSS shall come into effect on 1 April 2007 (WARC Res 525)
S5.533	The use of the band 21.4-22 GHz by the ISS is limited to Regions 1 and 3
	The ISS shall not claim protection from interference from the RNSS.
S5.535 (Add99)	The use of the band 24.65-24.75 GHz by the ISS is limited to Regions 1 and 3
	In this band 24.75-25.25 GHz, the BSS feeder links shall have priority over uses in the FSS (Earth-to-space) (WARC92)
S5.535A (Add95+Mod97)	The use of the band 29.1-29.5 GHz (Earth-to-space) is limited to the GSO-FSS and the NGSO-SMS feeder links and is subject to the application of the provisions of Rev97 WRC95 Res 46 / No.S9.11A bu not subject to the provisions of No S22.2.
S5.536 (Add92)	The use of the band 25.25-27.50 GHz by the ISS is limited to SRe and EESS applications and data transmitting from Industrial and Medical Activities in Space.
S5.537 (Add92)	Space Services using non-GEO satellites operating in the ISS in the band 27-27.5 GHz are exempt from the provisions of No S22.2.
S5.538	The bands 27.500-27.501 and 29.999-30.000 GHz (space-to-Earth) are also allocated to the FSS on a primary basis for the beacons.
S5.539 (Add92)	The band 27.5-30.0 GHz (Earth-to-space) may be used by the FSS for the BSS feeder links (WARC92)
S5.540 (Add92)	The band 27.501-29.999 GHz is also allocated to the FSS (space-to-Earth) on a secondary basis for beacon transmissions intended for up-link power control.
S5.541 (Add92)	In the band 28.5-30 GHz (Earth-to-space), the EESS is limited to the transfer of data.
S5.541A (Add95+Mod97)	In the band 29.1-29.5 GHz (Earth-to-space) , the feeder links of the GSO-FSS and the non-GSO-SMS networks operating shall employ uplink adaptative power control or others methods of fade compensation.
	These methods shall apply to neworks for which APS4 coordination information is considered as having been received by the BR after 17 May 1996. These methods are also subject to review by the ITU-R (see Rev97 WRC95 Resolution 121)
S5.543 (Mod92)	The band 29.95-30 GHz may be used for space-to-space links for the EESS TTC on a secondary basis.
S5.547 (Add97)	The bands 31.8-33.4 GHz, 51.4-52.6 GHz, 55.78-59 GHz and 64-66 GHz are available for HD applications in the F service (see WRC97 Resolution 726)
S5.547A (Add97)	The use of the band 31.8-33.4 GHz by the F service shall be in accordance with WRC97 Resolution 126.
S5.548 (Mod92)	All necessary measures shall be taken to prevent interference between the ISS and RNSS in the band 32-33 GHz and the SR (deep space) in the 31.8-32.3 GHz.
S5.551B (Add97)	The use of the band 41.5-42.5 GHz by the FSS (space-to-Earth) is subject o WRC97 Res 128.
S5.551D (Add97)	Additional Allocation: The use of the band 40.5-42.5 GHz by the FSS (space-to-Earth) in certain countries in Region 1 shall be in accordance with WRC97 Res 134.
S5.551E (Add97)	The use of the band 40.5-42.5 GHz by the FSS (space-to-Earth) IN Region 3 shall be in accordance with WRC97 Res 134.
S5.552	To take all practible steps to reserve the band 47.2-49.2 GHz for FL for the BSS operating in the band 40.5-42.5 GHz
S5.552A (Add97)	The use of the bands 47.2-47.5 GHz and 47.9-48.2 GHz is designated for use by high altitude platform stations and is subject to the provisions of WRC97 Resolution 122
S5.553	In the bands 43.5-47 GHz, 66-71 GHz, 95-100 GHz, 134-142 GH z, 190-200 GHz and 252-265 GHz, stations in the LM Service may be operated but shall not cause interference to the Space Radiocom Services (see S5.43)
S5.554	Satellite links connecting land stations at specified fixed points are also authorized when used with the MSS or the RNSS in the bands 43.5-47 GHz, 66-71 GHz, 95-100 GHz, 134-142 GHz, 190-200 GHz & 252-265 GHz.
S5.556A (Add97)	The use of he bands 54.25-56.9 GHz, 57-58.2 GHz and 59-59.3 GHz by the ISS is limited to satellites in the GSO. (the pfd shall not exceed -147 dBW/m²/100 MHz)

920

S5.558A (Add97)	The use of the band 56.9-57 GHz by the ISS is limited to links between satellites in GSO and to transmissions from NGSO systems in HEO to those in LEO. (for links between satellites in the GSO, the pfd shall not exceed -147 dBW/m²/100MHz).
S5.558 (Mod97)	In the bands 55.78-58.20 GHz, 59-64 GHz, 66-71 GHz, 116-134 GHz, 170-182 GHz and 185-190 GHz , stations in the ANM Service may be operated but shall not cause interference to the ISS (see S5.43)
S5.559	In the bands 59-64 GHz and 126-134 GHz, airborne radars in the RL Service may be operated but shall not cause interference to the ISS (see S5.43)
S5.561	In the band 84-86 GHz, stations in the F, M, and B services shall not cause harmful interference to BSS
S5.565	The band 275-400 GHz may be used for experimentation with various active and passive services.

footnotes: category of frequency allocation

Additional allocation (country or service): (the band is also allocated to a service in a particular country or in an area smaller than a Region)

S5.207/218/262/263/271/318/339/341/344/343/342/350/355/357/358/359/373/381/384/385/393

S5.404/405/408/431/437/451/453/455/456/458/489/491/494/495/496/499/

S5.500/501/505/508/509/514/519/521/524/531/534/540/542/553/555/557&557A/564

Alternative allocation : (the band is allocated to one or more services in a particular country or in an area smaller than a Region)

S5.344/361/362/412/417/520/547C&D

Different category of service: (the allocation of the band is to another service in a particular country)

S5.204/205/206/347/349/369/370/382/397/400/432/464/518

Other use of bands and Constraints :

(the use of the service in the band may also or is limited to or is authorized or shall not cause interference to, shall be subject to the condition that, or is exempt from the provisions of ....)

S5.149/222/223/333/372/376/389B/E/F/398/399/413/416/417/418/434/435/440/462/463/

504/511B/525/526/528/529/533/534/535/538/543/553/554/558/559/561/565

Symbols and abbreviations

Articles, Appendices, Resolutions or Recommendations of RR

Article S4 of RR   ex-Articles 6 & 9 of RR: Assignment and Use of frequencies

S4.4   Any frequency in derogation shall not cause harmful interference to, and shall not claim protection from harmful interference caused by a station operating in accordance with the provisions of the Regulations.

u	Earth-to-space		F	Fixed Service
d	space-to-Earth		B	Broadcasting Service
b	both-way			
s	space-to-space (in the SIS)			
x	link in the FSS limited to FL for the BSS or MSS (up link for the BSS and up or down links for the MSS)		M	Mobile Service
			LM	Land Mobile Service
			MxA	Mobile except AeroNautical Mobile Service
R1;R2;R3	Region 1; Region 2; Region 3		AM	AeroNautical Mobile Service
IMT	International Mobile Telecommunications			
FPLMTS	Future Public Land Mobile Telecom Systems			
(A)	Allotment Plan		ARN	AeroNautical RadioNavigation
FL	Feeder Link		RN	RadioNavigation
GSO	Geostationary Satellite Orbit		RL	RadioLocation
NGSO	Non Geostationary Satellite Orbit			
pfd	power flux density		RAst	RadioAstronomy
eirp	equivalent isotropically radiated power		SR	Space Research
e/s & s/s	earth station & space station		SOp	Space Operation (TTC)
DNR	Draft New Recommendation		TTC	Telemetry, Tracking and Control
PDNR	Preliminary DNR			

FSS	Fixed Satellite Service
BSS	Broadcasting Satellite Service
MSS	Mobile Satellite Service (MobSat)
MxASS	MSS except AeroNautical Mobile Satellite Service
AMSS	AeroNautical Mobile Satellite Service
LMSS	Land Mobile Satellite Service
MMSS	Maritime Mobile Satellite Service
ISS	Inter Satellite Service
EESS	Earth Exploration Satellite Service
MeteoSat	Meteorological Satellite Service
RNSS	RadioNavigation Satellite Service
RLSS	RadioLocation Satellite Service
RDSS	RadioDetermination Satellite Service

HEO	high-Earth Orbit
LEO	low-Earth Orbit
MEO	Middle-Earth Orbit

Sup : Deletion of a provision		Sup97 : text deleted by WRC97	
Mod : Change to the substance of the text		Mod97 : text modified by WRC97	
Add : Addition of a new provision		Add97 : text added by WRC97	
Rev : Revised text		Rev97 : text revised by WRC97	

MIFR	Master International Frequency Register
IFL	International Frequency List
(&List VIIIA)	
API	Advance Publication of Information

Article S5 of RR

S4.10   ex-N°953 of ex-Article 6 of RR / N° S4.10 of Article S4 of RR: Protection of security services

      ex-Article 8 of RR : Frequency Allocations

S5.33   When a band is allocated to a service on a primary basis in a particular country, this is a primary service only in that country.

Article S9 of RR	ex-Articles 11/14/14A/15 of RR : Procedure for effecting coordination with or obtaining agreement of other Administration
S9.11A	n° S9.11A of Article S9 of RR : footnotes of the table of Frequency Allocations referring to the provision under n°S9.27 (Appendix S5)
S9.13	n° S9.13 of Article S9 of RR : referring to Res46
S9.21	n° S9.11A of Article S9 of RR: footnotes of the table of Frequency Allocations referring to the provision under n°S9.27 (Appendix S5)
S9.27	n° S9.2 of Article S9 of RR : for coordination using Appendix S5
Article S11 of RR	ex-Articles 12/13/14A/15/15A of RR: Notifying and Recording of Frequency Assignments
Article S15 of RR	ex-Article 18 of RR : Interference from Radio Stations
S15.13	ex-n°1815 of ex-Article 18 of RR : n° S15.13 of Article S15 of RR
Article S21 of RR	ex-Article 27 &28 of RR : Terrestrial and Space Services Sharing Frequency Bands Above 1GHz
Article S22 of RR	ex-Articles 28 & 29 of RR : Space Services
S22.2	ex-n°2613 of ex-Article 29 of RR : n° S22.2 of Article S22 of RR: NGSO systems shall not cause unacceptable interference to GSO systems in the FSS and the BSS operating in accordance with these Regulations.
S22.5A (Add 95)	In the frequency 6700-7075 MHz, the pfd produced at the GSO by a NGSO in the FSS shall not exceed -168 dBW/m2 in any 4 kHz band.
Article S31 of RR	ex-Article N38 of RR: frequencies for the GMDSS 5Global Maritime Distress and Safety System) (see also Appendix S15)

Appendix S4 of RR	Appendices 1, 2, 3, 4 and 5 are replaced by Appendix S4 as from 1 January 97
Appendix S5 of RR	ex-Article 29 and Res46 of RR : Identification of Administrations with which Coordination is to be effected or agreement sought uner the provisions of Article S9 (see tables S5)
Appendix S13 of RR	ex-Articles 37/38/39/40/41/42 & Articles 55/56 of RR: Distress and Safety Communications (non-GMDSS)
Appendix S30 of RR	ex-Appendix 30 of RR : BSS Plan
Appendix S30A of RR	ex-Appendix 30A of RR : BSS FL Plan

RR Resolutions	*(relating to the FSS or BSS services and to certain space services or radio regulations)*
Resolution 1 (WARC79+Add97)	Notification of frequency assignments
Resolution 2 (WARC79)	Equitable use, by all countries, with equal rights, of the GSO and of frequency bands for Space Radiocommunication Services
SupOrb88 Resolution 3 (WARC79)	Use of the GSO and Planning of Space Stations using this Orbit
Resolution 4 (WARC79+Orb88)	Period of validity of frequency assignments to space stations using the GSO
Resolution 5 (WARC79)	Technical cooperation with the developing countries in the study of propagation in tropical areas
Resolution 7 (WARC79)	Development of National Radio Frequency Management
Resolution 13 (WARC79+WRC97)	Formation of Call Signs and allocation of New International Series
Resolution 14 (WARC79)	Transfer of Technology
Resolution 15 (WARC79)	International cooperation and technical assistance in the field of Space Radiocommunications
Resolution 26 (WRC95)	Footnotes to the Table of Frequency Allocations
Resolution 28 (WRC95)	References to ITU-R Rec in the RR (Principles of Incorporation by Reference)
Resolution 29 (WRC95)	Revision of References to ITU-R Rec incorporated by Reference in the RR
Resolution 33 (Add79+Mod97)	Bringing into use of space stations in the BSS, prior to the entry into force of agreements and associated Plans for the BSS
Resolution 34 (Add79)	Relating to the Establishment of the BSS in the band 12.5-12.75 GHz in Region 3 and to Sharing with Space and Terrestrial Services in Regions 1, 2 and 3.
Sup97 Resolution 37 (WARC79)	Automated Frequency Management
Sup97 Resolution 39 (MOB83)	Use of international monitoring facilities in applying decisions of WARCs
Resolution 42 (Orb85&88)	Use of Interim systems in R2 (BSS and FSS FL) in bands governed by RR APP30&30A (Orb85&88)

SupOrb88 Resolution 43 (Orb85)	limits to the orbital position for the BSS in Regions 1&3 in the bands12.2-12.5 GHz and for the FSS FL in Region 2 in the band 17.3-17.8 GHz
Sup97 Resolution 45 (Orb88)	Improvement of the Accuracy of the MIFR, the IFL and List VIIIA
Res 46 (WARC92+Mod95&97)	Rev WRC95&97 Res of RR: Interim procedures for the coordination and notification of frequency assignments of satellite networks in certain space and the other services to which certain bands are allocated (Need for procedures to regulate the frequency assignments of NGSO satellite networks.
	Immediate application of RR Resoluion 46 in some bands
Sup97 Resolution 47 (WRC95)	Conditions for recommencing the procedures for API
Sup97 Resolution 48 (WRC95)	Relating to Information on the Propagation of Radio Waves used in the Determination of the Coordination Area
Resolution 60 (WARC79)	Relating to the Division of the World into Climatic Zones for the purpose of Calculation of Propagation Parameters
Sup97 Resolution 61 (WARC79)	Relating to the circulation of current information on ITU-R Recommendations referred to in the RR (Cross-referencing of ITU-R Recommendations in the RR)
Sup97 Resolution 65 (WARC79)	Division of the World into 3 Regions for the allocation of frequencies
Sup92 Resolution 66 (WARC79)	Improvements in the Design and use of Radio Equipment
Sup92 Resolution 67 (WARC79)	Redefinition of certain terms applicable to the RR
Sup92 Resolution 68 (WARC79)	Estimation of interference between Satellite Networks using Simplified Methods (cross-interfering simplified methods)
Sup97 Resolution 69 (Orb88)	Establishment of standards for the operation of LEOs
Resolution 70 (WARC92)	Revision,or Replacement or Abrogation of RR Resolutions/Recommendations
Sup92 Resolution 90 (WARC79)	Review and treatment of certain RR Resolutions/Recommendations
Sup97 Resolution 93 (WARC92)	Review of RRResolutions/Recommendations
Sup97 Resolution 94 (WARC92)	
SupOrb88 Resolution 100 (WRC79)	Coordination,Notification and Recording in the MIFR of Assignments to stations in the FSS with respect to stations in the SRS in Region2
SupOrb88 Resolution 101 (WRC79)	Agreements and Establishment Associated Plans for the FL to space stations in the SRS operating in the band 12GHz, in conformity with the Plan adopted by the WRC77 in Regions 1&3
SupOrb88 Resolution 102 (WRC79)	Coordination between administrations concerning technic characteristics of the FL to space stations in the SRS operating in the band 11.7-12.5 GHz (Region 1) and 11.7-12:2 GHz (Region 3) during the period from the entry into force WRC79 Final Acts until the entry into force WARC Orb85&88 Final Acts concerning the planning of FL to such space stations
Sup97 Resolution 104 (Orb88)	Application of the provisions of RR No 1550 as modified by WARC Orb88
Resolution 105 (Orb88)	Improvement of the quality of certains allotments in Part A of the FSS Plan
Sup97 Resolution 106 (Orb88)	Provisional application of RR AP30A as contained in the Final Acts of the WARC Orb88
Sup97 Resolution 107 (Orb88)	Satellite Networks intended for Use in the FSS Plan 88 (RR Ap 30B) before the WARC Orb88 (Existing networks RR AP30B)
Sup92 Resolution 108 (Orb88)	Use of the bands 4500-4800 and 6725-7025 MHz , 10.70-10.95, 11.2-11.45 and 12.75-13.25 GHz prior to the date of entry into force of Appendix 30B.
Sup97 Resolution 109 (Orb88)	Recording in the Master International Frequency Register MIFR of the Assignments for Regions 1 & 3 contained in AP30A (RR AP30A of the WARC Orb88 in the MIFR)
Sup97 Resolution 110 (Orb88)	Improved Procedures for Certain bands of the FSS (Concept of Multilateral Planning Meetings MPM)
Resolution 114 (WRC95)	WRC95 Res of RR : Use of the band 5091-5150 MHz by the FL of the NON-GSO MSS (Earth-to-space) (the band 5000-5250 MHz to the ARNS)
Sup97 Resolution 115 (WRC95)	WRC95 Res of RR : Calculation of the PFD at the GSO in the band 6700-7075 MHz used for FL of NGEO Satellite Systems in the MSS (Space-to-Earth)
Resolution 116 (WRC95)	WRC95 Res of RR : Allocation of Frequencies to the FSS (Space-to-Earth) in the band 15;4-15.7 GHz for FL of NON-GEO Satellite networks in the MSS
Resolution 117 (WRC95)	WRC95 Res of RR: Allocation of Frequencies to the FSS (Earth-to-Space) in the band 15.45-15.65 GHz for use by FL of NON-GSO satellite networks operating in the MSS.
Sup97 Resolution 118 (WRC95)	WRC95 Res of RR: Use of the bands 18.8-19.3 GHz and 28.6-29.1 GHz by the NON-GSO FSS systems
Sup97 Resolution 119 (WRC95)	WRC95 Res of RR: Sharing between the FSS and the FS in the 19.3-19.6 GHz band when used by the FSS to provide
Sup97 Resolution 120 (WRC95)	WRC95 Res of RR: Use of FL for NON-GEO satellite systems in the MSS by Feer Links for NON-GSO MSS networks
Resolution 121 (WRC95)	WRC95 Res of RR: Development of Interference criteria and Methodologies for coordination between FL for NON-GSO satellite systems in the MSS and GSO satellite systems in the FSS FL in the bands 19.3-19.6 GHz and 29.1-29.4 GHz
Resolution 212 (WARC92+Mod95)	Rev WRC95 Res of RR: implementation of future public land mobile telecommunication systems (FPLMTS)
Resolution 213 (WARC92+Mod95)	Rev WRC95 Res of RR: Sharing studies concerning possible use of the band 1675-1710 MHz by the MSS

Resolution 214 (WRC95)	WRC95 Res of RR : Sharing studies relating to consideration of the allocation of bands below 1GHz to the Non-GSO MSS
Resolution 215 (WRC95)	WRC95 Res of RR : Coordination process among Non-GSO MSS systems
Sup97 Resolution 505 (WARC79)	Relating to the BSS (sound) in 1.5 GHz (in the frequency range 0,5 to 2 GHz)
Resolution 506 (Orb88+Mod97)	Use by Space Stations operating in the 12 GHz frequency bands allocated to the BSS of the GSO and no other
Resolution 507 (WARC79)	Relating to the establishment of Agreements and associated Plans for the BSS (Agreements/Plans for BSS)
Resolution 518 (Orb88)	Area/Country symbols in RR AP30/30A
Resolution 519 (Orb88)	Possible Extensions to Regions 1 &3 of Provisions for Interim systems
Sup97 Resolution 522 (WARC92)	Further work by ITU-R Study Groups concerning the BSS (sound)
Resolution 524 (WARC92+Mod95)	Rev WRC95 Res of RR: future consideraration of the Plans for the BSS in the bands 11.7-12.5 GHz (R1) and 11.7-12.2 GHz (R3) in Appendix S30 and the associated FL Plans in Appendix S30A.
Resolution 525 (WARC92+Mod95)	Rev WRC95 Res of RR: Introduction of HDTV systems of the BSS in the band 21.4-22 GHz in R1 & 3
Resolution 526 (WARC92+Mod95)	Rev WRC95 Res of RR: future adoption of procedures to ensure flexibility in the use of the frequency band allocated to the BSS for wide RF-band HDTV and the associated FL
Resolution 528 (WARC92+Mod95)	Rev WARC95 Res of RR: Introduction of the BSS(sound) systems in the range 1-3 GHz
Resolution 531 (WRC95)	WRC95 Res of RR: Review of Appendices S30 and S30A of the RR
Sup97 Resolution 643 (WRC95)	WRC95 Res of RR: Inter Satellite Links between 50 and 70 GHz
Resolution 703 (WARC79+Mod92)	WARC79 & WRC92 Res of RR: Calculation metods and Interference criteria recommended by ITU-R for sharing frequency bands between Space and Terrestrial Radiocommunication Services or between Space Radiocommunication Services.
Sup97 Resolution 711 (WRC92)	Possible Relocation of Frequency Assignments to certain space Missions from the 2 GHz band to bands above 20 GHz
Sup97 Resolution 712 (WRC92)	Consideration after WARC92 by a future competent WARC of issues dealing with allocations to Space Services
Resolution 714 (WRC95)	WRC95 Res of RR: pfd level applicable in the frequency band 137-138 MHz shared by the MSS and Terrestrial Services
Resolution 715 (WRC95+Mod97)	Rev WARC95 Res of RR: Studies concerning sharing between the RNSS and the MSS in the bands 149.9-150.05 MHz and 399.9-400.05 MHz
Resolution 716 (WRC95)	WRC95 Res of RR: Use of the frequency bands 1980-2010 MHz and 2170-2200 MHz in all 3 Regions and 2010-2025 MHz and 2160-2170 MHz in Region 2 by the FS and MSS
Resolution 717 (WRC95)	WRC95 Res of RR: revew of allocations to the MSS in the 2 GHz range
Sup97 Resolution 718 (WRC95)	WRC95 Res of RR: Agenda for the 1997 World Radiocommunication Conference
Sup97 Resolution 719 (WRC95)	WRC95 Res of RR: urgent studies required in preparation for the for the 1997 World Radiocommunication Conference
Sup97 Resolution 720 (WRC95)	WRC95 Res of RR: Preliminary agenda for the 1999 for the 1997 World Radiocommunication Conference

New Resolutions (relating to FSS or BSS services) adopted by WRC97 (see Finals Acts of the WRC97)

WRC97 Res 72 / PLEN2	Regional preparations for WRCs
WRC97 Res 532 / PLEN3	Review and possible revision of the 1997 BSS Plans for Regions 1 &3
WRC97 Res 533 / PLEN4	Implementation of the decisions of WRC97 relating to RR Appendices S30 & S30A
WRC97 Res 534 / PLEN5	Implementation of Annex5 to RR Rev97AppendixS30 and Annex3 to RR Rev97AppendixS30A
WRC97 Res 95 / GTPLEN1-1	General review of the Resolutions and Recommendations of WARCs and WRCs
WRC97 Res 50 / GTPLEN1-2	Interval between WRCs
WRC97 Res 721 / GTPLEN1-3	Agenda for the 1999 WRC
WRC97 Res 722 / GTPLEN1-4	Preliminary agenda for the 2001 WRC
WRC97 Res 49 / GTPLEN2-1	Administrative due diligence applicable to some satellite communication services
WRC97 Res 80 /CLM-CTR-EQA1	Due dilidence in applying the principles embodied in the Constitution.
WRC97 Res 73 / COM4-15	Measures to solve the incompatibility between BSS in Region 1 and FSS in Region 3 in the frequency band 12.2 - 12.5 GHz
WRC97 Res 30 / COM4-17	Publication of the Weekly Circular including special sections.

WRC97 Res 51 / COM4-18	Provisional application of certain provisions of the RR as modified by WRC97 and transitional arrangements
WRC97 Res 52 / COM4-19	Provisional application of Nos S11.24 and S11.26 of the RR adopted by WRC97 with regard to high altitude platform stations
WRC97 Res 53 / COM4-20	Updating of the "remarks" comumns in the tables of Article 9A of RR Appendix S30A and Article 11 of RR Appendix S30
WRC97 Res 536 / COM4-23	Operation of broadcasting satellites serving other countries
WRC97 Res 216 / COM5-2	Possible broadening of the secondary allocations to MSS (Earth-to-space) in the band 14-14:5 GHz to cover aeronautical applications
WRC97 Res 122 / COM5-7	Use of the bands 47.2-47.5 GHz and 47.9-48.2 GHz by high altitude platform stations in the Fixed Service and by other services
WRC97 Res 123 / COM5-8	Feasibility of implementation NGSO MSS FL in the band 15.43-15.63 GHz (space-to-Earth)
	while taking into account the protection of the RA, the ESSS & Space Research passive services in the band 15.35-15.4 GHz
WRC97 Res 124 / COM5-9	Protection of the Fixed Service in the frequency band 8025-8400 MHz sharing with the GSO EESS (space-to-Earth)
WRC97 Res 126 / COM5-11	Use of the frequency band 31.8-33.4 GHz for high density systems in the Fixed Service.
WRC97 Res 726 / COM5-12	Frequency bands above 30GHz available for high-density applications in the Fised Service
WRC97 Res 128 / COM5-16	Allocation to the FSS (space-to-Earth) in the 41.5-42.5 GHz band and protection of the RA service in the 42.5-43.5 GHz band
WRC97 Res 129 / COM5-17	Criteria and methodologies for sharing between the FSS and other services in the band 40.5-42.5 GHz
WRC97 Res 130 / COM5-18	Use of the NGSO FSS in certain frequency bands
WRC97 Res 538 / COM5-19	Use of the frequency bands covered by RR Appendices S30 &S30A by the NGSO FSS
WRC97 Res 131 / COM5-23	Pfd limits applicable to the NGSO FSS for protection of terrestrial services in the bands 10.7-12,75 GHz and 17.7-19.3 GHz
WRC97 Res 132 / COM5-27	Use of the bands 18.8-19.3 GHZ and 28.6-29.1 GHz by networks operating in the FSS
WRC97 Res 133 / COM5-28	Sharing between the Fixed Service and other services in the band 37-40 GHz
WRC97 Res 134 / COM5-29	Use of the frequency band 40.5-42.5 GHz by the FSS
WRC97 Res 54 / COM5-30	Implementation of Rev WRC97 Resolution 46

RR Recommendations

(relating to the FSS or BSS services and to certain space services or radio regulations)

Sup97 Recommendation 1 (WARC79)	Use of space systems in disasters....
Sup97 Recommendation 2 (WARC79)	Relating to the Examination by WRC of the situation with regard to occupation of the Frequency Spectrum in Space Radiocommunications
Sup97 Recommendation 6 (WARC79)	Assistance to developing countries
Recommendation 8 (WARC79)	Relating to Automatic Identification of Stations
Sup97 Recommendation 10 (WARC79)	Relating o the presentation of Draft Amendments to the RR
Sup97 Recommendation 11 (WARC79)	Relating to the Marginal Numbering of the RR
Sup92 Recommendation 12 (WARC79)	Relating to the convening of future WARCs to deal with specific services
Sup97 Recommendation 13 (WARC79)	Relating to a WARC to carry out a general or partial revision of the RR
Sup97 Recommendation 15 (Orb88)	Review of RR Article 14 and further development of technical criteria for its application
Sup97 Recommendation 30 (WARC79)	Relating to International Monitoring
Sup97 Recommendation 31 (WARC79)	Handbook for computer-aided techniques in Radio Frequency Management
Recommendation 32 (Orb88)	International monitoring of emissions originating from space stations
Recommendation 34 (WRC95)	Principles for the Allocation of Frequency Bands
Recommendation 35 (WRC95)	WRC95 Rec of RR: Procedures for Modification of a Frequency Allotment or Assignment Plan
Sup97 Rec60 (WARC79)	Technical standards of the IFRB/RB
Recommendation 61 (WARC79)	Relating to Technical Standards for the Assessment of Harmful Interference in the frequency bands above 28 MHz
Sup97 Rec62 (WARC79)	Additional characteristics for classifying Emissions
Recommendation 63 (WARC79)	Relating to the provision of formulae and examples for the calculation of necessary bandwidths

Recommendation 64 (WARC79)	Relating to Protection Ratios and minimum Field Strengths required
Sup97 Rec65 (WARC79)	Relating to the Technology for New Spectrum Sharing and Band Utilisation schemes
Recommendation 66 (WARC92)	Studies of the maximum permitted levels of spurious emissions
Sup92 Rec67 (WARC79)	Relating to the definitions of "service area" and "coverage area"
Sup97 Rec68 (WARC79)	Relating to studies and prediction of radio propagation and radio noise
Sup92 Rec69 (WARC79)	Relating to the Frequency tolerances of transmitters
Sup92 Rec70 (WARC79)	Relating to Studies of the Technical characteristics of radio equipment
Recommendation 71 (WARC79)	Standardisation of the technical and operational characteristics of radioequipment
Sup97 Rec72 (WARC79)	Relating to Terminology
Sup97 Rec73 (WARC79)	Relating to the use of the term "channel" in the RR
Sup97 Rec74 (WARC79)	Relating to the use of the SI system (Système International d'Unités)
Rec 100 (WARC79+Add95)	Relating to preferred frequency bands for systems using tropospheric scatter
Sup92 Rec101 (WARC79)	Relating to feeder links for the BSS
Sup92 Rec102 (WARC79)	Relating to studies of modulation methods for Radio relay systems for frequency sharing with FSS systems
Sup97 Rec 103 (WARC79)	Relating to carrier energy dispersal ED in FSS systems
Recommendation 104 (Add95)	WRC95 Rec of RR: Development of PFD and EIRP limits to be met by FL of NGSO satellite networks in the FSS in bands where N° 2613/S22.2 of the RR applies.
Recommendation 105 (Add95)	WRC95 Rec of RR: Determination of the coordination area around e/s operating with GSO satellite networks in the FSS and e/s providing FL to NGSO satellite networks in the MSS operating in opposite directions of transmissions.
Sup97 Rec 505 (WARC79)	Relating to Studies of Propagation at 12 GHz for the BSS
Sup97 Rec 506 (WARC79)	Relating to the Harmonics of the fundamental frequency of BSS Stations
Sup97 Rec 507 (WARC79)	Relating to Spurious Emissions in the BSS
Sup97 Rec 508 (WARC79)	Relating to transmitting Antennae for the BSS
Recommendation 521 (Add95)	WRC95 Rec of RR: Technical parameters for use in the revision of Appendices 30/S30 and 30A/S30A in response to Resolution 524 of the RR
Recommendation 705 (WARC79)	Criteria to be applied for frequency sharing between the BSS and the terrestrial BS in the band 620-790 MHz
Recommendation 706 (WARC79)	Frequency sharing by the EESS (passive sensors) and the Space Research (passive sensors) with the FS, MxAS and FSS in the band 18.6-18.8 GHz
Recommendation 707 (WARC79)	Relating to the use of the frequency band 32-33 GHz shared between the ISS and the RadioNavigation.
Recommendation 708 (WARC79)	Space and Terrestrial services
Recommendation 709 (WARC79)	Relating to sharing frequency bands between the AeroNautical Mobile AMS and the ISS
Recommendation 710 (WARC79)	Relating to the use of Airborne Radars in the frequency bands shared between the ISS and the Radiolocation RL.
Recommendation 711 (WARC79)	Relating to the Coordination of Earth stations
Sup97 Rec 712 (WARC79)	Relating to the Interdependence of receiverdesign, channel grouping and sharing criteria in the BSS (Design characteristics for BSS)
Recommendation 715 (Orb88)	Multiband and/or Multiservice Satellite Networks using the GSO
Sup97 Rec 717 (WARC92+Mod95)	WRC95 Rec of RR: Frequency sharing in bands shared by the MSS and the FS, MS and other terrestrial services below 3 GHz
Recommendation 719 (WARC92)	Multiservice satellite networks using the GSO
Sup97 Rec 721 (WRC95)	WRC95 Rec of RR: Frequency sharing in the bands 1610.6 - 1613.8 MHz and 1660 - 1660.5 MHz between the MSS and the FS and the Radio Astronomy service.

New Recommendations of WRC97 (see Finals Acts of WRC97)

WRC97 Rec 36 / GT PLEN2A	Role of International monitoring in reducing apparent congestionin the use of orbital and spectrum resources
WRC97 Rec 622 / COM5A	Use of the frequency bands 2025-2110 MHz and 2200-2290 MHz by he Space Research , Space Operation, EESS, FS and MS.

APPENDIX 9.2

Example of determination of coordination distance around an earth station

AP9.2-1 Introduction

The example included in this appendix is based on an earth station working to the INTELSAT Pacific Ocean satellites at 174° E and 179° E longitude. The particular earth station is the one at Honiara, in the Solomon Islands. Details on this earth station are given in Special Section SPA-AJ/234*, as corrected by Special Section SPA-AJ/284**.

The particular situation for which the contour has been determined is for reception of analogue modulated signals in the 4 GHz band. The assumed characteristics of the "interfering" terrestrial station are those set by the limits of RR S21.3 and RR S21.5 of Article S21.

AP9.2-2 Basic parameters

From SPA-AJ/234* and 284**, the following data is obtained:

earth station location:	159° 57' E, 9° 26' S
space station location:	in geostationary-satellite orbit at 174° E, 179° E
antenna radiation diagram:	in accordance with Recommendation ITU-R S.465
noise temperature at receiving system:	80 K

horizon elevation angle is obtained from Figure 6.4.

For the terrestrial station:

e.i.r.p.: 55 dBW	
transmitter power:	13 dBW

Table II of RR Appendix S7 lists the other parameters which are required. In this case:

p_0 0.03%

n 3

---

*   Special Section SPA-AJ (now AR11A)/234/1430, annexed to ex-IFRB Circular No. 1430, 5 August 1980.

** Special Section SPA-AJ (now AR11A)/284/1446, annexed to (ex-IFRB) Circular No. 1446, 25 November 1980.

p   0.01%

J   −8 dB

$M_0(p_0)$ 17 dB

W   4 dB

E   55 dBW

P_t 13 dBW as stated above

ΔG 0 dB

B 10^6 Hz

For this calculation, transmission loss values are calculated based on p = 0.01%, i.e. the calculated contour allows for the required performance for 99.99% of the time.

AP9.2-3 Earth station antenna gain

The gain of the earth station antenna towards the physical horizon (for all values of azimuth) can be determined by first finding the off-beam angle to the horizon and then applying the equations of Recommendation ITU-R S.465 to determine the antenna gain. Figure II-1 of Annex II to Appendix S7 can be used to obtain the various values of the off-beam angle.

Figure AP9.2-1 of this appendix shows the necessary graphical construction. In this case, the two satellites are located 14° 3' and 19° 3' East of the earth-station longitude. Hence the arc at the top left of Figure AP9.2-1 can be drawn to represent the orbital arc between 174° E and 179° E. The horizon elevation angle for azimuths of 0° - 90° and 270° - 360° can also be plotted at the base of the diagram.

It is then found that the distance from the horizon elevation angle line to the arc is always greater than 48°; hence the earth station antenna gain, to the horizon, for all angles of azimuth is −10 dB.

AP9.2-4 Permissible level of the interfering emission

Equation (3) of Appendix S7 gives the formula for $P_r(p)$, so that using the values listed in § 2 of this appendix:

$$P_r(0.01) = -228.6 + 60 + 19 - 8 + 17 - 4 = -144.6 \quad \text{dB}$$

AP9.2-5 Minimum permissible basic transmission loss (propagation model (1))

The minimum permissible basic transmission loss, for propagation mode (1), can be found from equation (2) of RR Appendix S7 as follows:

$$L_b(0.01) = 55 - 10 - (-144.6) = 189.6 \quad \text{dB}$$

AP9.2-6 Normalized transmission loss (propagation model (2))

The normalized transmission loss for propagation mode (2) (rain-scatter) can be found from equation (20) of Appendix S7 as follows, noting that ΔG and $F(p,f)$ are both zero:

$$L_2\,(0.01) = 13 - (-144.6) = 157.6 \quad \text{dB}$$

AP9.2-7 Coordination distance for propagation mode (1)

The coordination loss L_1, can be found from equation (14) of Appendix S7. The horizon angle correction A_h needs to be found first. Considering Figure 1 of SPA-AJ/234 and Figure 1 of Appendix S7, as well as equation (7a) of Appendix S7, the following data is obtained:

Azimuth (degrees from true North)	Horizon elevation angle (degrees)	Horizon angle correction, A_h (dB)
0 - 26	15	66
26 - 32	10	55
32 - 80	15	66
80 - 112	10	55
112 - 180	15	66
180 - 342	5	42
350	9	53
360	15	66

NOTE – For $\varepsilon = 15°$, $f = 4$ GHz, equation (7a) of Appendix S7 gives $A_h = 66$ dB.

The coordination loss, L_1, is therefore:

Azimuth (degrees)	A_h (dB)	L_1 (dB)
0 - 26	66	123.6
26 - 32	55	134.6
32 - 80	66	123.6
80 - 112	55	134.6
112 - 180	66	123.6
180 - 342	42	147.6
350	53	136.6
360	66	123.6

APPENDIX 9.2   Example of determination of coordination distance around an earth station

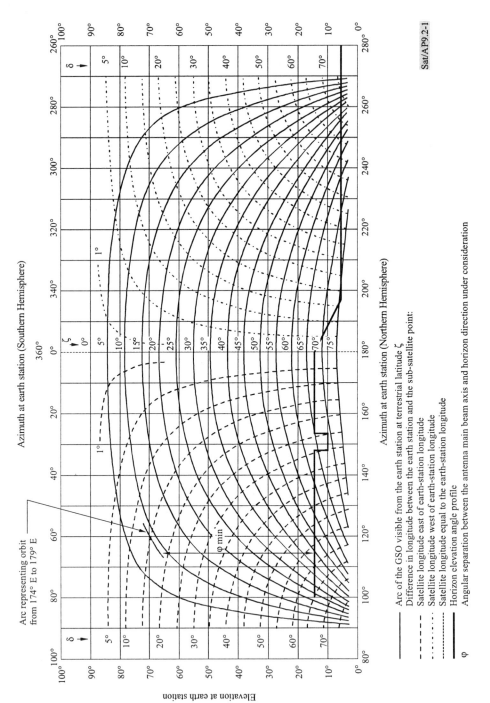

Sat/AP9.2-1

FIGURE AP9.2-1

Determination of angle between the main beam axis of the earth station antenna and the horizon

The area surrounding the earth station (see map in Figure AP9.2-2), is composed of both radio-climatic zones A and C. Considering Figures 2 (Zone A) and 5 (Zone C) of Appendix S7, it is found that for all values of L_1, other than $L_1 = 147.6$ dB, the coordination distance, d(0.01), is less than 100 km. Hence, noting § 5 of Appendix S7, d_1 is taken in these cases as 100 km.

For azimuth angles of 180° to 342° this result does not apply. For Zone A, and $L_1 = 146.6$ dB, $d_A = 100$ km. For Zone C, $L_1 = 147.6$ dB, $d_c = 400$ km.

In these cases, the method of Annex III to Appendix S7 has to be used to determine the coordination distance. For the geographical situation around Honiara this method gives the following distances:

Azimuth (degrees)	Distance in Zone A (km)	Distance in Zone C (km)	Distance in Zone A (km)
180	40	>400	
190	38	>400	
200	42	>400	
210	39	>400	
220	40	>400	
230	37	>400	
240	36	>400	
250	39	>400	
260	40	>400	
270	38	>400	
280	39	>400	
290	41	165	30
300	38	242	18
310	29	>400	
320	1	>400	
330	1	154	39
340	1	105	56

Figure AP9.2-3 has been drawn using the method given in Annex III to Appendix S7, and can be used for determining the overall values of the coordination distance. For most azimuth angles, there exists a simple mixed propagation path passing first through Zone A and then Zone C. In these cases, the procedure for calculating the required coordination distance is simple. For example, at an azimuth of 180°, one enters the graph on the Zone A axis at 40 km which gives a distance of 240 km on the Zone C axis, thus obtaining an overall distance of 280 km.

For the more complex paths, e.g. at 290°, 300°, 330° and 340°, it is necessary to use a different construction. For example, at 290°, the construction OA_1, AC_1, C_1X_1 is drawn up to give an overall distance of $41 + 165 + 18 = 224$ km. Again at 340°, the construction OA_2, A_2C_2, C_2A_3, A_3X_3 is drawn to give an overall distance of $1 + 105 + 56 + 67 = 229$ km.

Using these methods, the following distances are obtained:

Azimuth (degrees)	Distance d_1 (km)	Azimuth (degrees)	Distance d_1 (km)
180	280	270	286
190	286	280	283
200	274	290	224
210	283	300	282
220	280	310	313
230	289	320	397
240	292	330	280
250	283	340	229
260	280		

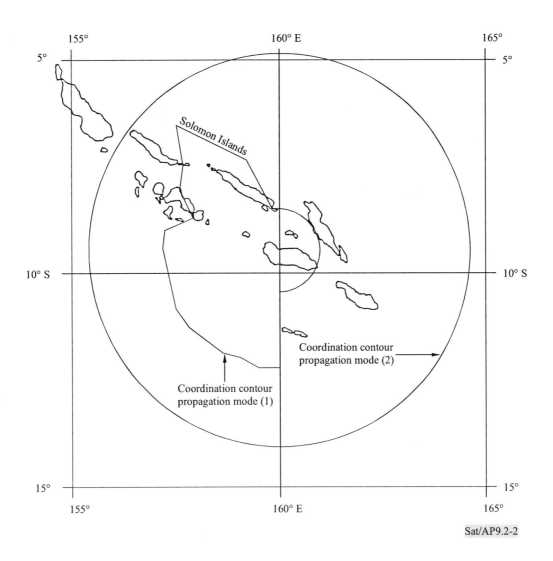

FIGURE AP9.2-2

Coordination contours for propagation modes (1) and (2)

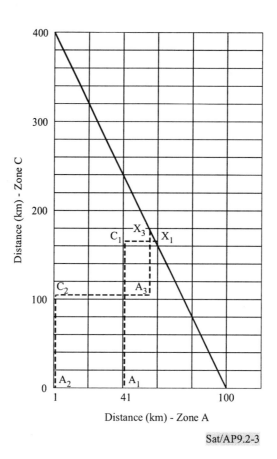

Sat/AP9.2-3

FIGURE AP9.2-3

Determination of coordination distance for mixed paths involving Zones A and B

APPENDIX 9.3

C/I calculation for digital carriers

The calculation of the interference between two satellite systems working in the same frequency bands using digital signals is based on Recommendation ITU-R S.523, which states that the maximum single entry interference level should not exceed 6% of the total noise power level. This can be expressed by the following relation for the single entry criterion:

$$(C/I) = (C/N) + 12.2 \quad dB$$

where C/N is the total operating carrier-to-noise ratio corresponding to a bit error ratio of 1 in 10^6 and I is the interference power in the occupied bandwidth of the wanted carrier.

The equation for the total C/I numerical ratio is expressed as:

$$C/I = \left[(C/I)_u^{-1} + (C/I)_d^{-1} \right]^{-1}$$

where $[C/I]_u$ and $[C/I]_d$ are numerical ratios.

The uplink and downlink C/I can be computed by the following relations:

$$C/I_u = P_{ei} + G_{eti} + G_{sri} - L_{ui} - (P_{ej} + G_{etj}(\theta_{ji}) + G_{sri}(\theta_{ji}) - L_{uj}) \quad dB \tag{1}$$

$$C/I_d = P_{si} + G_{sti} + G_{eri} - L_{di} - (P_{sj} + G_{stj}(\theta_{ji}) + G_{eri}(\theta_{ji}) - L_{dj}) \quad dB \tag{2}$$

where I_u and I_d represent the uplink and downlink interference power in the bandwidth of the wanted carrier, and

e   :   subscript for earth station,

s   :   subscript for space station,

u   :   subscript for uplink,

d   :   subscript for downlink,

r   :   subscript for receiving,

t   :   subscript for transmitting,

i   :   subscript for wanted system,

j   :   subscript for interfering system,

P   :   transmitting power,

G   :   maximum antenna gain,

$G(\theta_{ji})$:   antenna gain towards interfering/interfered-with direction,

L   :   free-space loss.

The total C/N value is given by:

$$C/N = E_b/N_0 + m + 10 \log R - 10 \log B \quad dB$$

where

E_b/N_0 :   energy per bit to noise ratio relative to the required BER (dB),

m   :   implementation margin (dB),

R   :   bit rate (bit/s),

B   :   noise bandwidth (Hz).

As an example, two satellite systems A and B are considered which work in the 11/14 GHz frequency bands, with the link parameters given in Table AP9.3-1. This example considers system A as the wanted one and system B as the interfering one. The first step is to compute the total C/I. This can be done by substituting the parameter values in equations (1) and (2). Assuming:

• overlapping service areas,

• free-space loss of:

206 dB at 14 GHz,

204 dB at 11 GHz,

• satellite separation angle of $6° = 7° - 1°$,

• earth-station side-lobe pattern given by $32 - 25 \log 6° = 125.5$ dB,

• satellite beam edge gain loss of 3 dB for the wanted signal,

then:

$$(C/I_u) = 17 + 59.5 + 33.5 - 206 - (24 + 12.5 + 36.5 - 206) + 4.9 = 41.9 \quad \text{dB}$$

$$(C/I_d) = 15.5 + 34.5 + 46.7 - 204 - (12.0 + 34.0 + 12.5 - 204) + 4.9 = 43.1 \quad \text{dB}$$

The last term of 4.9 dB in the equations above is the ratio of the occupied bandwidths of the interfering and wanted carrier ($10 \log 80/26 = 4.9$ dB).

$$(C/I)_{tot} = -10 \log \left(10^{-\frac{41.9}{10}} + 10^{-\frac{43.1}{10}} \right) = 39.4 \quad \text{dB}$$

For the computation of the total C/N, assuming:

• E_b/N_0 =   10.8 dB for a 10^{-6} BER,

• m   =   2 dB,

it follows:

$$C/N = 10.8 + 2 + 10 \log 40 \times 10^6 - 10 \log 26 \times 10^6 = 14.7 \quad \text{dB}$$

The resulting margin will then be given by:

$$M = (C/I)_{tot} - C/N - 12.2 = 39.4 - 14.7 - 12.2 = 12.5 \quad \text{dB}$$

TABLE AP9.3-1

Satellite system parameters	A	B
Satellite transmitting antenna gain (G_{st})	37.5 dB	34.0 dB
Satellite transmission power (P_s)	15.5 dBW	12.0 dBW
Satellite receiving antenna gain (G_{sr})	36.5 dB	28 dB
Earth station transmitting antenna gain (G_{et})	59.5 dB	65.6 dB
Earth station transmitting power (P_{et})	17.0 dBW	24 dBW
Earth station receiving antenna gain (G_{er})	46.7 dB	63.5 dB
Satellite longitude	1° E	7° E
Noise bandwidth (B)	26 MHz	80 MHz
Modulation	4-PSK	4-PSK
Bit rate (R)	40 Mbit/s	130 Mbit/s

ANNEX 1

Propagation

AN1 Introduction

This annex summarizes the different effects that contribute to the overall propagation loss along Earth-space paths. The intention is to outline the significant factors that impair transmissions for frequencies below 30 GHz. The information is a summary of the material that is included in, or referred to by, the ITU-R Handbook entitled "Radiowave Propagation Information for Predictions for Earth-to-Space Path Communications". In determining propagation loss, the latest version of the appropriate ITU-R Study Group 3 Recommendations and the Handbook on Earth-space propagation should be consulted.

The procedures presented in this annex are for predicting impairments along a fixed Earth-space path (i.e. a GSO satellite link). Where users plan to install systems that operate to non-geostationary satellite orbit (non-GSO) constellations, they should bear in mind the elevation angles will change with time, and sometimes very rapidly.

AN1.1 Overview of atmospheric effects

Earth-space transmissions propagate through two important atmospheric regions: the troposphere and the ionosphere. The troposphere is a non-ionized part of the atmosphere that extends from the ground up to a height of about 15 km and contains most of the Earth's weather effects in terms of clouds, rainfall, and snow. The ionosphere is an ionized region of the atmosphere extending from a height of around 30 km to around 1 000 km.

The dual effects of the troposphere and the ionosphere can give rise to the following signal impairments:

a)   signal attenuation caused by atmospheric gases, clouds, precipitation, sand and dust;

b)   emission noise (sky noise increases) from absorbing media;

c)   signal depolarization by raindrops, ice crystals, and Faraday rotation;

d)   signal scintillations by refractivity variations and low elevation angles;

e)   refraction and atmospheric multipath by atmospheric gases and ionospheric electron distributions.

Each of these contributions has its own characteristics as a function of frequency, geographic location and elevation angle. Signal impairments from ionospheric effects arise from refraction, Faraday rotation, scintillations and excess time delay, while signal impairments from tropospheric effects arise from gaseous absorption, precipitation, depolarization, refraction and scintillations.

In general, tropospheric effects on Earth-space paths only start to become significant for frequencies above 1 GHz. Transmissions for elevation angles greater than 10° and below 6 GHz are relatively unaffected, while transmissions of low elevation angles (<3°), or those above 10 GHz, suffer significant effects. Ionospheric impairments on propagation are only dominant for lower frequencies (normally <1 GHz), although severe scintillation effects can occur at frequencies up to around 6 GHz at high latitudes or within ±20° of the geomagnetic equator.

As a rule, for Earth-space transmissions with elevation angles above 10°, attenuation exceeding a few tenths of a decibel will result only from rain attenuation, gaseous attenuation and possibly scintillation, depending on propagation conditions.

AN2 Signal attenuation

AN2.1 Method of predicting the long-term rainfall attenuation statistics from point rainfall rate

For satellite systems operating above 10 GHz, rainfall is the major effect causing loss of signal. The attenuation increases with rainfall rate, signal frequency and decreasing elevation angle. Rain attenuation can normally be neglected for frequencies below 6 GHz.

The procedure to calculate the mean long-term attenuation statistics due to rainfall along an Earth-space path at a given location for frequencies up to 30 GHz is presented below. It is taken from Recommendation ITU-R P.618-4 and consists of the estimation of the attenuation exceeded at 0.01% of the time ($A_{0.01}$) from the rainfall rate exceeded at the same time percentage ($R_{0.01}$). The method consists of following steps 1 to 7 to predict the attenuation exceeded for 0.01% of the time, and step 8 to convert this value to other time percentages. Note, if reliable long-term statistical attenuation data measured at one frequency is available for the site, it is recommended that the attenuation be derived by using the frequency-scaling method discussed in § 2.2.1.2 of Recommendation ITU-R P.618, rather than the prediction method of using rain rate data presented below.

The procedure to calculate the attenuation exceeded for 0.01% of the time ($A_{0.01}$) requires the following parameters to be used in the calculation:

$R_{0.01}$: point rainfall rate for the location for 0.01% of an average year (mm/hour)

h_s: height above mean sea level of the earth station (km)

θ: elevation angle (degrees)

φ: latitude of the earth station (degrees)

f: frequency (GHz)

The geometry of the path is illustrated in Figure AN1-1 of this annex. The attenuation exceeded for 0.01%, $A_{0.01}$, is calculated as the product of the path reduction factor ($r_{0.01}$), the slant path length (L_s) and the value of specific attenuation (γ_R) corresponding to a given value of rain rate $R_{0.01}$.

Step 1: Calculate the effective rain height[1], h_R, for the latitude of the station φ:

$$h_R \text{ (km)} = \begin{cases} 5 - 0.075\,(\varphi - 23.0) & \varphi > 23° & \text{Northern hemisphere} \\ 5 & 0 \leq \varphi \leq 23° & \text{Northern hemisphere} \\ 5 & 0 \geq \varphi \geq -21° & \text{Southern hemisphere} \\ 5.0 + 0.1\,(\varphi + 21) & -71 \leq \varphi < -21° & \text{Southern hemisphere} \\ 0 & \varphi < -71° & \text{Southern hemisphere} \end{cases} \tag{1}$$

Step 2: Compute the slant-path length, L_s, below the rain height. For $\theta \geq 5°$, use the following formula:

$$L_s = \frac{(h_R - h_s)}{\sin\theta} \quad \text{km} \tag{2}$$

For $\theta < 5°$, use:

$$L_s = \frac{2\,(h_R - h_s)}{\left(\sin^2\theta + \dfrac{2\,(h_R - h_s)}{R_e}\right)^{1/2} + \sin\theta} \quad \text{km} \tag{3}$$

Where R_e is the effective earth radius after accounting for refraction. Typically, a value of $R_e = 8\,500$ km is appropriate for $h_s \leq 1$ km. For $h_s > 1$ km, Recommendation ITU-R P.676 should be consulted.

Step 3: Calculate the horizontal projection, L_G, of the slant-path length using the formula (see Figure AN1-1 of this annex):

$$L_G = L_s \cos\theta \quad \text{km} \tag{4}$$

Step 4: Obtain the rain intensity, $R_{0.01}$, exceeded for 0.01% of an average year (with an integration time of 1 min). If this information cannot be obtained from local data sources, an estimate can be obtained from the rain climatic zone maps contained in Figure AN1-3 and Table AN1-1 of this annex (reprinted from Recommendation ITU-R P.837-1).

Step 5: Calculate the reduction factor, $r_{0.01}$, for 0.01% of the time:

$$r_{0.01} = \frac{1}{1 + L_G / L_0} \tag{5}$$

where $L_0 = 35 e^{-0.015\,R_{0.01}}$ (for $R_{0.01} \leq 100$ mm/hour).

---

[1] Equations for the effective rain height, h_R, have been updated in accordance with the proposed revision to Recommendation ITU-R P.618-4.

For $R_{0.01} > 100$ mm/hour, use the value of 100 mm/hour in place of $R_{0.01}$ in the calculation of L_0.

Step 6: Obtain the specific attenuation, γ_R, using the frequency-dependent coefficients k and α and the rainfall rate, $R_{0.01}$, determined from step 4, by the equation:

$$\gamma_R = k(R_{0.01})^\alpha \quad \text{dB/km} \tag{6}$$

The frequency-dependent coefficients k and α are given in Table AN1-2 of this annex for frequencies ≤ 35 GHz and in terms of horizontal (H) and vertical (V) polarizations. The values have been reproduced from Table 1 of Recommendation ITU-R P.838 which contains values for frequencies ≤ 400 GHz. Values of k and α at frequencies in between those tabulated in Table AN1-2 of this annex can be obtained by interpolation using a logarithmic scale for frequency, a logarithmic scale for k and a linear scale for α.

The specific attenuation due to rain can also be determined using the nomogram in Figure AN1-4 of this annex.

To determine the specific attenuation with circular polarization, a good approximation is obtained by taking the arithmetic mean of the attenuation determined for horizontal and vertical polarization. For a more accurate calculation for circular polarization, or for a calculation of linear polarization with a particular polarization tilt angle, reference should be made to equations (2) and (3) of Recommendation ITU-R P.838.

Step 7: The predicted attenuation exceeded for 0.01% of an average year is obtained from:

$$A_{0.01} = \gamma_R \cdot L_S \cdot r_{0.01} \quad \text{dB} \tag{7}$$

Step 8: The estimated attenuation to be exceeded for other percentages of an average year (A_P), in the range 0.001% to 1%, is determined from the attenuation to be exceeded for 0.01% for an average year by using:

$$A_p = 0.12 \cdot A_{0.01} \cdot P^{-(0.546 + 0.043 \log (P))} \tag{8}$$

This interpolation formula gives factors of 0.12, 0.38, 1.0 and 2.14 for 1%, 0.1%, 0.01% and 0.001%, respectively.

Seasonal variations – Worst month

System planning often requires the attenuation exceeded for a time percentage P_w of the "worst month". The conversion from annual statistics to "worst-month" statistics is discussed in the ITU-R Handbook on Radiometeorology. The relationship between P_w and the annual time percentage, P, can be expressed as:

$$P = Q_1^{\frac{-1}{1-\beta}} \cdot P_w^{\frac{1}{1-\beta}} \tag{9}$$

The above expression is applicable for the time percentage range (0.001%< P < 3%). Values of Q_1 and β measured in various locations and for several propagation effects are given in the Handbook on Radiometeorology and Table 1 of Recommendation ITU-R P.841. For global planning purposes, in the absence of precise information, a single "average" relationship between annual attenuation and worst-month attenuation may be used. For climates with relatively small seasonal variations in rainfall intensity, a value of $Q_1 = 2.85$ and $\beta = 0.13$ is appropriate giving:

$$P = 0.3 \, P_w^{1.15} \tag{10}$$

for $1.9 \cdot 10^{-4} < P_w\% < 7.8$.

AN2.2 Attenuation due to atmospheric gases, clouds and other precipitation

Attenuation due to rain, cloud and fog on Earth-space paths is not usually a significant factor for high availability services (99.96% link up-time in an average year, 0.2% down-time in an average worst month for a one-way connection) where the additional margin required for clouds over that necessary to counteract rain attenuation is relatively small. However, some additional attenuation can result from clouds of high water content. This is not usual, but for low-fade margin systems, the additional margin for attenuation due to clouds can be significant at frequencies above 10 GHz and low elevation angles – where more than 2 dB of attenuation can occur. In such cases, the value of the attenuation can be determined by reference to Recommendation ITU-R P.840. At very low elevation angles, it is most likely that tropospheric scintillation will be of greater significance than cloud losses.

The effects of ice cloud, dry hail and dry snow can generally be neglected for frequencies below 30 GHz.

Attenuation due to atmospheric gases depends mainly on frequency, elevation angle, altitude above sea level and water vapour concentration (absolute humidity). It may normally be neglected at frequencies below 10 GHz. For frequencies above 10 GHz, its importance increases especially for low elevation angles. At a frequency of 22 GHz (which corresponds to a water vapour absorption band), for average water vapour density and for elevation angles greater than 10°, the absorption does not exceed 2 dB.

The effect of attenuation by atmospheric gases is discussed in detail in Recommendation ITU-R P.676.

AN2.3 Signal scintillation and multipath effects

Scintillations are rapid level fluctuations in the received signal's amplitude, phase and apparent angle-of-arrival. They arise due to small-scale irregularities in the atmospheric refractive index and affect low margin satellite systems and antenna tracking systems.

There are two types of scintillation mechanism: tropospheric effects and ionospheric effects. Tropospheric scintillations can, on occasion, be severe at low elevation angles ($\leq 10°$) and at frequencies above about 10 GHz. They can be very severe (>10 dB attenuation) for very small time percentages and for very low elevation angles ($\leq 4°$ on inland paths or $\leq 5°$ on coastal paths) where multipath effects due to large-scale tropospheric stratification can also be encountered. Ionospheric scintillations can, at some locations, be severe at frequencies below about 6 GHz.

AN2.3.1 Tropospheric effects

The extent of tropospheric scintillations depends on the magnitude and structure of the refractive index variations. The magnitude of the scintillations increases with frequency and path length through the medium and decreases as the antenna beamwidth decreases because of aperture averaging.

In temperate countries, for an elevation angle greater than 10° and for a frequency of 20 GHz, the peak-to-peak amplitude of the fluctuations is generally less than 1 dB. Tropospheric scintillations may exceptionally exceed fades of 2 dB for small percentages of time, and periods of between a few seconds to tens of seconds. The technique of predicting the magnitude of tropospheric scintillations is documented in § 2.4 of Recommendation ITU-R P.618-4.

AN2.3.2 Ionospheric effects

Ionospheric scintillations are caused by small-scale electron density fluctuations and are known to occur under certain ionospheric disturbances such as solar and geomagnetic conditions. Geographically, there are two intense zones of scintillations, one at high latitudes and the other centred around ±20° of the magnetic equator. In these sectors, there is a pronounced night-time maximum of the activity.

Studies have reported that ionospheric scintillations at 4 GHz can have fades of several dBs and fading periods of between 1 to 10 seconds. The ionospheric events can last from 30 minutes to many hours. For an equatorial station at years of solar maximum, ionospheric scintillation occurs almost every evening after sunset. Further information on ionospheric scintillation effects is provided in Recommendation ITU-R P.531-3.

AN3 Effects of the atmosphere on polarization

Use of orthogonally polarized transmissions (linear or circular) is made in some satellite communications systems to increase channel utilization without increasing the bandwidth requirements. The technique is, however, restricted by depolarization (cross-polarization) effects arising within the atmosphere. This alters the polarization properties of a transmitted wave by causing some of the energy that is transmitted in one polarization state to be transferred to the orthogonal polarization state, thus resulting in interference between the two channels.

The effect of depolarization on a telecommunication system depends on several factors:

- frequency of operation;

- path geometry (e.g. elevation angle and tilt angle of the received polarization);

- local climatic factors (e.g. severity of the rain climate);

- sensitivity to cross-polar interference (e.g. whether the system employs frequency reuse).

For dual-orthogonal polarization systems, depolarization is often the most significant path impairment for 6/4 GHz satellite systems and is the limiting performance factor for some 14/11 GHz satellite propagation paths, especially at lower path elevation angles in moderate rain climates. For frequencies of 18 GHz and above, performance will in general be limited by fading and not by path depolarization, at least for fade margins up to 10-15 dB.

Cross-polarization may also be caused by the characteristics of the antenna systems at each terminal. This type of cross-polarized component will give rise to a basic level of interference prior to the inclusion of any propagation-induced depolarization effects.

AN3.1 Rotation of the plane of polarization due to the ionosphere

When a linearly polarized wave propagates through the ionosphere it is split into two circularly polarized waves due to the effect of the Earth's magnetic field on the ionosphere. These two rays do not travel at the same speed and on leaving the ionosphere, the rays recombine to form a linearly polarized wave which has its plane of polarization rotated in relation to that of the incident wave (referred to as Faraday rotation). The rotation exhibits fairly predictable temporal variations which may be compensated for by adjustment of the polarization tilt angle of the earth station antenna. However, few FSS systems try to compensate for Faraday rotation effects because:

- most Faraday rotation effects do not cause polarization isolation to fall below 20 dB at 6/4 GHz (with the isolation greater for higher frequencies) and most analogue systems can handle cross-polar discriminations down to 12 dB;

- large deviations from the regular behaviour can arise for small percentages of the time and cannot be predicted in advance;

- as viewed from the earth station, the linear polarization planes of the wave rotate in the same direction (e.g. clockwise) on the uplinks and downlinks. To compensate, the polarization of the antenna at the earth station would have to be rotated in opposite directions for transmission and reception. It is therefore not possible to compensate for Faraday rotation by rotating the feed system of the antenna, if the same antenna is used for both transmitting and receiving.

The magnitude of Faraday rotation is proportional to the geomagnetic field strength and the electron density of the ionosphere, and inversely proportional to the square of the frequency. It is maximum at lower frequencies, when the direction of propagation is parallel to the Earth's magnetic field, and during the day when peak ionization is encountered. Figure AN1-2 of this annex shows typical values of the angle of rotation as a function of frequency for representative Total Electron Content (TEC) values. Note, the precise value for TEC along a particular Earth-space path is difficult to predict because the electron density of the plasma along the path is highly variable. Normally, values for TEC are between 10^{16} (electrons/m^2) for low solar activity to 10^{18} (electrons/m^2) for high solar activity.

High values of rotation can arise for small time percentages for frequencies around 1 GHz or below. This has led to the use of circularly polarized antennas at these frequencies. At frequencies of 6/4 GHz the Faraday rotation is much lower giving rise to angles of several degrees even in regions where ionospheric impairments are strong (magnetic equator and auroral zones). The smaller rotational angle means that linearly polarized antenna feed systems can be used at these frequencies. Above 10 GHz, the Faraday rotation will seldom exceed 1° and can be disregarded.

AN3.2 Wave depolarization due to the troposphere

Various wave depolarization (cross-polarization) mechanisms occur in the troposphere. Above 6 GHz, the major cross-polarization effects for Earth-space paths are caused by hydrometeors. The method of predicting the long-term cross-polarization statistics is described below and is taken from § 4.1 of Recommendation ITU-R P.618-4 and the Handbook "Radiowave Propagation Information For Predictions For Earth-to-Space Path Communications".

AN3.2.1   Prediction of statistics of hydrometeor-induced cross-polarization

Long-term statistics of depolarization are calculated from rain attenuation statistics using the following parameters:

A_p: rain attenuation (dB) exceeded for the required percentage of time, p, for the path in question. This is referred to as co-polar attenuation (CPA). The procedure to calculate A_p is detailed in § AN2.1 of this annex

τ: tilt angle of the linearly polarized electric field vector with respect to the horizontal (for circular polarization use $\tau = 45°$)

f: frequency (GHz)

θ: path elevation angle (degrees)

The method calculates depolarization statistics in terms of the cross-polarization discrimination (XPD). This is expressed as the ratio of the co-polarized received signal to the cross-polarized received signal when only one polarization is transmitted. The method is valid for frequencies between 8 GHz and 35 GHz and for elevation angles ≤60°. Statistics for frequencies below 8 GHz down to 4 GHz require scaling of statistics obtained at frequencies ≥8 GHz as indicated in step 9.

Step 1: Calculate the frequency dependent term:

$$C_f = 30 \log f \qquad\qquad \text{for } 8 \leq f \leq 35 \text{ GHz} \tag{11}$$

Step 2: Calculate the rain attenuation dependent term:

$$C_A = V(f) \log A_P \tag{12}$$

where:

$$V(f) = 12.8\, f^{0.19} \qquad\qquad \text{for } 8 \leq f \leq 20 \text{ GHz}$$
$$V(f) = 22.6 \qquad\qquad\qquad\ \text{for } 20 < f \leq 35 \text{ GHz} \tag{13}$$

Step 3: Calculate the polarization improvement factor:

$$C_\tau = -10 \log \left[1 - 0.484\left(1 + \cos 4\,\tau\right)\right] \quad \text{dB} \tag{14}$$

The improvement factor $C_\tau = 0$ for $\tau = 45°$ and reaches a maximum value of 15 dB for $\tau = 0°$ or $90°$.

Step 4: Calculate the elevation angle-dependent term:

$$C_\theta = -40 \log \left(\cos \theta\right) \quad \text{dB} \quad \text{for } \theta \leq 60° \tag{15}$$

Step 5: Calculate the canting angle-dependent term:

$$C_\sigma = 0.0052\,\sigma^2 \tag{16}$$

σ is the effective standard deviation of the raindrop canting angle distribution, expressed in degrees; σ takes the value 0°, 5°, 10° and 15° for 1%, 0.1%, 0.01%, 0.001% of the time, respectively.

Step 6: Calculate rain XPD not exceeded for p% of the time:

$$(\text{XPD})_{\text{rain}} = C_f - C_A + C_\tau + C_\theta + C_\sigma \quad \text{dB} \tag{17}$$

Step 7: Calculate the ice crystal dependent term:

$$C_{\text{ice}} = (\text{XPD})_{\text{rain}} \times (0.3 + 0.1 \log p)/2 \quad \text{dB} \tag{18}$$

Step 8: Calculate the XPD not exceeded for p% of the time, including the effects of ice:

$$\text{XPD}_p = \text{XPD}_{\text{rain}} - C_{\text{ice}} \quad \text{dB} \tag{19}$$

Step 9: If required, scale XPD statistics obtained at frequencies greater than 8 GHz to give statistics for frequencies between 4 and 8 GHz.

Long-term XPD statistics obtained at one frequency and polarization tilt angle can be scaled to another frequency and tilt angle using the formula:

$$\text{XPD}_2 = \text{XPD}_1 - 20 \log \left[\frac{f_2 \sqrt{1 - 0.484 \left(1 + \cos 4\,\tau_2\right)}}{f_1 \sqrt{1 - 0.484 \left(1 + \cos 4\,\tau_1\right)}} \right] \quad \text{for } 4\,\text{GHz} \le f_1, f_2 \le 30\,\text{GHz} \tag{20}$$

where XPD_1 and XPD_2 are the XPD values not exceeded for the same percentage of time at frequencies f_1 and f_2 and polarization tilt angles, τ_1 and τ_2, respectively.

AN4 Increases in antenna noise temperature

AN4.1 Sources of radio noise

Radio noise is important in system design because it sets a limit to the performance of radio systems. For satellite systems, there are many sources of radio noise that must be considered which are both internal and external to the radio receiving system. The external noise sources arriving at a receiving antenna may be due to:

- signal absorption by atmospheric gases and hydrometeors;
- radiation from celestial radio sources such as from the galaxy or from the sun;
- the ground or other obstructions within the antenna beam;
- atmospheric noise due to lightning discharges;
- man-made noise due to electrical machinery, power transmission lines, electronic equipment, engine ignition systems, etc.

Noise may also be received as interference from unwanted transmission. Generally, only one type of external noise will predominate.

AN4.2 Atmospheric noise temperature effects on Earth-space paths

The noise temperature of satellite antennas is dominated by the high temperature emitted from the Earth, with any additional noise from precipitation or other sources being insignificant. The noise temperature is essentially constant, except in the case of a global beam where the noise temperature is dependent both on frequency and on the position of the satellite with relation to the major land masses of the Earth. Figure 9 of Recommendation ITU-R P.372 shows this dependence.

Ground-based antennas, however, observe the relatively cool sky and therefore contributions from galactic and cosmic sources, rain and the ground (at low elevation angles) can significantly raise the noise temperature of the antenna. Furthermore, for earth stations with low receiver front ends, increases in antenna noise temperature from signal absorption by rain, may have a greater impact on the resulting decrease in the signal-to-noise ratio than that due to the attenuation itself.

Additional information on the effect of radio noise on an antenna's noise temperature is provided in § AN2.1.4.3 and § AN7.2.1.2 of this handbook.

AN5 Other propagation effects

The most commonly encountered propagation factors have been presented in § AN2 to AN4 of this annex. A summary of other additional impairments are:

* attenuation due to defocusing (or beam spreading). This is signal dispersion caused by variations in the atmospheric refractive index with height creating a ray-bending effect of the signal. The magnitude of the defocusing loss is given in Recommendation ITU-R P.834. The effect can be ignored at elevation angles above about 3° at latitudes less than 53° and elevations above about 6° at higher latitudes;

* attenuation due to wave-front incoherence. This effect is caused by small scale irregularities in the refractive index structure of the atmosphere which results in an apparent decrease in an antenna's gain. The effect is unlikely to be of importance in system design in comparison with attenuation due to other causes;

* attenuation by sand and dust storms. Attenuation due to sand/dust is negligible except at frequencies well above 10 GHz, and then only if there is significant water content. Depolarization may also result from sand or dust, even when dry, at frequencies above 10 GHz, especially at elevation angles below about 10°, when the optical visibility is very low. This depolarization effect becomes even more severe in humid conditions. (See § 3.5.4 of the Handbook on Earth-space propagation.) Dry dust and sand can also be a problem if they physically interfere with the operation of the antenna system. In many cases, the strong winds that generate a dust storm may cause antenna depointing that will lead to a loss of signal strength which is difficult to distinguish from the attenuating effect of dust particles;

* attenuation due to snow and dry ice accumulations on the surfaces of antenna reflectors and feeds. This is not normally a serious problem unless the weight of the snow and ice causes the antenna to distort from a parabolic shape. Ice and snow accumulations are only a problem when

they start to melt. It has been recommended that should de-icing equipment be installed to mitigate this effect, then it only be used when the temperature is between about +4°C and -4°C;

- attenuation by the local environment of the ground terminal (buildings, trees, etc.). This is not normally a factor in fixed-satellite service systems.

AN6 Site diversity

Path diversity is a method of overcoming the high attenuation and depolarization effects by rain which are encountered at frequencies above 10 GHz. The method requires the provision of alternate propagation paths for signal transmissions, with the capability to select the least-impaired path when conditions warrant. For satellite communications systems, implementation of path diversity involves the deployment of two or more interconnected earth stations at spatially separated sites. (Note, studies have shown that the additional availability obtained by operating at more than two diversity sites is small. Furthermore, extra costs and complexity are required to deploy more than two interconnected earth stations. It is therefore normal to assume that a site diversity scheme comprises two spatially separated earth stations and the emphasis of investigations into site diversity has concentrated on this.)

The separation between the sites must be sufficient to ensure that the propagation impairments on different paths are essentially uncorrelated. This distance will depend on certain factors which include path geometry (elevation and azimuth angles), local meteorological characteristics (rain rate statistics, rain cell dimensions, etc.), transmission frequency, orientation of the baseline joining the sites, and local topographical features. The performance of a particular diversity system will also depend on the procedure for selecting and switching to the least-impaired path. A detailed discussion of how these factors affect site diversity performance is given in § 3.3.1.7 of the ITU-R Handbook "Radiowave Propagation Information For Predictions For Earth-to-Space Path Communications".

Two concepts exist for characterizing diversity performance: the "diversity improvement factor", which is defined as the ratio of the single-path time percentage to the diversity time percentage for a specified rain attenuation level; and the "diversity gain", defined as the difference (in dB) between the single-path rain attenuation to the diversity rain attenuation for a given time percentage. Both parameters are important, depending on the system design approach chosen, although the diversity gain is commonly used as the parameter for specifying diversity performance.

The procedures to predict the diversity improvement factor and the diversity gain are given below and can be applied to frequencies between 10 and 30 GHz. In broad terms, the majority of the available diversity performance is achieved when the site separation is at the minimum distance of between about 15 to 30 km. For time percentages above 0.1%, the rainfall rate is generally small and the corresponding site improvement is not significant.

AN6.1 Diversity improvement factor

For two interconnected earth-station sites, the diversity improvement factor, I, is given by (from § 2.2.4.1 of Recommendation ITU-R P.618-4):

$$I = \frac{p_1}{p_2} = \frac{1}{(1+\beta^2)}\left(1 + \frac{100\,\beta^2}{p_1}\right) \approx 1 + \frac{100\,\beta^2}{p_1} \qquad (21)$$

where p_1 and p_2 are the respective single-site and diversity time percentages, and β is the parameter depending on link characteristics. The approximation on the right-hand side of equation (21) is acceptable since β^2 is generally small.

From a large number of measurements, it has been found that β^2 can be expressed by the following empirical relationship:

$$\beta^2 = 10^{-4}\,d^{1.33} \qquad (22)$$

where d is the distance between the earth stations in km.

Figure AN1-5 of this annex shows p_2 versus p_1 on the basis of equations (21) and (22).

AN6.2 Diversity gain

For two interconnected earth-stations sites, the calculation of the diversity gain, G, is given by § 2.2.4.2 of Recommendation ITU-R P.618-4. The procedure is given below and requires the following parameters:

        d: separation between the two sites (km)

        A: path attenuation due to rain for a single site (dB)

        f: frequency (GHz)

        θ: path elevation angle (degrees)

        ψ: angle made by the azimuth of the propagation path with respect to the baseline between sites, chosen such that $\psi \leq 90°$ (degrees)

Step 1: Calculate the gain contributed by the spatial separation of the earth station sites at a separation d (km) from:

$$G_d = a\left(1 - e^{-bd}\right) \qquad (23)$$

where:

$$a = 0.78\,A - 1.94\left(1 - e^{-0.11A}\right)$$
$$b = 0.59\left(1 - e^{-0.1A}\right) \qquad (24)$$

Step 2: Calculate the gain contributed by the frequency-dependent term from:

$$G_f = e^{-0.025f} \qquad (25)$$

Step 3: Calculate the gain contributed by the elevation angle element from:

$$G_\theta = 1 + 0.006\,\theta \qquad (26)$$

Step 4: Calculate the gain contributed by the baseline-dependent element from:

$$G_\psi = 1 + 0.002\,\psi \qquad (27)$$

Step 5: Compute the net diversity gain as the product of the four diversity gain elements:

$$G = G_d G_f G_\theta G_\psi \qquad (28)$$

BIBLIOGRAPHY

[1]   ITU-R Handbook – 1996 – "Radiowave Propagation Information for Predictions for Earth-to-Space Path Communications". International Telecommunication Union.

[2]   Recommendations ITU-R P.372, P.531, P.618, TF.376, P.837, P.838, P.840 and P.841.

[3]   Satellite-to-ground radiowave propagation – 1989 – A.E. Allnutt, Institution of Electrical Engineers/Peter Peregrinus Ltd.

[4]   Propagation of radiowaves – 1996 – M. Hall, L. Barclay and M. Hewitt, Institution of Electrical Engineers/Peter Peregrinus Ltd, pages 173-195.

[5]   Satellite Communications Systems (2nd Edition) – 1991 – B.G. Evans, Institution of Electrical Engineers/Peter Peregrinus Ltd, pages 113-127.

[6]   Satellite Communications Systems (2nd Edition) – 1993 – G. Maral and M. Bousquet, John Wiley & Sons, pages 45-54.

[7]   Radiowave Propagation in Satellite Communications – 1986 – L. Ippolito, Van Nostrand Reinhold.

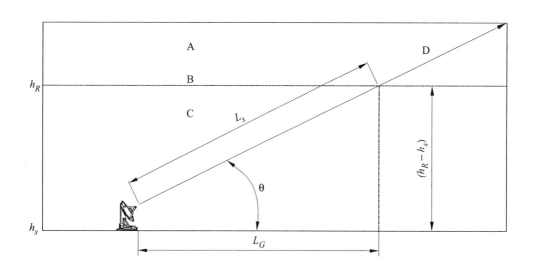

A: frozen precipitation
B: rain height
C: liquid precipitation
D: Earth-space path

Sat/C7-A1

FIGURE AN1-1

Schematic presentation of an Earth-space path giving the parameters to be input into the attenuation prediction process

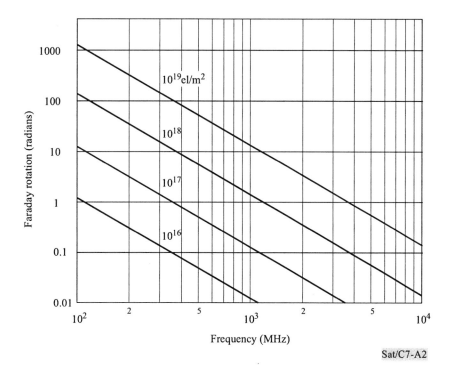

FIGURE AN1-2

Faraday rotation as a function of a Total Electron Content (TEC) and frequency

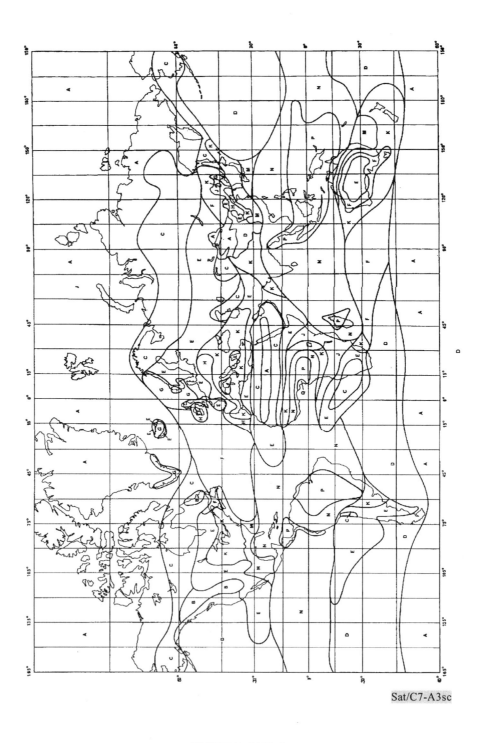

Sat/C7-A3sc

FIGURE AN1-3

Rain climatic zones (Reference to Table 1 of the Propagation Annex)

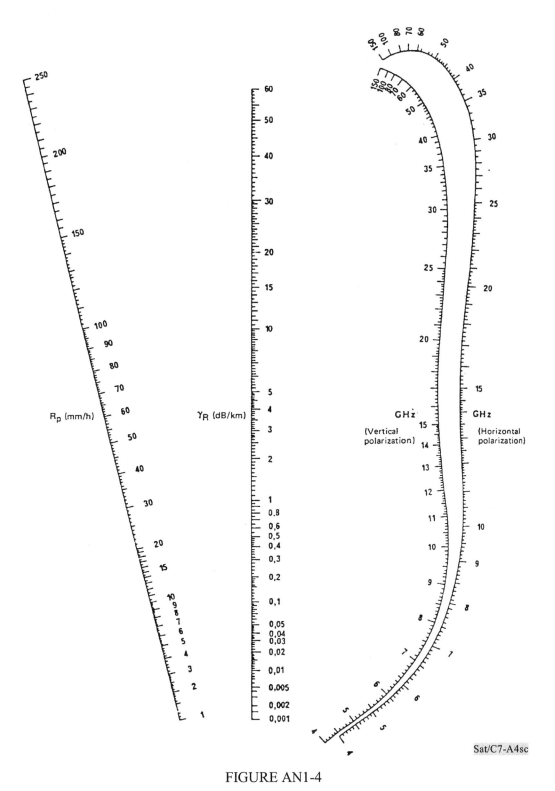

Sat/C7-A4sc

FIGURE AN1-4

Nomogram to determine the specific rainfall attenuation coefficient, γR'
as a function of the frequency (GHz) and rainfall rate density (mm/h)

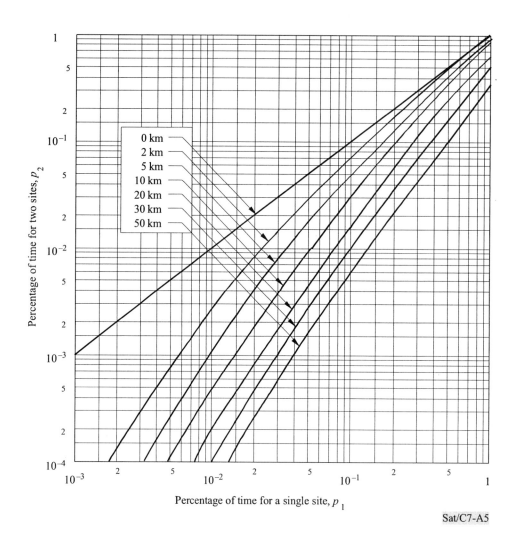

Sat/C7-A5

FIGURE AN1-5

**Relationship between percentages of time with and without diversity
for the same attenuation (Earth-satellite paths)**

TABLE AN1-1

Rainfall intensity exceeded (mm/hour)
(Reference to Figure 3 of the Propagation Annex)

Percentage of time (%)	A	B	C	D	E	F	G	H	J	K	L	M	N	P	Q
1.0	<0.1	0.5	0.7	2.1	0.6	1.7	3	2	8	1.5	2	4	5	12	24
0.3	0.8	2	2.8	4.5	2.4	4.5	7	4	13	4.2	7	11	15	34	49
0.1	2	3	5	8	6	8	12	10	20	12	15	22	35	65	72
0.03	5	6	9	13	12	15	20	18	28	23	33	40	65	105	96
0.01	8	12	15	19	22	28	30	32	35	42	60	63	95	145	115
0.003	14	21	26	29	41	54	45	55	45	70	105	95	140	200	142
0.001	22	32	42	42	70	78	65	83	55	100	150	120	180	250	170

TABLE AN1-2

Regression coefficients for estimating specific attenuation
as required in equation (6) of the Propagation Annex

Frequency (GHz)	k_H	k_V	α_H	α_V
1	0.0000387	0.0000352	0.912	0.880
2	0.000154	0.000138	0.963	0.923
4	0.000650	0.000591	1.121	1.075
6	0.00175	0.00155	1.308	1.265
7	0.00301	0.00265	1.332	1.312
8	0.00454	0.00395	1.327	1.310
10	0.0101	0.00887	1.276	1.264
12	0.0188	0.0168	1.217	1.200
15	0.0367	0.0335	1.154	1.128
20	0.0751	0.0691	1.099	1.065
25	0.124	0.113	1.061	1.030
30	0.187	0.167	1.021	1.000
35	0.263	0.233	0.979	0.963

ANNEX 2

Typical examples of link budget calculations

In order to illustrate the link budget calculation method described in § 2.3 of Chapter 2, a few typical examples are given in this annex. It should be noted that these examples being taken from actual cases, each one has its own presentation and precise procedure and calculations.

AN2.1 Analogue multiplexed telephony (FDM-FM-FDMA)

Although analogue transmissions are progressively becoming obsolete, the following calculations relate to frequency modulated transmission of frequency multiplexed telephone channels, with FDMA access. The reference parameters of an INTELSAT VI transponder (6/4 GHz, 72 MHz hemi-zone) are used. Note that the earth station implemented is an "old" high-capacity station with a 32 m antenna compliant with the previous INTELSAT "Standard A" specifications.

a)   Satellite transponder and earth station parameters

Satellite

Frequency bands	transmit	3 625-4 095 MHz
	receive	5 850-6 320 MHz

Minimum satellite G/T	-9.2 dB(K^{-1})

Power flux-density (saturation) (low gain mode)	-67.1 dB(W/m^2)

Input level to saturate TWT	-104.5 dBW

Earth stations

Antenna gain	transmit	64.5 dBi
	receive	61.0 dBi

Minimum G/T (clear sky)	40.7 dB(K^{-1})

Table AN2-1 gives the transmission parameters of the regular FDM-FM carriers in the INTELSAT system, corresponding to a quality objective of 8 200 pW0p channel noise.

b)   Reference link budget

Uplink

Uplink frequency	6.0 GHz
Earth station e.i.r.p. at satellite input back-off	85.5 dBW
Input power flux-density to saturate the transponder	-67.1 dB(W/m^2)
Transponder input back-off	-11.0 dB
$10 \log_{10}(\lambda^2/4\pi)$ (effective area of an isotropic antenna)	-37.0 dB(m^2)
Satellite G/T	-9.2 dB(K^{-1})
Uplink C/T (thermal noise)	-124.3 dB(W/K)
$10 \log_{10} b$ (b = 90% of 72 MHz)	78.1 dB/Hz
$10 \log_{10} k$	-228.6 dB(W/kHz)
Uplink C/N (thermal noise)	26.2 dB
Uplink C/I (frequency reuse within the spacecraft)	21.6 dB

Intermodulation

(C/I)$_i$	21.2 dB

This value was calculated using a computer program to optimize the performance of the transponder based upon the principle which is explained in Figure 2.11 of Chapter 2.

Downlink

Downlink frequency	4.0 GHz
Saturation e.i.r.p. of satellite (beam edge)	31.0 dBW
Output back-off	-6.6 dB
Geographical advantage	0.6 dB
Downlink attenuation	196.7 dB
Earth station G/T	40.7 dB(K^{-1})
Downlink C/T	-131.0 dB(W/K)
$10 \log_{10} b$ (b = 90% of 72 MHz)	78.1 dB/Hz
$10 \log_{10} k$	-228.6 dB(W/kHz)
Downlink C/N (thermal noise)	19.5 dB
Downlink C/I (frequency reuse within the spacecraft)	21.6 dB

The clear sky (C/N) is obtained as follows:

$$(C/N)_{\text{clear sky}} = \frac{1}{\dfrac{1}{(C/N)_u} + \dfrac{1}{(C/I)_u} + \dfrac{1}{(C/I)_i} + \dfrac{1}{(C/N)_d} + \dfrac{1}{(C/I)_u}} \tag{1}$$

$$(C/N)_{\text{clear sky}} = 14.5 \text{ dB}$$

This result indicates that an INTELSAT-VI 72 MHz hemi-zone transponder can accommodate, as an example, up to approximately 1 000 channels:

    4   (252 channel/15 MHz) carriers, or

    8   (132 channel 7.5 MHz) carriers, or

   14 (72 channel/5.0 MHz) carriers,

with adequate system margin, while meeting the quality objective stated in Table 2-1 of Chapter 2.

TABLE AN2-1

INTELSAT transmission parameters for regular FDM-FM carriers

Carrier capacity (number of channels)	Top base-band frequency (kHz)	Allo-cated satellite BW unit (MHz)	Occu-pied band-width (MHz)	Devia-tion (r.m.s.) for 0 dBm0 test tone (kHz)	Multi-channel r.m.s. devia-tion (kHz)	Carrier-to-total noise temperature ratio at operating point 8 000 + 200 pW0p from RF sources (dB(W/K))	Carrier-to-noise ratio in occupied BW (dB)	Ratio of unmodulated carrier power to maximum carrier power density under full load conditions
n	f_m	B_a	B_o	f_r	f_{mc}	C/T	C/N	(dB/4 kHz)
12	60	1.25	1.125	109	159	−154.7	13.4	20.0
24	108	2.5	2.00	164	275	−153.0	12.7	22.3
36	156	2.5	2.25	168	307	−150.0	15.1	22.8
48	204	2.5	2.25	151	292	−146.7	18.4	22.6
60	252	2.5	2.25	136	296	−144.0	21.1	22.4
60	252	5.0	4.0	270	546	−149.0	12.7	25.3
72	300	5.0	4.5	294	616	−149.1	13.0	25.8
96	408	5.0	4.5	263	584	−145.5	16.6	25.6
132	552	5.0	4.4	223	529	−141.4	20.7	24.2* (X = 1)
96	408	7.5	5.9	360	799	−148.2	12.7	27.0
132	552	7.5	6.75	376	891	−145.9	14.4	27.5
192	804	7.5	6.4	297	758	−140.6	19.9	25.8* (X = 1)
132	552	10.0	7.5	430	1 020	−147.1	12.7	28.0
192	804	10.0	9.0	457	1 167	−144.4	14.7	28.6
252	1 052	10.0	8.5	358	1 009	−139.9	19.4	27.0* (X = 1)
252	1 052	15.0	12.4	577	1 627	−144.1	13.6	30.0
312	1 300	15.0	13.5	546	1 716	−141.7	15.6	30.2
372	1 548	15.0	13.5	480	1 646	−138.9	18.4	30.1
432	1 796	15.0	13.0	401	1 479	−136.2	21.2	27.6* (X =1)
312	1 300	17.5	15.75	663	2 081	−143.3	13.2	31.2
372	1 548	·17.5	15.75	583	1 999	−140.8	15.9	31.0
432	1 796	17.5	15.75	517	1 919	−138.5	18.2	30.8
432	1 796	20.0	18.0	616	2 276	−139.9	16.1	31.5
492	2 044	20.0	18.0	558	2 200	−137.8	18.2	31.4
552	2 292	20.0	18.0	508	2 121	−136.0	20.0	30.2* (X =1)
432	1 796	25.0	20.7	729	2 688	−141.1	14.1	32.2
492	2 044	25.0	22.5	738	2 911	−140.3	14.8	32.6
552	2 292	25.0	22.5	678	2 833	−138.5	16.6	32.5
612	2 540	25.0	22.5	626	2 755	−136.9	18.1	32.4
612	2 540	36.0	32.4	983	4 325	−141.0	12.5	34.3
792	3 284	36.0	32.4	816	4 085	−137.0	16.5	34.1
972	4 028	36.0	32.4	694	3 849	−133.8	19.7	32.8* (X = 1)
792	3 284	36.0	36.0	930	4 653	−138.3	14.8	34.7
972	4 028	36.0	36.0	802	4 417	−135.2	17.8 ·	34.5

* This value is X dB lower than the value calculated according to the normal formula used to derive this ratio:

$$10 \log_{10} (f_{mc}\sqrt{2\pi / 4}$$

where: X is the value in brackets in the last column and f_{mc} is the r.m.s. multichannel derivation in kHz. This factor is necessary in order to compensate for low modulation index carriers which are not considered to have a Gaussian power density distribution.

AN2.2 Digital transmission: rural communications networks (FDMA/SCPC/PSK)

AN2.2.1 Star network

This section gives a typical example of link budget calculation for a rural communications network, e.g. for an isolated area in Africa using a star configuration satellite system, i.e. with small remote earth stations communicating through a central earth station (the hub). Digital communications, with an 8 kbit/s bit-rate are implemented in this system. This bit rate can be implemented with currently available low bit-rate encoding (LRE) codecs and may be considered as quite efficient and attractive for rural telephony systems although lower bit-rate codecs (e.g. 4.8 kbit/s or even lower) are becoming available.

The main features of star network architecture are described in Chapter 5 (§ 5.6) of this Handbook. It should be noted, however, that in the case of rural telephony, such systems are optimized when most of the traffic is operated between the local (remote) stations and the hub. If, on the contrary, an important part of the traffic is exchanged between the local stations themselves, then a mesh architecture, which does not suffer double hop communications, is preferable (see below § AN2.2.2).

Through a computer program, the calculation delivers optimized conditions, i.e. a $(E_b/N_0)_{Total}$ (including all possible interference noise and rain conditions) which correspond to quasi-error-free telephony on both outbound (hub-to-remote) and inbound (remote-to-hub) links (BER better than 10^{-7}).

The various parameters which are accounted for are summarized in Tables AN2-2a to d below. The general results of the link budget are given in Table AN2-2e. The importance of interference noise in the link budgets needs to be emphasized. This is, in fact, a very common feature in the case of GSO satellite communications with small earth stations.

Traffic capacities: The traffic capacities through the transponder can be easily calculated by referring to Table AN2-2a (satellite e.i.r.p. = 28 dBW, EOC, clear sky, accounting for the back-off) and to Table AN2-2e (e.i.r.p. per carrier = –10.6 dBW, remote-to-hub and –2.8 dBW, hub-to-remote, EOC, clear sky). The number of continuous carriers which can be transmitted through the transponder is given below. It is very important to note that the actual number of circuits can be multiplied by a factor 2.5 in case of voice activated transmissions[1].

- Transponder fully used for remote-to-hub transmissions $\cong$ 7 230 carriers

- Transponder fully used for hub-to-remote transmissions $\cong$ 1 200 carriers

- Transponder fully used for telephony two-way transmissions $\cong$ 2 048 carriers (or half-circuits) i.e. $\cong$ 1 024 circuits (or up to $\cong$ 2 560 circuits with voice activation)

---

[1] Note that the traffic capacity, although generally limited by the available transponder power (power limitation), may be, in some instances where voice activation is used, limited by the available transponder bandwidth (frequency band limitation). In fact, in the example dealt with in this section, the capacity of the transponder is limited to about 4 800 frequency slots.

TABLE AN2-2a

Rural communications star network (FDMA/SCPC/PSK): main characteristics

General
- Satellite: INTELSAT VII (at 359° E)
- Frequency bands: 6/4 GHz
- Hub earth station: 7.3 m antenna
- Remote earth stations: 1.8 m antenna
- Multiple Access: FDMA/SCPC

Carrier transmission characteristics
- Digital, 8 kbit/s
- Possible voice activation for telephony (average carrier activity: 40%)
- Composite information bit rate: 8 kbit/s
- Error correction: FEC (1/2 rate)
- QPSK modulation
- Bandwidth (Nyquist): 8 kHz

Satellite transponder parameters
- Transponder bandwidth: 72 MHz
- e.i.r.p. (EOC: edge of coverage): 33.0 dBW
- G/T (EOC): −7.5 dB/K
- Saturation Flux-Density (SFD)(EOC): −80 dBW/m^2
- Geographical advantage (GA), uplinks and downlinks: +2 dB

  (NOTE – GA is the increase of the transponder parameters due to the actual location of the earth stations in the coverage, compared to EOC.)
- Transponder Back-Off:
  - Input (BO)i : 6.8 dB (minimum) to 7.2 dB (with uplink rain)
  - Output (BO)o: 5 dB (minimum) to 5.4 dB (with uplink rain)

TABLE AN2-2b

**Rural communications star network (FDMA/SCPC/PSK):
earth station characteristics**

Earth station (ES) characteristics	Remote ES	Hub ES	Units
Antenna diameter	1.8	7.3	m
Transmit (TX) characteristics			
• Antenna gain	38.9 (0.6 efficiency)	51.0 (0.6 efficiency)	dBi
• antenna depointing loss	0.1	0.2	dB
• elevation	54.3	70.2	degree
• Antenna side lobes	$29 - 25 \log \theta$	$29 - 25 \log \theta$	dBi
• Power per carrier	0.45	0.25	W
• feeder loss	0.5	2.5	dB
• e.i.r.p. per carrier (clear sky)	34.9	42.6	dBW
• e.i.r.p. density (EOC, clear sky)	33.9	41.6	dBW/4 kHz
• Transmit off-axis emission	10.1	5.6	dBW/4 kHz
Receive (RX) characteristics			
Antenna gain	35.7 (0.65 efficiency)	35.7 (0.65 efficiency)	dBi
• antenna depointing loss	0.1	0.1	dB
• Antenna side lobes	$29 - 25 \log \theta$	$29 - 25 \log \theta$	dBi
• Noise			
• antenna noise	40	30	K
• receiver noise	50	60	K
• G/T			
Clear sky	16.1	28.3	$dB(K^{-1})$
• degraded conditions	15.8	27.6	$dB(K^{-1})$
• degraded and including rain loss	15.7	27.3	$dB(K^{-1})$

TABLE AN2-2c

**Propagation conditions: free-space attenuations, rain conditions
and availability** (assumed in the calculations)

NOTE – The link calculation is effected in view of an availability of 95% of the worst month. Taking into consideration realistic rain conditions on the uplinks and downlinks, a detailed balance calculation shows that this corresponds to include the rain attenuations shown below.

Attenuations	Remote-to-hub	Hub-to-remote	Units
Free-space attenuation			
• Uplink	199.3	199.2	dB
• Downlink	195.6	195.8	dB
Rain attenuation			
• Uplink	0.3	0.3	dB
• Downlink	0.3	0.1	dB

TABLE AN2-2d

Basic parameters for interference and intermodulation noise calculations

Basic parameters	Remote-to-hub	Hub-to-remote	Units
Uplink interference (referred to ES output)			
• Off-axis e.i.r.p. density (EOC) (NOTE 1)	10		dBW/4 kHz
• Cross-polarization (EOC) (NOTE 2)	15.2		dBW/4 kHz
• Co-polarization (EOC) (NOTE 3)	17.0		dBW/4 kHz
• Total uplink interference noise (EOC)	19.7		dBW/4 kHz
(equivalent e.i.r.p. in Nyquist bandwidth)	22.7		dBW
Downlink interference (referred to satellite output)			
• Adjacent Satellite Interference (ASI):			
• ASI on axis e.i.r.p. density (EOC) (NOTE 4)	−35.8	−23.6	dBW/4 kHz
• Cross-polarization (EOC, on-axis) (NOTE 2)	−30	−30	dBW/4 kHz
• Co-polarization (EOC) (NOTE 5)	−31	−31	dBW/4 kHz
• Total downlink interference noise (EOC)	−26.9	−22.1	dBW/4 kHz
(equivalent e.i.r.p. in Nyquist bandwidth)	−23.9	−19.1	dBW
Satellite transponder intermodulation (referred to satellite output)	−30		dBW/4 kHz

NOTE 1 – Estimated value of the off-axis emission level from an earth station transmitting to another nearby satellite (may be compared to the transmit off-axis emission quoted in Table AN2-2b).
NOTE 2 – Based on 25 dB cross-polarization (X-Polar) isolation.
NOTE 3 – Estimated value of the worst case contribution of the earth station power amplifier intermodulation.
NOTE 4 – The adjacent satellite is assumed similar to the operating satellite and separated by 3°. The ASI e.i.r.p. is corrected by the assumed attenuation of the adjacent satellite antenna pattern in the direction of the operating zone.
NOTE 5 – Estimated value of interference from other nearby satellite carriers.

TABLE AN2-2e

Rural communications star network (FDMA/SCPC/PSK): main results

	Remote-to-hub	Hub-to-remote	Units
Satellite carrier power levels			
• e.i.r.p. per carrier, maximum (EOC, clear sky)	−10.6	−2.8	dBW
• e.i.r.p. per carrier (EOC, rain on uplink)	−11	−3.2	dBW
Link budget summary			
• $(E_b/N_0)_U$ Uplink, clear sky	19.6	27.3	dB
$(E_b/I_0)_U$ Uplink, clear sky (refer to Table AN2-2d)	14.1	21.7	dB
• $(E_b/N_0)_D$ Downlink, clear sky	13.5	8.9	dB
• $(E_b/I_0)_D$ Downlink, clear sky (refer to Table AN2-2d)	13.2	16.2	dB
• $(E_b/I_0)_{IM}$ Satellite intermodulation interference noise (refer to Table AN2-2d)	16.3	24.1	dB
• $(E_b/N_0)_{Total}$ Clear sky, no interference	11.0	8.7	dB
• $(E_b/N_0)_{Total}$ Clear sky, with interference	7.8	7.8	dB
• $(E_b/N_0)_{Total}$ Rain (worst case, downlink or uplink), with interference	7.5	7.5	dB

AN2.2.2 Mesh network

As pointed out above, if an important part of the traffic is exchanged between local stations a mesh architecture should be the good choice. Demand assignment (DAMA, see Chapter 5, § 5.5) is, in general, needed to implement the various interconnections in such a network. The DAMA system is usually operated through a control station, i.e. in a centralized mode, during the call establishment and the disconnection phases.

In the case of a mesh network, there is only one type of links, i.e. remote-to-remote links. Of course, the price to be paid, in such a system, for operating all communications from (uplink) and towards (downlink) earth stations with small antennas is either a reduction in the traffic capacity and/or an increase in the remote stations transmit power and antenna size.

Tables AN2-2f and AN2-2g below give a short summary of the earth station characteristics and of a link budget calculation (similar to the one above and with the same satellite, carrier characteristics and propagation conditions) for a typical rural communications mesh network. Note that the downlink interference parameters, although not recalled here, are slightly different from those in Table AN2-2d.

Traffic capacity: Under these conditions, the traffic capacity through the transponder can again be easily calculated by referring to Table AN2-2a (satellite e.i.r.p. = 28 dBW, EOC, clear sky, accounting for the back-off) and to Table AN2-2g (e.i.r.p. per carrier = −5 dBW, EOC, clear sky). The number of continuous carriers which can be transmitted through the transponder (assumed to be fully used for telephony two-way transmissions) is about 2 000 carriers (or half-circuits) i.e. $\cong$ 1 000 circuits (or up to $\cong$ 2 500 circuits with voice activation).

TABLE AN2-2f

Rural communications mesh network (FDMA/SCPC/PSK): earth station characteristics
(Characteristics not mentioned are the same as in Table AN2-2b)

Earth station (ES) characteristics	Remote ES	Units
Antenna diameter	**2.4**	**M**
Transmit (TX) characteristics		
• Antenna gain	41.3 (0.6 efficiency)	dBi
• Power per carrier	0.9	W
• • feeder loss	0.5	dB
• e.i.r.p. per carrier (clear sky)	40.3	dBW
• e.i.r.p. density (EOC, clear sky)	39.3	dBW/4 kHz
• Transmit off-axis emission	13.1	dBW/4 kHz
Receive (RX) characteristics		
• Antenna gain	38.2 (0.65 efficiency)	dBi
• G/T		
Clear sky	18.6	$dB(K^{-1})$
• degraded conditions	18.3	$dB(K^{-1})$
• degraded and including rain loss	18.2	$dB(K^{-1})$

TABLE AN2-2g

Rural communications mesh network (FDMA/SCPC/PSK): main results

	Remote to-remote	Units
Satellite carrier power levels		
• e.i.r.p. per carrier, maximum (EOC, clear sky)	−5	dBW
• e.i.r.p. per carrier (EOC, rain on uplink)	−5.3	dBW
Link budget summary		
• $(E_b/N_0)_U$ Uplink, clear sky	25.2	dB
• $(E_b/I_0)_U$ Uplink, clear sky (refer to Table AN2-2d)	19.6	dB
• $(E_b/N_0)_D$ Downlink, clear sky	9.3	dB
• $(E_b/I_0)_D$ Downlink, clear sky (refer to Table AN2-2d)	15.7	dB
• $(E_b/I_0)_{IM}$ Satellite intermodulation interference noise (refer to Table AN2-2d)	22	dB
• $(E_b/N_0)_{Total}$ clear sky, no interference	8.9	dB
• $(E_b/N_0)_{Total}$ clear sky, with interference	7.8	dB
• $(E_b/N_0)_{Total}$ Rain (worst case, downlink or uplink), with interference	7.5	dB

AN2.3 Multiplexed digital transmission (FDMA-TDM-PSK)

This section gives a typical example of a link budget calculation for transmitting multiplexed, medium bit-rate, digital transmission continuous carriers, such as the IDR and IBS links (standardized by INTELSAT) or the SMS links (EUTELSAT) between medium-sized earth stations. Moreover, in order to diversify the examples, this particular calculation refers to a particular arrangement of INTELSAT satellite transponders, i.e. a cross-connection between 6/4 GHz and 14/12 GHz transponders.

The calculation is presented similarly and under the same format as the previous example in § AN2.2.1.

Traffic capacities: The traffic capacity resulting from this calculation is of about six carriers (8 544 kbit/s or 120 64 kbit/s channels), i.e. about 720 channels as concerns the links between a 14/12 GHz earth station and a 6/4 GHz earth station. It could theoretically be much higher for the reverse links (between a 6/4 GHz earth station and a 14/12 GHz earth station). This is due to the much higher e.i.r.p. of the 14 GHz transponder.

TABLE AN2-3a

Multiplexed digital transmission (FDMA/TDM/PSK): main characteristics

General
• Satellite: INTELSAT VII (at 307° E)
• Frequency bands: Cross-connection of 6/4 GHz and 14/12 GHz transponders
• 6/4 GHz earth station: 6.1 m antenna
• 14/12 GHz earth station: 4.5 m antenna
• Multiple Access/Multiplexing: FDMA/TDM

Carrier transmission characteristics
• Composite information bit rate: 8 544 kbit/s
• Error correction: FEC (3/4 rate)
• QPSK modulation
• Bandwidth (Nyquist): 5 696 kHz

Satellite transponder parameters

	14/12 GHz ES →6/4 GHz ES	6/4 GHz ES→ 14/12 GHz ES	Units
• e.i.r.p. (EOC)	32.6	47	dBW
• G/T (EOC)	0.9	−8.0	dB/K
• Saturation Flux-Density (SFD) (EOC)	−80	−80	dBW/m^2
• Geographical Advantage (GA) Uplink (U/L)	0	0	dB
• Geographical Advantage (GA) Downlink (D/L)	2	2	dB
• Transponder Back-Off:			
• Input (minimum/with U/L rain)	5.5/7.5	3.7/4.6	dB
• Output (minimum/with U/L rain)	4.0/6.0	3.0/3.9	dB

TABLE AN2-3b

Multiplexed digital transmission (FDMA/TDM/PSK): earth station characteristics

Earth station (ES) characteristics	14/12 GHz ES	6/4 GHz ES	Units
• Antenna diameter	4.5	6.1	m
• elevation	8	43.5	degree
Transmit (TX) characteristics			
• Antenna gain	54.5 (0.65 efficiency)	50.4 (0.65 efficiency)	dBi
• antenna depointing loss	1	0.5	dB
• Antenna side lobes	$29 - 25 \log \theta$	$29 - 25 \log \theta$	dBi
• Power per carrier	68.7	68.6	W
• feeder loss	3	3	dB
• e.i.r.p. per carrier (clear sky)	69.9	65.7	dBW
• e.i.r.p. density (EOC, clear sky)	38.4	34.2	dBW/4 kHz
• Transmit off-axis emission	0.9	0.9	dBW/4 kHz
Receive (RX) characteristics			
• Antenna gain	53.2 (0.65 efficiency)	46.3 (0.65 efficiency)	dBi
• antenna depointing loss	0.8	0.2	dB
• Antenna side lobes	$29 - 25 \log \theta$	$29 - 25 \log \theta$	dBi
• Noise			
• antenna noise	30	30	K
• receiver noise	110	50	K
• G/T			
• Clear sky	31.7	27.2	$dB(K^{-1})$
• degraded conditions	30.6	26.1	$dB(K^{-1})$
• degraded and including rain loss	29.9	25.6	$dB(K^{-1})$

TABLE AN2-3c

**Propagation conditions: free-space attenuations, rain conditions
and availability** (assumed in the calculations)

NOTE – The link calculation is effected in view of an availability of 90% of the worst month. Taking into consideration realistic rain conditions on the uplinks and downlinks, a detailed balance calculation shows that this corresponds to include the rain attenuations shown below.

Attenuations	14/12 GHz ES→ 6/4 GHz ES	6/4 GHz ES→ 14/12 GHz ES	Units
Free-space attenuation			
• Uplink	207.2	200.1	dB
• Downlink	196.0	206.2	dB
Rain attenuation			
• Uplink	14.3	0.4	dB
• Downlink	0.1	10.4	dB

TABLE AN2-3d

Basic parameters for interference and intermodulation noise calculations

Basic parameters	14/12 GHz ES →6/4 GHz ES R	6/4 GHz ES→ 14/12 GHz ES	Units
Uplink interference (referred to ES output)			
• Off-axis e.i.r.p. density (EOC) (NOTE 1)	1.0	1.0	dBW/4 kHz
• Cross-polarization (EOC) (NOTE 2)	−100	9.0	dBW/4 kHz
• Co-polarization (EOC) (NOTE 3)	16.0	15	dBW/kHz
• Total uplink interference noise (EOC) (equivalent e.i.r.p. in Nyquist bandwidth)	16.1 47.7	16.1 47.6	dBW/4 kHz dBW
Downlink interference (referred to satellite output)			
• Adjacent Satellite Interference (ASI):			
• ASI on axis e.i.r.p. density (EOC) (NOTE 4)	−39.2	−37.1	dBW/4 kHz
• Cross-polarization (EOC, on-axis) (NOTE 2)	−35	−100	dBW/4 kHz
• Co-polarization (EOC) (NOTE 5)	−36	−27	dBW/4 kHz
• Total downlink interference noise (EOC) (equivalent e.i.r.p. in Nyquist bandwidth)	−31.6 −0.1	−26.6 4.9	dBW/kHz dBW
Satellite transponder intermodulation (referred to satellite output)	−33.0	−16.0	dBW/4 kHz

NOTE 1 – Estimated value of the off-axis emission level from an earth station transmitting to another nearby satellite (may be compared to the transmit off-axis emission quoted in Table AN2-3b).

NOTE 2 – The 6/4 GHz satellite and earth station antennas operate traditionally in circular polarization and the cross-polarization (X-Polar) figures are based on a 25 dB isolation. On the contrary, the 14/12 GHz satellite and earth station antennas use linear polarization, which gives very high cross-polarization isolation figures.

NOTE 3 – Estimated value of the worst-case contribution of the earth station power amplifier intermodulation.

NOTE 4 – The adjacent satellite is assumed similar to the operating satellite and separated by 3°. The ASI e.i.r.p. is corrected by the assumed attenuation of the adjacent satellite antenna pattern in the direction of the operating zone.

NOTE 5 – Estimated value of interference from other nearby satellite carriers.

TABLE AN2-3e

Multiplexed digital transmission (FDMA/TDM/PSK): main results

	14/12 GHz ES →6/4 GHz ES	6/4 GHz ES→ 14/12 GHz ES	Units
Satellite carrier power levels			
• e.i.r.p. per carrier, maximum (EOC, clear sky)	20.8	30.9	dBW
• e.i.r.p. per carrier (EOC, rain on uplink)	18.8	30.1	dBW
Link budget summary			
• $(E_b/N_0)_U$ Uplink, clear sky	21.5	16.5	dB
• $(E_b/I_0)_U$ Uplink, clear sky (refer to Table AN2-2d)	19.2	15.9	dB
• $(E_b/N_0)_D$ Downlink, clear sky	12.1	16.5	dB
• $(E_b/I_0)_D$ Downlink, clear sky (refer to Table AN2-2d)	18.1	23.8	dB
• $(E_b/I_0)_{IM}$ Satellite intermodulation interference noise (refer to Table AN2-2d)	19.5	13.2	dB
• $(E_b/N_0)_{Total}$ Clear sky, no interference	11.0	10.3	dB
• $(E_b/N_0)_{Total}$ Clear sky, with interference	9.7	9.1	dB
• $(E_b/N_0)_{Total}$ Rain (worst case, downlink or uplink), with interference	8.7	8.7	dB

AN2.4 High bit-rate digital transmission (120 Mbit/s TDMA)

The calculations given here relate to time division multiple access transmission of pulse coded modulated telephone channels via the EUTELSAT II satellites.

AN2.4.1 Characteristics of the TDMA system

Voice channel coding	64 kbit/s PCM-DSI
Data channel coding	64 kbit/s non-interpolated
Multiple access	TDMA
Bit rate	120.832 Mbit/s
Modulation	QPSK
Demodulation	Coherent
Resolution of phase ambiguity	Unique word detection

AN2.4.2 Characteristics of the EUTELSAT II capacity used for the TDMA system

Satellite

Frequency band … uplink	14 166 to 14 500 MHz
… downlink	10 950 to 11 200 MHz
	11 616 to 11 700 MHz

Transponder useful bandwidth	72 MHz
Minimum G/T (edge of coverage)	-0.5 dB(K^{-1})
Minimum e.i.r.p. (edge of coverage)	42.5 dBW

Earth stations

Minimum G/T in the direction of the satellite	37 dB(K^{-1})
Nominal on-axis e.i.r.p.	83 dBW
Loss due to e.i.r.p. instability	0.5 dB
Noise temperature at receiver input	200 K

AN2.4.3 Performance objectives

The EUTELSAT II satellites can accommodate the EUTELSAT TDMA service by providing a BER performance quality compliant with both Recommendation ITU-R S.522 dealing with performance objectives for PCM telephony applications and Recommendation ITU-R 614 dealing with performance objectives for ISDN compatibility.

The performance objectives of Recommendations ITU-R S.522 and ITU-R S.614, expressed in terms of percentage of the available time during which given BER values can be exceeded, are given in Table AN2-4a. Additionally, the third column of the table gives a set of performance objectives expressed in terms of the total time which allows both recommendations to be met. In the case of the objectives at 10^{-3} and 10^{-6} BER the more stringent of the two recommendations, namely Recommendation ITU-R S.614, has been retained. At 10^{-3}, the required percentage of the total time (0.2%) has the value fixed in Recommendation ITU-R S.614 as a design objective. At 10^{-4}, 10^{-6} and 10^{-7}, the percentages of the total time have the same values fixed in the recommendations for the available time; this represents a conservative assumption.

Table AN2-4b gives the contributions which make up the ratios E_b/N_0 and C/T required to obtain the bit-error ratios specified in the performance objectives. It should be noted that the required E_b/N_0 is the sum of the margins plus the theoretical E_b/N_0 ratio.

It has to be noted that, if the system had to be used for applications other than telephony, e.g. broadcast, ISDN or video contribution, the performance objectives should be in line with Recommendation ITU-R S.1062 (i.e. 10^{-6} during 0.2% of the month) or the DVB Recommendation as per ETSI ETS 300421.

TABLE AN2-4a

Performance objectives to meet Recommendations ITU-R S.522 and ITU-R S.614

Recommendation ITU-R S.522		Recommendation ITU-R S.614		Recommendations ITU-R S.522 and ITU-R S.614	
BER	**% of any month available time**	**BER**	**% of any month available time**	**BER**	**% of any month total time**
10^{-3}	0.05	10^{-3}	0.03	10^{-3}	≈ 0.03
10^{-4}	0.3			10^{-4}	≈ 0.3
10^{-6}	20.0	10^{-6}	2.0	10^{-6}	≈ 2.0
		10^{-7}	10.0	10^{-7}	≈ 10.0

TABLE AN2-4b

Calculation of the carrier-to-noise temperature ratios (C/T) required for a given BER for a TDMA link via the EUTELSAT II satellites

BER	10^{-3}	10^{-4}	10^{-6}	10^{-7}
Theoretical E_b/N_0 required (dB)	6.8	8.4	10.5	11.3
Margins (dB):				
• modem implementation	0.5	0.8	1.1	1.2
• channel distortion (linear and non-linear)	1.3	1.5	2.1	2.4
• interference from co-frequency and adjacent frequency channels	2.2	2.2	1.9	1.9
• interference from other networks	1.5	1.5	1.5	1.5
E_b/N_0 required (dB)	12.3	14.4	17.1	18.3
10 log R (R: bit rate) (dB)	80.8	80.8	80.8	80.8
10 log K (dB(W/kHz))	−228.6	−228.6	−228.6	−228.6
C/T required (dB(W/K))	−135.5	−133.4	−130.7	−129.5

AN2.4.4 Link budget calculation

For each performance objective, which consists of a BER level not to be exceeded for a given percentage of time p%, link budgets are used to evaluate the available overall carrier-to-noise $(C/T)_{total}$ at the earth-station receiver so as to verify that the required C/T corresponding to the performance objective can be provided.

The carrier-to-noise ratios in the uplink $(C/T)_u$ and in the downlink $(C/T)_d$ are evaluated separately. The total C/T available is given by the equation:

$$(C/T)_{total}^{1} = (C/T)_{u}^{1} + (C/T)_{d}^{1} \tag{2}$$

It should be noted that C/T ratios in the above formula are in numerical value.

It is assumed that atmospheric attenuation does not occur simultaneously on both the uplink and the downlink; consequently, two link budgets have to be performed separately for the following configurations:

i)   atmospheric attenuation in the uplink, corresponding to a percentage p_{up}% of the total time, and clear weather in the downlink;

ii) clear weather in the uplink and atmospheric attenuation in the downlink, corresponding to a percentage p_{dw}% of the total time.

The percentage p_{up} and p_{dw} are such that their sum equals the percentage of time fixed by the performance objective:

$$p_{up}\% + p_{dw}\% = p\%$$

The correct split between p_{up} and p_{dw} is that which makes the value of $(C/T)_{total}$ evaluated under configuration i) (attenuation in the uplink) equal to the value of $(C/T)_{total}$ evaluated under configuration ii) (attenuation in the downlink). The assessment of this split is usually performed by iteration on the computer letting p_{up} vary from 0 to p, with p_{dw} given by $p - p_{up}$, until the values of $(C/T)_{total}$ in the two configurations are equal.

For pessimistic purposes, the TDMA link budgets for EUTELSAT II given here have been calculated for the case of the worst climatic conditions experienced in Europe.

Table AN2-4c gives the values of p_{up} and p_{dw} and the values of the atmospheric attenuations corresponding to these percentages of time.

TABLE 2-4C

Percentages of time and attenuations used in the link budgets of TDMA via EUTELSAT II*

Performance objective

BER	10^{-3}	10^{-4}	10^{-6}	10^{-7}
p%	0.2	0.3	2	10

Fading in the uplink

p_{up}%	0.12	0.17	1	5
Attenuation corresponding to P_{up}% of the month (dB)	7.2	5.7	1.8	0.8

Fading in the downlink

p_{dw}%	0.08	0.13	1	5
Attenuation corresponding to P_{dw}% of the month (dB)	5.6	4.2	1.3	0.4
G/T degradation of the receive earth station (dB)	2.2	2.0	0.9	0.3

AN2.4.4.1   Fading in the uplink

BER objective	10^{-3}	10^{-4}	10^{-6}	10^{-7}
Percentage of the month	0.2	0.3	2	10

a)   Calculation of the carrier-to-noise temperature ratio in the uplink

Earth station

e.i.r.p. (dBW)	83	83	83	83
Loss due to e.i.r.p. instability (dB)	−0.5	−0.5	−0.5	−0.5

Path

Free-space attenuation (dB)	−207.6	−207.6	−207.6	−207.6
Atmospheric absorption (dB)	−0.3	−0.3	−0.3	−0.3
Rain attenuation (dB)	−7.2	−5.7	−1.8	−0.8

---

*   The values have been evaluated by computed iteration.

Satellite

Minimum G/T (dB(K^{-1}))	−0.5	−0.5	−0.5	−0.5
Available (C/T)$_u$ (dB(W/K))	−133.1	−131.6	−127.7	−126.7

b)   Calculation of the carrier-to-noise temperature in the downlink

Satellite

Minimum e.i.r.p. (dBW)	42.5	42.5	42.5	42.5
Output power reduction caused by the attenuation in the uplink (dB)	−2.7	−1.9	−0.6	0

Path

Free-space attenuation (dB)	−205.3	−205.3	−205.3	−205.3
Atmospheric absorption (dB)	−0.2	−0.2	−0.2	−0.2

Earth station

Clear sky G/T (dB(K^{-1}))	37	37	37	37
Available (C/T)$_d$ (dB(W/K))	−128.7	−127.9	−126.6	−126.0

c)   Calculation of the total carrier-to-noise temperature ratio

Available (C/T)$_{total}$ (dB(W/K))	−134.4	−133.1	−130.2	−129.3

AN2.4.4.2   Fading in the downlink

BER objective	10^{-3}	10^{-4}	10^{-6}	10^{-7}
Percentage of the month	0.2	0.3	2	10

a)   Calculation of the carrier-to-noise temperature ratio in the uplink

Earth station

e.i.r.p. (dBW)	83	83	83	83
Loss due to e.i.r.p. instability (dB)	−0.5	−0.5	−0.5	−0.5

Path

Free-space attenuation (dB)	−207.6	−207.6	−207.6	−207.6
Atmospheric absorption (dB)	−0.3	−0.3	−0.3	−0.3

Satellite

Minimum G/T (dB(K^{-1}))	−0.5	−0.5	−0.5	−0.5
Available (C/T)$_u$ (dB(W/K))	−125.9	−125.9	−125.9	−125.9

b)   Calculation of the carrier-to-noise temperature in the downlink

Satellite

Minimum e.i.r.p. (dBW)	42.5	42.5	42.5	42.5

Path

Free-space attenuation (dB)	−205.3	−205.3	−205.3	−205.3
Atmospheric absorption (dB)	−0.2	−0.2	−0.2	−0.2
Rain attenuation (dB)	−5.6	−4.2	−1.3	−0.4

Earth station

Clear sky G/T (dB(K^{-1}))	37	37	37	37
G/T degradation (dB)	−2.2	−2.0	−0.9	−0.3
Available $(C/T)_d$ (dB(W/K))	−133.8	−132.2	−128.2	−126.7

c)   Calculation of the total carrier-to-noise temperature ratio

Available $(C/T)_{total}$ (dB(W/K))	−134.4	−133.1	−130.2	−129.3

AN2.4.4.3 Comparison of available and required C/T

The link model used in this example is rather pessimistic (receive and transmit earth stations at edge of coverage, worst statistical models of rain attenuation and worst interference situation) and the link budgets provide performance levels worse than those expected by real TDMA links on EUTELSAT II.

Nevertheless, even under such assumptions the available C/T provides the following positive margins over the required C/T:

BER objective	10^{-3}	10^{-4}	10^{-6}	10^{-7}
Percentage of the month	0.2	0.3	2	10
Available $(C/T)_{total}$ (dB(W/K))	−134.4	−133.1	−130.2	−129.3
Required C/T (dB(W/K))	−135.5	−133.4	−130.7	−129.5
Margin* (dB)	+1.1	+0.3	+0.5	+0.2

* Available (C/T) minus required (C/T).

AN2.5 Digital TV broadcasting (multiple programmes per transponder)

Several programmes can be carried on one transponder. They are time division multiplexed (TDM) according to MPEG 2 format specified in the ISO 13818-1 standard. The transmission over the satellite of such an MCPC (multiple channel per carrier) digital stream follows the DVB Forum recommendation which is also a European (ETSI) standard (ETS 300 421).

Table AN2-5 below shows two examples of link budgets for the EUTELSAT Hot Bird 2 satellite, one for reception with a 45 cm antenna dish diameter located on a downlink e.i.r.p. contour of 53 dBW (Case 1) and the other one for a receive antenna of 70 cm located on a downlink e.i.r.p. contour of 49 dBW (Case 2).

In both cases, the performance objective is the one defined in ETS 300 421, i.e. quasi-error free (less than one error event over one hour transmission).

The margins shown at the end of the table are the downlink rain attenuations that can be accepted with the selected FEC scheme specified in ETS 300 421 and include the receiver system noise temperature increase.

TABLE AN2-5

Reference link budget for digital TV carriers

General						Units
Satellite Service type Carrier modulation Symbol rate Outer code	EUTELSAT Hot Bird 2 Digital TV broadcasting QPSK 27.5 Shortened Reed Solomon (RS 204,188)					 MBaud
FEC Useful bit rate	1/2 25.343	2/3 33.791	3/4 38.015	5/6 42.239	7/8 44.350	Mbit/s
Uplink						
Frequency	17.5					GHz
Transmit e.i.r.p.	80.0					dBW
e.i.r.p. instability	0.5					dB
HPA intermodulation loss	0.10					dB
HPA intermodulation equivalent C/T	−122.3					dBW/K
Earth station elevation angle	33					degree
Gaseous absorption	0.47					dB
Rain attenuation	0.00					dB
Spreading loss (i.e. $10 \log (4 \pi D^2)$ with $D \approx$ 38 500 km)	162.7					dB(1/m²)
Effective area of an isotropic antenna	−46.3					dB m²
Free space loss	209.0					dB
Satellite G/T	0.0					dB/K
Uplink C/T (thermal noise)	−129.9					dBW/K
Uplink C/T (co-channel)	−127.2					dBW/K
Uplink C/T (adjacent channels)	−125.7					dBW/K
Uplink (adjacent satellites)	−116.6					dBW/K
Uplink C/T (total)	**−133.3**					dBW/K

TABLE AN2-5

(continued)

Downlink	Case 1					Case 2					
Satellite											
Operating flux-density at the GSO level	−83.64										dBW/m²
Input power flux-density at transponder saturation	−83.00										dBW/m²
Input back-off at transponder operating point	0.64										dB
Output back-off at transponder operating point	0.02										dB
TWTA intermodulation equivalent loss	0.50										dB
TWTA intermodulation equivalent C/T	**−133.1**										dBW/K
Frequency	11.9					11.9					GHz
Saturation transponder e.i.r.p. (edge of coverage)	**53.0**					**49.0**					**dBW**
Output back-off at transponder operating point	0.02					0.02					dB
Earth station elevation angle	30.0					30.0					degree
Gaseous absorption	0.17					0.17					dB
Rain attenuation	0.00					0.00					dB
Spreading loss	162.7					162.7					dB(1/m²)
Effective area of an isotropic antenna	−43.0					−43.0					dB m²
Free space loss	205.7					205.7					dB
Receive antenna diameter	**0.45**					**0.70**					**m**
Receive equivalent noise temperature (clear sky)	127.0					127.0					K
Receive earth station G/T (clear sky)	12.4					16.2					dB/K
Depointing loss	0.5					0.5					dB
Receive earth station G/T to satellite (clear sky)	11.9					15.7					dB/K
Downlink C/T (thermal noise)	−141.0					−141.2					dBW/K
Downlink C/T (co-channel)	−133.9					−134.9					dBW/K
Downlink C/T (adjacent channels)	−114.2					−115.2					dBW/K
Downlink C/T (adjacent satellites)	−126.4					−126.6					dBW/K
Downlink C/T (total)	**−141.9**					**−142.2**					dBW/K
Total											
C/T	−143.0					−143.2					dBW/K
C/N₀	85.6					85.4					dB(Hz)
C/N (in Nyquist bandwidth)	11.2					11.0					dB

FEC	1/2	2/3	3/4	5/6	7/8	1/2	2/3	3/4	5/6	7/8	
E_b/N_0	11.6	10.3	9.8	9.4	9.2	11.4	10.1	9.6	9.1	8.9	dB
E_b/N_0 for BER = 2.10⁻⁴	4.5	5.0	5.5	6.0	6.4	4.5	5.0	5.5	6.0	6.4	dB
Downlink rain attenuation margin (including receive noise increase)	4.7	3.4	2.6	1.9	1.6	4.6	3.2	2.5	1.8	1.4	dB

AN2.6 Satellite news gathering

This section gives two typical examples of link budget calculations for satellite news gathering (SNG) operation in the framework of the INTELSAT's Lease and Sales Transmission Plan programme (LST). Multi-carrier transponder operation is assumed.

The first example (Table AN2-6a) is based on the following assumptions:

- Frequency band: 6/4 GHz

- Hemispheric beam of an INTELSAT VI satellite with 1 dB antenna pattern advantage; transponder set in extra high gain operation

- SNG antenna (transmit): 4.5 m diameter

- Receive earth station G/T = 35 dB/K (antenna diameter ≈ 15 m)

- Analogue transmission by a FM carrier; occupied bandwidth ≈ 30 MHz

- Typical medium rain region in the United States.

The second example (Table AN2-6b) is based on the following assumptions:

- Frequency band: 14/11-12 GHz

- Spot beam of an INTELSAT VI satellite at edge of coverage; Transponder set in High Gain operation

- SNG antenna (transmit): 1.2 m diameter

- Receive earth station G/T = 34 dB/K (antenna diameter ≈ 9 m)

- Digital transmission by an 8 448 kbit/s QPSK modulated carrier with 3/4 FEC; occupied bandwidth ≈ 6.8 MHz

- Typical medium rain region in the United States.

TABLE AN2-6A

Typical link budget for SNG (analogue TV carriers, 6/4 GHz)

Item			Units
1	**Uplink**		
a	TV carrier e.i.r.p.	74.1	dBW
b	Path loss	199.9	dB
c	Satellite G/T (EOC: edge of coverage)	−9.2	dB/K
d	Antenna pattern advantage (geographical advantage)	1.0	dB
e	Margin for tracking error, rain, etc.	0.5	dB
	$(C/T)_{up} = a–b+c+d–e$	−134.5	dBW/K
2	**Earth station (ES) HPA intermodulation**		
a	HPA intermodulation limit toward ES	19.8	dBW/4 kHz
b	$(C/T)_{HPA\ IM}$ (limit per carrier)	−138.3	dBW/K
3	**Satellite TWT intermodulation**		
a	TWT intermodulation e.i.r.p. limit (EOC)	−37.0	dBW/4 kHz
b	$(C/T)_{TWT\ IM}$ (per carrier)	−130.8	dBW/K
4	**Downlink**		
a	Earth station (ES) elevation angle	23.3	degree
b	Downlink e.i.r.p. toward smallest ES	26.2	dBW
c	Path loss	196.4	dB
d	Smallest ES G/T	35.0	dB/K
e	Margin for tracking error, rain, etc.	0.5	dB
f	$(C/T)_{down} = b–c+d–e$	−135.7	dBW/K
5	**Co-channel interference**		
a	C/I co-channel interference, total	17.0	dB
b	$(C/T)_{co-ch\ INT}$ (total)	−136.8	dBW/K
6	**Total**		
a	$(C/T)_{Total}$	**−142.9**	dBW/K
b	Boltzmann's constant	−228.6	dBW/K-Hz
c	Receive noise bandwidth	74.8	dB-Hz
d	$(C/N)_{Total} = a–b–c$	**11.0**	dB
7	**Video signal/noise**		
a	Peak frequency deviation at cross-over frequency of the pre-emphasis network	10.75	MHz
b	Weighting factor + de-emphasis	14.8	dB
c	**(S/N) at operating C/N**	**48**	dB
	(S/N) at threshold	**46**	dB
	Availability	**99.6**	%

TABLE AN2-6b

Typical link budget for SNG (digital TV carriers, 14/11-12 GHz)

Item			Units
1	**Uplink**		
a	e.i.r.p. (per TV carrier)	62.0	dBW
b	Path loss	207.3	dB
c	Satellite G/T (EOC: edge of coverage)	1.0	dB/K
d	Antenna pattern advantage (geographical advantage)	0.0	dB
e	Margin for tracking error, rain, etc.	0.5	dB
	$(C/T)_{up}$ = a–b+c+d–e	**–144.8**	dBW/K
2	**Earth station (ES) HPA intermodulation**		
a	HPA intermodulation limit toward ES	9.8	dBW/4 kHz
b	$(C/T)_{HPA\,IM}$ (limit per carrier)	**–140.4**	dBW/K
3	**Satellite TWT intermodulation**		
a	TWT intermodulation e.i.r.p. limit (EOC)	–22.0	dBW/4 kHz
b	$(C/T)_{TWT\,IM}$ (per carrier)	**–143.3**	dBW/K
4	**Downlink**	b	
a	Earth station (ES) elevation angle	23.3	degree
b	Downlink e.i.r.p. toward smallest ES	27.3	dBW
c	Path loss	205.1	dB
d	Smallest ES G/T	34.0	dB/K
e	Margin for tracking error, rain, etc.	0.5	dB
f	$(C/T)_{down}$ = b–c+d–e	**–144.4**	dBW/K
5	**Co-channel interference**		
a	C/I co-channel interference, total	30.0	dB
b	$(C/T)_{co\text{-}ch\,INT}$ (total)	**–130.3**	dBW/K
6	**Total**		
a	$(C/T)_{Total}$ (per carrier)	**–149.6**	dBW/K
b	Boltzmann's constant	**–228.6**	dBW/K-Hz
c	Receive noise bandwidth	**68.3**	dB-Hz
d	$(C/N)_{Total}$ = a–b–c	**10.7**	dB
e	(E_b/N_0) (Information signal + Overhead)	**9.7**	dB
	(E_b/N_0) at threshold	**5.7**	dB
	Availability	**99.6**	%

AN2.7 30/20 GHz system with on-board processing (OBP)

As a typical example of an advanced system using on-board regeneration and processing and very high RF frequencies (30/20 GHz), this section gives a generic format for calculating the link budget of US GSO network using FDMA uplinks and a TDMA downlink and implementing very small VSAT earth stations.

TABLE AN2-7

Generic link budget for a FDMA/TDMA system with on-board regeneration and processing

Parameters	USA VSAT "T5"	USA VSAT "T1"	Units
Uplink			
Access type	FDMA	FDMA	
Modulation type	QPSK	QPSK	
Variable coding rate	No	No	
Noise bandwidth per carrier	0.68	0.68	MHz
Threshold C/(N+I)	8	8	dB
for % of the year where it should be exceeded	99.3	99.8	%
Downlink			
Access type	TDMA	TDMA	
Modulation type	QPSK	QPSK	
Variable coding rate	No	No	
Noise bandwidth per carrier	100	100	MHz
Threshold C/(N+I)	6	6	dB
for % of the year where it should be exceeded	99.3	99.8	%
Earth station (ES) transmit characteristics			
Elevation angle	20	20	
Rain model/zone	ITU-R/K	ITU-R/K	
On-axis e.i.r.p.	45.2	49.1	
Antenna pointing loss	0.5	0.5	
Intermodulation ES C/I (N/A if not applicable)	N/A	N/A	
Power control range	3	3	
Polarization isolation (C/I: wanted/unwanted polarization ratio)	12	12	
Earth station (ES) receive characteristics			
Elevation angle	20	20	degree
Rain model/zone	ITU-R/K	ITU-R/K	
ES receive noise temperature	275	275	K
Antenna diameter	0.7	1.1	m
On-axis antenna gain	41.8	45.7	dBi
Antenna pointing loss	0.5	0.5	dB
Polarization isolation (C/I: wanted/unwanted polarization ratio)	12	12	dB
Space station receive characteristics			
Receive frequency	29.7	29.7	GHz
Receive polarization	Circular	Circular	
Maximum receive antenna gain	45	45	dBi
Receive antenna gain in the direction of the transmitting ES	41	41	dBi
Satellite noise temperature	600	600	K
Receive cross-polarization isolation	8	8	dB
Receive frequency reuse isolation	14	14	dB

TABLE AN2-7

(continued)

Space station transmit characteristics			
Transmit frequency	19.95	19.95	GHz
Transmit polarization	Circular	Circular	
Transmit total output back-off	N/A	N/A	dB
Satellite e.i.r.p. in the direction of the receiving ES	59	59	dBW
Transmit cross-polarization isolation	N/A	N/A	dB
Transmit frequency reuse isolation	14	14	dB
Satellite adjacent transponder isolation	N/A	N/A	dB
Transponder intermodulation C/I	N/A	N/A	dB
Interference from other GSO networks and terrestrial services			
	15.5	19.4	dB
Uplink clear-sky C/I due to other GSO networks	N/A	N/A	dB
Uplink clear-sky C/I due to sharing with fixed services	13.3	17.2	dB
Downlink clear-sky C/I due to other GSO networks	N/A	N/A	dB
Downlink clear-sky C/I due to sharing with fixed services			

ANNEX 3

General overview of existing systems

This Annex provides information on the main characteristics of existing satellite systems. It should be noted that the included information has been supplied by the relevant administrations/organizations and is limited to those administrations/organizations which have answered the questionnaire sent by ITU and which have actually supplied updated information. In consequence, this Annex gives important examples but makes no claim for completeness.

AN3.1 International satellite systems

AN3.1.1 INTELSAT

AN3.1.1.1 General

INTELSAT, the world's largest satellite communication services provider, operates a global satellite system which brings television, telephone and data distribution services to users in more than 200 nations, territories and dependencies all over the world.

Founded in 1964, INTELSAT was the first organization to provide global satellite coverage and connectivity and continues to be the communications provider with the broadest reach and the most comprehensive range of services.

INTELSAT owns and operates a global satellite system implementing a fleet of 20 high-powered, technically advanced spacecraft in the geostationary orbit: the INTELSAT V/V-A series, the INTELSAT VI series and the INTELSAT VII/VII-A series. Four of the new INTELSAT VIII/VIII-A satellites are in service. The newest generation of spacecraft, the INTELSAT IX series is in production.

INTELSAT establishes the technical and operating standards for earth stations with which any INTELSAT user must comply. Thousands of earth stations, ranging in size from as large as 30 m to as small as 0.5 m or less, access the system. For the main characteristics of the standardized INTELSAT earth stations, the reader is referred to Chapter 7 (Appendix 7.1).

INTELSAT is an international not for profit cooperative of more than 140 member nations. The owners contribute capital in proportion to their relative use of the system and receive a return on their investment. Users pay a charge for all INTELSAT services, depending on the type, amount and duration of the service. Any nation may use the INTELSAT system, whether or not it is a member. Some member nations authorize several organizations to provide INTELSAT services within their countries; therefore there are currently more than 300 authorized customers.

INTELSAT customers include the major telecommunication operators throughout the world, such as:

- the world's major broadcasters (DBS, BBC, CNN, the European Broadcasting Union, the Asian Broadcasting Union, etc.) for transmitting news, sports and entertainment programming;

- basic long-distance telephone service providers (BT, Cable and Wireless, France Telecom, Deutsche Telecom, etc.);

- airlines, for transcontinental booking;

- international banks for credit verification and authorization;

- multinational manufacturers, petroleum companies, news and financial information agencies (Reuters, Agence France Presse, ITAR Tass of Russia, etc.) for facilitating their global operations;

- international newspaper distributors (the International Herald Tribune, the Financial Times, the Wall Street Journal, etc.) for simultaneous remote printing of daily editions;

- disaster relief and health care agencies and organizations, regional economic organizations, national governments and the United Nations, for fostering development and global interaction.

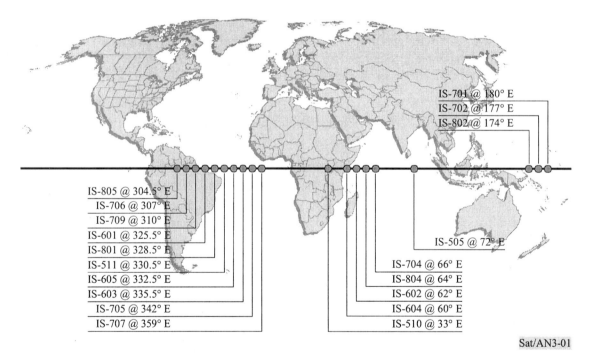

FIGURE AN3-1

The INTELSAT satellite system

AN3.1.1.2 The INTELSAT satellite system

The main technical characteristics of the INTELSAT satellites are summarized in Table AN3-1.

The currently in-orbit INTELSAT satellites, represented by Figure AN3-1, as well as the future planned spacecraft and satellite locations, are distributed as follows in the INTELSAT operating regions:

- the Atlantic Ocean Region (AOR) serving the Americas, the Caribbean, Europe, the Middle East, India and Africa with satellites at orbital locations ranging from 304.5° E to 359° E;

- the Indian Ocean Region (IOR) serving Europe, Africa, Asia, the Middle East, India and Australia with satellites at orbital locations ranging from 33° E to 66° E;

- the Asia Pacific Region (APR) serving Europe, Africa, Asia, the Middle East, India and Australia with two satellites at orbital locations 72° E and 83° E;

- the Pacific Ocean Region (POR) with coverage of Asia, Australia, the Pacific and the western part of North America with satellites at orbital locations ranging from 174° E to 180° E.

TABLE AN3-1

Evolution of INTELSAT satellites (1/2)

INTELSAT satellite series	Year of launch	Capacity (circuits/TV channel)	Total RF BW (MHz)	RF band (C: 6/4 GHz Ku: 14/10-12 GHz)	Solar array power (Watts)	Weight (kg)	Design life (years)	Features/ new technologies introduced
I	1965	240 or 1 TV channel	50	C	45	45	1.5	• First commercial communications satellite for international telephony.
II	1966	240 or 1 TV channel	125	C	75	45	3	• Multipoint communications capability between earth stations.
III	1968	1 500 and 4 TV channels	450	C	120	300	5	• Equipped with mechanically despun antenna. • Expanded multipoint communications simultaneously: telephone, telegraph, television, high-speed data and facsimile. TV service provided without interrupting telephone service.
IV	1971	4 000 and 2 TV channels	480	C	460	720	7	• SCPC service introduced. • Two steerable spot-beam transmit antennas.
IV-A	1975	6 000 and 2 TV channels	800	C	595	795	7	• Frequency reuse by spatial isolation yielding equivalent of nearly twice the number of transponders.
V	1980	12 000 and 2 TV channels	2 144	C and Ku	1 175	970	7	• Introduction of frequency reuse by polarization isolation. • Ku-band package and cross-strapped operation. • The 6/4 GHz frequencies are used four times by orthogonally polarized east and west hemispheric and zone beams. • Also carried L-band maritime communications subsystems.
V-A	1985	15 000 and 2 TV channels	2 250	C and Ku	1 475	970	7	• Frequency reuse for global beams and steerable spot beams. • Three cross polarized spot beams at 6/4 GHz for domestic leased services. • Use of nickel hydrogen batteries.
VI	1989	24 000 and 3 TV channels (up to 120 000 with DCME)	3 300	C and Ku	2 100	1 800	10	• Introduction of SS/TDMA. • The 6/4 GHz frequencies are reused six times. The 14/11 GHz frequencies are reused twice. • Higher power in the 14/11 GHz, promotes access by smaller stations.
K	1992	16 54 MHz Ku-band transponders can be configured to provide up to 32 high-quality TV channels	860	Ku	3 690	1 200	10	• First Ku-band only satellite and first Ku-band downlink services to Latin America. • Compatible with all international TV transmission formats (NTSC, PAL, SECAM, B-MAC, D-MAC and D2-MAC) and with encryption technologies. Access from earth station 1.2 m and smaller. • Made possible the first transatlantic direct-to-home broadcasting service.
VII	1993	18 000 and 3 TV channels (up to 90 000 with DCME)	2 432	C and Ku	4 000	1 437	10.9	• Independently steerable C- and Ku-band spot beams. • Four times frequency reuse at C-band and two times at Ku-band. • Solid state power amplifiers (SSPAs) at C-band and linearized travelling wave tube amplifiers (TWTAs). • Real time ability to reconfigure the satellite's coverage capabilities in-orbit.
VII-A	1995	22 500 and 3 TV channels (up to 112 500 with DCME)	3 160	C and Ku	5 000	1 823	10.9	• Same as VII. Linearized TWTAs and paralleled LTWTAS in Ku-band for a high-power mode.

AN3.1 International satellite systems

TABLE AN3-1

Evolution of INTELSAT satellites (2/2)

INTELSAT satellite series	Year of launch	Capacity (circuits/TV channel)	Total RF BW (MHz)	RF band (C: 6/4 GHz Ku: 14/10-12 GHz)	Solar array power (Watts)	Weight (kg)	Design life (Years)	Features/ new technologies introduced
VIII	1997	22 500 and 3 TV (up to 112 500 with DCME)	2 550	C and Ku	5 100	1 587	10	• Predominantly C-band spacecraft. • Highest C-band power levels ever. • Better optimization of transponder capacity via adjustable transponder gain settings from ground command. • Flexibility to optimize coverage for a specific demand through operation in either normal or inverted attitude mode. • Expanded SNG service provided by capability to connect spot beams to global beams. • New broadcast mode capability which is switchable between zone beams. • Intended for one-way broadcast TV operation and switchable on a channel-by-channel basis. • Compatibility with INTELSAT VII/VII-A satellites to facilitate transition. • Sixfold hemi/zone coverage frequency reuse in six/five channel banks in C-band. • Twofold global coverage frequency reuse of C-band expanded capacity in three channel banks. • Channel-bank (9) switchable between hemi and the global beams. • In-orbit/activation of up to a maximum of six out of 10 Ku-band transponders.
VIII-A	1998	325 digital TV channels at 5 Mbps	1 512	C and Ku	5 410	1 530	10	• Predominantly C-band spacecraft. • Higher e.i.r.p. than all previous INTELSAT satellites. • Linearizers at C-band and at Ku-band. • Increased TWTA and bus spacecraft power. • Complex coverages provided utilizing reflector antenna technology. • Twofold frequency reuse at C-band with high antenna isolation. • C-band coverage available for either dual linear or dual circular polar mode.
KTV	1999 (Plan)	210 SCPC digital TV channels at 5 Mbps	1 080	Ku	7 502	1 441	12	• A high-power Ku-band satellite specifically for direct-to-home TV service and for VSAT network applications, including Internet/intranet. • Common repeater configuration with alternative "orbit specific" beam coverages, and a transponder frequency plan for reduced complexity and increased utilization flexibility. • Switchable "Automatic Level Control" (ALC) feature to maintain the transponder output power within 0.3 dB for an uplink rain fade of up to 12 dB. • Ability to operate in collocation with up to three satellites, such that all satellites are within the nominal N-S and E-W station-keeping limits.
IX	2000 (Plan)	32 000 circuits assuming 50% IDR/ITEM implementation and 8 digital TV carriers (up to 160 000 circuits with DCME)	3 456	C and Ku	8 085	1 900	13	• Designed to replace INTELSAT VI satellites in the IOR and AOR, with an "ocean region" specific antenna coverage and a common repeater configuration for reduced complexity. • Ability to correct beam deviations in inclined orbit operation of up to 3°. • INTELSAT's largest capacity satellite with superior RF performance. • Like in the INTELSAT VII/VIIA satellites, ability to match east-west traffic imbalances by using a sixfold frequency reuse transponder configuration on a sevenfold reuse hemi/zone antenna design.

AN3.1.1.3   INTELSAT earth stations

The main characteristics of the INTELSAT standard earth stations are summarized in Chapter 7 (Appendix 7.1).

AN3.1.1.4   INTELSAT services

- Voice/Data: INTELSAT international and domestic communications voice and data transmission services – either for private, business or specialized networks – can be offered in a variety of ways in order to give customers maximum flexibility: they include on-demand applications providing service on a per-minute basis, as well as long-term capacity options. In particular:

  - in order to meet needs for thin-route, public switched, communication requirements, INTELSAT has developed a digital, on-demand, satellite service for use in domestic and international rural networks, which provides thin-route operators instant dial-up global connectivity at a very minimal cost;

  - INTELSAT's TDMA service offers maximum digital connectivity and flexibility to customers with large traffic needs;

  - intermediate data rate service – IDR – is the foundation service to INTELSAT's public switched network system. It provides ISDN quality international and domestic public switched services and is the ideal solution for PSN operators with point-to-point or low connectivity requirements. IDR carriers may be configured for different capacity rates, from 64 kbit/s to 155 Mbit/s, the most popular user rates being 2.048 or 1.544 Mbit/s. They can be used for the full range of international and domestic PSN applications, including: voice, data, videoconferencing, digital television, audio, printed material distribution and, more generally, for all ISDN applications or for dedicated private digital networks (virtual private networks);

  - INTELSAT also provides private business networks for a growing number of customers worldwide. For example, Venezuela and Colombia recently implemented a new 14/11 GHz private business service on the INTELSAT VII satellite at 310° E, incorporating advanced digital voice, data and television applications through small earth stations with 1.8 m antennas;

  - looking toward the future, INTELSAT is positioning itself to offer a range of Internet and multimedia applications. Satellites are actually offering promising opportunities for serving these evolving markets.

- Video services: with the largest distribution network in the world, INTELSAT can deliver a global audience or the most targeted narrowcast viewing programme, from the Olympic Games and World Cup football to local sitcoms. The INTELSAT Video Services provide technologically advanced features – including, e.g. digital compression technology – and offer comprehensive experience in satellite, video and customer support.

AN3.1.2   Intersputnik

AN3.1.2.1 General

Intersputnik is an international intergovernmental organization with its principal business in the field of operation of the global satellite communication system. The organization was founded in 1971 in accordance with the intergovernmental Agreement on the Establishment of the Intersputnik International System and Organization of Space Communications signed on 15 November.

The legal status of Intersputnik was also defined by the Agreement on the Legal Capacity, Privileges and Immunities of the Intersputnik International Organization of Space Communications of 20 September 1976; Organization's Charter approved by the XVIInd session of the Intersputnik Board on 16 December 1977 and by a number of other international legal documents. Owing to the transition of the Organization to the commercial operation of the satellite system, new regulatory documents were developed, such as the Protocol on Amendments to the Agreement on the Establishment of the Intersputnik International System and Organization of Space Communications dated 15 November 1971 and Intersputnik's Operation Agreement. These documents will come into force after the Depositary of the Agreement, the Russian Federation Government, receives the approval of the above documents from the two thirds of the governments, members of the Organization.

At present Intersputnik has 23 member countries: the Islamic State of Afghanistan, the Republic of Bulgaria, the Republic of Belarus, the Republic of Hungary, the Socialist Republic of Viet Nam, the Federal Republic of Germany, Georgia, the Republic of Yemen, the Democratic People's Republic of Korea, the Republic of Kazakstan, the Kyrgyz Republic, Cuba, Lao People's Democratic Republic, Mongolia, Nicaragua, the Republic of Poland, Romania, the Russian Federation, the Syrian Arab Republic, the Republic of Tajikistan, Turkmenistan, Ukraine and the Czech Republic.

Intersputnik benefits from a status of observer at the United Nations Committee on the Peaceful Uses for Outer Space; International Telecommunication Union and UNESCO and actively participates in the activity of the above organizations. Intersputnik is a member and is on the Asia-Pacific Satellite Communication Council (headquartered in Seoul) and Telecom Forum (Moscow). Intersputnik maintains relationships and develops cooperation with other global, regional and private organizations of satellite communications.

There are more than 100 state and private run companies from all over the world among Intersputnik's customers.

AN3.1.2.2 The Intersputnik satellite system

The Intersputnik space segment comprises the communication satellites owned by the Russian Federation (Gorizont, Express and Gals type) and LMI-series satellites designed within the framework of the Lockheed Martin Intersputnik joint venture. The project calls for the launch of up to four state-of-the-art communication satellites by the year 2000, based on the A2100 platform (developed by Lockheed Martin). The first LMI satellite will be launched by the Proton launch vehicle in the second half of 1999.

At present, Intersputnik uses the satellites located at seven orbital slots. Fifteen GEO slots were filed by Intersputnik for its planned networks. All of the 15 slots are at the stage of coordination.

The Intersputnik satellites are operated and controlled from the ASOC Sunnyvale (United States), Dubna (Russia), the technical characteristics of satellites are monitored by TT&C monitoring stations Duban (Russia), Shipka (Bulgaria) and Subic Bay (the Philippines).

The main technical characteristics of the Intersputnik satellites as of the middle of 1999 are shown in Table AN3-2.

AN3.1 International satellite systems

TABLE AN3-2

Main technical characteristics of the Intersputnik satellites (mid-1999)

Satellite	Gorizont	Gals	Gals-R16	Express	Express-A	LMI-1	LMI-2	Yamal-90E
First launch	1981	1994	1999	1994	1999	1999	2001	1999
Prime contractor	NPO PM* Russia	Inform-Cosmos Russia	NPO PM* Russia	Inform-cosmos Russia	NPO PM* Russia	Lockheed Martin	Lockheed Martin	RKK (Energia/Space Systems/Loral)
Lifetime, years	5	5-7	10	5-7	10	15	15	10-15
Launcher	Proton	Proton	Proton	Proton	Proton	Proton	Proton	Proton
Number of transponders (bandwidth, MHz)	– 6/4 GHz band: 5 (36) – 14/10-12 GHz band: 1 (36)	– 14/10-12 GHz band: 3 (27)	– 14/10-12 GHz band: 16 (33)	– 6/4 GHz band: 8 (36) + 1 (40) – 14/10-12 GHz band: 2 (36)	– 6/4 GHz band: 11 (36) + 1 (40) – 14/10-12 GHz band: 5 (36)	– 6/4 GHz band: 28 (36) – 14/10-12 GHz band: 16 (27)	– 6/4 GHz band: 24 (36) – 14/10-12 GHz band: 24 (36, 72)	– 6/4 GHz band: 10 (36)
e.i.r.p. (max), (dBW)***	– 6/4 GHz band: 31-49.3 – 14/12 GHz band: 42.3	– 14/10-12 GHz band: 55	– 14/10-12 GHz band: 50-54	– 6/4 GHz band: 31-48 – 14/10-12 GHz band: 42	– 6/4 GHz band: 36.5-48 – 14/10-12 GHz band: 43.5	– 6/4 GHz band: 35-37 – 14/10-12 GHz band: 47	– 6/4 GHz band: 42-45 – 14/10-12 GHz band: 50-54	– 6/4 GHz band: 39-40
Mass in orbit, (kg)	2 120	2 500	2 570*	2 500	2 600	1 730**		1 300
Station-keeping accuracy, (degrees)	±2.0 N and S ±0.2 E and W	±0.2 E and W; N and S	±0.1 E and W; N and S	±0.2 E and W; N and S	±0.2 E and W; N and S	±0.05 E and W; N and S	±0.05 E and W; N and S	±0.1 E and W; N and S
Power (W)	1 300	2 400	5 000	2 400	2 540	7 560		2 500

* NPO PM – Applied Mechanics Research and Production Association.
** Satellite weight (dry), kg.
*** Depending on the used antennas and service areas.

AN3.1.2.3 Intersputnik earth stations

Intersputnik ISSC also includes the ground segment comprising earth stations, which are the property of the state's duly authorized operations organizations. These stations are granted access to the system provided they meet the technical requirements set forth in the Intersputnik Regulations and approved by the Board. At present, there are more than 80 stations in the Intersputnik system all over the world.

The main technical characteristics of the Intersputnik standard earth stations (ES) are summarized in Table AN3-3: the 6/4 GHz (or C-band) ES standards are designated with a capital "C" followed by a number and the 14/11 GHz and 13/12 GHz (or Ku-band) ES standards are designated with a capital "K", also followed by a number.

TABLE AN3-3

Intersputnik standard earth station characteristics

Earth station standard	G/T* $(dB(K^{-1}))$	Typical antenna diameter** (m)	Basic services and functions
C1 K1	≥31.0 ≥38.0	9-12 9-10	Exchange of any traffic including video and audio programmes, voice, fax and telex messages, data services, videoconferencing, etc.
C2 K2	≥28.0 ≥35.0	6.5-7.5 6.5-7.5	
C3 K3	≥23.5 ≥30.0	3.5-5 3.5-5	Voice, fax and telex messages, data services, videoconferencing. Reception of television and audio programmes.
C4 K4	≥19.3 ≥25.3	2-3 2-3	Voice, fax and telex messages, data services. Reception of television and audio programmes.

NOTES:

* G/T is given for clear-sky conditions, for an antenna elevation $\beta = 5°$ (C Standards) or $\beta = 10°$ (K Standards) and for centre frequency of receive operating band.

** Small antenna diameter corresponds to a LNA with a low noise temperature ≤40° K (C Standards) or ≤ 50° K (K Standards).

AN3.1.2.4 Services provided by the Intersputnik system

Based on the alliances on integrated services, the transponders leased from Intersputnik can be used to establish all kinds of networks:

- Operator networks:
  - administrative connection;
  - second type operators;
    - international syndicates;
    - regional and local telephone stations.

- Television networks:
  - broadcasting companies;
  - cable broadcasting;
  - occasional programmes;
  - videoconferencing;
  - tele-education;
  - telemedicine;
  - educational programmes;
  - radio broadcasting.
- Internet networks:
  - national providers;
  - Internet providers;
  - interactive multimedia.
- Corporate networks;
  - financial and bank establishments;
  - oil and gas companies;
  - automobile companies;
  - retail networks;
  - direct PC information services;
  - tele-education networks;
  - medical and pharmaceutical establishments;
  - emergency networks.
- Time payment services:
  - data transmission;
  - public switched telephone network;
  - open-end lines.

AN3.1.3 Inmarsat

AN3.1.3.1 The formation and organization of Inmarsat

The International Mobile Satellite Organization or Inmarsat (previously named International Maritime Satellite Organization) is an internationally owned cooperative that provides mobile satellite communications worldwide. Established in 1979 to serve the maritime community, Inmarsat has since evolved to become the only provider of global mobile satellite communications for commercial and distress and safety applications, at sea, in the air and on land. Headquartered in London, Inmarsat has 86 member countries.

On 24 September 1998 Inmarsat's Assembly of member governments agreed by consensus that Inmarsat will become a privatized company from April 1999. The new structure will comprise two entities:

- a public limited company that will seek an initial public offering (IPO) within approximately two years of formation; and

- an intergovernmental body to ensure Inmarsat meets its public service obligations, including the Global Maritime Distress and Safety System (GMDSS).

The new company will be governed by a 15-member fiduciary Board of Directors comprising the Chief Executive Officer and 14 non-executive directors, three of which will represent developing countries.

AN3.1.3.2 Current Inmarsat services

The services supported by Inmarsat satellite networks include direct-dial phone, telex, fax, electronic mail and data connections for maritime applications; flight deck voice and data, automatic position and status reporting, and direct-dial passenger telephone, fax and data communications for aircraft and in-vehicle and portable phone, fax and two-way data communications, position reporting, electronic mail and fleet management for land transport. Inmarsat is also used for disaster and emergency communications and by media for news reporting from areas where communications would otherwise be difficult or impossible. Systems are also available for temporary or fixed communications to areas beyond the reach of normal communications.

Inmarsat offers several different mobile communications systems designed to provide users with a wide variety of mobile terminals and services:

- Inmarsat-phone is a digital phone, fax and data system using very compact terminals. Portable Inmarsat-phone terminals range in size from a small laptop computer upwards and can provide direct-dial phone, fax or 2.4 kbit/s data connections. Versions for mounting on cars and other vehicles are also available, along with maritime models. User charges are around USD 3 per minute.

- Inmarsat-B is the digital successor to the Inmarsat-A. The maritime version of these terminals features a dynamically-driven parabolic antenna less than one metre in diameter, generally housed in a radome and mounted high on the ship's superstructure. Inmarsat-B can support high-quality direct-dial phone, telex, fax and data up to 64 kbit/s. Inmarsat-B terminals are also produced in the transportable form; a system and its folding antenna can fit into one or two suitcase-size containers.

- Inmarsat-A is the predecessor to Inmarsat-B. It supports a similar range of services, but is more expensive to use because it is based on analogue telecommunications technologies and is less efficient in its use of satellite resources.

- The Inmarsat-C system provides data messaging communications through small, lightweight terminals. These come in fixed, mobile, transportable, maritime and aeronautical forms all with omnidirectional antennas. Inmarsat-C supports two-way, store-and-forward messaging, text and data-reporting communications at a data rate of 600 bit/s.

- Four systems are available for aeronautical services:

  - Aero-C, which allows store-and-forward text and data messages – flight safety communications excluded – to be sent and received by aircraft operating anywhere in the world;

  - Aero-L (low-gain antenna terminal) and low-speed (600 bit/s) real-time data communications, mainly for airline operational and administrative purposes;

  - Aero-I (intermediate gain antenna terminal) for multiple channel passenger and operational voice and data services for national and regional aircraft;

  - Aero-H (high-gain antenna terminal), a high-speed (up to 10.5 kbit/s) service supporting multichannel voice, fax and data communications for passenger and aircraft operational and administrative applications, for larger, intercontinental aircraft.

AN3.1.3.3 The Inmarsat space segment

The Inmarsat satellite network provides global mobile satellite communications services via its own INMARSAT-2 (2nd-generation satellites) and INMARSAT-3 (3rd-generation) satellites.

Inmarsat operates a total of four INMARSAT-2 satellites. Launched in 1990-92, they each have a capacity equivalent to about 250 INMARSAT-A voice circuits.

In 1996 Inmarsat launched the first of five of the INMARSAT-3 satellites into orbit. These use the latest spot-beam technology and higher power to supply voice and data communications services worldwide to mobile terminals as small as pocket-size messaging units. The simultaneous voice channel capacity of INMARSAT-3 is up to eight times that of INMARSAT-2. Furthermore, INMARSAT-3 also carries a navigation transponder payload designed to enhance the accuracy, availability and integrity of the GPS and Glonass satellite navigation systems.

The main characteristics of the Inmarsat satellites are summarized in Table AN3-4 below.

A decision on Inmarsat's third generation follow-on system is shortly expected.

AN3.1.3.4 The Inmarsat ground segment

The Inmarsat ground segment comprises the mobile earth stations (MES), the network control stations (NCS) and the land earth stations (LES).

The MESs are the portable and transportable terminals, which provide the various types of service to the user. As of April 1999, more than 140 000 terminals of different types (Inmarsat-A, -B, -C, -D, -phone (M and mini-M) and aeronautical terminals) have been commissioned for use with the Inmarsat system.

TABLE AN3-4

Main characteristics of the Inmarsat satellites

	INMARSAT-3	INMARSAT-2
Launch period	1996-1998	1990-1992
Launch mass	2 066 kg	1 300 kg
Power	2.8 kW (end of life)	1.2 kW (initial)
1.5 GHz band service link e.i.r.p. (edge of coverage - eoc)	49 dBW (spot) or 44 dBW (global)	39.0 dBW
4 GHz band feeder link e.i.r.p. (eoc)	30 dBW	24 dBW
6 GHz band feeder link G/T (eoc)	-13 dB(K^{-1})	-14 dB(K^{-1})
Launchers	Atlas IIA, Proton, Ariane 4	Delta, Ariane 4
Constellation (INMARSAT-2/3 located within:)	142° W, 98° W, 55° W, 54° W, 17° W, 15.5° W, 25° W, 64° E, 65° E, 109° E, 178° E, 179° E.	

The LES – sometimes referred to as coast earth stations (CES) in the maritime environment and ground earth stations (GES) in the aeronautical environment – link the Inmarsat's satellites with the national and international telecommunication network. The LES are owned independently by telecommunication operators. These are often, but not always, the national signatory – the organization nominated by the government to invest in and work with Inmarsat – of the country in which the LES is located. One LES can support multiple services providers. There are more than 30 LESs distributed around the globe, with at least one on every continent.

The NCSs control the flow of traffic through each operational satellite, using demand assigned multiple access techniques.

AN3.1.3.5 INMARSAT-3 frequency bands

The Inmarsat satellite network operates in two separate frequency bands, the 1.6/1.5 GHz band for the mobile services (MSS) and the 6/4 GHz band for the feeder links and TT&C links (FSS).

The precise frequency bands used by INMARSAT-3 are shown in Table AN3-5.

TABLE AN3-5

Inmarsat frequency bands

	Downlink frequencies (MHz)	Uplink frequencies (MHz)
Service links (used by the MESs)	1 525.0-1 544.0 and 1 545.0-1 559.0	1 626.5-1 645.5 and 1 646.5-1 660.5
Feeder links (used by the LESs)	3 599.0-3 614.0 and 3 615.0-3 629.0	6 424.0-6 439.0 and 6 440.0-6 454.0
Telecommand, telemetry and ranging: • **telemetry** • **telecommand and ranging (INMARSAT-3 normal operation)** • **telecommand and ranging (INMARSAT-3 emergency)** • **telecommand and ranging (INMARSAT-2)**	3 945.0-3 955.0	6 338.0-6 342.0 6 420.0-6 425.0 6 170.0-6 180.0
Navigation payload (INMARSAT-3)	1 574.4-1 576.6 and 3 629.4-3 631.6	6 454.4-6 456.6

AN3.2 Regional satellite systems

AN3.2.1   APSTAR satellite system (APT)

AN3.2.1.1   General

APT Satellite Holdings Limited ("APT"), is a private company listed both in Hong Kong and New York. Its wholly-owned subsidiary, APT Satellite Company Limited, was established in 1992. It primarily provides high-quality satellite transponder services for international and Asia-Pacific broadcasting and telecommunication sectors and has achieved remarkable performance.

AN3.2.1.2 APSTAR space segment

APT currently owns and operates three in-orbit satellites, APSTAR-I, APSTAR-IA and APSTAR-IIR, through its own satellite control centre located in Hong Kong.

The APSTAR-I satellite operates in the 6/4 GHz band with 24 transponders. It was launched in July 1994 by a Long March 3 launch vehicle from Xichang, People's Republic of China (PRC) with a design lifetime of at least ten years. It was placed at 138° E and has an extensive footprint that covers the PRC, Japan and South-East Asia.

The APSTAR-IA satellite is similar to APSTAR-I. It was launched in July 1996, also by a Long March 3 launch vehicle from Xichang with a design lifetime of at least ten years. APSTAR-IA was placed at 134° E and has a similar footprint to APSTAR-I that covers the PRC, Japan and South-East Asia, but with an enhanced coverage in Southern Asia.

The APSTAR-IIR, the third satellite of APT operates both in the 6/4 GHz band with 28 transponders and in the 14/12 GHz band with 16 Ku-band high power transponders. APSTAR-IIR was launched in November 1997 by a Long March 3B launch vehicle from Xichang and was located at 76.5° E. It has a very broad 6/4 GHz band footprint that covers over 100 countries across Asia, Europe, Africa and Australia, or approximately 75% of the world's population.

The 14/12 GHz band transponders mainly cover the PRC including Hong Kong, Macau and Taiwan, with two different, switchable antenna beams. Beam 1 is mainly for VSAT telecommunication networks whereas Beam 2 is mainly for direct satellite TV broadcasting, capable of being received by 0.5 m antennas.

The main technical characteristics of the APSTAR satellites communication subsystems are given in Table AN3-6 below.

AN3.2.1.3 APSTAR satellite services

APT provides transponder leasing services to leading international and PRC broadcasting customers and telecommunication networks operators. APT has quickly built up a well-established and diversified base of prominent international and PRC customers such as China Central Television, ChinaSat, Walt Disney, ESPN, HBO, Sony Pictures Entertainment, WTCI, Turner, Reuters and Viacom.

All satellites are monitored and controlled by APT's Satellite Control Centre ("SCC"), which is equipped with an 11 m antenna (full motion, monopulse tracking) and operated by a team of about 40 specialists and engineers. The ground segment currently comprises seven antennas (up to 13 m for C-band and 3 m for Ku-band) and nine transmission facilities as well as associated equipment.

TABLE AN3-6

APSTAR satellites communication subsystems

Satellite	Frequency bands (GHz): uplink/ downlink (polarization)	Number of operational transponders (bandwidth, B, MHz) (transmitter power P, W)	Coverage	e.i.r.p. (edge of coverage - eoc) (dBW)	G/T (eoc) $(dB(K^{-1}))$	Saturating power flux-density $dB(W/m^{-2})$
APSTAR- 1	6/4 (Linear)	20 (B = 36) (TWTA, P = 15)	PRC, Japan, South-East Asia	34	–4	–90 to –76
		4 (B = 72) (TWTA, P = 15)				
APSTAR- 1A	6/4 (Linear)	20 (B = 36) (TWTA, P = 16)	PRC, Japan, South-East Asia, Southern Asia (enhanced)	34	–4	–88 to –74
		4 (B = 72) (TWTA, P = 16)				
APSTAR- 2R	6/4 (Linear)	1 (B = 30) (TWTA, P = 15)	100 countries over: Asia, Europe, Africa, Australia	33	–10.4	–98 to –77
		27 (B = 36) (TWTA, P = 60)				
	14/12 (linear)	1 (B = 36) (TWTA, P = 110)	PRC incl. Hong Kong, Macau, Taiwan	Beam 1 (VSAT): 46 Beam 2 (TVRO): 45	+0.6	
		15 (B = 54) (TWTA, P = 110)				
NOTE – The 36 MHz transponder and three of the 54 MHz transponders can be switched between Beam 1 and Beam 2.						

AN3.2.2   AsiaSat

AN3.2.2.1 General

Asia Satellite Telecommunications Co. Ltd. (AsiaSat) was formed in 1988 as Asia's first privately-owned regional satellite operator. AsiaSat has been dedicated to providing satellite telecommunications and broadcasting services throughout the Asia Pacific region. Its ultimate holding company, Asia Satellite Telecommunications Holdings Limited, was listed on the Hong Kong and New York stock exchanges in June 1996.

AsiaSat's satellite footprints cover more than 50 countries, serving over 60% of the world's population.

AsiaSat provides telecommunications and broadcast facilities to governments, telecommunication service operators, wholesale news agencies, financial institutions, securities exchanges, television and radio broadcasters, Internet service providers and industries such as aviation, print media and petroleum.

AN3.2.2.2   AsiaSat space segment

AsiaSat presently owns and operates two satellites, ASIASAT 2 and ASIASAT 3S. Both ASIASAT 2 and ASIASAT 3S are 3-axis body stabilized satellites. ASIASAT 2, operating at 100.5° E, was launched in November 1995 with a 13-year operational life. ASIASAT 3S was launched on 21 March 1999 and started commercial service on 8 May. It replaces ASIASAT 1 at 105.5° E and has a 15-year operational life. ASIASAT 1 has been drifted away from 105.5° E and will reach 122° E orbital location in early August.

The main characteristics of the satellite communication subsystems are summarized in Table AN3-7 (A).

TABLE AN3-7 (A)

AsiaSat satellites communication subsystems

Satellite	Frequency bands (GHz): uplink/ downlink (polarization)	Number of operational transponders (bandwidth, B, MHz) (transmitter power P, W)	Coverage	e.i.r.p. (dBW)	G/T $(dB(K^{-1}))$ (max.)	Saturating power flux-density $dB(W/m^{-2})$
ASIASAT 2	6/4 (linear)	20 (B = 36) (TWTA, P = 55)	Asia, Middle East, CIS, Australasia	40	0	-97 ± 1
		4 (B = 72) (TWTA, P = 55)				
	14/12 (linear)	9 (B = 54) (TWTA, P = 115)	Greater China Region, Korea, Japan	53	6	
ASIASAT 3S	6/4 (linear)	28 (B = 36) (TWTA, P = 55)	Asia, Middle East, CIS, Australasia	41	-1	-98 @G/T = -1 dB(K^{-1})
	14/12 (linear)	16 (B = 54) (TWTA, P = 140)	3 beams: East Asia, South Asia + 1 steerable beam	54	7	-100 @G/T = 7 dB(K^{-1})

AN3.2.2.3 AsiaSat earth segment

The main characteristics of the earth stations in service on AsiaSat are summarized in Table AN3-7 (B).

TABLE AN3-7 (B)

AsiaSat earth stations

Satellite	Frequency bands (GHz): uplink/downlink	Antenna diameter (m)	Antenna type
ASIASAT 1	6/4	11	Step track
ASIASAT 2	6/4	9	Full motion
	14/12	6	Step track
ASIASAT 3S	6/4	11	Full motion
		5	Step track
	14/12	2 x 6 m	Step track

AN3.2.3   EUTELSAT

EUTELSAT, the European Telecommunications Satellite Organization operates the EUTELSAT space segment. EUTELSAT is an international organization created in September 1985 which currently comprises 47 member countries. The investors are the signatory entities, private or public telecommunication operators of these countries, designated by their respective governments which are parties to the EUTELSAT Convention.

AN3.2.3.1 Space segment

All EUTELSAT satellites operate in the 14/11-12 GHz bands.

Two satellites of the first generation (EUTELSAT I-F4 and -F5) currently remain in service. As of June 1999, a total of seven transponders with 83 MHz bandwidth and dual orthogonal linear polarizations remain available for service (four transponders on EUTELSAT I-F4 and three transponders on EUTELSAT I-F5). The EUTELSAT II satellites, introduced between 1990 and 1995, offer more capacity (16 transponders 72 MHz and 36 MHz bandwidth wide) and improved European coverage, including a high gain beam for TV direct-to-home (DTH) distribution.

In 1996, EUTELSAT opened service with its third-generation satellites, which comprises two series: the "HOT BIRD™" family (EUTELSAT II-F6, renamed HOT BIRD™ 1 in 1995), dedicated to entertainment TV services at 13° E and the "W" series for telecommunication services on the 7° E to 48° E orbital arc.

HOT BIRD™s 2, 3 and 4 (1996/97/98) provide 20 high-power transponders (33, 36, 47 or 50 MHz bandwidth) with a design lifetime of 12 years. There are two transmit coverages, the higher power "Superbeam" coverage and the "Widebeam". All channels can be connected to either coverage on a channel-to-channel basis. HOT BIRD™s 4 and 5 (1998) feature the "Skyplex" on-board digital multiplexing package, enabling the uplinking of different programmes independently from different sites using low-cost transmitting earth stations, while maintaining a single downlink multiplexed signal (see Chapter 7, § 7.10.1.1). An additional satellite of the HOT BIRD™ design, provisionally named "RESSAT" (for "REServe SATellite") will be delivered in early 2000 to provide on-ground back-up capacity for the W series.

The W1R, W2 and W3 satellites provide 24 high-power transponders (72 MHz bandwidth on W1R, 36 and 72 MHz on W2 and W3) also with a design lifetime of 14 years. The W4 satellite will feature 31 transponders of 33 MHz.

Twelve of these transponders (uplink in the 13.75-14.25 GHz band) can be downlinked by on-board switching, on a channel-per-channel basis, either in the 11.45-11.70 GHz or 12.50-12.75 GHz bands, depending on the service provided. In addition to fixed beam, all W satellites have steerable elliptical beams (with the possibility of rotating the ellipse by 90°).

Finally, a satellite called SESAT is under construction. SESAT will be launched in the summer of 1999 by a Russian PROTON launcher and located at 36° E. It will provide coverage up to Siberia, the Middle East and India with fixed and steerable antennas.

EUTELSAT first- and second-generation capacity is being made available at 12.5° W for services between Europe and North America and there are plans to put a new satellite into service at this orbital location in the next two years.

Table AN3-8 below summarizes the main technical characteristics of most EUTELSAT satellites communication subsystems.

Examples of EUTELSAT satellite coverages are given in Chapter 2 (Figure 2.20 for EUTELSAT II) and below (Figure AN3-2 for HOT BIRD™).

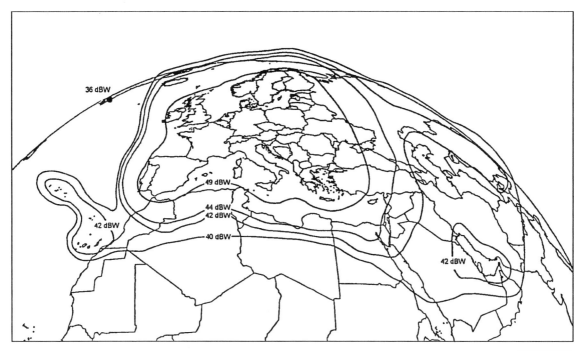

Sat/AN3-02

FIGURE AN3-2

HOT BIRD^{TM} **widebeam coverage**

AN3.2.3.2   Telecommunication services

AN3.2.3.2.1   Public telephony services

a)   Trunk telephony

Voice and data digitally encoded signals are transmitted in TDMA at 120 Mbit/s. DSI is applied to voice signals for enhancing transponder capacity. The EUTELSAT TDMA/DSI is similar to the system operated by INTELSAT (see Chapter 3, § 5.3.3 and Chapter 7, § 7.6.3.2). The earth stations are generally similar to the INTELSAT Standard C (see Chapter 7, Appendix 7.1, Table AP7.1-2). There are at present 11 stations in operation, serving 12 countries.

b)   DAMA-STS

The EUTELSAT demand assigned multiple access suitcase telephony system (DAMA-STS) is a state-of-the-art spread spectrum 14/11 GHz VSAT network designed to provide the end user with a two-way data, fax and toll quality telephone communication using either fixed or portable (suitcase) terminals.

The DAMA-STS network can be connected to external networks (e.g. PSTN, PABX) via the gateway hub earth station, enabling any remote terminal user to communicate with an end user outside the DAMA-STS network.

AN3.2.3.2.2   Audiovisual services

a)   Contribution

EUTELSAT satellites provide capacity for the Eurovision and Euroradio systems transmission, i.e. for exchanging programmes between broadcasting entities which are members of the European Broadcasting Union (EBU). These contribution links currently use FM transmission (19 MHz/V or 25 MHz/V frequency deviation). The earth stations are the same as those established for public telephony traffic (standard TDMA/TV earth stations). Earth stations with smaller antennas, to be used solely for television, can be established in other areas covered by EUTELSAT, such as North Africa or the Middle East.

b)   Broadcasting and distribution

The largest demand at present is for analogue and digital satellite distribution and direct broadcasting. In this case, high e.i.r.p. transponders, such as those of the "HOT BIRD™s", are generally implemented to permit reception by low-cost, small-dish terminals. Depending on the location in the coverage and of the type of modulation, good quality is achievable with antenna diameters ranging from 0.8 to 0.5 m for individual, direct-to-home (DTH) reception. For collective applications (cable head-end, SMATV) where requirements are often more stringent, 1.5 m to 3 m antennas can be used. The EUTELSAT satellites provide also digitally multiplexed TV programmes "bouquets", some of which are encrypted, for DTH reception through specialized TV network operators (see Chapter 7, § 7.10).

AN3.2.3.2.3   Business services (SMS and IDC)

The EUTELSAT satellite multiservices system (SMS) supports a range of digital telecommunications, including point-to-point, point-to-multipoint links (unidirectional and bidirectional) and VSAT networks throughout Europe. The system is highly flexible in its utilization and possible network configurations. Applications span a range of business sectors, including telephony, videoconferencing, data distribution, remote printing, computer communications and digital video/audio distribution (non-exhaustive list) with data rates from 1.2 kbit/s to 8 Mbit/s. Earth stations typically range from small VSATs with 0.6 m antenna to large ones with 9 m antenna or more.

In the SMS service, EUTELSAT either "retails" (on a full-time or occasional use basis) carriers or groups of carriers (power and bandwidth) within the SMS transponders or "wholesales" (on a full-time basis) complete or partial (down to 1.125 MHz) transponders (DigiLease).

In the SMS open network, terminals must operate in accordance with well-defined EUTELSAT specifications, thus guaranteeing interoperability between systems. In an SMS closed network, different types of carriers and earth stations can be used to suit the customers' requirements. Such transmissions are subject to less stringent specifications than the SMS open network.

The EUTELSAT intermediate rate data carrier (IDC) service supports high-quality telephony circuits at 2 Mbit/s and 8 Mbit/s.

AN3.2.3.2.4 Mobile communications (EUTELTRACS)

EUTELTRACS is a two-way message exchange and position reporting service for land and maritime mobiles operating in Europe, North Africa and the Middle East. Up to 45 000 mobiles can be accommodated through capacity allocated on two EUTELSAT satellites. The traffic is managed by EUTELSAT, through a dedicated EUTELSAT hub earth station near Paris.

EUTELSAT is now introducing the *emsat* mobile service, which makes use of the L-band EMS payload on board the ITALSAT F2 satellite. This service provides voice communications, as well as fax, data and SMS, connected to the PSTN. Like EUTELTRACS, *emsat* also provides positioning services.

AN3.2.3.2.5 Transponder leasing

A variety of transponder lease possibilities are offered on available EUTELSAT transponders: partial or full transponder, full-time, part-time or occasional (e.g. for SNG), pre-emptible or non-pre-emptible.

AN3.2.3.2.6   Other services

Other communication services, such as Internet or multimedia applications, provision of ATM links or networks, etc. are being introduced in the EUTELSAT system.

TABLE AN3-8

EUTELSAT satellites communication subsystems

Satellite	Frequency bands (GHz): uplink Rx/ downlink Tx	Number of operational transponders (bandwidth, B, MHz)	Coverage	e.i.r.p. (dBW)	G/T (dB(K^{-1}))	Saturating power flux-density dB(W/m^{-2})
EUTELSAT II F1 to F4	14/11-12	16 (B = 36 and 72)	Widebeam Rx	N/A	−3.5 to +6	−92 to −77 (from −0.5 dB(K^{-1}))
			Widebeam Tx (all Europe, shaped)	39 to 47	N/A	N/A
			Superbeam Tx (shaped)	40 to 52	N/A	N/A
HOT BIRD^{TM} (HB) 1 to 5	– HB 1: 13/11 – HB 2, 3, 4: 13-14-18/11-12 – HB 5: 14/11-12	– HB 1: 16 (B = 36) – HB 2, 3, 4: 20 (B = 33, 36, 72) – HB 5: 22 (B = 33, 36, 72)	Widebeam Rx (all Europe, shaped)	N/A	– HB 1: −7 to +2 – HB 2, 3, −4.5 to +2 – HB 5: −4 to +4	– HB 1: −92 to −77 (from −0.5 dB(K^{-1})) – HB 2, 3, 4, 5: −90 to −70 (from 0 dB(K^{-1}))
			Widebeam Tx (all Europe, shaped)	– HB 1: 40 to 49 – HB 2, 3, 4, 5: 40 to 50	N/A	N/A
			Superbeam Tx (shaped) (HB 2, 3, 4, 5)	42 to 53	N/A	N/A
			Steerable spot beam (circular) Tx (HB 2, 3, 4)	42 to 50	N/A	N/A
W1, W2, W3	13-14/11-12	– W1: 24 (B = 72) – W2, W3: 24 (B = 36 and 72) – W4: 31 (B = 33)	Widebeam Rx (all Europe, shaped)	N/A	−3 to +4	−92 to −77 (from dB(K^{-1}))
			Widebeam Tx (all Europe, shaped)	– W1, W3: 40 to 47 – W2: 40 to 49	N/A	N/A
			Steerable spot beam Tx and Rx (elliptical)	42 to 52	−2 to +8	−92 to −77 (from 0 dB(K^{-1}))
SESAT	13-14/11-12	18 (B = 72)	Widebeam Rx (all Europe and Asia, shaped)	N/A	−3 to +4	−92 to −77 (from 0 dB(K^{-1}))
			Widebeam Tx (all Europe, shaped)	40 to 47	N/A	N/A
			Steerable spot beam Tx and Rx (circular)	40 to 45	−4 to +6	−92 to −77 (from 0 dB(K^{-1}))

AN3.3 National satellite systems

AN3.3.1 Argentina (NAHUELSAT)

The Nahuel C telecommunication system has been established to provide an operational satellite system from the south of the United States down to the south of Argentina. Three beams are implemented: Beam A over Argentina, Beam B over Brazil, Beam C over Latin America and south of the United States. The NAHUELSAT C satellite, launched in 1997 with a design lifetime of 12 years, is located at 72° W. It is a body (3-axis) stabilized spacecraft with a mass of 830 kg in orbit and a power (end of life) of 3 000 W.

The main characteristics of the satellite communication subsystems are summarized in Table AN3-9.

There are various types of earth stations, with antennas of 9.3 m (G/T = 35.3 dB(K^{-1}), e.i.r.p. = 88 dBW) and 4.8 m diameter (G/T = 30 dB(K^{-1}), e.i.r.p. = 82 dBW).

The access types are FDMA and TDMA.

TABLE AN3-9

NAHUELSAT satellite communication subsystem

Satellite	Number of transponders (bandwidth, B, MHz)	Coverage (14/12 GHz band, linear polarization)	e.i.r.p. (edge of coverage) (dBW)	G/T (edge of coverage) (dB(K^{-1}))	Saturating power flux-density dB(W/m^{-2})
NAHUEL C	20 (18 simultaneous) + 4 back-up (B = 54)	A: Argentina	A1: 50 A2: 48	A1: 5 A2: 3.5	A1: −94 A2: −92
		B: Brazil	B1: 51 B2: 48 B3: 38	B1: 5 B2: 2 B3: −5	B1: −94 B2: −91 B3: −84
		C: Latin America and south of the United States	41	−3	−86

AN3.3.2 Australia

AN3.3.2.1 General

The telecommunications regulatory situation in Australia has undergone many changes in recent years. Since 1 July 1997, the Australian telecommunications market has been deregulated and any service may be provided by any carrier organization which satisfies the government criteria.

Currently, two carrier organizations are operating:

• The Telstra corporation (formerly Australian and Overseas Telecommunications Corp., formed by the merger, in 1992 of Telecom Australia and OTC) which dominates international satellite communications business through its involvement with INTELSAT.

• Cable and Wireless Optus (which took over AUSSAT, the previous government satellite operator) which is the main domestic satellite operator.

AN3.3.2.1.1 Telstra

Telstra offers business telecommunication services (VSATs) in Australia and internationally.

In Australia, Telstra has provided "Iterra" ("be quick" in aboriginal language), a network (operating on a Panamsat satellite) for mining and resources exploration in remote outback and offshore rigs.

Internationally, Telstra has provided, since 1990, the Pacific Area Cooperative Telecommunications (PACT) services (with DAMA), to 11 Pacific Island nations. These services currently operate on the INTELSAT satellites at 174° E (POR) and 66° E (IOR). These domestic and international services have been recently augmented by a new digital system which features full mesh, single hop capability for telephony (16 kbit/s) fax and data (14.4 kbit/s) with seamless connectivity to the PSTN. Such services are currently operational in Fiji, Tokelau in the Pacific. In the Indian Ocean region, they are being provided in Russia, Kazakstan and Somalia.

AN3.3.2.1.2 Cable and Wireless Optus

Cable and Wireless Optus operated an interactive VSAT service between 1987 and 1996. The main users were banks for connecting their widely dispersed branches to their computer hubs. At its peak, some 1 000 terminals were in service, in a star network architecture ("Optus STARNET"). The service was decommissioned since it was only of low capacity and as it was easier to supply the customers' evolving needs via terrestrial services.

A growing number of private VSAT networks use the Optus satellite, but these have been limited by the regulations which restrict interconnection with public networks.

One of the larger private networks is AAP Communications which operates a two-way VSAT network comprising over 200 sites for news, radio paging services, etc.

Another VSAT network is operated by Airservices Australia to provide a relay via satellite of remote VHF air-to-ground relay sites.

For security reasons, this Airservices system features full duplication through two separate satellites. About 120 terminals are used on this network which is not available for public traffic.

AN3.3.2.2 The Australian Domestic Satellite System

The Australian Domestic Satellite System (formerly AUSSAT), now owned by C&W Optus, currently consists in two first-generation (A series) and two second-generation (B series) satellites located at 152° E, 156° E, 160° E and 164° E.

These satellites provide:

- domestic fixed communication services (14.0-14.5/12.25-12.75 GHz bands);

- community broadcasting services (same bands);

- mobile-satellite services (on the B spacecraft; 1.5/1.6 GHz with feeder links in the same bands as above).

The service covers Australia, Papua New Guinea (separate beam on A series), South West Pacific (A3 only) and New Zealand (separate beam on B series). The downlinks are configured to implement either a single national beam or five separate spot beams for broadcasting services.

The A series are spin-stabilized satellites with a mass of 550 kg in orbit and a power (end of life) of 860 W. The first one, A1, was launched in 1985 with a design life of seven years. It was taken out of service in 1992. A2 and A3 will continue to operate in inclined orbit until 2000 (A2) and 2005 (A3).

The B series are body-stabilized satellites with a mass of 1 300 kg in orbit and a power (end of life) of 3 500 W. The first one, B1, was launched in 1992 with a design life of 13.5 years. Because B2 suffered a launch failure, it was replaced by B3.

The main characteristics of the satellite communication subsystems are summarized in Table AN3-10.

There are various types of earth stations, ranging from the major stations with 13 m antenna diameter to smaller ones with 1.8 m antenna and the TVROs with antennas from 1.5 m to 0.5 m.

The access types are FDMA (preassigned or demand assigned – DAMA), TDMA or CDMA.

TABLE AN3-10

C&W OPTUS satellites communication subsystems

Satellite	Number of operational transponders (bandwidth, MHz)	Coverage (14/12 GHz band, linear polarization)	e.i.r.p. (edge of coverage) (dBW)	G/T (edge of coverage) $(dB(K^{-1}))$	Saturating power flux-density $dB(W/m^{-2})$
A3	15 (45)	Australian national Tx and Rx	36	−3	−90 to −80
		Spot Tx	47	N/A	N/A
	3 shared (45)	South West Pacific Tx and Rx	34	−7	−90 to −80
B1, B3	15 (54)	Australian national Tx and Rx	40	−2	−95.5 to −74.5
		Spot Tx	50	N/A	N/A
	Mobile Service: Mobile link transponder (Circular polarization): 1.5/1.6 GHz, 150 W Feeder link transponder (Linear polarization): 12/14 GHz, 50 W				

AN3.3.3   China

AN3.3.3.1 General

The planning of a public satellite communication network in China started in the mid-1970s. This network was formed by using leased INTELSAT transponders in the early 1980s and completed mainly by using its own satellites (CHINASAT series) in the late 1980s. Transponders of ASIASAT and APSTAR satellites are also employed in Chinese satellite communication networks now.

At the present time, there are three satellite operators in China. These are: China Telecommunications Broadcast Satellite Corporation (ChinaSat), China Orient Telecomm Satellite Company, Ltd. ("China Orient") and SINO Satellite Communication Company Ltd. (SINOSATCOM). Earth stations providing public services are operated by China Telecom now. There are many private satellite communication networks providing services such as telephony, data, TV distribution, etc.

VSAT systems are widely established in China with a few thousand earth stations in operation. Tens of thousands of TVROs are employed.

AN3.3.3.2 ChinaSat

The first ChinaSat satellite was launched in 1988. At present, ChinaSat owns three satellites, CHINASAT-5 (located at 115.5° E), CHINASAT-6 (at 125° E) and CHINASAT-8 (at 115.5° E). CHINASAT-8 will be launched in the second quarter of 1999.

The main characteristics of the communication subsystems of these satellites are listed in Table AN3-11.

Besides traffic of public services, ChinaSat operates a VSAT system with 2 000 earth stations providing one-way and two-way services in civil air tickets booking, ocean weather forecast, seismological observation, financial information, securities and high-speed data.

TABLE AN3-11

ChinaSat satellites communication subsystems

Satellite	Frequency band (GHz)	Number of operational transponders (bandwidth, B, MHz)	Coverage	e.i.r.p. (edge of coverage) (dBW)	G/T (dB(K^{-1}))	Saturating power flux-density dB(Wm^{-2})
CHINASAT-5	6/4	10 (B = 36)	China and adjacent areas	32	−1	−87 to −81
		6 (B = 72)		33	−6	
CHINASAT-6	6/4	24 (B = 36)		38 or 37	−1	−88 to −76 −86 to −78
CHINASAT-8	6/4	34 (B = 36)		33	−6	−96 to −74
		2 (B = 72)				
	14/11	10 (B = 36)		46	−2	
		6 (B = 72)				

AN3.3.3.3 China Orient

China Orient was founded in 1995. It operates the CHINASTAR-1 satellite, launched in May 1998 and located at 87.5° E. The main characteristics of the communication subsystem of the CHINASTAR-1 satellite are listed in Table AN3-12.

At present, China Orient mainly provides two-way long-distance telephone and data circuits nationwide. Most of them are part of the public telecommunication networks. The others are various types of private satellite communication service networks providing paging, occasional TV distribution, two-way data, etc.

TABLE AN3-12

CHINASTAR-1 satellite communication subsystem

Satellite	Frequency band (GHz)	Number of operational transponders (bandwidth, B, MHz)	Coverage	e.i.r.p. (edge of coverage) (dBW)	G/T (dB(K^{-1}))	Saturating power flux-density dB(W/m^{-2})
CHINASTAR-1	6/4	12 (B = 36) 6 (B = 72)	China and adjacent areas	41	1	−97 to −75
	14/11	16 (B = 36)		52	5	−92 to −70
		4 (B = 72)		54		

AN3.3.3.4 SINOSATCOM

SINOSATCOM currently operates one satellite. It was launched in July of 1998 and located at 110.5° E. The main characteristics of its communication subsystem are shown in Table AN3-13.

TABLE AN3-13

SINOSATCOM satellite communication subsystem

Satellite	Frequency band (GHz)	Number of operational transponders (bandwidth, B, MHz)	Coverage	e.i.r.p. (edge of coverage) (dBW)	G/T (dB(K^{-1}))	Saturating power flux-density dB(W/m^{-2})
SINOSATCOM	6/4	23 (B = 36) 1 (B = 54)	China and adjacent areas	36	−3	−98 to −80
	14/11	14 (B = 54)	China and neighbouring area	46	0	−90 to −70 or −90 to −78

AN3.3.4 France

AN3.3.4.1 General

Since the beginning of satellite communications, France always played an important role in their development, both in being a forerunner in all major satellite international programmes and organizations (INTELSAT, EUTELSAT, ESRO, etc.) and in being active in the development of new satellite systems, applications and techniques. For example, France launched, in 1974, in cooperation with Germany, Symphony-1, the first operational 3-axis, body stabilized geostationary communication satellite.

In 1979, France decided to establish its own telecommunication satellite network. Since 1984, France Telecom has operated a first-generation system comprising three satellites: TELECOM 1A, 1B and 1C. The second-generation was launched between 1991 and 1996, comprising four satellites TELECOM 2A (located at 8° W), 2B (5° W), 2C (3° W) and 2D (5° W).

The TELECOM 2 system features two FSS payloads operating simultaneously on each spacecraft for the following missions:

- Telephony, data transmission and audiovisual services between mainland France and its overseas departments and between these departments (6/4 GHz C-band).

- A wide range of domestic services (14/12 and 14/11 GHz Ku-band), including:

    - radio and TV broadcasting;

    - customized business services at high and variable data rates;

    - fixed or temporary links for business television, live TV reporting or closed-circuit video transmissions.

AN3.3.4.2 The TELECOM 2 space segment

The main characteristics of the TELECOM 2 satellite communication subsystems are summarized in Table AN3-14 and the main characteristics of the earth stations in Table AN3-15.

TABLE AN3-14

TELECOM 2 satellite communication subsystem

Frequency band (GHz)/ (polarization)	Number of transponders (bandwidth, B, MHz) (transmitter type and power P, W)	Coverage	e.i.r.p. (edge of coverage) (dBW)	G/T $dB(K^{-1})$	Saturating power flux-density $dB(W/m^2)$
6/4 (Circular)	6 (2, B = 50 and 4, B = 92) (SSPA, P = 11)	Spot beams : • French Antillas • French Guyana • Saint-Pierre et Miquelon • Mainland France	39 36	−11.3	−70
	4 (B = 50) (SSPA, P = 11)	Hemispheric beam: • Same zones - + Reunion island	32.5		
Telecom 2A: 2B, 2C: 14/12 Telecom 2D: 14/11 (Linear)	11 (B = 36) (TWTA, P = 55)	Mainland France	52	7.5	−91

TABLE AN3-15

TELECOM 2 earth stations

Station type	Frequency band (GHz)/ (polarization)	Transmitting antenna gain (dBi)	e.i.r.p. (dBW)	Receiving antenna gain (dBi)	G/T (dB(K⁻¹))
VSAT data networks	14/12 or 14/11 (linear)	41 to 53.1	36.8 to 67	39.2 to 52	14.9 to 28.7
Radio and TV broadcasting and transmission	14/12 or 14/11 (linear)	52	71 to 73	36.4	13.5
Others (Ku-band)	14/12 or 14/11 (linear)	52	80	40 to 51	15.7 to 27.2
Links with overseas departments	6/4 (circular)	47 to 64.5	86.2 to 90.2	45 to 61	21.7 to 40.4

AN3.3.5   India

AN3.3.5.1   General

The Indian National Satellite System (INSAT) is a domestic multipurpose satellite system which provides the following services:

- fixed and mobile telecommunications;

- direct TV broadcasting;

- networking of sound broadcast transmitters;

- mobile services;

- synoptic Earth observation and relay of meteorological, hydrological and oceanographic data for weather forecast and disaster warning;

- research and rescue functions.

AN3.3.5.2   The INSAT satellites

The multipurpose spacecraft combines the capacity for providing the fixed-satellite service (FSS), the broadcasting-satellite service (BSS), and the meteorological-satellite service. To a considerable extent, the earth segment facilities are also integrated to achieve significant economies.

The first-generation INSAT-1 series of satellites were launched during the eighties. The INSAT-2 series launched in the nineties utilize additional higher frequency bands and provide enhanced fixed and mobile communication services

At the moment, the space segment consists in the following satellites:

- INSAT-1D at 83° E;

- INSAT-2A at 74° E;

- INSAT-2B and INSAT-2C co-located at 93.5° E;

- INSAT-2DT at 55°.

This space segment is controlled and monitored by a master control facility (MCF).

The INSAT-1 series are three axis stabilized satellites with a mass of 650 kg in orbit and a power (end of life) of 900 W. The first one was launched in 1982 with a design life of seven years.

The INSAT-2 series, manufactured by ISRO (India) are also three axis stabilized satellites with a mass of 1 040 kg in orbit and a power (end of life) of 1 150 W. The first one was launched in 1992 with a design life of seven years. The INSAT-2DT has been purchased from the ARABSAT organization and relocated at 55° E.

The currently operated INSAT-2 series satellites are as per Table AN3-16 and the typical characteristics of the INSAT-2C satellite communication subsystem are summarized in Table AN3-17.

TABLE AN3-16

INSAT operational satellites

Satellite type	Location launch date	Transponders (number per band)						Other functions
		6/4 GHz band	Extended 6/4 GHz band	2.5-3 GHz band	Search and rescue	Direct TV	14/11 GHz band	MSS/VHRR /CCD
INSAT-1D	83° E June 1990	12		2		1		VHRR
INSAT-2A	74° July 1992	12	6	2	1	1		VHRR
INSAT-2B	93.5° E July 1993	12	6	2	1	1		VHRR
INSAT-2C	93.5° Dec. 1995	12	6	1			3	MSS
INSAT-2E	83° E April 1999	12	5					CCD
INSAT-2DT	55° E	25 (circular polarization)		1				

TABLE AN3-17

INSAT-2C satellite communication subsystem

Service and transponder	Frequency band: uplink/ downlink (GHz)	Bandwidth (MHz)	Transmitter type and e.i.r.p. (edge of coverage) (dBW)	G/T (edge of coverage) (dB(K^{-1}))	Saturating power flux-density dB(W/m^{-2})
FSS 6/4 band	5.9-6.4/ 3.7-4.2	36	SSPA/TWTA 32-36	−5	−92 to −70
FSS extended 6/4 band	6.735-6.975/ 4.53-4.73	36	SSPA 32-35	−5	−92 to −70
FSS 14/11 band	14.2-14.5/ 10.7-11.7	72-77	TWTA 41	−2	−90
MSS 6/2.5 band	6.45-6.47/ 2.50-2.52	20	TWTA 35	−5	−92 ± 2
MSS 2.6/3.7 band	2.67-2.69/ 3.68-3.70	9	SSPA 30	−9	−109 ± 2
Search and rescue	0.406				
BSS 5.9/2.6 band	5.890-5.930 2.590-2.630	36	TWTA 42	−5	−95 ± 2

AN3.3.5.3 The INSAT earth segment

The earth stations are characterized by different types of antenna as given below in Table AN3-18. (The e.i.r.p. values given are typical values. It varies from carrier to carrier type.)

TABLE AN3-18

INSAT typical earth stations characteristics

Number of stations and type		Antenna diameter (m)	G/T (dB(K^{-1}))	e.i.r.p. (typical values) (dBW)
13 main stations		11	31.7	75-83
44 primary stations		7	25.7	64-71
33 remote stations		4.5	19.7	38
20 transportable		6.1	22	76
5 700 VSAT stations	**Mesh type**	3.8	22.8	42
	Star type	1.8	15	40

AN3.3.5.4 Services

The telecommunication network is digitalized; digital, IDR-type carriers up to 8 Mbit/s capacity in the 6/4 and extended 6/4 bands are used for connecting main-to-main and main-to-primary stations. Thirty five IDR links are currently in operation. An IDR-type carrier at 34 Mbit/s in the 14/11 GHz band of INSAT-2C is also in operation. The thin route traffic for main-to-primary is based on FDM/FM which is likely to be phased out.

The National Informatics Centre Network (NICNET) was the first VSAT-based network in India. NICNET became operational in 1989 and provides services to government departments for computer communications connecting district headquarters.

The remote area business message network (RABM) of the Department of Telecommunications (DOT), which provides data communications to private users through leased VSATs became operational in July 1991. The network uses code division multiple access (CDMA) similar to NICNET and each network has above 500 terminals at present.

The network (I-NET) has connectivity to the telex network, to the X.25 public switched data network (PSDN) and to the international gateway packet switch. The primary customers are from industry (steel, automobile, cement, petrochemical), mining, civil contractors, power generation, banks and the postal department.

Both NICNET and RABM are based on spread spectrum technique (CDMA) as already mentioned. These can support terminals with transmit data rates of 1 200 baud and 9 600 baud at the customer premises, either with asynchronous or X.25 interfaces. A 14/11 GHz overlay network is also operational under NICNET.

HVNET (high-speed VSAT data network), using TDM/TDMA techniques is operational, providing data communication with bit rates up to 64 kbit/s. It provides high-speed data communications between computers and data terminals with star connectivity. HVNET offers access to the PSTN for voice communications, to the PSDN (I-NET/RABM) and to international data networks.

Multiple channel-per-carrier (MCPC) VSAT networks of DOT with 128 kbit/s digital PSK-FDMA carriers are used for linking inaccessible places in various states. These MCPC VSATs connect spur routes for 8 to 10 channel connectivity to small exchanges in rural areas. There are 191 stations in operation. The introduction of DCME equipment is started in the network for capacity enhancement.

Access to isolated villages is provided by long-distance satellite telephone. Remote inaccessible locations and islands are connected via the SCPC and MCPC VSAT networks. The possibility of providing public calling offices (PCO) in villages via satellite is planned.

Initially, low bit-rate data communications services through VSATs were provided using CDMA with small terminals equipped with a 1.8 x 1.2 m reflector antenna operating in the 6/4 GHz bands. Private business communication for closed user groups is provided in TDMA or digital SCPC-DAMA, using two types of earth stations, operating in the "upper" 6/4 GHz bands.

The broadcast service of television is provided by more than 700 TV broadcast transmitters and around 200 audio programme transmitters are linked up via INSAT satellites. Direct TV is also provided to TVROs all over India, through uplinks transmitted by some of the large and medium earth stations.

The meteorological services are provided by about 100 data collection platforms (DCP) located on land and ten in ocean areas to gather meteorological information and relay it through the INSAT DCP transponder towards a main earth station and from there to a meteorological data utilization centre (MDUC) at New Delhi.

A network operations control centre (NOCC) for monitoring and control of the satellite transmissions is co-located with the Delhi earth station.

AN3.3.5.5 VSAT network and village telephone

Consequent to the liberalization of telecommunication services in India, private operators have been permitted to provide closed user group, VSAT-based services in the upper 6/4 GHz (6.7-7/4.5-4.8) band switch star and mesh connectivity.

Several private companies are setting up such VSAT networks. A VSAT network with more than 5 000 VSATs has been commissioned. The National Stock Exchange, several banks, national and multinational companies have linked their headquarters with regional and local centres or factories, etc. utilizing VSATs for voice, data and fax communication. Generally data communications are implemented through a star network operating in TDMA, while voice communications use mesh connectivity with LRE encoding on low bit-rate SCPC-DAMA.

Access to the villages' satellite is provided by long-distance satellite telephone. Remote inaccessible places and islands are connected via SCPC and MCPC VSAT network. The possibility of providing public calling office (PCO) in villages is planned via satellite.

AN3.3.6   Italy

AN3.3.6.1 Introduction

Independently of its participation in all major satellite communications organizations (INTELSAT, EUTELSAT, ESRO, etc.), Italy has always been very active in the development of new satellite systems, applications and techniques.

In the field of new applications and services, it is worth mentioning that Italy has been the first country in Europe to implement a two-way, closed user group TDM/TDMA VSAT network, the ARGO system, installed in 1990, with 122 fixed or transportable remote earth stations.

Italy remains one of the most important European players in this field with various other networks approaching 10 000 VSATs. This includes a network for the Central Bank of Italy (100 sites), one for the civil aviation (sites in 43 airports), one for the automotive industry, etc.

In the field of new techniques Italy started, in 1977, its domestic satellite communication programme with an advanced experimental SHF satellite, SIRIO-1. As a consequence of the successful results of this satellite, ITALSAT-1, Europe's first 30/20 GHz satellite, was launched in January 1991, followed, in August 1996, by the launching of ITALSAT-2.

AN3.3.6.2 The ITALSAT system

The ITALSAT system was conceived as a pre-operational programme to implement a large traffic capacity satellite switched network, providing digital circuits for both telephony and business services with high flexibility and connectivity with the terrestrial network. The satellite system incorporates both:

- real-time demand assignment; and

- non-real-time traffic rearrangement for re-allocating satellite capacity between the earth stations to match daily, weekly, seasonal and unforeseeable variations in traffic demand.

AN3.3.6.3 The ITALSAT space segment

The ITALSAT satellites are 3-axis body stabilized satellites. The ITALSAT-1, with a design life of five years (minimum) has a mass of 1 120 kg in orbit and a power (end of life) of 1 760 W and ITALSAT-2 (design life of eight years) has a mass of 1 200 kg in orbit and a power (end of life) of 2 200 W.

The telecommunications payload of the ITALSAT satellites, the characteristics of which are summarized in Table AN3-19, is composed as follows:

- A multibeam Italian coverage 30/20 GHz subsystem. This subsystem features on-board processing (OBP) with six regenerative transponders connectable, through a 6 x 6 switching matrix, to two deployable 2 m circular reflector antennas.

  Each antenna generates three circular spot beams 0.435° wide. The six spot beams, which cover most of Italy, are operating in SS-TDMA at 147.456 Mbit/s.

- A whole Italy coverage 30/20 GHz subsystem. This subsystem is composed of three transparent transponders connected to an elliptical reflector antenna covering the whole of Italy. These transponders operate either in TDMA, at a maximum bit rate of 24.576 Mbit/s, or in FDMA (digital SCPC, IDR, etc.). They are not interconnected on board with the multibeam subsystem.

- A European Mobile System (EMS, only ITALSAT-2) subsystem. This subsystem, which provides coverage for mobiles extending to Europe, North Africa and Turkey, is composed of:

  - one 12 MHz bidirectional channelized transponder (3 x 4 MHz forward and 12 x 1 MHz return channels) operating in the 1.6/1.5 GHz band. This transponder is connected to an antenna made up of a radiator illuminating the two reflectors of the multibeam subsystem;

  - one bidirectional transponder in the 14/12 GHz band for the feeder link (satellite to gateway station). This transponder is connected to a dedicated dual-offset antenna.

- A propagation experiment subsystem (only ITALSAT-1) which consists of two (redundant) 1 W 50-40 GHz beacons covering a large part of western Europe by means of two small dedicated antennas.

TABLE AN3-19

ITALSAT satellite communication subsystem

Frequency band (GHz) (polarization)	Coverage	Number of transponders (type) (bandwidth, B, MHz) (transmitter type and power P, W)	e.i.r.p. (edge of coverage) (dBW)	G/T (edge of coverage) (dB(K^{-1}))	Saturating power flux-density dB(W/m^{-2})
30/20 (linear)	Italy (spots)	6 (regenerative) (B = 110) (TWTA, P = 20)	55.5	18.8	−84.7
30/20 (linear)	Italy (global)	3 (transparent) (B = 36) (TWTA, P = 20)	47	6.8	−77.2
1.6/1.5 (circular) (payload not carried on ITALSAT-1)	Europe (European Mobile System)	1 (transparent) (B = 12, i.e. 3 x 4 MHz, channelized) (SSPA, P = 22)	42.5	−20	−123
14/12 (linear) (payload not carried on ITALSAT-1)	Europe	1 (transparent) (B = 12) (SSPA, P = 4.5)	32	−1.4	−97.0
50-40 (circular) (payload not carried on ITALSAT-2)	Europe	2 (beacons) (P = 1)	28.5	N/A	N/A

AN3.3.6.4   The ITALSAT earth stations

The FSS earth station of the ITALSAT system utilizes 3.5 m circular antennas or 3.5 x 7 m elliptical antennas, depending on climatic zones. The selected antenna configuration is a multi-reflector offset type with shaped sub-reflectors.

AN3.3.7   Japan

AN3.3.7.1   CS-2, CS-3 and N-STAR systems and space segment

The CS-2 satellite programme established the first operational domestic satellite communication system in Japan, with the following objectives:

- to secure the important communication services in the case of natural disasters and emergencies;

- to establish a public communication network between regional centres;

- to establish a public communication network between the mainland and remote islands;

- to provide temporary communication links;

- to provide satellite digital communication services;

- to provide satellite video communication services;

- to provide small capacity satellite communications in remote areas or situations where telecommunications do not exist.

The CS-2 (CS-2a and CS-2b) was used by Nippon Telegraph and Telephone Corporation (NTT) and other users, e.g. National Police Agency, Ministry of Construction, Japan railways, etc. in order to promote the objectives listed above.

The CS series, developed and manufactured in Japan, are spin-stabilized satellites with an in-orbit mass of 350/550 kg (CS-2/CS-3) and a power (end of life) of 480/840 W (CS-2/CS-3). The first one (CS-2a) was launched in early 1983 and the second one (CS-2b) six months later by the National Space Development Agency (NASDA), then handed over to the Telecommunication Satellite Corporation of Japan (TSCJ) for operational use.

The successors of CS-2a and CS-2b are CS-3a, launched in February and September 1988 with a design lifetime of seven years.

The CS-3 satellite communication system utilized the 6/4 GHz and 30/20 GHz bands for the transmission of telephone, fax, data and television signals. The CS-3 satellites finished operation in 1996.

The N-STAR system (N-STAR a and N-STAR b) was developed to take over the services provided by the CS-3 system and to enable new services in the multimedia world.

The N-STAR series are three axis, body-stabilized satellites with an in-orbit mass of 2 000 kg and a power (end of life) of 5 000 W. Their design lifetime is more than ten years. The first one (N-STAR a) was launched in August 1995 and located at 132° E and the second one (N-STAR b) was launched in February 1996 and located at 136° E.

Each N-STAR satellite carries five telecommunication payloads in four frequency bands: 3 GHz band (multibeam), 6/4 GHz band (shaped beam), 14/11-12 GHz band (shaped beam), 30/20 GHz band (shaped beam) and 30/20 GHz (multibeam).

AN3.3.7.2 JCSAT and Superbird systems and space segments

In addition to the N-STAR system, two separate commercial satellite systems provide FSS communication services in Japan.

Besides conventional telephone trunking, both systems mainly find users in such applications as: corporate data networks, teleconferencing, television programme transmissions for broadcasters and CATV, satellite news gathering (SNG), etc.

The JCSAT satellite system, owned by Japan Satellite Systems, Inc., comprises three satellites (JCSAT-2, JCSAT-3, JCSAT-4, JCSAT-1B) located at 154° E and 128° E, 150° E and 150° E, respectively.

JCSAT-2 is a spin-stabilized satellite with an in-orbit mass of 1 370 kg and a power (end of life) of 2 200 W (design lifetime: more than ten years). JCSAT-2 was launched in 1990. It carries 32 transponders at 14/12 GHz.

JCSAT-3, JCSAT-4 and JCSAT-1B are three axis, body-stabilized satellites with an in-orbit mass of 1 800 kg and a power (end of life) of 5 200 W (design lifetime: 12 years (3, 4), 13 years (1B)). JCSAT-3 was launched in 1995 and both JCSAT-4 and JCSAT-1B in 1997. Each JCSAT-3 or JCSAT-4 carries 28 transponders at 14/12 GHz and 12 transponders at 6/4 GHz. JCSAT-1B carries 32 transponders at 14/12 GHz.

The Superbird satellite system, owned by the Space Communications Corporation, comprises three satellites (SUPERBIRD-A, SUPERBIRD-B and SUPERBIRD-C) located at 158° E, 162° E and 144° E.

The SUPERBIRD-A and SUPERBIRD-B are three axis, body-stabilized satellites with an in-orbit mass of 2 550 kg and a power (end of life) of 3 800 W (design lifetime: more than ten years). The SUPERBIRD-C is also a three axis, body-stabilized satellite with an in-orbit mass of 3 100 kg and a power (end of life) of 4 300 W (design lifetime: more than 13 years).

The SUPERBIRD-A and SUPERBIRD-B were launched in 1992 and each carries 23 transponders at 14/12 GHz and 3 transponders at 30/20 GHz. The SUPERBIRD-C was launched in 1997 and carries 24 transponders at 14/12 GHz.

AN3.3.7.3 VSAT networks in Japan

There are several VSAT network carriers operating hub earth stations and providing services to customers leasing VSATs. These services include:

- packet switched data communication services;
- circuit switched data communication services;
- dedicated data circuit services;
- data broadcast services;
- video broadcast services.

There are also corporations which lease satellite capacity directly from the satellite operator and operate their own private VSAT networks.

A particular VSAT network is operated for local authorities, such as prefectures, cities and towns with the purpose of improving the exchange of information between local authorities, especially to secure means of communication in the event of disasters.

AN3.3.8 Korea

AN3.3.8.1 General

With the launching in 1995 of the KOREASAT-1 satellite, located at 116° E, Korea Telecom put into operation a multipurpose satellite system providing the following services:

- direct broadcasting of digital TV and HDTV;
- video distribution and CATV;
- high- and low-speed data transmission;

- inter-city trunking;

- narrow-band digital data services (VSAT);

- wideband digital data services;

- rural area communication services (SCPC/PAMA or DAMA).

At present, Korea Telecom is operating two satellites, KOREASAT-1 and KOREASAT-2 and is planning to launch KOREASAT-3 in mid-1999. This advanced type satellite, which will take over the KOREASAT-1 mission, will feature greater traffic and broadcasting capacity. A 30/20 GHz payload with large bandwidth transponders is also provided on this satellite.

AN3.3.8.2 KOREASAT-1 space segment

The two operating KOREASAT satellites are three axis, body-stabilized satellites with an in-orbit mass of 711 kg (KOREASAT-2: 834 kg) and a power (end of life) of 1 600 W (design lifetime: ten years).

The KOREASAT-3 satellite will have an in-orbit mass of 1 690 kg and a power (end of life) of 5 200 W (design lifetime: 15 years).

The main characteristics of the KOREASAT satellite communication subsystems are shown in Table AN3-20.

Two TT&C stations, primary and back-up, located at Yongin and Teajon respectively, are controlling the satellite and system operation.

The main characteristics of the most typical earth station types implemented in the framework of the KOREASAT satellite system are shown in Table AN3-21.

TABLE AN3-20

KOREASAT satellites communication subsystems

Satellite type	Frequency band (GHz) (polarization)	Coverage	Number of transponders (bandwidth, B, MHz) (transmitter type and power P, W)	e.i.r.p. (edge of coverage) (dBW)	G/T (edge of coverage) $(dB(K^{-1}))$	Saturating power flux-density $dB(W/m^{-2})$
KOREASAT-1 and -2	14.0-14.5/ 12.25-12.75 (linear)	Republic of Korea	12 (FSS) (B = 36) (TWTA, P = 14)	50.2	13.5	−90 (nominal)
	14.5-14.8/ 11.7-12.0 (circular)		3 (DBS) (B = 27) (TWTA, P = 120)	59.4	13.0	−82 (nominal)
KOREASAT-3	14.0-14.5/ 12.25-12.75 (linear)	Republic of Korea and steerable	24 (FSS) (B = 36) (TWTA, P = 45)	54.7 (domestic) 49.5 (steerable)	13.5 (domestic) 7.0 (steerable)	−90 (domestic) −84 (steerable)
	14.5-14.8/ 11.7-12.0 (circular)	Republic of Korea	6 (DBS) (B = 27) (TWTA, P = 120)	59.4	13.0	−82
	30.085-30.885/ 20.355-21.155 (circular)	Korean peninsula	3 (FSS) (B = 200) (TWTA, P = 85)	55.0	9.3	−86

TABLE AN3-21

KOREASAT-1 satellite communication system – main characteristics of the earth stations

Station type	Antenna diameter (m)	e.i.r.p. (dBW)	G/T $(dB(K^{-1}))$	Multiple access/ modulation
CATV	9 (Tx)/3.7 (Rx)	56	30	MCPC/QPSK
Trunking	6.0	50	35	TDM/QPSK
TVRO	9 (Tx)/1.8 (Rx)	58.5	20	MCPC/QPSK
SNG	1.8 (Rx)/9 (Tx)	51	37	SCPC/QPSK
VSAT	1.8	34	20	TDMA/TDM
Multimedia	9 (Tx)/0.75 (Rx)	64	13	MCPC/QPSK
DBS	9 (Tx)/0.45 (Rx)	74	11	MCPC/QPSK

AN3.3.9   Spain

AN3.3.9.1 General

HISPASAT is the Spanish satellite communication system. It has been established as a multi-mission system providing domestic and international services in Western Europe and in the Americas through two high-power satellites (HISPASAT-1A and HISPASAT-1B).

AN3.3.9.2 HISPASAT-1 satellite system and space segment

The various missions assigned to the system are:

- Broadcasting-satellite service mission: five high-power transponders allow BSS towards the whole Spanish territory according to the ITU BSS Plan.

- Fixed-satellite service mission: 16 transponders of different bandwidths (36 to 72 MHz) and with frequency reuse implementation are offered for carrying all types of traffic over a wide area in Europe.

- America mission: service is offered between Spain (or anywhere within the FSS coverage) and a great part of the Americas, from Canada to Terra di Fuego (Argentina) and especially towards/from Spanish speaking states. This mission comprises two sub-missions:

  - TV America (TVA): two high-power transponders (one in each satellite) allow transmission of television (and/or radio) programmes towards the Americas. High-quality/availability reception is possible in all climatic zones with small TVRO type terminals.

  - America Return (TVR): the HISPASAT-1B satellite is also equipped with two of the FSS channels (54 and 72 MHz bandwidth) which permit to uplink from any place in the America coverage area. This allows to implement TV contribution or distribution links towards Spain. It also permits the implementation of data distribution networks.

  - Government mission: two transponders in the 8/7 GHz band allow the development of government telecommunication networks of various types.

The two operating HISPASAT-1 satellites (HISPASAT-1A and HISPASAT-1B) are three axis, body-stabilized satellites with an in-orbit mass of 2 150 kg and a power (end of life) of 3 500 W (design lifetime: ten years).

The main characteristics of the HISPASAT-1 satellites communication subsystems are summarized in Table AN3-22.

A satellite control centre (SCC), located at Arganda, near Madrid and two payload monitoring centres are controlling the satellite and system operation.

AN3.3.9.3 HISPASAT earth stations and operational support

There are various types of earth stations. HISPASAT has made available CTETH documents (recomendaciones para las caracteristicas téchnicas de las estaciones terrenas de Hispasat) which provide technical recommendations for the stations operating in the system as well as the description of the various services: TV/FM (contribution, distribution, SNG, etc.), digital carriers for business applications (data rates from 64 kbit/s to 34 Mbit/s, VSATs, etc.), etc.

Moreover, in order to provide operational support to the users, MUSSH documents have been elaborated (manual de utilizacion del sistema de satélites HISPASAT). These are the HISPASAT user operational guide, providing user friendly information about the procedures to access the HISPASAT space segment.

TABLE AN3-22

HISPASAT-1 satellite communication subsystem

Service	Frequency band (GHz) (polarization)	Coverage	Number of transponders (type) (bandwidth, B, MHz) (transmitter type and power P, W)	e.i.r.p. (edge of coverage) (dBW)
Direct TV Broadcasting (BSS)	17.0-17.7/ 12.0-12.25 (circular)	Spain national	5 (TWTA, P = 110)	56
FSS	14.0-14.5/ 11.45-11.7 and 12.5-12.75 (linear, V and H, frequency reuse)	Spain and Western Europe	16 (8 x B = 36, 2 x B = 46, 2 x B = 54, 4 x B = 72) (TWTA, P = 55)	50
TV America (TVA)	14/10-12 (linear)	Uplink: from Spain Downlink: to the Americas (shaped beam)	1 in each satellite (TWTA, P = 110)	49.5 to 41
America Return (TVR)	14/10-12 (linear)	Uplink: from Spanish speaking America up to Canada Downlink: to Spain	In HISPASAT-1B: 2 (two of the FSS transponders) (B = 54 and 72) (TWTA, P = 55)	
Government	8/7	Spain		

AN3.3.10   United States

AN3.3.10.1 Background

The United States domestic satellite systems provide a wide range of telephone, programme distribution and specialized communication services. They are used to supplement the terrestrial microwave network by providing long-haul, high-usage trunk facilities. A significant portion of traffic to Alaska and the offshore areas of Hawaii, Puerto Rico and the Virgin Islands is carried over domestic facilities.

Services range from 4 kHz voice channels, digital signals with various bit rates for commercial uses (64 kbit/s to 1.5 Mbit/s), special bit rates (3-15 Mbit/s) for special government and industry applications (relay of weather data, facsimile transmissions for automated publications, financial record distribution, etc.), high-speed TDMA systems (40-60 Mbit/s) for multimode transmissions and wideband applications where the transmission distances are large and/or varied.

Customer premises earth stations for private and government networks are used for many of these applications.

Early in the implementation of satellite communication services, it was assumed that two-way long-distance voice traffic would be a major source of business. However, the time delay associated with the relay of radio signals over geostationary satellites was a significant obstacle, especially after the introduction of fibre-optic cables. Over the last decade other applications, such as video distribution and VSAT services, more than offset the lack of demand for voice. Communication satellites have grown to become a major industry in the interim.

AN3.3.10.2 Communication applications and growth

During the 1970s, when domestic satellite communications were being established in the United States, the 6/4 GHz band became the primary radio frequency for satellites. The standard earth station facility featured antennas of 10 m to 30 m diameter for long-distance trunk circuits. Smaller antennas, however, were feasible for the distribution of television and audio programme material to cable television systems, broadcast stations, hotels and other points for retransmission and reception. This application grew into a large business for satellite equipment suppliers. Thousands of earth terminals with 4.5 m to 6 m diameter antennas were installed within a few years. All public, non-commercial television and radio broadcasting stations became interconnected by satellites for distribution and interchange of programme material. Many commercial establishments followed this example, enhanced by the development and growth of satellite news gathering (SNG) for the immediate distribution of news, sports and other events of interest to the public.

Early in the 1980s, satellites began operations in the 14/12 GHz band. Unlike the 6/4 GHz band, which had to be shared with terrestrial radio services, the 14/12 GHz band was allocated by the United States Federal Communication Commission (FCC) to the domestic FSS on a relatively exclusive basis.

The objective was to encourage the growth of satellite communication applications. Another attractive feature was that earth station antennas one-half the size of 6/4 GHz systems were feasible in the 14/12 GHz band. These conditions were instrumental in the phenomenal growth of satellite communications during the last decade. Narrow-band digital applications (SCPC) expanded when it was demonstrated that user premises earth stations with 1.2 m to 2.4 m diameter antennas were feasible.

AN3.3.10.3 VSAT applications

The first application for the use of small aperture antennas in the United States was initiated by the need for the transmission of data from remote sites (oil drilling, pumping platforms, etc.) to a central office facility. The first operational satellite systems available in the 1970s used the 6/4 GHz band. In order to employ small transportable earth station antennas (2 m to 4 m diameter) and avoid excessive interference between satellite networks using the GSO, spread spectrum or CDMA modulation was used. Since the 6/4 GHz band is shared with the fixed service (FS), it was necessary to coordinate all earth installations with the FS networks. Also, it became apparent that the use of many of these small earth stations resulted in self-interference as well as raising the interference threshold to other satellite networks.

The solution, early in the 1980s, was the introduction of 14/12 GHz satellite operations. Increasing demand for very small satellite terminals (VSATs), by large commercial enterprises, bank, stock and bond brokers, educational institutions, video distributors, and many others, resulted in the development of the classic VSAT network. This has been generally described as a) many widely dispersed user terminals employing 1.2 m to 2.4 m diameter antennas linked to b) a central facility (hub) employing a more conventional earth station with a 5 m to 7 m diameter antenna.

The FCC in 1986, faced with the burden of licensing thousands of VSAT networks, established a set of technical standards by which conforming applicants could receive a blanket authorization for their network(s), thus eliminating the need for licensing each individual earth station. The result has been that the United States now has many thousands of VSATs distributed throughout the country.

AN3.3.10.4 Space segment

The United States domestic satellites (DOMSATs) are located in the GSO from approximately 60° W to 135° W longitude. The first commercial GSO satellite was the Westar (1974), which employed the 6/4 GHz band (500 MHz) and covered the continental United States, Hawaii and Puerto Rico with 12 transponder channels each 36 MHz wide. Subsequently, all 6/4 GHz satellites used cross-polarization to double the number of transponders (frequency reuse) to 24, each with 36 MHz of bandwidth. The first commercial 14/12 GHz (500 MHz) satellite network was operated by SBS (1980) with ten transponders each 43 MHz wide. The next generation of 14/12 GHz satellites used cross-polarization and essentially doubled the number of transponders. By the mid-1980s satellites were developed to employ both 6/4 GHz and 14/12 GHz bands thus expanding the communication capacity of a single satellite to 2 000 MHz of bandwidth. Another FCC regulation, which is designed to maintain a high capacity of usage for the GSO arc, is an orbital separation (geocentric) between domestic satellites of 2°.

Currently, the domestic satellite industry is completing its third generation, featuring very large satellites employing high-power transponders to service smaller and smaller earth station antennas. This legacy has been enhanced by the developments in the broadcasting-satellite service (BSS) industry. In addition, there are plans for using approximately 1 000 MHz of the 30/20 GHz bands for services to earth stations employing 0.7 m antennas. The satellites will employ spot beam antennas to ensure the necessary pfds on the surface of the Earth for good performance. The 0.7 m earth station antennas are equivalent to the VSAT antenna applications at 14/12 GHz described above.

The main characteristics of United States satellites are summarized in Table AN3-23.

AN3.3.10.5 Earth segment

The United States domestic commercial satellite systems interconnect several thousands of earth terminals in the country and across local borders. The 6/4 GHz satellite networks employ earth station antennas ranging from approximately 4 m to 30 m. Their applications were reviewed in the sections above. The 14/12 GHz satellite networks employ earth station antennas ranging from approximately 1 m to 7 m, with the former associated with VSAT applications as reviewed above.

TABLE AN3-23

Main characteristics of the United States satellites

Satellite series name	Designation	Frequency band (GHz) (H: hybrid: 6/4 +14/12)	Launch period (year)	Stabil. type (S: spin, F: 3-axis)	Design lifetime (years)	Mass in orbit (kg)	Power (end of life) (W)
Aurora	II	6/4	1991	F	10.5	736	1 100
Galaxy	I-R	6/4	1994	S	10	788	1 050
	V-W	6/4	1992	S	10	788	950
	VI	6/4	1990	S	10	584	950
	III/H	H	1995	F	12	1 700	4 300
	IV/H	H	1993	F	12	1 700	4 300
	VII/H	H	1992	F	12	1 700	4 300
	IX	6/4	1996				
General Electric	GE-1	H	1996				
GSTAR	I	14/12	1985	F	10	715	1 900
	II	14/12	1986	F	10	715	1 900
	III	14/12	1988	F	10	715	1 900
	IV	14/12	1990	F	10	715	1 900
SATCOM	C-1	6/4	1991	F	10	510	1 040
	C-3	6/4	1992	F	12	620	1 400
	C-4	6/4	1992	F	12	620	1 400
	C-5	6/4	1991	F	10	736	1 100
	SN2	H	1984	F	10	692	1 300
	SN3	H	1988	F	10	692	1 300
	SN4	H	1991	F	10	692	1 300
	IIR	6/4	1983	F	10	610	1 035
	VIR	6/4	1991	F	12	620	1 400
	Ku-1	14/12	1986	F	10	780	2 490
	Ku-2	14/12	1985	F	10	780	2 490
SBS	3 (Inc)	14/12	1983	S	10	560	900
	4 (Inc)	14/12	1984	S	10	590	1 000
	5	14/12	1988	S	10	600	1 200
	6	14/12	1990	S	10	1 160	2 000
Telstar	302	6/4	1984	S	10	650	670
	303	6/4	1985	S	10	650	670
	401	H	1993	F	12	1 912	6 000
	402R	H	1995	F	12	1 912	6 000
International services							
Panamsat	PAS-1	H	1988	F	13	1 560	
	PAS-2/4	H	1994	F	15		
	PAS-3/2R	H	1996	F	15		
	PAS-4/6	H	1995	F	15		
TDRS	TDRS-AOR	H	1989	F	10	2 400	1 700
	TDRS-POR		1991	F	10	2 400	1 700
		H					

ANNEX 4

ITU Bibliography

AN 4.1 Introduction

Communication satellites have provided an important dimension for national, regional and international telecommunications. As they have become part of telecommunication networks, they must comply with performance specifications of established networks or those still being developed, which are based on generally agreed end-to-end transmission characteristics of all types of radiocommunication modes.

A list of the main ITU publications relevant to the subject is given in this annex.

AN 4.2 Constitution of the International Telecommunication Union, Geneva, 1992

Chapter II	Radiocommunication Sector
Chapter VI	General provisions relating to telecommunications
Chapter VII	Special provisions for radio

AN 4.3 Radio Regulations, Geneva, 1998 Edition

Article S1	Terms and definitions
Article S2	Nomenclature of the frequency and wavelength bands used in radiocommunication
Article S5	Frequency allocation
Article S9	Procedure for effecting coordination with or obtaining agreement of other Administrations
Article S11	Notification, and recording in Master International Frequency Register of frequency assignments to radio astronomy and space radiocommunication stations (except stations in the broadcasting-satellite service)
Article S15	Interference
Article S21	Terrestrial and space services sharing frequency bands above 1 GHz
Article S22	Space services
Appendix S1	Classification of Emissions and Necessary Bandwidths
Appendix S2	Table of Transmitter Frequency Tolerances
Appendix S3	Table of Maximum Permitted Spurious Emission Power Levels

Appendix S4	Consolidated list and tables of characteristics for use in the application of procedures of Chapter SIII (Coordination, Notification and Recording of Frequency Assignments and Plan Modification)
Appendix S5	Identification of Administrations with which Coordination is to be Effected or Agreement sought under the Provisions of Article S9
Appendix S7	Method for the Determination of the Coordination Area Around an Earth Station in Frequency Bands Between 1 GHz and 40 GHz Shared Between Space and Terrestrial Radiocommunication Services
Appendix S8	Method of Calculation for Determining if Coordination is Required Between Geostationary-Satellite Networks Sharing the Same Frequency Bands
Appendix S9	Report of an Irregularity or of an Infringement of the Convention or the Radio Regulations
Appendix S30	Provisions for All Services and Associated Plans for the Broadcasting-Satellite Service in the Frequency Bands 11.7-12.2 GHz (in Region 3), 11.7-12.5 GHz (in Region 1) and 12.2-12.7 GHz (in Region 2)
Appendix S30A	Provisions and Associated Plans for Feeder Links for the Broadcasting-Satellite Service (11.7-12.5 GHz in Region 1, 12.2-12.7 GHz in Region 2 and 11.7-12.2 GHz in Region 3) in the Frequency Bands 14.5-14.8 GHz and 17.3-18.1 GHz in Regions 1 and 3, and 17.3-17.8 GHz in Region 2
Appendix S30B	Provisions and Associated Plans for Fixed-Satellite Service in the Frequency Bands 4 500-4 800 MHz, 6 725-7 025 MHz, 10.70-10.95 GHz, 11.20-11.45 GHz and 12.75-13.25 GHz

AN 4.4 ITU-R S Series Recommendations (Fixed-satellite service)

Section 4A – Definitions

Rec. ITU-R S.673   Terms and definitions relating to space radiocommunications

Section 4B – Systems aspects – Performance and availability

Section 4B1 – System aspects

Rec. ITU-R S.725	Technical characteristics for very small aperture terminals (VSATs)
Rec. ITU-R S.726-1	Maximum permissible level of spurious emissions from very small aperture terminals (VSATs)
Rec. ITU-R S.727	Cross-polarization isolation from very small aperture terminals (VSATs)
Rec. ITU-R S.728-1	Maximum permissible level of off-axis e.i.r.p. density from very small aperture terminals (VSATs)
Rec. ITU-R S.729	Control and monitoring function of very small aperture terminals (VSATs)
Rec. ITU-R S.1001	Use of systems in the fixed-satellite service in the event of natural disasters and similar emergencies for warning and relief operations
Rec. ITU-R S.1061	Utilization of fade countermeasures strategies and techniques in the fixed-satellite service

Rec. ITU-R S.1149-1   Network architecture and equipment functional aspects of digital satellite systems in the fixed-satellite service forming part of synchronous digital hierarchy transport networks

Rec. ITU-R S.1250   Network management architecture for digital satellite systems forming part of SDH transport networks in the fixed-satellite service

Rec. ITU-R S.1251   Network management – Performance management object class definitions for satellite systems network elements forming part of SDH transport networks in the fixed-satellite service

Rec. ITU-R S.1252   Network management – Payload configuration object class definitions for satellite system network elements forming part of SDH transport networks in the fixed-satellite service

Section 4B2 – Performance and availability

Rec. ITU-R S.352-4   Hypothetical reference circuit for systems using analogue transmission in the fixed-satellite service

Rec. ITU-R S.353-8   Allowable noise power in the hypothetical reference circuit for frequency-division multiplex telephony in the fixed-satellite service

Rec. ITU-R S.354-2   Video bandwidth and permissible noise level in the hypothetical reference circuit for the fixed-satellite service

Rec. ITU-R S.521-3   Hypothetical reference digital paths for systems using digital transmission in the fixed-satellite service

Rec. ITU-R S.522-5   Allowable bit error ratios at the output of the hypothetical reference digital path for systems in the fixed-satellite service using pulse-code modulation for telephony

Rec. ITU-R S.614-3   Allowable error performance for a hypothetical reference digital path in the fixed-satellite service operating below 15 GHz when forming part of an international connection in an integrated services digital network

Rec. ITU-R S.1062-1   Allowable error performance for a hypothetical reference digital path operating at or above the primary rate

Rec. ITU-R S.579-4   Availability objectives for a hypothetical reference circuit and a hypothetical reference digital path when used for telephony using pulse code modulation, or as part of an integrated services digital network hypothetical reference connection, in the fixed-satellite service

Rec. ITU-R S.730   Compensation of the effects of switching discontinuities for voiceband data and of Doppler frequency-shifts in the fixed-satellite service

Rec. ITU-R S.1420   Allowable error performance for a hypothetical reference digital path based on the SDH

Rec. ITU-R S.1424   Availability objectives for a hypothetical reference digital path when used for transmission of B-ISDN ATM mode in the FSS by GSO satellite systems using frequencies below 15 GHz

Rec. ITU-R S.1425   Transmission considerations for digital carriers using higher levels of modulation on satellite circuits

Rec. ITU-R S.1429 — Error performance objectives due to interference between GSO and non-GSO FSS systems for hypothetical reference digital paths operating at or above the primary rate carried by systems using frequencies below 15 GHz

Rec. ITU-R S.1432 — Apportionment of the allowable error performance degradations to FSS hypothetical reference digital paths arising from time invariant interference for systems operating below 15 GHz

Section 4C – Earth station and baseband characteristics – Earth station antennas – Maintenance of earth stations

Rec. ITU-R S.465.5 — Reference earth-station radiation pattern for use in coordination and interference assessment in the frequency range from 2 to about 30 GHz

Rec. ITU-R S.731 — Reference earth-station cross-polarized radiation pattern for use in frequency coordination and interference assessment in the frequency range from 2 to about 30 GHz

Rec. ITU-R S.580-5 — Radiation diagrams for use as design objectives for antennas of earth stations operating with geostationary satellites

Rec. ITU-R S.732 — Method for statistical processing of earth-station antenna side-lobe peaks

Rec. ITU-R S.733-1 — Determination of the G/T ratio for earth stations operating in the fixed-satellite service

Rec. ITU-R S.734 — The application of interference cancellers in the fixed-satellite service

Rec. ITU-R S.464-2 — Pre-emphasis characteristics for frequency-modulation systems for frequency-division multiplex telephony in the fixed-satellite service

Rec. ITU-R S.446-4 — Carrier energy dispersal for systems employing angle modulation by analogue signals or digital modulation in the fixed-satellite service

Rec. ITU-R S.481-2 — Measurement of noise in actual traffic for systems in the fixed-satellite service for telephony using frequency-division multiplex

Rec. ITU-R S.482-2 — Measurement of performance by means of a signal of a uniform spectrum for systems using frequency-division multiplex telephony in the fixed-satellite service

Section 4D – Frequency sharing between networks of the fixed-satellite service – Efficient use of the spectrum and geostationary-satellite orbit

Section 4D1 – Permissible levels of interference

Rec. ITU-R S.466-6 — Maximum permissible level of interference in a telephone channel of a geostationary-satellite network in the fixed-satellite service employing frequency modulation with frequency-division multiplex, caused by other networks of this service

Rec. ITU-R S.483-3 — Maximum permissible level of interference in a television channel of a geostationary-satellite network in the fixed-satellite service employing frequency modulation, caused by other networks of this service

Rec. ITU-R S.523-4   Maximum permissible levels of interference in a geostationary-satellite network in the fixed-satellite service using 8-bit PCM encoded telephony, caused by other networks of this service

Rec. ITU-R S.735-1   Maximum permissible levels of interference in a geostationary-satellite network for an HRDP when forming part of the ISDN in the fixed-satellite service caused by other networks of this service below 15 GHz

Rec. ITU-R S.1323   Maximum permissible levels of interference in a satellite network (GSO/FSS; non-GSO/FSS; non-GSO/MSS feeder links) for a hypothetical reference digital path in the fixed-satellite service caused by other codirectional networks below 30 GHz

Rec. ITU-R S.1325   Simulation methodology for assessing short-term interference between co-frequency, codirectional non-geostationary-satellite orbit (GSO) fixed-satellite service (FSS) networks and other non-GSO FSS or GSO FSS networks

Rec. ITU-R S.1324   Analytical method for estimating interference between non-geostationary mobile-satellite feeder links and geostationary fixed-satellite networks operating co-frequency and codirectionally

Rec. ITU-R S.671-3   Necessary protection ratios for narrow-band single channel-per-carrier transmissions interfered with by analogue television carriers

Rec. ITU-R S.524-5   Maximum permissible levels of off-axis e.i.r.p. density from earth stations in the fixed-satellite service transmitting in the 6 and 14 GHz frequency bands

Rec. ITU-R S.736-3   Estimation of polarization discrimination in calculations of interference between geostationary-satellite networks in the fixed-satellite service

Rec. ITU-R S.1063   Criteria for sharing between BSS feeder links and other Earth-to-space or space-to-Earth links of the FSS

Rec. ITU-R S.1150   Technical criteria to be used in examinations relating to the probability of harmful interference between frequency assignments in the fixed-satellite service as required in No. 1506 of the Radio Regulations

Rec. ITU-R S.1328   Satellite system characteristics to be considered in frequency sharing analyses between geostationary-satellite orbit (GSO) and non-GSO satellite systems in the fixed-satellite service (FSS) including feeder links for the mobile-satellite service (MSS)

Rec. ITU-R S.1418   Method for calculating single entry carrier-to-interference ratios in inter-satellite service using GSO

Rec. ITU-R S.1427   Methodology and criterion to assess interference from RLAN transmitters to non-GSO MSS feeder links in the band 5 150-5 250 MHz

Rec. ITU-R S.1430   Determination of the coordination area for earth stations operating with non-GSO space stations with respect to earth stations operating in the reverse direction in frequency bands allocated bidirectionally to the FSS

Rec. ITU-R S.1433   Equivalent power flux-density EPFDup and EPFDis

Section 4D2 – Coordination methods

Rec. ITU-R S.737	Relationship of technical coordination methods within the fixed-satellite service
Rec. ITU-R S.738	Procedure for determining if coordination is required between geostationary-satellite networks sharing the same frequency bands
Rec. ITU-R S.739	Additional methods for determining if detailed coordination is necessary between geostationary-satellite networks in the fixed-satellite service sharing the same frequency bands
Rec. ITU-R S.740	Technical coordination methods for fixed-satellite networks
Rec. ITU-R S.741-2	Carrier-to-interference calculations between networks in the fixed-satellite service
Rec. ITU-R S.742-1	Spectrum utilization methodologies
Rec. ITU-R S.743-1	The coordination between satellite networks using slightly inclined geostationary-satellite orbits (GSOs) and between such networks and satellite networks using non-inclined GSO satellites
Rec. ITU-R S.744	Orbit/spectrum improvement measures for satellite networks having more than one service in one or more frequency bands
Rec. ITU-R S.1002	Orbit management techniques for the fixed-satellite service
Rec. ITU-R S.1253	Technical options to facilitate coordination of fixed-satellite service networks in certain orbital arc segments and frequency bands
Rec. ITU-R S.1254	Best practices to facilitate the coordination process of fixed-satellite service satellite networks
Rec. ITU-R S.1003	Environmental protection of the geostationary-satellite orbit
Rec. ITU-R S.1255	Use of adaptive uplink power control to mitigate codirectional interference between geostationary satellite orbit/fixed-satellite service (GSO/FSS) networks and feeder links of non-geostationary satellite orbit/mobile satellite service (non-GSO/MSS) networks and between GSO/FSS networks and non-GSO/FSS networks
Rec. ITU-R S.1256	Methodology for determining the maximum aggregate power flux-density at the geostationary-satellite orbit in the band 6 700-7 075 MHz from feeder links of non-geostationary satellite systems in the mobile-satellite service in the space-to-Earth direction
Rec. ITU-R S.1257	Analytical method to calculate visibility statistics for non-geostationary satellite orbit satellites as seen from a point on the Earth's surface
Rec. ITU-R S.1419	Interference mitigation techniques to facilitate coordination between non-GSO MSS feeder links and GSO FSS networks in the bands 19.3-19.7 GHz and 29.1-29.5 GHz
Rec. ITU-R S.1428	Reference FSS earth station radiation pattern for use in interference assessment involving non-GSO satellite in frequency bands between 10.7 GHz and 30 GHz

Rec. ITU-R S.1431   Methods to enhance sharing between non-GSO FSS systems (except MSS feeder links) in the frequency bands between 10-30 GHz

Section 4D3 – Spacecraft station-keeping – Satellite antenna radiation pattern – Pointing accuracy

Rec. ITU-R S.484-3   Station-keeping in longitude of geostationary satellites in the fixed-satellite service

Rec. ITU-R S.670-1   Flexibility in the positioning of satellites as a design objective

Rec. ITU-R S.672-4   Satellite antenna radiation pattern for use as a design objective in the fixed-satellite service employing geostationary satellites

Rec. ITU-R S.1064-1   Pointing accuracy as a design objective for earthward antennas on board geostationary satellites in the fixed-satellite service

Section 4E – Frequency sharing between networks of the fixed-satellite service and those of other space radiocommunication systems

Rec. ITU-R S.1065   Power flux-density values to facilitate the application of RR Article 14 for the FSS in Region 2 in relation to the BSS in the band 11.7-12.2 GHz

Rec. ITU-R S.1066   Ways of reducing the interference from the broadcasting-satellite service of one Region into the fixed-satellite service of another Region around 12 GHz

Rec. ITU-R S.1067   Ways of reducing the interference from the broadcasting-satellite service into the fixed-satellite service in adjacent frequency bands around 12 GHz

Rec. ITU-R S.1068   Fixed-satellite and radiolocation/radionavigation services sharing in the band 13.75-14 GHz

Rec. ITU-R S.1069   Compatibility between the fixed-satellite service and the space science services in the band 13.75-14 GHz

Rec. ITU-R S.1151   Sharing between the inter-satellite service involving geostationary satellites in the fixed-satellite service and the radionavigation service at 33 GHz

Rec. ITU-R S.1340   Sharing between feeder links for the mobile-satellite service and the aeronautical radionavigation service in the Earth-to-space direction in the band 15.4-15.7 GHz

Rec. ITU-R S.1341   Sharing between feeder links for the mobile-satellite service and the aeronautical radionavigation service in the space-to-Earth direction in the band 15.4-15.7 GHz and the protection of the radio astronomy service in the band 15.35-15.4 GHz

Rec. ITU-R S.1339   Feasibility of sharing between spaceborne passive sensors of the earth exploration-satellite service and inter-satellite links of geostationary-satellite networks in the range 50 to 65 GHz

Rec. ITU-R S.1326   Feasibility of sharing between the inter-satellite service and the fixed-satellite service in the frequency band 50.4-51.4 GHz

Rec. ITU-R S.1327   Requirements and suitable bands for operation of the inter-satellite service within the range 50.2-71 GHz

Rec. ITU-R S.1342	Method for determining coordination distances, in the 5 GHz band, between the international standard microwave landing system stations operating in the aeronautical radionavigation service and non-geostationary mobile satellite service stations providing feeder uplink services
Rec. ITU-R S.1329	Frequency sharing of the bands 19.7-20.2 GHz and 29.5-30.0 GHz between systems in the mobile-satellite service and systems in the fixed-satellite service
Rec. ITU-R S.1426	Aggregate pfd limits, at the FSS satellite orbit for RLAN transmitters sharing frequencies with the FSS

AN 4.5 ITU-R SNG Series Recommendations (Satellite news gathering)

Rec. ITU-R SNG.722-1	Uniform technical standards (analogue) for Satellite News Gathering (SNG)
Rec. ITU-R SNG.770-1	Uniform operational procedures for Satellite News Gathering (SNG)
Rec. ITU-R SNG.1007-1	Uniform technical standards (digital) for Satellite News Gathering (SNG)
Rec. ITU-R SNG.1421	Common operating parameters to ensure interoperability for transmission of digital TV news gathering

AN 4.6 ITU-R SF Series Recommendations

Section 4/9A – Sharing conditions

Rec. ITU-R SF.355-4	Frequency sharing between systems in the fixed-satellite service and radio-relay systems in the same frequency bands
Rec. ITU-R SF.356-4	Maximum allowable values of interference from line-of-sight radio-relay systems in a telephone channel of a system in the fixed-satellite service employing frequency modulation, when the same frequency bands are shared by both systems
Rec. ITU-R SF.357-4	Maximum allowable values of interference in a telephone channel of an analogue angle-modulated radio-relay system sharing the same frequency bands as systems in the fixed-satellite service
Rec. ITU-R SF.558-2	Maximum allowable values of interference from terrestrial radio links to systems in the fixed-satellite service employing 8-bit PCM encoded telephony and sharing the same frequency bands
Rec. ITU-R SF.615-1	Maximum allowable values of interference from the fixed-satellite service into terrestrial radio-relay systems which may form part of an ISDN and share the same frequency band below 15 GHz
Rec. ITU-R SF.358-5	Maximum permissible values of power flux-density at the surface of the Earth produced by satellites in the fixed-satellite service using the same frequency bands above 1 GHz as line-of-sight radio-relay systems
Rec. ITU-R SF.674-1	Power flux-density values to facilitate the application of Article 14 of the Radio Regulations for FSS in relation to the fixed-satellite service in the 11.7-12.2 GHz band in Region 2

Rec. ITU-R SF.1320 | Maximum allowable values of power flux-density at the surface of the Earth produced by non-geostationary satellites in the fixed-satellite service used in feeder links for the mobile-satellite service and sharing the same frequency bands with radio-relay systems

Rec. ITU-R SF.1004 | Maximum equivalent isotropically radiated power transmitted towards the horizon by earth stations of the fixed-satellite service sharing frequency bands with the fixed service

Rec. ITU-R SF.1005 | Sharing between the fixed service and the fixed-satellite service with bidirectional usage in bands above 10 GHz currently unidirectionally allocated

Rec. ITU-R SF.1008-1 | Possible use by space stations in the fixed-satellite service of orbits slightly inclined with respect to the geostationary-satellite orbit in bands shared with the fixed service

Section 4/9B – Coordination and interference calculations

Rec. ITU-R SF.1006 | Determination of the interference potential between earth stations of the fixed-satellite service and stations in the fixed service

Rec. ITU-R SF.766 | Methods for determining the effects of interference on the performance and the availability of terrestrial radio-relay systems and systems in the fixed-satellite service

Rec. ITU-R SF.1193 | Carrier-to-interference calculations between earth stations in the fixed-satellite service and radio-relay systems

AN 4.7 ITU-R IS Series Recommendations (Inter-service sharing and compatibility)

Rec. ITU-R IS.847-1 | Determination of the coordination area of an earth station operating with a geostationary space station and using the same frequency band as a system in a terrestrial service

Rec. ITU-R IS.848-1 | Determination of the coordination area of a transmitting earth station using the same frequency band as receiving earth station in bidirectionally allocated frequency bands

Rec. ITU-R IS.849-1 | Determination of the coordination area for earth stations operating with non-geostationary spacecraft in bands shared with terrestrial services

Rec. ITU-R IS.850-1 | Coordination areas using predetermined coordination distances

AN 4.8 ITU-R SA Series Recommendations (Space applications and meteorology)

Rec. ITU-R SA.1071 | Use of the 13.75 to 14.0 GHz band by the space science services and the fixed-satellite service

Rec. ITU-R SA.1277 | Sharing in the 8 025-8 400 MHz frequency band between the earth exploration-satellite service and the fixed, fixed-satellite, meteorological-satellite and mobile services in Regions 1, 2 and 3

Rec. ITU-R SA.1278   Feasibility of sharing between the earth exploration-satellite service (space-to-Earth) and the fixed, inter-satellite and mobile services in the band 25.5-27.0 GHz

Rec. ITU-R SA.1156   Method of calculating low-orbit satellite visibility statistics

AN 4.9 ITU-R P Series Recommendations (Radiowave propagation)

Rec. ITU-R P.1144   Guide to the application of the propagation methods of Radiocommunication Study Group 3

Rec. ITU-R P.341-4   The concept of transmission loss for radio links

Rec. ITU-R P.581-2   The concept of "worst-month"

Rec. ITU-R P.841   Conversion of annual statistics to worst-month statistics

Rec. ITU-R P.1058-1   Digital topographic databases for propagation studies

Rec. ITU-R P.453-6   The radio refractive index: its formula and refractivity data

Rec. ITU-R P.676-3   Attenuation by atmospheric gases

Rec. ITU-R P.838   Specific attenuation model for rain for use in prediction methods

Rec. ITU-R P.840-2   Attenuation due to clouds and fog

Rec. ITU-R P.618-5   Propagation data and prediction methods required for the design of Earth-space telecommunications systems

Rec. ITU-R P.531-4   Ionospheric propagation data and prediction methods required for the design of satellite services and systems

AN 4.10 ITU-R BO Series Recommendations (Broadcasting-satellite service)

Section 10/11S-A – Terminology

Rec. ITU-R BO.566-3   Terminology relating to the use of space communication techniques for broadcasting

Section 10/11S-B – Systems

Rec. ITU-R BO.650-2   Standards for conventional television systems for satellite broadcasting in the channels defined by Appendix 30 of the Radio Regulations

Rec. ITU-R BO.651   Digital PCM coding for the emission of high-quality sound signals in satellite broadcasting (15 kHz nominal bandwidth)

Rec. ITU-R BO.712-1   High-quality sound/data standards for the broadcasting-satellite service in the 12 GHz band

Rec. ITU-R BO.786   MUSE system for HDTV broadcasting-satellite services

Rec. ITU-R BO.787   MAC/packet based system for HDTV broadcasting-satellite services

Rec. ITU-R BO.788-1   Coding rate for virtually transparent studio quality HDTV emissions in the broadcasting-satellite service

Rec. ITU-R BO.789-2   Service for digital sound broadcasting to vehicular portable and fixed receivers for broadcasting-satellite service (sound) in the frequency range 1 400-2 700 MHz

Rec. ITU-R BO.1130-1   Systems for digital sound broadcasting to vehicular, portable and fixed receivers for broadcasting-satellite service (sound) bands in the frequency range 1 400-2 700 MHz

Rec. ITU-R BO.1211   Digital multiprogramme emission systems for television, sound and data services for satellites operating in the 11/12 GHz frequency range

Rec. ITU-R BO.1294   Common functional requirements for the reception of digital multiprogramme television emissions by satellites operating in the 11/12 GHz frequency range

Section 10/11S-C – Technology

Rec. ITU-R BO.652-1   Reference patterns for earth-station and satellite antennas for the broadcasting-satellite service in the 12 GHz band and for the associated feeder links in the 14 GHz and 17 GHz bands

Rec. ITU-R BO.790   Characteristics of receiving equipment and calculation of receiver figure-of-merit (G/T) for the broadcasting-satellite service

Section 10/11S-D – Planning and sharing

Rec. ITU-R BO.600-1   Standardized set of test conditions and measurement procedures for the subjective and objective determination of protection ratios for television in the terrestrial broadcasting and the broadcasting-satellite services

Rec. ITU-R BO.791   Choice of polarization for the broadcasting-satellite service

Rec. ITU-R BO.792   Interference protection ratios for the broadcasting-satellite service (television) in the 12 GHz band

Rec. ITU-R BO.793   Partitioning of noise between feeder links for the broadcasting-satellite service (BSS) and BSS down links

Rec. ITU-R BO.794   Techniques for minimizing the impact on the overall BSS system performance due to rain along the feeder-link path

Rec. ITU-R BO.795   Techniques for alleviating mutual interference between feeder links to the BSS

Rec. ITU-R BO.1212   Calculation of total interference between geostationary-satellite networks in the broadcasting-satellite service

Rec. ITU-R BO.1213   Reference receiving earth station antenna patterns for replanning purposes to be used in the revision of the WARC-77 BSS plans for Regions 1 and 3

Rec. ITU-R BO.1295   Reference transmit Earth station antenna off-axis e.i.r.p. patterns for planning purposes to be used in the revision of the Appendix 30A (Orb-88) Plans of the Radio Regulations at 14 GHz and 17 GHz in Regions 1 and 3

Rec. ITU-R BO.1296   Reference receive space station antenna patterns for planning purposes to be used for elliptical beams in the revision of the Appendix 30A (Orb-88) Plans of the Radio Regulations at 14 GHz and 17 GHz in Regions 1 and 3

Rec. ITU-R BO.1297   Protection ratios to be used for planning purposes in the revision of the Appendices 30 (Orb-85) and 30A (Orb-88) Plans of the Radio Regulations in Regions 1 and 3

Rec. ITU-R BO.1293   Protection masks and associated calculation methods for interference into broadcast satellite systems involving digital emissions

AN 4.11 ITU-R BT Series Recommendations (Broadcasting service (television))

Rec. ITU-R BT.470-6   Conventional television systems

Rec. ITU-R BT.1117-2   Studio format parameters for enhanced 16:9 aspect ratio 625-line television systems (D- and D2-MAC, PALplus, enhanced SECAM)

Rec. ITU-R BT.1299   The basic elements of a worldwide common family of systems for digital terrestrial television broadcasting

Rec. ITU-R BT.601-5   Studio encoding parameters of digital television for standard 4:3 and wide-screen 16:9 aspect ratios

Rec. ITU-R BT.1127   Relative quality requirements of television broadcast systems

Rec. ITU-R BT.500-9   Methodology for the subjective assessment of the quality of television pictures

Rec. ITU-R BT.654   Subjective quality of television pictures in relation to the main impairments of the analogue composite television signal

Rec. ITU-R BT.1128-2   Subjective assessment of conventional television systems

Rec. ITU-R BT.813   Methods for objective picture quality assessment in relation to impairments from digital coding of television signals

Rec. ITU-R BT.1210-1   Test materials to be used in subjective assessment

AN 4.12 ITU-R V Series Recommendations (Vocabulary and related subjects)

Rec. ITU-R V.573-3   Radiocommunication vocabulary

Rec. ITU-R V.431-6   Nomenclature of the frequency and wavelength bands used in telecommunications

Rec. ITU-R V.666-2   Abbreviations and initials used in telecommunications

Rec. ITU-R V.574-3   Use of the decibel and the neper in telecommunications

Rec. ITU-R V.662-2   Terms and definitions

AN 4.13 ITU-T Recommendations

D.186 (10/96)   General tariff and accounting principles for the international two-way multipoint telecommunication service via satellite

E.170 (10/92)   Traffic routing

E.430 (6/92)   Quality of service framework

E.800 (8/94)	Terms and definitions related to quality of service and network performance including dependability
E.862 (6/92)	Dependability planning of telecommunication networks
F.140 (3/93)	Point-to-multipoint telecommunication service via satellite
F.141 (6/94)	International two-way multipoint telecommunication service via satellite
G.101 (8/96)	The transmission plan
G.102 (11/88)	Transmission performance objectives and Recommendations
G.103 (12/98)	Hypothetical reference connections
G.111 (3/93)	Loudness ratings (LRs) in an international connection
G.113 (2/96)	Transmission impairments
G.114 (2/96)	One-way transmission time
G.131 (8/96)	Control of talker echo
G.164 (11/88)	Echo suppressors
G.165 (3/93)	Echo cancellers
G.180 (3/93)	Characteristics of N + M type direct transmission restoration systems for use on digital and analogue sections, links or equipment
G.181 (3/93)	Characteristics of 1 + 1 type restoration systems for use on digital transmission links
G.211 (11/88)	Make-up of a carrier link
G.212 (11/88)	Hypothetical reference circuits for analogue systems
G.223 (11/88)	Assumptions for the calculation of noise on hypothetical reference circuits for telephony
G.233 (11/88)	Recommendations concerning translating equipments
G.241 (11/88)	Pilots on groups, supergroups, etc.
G.322 (11/88)	General characteristics recommended for systems on symmetric pair cables
G.423 (11/88)	Interconnection at the baseband frequencies of frequency-division multiplex radio-relay systems
G.701 (3/93)	Vocabulary of digital transmission and multiplexing, and pulse code modulation (PCM) terms
G.702 (11/88)	Digital hierarchy bit rates
G.703 (10/98)	Physical/electrical characteristics of hierarchical digital interfaces
G.707 (3/96)	Network node interface for the synchronous digital hierarchy (SDH)
G.711 (11/88)	Pulse code modulation (PCM) of voice frequencies
G.712 (11/96)	Transmission performance characteristics of pulse code modulation channels
G.722 (11/88)	7 kHz audio-coding within 64 kbit/s

G.723 (3/96)	Speech coders
G.726 (12/90)	40, 32, 24, 16 kbit/s adaptive differential pulse code modulation (ADPCM)
G.728 (9/92)	Coding of speech at 16 kbit/s using low-delay code excited linear prediction
G.729 (3/96)	Coding of speech at 8 kbit/s using conjugate structure algebraic-code-excited linear-prediction (CS-ACELP)
G.732 (11/88)	Characteristics of primary PCM multiplex equipment operating at 2 048 kbit/s
G.733 (11/88)	Characteristics of primary PCM multiplex equipment operating at 1 544 kbit/s
G.763	Digital circuit multiplication equipment using ADPCM (Recommendation G.726) and digital speech interpolation
G.764 (12/90)	Voice packetization – Packetized voice protocols
G.765 (9/92)	Packet circuit multiplication equipment
G.783 (4/97)	Characteristics of synchronous digital hierarchy (SDH) equipment functional blocks
G.792 (11/88)	Characteristics common to all transmultiplexing equipments
G.801 (11/88)	Digital transmission models
G.803 (6/97)	Architecture of transport networks based on the synchronous digital hierarchy (SDH)
G.811 (9/97)	Timing characteristics of primary reference clocks
G.821 (8/96)	Error performance of an international digital connection operating at a bit rate below the primary rate and forming part of an integrated services digital network
G.823 (3/93)	The control of jitter and wander within digital networks which are based on the 2 048 kbit/s hierarchy
G.824 (3/93)	The control of jitter and wander within digital networks which are based on the 1 544 kbit/s hierarchy
G.826 (2/99)	Error performance parameters and objectives for international, constant bit rate digital paths at or above the primary rate
G.827 (8/96)	Availability parameters and objectives for path elements of international constant bit rate digital paths at or above the primary rate
G.831 (8/96)	Management capabilities of transport networks based on the synchronous digital hierarchy (SDH)
G.921 (11/88)	Digital sections based on the 2 048 kbit/s hierarchy
G.957 (7/95)	Optical interfaces for equipments and systems relating to the synchronous digital hierarchy
G.958 (11/94)	Digital line systems based on the synchronous digital hierarchy for use on optical fibre cables

H.100 (11/88)	Visual telephone systems
H.120 (3/93)	Codecs for videoconferencing using primary digital group transmission
I.112 (3/93)	Vocabulary of terms for ISDNs
I.120 (3/93)	Integrated services digital networks (ISDNs)
I.121 (4/91)	Broadband aspects of ISDN
I.210 (3/93)	Principles of telecommunication services supported by an ISDN and the means to describe them
I.310 (3/93)	ISDN – Network functional principles
I.311 (8/96)	B-ISDN general network aspects
I.340 (11/88)	ISDN connection types
I.356 (10/96)	On B-ISDN ATM layer cell transfer performance
I.357 (8/96)	On B-ISDN semi-permanent connection availability
I.363 (3/93)	B-ISDN ATM adaptation layer specification
I.375 (6/98)	Network capabilities to support multimedia services
I.411 (3/93)	ISDN user-network interfaces – Reference configurations
I.413 (3/93)	B-ISDN user-network interface
I.414 (9/97)	Overview of Recommendations on layer 1 for ISDN and B-ISDN customer accesses
I.430 (11/95)	Basic user-network interface – Layer 1 specification
I.500 (3/93)	General structure of the ISDN interworking Recommendations
I.501 (3/93)	Service interworking
I.525 (8/96)	Interworking between networks operating at bit rates less than 64 kbit/s with 64 kbit/s-based ISDN and B-ISDN
I.530 (3/93)	Network interworking between an ISDN and a public switched telephone network (PSTN)
I.580 (11/95)	General arrangements for interworking between B-ISDN and 64 kbit/s based ISDN
J.11 (11/88)	Hypothetical reference circuits for sound-programme transmissions
J.12 (11/88)	Types of sound-programme circuits established over the international telephone network
J.51 (8/94)	General principles and user requirements for the digital transmission of high quality sound programmes
J.61 (6/90)	Transmission performance of television circuits designed for use in international connections
J.62 (2/78)	Single value of the signal-to-noise ratio for all television systems
J.85 (6/90)	Digital television transmission over long distances – General principles

M.20 (10/92)	Maintenance philosophy for telecommunications networks
M.21 (10/92)	Maintenance philosophy for telecommunication services
M.460 (11/88)	Bringing international group, supergroup, etc., links into service
M.580 (11/88)	Setting up and lining up an international circuit for public telephony
M.2100 (7/95)	Performance limits for bringing-into-service and maintenance of international PDH paths, sections and transmission systems
M.2110 (4/97)	Bringing-into-service of international PDH paths, sections and transmission systems and SDH paths and multiplex sections
M.3010 (5/96)	Principles for a telecommunications management network
N.1 (3/93)	Definitions for application to international sound-programme and television-sound transmission
N.51 (11/88)	Definitions for application to international television transmissions
N.64 (11/88)	Quality and impairment assessment
Q.7 (11/88)	Signalling systems to be used for international automatic and semi-automatic telephone working
Q.48 (11/88)	Demand assignment signalling systems
Q.50 (6/97)	Signalling between circuit multiplication equipments (CME) and international switching centres (ISC)
Q.115 (6/97)	Logic for the control of echo control devices
Q.601 (3/93)	Interworking of signalling systems – General
V.24 (10/96)	List of definitions for interchange circuits between data terminal equipment (DTE) and data circuit-terminating equipment (DCE)
V.32 (3/93)	A family of 2-wire, duplex modems operating at data signalling rates of up to 9 600 bit/s for use on the general switched telephone network and on leased telephone-type circuits
V.230 (11/88)	General data communications interface layer 1 specification
X.1 (10/96)	International user classes of service in, and categories of access to, public data networks and integrated services digital networks (ISDNs)
X.3 (3/93)	Packet assembly/disassembly facility (PAD) in a public data network
X.21 (9/92)	Interface between data terminal equipment and data circuit-terminating equipment for synchronous operation on public data networks
X.24 (11/88)	List of definitions for interchange circuits between data terminal equipment (DTE) and data circuit-terminating equipment (DCE) on public data networks
X.25 (10/96)	Interface between data terminal equipment (DTE) and data circuit-terminating equipment (DCE) for terminals operating in the packet mode and connected to public data networks by dedicated circuit

X.28 (12/97)	DTE/DCE interface for a start-stop mode data terminal equipment accessing the packet assembly/disassembly facility (PAD) in a public data network situated in the same country
X.29 (12/97)	Procedures for the exchange of control information and user data between a packet assembly/disassembly (PAD) facility and a packet mode DTE or another PAD
X.50 (11/88)	Fundamental parameters of a multiplexing scheme for the international interface between synchronous data networks
X.55 (11/88)	Interface between synchronous data networks using a 6 + 2 envelope structure and single channel per carrier (SCPC) satellite channels
X.56 (11/88)	Interface between synchronous data networks using an 8 + 2 envelope structure and single channel per carrier (SCPC) satellite channels
X.200 (7/94)	Information technology – Open systems interconnection – Basic reference model: The basic model
X.300 (10/96)	General principles for interworking between public networks and between public networks and other networks for the provision of data transmission services
X.361 (10/96)	Connection of VSAT systems with packet-switched public data networks based on X.25 procedures
X.637 (10/96)	Basic connection-oriented common upper layer requirements
X.638 (10/96)	Minimal OSI facilities to support basic communications applications
X.641 (12/97)	Information technology – Quality of service: framework

AN 4.14 ITU Special Autonomous Group (GAS) Handbooks

GAS-7	Handbook on economics and financing of telecommunications projects in developing countries
GAS-7	Rural telecommunications, Geneva, 1994

Concluding remarks

Satellite communications are nowadays faced with rapidly changing perspectives. In terms of market (in a more and more deregulated world), some keywords underlying these changes are: Global Information Infrastructure (GII), telecommunication needs in developing countries, mobility and Personal Communications, multimedia, Internet and multimedia, information broadcasting (including TV programmes), competition with fibre optic cables.

In terms of the current status of the technical environment and of the space and earth segment equipment development, some keywords (listed here in a very mixed style) are: digitization, non-Geostationary Orbit (non-GSO) satellite systems and networks, orbit/frequency spectrum allotments and utilization of higher frequency bands, transparent vs. regenerative transponders, B-ISDN and ATM operation, very small earth stations for business networks and rural communications, compressed digital TV broadcasting. It is the hope of the authors of this new Edition of the Handbook that the reader will be able to find his way towards these different subjects that are shortly reviewed below.

Digitization: It should be evident to the reader that a major difference between this edition and the previous one is the accent put on the transmission of digital signals and on the correlative signal processing techniques. In fact, reference to analogue signals has not been completely omitted, because some analogue satellite links and equipment remain in current operation. However, due to their progressive obsolescence, their part in the text has been considerably shortened.

The reasons for the very rapid development of digital transmission in all domains of communications are well known. They are particularly valid in the case of satellite communications: very high-quality transmission through the utilization of powerful error correction schemes, low sensitivity to non-linearities, better bandwidth utilization by signal processing (notably by bit-rate reduction techniques), operation control by software. For all types of integrated circuits (including the microwave analogue units), the progress in the performance and technology (compactness and level of integration) is such that complete functions and equipment sets can currently be mass produced at low cost, which permits to envisage new classes of applications which were unthinkable before[1].

Non-Geostationary orbit (non-GSO) satellite systems and networks

Current projects in the field of mobile satellite communications, i.e. via the mobile-satellite service (MSS), such as Iridium, ICO, Globalstar, etc. deserve mentioning here although they are outside the scope of this Handbook (which is devoted to FSS)[2]. This is because the current developments in mobile satellite systems, networks and terminals should induce cross-fertilization effects on FSS

---

[1] Much experience has been gained from the development of the currently operating terrestrial cellular telephony systems (GSM and others) and, in particular from the realization of good quality, very low bit-rate (LRE) voice codecs.

[2] Note that ITU-R Study Group 8 is in the process of finalizing a mobile-satellite service Handbook.

systems. This should be the case, notably, for the new FSS projects based on Low or Medium earth Orbiting (LEO/MEO) satellites.

In fact, communication satellite systems, generally based on constellations of LEOs/MEOs, (so-called non-GSO/FSS systems), allowing direct access from users, are currently under development or construction. Their main features are:

- they will be implemented in the FSS 14/10-12 GHz and 30/20 GHz bands (see Chapter 9, in particular § 9.5);

- they should, ultimately, operate on a global basis;

- they should provide on demand, dynamic, large bandwidth allocations, in particular for multimedia applications and should, in consequence provide wideband, Internet type, connections.

Detailed information on two different systems of this type (SkyBridge and Teledesic) is given in Chapter 6, Appendix 6.1. Other projects are in course.

Orbit/Frequency spectrum allotments and utilization of higher frequency bands

The ever-increasing demands for orbital positions and frequency bands assignments have constrained the various authorities in charge of these matters to envisage a complete revision of their previous rules and regulations. This is particularly due to the advent of the above-mentioned non-GSO systems and networks, which makes considerably more complex the resolution of interference and frequency coordination problems and which makes it necessary to augment the frequency resources by calling for the utilization of higher frequency bands (30/20 GHz and higher in the future). The Teledesic system actually uses the 30/20 GHz band[3].

In the first rank of actors in this field is the ITU through world radiocommunication conferences (WRCs). The relevant work in these conferences is technically prepared by ITU-R Study Groups (and notably by Study Group 4). Considerable work has already been accomplished by WRC-97, which began a complete revision of the relevant parts (articles and annexes) of the Radio Regulations (RR). This work should be completed by the incoming WRC-2000.

These aspects are discussed in detail in Chapter 9.

Transparent vs. regenerative transponders

On-Board Processing (OBP) transponders, i.e. capable of performing one or more of the following functions: switching (in frequency, space and/or time), regeneration and baseband processing, have been the subject of much R&D (See Chapter 6, § 6.3.4[4]). Although only a few realizations are actually in operation, it is to be expected that future projects, especially those operating in the higher frequency bands, will be based on OBP satellites. This is already the case for the Teledesic system,

---

[3] The main trouble with higher frequencies is the increased influence of atmospheric propagation effects (increased path loss, depolarization by rain, etc.). These effects are, of course reduced in the case of high operating elevation angles (e.g. in the case of LEOs).

[4] The implementation of satellites with multiple spot beams and on-board processing is also dealt with in Supplement 3 to Edition 2 of the Handbook ("VSAT Systems and Earth Stations", ITU, Geneva, 1995, Chapter 6, § 6.3).

which associates OBP with inter-satellite links (ISLs) between the satellites of the constellation. Of course, systems based on transparent satellites, due to their high reliability and simplicity of implementation and system operation will remain active contenders of OBP satellite systems.

B-ISDN and ATM operation

At first glance, the implementation of high-capacity trunk satellite links appears to be slowing down due to the growing worldwide installation of optical cables. Whenever advisable in terms of cost and of return on investments, optical cables may offer advantages over satellite links, in terms of intrinsic bandwidth and of reduced transmission delay.

However, even in this market segment, satellites will continue to be largely used wherever the geographical and environmental conditions render cable installation impractical or uneconomical with regard to the expected traffic needs.

In consequence, satellite communications are and will remain an integral part of the worldwide telecommunication system, or, to use a new terminology, of the Global Information Infrastructure (GII). This is the reason why ITU devotes so much effort to ensure full compatibility of satellite links with the most recent telecommunications standards and Recommendations, notably in the framework of the Broadband Integrated Services Digital Network (B-ISDN, see Chapter 3, § 3.5.5) and also for operation with the Asynchronous Transfer Mode (ATM).

Specifically, satellite ATM networks, i.e. satellite links with ATM networks or even satellite links incorporating ATM switches (possibly on-board), are receiving particular attention at the moment (see Chapter 3, § 3.5.4 and Appendix 3.4 and also Chapter 8, § 8.6). This is because satellite links can provide the advantages of ATM and the ATM services, not only through point-to-point trunk links, but also through point-to-multipoint links, in particular towards extending services to remote areas. ATM technology by itself offers common transport for voice, video and data services (an ISDN specificity) and also the flexibility for providing bandwidth on demand. The latter is ideally matched with the specific advantages of satellite communications (multipoint networking including network reconfiguration capabilities, etc.).

Business communications networks: Such private systems are usually known as "VSAT networks", the acronym "VSAT" designating in fact the very small earth stations (also called simply user terminals) directly installed in the user premises. A complete ITU-R Handbook has already been devoted to the subject[5] . New applications are likely to occur in this area, e.g. for providing broadband interconnections (e.g. up to 155.52 Mbit/s) with direct access by professional or even individual users to Internet or other data networks. It is actually one of the main objectives of the new LEO constellation projects to provide such broadband Internet direct connections throughout the world.

Rural communications systems: At the eve of the 21st century, there remain in our world millions of villages and other isolated areas, and therefore billions of humans without any type of electrical communication. There are currently good expectations that satellite systems can afford economically realistic solutions to break this insufferable barrier to the development of these parts of the world.

---

[5] "VSAT Systems and Earth Stations", see Footnote 3.

This is because, as explained above, technical and technological progress should now permit to offer low cost remote earth stations and to also reduce the recurrent costs of the operational space segment (reduction of the needed bandwidth through low bit rate encoding and demand assignment).

Broadcast systems: One unsurpassed specificity of satellites is their capability for distributing information directly to a part or to the totality of end users located in the satellite downlink coverage. Throughout the world, millions of homes are nowadays receiving directly TV programmes via FSS or BSS satellites (see Chapter 7, § 7.10). The new, very sophisticated possibilities offered by digital TV transmission, associated with bit-rate compression (notably MPEG protocols, see Chapter 3, Appendix 3.3) are in the course of drastically increasing this audience. Numbers of satellites are already or will be entirely dedicated to broadcasting TV programmes "bouquets". Systems are also projected for Digital Audio Broadcasting (DAB), with the purpose of distributing audio programmes towards very low cost receivers, particularly in developing countries.

A few final remarks:

Mobile communications and Personal Communications networks: Although mobile communications are, in principle, outside the framework of this FSS Handbook, mention should be made here to the studies, research and development efforts and standardization processes which are currently conducted throughout the world, in the field of next generations of mobile communications. In ITU-R, in particular, IMT-2000 Task Group 8/1 (under Study Group 8) is working on the third generation of mobile telecommunication, i.e. IMT-2000 (International Mobile Telecommunication). The reasons for this interest in mobile communications are, on the one hand and as explained above, their cross-fertilization effects on FSS systems (e.g. LEO/MEO systems) and, on the other hand, the perspectives offered by the concept of Universal Personal Communications (UPC) in which the service is provided to a user, either fixed or mobile, identified by a unique personal identity number independently of his geographical location.

Of course, to be available all over the world, these services will necessarily need a satellite component.

Satellite access systems: Many of the new satellite systems (in particular business and rural systems) are characterized by their direct connection to the end users and can therefore be classified as access systems, in contrast with the traditional satellite links which are mostly trunked parts of public networks and which need local connections through switches. Of course, this does not mean that satellite systems cannot also be connected to public networks (this is the case, in particular, for the mobile systems).

In conclusion, satellites, in the future, while continuing to exploit the geostationary orbit, should take profit of the opportunities offered by the non-GSO satellite constellation systems and should consolidate their specific advantages in the challenging applications which are currently developing or opening, such as wideband business communication systems, rural networks and TV multiprogramme broadcasting.

General index

A

D

E

International switching centre (ISC): § 8.1.1.

International Telecommunication Union
(ITU)*: § 1.5, § 8.5, Chap. 9.

International transmission maintenance centre
(ITMC): § 8.1.1.

Internet, Internet protocol (TCP/IP):
AP6.1-2.1, § 8.6.6.1.

Inter-satellite link (ISL): § 1.3, § 6.3.2.4,
§ 6.4, AP6.1-1, see also inter-satellite
service (ISS).

Inter-satellite service (ISS): § 1.3, § 9.7.

Intersputnik earth stations: AN3.1.2.3.

Intersputnik satellites: AN3.1.2.2.

Intersputnik: § 1.2, § 6.1.4.4, AN3.1.2.

Inter-symbol interference (ISI): § 4.2.6.1.

Ionosphere, Ionospheric propagation: AN1.1,
AN1.2.3.2.

Iridium: § 6.1.4.4, § 6.5.

Italsat satellite: See Italy satellite system.

Italy satellite system: AN3.3.6.

J

Jamming margin: § 5.4.4.

Japan satellite system: AN3.3.7.

JCSat satellite (Japan): AN3.3.7.2.

Joint photographic experts group (JPEG):
§ 3.3.4, AP3.3-2.2.

K

Keplerian orbit: § 6.1.3.1.

Kepler's laws: § 6.1.

King post (Az-El antenna mount): § 7.2.3.3.

Klystron amplifier: § 7.4.1.1.3 b).

Klystron: § 7.4.2.2.

Korea satellite system: AN3-3.8.

Koreasat satellite: see Korea satellite system.

L

Launcher, Launching (satellite): § 6.6.1, § 6.9.

Layer: see Open system interconnection.

Left-hand circular polarization (LHCP): see
Polarization, Polarizer.

Lifetime (satellite): § 6.7.

Line terminating equipment (LTE): § 8.1.1.

Linear polarization (LP): § 7.2.2.2.3.

Linear prediction coding (LPC): AP3.1-2.1.

Linearizer: § 2.1.5.1, § 7.4.5.2.3.

Link budget*: § 2.3, AN2.

LLV Series (US launcher): § 6.9.5.

Local oscillator (LO)*: § 7.5.2.1, § 7.5.2.3.

Long March (Chinese launcher): § 6.9.3.
§ 6.9.5.

Low bit-rate encoding (LRE)*: § 3.3.2,
AP3.1, § 7.6.2.1 a).

Low earth orbit (LEO) satellite or system*:
§ 1.1, § 1.4.1.1 i), § 2.1.3.1, § 5.6.3,
§ 6.1, § 6.2.4.3, § 6.5, AP6.1-1,
AP6.1-2.

Low noise converter (LNC): § 7.3.4.

Low-noise amplifier (LNA)*: § 2.3, AP5.1-2,
§ 7.1.1.2, § 7.3.

Lunisolar attraction: see Sun, Moon.

M

MAC TV signal: § 3.2.2.2.

Management (satellite operation): § 6.8.

Mapping: AP3.2-2, § 4.2.3.2.

Maritime earth station: § 7.1.2.4.2, AN3.1.3.4,
see also: Inmarsat and Mobile satellite
service (MSS).

Maritime mobile-satellite service: See mobile-
satellite service.

M-ary phase shift keying (MPSK)*: § 4.1.2.

Master International Frequency Register
(MIFR): § 1.5.2.2, Chap. 9.

Measurement, measuring equipment:
§ 7.1.1.7.2, § 7.2.1.3.

Mechanical antenna system (earth station):
§ 7.1.1.1.

Medium earth orbit (MEO) satellite or system:
§ 1.1, § 1.4.1.1 i), § 2.1.3.1, § 6.1,
§ 6.2.4.3, § 6.5.

Mesh network: § 5.5.7, § 5.6.2, AP5.1-2,
AN2.2.2.

Message switching: § 3.3.3.

Microwave integrated circuit (MIC)*:
§ 6.3.2.4, § 6.3.3.5.

Microwave tube amplifier: § 7.4.3.

Microwave tube: § 7.4.2.

Self-interference: see Self-noise.

Self-noise: § 5.4.1.

Sequential decoding: § 3.3.5, AP3.2.

Servo subsystem (earth station): § 7.2.3.4.

Shaped beam: see Antenna beam.

Shaped reflector (antenna): § 7.2.2.1.1.

Side lobes (antenna)*: AP2.1-3, § 7.2.1.1,
§ 7.2.2.1.1.

Signal constellation: § 4.2.3.2.

Signal processing* (digital TV): § 3.3.4,
AP3.3.

Signal processing*: see Baseband processing.

Signalling system No. 5: § 8.1.7.1.

Signalling system No. 7: § 8.1.7.3, AP8.1-5.

Signalling system R1: § 8.1.7.1, AP8.1-2.

Signalling system R2: § 8.1.7.1, AP8.1-3.

Signalling: § 8.1.7, AP8.1.

Signalling system No. 6: § 8.1.7.3, AP8.1-4.

Signal-to-Noise ratio (S/N or SNR)*: § 2.1.1,
§ 2.3.1, § 3.3.2, § 4.1.1.

Single channel per carrier (SCPC)*: § 3.1,
§ 3.2, § 4.1.1.2, § 5.2.2, § 5.5.4.3,
AP5.1-2, § 6.3.4.5, § 7.5.4, § 7.6.2.1,
§ 7.5.4, § 7.6.1, § 7.6.2.1, § 8.1.3.3,
AN2.2.

Single entry interference: § 9.3.1.4,
§ 9.5.1.2.3.

Sinosatcom satellite (China): AN3.3.3.4.

Skybridge satellite system: § 6.1.4.4, § 6.5,
AP6.1-2.

Slotted Aloha: see Random access time
division multiple access
(RA-TDMA).

Small earth stations (see also: VSAT): § 7.1.3.

Soft decision: § 3.3.5, AP3.2.

Solar array: § 6.2.5 a).

Solar cell: § 6.2.2.

Solar interference (outage): § 7.2.1.2.

Solar panel: see Solar cell.

Solid state power amplifier (SSPA)*:
§ 6.3.3.1, § 6.3.3.3, § 6.3.3.5,
§ 7.1.1.3, § 7.4.4.

Sound-in-sync (SIS): see Audio programme.

Soyuz (Russian launcher): § 6.9.3.2.

Space segment*: § 1.4.1.1 i), Chap. 6.

Space shuttle (launcher): § 6.7, § 6.9.3.1.

Space stations*: 1.3, Chap. 6.

Space transportation system (STS): See Space
shuttle.

Spacecraft control centre (SCC): § 6.8.1.2.

SPADE: § 5.5.4.2.

Spain satellite system: AN3.3.9.

Speech interpolation: see Digital speech
interpolation.

Spin stabilization (satellite): § 6.2.2, § 6.2.4.2,
§ 6.3.2.2.

Spot beam: see antenna beam.

Spread spectrum multiple access (SSMA):
§ 1.4.4.2, § 5.4.

Spreading factor: § 5.4.4.

Star network*: § 5.5.7, § 5.6.2, AP5.1-2,
§ 8.2.2.4, AN2.2.1.

Station keeping (satellite): § 6.2.4.3, § 6.6.1.

Step-track (or Step-tracking): see Tracking
system.

Structure (satellite): § 6.2.2, § 6.3.2.2.

Sub-burst: § 5.3.1.

Sub-carrier (Television): § 7.5.5.2.

Sub-reflector: see Cassegrain.

Sun: § 6.1.3.2, § 6.2.3, § 6.2.4.3, § 6.2.5,
§ 6.6.3, § 7.2.1.2.

SuperBird satellite (Japan): AN3.3.7.2.

Switching matrix (SS-TDMA): § 6.3.3.4,
§ 6.3.4, § 6.3.4.5.

Syllabic companding: § 3.2.1.1.

Symbol (rate, etc.): § 4.2, § 5.3.3.

Symbol rate: § 3.3.5.1, AP3.2.

Synchronization*: § 4.2.4.2, § 5.3.2.

Synchronous code division multiple access
(S-CDMA): § 5.4.5.

Synchronous data link control (SDLC):
§ 8.3.2, § 8.4.1, see also: Packet
switching.

Synchronous digital hierarchy (SDH)*:
§ 1.4.5.4, § 3.5.3, AP3.5.

Synchronous transport module (STM):
§ 3.5.3, AP3.5.

Synthesizer (frequency synthesizer): § 7.5.4, §
7.6.4.4.

Systematic feedback encoder: § 4.2.3.2.

T

TDRS satellite (USA): AN3.3.10.

U

Nomenclature of main abbreviations

A

AAL: ATM Adaptation Layer.

A-CDMA: Asynchronous CDMA.

ADC (or A/D):   Analogue-to-digital converter.

ADPCM: Adaptive differential pulse code modulation (ADPCM). A voice coding process.

AFC: Automatic frequency control.

AGC: Automatic gain control.

AM-AM: Amplitude modulation to amplitude modulation conversion. An effect in non-linear circuits.

AM-PM: Amplitude modulation to phase modulation conversion. An effect in non-linear circuits.

AOR: Atlantic Ocean region (Intelsat).

APFD: Aggregate power flux-density.

API: Advanced publication of information.

AR: Axial ratio (polarization).

ARM: Apogee rocket motor.

ARNS: Aeronautical radionavigation service.

ARQ: Automatic repeat request; A type of error correction in which errored data are automatically repeated.

ASK: Amplitude-shift keying.

AVD: Alternate voice data.

AWGN: Additive white Gaussian noise.

Az-El: Azimuth-Elevation. A type of antenna mount (earth station).

B

BCH: Bose-Chaudhuri-Hocquenghem code. A class of block codes for error correction.

BEP:	Bit error probability.
BER:	Bit error ratio.
BFN:	Beam-forming network. A complex microwave circuit used in antennas with shaped beams.
B-ISDN:	Broadband Integrated Services Digital Network. An ISDN with capabilities up to Gigabit transmission standardized by ITU since 1985.
BO:	Back-off. The level below saturation for non-linear operation of amplifiers.
BPSK:	Binary (2-phase) phase shift keying.
BSS:	Broadcasting-satellite service.
BWG:	Beam waveguide (antenna).

C

C/N:	Carrier-to-noise ratio.
C/T:	Carrier-to-noise temperature ratio.
CATV:	Cable television.
CCIR:	International Radio Consultative Committee: now ITU-R.
CCITT:	International Telegraph and Telephone Consultative Committee: now ITU-T.
CCS:	Common channel signalling (CCS).
CDMA:	Code-division multiple access. The most usual type of Spread spectrum multiple access (SSMA).
CDV:	Cell delay variation.
CELP:	Code excited linear prediction. A voice low bit-rate encoding process.
CER:	Cell error ratio.
CFM:	Companded frequency modulation.
CLR:	Cell loss ratio.
CP:	Circular polarization.
CPM:	Continuous phase modulation.
CPSK:	Coherent PSK.
CSC:	Common signalling channel.

CSM:	Communications system monitoring.
CTD:	Cell transfer delay.
CU:	Channel unit (SCPC).
CUG:	Closed user group.

D

DAB:	Digital audio broadcasting.
DAMA:	Demand assigned multiple access.
DC (or D/C):	Down-converter
DCE:	Data circuit terminating equipment.
DCME:	Digital circuit multiplication equipment. An equipment implementing circuit concentration techniques.
DCT:	Discrete cosine transform. A technique used in video compression.
DM:	Delta modulation.
DPCM:	Differential pulse code modulation. A voice coding process.
DS:	Direct sequence modulation. A CDMA modulation technique.
DSI:	Digital speech interpolation. A circuit concentration technique based on silent gaps in a two-way telephone conversation.
DSP:	Digital signal processing.
DTE:	Data terminal equipment.
DTH:	Direct-to-home television distribution.
DVB:	Direct video broadcasting.

E

e.i.r.p.:	Equivalent isotropically radiated power. The product of the power at the input of a transmitting antenna by the antenna gain (usually expressed in dBW).
EESS:	Earth exploration-satellite service.
EMC:	Electro magnetic compatibility.
EPFD:	Equivalent power flux-density.

ESA:	European Space Agency
ESC:	Engineering service circuits.
ETSI:	European Telecommunications Standards Institute.

F

FBW:	Forward band working (co-directional, co-frequency sharing).
FDM:	Frequency division multiplex.
FDMA:	Frequency division multiple access.
FDMA-TDMA:	Frequency and time division multiple access.
FDP:	Fractional degradation of performance.
FEC:	Forward error correction; A data coding process in which supplementary bits are included allowing error correction at receive.
FET:	Field effect transistor (generally GaAS: Gallium, Arsenide).
FFSK:	Fast frequency shift keying: same as MSK.
FFT:	Fast Fourier transform.
FH:	Frequency hopping modulation. A CDMA modulation technique.
FIR:	Finite impulse response (filter).
FM:	Frequency modulation.
FS:	Fixed service.
FSK:	Frequency shift keying.
FSS:	Fixed-satellite service.

G

G/T:	Gain-to-noise temperature ratio. Also called Figure of merit (of a receiving station).
GCE:	Ground communication equipment: a (rather obsolete) denomination for the communication equipment of an earth station.
GDE:	Group delay equalizer.
GPS:	Global positioning system.

GSO:	Geostationary satellite orbit.

H

HDLC:	High level data link control.
HDTV:	High definition television.
HEMT:	High electron mobility transistor.
HPA:	High power amplifier.
HRC:	Hypothetical reference circuit.
HRDP:	Hypothetical reference digital path.
HRX (ISDN):	Hypothetical reference connection, ISDN.

I

IBS:	Intelsat business system. An Intelsat standard for business communications.
IDR:	Intermediate data rate. An Intelsat standard for medium bit-rate digital transmissions.
IDU:	Indoor unit (VSAT).
IF:	Intermediate frequency.
IFRB:	International Frequency Registration Board, now in ITU-R.
IOR:	Indian Ocean region (Intelsat).
IRD:	Integrated receiver decoder.
ISC:	International switching centre.
ISDN:	Integrated services digital network.
ISI:	Inter-symbol interference.
ISL:	Inter-satellite link.
ISO:	International Organization for Standardization.
ISS:	Inter-satellite service.
ITMC:	International transmission maintenance centre.
ITU:	International Telecommunication Union.

J

JPEG: Joint photographic experts group. Term used for a fixed image data compression standard.

L

LEO: Low-earth orbit.

LHCP: Left-hand circular polarization.

LNA: Low-noise amplifier.

LNC: Low noise converter. A unit including LNA + D/C.

LO: Local oscillator.

LP: Linear polarization.

LPC: Linear prediction coding. A voice low bit rate encoding process.

LRE: Low bit rate encoding: § 3.3.2 A voice compression coding process.

LSI: Large scale integrated circuit.

LTE: Line terminating equipment.

M

MAC: Monitoring, alarm and control.

MAI: Multiple access interference (MAI). Also called self-noise or self-interference in CDMA.

MCPC: Multiple channel per carrier.

MEO: Medium-earth orbit.

MIC: Microwave integrated circuit. See also MMIC.

MIFR: Master International Frequency Register (MIFR). The ITU-R Master book for registration of frequency assignments.

MMIC: Monolithic microwave integrated circuits.

MMSS: Maritime mobile-satellite service.

MPEG: Moving picture experts group. A moving picture data compression standard.

MPSK: M-ary phase shift keying. E.g. M = 2: BPSK, M = 4, QPSK.

MSS:	Mobile-satellite service.
MUX:	Multiplexer/Demultiplexer.

N

NASA:	National Aeronautical and Space Agency (USA).
NCC:	Network control centre.
NGSO:	see non-GSO.
NIC:	Nearly instantaneous companding.
N-ISDN:	Narrow-band Integrated Services Digital Network (often called simply ISDN and opposed to B-ISDN).
NNI:	Network node interface.
NOCC:	Network operations control centre.
Non-GSO:	Non-geostationary satellite orbit.
NT:	Network termination (equipment) at reference point T (NT1) or S (NT2).

O

OB:	Outside broadcasting (OB).
OBP:	On-board processing. Data processing on board of a satellite.
ODU:	Outdoor unit (VSAT).
OMJ:	Orthomode junction: see OMT.
OMT:	Orthomode transducer: A microwave device for separating orthogonal polarizations.
OPSK:	Octal-Phase shift keying (or 8-PSK). 8-phase shift keying.
OQPSK:	Offset quadrature phase shift keying.
OSI:	Open systems interconnection.

P

PA:	Power amplifier.
PABX:	Private automatic branch exchange. A private switch.

PAD:	Packet assembly/disassembly.
PAMA:	Pre-assigned multiple access.
PCM:	Pulse code modulation.
PDH:	Plesiochronous digital hierarchy.
PFD (or pfd):	Power flux-density.
PLL:	Phase locked loop.
PN:	Pseudo-noise modulation. See DS.
POR:	Pacific Ocean region (Intelsat).
PPM:	Periodic permanent magnet. A magnetic focusing technique used in microwave tubes.
PSD:	Power spectrum density.
PSDN:	Packet switched data network.
PSK:	Phase shift keying. The most usual digital modulation method. Commonly implemented with 2 or 4 phases (BPSK, QPSK).
PSPDN:	Packet switched public data network.
PSTN:	Public switched telephone network.
PTCM:	Pragmatic trellis codes.
PvtDN:	Private data network.

Q

| QO-CDMA: | Quasi-orthogonal codes. The most common codes for CDMA. |
| QPSK: | Quadrature (4-phase) phase shift keying. |

R

RA-TDMA:	Random access time division multiple access.
RBW:	Reverse band working.
RDSS:	Radiodetermination satellite service (RDSS).
RF:	Radio frequency.
RHCP:	Right-hand circular polarization.

RO:	Receive only (earth stations).
RR:	Radio Regulations (of ITU).
RS:	Reed-Solomon codes. A powerful error correction block coding technique.
RX:	An abbreviation for Receive (or Reception).

S

S/N (or SNR):	Signal-to-Noise ratio.
SCC:	Spacecraft control centre.
S-CDMA:	Synchronous CDMA.
SCPC:	Single channel per carrier. A particular, non-multiplexed form of FDMA.
SDH:	Synchronous digital hierarchy.
SDLC:	Synchronous data link control. An IBM protocol allowing multiple unacknowledged data frames.
SIS:	Sound-in-sync. An audio programme transmission method.
SMS:	Satellite multiservice: In Eutelsat, similar to IBS.
SNG:	Satellite news gathering.
SNR:	see S/N.
SPADE:	SCPC PCM multiple access demand assignment equipment. A – now obsolete - Intelsat DAMA system.
SSMA:	Spread spectrum multiple access.
SSOG:	Satellite system operations guide (Intelsat).
SSPA:	Solid state power amplifier.
SS-TDMA:	Satellite switched TDMA.
STM:	Synchronous transport module. The block frame structure in SDH.
STS:	Space transportation system: the USA space shuttle.

T

| TCM: | Trellis coded modulation. A technique combining error correction coding with carrier modulation. |

TCP/IP:	Transmission control protocol over Internet protocol.
TCR:	Telemetry, command and ranging.
TDE:	Time delay equalizer.
TDM:	Time division multiplexing.
TDMA:	Time division multiple access.
TDRSS:	Tracking and data relay satellite system. A USA/NASA advanced satellite system.
TED:	Threshold extension demodulator.
TIM:	Terrestrial interface module.
TMC:	Tracking mode coupler (antenna, earth station).
TOCC:	Technical and Operational Control Centre.
TST:	Time-space-time switching.
TT&C:	Tracking, telemetry and telecommand.
TV:	Television.
TVRO:	Television receive only (earth stations).
TWT:	Travelling wave tube.
TWTA:	Travelling wave tube amplifier.
TX:	An abbreviation for Transmit (or Transmission).

U

UC (or U/C):	Up-converter
UNI:	User network interface.
UPC:	Uplink power control.
UPS:	Uninterrupted power supply.

V

| VC: | Virtual container (SDH); Virtual channel (ATM). |
| VCO: | Voltage controlled oscillator. |

VCXO:	Voltage controlled crystal oscillator. See also VCO.
VLSI:	Very large scale integrated circuit.
VP:	Virtual path.
VSAT:	Very small aperture terminal. This term designates small earth stations, generally directly installed on the user premises. It also designates satellite systems and networks implementing VSAT earth stations.

W

WARC:	World administrative radio conference.
WRC:	World radiocommunication conference.

X

X.25:	An ITU-T recommendation specifying interfaces for terminals operating in the packet mode.
XPD:	Cross-polarization discrimination.
XPI:	Cross-polarization isolation.
X-Y:	A type of antenna mount (earth station).